LEWIS' DICTIONARY OF OCCUPATIONAL *and* ENVIRONMENTAL SAFETY *and* HEALTH

Jeffrey W. Vincoli

LEWIS PUBLISHERS
Boca Raton London New York Washington, D.C.

Library of Congress Cataloging-in-Publication Data

Lewis dictionary of occupational and environmental safety and health /
edited by Jeffrey W. Vincoli.
p. cm.
ISBN 1-56670-399-9 (alk. paper)
1. Industrial safety Dictionaries. 2. Environmental health
Dictionaries. I. Vincoli, Jeffrey W.
T55.L468 1999
363.11′03—dc21

99-32640
CIP

International Standard Book Number 1-56670-399-9
Library of Congress Card Number 99-32640
Printed in the United States of America 1 2 3 4 5 6 7 8 9 0
Printed on acid-free paper

Preface

The occupational and environmental safety and health professions have been on merging paths for several years now. Corporate "down-sizing" or "right-sizing" has resulted in a more streamlined approach to these once very diverse and quite separate disciplines. Although they both may now be practiced in tandem, often by the same individuals, each has evolved and developed as a separate area of study. As such, there are literally thousands of words, terms, and phrases that have specific meanings within their respective disciplines that may not always be clear and simple.

The practicing professional who has responsibilities in both occupational *and* environmental safety and health must be familiar with the "language of the profession" to successfully maneuver through the maze of compliance, regulatory, management, administrative, legal, technical, scientific, and even industry-specific slang terminology that are encountered every working day.

The *Lewis Dictionary of Occupational and Environmental Safety and Health* is the most comprehensive reference source of its kind available to today's diversified professional. Words, terms, and phrases from the following specific areas have been included in this publication. In total, there are approximately 25,000 definitions from the various listed areas of study.

Anatomy and Anthropometrics
Accident Investigation and Prevention
Aviation and Aerospace Safety
Biological and Medical Waste Management
Biology and Microbiology
Biostatistics
Chemistry
Clean Air Act
Clean Water Act
Collective Bargaining and Union Agreements
Computing and Computer Science
Ecology
Emergency/Disaster Preparedness and Response
Epidemiology
Environmental Compliance
Environmental Protection Agency Terms
Environmental Sanitation and Pollution Control
Ergonomics
Expert Witnessing
Fire Protection and Prevention
Fire Science and Fire Engineering
Geology and Hydrogeology
Hazardous Waste Management
Healthcare
Health Physics (Radiation)
Human Resources and Labor Management
Industrial Hygiene and Occupational Health
Industrial Security
Industrial Toxicology
Inspections and Audits
Insurance and Loss Control
Internet
Law and Litigation
Manufacturing
National Institute for Occupational Safety and Health
Occupational Medicine
Occupational Safety and Health Act
Occupational Safety and Health Administration Terms
Occupational Safety and Health Review Commission
Pollution Prevention Act
Product Liability
Public Health
Regulations and Standards
Risk Management
Robotics
Safety and Health Training
Safety Engineering
Safety Management and Administration
Site Assessments and Audits
Superfund (CERCLA)
System Safety Engineering
Transportation (Air, Road, Rail, Water)
Wet Lands Management
Workers' Compensation

Occupational and environmental safety and health disciplines are, indeed, separate functions. However, changes in the way corporate America does business has forced a continued divergence of the two professions. Those stuck in the middle, the practicing safety and environmental professionals, are forced to contend with an increasing number of responsibilities in areas where they may only possess cursory knowledge. This development has created a drastic need for new, quick-reference sources of knowledge and information. The more complete and comprehensive the source, the more beneficial it will be to the user.

This *Lewis Dictionary of Occupational and Environmental Safety and Health* is an attempt to fill this need and provide the professional with the single-source of reference for defining the thousands of words, terms, and phrases they are faced with literally every working day.

Jeffrey W. Vincoli
Editor

THE EDITOR

Jeffrey W. Vincoli, CSP, CHCM has worked in the field of occupational safety and environmental health for more than eighteen years. This experience has included the development, implementation, administration, and management of occupational and environmental safety and health programs for a number of Fortune 500 companies. Currently, Mr. Vincoli is President and Principal Consultant of J.W. Vincoli & Associates specializing in providing occupational safety and environmental training and consulting services to a number of domestic and international clients.

Mr. Vincoli has provided safety, health, and environmental training and management consulting services for literally thousands of professionals across the United States and in more than 15 countries. He also specializes in providing expert testimony on matters of fact pertaining to occupational safety, health, and the environment. His consulting practice has extended across a wide range of industries including aerospace and aviation, military, mass transit, nuclear, chemical, manufacturing, and many others. This experience has led to an appreciation for the specialized terminology that seems to be somewhat unique to the various industries. This appreciation has subsequently resulted in the compilation of the *Lewis Dictionary of Occupational and Environmental Safety and Health.*

Prior to beginning his own consulting practice, Mr. Vincoli spent 14 years working in our nation's missile, space, and strategic defense programs. With more than 10 years working for the former McDonnell Douglas Corporation (now Boeing), he worked first as a Safety Engineer and then as Manager of occupational safety and health, system safety engineering, industrial hygiene, hazardous waste management, and environmental compliance programs. Mr. Vincoli has worked on such programs as the Space Shuttle, Space Station, unmanned launch vehicle operations, Tomahawk Cruise Missile, and other specialized weapon systems for the United States and allied governments. Mr. Vincoli also worked for companies such as EG&G, Inc., and United Technologies Corporation, always with a focus on ensuring a safe and healthy work environment for several thousand employees.

Mr. Vincoli received his undergraduate degrees from the Florida Institute of Technology and completed his Master of Science and Master of Business Administration from Embry-Riddle Aeronautical University. He is a Certified Safety Professional, a Certified Hazard Control Manager, and a Registered Environmental Professional.

Mr. Vincoli is a member of many recognized organizations, including the American Society of Safety Engineers, the System Safety Society, the National Environmental Health Association, and the Veterans of Safety. He has published more than two dozen articles in professional trade journals such as *Professional Safety, Occupational Health and Safety, Hazard Prevention, Green Cross* (Hong Kong), and *Noticias de Seguridad* (Mexico). The *Lewis Dictionary of Occupational and Environmental Safety and Health,* his second work for Lewis Publishers, is his seventh published text in the field of safety, health, and the environment. He has served on the Editorial Board for *Occupational Hazards* magazine (1995-1998) and on the Advisory Committee for the Bureau of Business Practice (1995-1997).

Mr. Vincoli is an active member of the American Society of Safety Engineers. He has held office on the Chapter level, including President, and has served on numerous Regional and National Committees and Special Task Force assignments focusing on the Society's service to its members and to the safety profession. He has received numerous awards from professional societies, including the System Safety Society's Manager of the Year (1994) and the American Society of Safety Engineer's Regional Safety Professional of the Year (1987 and 1993). In 1998, he received the Charles V. Culbertson Outstanding Volunteer Service Award from the American Society of Safety Engineers for his contributions to the Society and its members.

Mr. Vincoli is a noted speaker, lecturer, trainer, and published author on subjects that extend across the broad scope of the occupational safety, health, and environmental industries.

Acknowledgments

This work was developed with the help of a number of organizations, contributors, and specialists representing the various areas of study that are of interest to the practicing safety and environmental professional. Specifically, I am particularly grateful to the following individuals/organizations and/or quoted sources for their contributions without which this publication would not have been possible:

American Association of Railroads
American Medical Association
American National Standards Institute
American Public Transit Association
Black's Law Dictionary (6th Edition)
Environmental Protection Agency
Federal Aviation Administration
Federal Railroad Administration
National Aeronautics and Space Administration
National Institute for Occupational Safety and Health
Occupational Safety and Health Administration
United States Air Force
United States Coast Guard
United States Department of Labor
United States Department of Transportation

Thomas M. Pankratz
James H. Stramler, Jr.
John Voorhees
Anton Cammarota
Bob Woellner
Benjamin F. Miller, M.D.
Claire Brackman Keane, R.N., B.S.

Finally, I would like to thank the many professionals at Lewis Publishers who have made the publication of this text as painless as possible. Specifically, I am grateful to Kenneth P. McCombs, Bob Hauserman, Suzanne Lassandro, and Mimi Williams for their efforts in making this publication a reality.

It is appropriate that I dedicate this "book of terms" to the two people who taught me the meaning of some of the most important things in life long before I could even read.
To my mother, Carmela Vincoli,
whose courageous battle against cancer is an inspiration to all who know her.
To my father, Joseph Vincoli,
a lifelong example of dependability and support, always putting the interests of others before his own.

A

A1 carcinogen
A confirmed human carcinogen as classified by the ACGIH TLV Committee. Substances associated with industrial processes, recognized to have carcinogenic potential.

A2 carcinogen
A suspected human carcinogen as classified by the ACGIH TLV Committee. Chemical substances, or substances associated with industrial processes, which are suspect of inducing cancer, based on either limited epidemiological evidence or demonstration of carcinogenesis on one or more animal species by appropriate methods.

"A" basis allowables
The minimum mechanical strength values guaranteed by the material producers or suppliers such that at least 99 percent of the material they produce or supply will meet or exceed the specified values with a 95 percent confidence level.

AAEE
American Academy of Environmental Engineers.

AAI
*See **arrival aircraft interval**.*

AAIH
*See **American Academy of Industrial Hygiene**.*

AALACS
*See **ambient aquatic life advisory concentrations**.*

AAOHN
*See **American Association of Occupational Health Nurses**.*

AAOO
American Academy of Ophthalmology and Otolaryngology.

AAP
*See **acoustical assurance period**.*

AAQS
Ambient air quality standards.

AAR
*See **airport acceptance rate**.*

AAS
*See **atomic absorption spectroscopy**.*

ABIH
*See **American Board of Industrial Hygiene**.*

abs
Absolute.

A-scale sound pressure level
A measurement of sound approximating the sensitivity of the human ear, used to note the intensity or annoyance of sounds.

A-shift
*See **first shift**.*

A-weighted network
Weighing network that is present on sound level meters and octave band analyzers which mimics the human ear's response to sound. Represented as dB(A).

abaft
A point beyond the midpoint of a ship's length.

abandon
Law. To desert, surrender, forsake, or cede. To relinquish or give up with intent of never resuming one's right or interest. To cease to use. To give up absolutely; to forsake entirely; to renounce utterly; to relinquish all connection with or concern in; to desert. It includes the intention, and also the external act by which it is carried into effect.

abandoned property
Law. Property over which the owner has given up dominion and control with no intention of recovering it. *See also **abandonment**.*

abandoned runway
An airstrip that is intact but not maintained or intended for use as a runway.

abandoned well
A well whose use has been permanently discontinued or which is in a state of disrepair such that it cannot be used for its intended purpose.

abandonee
Law. A party to whom a right or property is abandoned or relinquished by another. Term is applied to the insurers of vessels and cargoes.

abandonment
Law. The surrender, relinquishment, disclaimer, or cession of property or of rights. Voluntary relinquishment of all right, title, claim, and possession, with the intention of not reclaiming it. Time is not an essential element of act, although the lapse of time may be evidence of an intention to abandon, and where it is accompanied by acts manifesting such an intention, it may be considered in determining whether there has been an abandonment. Abandonment differs from *surrender* in that the latter requires an agreement, and also from *forfeiture* in that forfeiture may be against the intention of the party alleged to have forfeited.

ab assuetis non fit injuria
Law (Latin). From things to which one is accustomed (or in which there has been long acquiescence) no legal injury or wrong arises. In other words, if a person neglects to insist on his/her right, he/she is deemed to have abandoned it.

abatable nuisance
Law. A nuisance which is practically susceptible of being suppressed, or extinguished, or rendered harmless, and whose continued existence is not authorized under the law.

abate
Law. To throw down, to beat down, destroy, quash. To do away with, nullify, lessen, or diminish.

abatement
(1) *Air Pollution.* The reduction in the intensity or concentration of an ambient air pollutant. (2) *Asbestos.* Control of the release of fibers from a source of asbestos-containing materials during removal, enclosure, or encapsulation. (3) *General.* The removal or elimination of a nuisance; the actions taken to effect same; reducing the degree or intensity of, or eliminating, pollution. (4) *Law.* A reduction, a decrease, or a diminution. The suspension or cessation, in whole or in part, of a continuing charge, such as rent.

abatement in action
Law. An entire overthrow or destruction of the suit so that it is quashed and ended.

abator
Law. (1) In real property law, a stranger who, having no right of entry, contrives to get possession of an estate of freehold, to the prejudice of the heir or divisee, before the latter can enter, after the ancestor's death. (2) In the law of torts, one who abates, prostrates, or destroys a nuisance.

abbreviated injury scale (AIS)
An integer scale developed by the Association for the Advancement of Automotive Medicine to rate the severity of individual injuries. A numerical rating system used in an attempt to quantify an automobile accident victim's severity of injuries, as follows:

Rating	Severity
1	minor
2	moderate
3	serious
4	severe
5	critical (survival uncertain)
6	maximum (virtually unsurvivable)
9	unknown

abbreviated instrument flight rules (IFR) flight plans
Federal Aviation Administration. An authorization by Air Traffic Control (ATC) requiring pilots to submit only that information needed for the purpose of ATC. It includes only a small portion of the usual Instrument Flight Rules (IFR) flight plan information. In certain instances, this may be only aircraft identification, location, and pilot request. Other information may be requested if needed by ATC for separation/control purposes. It is frequently used by aircraft which are airborne and desire an instrument approach or by aircraft which are on the ground and desire a climb to Visual Flight Rule (VFR)-on-top.

ABC test
Unemployment compensation law exclusion tests providing that employer is not covered if individuals he/she employs are free from his/her control, the services are performed outside employer's places of business, and employees are customarily engaged in independently established trades or professions.

ABC transaction
In mining and oil drilling operations, a transfer by which A, the owner, conveys the working interest to B, the operator and developer for cash consideration, reserving a pro-

duction payment usually larger than the cash consideration paid by B. Later, A sells the reserved production payment to C for cash. The tax advantages of this type of transaction were eliminated by the Tax Reform Act of 1969.

abdication
Renunciation of the privileges and prerogatives of an office. It differs from *resignation*, in that resignation is made by one who has received his/her office from another and restores it into his/her hands, as an inferior into the hands of a superior. Abdication is the relinquishment of an office which has devolved by act of law.

abdomen
That part of the human body which lies between the thorax and the pelvis, containing the stomach, liver, spleen, pancreas, kidneys, bladder, and intestines.

abdominal cavity
That cavity within the abdomen which contains such organs as the intestines, liver, and bladder.

abdominal circumference
The surface distance measure of the lower torso at the level of the maximum anterior protrusion of the abdomen. It is measured with the individual sitting erect.

abdominal depth, standing
The horizontal linear distance from the back to the front of the abdomen, at the level of the maximum anterior protrusion. Measured with the individual standing erect; note the level at which the measurement is taken.

abdominal extension circumference
See ***abdominal circumference***.

abdominal extension depth
See ***abdominal depth.***

abdominal extension height
The vertical distance from the floor or other reference surface to the level of the maximum anterior protrusion of the abdomen in the midsagittal plane. Measurement is taken with the person standing.

abdominal extension level
See ***abdominal extension height.***

abdominal extension to wall
The horizontal distance from the most laterally protruding point of the abdomen to a wall. Measured with the individual standing erect against the wall with minimal buttock compression.

abdominal skinfold
The thickness of a horizontal skinfold centered at 3 cm lateral from 1 cm inferior to the umbilicus. Measured with the person standing comfortably erect, the body weight equally distributed to both feet, and the abdominal muscles relaxed.

abdominal wall
The covering of the abdominal cavity, consisting of fibrous and fatty tissue, muscles, and skin.

abdominoscopy
Examination, especially by means of an instrument, of the abdomen or its contents.

abducens
The sixth cranial nerve, which supplies the external rectus muscle of the eye.

abducent muscles
Muscles that pull back certain parts of the body from the mesial line.

abduct
To draw away from the main axis of the body or from a part of the body.

abduction angle
That angle through which a joint is abducted from a specified reference position. Also referred to as the *angle of abduction.*

abductor
A muscle that moves certain parts from the axis of the body.

abeam
Federal Aviation Administration. An aircraft is "abeam" a fix, point, or object when that fix, point, or object is approximately 90 degrees to the right or left of the aircraft track. Abeam indicates a general position rather than a precise point.

Abel Test
A colorimetric test that involves the use of moist potassium iodide paper which turns violet in the presence of gasses evolved from nitroglycerin, nitrocellulose, and nitroglycol.

aberration
(1) The failure of light rays to converge at a focal point in an optical system, resulting in blur. *See also* ***spherical aberration, chromatic aberration***. (2) Deviation from the

right course. (3) Deviation of refracted light rays. (4) A mental disorder.

aberrant
(1) Straying from the usual or normal method or course of action. (2) An aberrant structure, especially in regard to variable chromosome numbers.

abet
Law. (1) To encourage, incite, or set another on to commit a crime. This word is usually applied to aiding in the commission of a crime. To abet another to commit a crime is to command, procure, counsel, encourage, induce, or assist. (2) To facilitate the commission of a crime, promote its accomplishment, or help in advancing or bringing it about. In relation to charge of aiding and abetting, term includes knowledge of the perpetrator's wrongful purpose, and encouragement, promotion, or counsel of another in the commission of the criminal offense.

abettor
Law. An instigator, or setter on; one who promotes or procures a crime to be committed. A person who, being present, incites another to commit a crime, and thus becomes a principal.

abeyance
Law. Lapse in succession during which there is no person in whom title is vested. In the law of estates, the condition of a freehold when there is no person in whom it is vested.

abide
To accept the consequences of; to rest satisfied with; to wait for. With reference to an order, judgment, or decree of a court, to perform or execute.

abiding conviction
Law. (1) A definite conviction of guilt derived from a thorough examination of the whole case. Used commonly to instruct juries on the frame of mind required for guilt proved beyond a reasonable doubt. (2) A settled or fixed conviction.

ability
Having the physical and/or mental capacity to perform a given task effectively.

ab inconvenienti
(Latin) *Law.* From hardship, or inconvenience. An argument founded upon the hardship of the case, and the inconvenience or disastrous consequences to which a different course of reasoning would lead.

ab initio
(Latin) *Law.* From the beginning; from the first act; from the inception. An agreement is said to be "void ab initio" if it has at no time had any legal validity.

abiotic
Indicating the absence of life; non-biological.

ablate
To remove surgically.

able-bodied
Not having a physical handicap.

ABLEDATA
A computerized database containing consumer product information on devices for disabled individuals. Available from the National Clearinghouse on Technology and Aging.

able to earn
Law. Ability to obtain and hold employment means that the person referred to is either able or unable to perform the usual duties of whatever employment may be under consideration, in the manner that such duties are customarily performed by the average person engaged in such employment.

abnormal reading
*See **abnormal time**.*

abnormal time
An observed elemental time value which is beyond typical statistical or policy limits. Also referred to as *abnormal reading*.

ABO
*See **aviator's breathing oxygen**.*

abode
One's home; habitation; place of dwelling; or residence. Ordinarily means *domicile*.

ABOHN
American Board of Occupational Health Nurses, Incorporated.

abort
(1) *General.* Terminate some ongoing process or activity prior to its scheduled or expected completion. (2) *Federal Aviation Administration.* To terminate a preplanned aircraft maneuver (e.g., an aborted takeoff).

above elbow (AE)
Pertaining to an amputation at some level of the upper arm or a prosthesis which is fitted over the upper arm.

above knee (AK)
Pertaining to an amputation at some level of the thigh or a prosthesis which is fitted over the thigh.

abrade
To rub or wear down skin, primarily through friction.

abrasion
(1) The act of abrading. (2) An injury of the skin by abrading the outer layer. (3) Any scrapped area.

abrasive
A collection of discrete, solid particles that, when impinged on a surface, cleans, removes surface coatings, or improves the quality of, or otherwise prepares to modify the characteristics of that surface, either by impact or friction.

abrasive blasting
*See **abrasive cleaning**.*

abrasive cleaning
Process of cleaning surfaces by use of materials such as sand, alumina, steel shot, walnut shells, etc. in a stream of high pressure air or water.

ABS
(1) Acrylonitrile-butadiene-styrene, a black plastic material used in the manufacture of pipes and other components. (2) Alkylbenzene-sulfonate, a surfactant formerly used in synthetic detergents that resisted biological breakdown.

abscess
A collection of purulent matter in the tissue of a body organ or part, with pain, heat, and swelling.

abscissa
The horizontal or independent axis on a two-dimensional graph.

abscond
Law. To go in a clandestine manner out of the jurisdiction of the courts, or to remain concealed, in order to avoid their process. To hide, conceal, or absent oneself clandestinely, with the intent to avoid legal process. Postponing limitations.

absconding debtor
Law. One who absconds from his/her creditors to avoid payment of debts. A debtor who has intentionally and purposely concealed himself/herself from his/her creditors, or withdrawn from the reach of their suits, with intent to frustrate their just demands.

absent
Not present at some location when one is normally expected to be there.

absentee
An individual who is not present at his/her workplace when he/she is supposed to be there.

absolute
(1) *General.* Perfect or pure, as absolute alcohol (ethyl alcohol containing not more than one percent by weight of water). (2) *Law.* Free from conditions, limitations, or qualifications; not dependent, modified, or affected by circumstances; that is, without any condition or restrictive provisions.

absolute block
Railroads. A block in which no train is permitted to enter while it is occupied by another train.

absolute deed
Law. A document of conveyance without restriction or defeasance; generally used in contradistinction to mortgaged deed.

absolute humidity
The weight of water vapor per unit volume of air (e.g., pounds per cubic foot or milligrams per cubic meter).

absolute law
Law. The true and proper law of nature, immutable in the abstract or principal, in theory, but not in application; very often the object, the reason, the situation, and other circumstances, may vary its exercise and obligation. *See also **natural law**.*

absolute liability
Law. Responsibility without fault or negligence.

absolute maximum
The highest value anywhere on the total extent of a curve. *See also **relative minimum**.*

absolute pitch
A skill or ability of a person to identify the pitch of a pure tone without the use of any external reference.

absolute pressure
Pressure measured with respect to zero pressure or a vacuum. It is equal to the sum of a pressure gauge reading and the atmospheric pressure at the measurement location.

absolute purity water
Water with a specific resistance of 18.3 megohms cm at 25°C.

absolute scale
A temperature scale based on absolute zero. *See also* ***Kelvin scale***.

absolute temperature
Temperature based on an absolute scale expressed in either degrees Kelvin or degrees Rankine corresponding, respectively, to the centigrade or Fahrenheit scales. Degrees Kelvin are obtained by adding 273 to the centigrade temperature or subtracting the centigrade temperature from 273 if below zero C. Degrees Rankine are obtained by algebraically adding the Fahrenheit reading to 460. Zero degrees K is equal to -273°C and zero R is equal to -459.69°F.

absolute threshold
That minimum stimulus intensity which represents the transition between a response and no response from an observer attending to a particular sensory/perceptual task under specified conditions. Also referred to as *lower threshold*. *See also* ***threshold,*** *and* ***threshold of audibility***.

absolute vorticity
See ***vorticity***.

absolute zero
The minimum point in the thermodynamic temperature scale, expressed as zero degrees Kelvin, -273.16 degrees centigrade, -459.69 degrees Fahrenheit, or zero Rankine. This is a hypothetical temperature at which there is a total absence of heat.

absolutely stable air
An atmospheric condition that exists when the environmental lapse rate is less than the moist adiabatic rate.

absolutely unstable air
An atmospheric condition that exists when the environmental lapse rate is greater than the dry adiabatic rate. Also referred to simply as *unstable air*.

absorb
The penetration of a substance into the body of another.

absorbance
Logarithm to the base 10 of the transmittance.

absorbed dose
For any ionizing radiation, the energy imparted to matter by ionizing particles per unit mass of irradiated materials at the point of exposure. *See also* ***RAD***.

absorbent
A substance that takes in and absorbs other materials; a substance applied to a wound to stanch or arrest the flow of blood.

absorbent gas mask
Any respirator which includes a container having some type of material to absorb toxic substance.

absorber
(1) Any material which is capable of taking up chemicals or radiation. (2) Any device which is capable of taking up chemicals or radiation.

absorption
(1) *Toxicological.* The ability of a substance to penetrate the body of another; the movement of a chemical from the site of exposure (oral, dermal, respiratory) across a biologic barrier and into the bloodstream or lymphatic system. (2) *Chemistry.* The process by which one material is pulled into and retains another to form a blended or homogeneous solution. (3) *Physiology.* The process by which porous tissues such as the skin and intestine walls permit passage of liquids and gases into the bloodstream. (4) *Radiation.* The process whereby the number of particles or quanta in a beam of radiation is reduced or degraded in energy as it passes through some medium. The absorbed radiation may be transformed into mass, other radiation, or energy by interaction with the electrons or nuclei of the atoms on which it impinges. (5) *Acoustics.* The conversion of acoustical energy to heat or another form of energy within the medium of

the sound-absorbing material. (6) *Environmental.* The adhesion of molecules of gas, liquid, or dissolved solids to a surface. Used as an advanced method of treating in which activated carbon removes organic matter from wastewater. (7) *Law.* Act or process of absorbing. Term used in collective bargaining agreements to provide seniority for union members if employer's business is merged with another.

absorption coefficient
In acoustics, the fraction of incident sound absorbed or otherwise not reflected by a surface.

absorptive muffler
A type of acoustic muffler that is designed to absorb sound energy as sound waves pass through it.

abstention doctrine
Law. Permits a federal court, in the exercise of its discretion, to relinquish jurisdiction where necessary to avoid needless conflict with the administration by a state of its own affairs.

abstract of record
Law. A complete history in short, abbreviated form of the case as found in the record, complete enough to show the appellate court that the questions presented for review have been properly reserved. An abbreviated, accurate, and authentic history of proceedings.

abut
To reach; to touch. To touch at the end; be contiguous; join at a border or boundary; terminate on; end at; border on; reach or touch with an end. The term *abutting* implies a closer proximity than the term *adjacent*. No intervening land.

abutting owner
An owner of land which abuts or adjoins. The term usually implies that the relative parts actually adjoin, but is sometimes loosely used without implying more than close proximity.

ac
Alternating current.

AC
See ***alcohol concentration***.

ACAIS
See ***Air Carrier Activity Information System***.

acantha
The spine; one of the acute processes of the vertebrae.

acariasis
A skin disease caused by mites. *See also* ***scabies***.

acaricide
Chemical used to kill ticks and mites.

ACBM
Asbestos-containing building material.

ACCC
See ***area control computer complex***.

accelerate stop distance available
Federal Aviation Administration. The runway plus stopway length declared available and suitable for the acceleration and deceleration of an airplane aborting a takeoff.

acceleration
A vector representing the rate of change of velocity with time.

acceleration illusion
Any perception of apparent motion or change in motion resulting from acceleration-induced stimulation of the vestibular apparatus, the visual system, or other mechanoreceptors.

acceleration loss
In ventilation, the energy required to accelerate air to a higher velocity.

acceleration power
Measured in kilowatts. Pulse power obtainable from a battery used to accelerate a vehicle. This is based on a constant current pulse for 30 seconds at no less than 2/3 of the maximum open-circuit voltage, at 80% depth-of-discharge relative to the battery's rated capacity and at 20 degrees Celsius ambient temperature.

acceleration syndrome
Any change in physiological and/or perceptual-motor-cognitive function due to the forces imposed on the body by changes in velocity. Also known as *g-force syndrome*. *See also* ***positive g***, ***negative g***, *and* ***transverse g***.

accelerator
A device for imparting kinetic energy to electrically charged particles such as electrons, protons, helium ions and other ions of elements of interest. Common types of accelerators include the Van der Graaf, Cockcroft-Walton, cyclotrons, betatrons, linear accelerators, and others.

accelerometer
A force transducer used in measuring acceleration. Also referred to as *acceleration transducer* and *acceleration pickup.*

accelerometry
The quantitative measurement of accelerations of a structure or its components.

accent lighting
Any form of directional or other unique illumination emphasis as an attempt to bring attention to a segment of the field of view or some object within the environment.

acceptability
With regard to the use of instruments, the willingness of personnel to use an instrument when considering its characteristics, such as weight, noise, response time, drift, portability, reliability, interference effects, etc.

acceptable daily intake (ADI)
An estimate similar in concept to the RfDs; however, derived using a less rigorously defined methodology. RfDs have replaced the ADI as the EPA's preferred value for use in evaluating potential non-carcinogenic health effects resulting from exposure to a chemical.

acceptable entry conditions
As pertains to confined space entry, the conditions that must exist in a permit space to allow entry and to ensure that employees involved with a permit-required confined space entry can safely enter into and work within the space.

acceptable indoor air quality
Indoor air in which there are no known contaminants at harmful levels and air with which 80% of the occupants of the indoor environment are satisfied with its quality.

acceptable intake for chronic exposure (AICs)
An estimate similar in concept to the RfDs; however, derived using a less rigorously defined methodology. RfDs have replaced AICs as the EPA's preferred value for use in evaluating potential non-carcinogenic health effects resulting from exposure to a chemical.

acceptable intake for subchronic exposure (AIS)
An estimate similar in concept to a subchronic RfDs; however, derived using a less rigorously defined methodology. Subchronic RfDs have replaced AICs as the EPA's preferred value for use in evaluating potential non-carcinogenic health effects resulting from exposure to a chemical.

acceptable lift
Ninety pounds multiplied by a series of factors related to the location of the object to be lifted, its distance from a specific position, and the lift frequency.

acceptable quality level (AQL)
The maximum allowable average percentage of vehicles or exhaust systems that can fail sampling inspection under a Selective Enforcement Audit under Chapter I, (Environmental Protection Agency), Subchapter G (Noise Abatement Programs) of Title 40 (Protection of Environment) of the Code of Federal Regulations.

acceptable risk
(1) That degree of risk which society is willing to take after societal, economic, and political factors are considered. (2) The residual risk that remains after all possible control measures have been implemented that is deemed acceptable by the party or parties that are exposed to the risk (e.g., management, employees, the public, the government, etc.).

accepted
With regard to electrical installations, accepted means it has been inspected and found by a nationally recognized testing laboratory to conform to specified plans or to procedures of applicable codes.

access, easement of
Law. An easement of access is the right which an abutting owner has of ingress to and egress from his/her premises, in addition to the public easement in the street.

access restrictions road gate
Constraints on use of a road.

access rights
This element identifies who has acquired legal access rights over a road segment.

access time
That temporal interval required to gain an opening to or achieve a certain position within a given structure.

access to counsel
Law. Right of one to consult with his/her attorney as guaranteed by the 6th Amendment of the U.S. Constitution.

accessibility
A measure of the ease with which a location may be reached, entered, or viewed.

accessibility score
A rating based on the cross-sectional area of the access path available for an individual or body part to reach the desired point.

accessible
(1) *Equipment.* Admitting close approach. Not guarded by locked doors, elevation, or other effective means. *See also* ***readily accessible***. (2) *Wiring Methods.* Capable of being removed or exposed without damaging the building structure or finish, or not permanently closed in by the structure or finish of the building. *See also* ***concealed*** *and* ***exposed***.

accessible environment
The atmosphere, land surfaces, surface waters, oceans and all of the lithosphere that is beyond the controlled area, according to Chapter I, (Environmental Protection Agency), Subchapter F (Radiation Protection Programs) of Title 40 (Protection of Environment) of the Code of Federal Regulations.

accessible stations
A public transportation passenger facility which provides ready access, is useable, and does not have physical barriers that prohibit and/or restrict access by individuals with disabilities, including individuals who use wheelchairs.

accessible vehicles
Public transportation revenue vehicles which do not restrict access, are usable, and provide allocated space and/or priority seating for individuals who use wheelchairs.

accessory
Criminal Law. (1) Contributing to or aiding in the commission of a crime. One who, without being present at the commission of a felonious offense, becomes guilty of such offense, not as a chief actor, but as a participator, as by command, advice, instigation, or concealment; either *before* or *after* the fact or commission. (2) One who is not the chief actor in the offense, nor present at its performance, but is in some way concerned therein, either before or after the act committed. (3) One who aids, abets, commands, or counsels another in the commission of a crime. *See also* ***abet, aid, accessory after the fact, accessory before the fact,*** and ***accessory during the fact.***

accessory after the fact
Law. Any person who, knowing a felony to have been committed by another, receives, relieves, comforts, or assists the felon, in order to enable him/her to escape from punishment, or the like.

accessory before the fact
Law. One who orders, counsels, encourages, or otherwise aids and abets another to commit a felony and who is not present at the commission of the offense. The primary distinction between the *accessory before the fact* and the *principle in the second degree* is presence.

accessory during the fact
Law. A person who stands by without interfering or giving such help as may be in his/her power to prevent the commission of a criminal offense.

accessory movement
See ***synkinesia.***

accessory or auxiliary equipment
A particular item of equipment added to a vehicle to aid or contribute to the vehicle's operation and/or mission.

accident
(1) *General.* An unplanned, unforeseen, and therefore unwanted or undesired event that may or may not result in physical harm and/or property damage; any unplanned event that interrupts or interferes with the orderly progress of a production activity or process. (2) *System Safety.* An unwanted event resulting from the occurrence of one or more fault incidents that have a negative impact on a system, product, equipment, or personnel. (3) *Worker's Compensation Law.* Any unforeseen, untoward happening which was not to be reasonably anticipated. An unlooked for and untoward event which is not expected or designed by injured employee; a result produced by a fortuitous cause. (4) *Insurance.* Includes continuous or repeated exposure to the same conditions resulting in public liability which the insured neither expected nor intended. (5) *Department of Transportation.* An occurrence involving a commercial motor vehicle operating on a public road which results in one of the following: a fatality; bodily injury to a

person, who as a result of the injury, immediately receives medical treatment away from the scene of the accident; or one or more motor vehicles incurring disabling damage as a result of the accident, requiring the vehicle to be transported away from the scene by a tow truck or other vehicle. It does not include an occurrence involving only boarding and alighting from a stationary motor vehicle; an occurrence involving only the loading or unloading of cargo; or an occurrence in the course of the operation of a passenger car or a multipurpose passenger vehicle, as defined in 49 CFR 571.3, by a motor carrier and is not transporting passengers for hire or hazardous materials of a type and quantity that require the motor vehicle to be marked or placarded in accordance with 49 CFR 177.823. (6) *American Gas Association.* (a) An event that involves the release of gas from a pipeline or of liquefied natural gas or gas from an LNG facility resulting in a death, or personal injury necessitating in-patient hospitalization; or estimated property damage, including cost of gas lost, of the operator or others, or both, of $50,000 or more; (b) An event that results in an emergency shutdown of an LNG facility; (c) An event that is significant, in the judgment of the operator, even though it did not meet the criteria of (a) or (b). (7) *Federal Transit Association.* An incident involving a moving vehicle. Includes collisions with a vehicle, object, or person (except suicides) and derailment/left roadway. (8) *National Safety Council.* Occurrence in a sequence of events that produces unintended injury, death, or property damage. Accident refers to the event, not the result of the event. (9) *Railroad Accident/Incident.* (a) Any impact between railroad on-track equipment and an automobile, bus, truck, motorcycle, bicycle, farm vehicle, or pedestrian at a rail-highway grade crossing; (b) Any collision, derailment, fire, explosion, act of God, or other event involving operation of railroad on-track equipment (standing or moving) that results in more than $6,300 in damages to railroad on-track equipment, signals, track, track structures, and road-bed; (c) Any event arising from the operation of a railroad which results in: i) Death of one or more persons; ii) Injury to one or more persons, other than railroad employees, that requires medical treatment; iii) Injury to one or more employees that requires medical treatment or results in restriction of work or motion for one or more days, one or more lost work days, transfer to another job, termination of employment, or loss of consciousness; or iv) Occupational illness of a railroad employee as diagnosed by a physician. (10) *Aviation.* An aircraft accident is defined by the National Transportation Safety Board (NTSB) as an occurrence associated with the operation of an aircraft which takes place between the time any person boards the aircraft with the intention of flight until all such persons have disembarked, and in which any person suffers death or serious injury as a result of being in or upon the aircraft or by direct contact with the aircraft or anything attached thereto, or in which the aircraft receives substantial damage.

accident analysis

A concerted, organized, methodical, planned process of examination and evaluation of all evidence and records identified during investigation of accidents.

accident classes

Transportation. Term used to categorize commercial vehicle accidents according to accident severity (i.e., fatal accidents, injury accidents, and property damage accidents).

accident consequences

Transportation. The physical results of motor vehicle accidents. Consequences include fatalities, injuries, and property damage.

accident frequency rate

An older term for the number of lost time accidents per 1,000,000 man-hours worked. Also known as *frequency rate,* it is represented by the following formula:

$$AFR = \frac{\textit{number of lost-time accidents}}{\textit{1,000,000 man-hours worked}}$$

accident insurance

Form of insurance which undertakes to indemnify the insured against expenses, loss of time, and suffering resulting from accidents causing him/her physical injury, usually by payment at a fixed rate per month while the consequent disability lasts, and sometimes including the payment of a fixed sum to his/her heirs in case of death by accident within the term of the policy. *See also* ***insurance*** *and* ***casualty insurance***.

accident investigation
A detailed and methodical effort to collect and interpret facts related to an individual accident, conducted to identify the causes and develop control measures to prevent recurrence; a systematic look at the nature and extent of the accident, the risks taken, and loss(es) involved; an inquiry as to how and why the accident event occurred.

accident phases
In an accident investigation, when evaluating the sequence of events that resulted in an accident, the events are divided into three phases or categories: *pre-contact* (before the accident), *contact* (the accident), and *post-contact* (after the accident). Analysis of the events occurring in each phase facilitates the identification of loss-inducing activities and conditions. Also referred to as the three stages of loss control.

accident potential
A situation comprised of human behaviors and/or physical conditions having a probability of resulting in an accident.

accident prevention
(1) Efforts or countermeasures that are taken to reduce the number and severity of accidents. (2) The design or application of countermeasures in an environment to reduce accidents or the accident potential.

accident proneness
A non-scientific determination or belief that a particular person may have a tendency toward being involved in or contributing to accidents.

accident rate
The accident experience relative to a base unit of measure (e.g., the number of disabling injuries per 1,000,000 person-hours worked). *See also* ***accident frequency rate***.

accident repeater
A person who has been principally involved, regardless of cause, in more than one accident within a predetermined and specified period of time, for example, one year.

accident risk
A measure of vulnerability to loss, damage, or injury caused by a dangerous element or factor (MIL-STD-1574A).

accident risk assessment
A written evaluation of those hazards associated with the operation of a given facility, including any equipment or hardware used in the facility. A determination of the accident potential and an explanation of control measures are also provided.

accident risk factor
A dangerous element of a system, event, process, or activity, including causal factors such as design or programming deficiency, component malfunction, human error, or environment, which can propagate a hazard into an accident if adequate controls are not effectively applied (MIL-STD-1574A).

accident severity rate
An older term for the number of lost workdays per 1,000,000 man-hours worked. Also known as severity rate, it is represented by the following formula:

$$ASR = \frac{\textit{number of lost workdays}}{\textit{1,000,000 man hours worked}}$$

accident site
The location of an unexpected occurrence, failure, or loss, either at a plant or along a transportation route, resulting in a release of hazardous materials, property damage, personnel injury or death, or some other combination of loss events.

accident sources
Accidents generally involve one or all of five elements: people, equipment, material, procedures, and the work environment, each of which must interact for successful business operations. However, when something unplanned and undesired occurs within either of these elements, there is usually some adverse effect on any one or all of the other elements, which if allowed to continue uncorrected, could lead to an incident or accident and subsequent loss.

accident type
Federal Highway Association. An accident type is classified as either "collision" or "non-collision."

accidental death
A death causally related to some accident.

accidental impact
An undesired, other than functional impact. Also referred to as a ***nuisance impact***.

accidental release
The unanticipated emission of a regulated substance or other extremely hazardous substance into ambient air from a stationary source.

acclimatization
An adaptive process which results in a reduction of the physiological response produced as a result of the application of a constant environmental stress, such as heat, on the body; the process of becoming accustomed to new conditions; the physiological and behavioral adjustments of an organism to changes in its environment.

accommodation
(1) *Physiology.* The ability of the eye to focus for varying distances; the adjustment of the eye lens whereby it is able to focus a clear image onto the retina. (2) *Law.* An arrangement or engagement made as a favor to another, not upon a consideration received. Something done to oblige, usually spoken of a loan of money or commercial paper, also a friendly agreement or composition of differences. The word implies no consideration.

accommodation of workers
See ***worker accommodation.***

accomplice
Law. One who knowingly, voluntarily, and without common intent unites with the principal offender in the commission of a crime. One who aids and assists, or is an accessory. One is liable as an accomplice to the crime of another if he/she gave assistance or encouragement or failed to perform a legal duty to prevent it with the intent thereby to promote or facilitate commission of the crime.

accomplice liability
Law. Criminal responsibility of one who acts with another before, during, or after the perpetration of a crime.

accord
An agreement between two persons, one of whom has a right of action against the other, to settle the dispute.

accouchement
Confinement; childbirth.

accountable
To be called upon to account for the accomplishments or non-accomplishments relative to an assigned function or task. Responsibility assigned by management to an individual to carry out an assignment.

accounting changes income (loss)
The difference between the amount of retained earnings at the beginning of the period in which a change in accounting has occurred and the amount of retained earnings that would have been reported, net of applicable taxes, at that date if the new accounting had been applied retroactively for all applicable periods.

accounts receivable insurance
Insurance coverage designed to protect against inability to collect because of damage to records which support the accounts.

accredited laboratory
Certification awarded to an analytical laboratory that has successfully participated in a proficiency testing program, such as that of the American Industrial Hygiene Association (AIHA).

accretion
The growth of a precipitation particle by the collision of an ice crystal or snowflake with a super-cooled liquid droplet that freezes upon impact. Also referred to as *riming*.

accumulation start date
That date when the first drop or piece of waste has been put into the container.

accumulative timing
A time-study technique in which multiple timers are used with electrical or mechanical linkage to obtain task or work cycle times. Also referred to as multiple watch timing.

accumulator
A tank installed in a circulating water system to allow for fluctuations in flow, temperature, pressure, or other variation in operation.

accuracy
The degree of agreement between a measured value and the accepted reference value, or the agreement of an instrument reading or analytical result to the true value. When referring to an instrument's accuracy it represents the ability of the device to indicate the true value of the measured quantity. For instruments, it is often expressed as a percentage of the full-scale range of the instrument.

ACD
Allergic contact dermatitis.

acetabulum
The cavity that receives the head of the thigh bone.

acetic
Having the properties of vinegar.

acetic acid
An acid, often prepared by the oxidation of alcohol, and with water forming the chief ingredient of vinegar. Acetic acid is used as a reagent and is sometimes taken internally.

acetone
(1). *Chemistry.* A chemical compound, CH_3COCH_3, with solvent properties and characteristic odor, obtained by fermentation or produced synthetically; it is a byproduct of acetoacetic acid. It can be produced synthetically. (2). *Physiology.* A colorless liquid found in minute amounts in the body and in larger amounts in the blood and urine in diabetes, faulty metabolism, and after lengthy fasting.

acetylcholine
A substance in the human body having important neurotransmitter effects on various internal systems; often used as a bronchoconstrictor.

acetylsalicylic acid
See ***aspirin***.

ACF
See ***area control facility***.

acfm
Actual cubic feet per minute.

ACGIH
See ***American Conference of Government Industrial Hygienists.***

achalasia
Failure of the sphincter or other muscular valves to relax normally and allow the gastrointestinal contents to pass.

ACh e enzyme
Acetylcholinesterase enzyme.

achievable duty
A term used to describe OSHA's approach to employer compliance with safety and health regulations and standards. The contention is that compliance must be achievable within the feasible bounds of economics and technology.

Achilles heel
Generally refers to the point of weakness that is most vulnerable or susceptible.

Achilles tendon
The tendon that joins the heel bone and the muscles of the calf.

achondroplasia
Defective development of cartilage causing dwarfism.

ac/hr
Air changes per hour. The movement of a volume of air in a given time; if a room has one air change per hour, it means that all of the air in the room will be replaced in a one-hour period.

achromatic
Without hue (color, chroma); appearing white, black, or gray.

achromatic lens
A lens corrected to have the same focal length for two or more specified wavelengths.

achromatic point
An equal energy white point on the CIE chromatic diagram with coordinates of $x = y = 0.33$. Also known as *white point.*

achromatin
That portion of the nucleus of a cell which is not stainable.

achromatopsia
Complete color blindness.

achromatous
Having no color; of a lighter color than is usual or normal.

ACI
American Concrete Institute.

acid
A compound consisting of hydrogen plus one or more other elements and which, in the presence of certain solvents or water, reacts with the production of hydrogen ions; a compound with pH between zero and seven. As pH decreases from seven to zero, acidity increases. An acid reacts with an alkali to form a salt and water; it turns litmus paper red.

acid-ash diet
A special diet prescribed for the purpose of lowering the urinary pH so that alkaline salts will remain in solution. The diet may be given to aid in the elimination of fluid in certain kinds of edema, in the treatment of some types of urinary tract infection, and to inhibit the formation of alkaline urinary calculi. Meat, fish, eggs, and cereals are emphasized; fruits, vegetables, and milk may be forbidden or restricted.

acid-base balance
The maintenance of a normal balance between the acidity and alkalinity of the body fluids located within the extracellular and intracellular compartments. Since most of the normal metabolic processes of the body produce acids as their end products, the body must work continuously to maintain this delicate balance. Chemical buffers, principally bicarbonates, phosphates, and salts of proteins, help in the neutralization process. The kidneys and lungs also participate in this mechanism because of their control of the availability of the electrolytes that are essential to proper functioning of the buffer system.

acid deposition
A complex chemical and atmospheric phenomenon that occurs when emissions of sulfur and nitrogen compounds and other substances are transformed by chemical processes in the atmosphere, often far from the original sources, and then deposited on earth in either a wet or dry form. The wet forms, popularly called "acid rain," can fall as rain, snow, or fog. The dry forms are acidic gases or particulates. *See also* ***acid rain.***

acid-fast
Not easily decolorized by acids when stained, as the tubercule bacillus.

acid gas
A gas that forms an acid when mixed with water.

acid mantle
The lipid (oily) outside layer of the skin structure, composed of oil and sweat, easily removed by washing. The acid mantle normally has a pH less than seven.

acid rain
The acidity in rain or snow (pH less than 5.6) that results from the oxidation of carbon, sulfur, or nitrogen compounds in the air, and their subsequent absorption into the precipitation, thereby making it acidic.

acidity
The capacity of an aqueous solution to neutralize a base.

acidophile
A tissue, organism, cell, or substance that shows an affinity toward an acidic environment.

acidophillic
(1) Having the quality of being easy to stain with acid. (2) Thriving or flourishing in an acid environment.

acidosis
A pathologic condition resulting from the accumulation of acid or depletion of the alkaline reserve (bicarbonate content) in the blood and body tissues, and characterized by an increase in hydrogen ion concentration (decrease in pH). The normal pH of the blood is approximately 7.4 (slightly alkaline) and is maintained at that level by chemical buffers and normal functioning of the kidneys and lungs. The opposite of acidosis is *alkalosis*.

acknowledging device
Railroad. A manually operated electric switch or pneumatic valve by means of which, on a locomotive equipped with an automatic train stop or train control device, an automatic brake application can be forestalled, or by means of which, on a locomotive equipped with an automatic cab signal device, the sounding of the cab indicator can be silenced.

acknowledging time
Railroad. As applied to an intermittent automatic train stop system, a predetermined time within which an automatic brake application may be forestalled by means of the acknowledging device.

acknowledgment circuit
Railroad. A circuit consisting of wire or other conducting material installed between the track rails at each signal in the territory where an automatic train stop system or cab signal system of the continuous inductive type with 2 indication cab signals is in service, to enforce acknowledgment by the engineer at each signal displaying an aspect requiring a stop.

ACLT
See ***actual calculated landing time.***

acne
(1) An eruption of hard, inflamed tubercles or pimples on the face, especially during adolescence but also resulting from exposure (for some individuals) to certain chemical substances. (2) An inflammatory disease of the skin, arising from the obstruction of the sebaceous glands.

acoustic
(1) The study of sound, including its generation, transmission, and effects. (2) The cause, nature, and phenomena of the vibrations of elastic bodies that affect the organ of hearing. (3) The properties determining audibility or fidelity of sound in an auditorium.

acoustic absorption coefficient ($_m$)
The ratio of energy absorbed by a material to the energy incident to it.

acoustic descriptor
The numeric, symbolic, or narrative information describing a product's acoustic properties as they are determined according to the EPA test methodology as per Chapter I (Environmental Protection Agency), Subchapter G (Noise Abatement Programs) of Title 40 (Protection of Environment) of the Code of Federal Regulations.

acoustic flanking
The structural transmission of vibrations to elements which re-radiate the sound in the acoustic range.

acoustic intensity
Represented as "I" it is the rate of flow of acoustic energy per specified cross-sectional area, as follows:

$$I = \frac{dW}{dA}$$

acoustic nerve
See ***auditory nerve***.

acoustic pressure
See ***sound pressure***.

acoustic reflex
The contraction of the tensor tympani and stapedius muscles attached to the conducting middle ear bones to increase acoustic impedance in response to a high intensity sound.

acoustic scattering
The irregular reflection, refraction, and/or diffraction of sound in many directions.

acoustic stimulus
Any varying pressure from air or other fluid having sufficient intensity within the transducing frequency range of the object or organism. *See also* ***auditory stimulus***.

acoustic trauma
A temporary or permanent hearing loss in one or both ears as a result of a sudden loud noise or blow to the head which caused injury or damage to the ear(s).

acoustical assurance period (AAP)
A specified period of time or miles driven after sale to the ultimate purchaser during which a newly manufactured vehicle or exhaust system, properly used and maintained, must continue in compliance with the federal standard; reference Chapter I (Environmental Protection Agency), Subchapter G (Noise Abatement Programs) of Title 40 (Protection of Environment) of the Code of Federal Regulations.

acoustical insulation
Material designed to absorb noise energy that is incident upon it.

acoustical treatment
The use of acoustical (sound) absorbents, acoustical isolation, or other changes or additions to a noise source to improve the acoustical environment.

acquired character
A biological change that results from use or environment rather than from heredity.

acquired immune deficiency syndrome (AIDS)
A severe (life-threatening) disease that represents the late clinical stage of infection with the *human immunodeficiency virus* (HIV). The HIV most often results in progressive damage to the immune system and various organ systems, especially the central nervous system. Body fluid-to-body fluid contact with an infected HIV carrier is required for transmission. HIV has been recovered from body fluids other than blood, such as tears, saliva, urine, bronchial secretions, spinal fluid, feces, vomitus, and others.

acquisition
Law. The act of becoming the owner of certain property; the act by which one acquires or procures the property.

acquisitive offense
Law. A generic term to describe all forms of larceny and offenses against the title or possession of property.

acquit
Law. To set free, release, or discharge as from an obligation, burden, or accusation. To absolve one from an obligation or a liability; or to legally certify the innocence of one charged with a crime.

acre
A quantity of land containing 160 square rods, 4,840 square yards, or 43,560 square feet of land, in whatever shape.

acre-foot
The volume of water that would cover a 1-acre area 1 foot deep. Equivalent to 1233.6 cubic meters or 325,850 gallons.

acrid
Sharp or biting to the taste or smell.

acrobatic flight
An intentional maneuver involving an abrupt change in an aircraft's attitude, an abnormal attitude, or abnormal acceleration not necessary for normal flight.

acrolein
An aldehyde compound used as a microbiocide and in the manufacture of organic chemicals.

acromegaly
A rare glandular disease associated with the overgrowth of bone, especially in the jaws, hands, and feet.

acromial
Pertaining to the acromion. The most lateral/superior point of the acromion.

acromial-biceps circumference-level length
The surface distance along the outer edge of the arm from acromial to the level at which the relaxed biceps circumference measure is taken. Also called the *acromion-biceps circumference-level length.* Measured with the individual standing erect with arms hanging naturally at the sides and the hands and fingers extended.

acromial-dactylion length
The vertical distance from the acromial to the tip of the middle finger. Also called *acromion-dactylion length* and *shoulder-fingertip length.* Measured with the individual standing erect with the arms hanging naturally at the sides.

acromial height, sitting
The vertical distance from the upper seat surface to acromial. Measured with the individual sitting erect and his arms hanging naturally at his sides.

acromial height, standing
The vertical distance from the floor or other reference surface to the acromial. Measure with the individual standing erect with the arms hanging naturally at the sides, and his/her weight equally distributed on both feet.

acromial-radial length
The vertical distance from acromial to radial. Also referred to as *acromion-radial length.* Measured with the individual standing erect and the arms hanging naturally at the sides.

acromion
(1) The flattened, expanded bony process at the lateral end of the spine of the scapula used as an anthropometric landmark. (2) The outward end of the spine of the scapula or shoulder blade.

acroosteolysis
A condition reported in workers exposed to vinyl chloride and manifested by ulcerating lesions on the hands and feet.

acrophase
The peak value in a biological rhythm cycle.

acropodium
The most posterior fleshy point on the heel.

act
(1) *General.* Denotes external manifestation of a person's will; expression of will or purpose; carrying of an idea into action. (2) *Criminal Law.* External manifestation of one's will which is prerequisite to criminal responsibility. There can be no crime without some act, affirmative or negative. An omission or failure to act may constitute an act for purpose of criminal law. (3) *Legislation.* An alternative name for statutory law. A bill which has been enacted by legislature into law, as the *Occupational Safety and Health Act* of 1970. When introduced into the first house of the legislature, a piece of proposed legislation is known as a bill. When passed to

the next house, it may then be referred to as an act. After enactment the terms *law* and *act* may be used interchangeably. An act has the same legislative force as a *joint resolution* but is technically distinguishable, being of a different form and introduced with the words "Be it enacted" instead of "Be it resolved."

act of God
An act occasioned by an unanticipated grave natural disaster or other natural phenomenon of exceptional, inevitable, and irresistible character the effects of which could not have been prevented or avoided by the exercise of due care or foresight.

actin
A protein important in the contraction of muscles. A globular protein involved in muscle contraction. *See also* ***actomyosin***.

actinic
Pertaining to that range of ultraviolet wavelengths within the electromagnetic spectrum which is capable of causing chemical changes, generally below about 315 nm.

actinic keratoconjunctivitis
An inflammatory condition of the corneal and/or conjunctival epithelium of the eye due to exposure to intense ultraviolet lights. *See also* ***welder's flash burn***.

actinodermatitis
See ***sunburn***.

actinomycin
One of the yellow-red or red polypeptide antibiotics separated from soil bacteria.

actinomycosis
A fungous disease in animals that is sometimes communicated to man. It most often invades the jaw.

action level
(1) Pesticides. Regulatory levels recommended by the EPA for enforcement by the FDA and the USDA when pesticide residues occur in food or feed commodities for reasons other than the direct application of the pesticide. (2) Environmental. In the Superfund Program, the existence of a contaminant concentration in the environment high enough to warrant action or trigger a response under SARA and the National Oil and Hazardous Substances Contingency Plan. The term can be used similarly in other regulatory programs. (3) Safety. An exposure limit usually set at 50% of the permissible exposure limit (PEL) as specified by the applicable Occupational Safety and Health Administration (OSHA) Standard. Exposures exceeding the action level typically require implementation of certain actions, such as medical surveillance, training, and monitoring programs, but not necessarily further controls (e.g., engineering controls) aimed at reducing exposures.

action limit
Ergonomics. A NIOSH guideline for the maximum load which should be lifted manually by a healthy person under given conditions to maintain acceptable injury incidence and severity rates.

action limit ratio
The ratio of average lift weight to the calculated action limit. *See also* ***action limit***.

action potential
A rapid change in electrical potential via the exchange of ions across the cell membrane in nerve and muscle tissues due to an initial depolarization beyond the threshold potential, followed by a return to the resting potential. Also referred to as *nerve impulse* and *spike*.

action spectrum
The spectral sensitivity curve for a given type of retinal photosensitive cell.

actionable fraud
Law. Deception practiced to induce another to part with property or surrender some legal right. A false representation made with an intention to deceive.

actionable misrepresentation
Law. A false statement respecting a fact material to the contract and which is influential in procuring it.

actionable negligence
Law. The breach or nonperformance of a legal duty, through neglect or carelessness, resulting in damage or injury to another. It is failure of duty, omission of something which ought to have been done, or doing of something which ought not to have been done. Essential elements are failure to exercise due care, injury, or damage, and proximate cause.

actionable nuisance
Law. Anything wrongfully done or permitted which injures or annoys another in the enjoyment of his/her legal rights.

activated alumina
A partially dehydrated form of aluminum oxide frequently used as an adsorbent, Chemical formula is Al_2O_3.

activated biofilter
Fixed-film biological wastewater treatment process with recycle of return sludge to reactor influent.

activated carbon (charcoal)
A highly adsorbent form of carbon used to remove odors and toxic substances from liquid or gaseous emissions. In waste treatment it is used to remove dissolved organic matter from wastewater. It is also used in motor vehicle evaporative control systems.

activated sludge
A material that results when primary effluent is mixed with bacteria-laden sludge and then agitated and aerated to promote biological treatment. This speeds breakdown of organic matter in raw sewage undergoing secondary waste treatment.

activated sludge process
A biological wastewater treatment process in which a mixture of wastewater and biologically enriched sludge is mixed and aerated to facilitate aerobic decomposition by microbes.

activation
In ionizing radiation, refers to the process of making a material radioactive by bombardment with neutrons, protons, or other nuclear radiation or simply the process of inducing radioactivity by irradiation.

activation analysis
A method of chemical analysis, especially for small traces of material, based on the detection of characteristic radionuclides.

activation energy
The energy required to initiate a process or reaction.

active
Resulting from internal causes and/or purposeful effort by an entity.

active aircraft
All legally registered civil aircraft which flew one or more hours.

active ingredient
Under the Federal Insecticide, Fungicide, and Rodenticide Act (FIFRA): (1) In the case of a pesticide other than a plant regulator, defoliant, or desiccant, an ingredient which will prevent, destroy, repel, or mitigate any pest. (2) In the case of a plant regulator, an ingredient which, through physiological action, will accelerate or retard the rate of growth or rate of maturation or otherwise alter the behavior of ornamental or crop plants or the product thereof. (3) In the case of a defoliant, an ingredient which will cause the leaves or foliage to drop from a plant. (4) In the case of a desiccant, an ingredient which will produce or artificially accelerate the drying of plant tissue.

active institutional control
Under Chapter I (Environmental Protection Agency), Subchapter F (Radiation Protection Programs) of Title 40 (Protection of Environment) of the Code of Federal Regulations: Controlling access to a disposal site by any means other than passive institutional controls; performing maintenance operations or remedial actions at a site; controlling or cleaning up releases from a site; or monitoring parameters related to disposal system performance.

active isolation
The energy attenuation or conversion to another form through the use of a system requiring its own energy to operate and acting near or within another system which is generating some undesired energy output.

active life
Environmental. The period of operation of a facility that begins with initial receipt of a solid waste and ends at completion of closure activities.

active movement
The process of moving a limb or other body part by an individual under one's own control. Also known as *volitional movement.*

active negligence
Law. A term of extensive meaning embracing many occurrences that would fall short of willful wrongdoing, or of crass negligence, for example, all inadvertent acts causing injury to others, resulting from failure to exer-

cise ordinary care; likewise, all acts the effects of which are misjudged or unforeseen, through want of proper attention, or reflection, and hence the term covers the acts of willful wrongdoing and also those which are not of that character.

active portion
Any area of a facility where treatment, storage, or disposal operations continue to be conducted.

active restraint
A restraining device which has a positive locking feature and requires no action by an individual to be held in place. An example would be a seat belt system in an automobile at the time of collision.

active safety measure
Any means of implementing safety precautions which requires an individual to take some action, such as reading or comprehending. An example would be a warning sign indicating an unsafe or hazardous condition.

active sampling
An air sampling method in which air is drawn into the sampler where it is exposed to a sensor which measures the concentration of the contaminant in the sampled air, or is absorbed/adsorbed by a sorbent for later analysis. Also referred to as a *pumped sample*. *See also* ***sample draw***.

active vehicle
American Public Transit Association. Transit passenger vehicles licensed, where required, and maintained for regular use, including spares and vehicles out of service for maintenance purposes, but excluding vehicles in "dead" storage, leased to other operators, in energy contingency reserve status, permanently not usable for transit service, and new vehicles not yet outfitted for active service.

active vehicles in fleet
Federal Transit Association. The vehicles in the year-end fleet that are available to operate in revenue service, including vehicles temporarily out of service for routine maintenance and minor repairs.

active window
That view on a display with which the user is currently interacting.

activities of daily living
Those functions normally performed on a daily or near daily basis that are involved in sustenance of the individual (e.g., eating, grooming, dressing, bathing, urination, defecation, etc.). *See also* ***daily living tasks***.

activity
In ionizing radiation, the rate of decay of radioactive material expressed as the number of nuclear disintegrations per second; the number of nuclear transformations occurring in a given quantity of material per unit time. The units of activity are the curie (Ci) and the becquerel (Bq). *See also* ***curie and becquerel***.

activity analysis
A study of the following set and any interrelationships within the set: a) involved individuals, b) the environment, c) the facilities or equipment present or available, and d) the actions required to perform the particular activity under study.

activity sampling
A sampling technique using many instantaneous observations of equipment or workers involved in an ongoing process to rate them as either functioning or non-functioning on each sample. Used for estimating the amount of time a machine or worker spends performing some function. Also referred to as *work sampling, snap reading technique, snap reading method,* and *random observation method*. *See also* ***work measurement***.

actomyosin
A combination of actin and myosin which is involved in muscle contraction.

actual 1985 emission rate
Under the Clean Air Act (CAA): For electric utility units it means the annual sulfur dioxide or nitrogen oxides emission rate in pounds per million Btu as reported in the NAPAP Emissions Inventory, Version 2, National Utility Reference File. For non-utility units, the term "actual 1985 emission rate" means the annual sulfur dioxide or nitrogen oxides emission rate in pounds per million Btu as reported in the NAPAP Emission Inventory, Version 2. *See* ***utility unit*** and ***non-utility unit***.

actual calculated landing time (ACLT)
A flight's frozen calculated landing time. An actual time determined at freeze calculated landing time (FCLT) or meter list display in-

terval (MLDI) for the adapted vertex for each arrival aircraft based upon runway configuration, airport acceptance rate, airport arrival delay period, and other metered arrival aircraft. This time is either the vertex time of arrival (VTA) of the aircraft or the tentative calculated landing time (TCLT)/actual calculated landing time (ACLT) of the previous aircraft plus the arrival aircraft interval (AAI), whichever is later. This time will not be updated in response to the aircraft's progress.

actual coverage
See *standard coverage.*

actual damages
Real, substantial, and just damages, or the amount awarded to a complainant in compensation for his/her actual and real loss or injury, as opposed on the one hand to *nominal* damages and on the other to *exemplary* or *punitive* damages. Synonymous with *compensatory damages* and *general damages. See also damages.*

actual service
Federal Transit Association. Total service operated during each time period. Actual service excludes missed trips and service interruptions (such as strikes, emergency shutdowns), but also includes deadheading. Actual service is measured by vehicles in service, in miles and/or hours.

actual severity
United States Coast Guard. On scene evaluation of the degree of danger that existed. An "after-the-fact" evaluation by the reporting unit.

actual time
See observed time.

actual vapor pressure
See vapor pressure.

actual vehicle miles/hours
The miles/hours a vehicle travels while in revenue service (actual vehicle revenue miles/hours) plus deadhead miles/hours. For rail vehicles, vehicle miles/hours refer to passenger car miles/hours. Actual vehicle miles/hours exclude miles and hours for charter services, school bus service, operator training, and maintenance testing.

actuation force
That force required to overcome static friction and begin the movement of a control or other mechanical device.

actuation torque
That force directed at some distance from the center of rotation which is required to overcome static friction and begin rotation of a control or other mechanical device.

acuity
Of or pertaining to the sensitivity of receptors used in hearing and vision; acuteness, or sharpness of the senses. *See also visual acuity, vernier acuity, Snellen acuity, stereoscopic acuity,* and *resolution acuity.*

acupuncture
An ancient Chinese technique of puncturing certain points in the body with long thin needles to treat painful conditions and to produce local anesthesia.

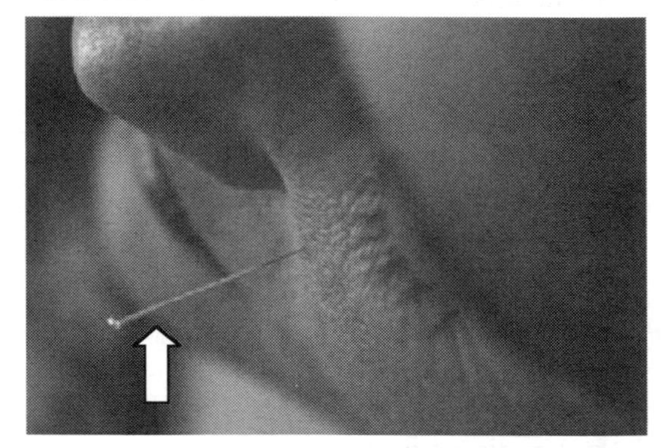

Acupuncture needle (at arrow) applied just under patient's eye

acute
Having a sudden onset and reaching a crisis rapidly. Effects are observed in a short period of time following exposure to an acute toxicant.

acute dermal LD_{50}
The single dermal dose of a substance, expressed as milligrams per kilogram of body weight, that is lethal to 50% of the test population of animals under specified test conditions.

acute effect
An effect that results following a brief exposure to a chemical, biological, or physical agent. For example, severe skin irritation and even corrosive damage can occur after brief exposure to acids or bases, depending upon the strength of the chemical, the duration of contact, and the size of the exposure area.

acute exposure
(1) *Chemical.* A sudden, short, rapid association with a chemical compound. (2) *Radia*

tion. Exposure of short duration, generally taken to be the total dose absorbed within 24 hours. (3) *Biologic.* A brief encounter with a pathogenic or nonpathogenic microorganism.

acute LC_{50}
A concentration of a substance, expressed as parts per million parts of medium (e.g., air, water, etc.), that is lethal to 50% of the test population of animals under specified test conditions.

acute oral LD_{50}
A single orally administered dose of a substance, expressed as milligrams per kilogram of body weight, that is lethal to 50% of the test population of animals under specified test conditions.

acute oxygen toxicity
A central nervous system disorder due to breathing pure oxygen at higher than normal pressures for several minutes to a few hours, depending on the pressure, and characterized by a range of symptoms from muscle twitching to convulsions. *See also* ***chronic oxygen toxicity***.

acute radiation effects
Any of one or more types of illnesses or other bodily disorders that follow exposure to relatively high doses of ionizing radiation resulting in deaths of significant numbers of cells.

acute radiation syndrome
A medical term for radiation sickness. *See also* ***acute radiation effects.***

acute reaction
A sudden physiologic response as a result of an exposure to a hazard (i.e., chemical, physical, biological, ergonomic, etc.).

acute toxicity
The ability of a substance to cause poisonous effects resulting in severe biological harm or death soon after a single exposure or dose. Also, any severe poisonous effect resulting from a single short-term exposure to a toxic substance.

ACV
See ***aquatic chronic value***.

adactylia
Absence of fingers or toes, or both, from birth.

Adam's apple
The prominence of the thyroid cartilage on the fore part of throat, predominantly in men.

adaptation
(1) A change in an organism's structure or habit that helps it to adjust to its surroundings. (2) A self-generated adjustment in a system in response to changes in the environment as an attempt to maintain functionality. Also referred to as *adaptive response. See also* ***sensory adaptation*** and ***perceptual adaptation.***

adaptive control
A form of automated control equipped with a self-contained decision-making capability for modifying its own operation based on previous experience.

adaptive equipment
Any type of equipment which enables a disabled or other individual to operate a machine or system.

adaptive response
See ***adaptation***.

adaptometer
An instrument designed to determine the degree of retinal adaptation or the time course over which adaptation occurs by measuring changes in an observer's threshold for light detection.

add on control device
An air pollution control device such as a carbon absorber or incinerator which reduces the pollution in an exhaust gas. The control device usually does not affect the process being controlled and thus is "add on" technology, as opposed to a scheme to control pollution through making some alteration to the basic process.

addict
One who is addicted to a practice or a habit, especially to narcotics.

addiction
The state of being addicted; habitual, compulsive use of narcotics.

addictive
Causing addiction; one of a class of drugs that are habit-forming in nature.

Addison's disease
A disease characterized by asthenia, digestive disturbances, and usually a brownish colora-

tion of the skin caused by disturbance of function of the adrenal glands.

additional capital invested
The difference between the price at which capital stock is sold and the par or stated value of the stock, gains or losses arising from the reacquisition and the resale or retirement of each class and series of capital stock, donations, the excess of retained earnings capitalized over par or stated value of capital stock issued, adjustments in capital resulting from reorganization or recapitalization, proceeds attributable to detachable stock purchase warrants related to debt issues, and contributions to the business enterprise by individual proprietors or partners.

additional insured
A person, other than the named insured, such as the insured person's spouse, who is protected under the terms of the contract. *See also* ***insurance***.

additional services
Advisory information provided by Air Traffic Control (ATC) which includes but is not limited to the following: 1) Traffic advisories. 2) Vectors, when requested by the pilot, to assist aircraft receiving traffic advisories to avoid observed traffic. 3) Altitude deviation information of 300 feet or more from an assigned altitude as observed on a verified (reading correctly) automatic altitude readout (Mode C). 4) Advisories that traffic is no longer a factor. 5) Weather and chaff information. 6) Weather assistance. 7) Bird activity information. 8) Holding pattern surveillance. Additional services are provided to the extent possible contingent only upon the controller's capability to fit them into the performance of higher priority duties and on the basis of limitations of the radar, volume of traffic, frequency congestion, and controller workload. The controller has complete discretion for determining if he is able to provide or continue to provide a service in a particular case. The controller's reason not to provide or continue to provide a service in a particular case is not subject to question by the pilot and need not be made known to him/her.

additional vehicle
Government Services Administration. A vehicle added to the inventory of a Fleet Management Center to fill a new program or to expand on an existing program of a participating agency.

additional work allowance
See ***excess work allowance***.

additive
A substance added in small amounts to another for improvement, as a drug added to a medicine.

additive color mixing
The addition of colored lights to an already illuminated surface or region, resulting in a change of apparent color. Also known as *light mixing.*

additivity
Interaction of a mixture of substances in which exposure results in a response equal to the sum of the responses expected from each component of the mixture.

adduct
(1) *Chemistry.* An un-bonded association of two molecules, in which a molecule of one component is either wholly or partly locked within the crystal lattice of the other. (2) *Physiology.* To draw (one's limb) toward the body's main axis.

adduction
The action by which a part of the body is drawn toward the body's axis.

adductor
In physiology, the muscle that draws toward the mesial line of the body.

adenalgia
A glandular pain.

adenitis
Inflammation of gland or lymph nodes.

adenocarcinoma
A malignant tumor that appears in glandular epithelium.

adenofibroma
A benign tumor of connective tissue frequently found in the uterus.

adenoid
An enlarged mass of lymphoid tissue in the upper pharynx that hinders nasal breathing.

adenoma
A benign epithelial tumor in which the cells form recognizable glandular structures or in

which the cells are clearly derived from glandular epithelium.

adenomatosis
The condition of multiple glandular enlargement.

adenosine (ATP)
A crystalline nucleoside, derived from the nucleic acid of yeast, which upon undergoing hydrolysis yields adenine and ribose.

ADF
*See **automatic direction finder**.*

adhesion
(1) *General.* Molecular attraction that holds surfaces of two substances in contact. (2) *Physiology.* A growth of scar tissue resulting from an incision; the abnormal union of adjacent tissues resulting from inflammation.

adhesive dirt
Any form of dirt which tends to remain attached to a surface through an inherent stickiness. *See also* ***dirt***.

adiabatic
Refers to a reaction that occurs without a gain or loss (or transfer) of heat between the system (such as an air parcel) and its surroundings. In an adiabatic process, compression always results in warming and expansion results in cooling.

adiabatic lapse rate
The constant rate at which temperatures decrease as altitude increases. In a dry atmosphere the dry adiabatic lapse rate (DALR) is approximately -1.00°C per 100 meter rise.

adipose tissue
That tissue composed primarily of fat cells with connective tissue for support.

ADIZ
*See **Air Defense Identification Zone**.*

adjudicate
To settle in the exercise of judicial authority.

adjudicated rights
Rights which have been recognized in a judicial or administrative proceeding.

adjudication
To hear and decide a case. Refers to the judge's decision. The legal right to resolve a dispute.

adjudicative facts
Factual matters concerning the parties to an administrative proceeding as contrasted with legislative facts which are general and usually do not touch individual questions of particular parties to a proceeding.

adjudicatory proceedings
A legal proceeding in a government agency wherein the rights and duties of specifically named parties are decided by applying law and policy to facts. Also known as a Trial Type proceeding. This usually carries with it a right to appeal to a higher level within the agency or court.

adjustable speed drives
Drives that save energy by ensuring the motor's speed is properly matched to the load placed on the motor. Terms used to describe this category include polyphase motors, motor oversizing, and motor rewinding.

adjutant general
An officer in charge of the National Guard of one of the States. The administrative head of a military unit having a general staff.

Adler's theory
An approach to psychology based on the hypothesis that behavior is governed by an effort to compensate for inferiority or deficiency.

administrative agency
A governmental body charged with administering and implementing particular legislation, such as the Occupational Safety and Health Administration (OSHA) that is charged with the administration of the Occupational Safety and Health Act.

administrative authority
The power of an agency or its head (the administrator) to carry out the terms of law creating that agency as well as to make regulations for the conduct of business before that agency; distinguishable from legislative authority to make laws.

administrative class V road
An administrative access road which consists of all public roads intended for access to administrative developments or structures such as offices, employee quarters, or utility areas.

administrative class VI road
A restricted road normally closed to the public, including patrol roads, truck trails, and other similar roads.

administrative control
A measure initiated to reduce worker exposure to various stresses in the work environment. An example is limiting the amount of time an employee can work around health hazards.

administrative law
Refers to that body of law that governs the methods by which administrative agencies make and implement decisions. Federal administrative law is based primarily on specific provisions of the U.S. Constitution, as well as various other federally mandated statutes. It is within this regulatory framework of administrative law that the basis for occupational safety and health legislation obtains the force of law.

administrative law judge
One who presides at an administrative hearing, with power to administer oaths, take testimony, rule on questions of evidence, regulate course of proceedings, and make agency determinations of fact.

administrative office of the United States courts
Created by the Administrative Office Act of 1939, it is responsible for administration of the federal court system, as a whole, including collection of statistics on court business, supervision of administrative personnel in the courts, and conducting of financial and management audits of the courts.

administrative officer
Politically, and as used in constitutional law, an officer of the executive department of government, and generally one of inferior rank; legally, a ministerial or executive officer, as distinguished from a *judicial officer.*

administrative order
The final disposition of a matter before an administrative agency, such as the Environmental Protection Agency (EPA), directing an individual, business, or other entity to take corrective action or refrain from an activity. It describes the violations and actions to be taken, and can be enforced in court. Such orders may be issued, for example, as a result of an administrative complaint whereby the respondent is ordered to pay a penalty for violations of a statute.

administrative procedure
Methods and processes before administrative agencies as distinguished from judicial procedure which applies to courts. Procedural rules and regulations of most federal agencies are set forth in the Code of Federal Regulations. *See also* ***Administrative Procedure Act.***

Administrative Procedures Act
A federal law enacted in 1946 (60 Stat. 237, 5 U.S.C.A) that provides procedures and requirements related to the promulgation of regulations by federal agencies, such as the EPA and OSHA. Individual states have also enacted variations of the federal act.

administrative process
In general, the procedure used before administrative agencies; in particular, the means of summoning witnesses before such agencies (e.g., the subpoena).

administrative record
The compilation of documents and exhibits by an agency to support or explain a decision. Often used to describe the written basis of a regulatory decision by EPA or a state environmental agency to promulgate a regulation or issue or deny a permit or license.

administrative review
Generally refers to judicial review of administrative proceedings; may also embrace appellate review within the administrative agency itself.

administrative road
Consists of all public and non-public roads intended to be used principally for administrative purposes. It includes roads servicing employee residential areas, maintenance areas and other administrative developments, as well as restricted patrol roads, truck trails, and similar service roads.

administrative rule
A regulation which implements a law. Since the implementation of a law often requires technical experience and familiarity with the area being regulated, the legislature often grants a state agency the authority to write rules. Before these rules are enforceable, the Administrative Procedures Act (1969, PA 306, as amended) requires that public hearings be conducted on the rules and approval is granted by the Legislative Service Bureau, the Attorney General, the Joint Committee on

Administrative Rules, and the agency which prepared the rules. The rules are then filed with the Secretary of State, after which they become enforceable.

administrative rule-making
Power of an administrative agency to make rules and regulations for proceedings before it.

administrative tribunal
A particular administrative agency before which a matter may be heard or tried as distinguished from a judicial forum.

admission
A granting of truth or a conceding of truth from the courts.

admixture
(1) A material or substance added in mixing. (2) A substance other than cement, aggregate, or water that is mixed with concrete.

ad nauseam
To the point of being sickening.

adolescent growth spurt
A phase in the maturing individual near puberty at which peak height and weight velocities occur, along with changes in body composition.

adrenal gland
In mammals, a gland adjacent to the kidney that produces the hormone adrenaline, or epinephrine. The hormone influences the heartbeat rate, dilates blood vessels, increases blood sugar, and plays a major role in other physiological activities of mammals.

adrenaline
*See **epinephrine**.*

adsorb
The condensation of a gas, liquid, or dissolved substance on the surface of a solid.

adsorbate
A material adsorbed on the surface of another.

adsorption
(1) Condensation of gases, liquids. or dissolved substances on the surface of solids. (2) The attraction and retention of atoms, molecules, or ions on the surface of a solid.

adult chorea
*See **Huntington's chorea**.*

adulterants
Chemical impurities or substances that by law do not belong in a food or in a pesticide.

adulterated pesticide
Applies to any pesticide if its strength or purity falls below the professed standard of quality as expressed on the labeling under which it is sold; any substance has been substituted wholly or in part for the pesticide; or any valuable constituent of the pesticide has been wholly or in part abstracted.

ad valorem
(1) A charge levied on persons or organizations based on the value of a transaction. It is normally a given percentage of the price at the retail or manufacturing stage and is a common form of sales tax, e.g., federal excise tax on new trucks and trailers. (2) A freight rate set at a certain percentage of the value of an article. 3) A set percentage of the value of taxable goods determined by the price at the port of shipment and calculated for duty assessment.

Advanced Dynamic Anthropometric Mannequin (ADAM)
An anthropometric dummy developed by the United States Air Force to represent human anthropometry and vertical dynamic response for ejection seat testing.

advanced oxidation processes (AOPs)
Processes using a combination of disinfectants, such as ozone and hydrogen peroxide, to mineralize toxic organic compounds to nontoxic form.

Advanced Notice of Proposed Rule-making (ANPRM)
A notice appearing in the Federal Register indicating the intention of a government agency to develop a regulation on the issue indicated in the notification.

advanced secondary treatment (AST)
Secondary wastewater treatment with enhanced solids separation.

advanced wastewater treatment (AWT)
Any treatment of sewage that goes beyond the secondary or biological water treatment stage and includes the removal of nutrients such as phosphorus or nitrogen and a high percentage of suspended solids.

advection
Movement caused by the motion of heat, air, water, or another fluid. It specifically refers to the horizontal movement by wind currents of chemical pollutants or heat.

advection fog
Occurs when warm, moist air moves over a cold surface and the air cools to below its dew point.

advection frost
*See **freeze**.*

adversary system
The jurisprudential network of laws, rules, and procedures characterized by opposing parties who contend against each other for a result favorable to themselves. In such system, the judge acts as an independent magistrate rather than a prosecutor; distinguished from inquisitorial system.

adverse effect
An effect which results in an impairment of the functioning of an organism or which reduces an organism's ability to respond to insult.

adverse environmental effect
Any significant and widespread adverse effect, which may reasonably be anticipated, to wildlife, aquatic life, or other natural resources, including adverse impacts on populations of endangered or threatened species or significant degradation of environmental quality over broad areas.

adverse weather
The weather conditions considered by the operator in identifying the response systems and equipment to be deployed in accordance with a response plan, including wave height, ice, temperature, visibility, and currents within the inland or Coastal Response Zone (defined in the National Contingency Plan (40 CFR 300)) in which those systems or equipment are intended to function.

adverse witness
A witness who gives evidence on a material that is prejudicial or unfavorable to the party that original called the witness.

advice of counsel
A defense used in actions for malicious prosecution which requires a finding that the defendant presented all facts to his/her counsel and the he/she honestly followed counsel's advice.

advise intentions
A spoken term, usually by an air traffic controller to a pilot or, sometimes, vice versa, meaning: "Tell me what you plan to do."

advising bank
A domestic bank which handles letters of credit for a foreign bank by notifying the exporter that the credit has been opened in his favor, fully informing him/her of the conditions and terms without responsibility on the part of the bank.

advisory
(1) A non-regulatory document that communicates risk information to persons who may have to make risk management decisions. (2) Counseling, suggesting, or advising, but not imperative or conclusive.

advisory committees
Groups of experts used by regulatory agencies such as OSHA to study and advise on certain regulatory issues. The committees consist of persons from outside the agency who may have certain expertise in a given area. These committees do not supersede the agency's regulatory powers or responsibilities. They only provide input on specific technical and/or policy issues that arise in the course of agency activities.

advisory counsel
Attorney retained to give advice as contrasted with trial counsel.

advisory frequency
Federal Aviation Administration. The appropriate frequency to be used for Airport Advisory Service.

advisory light
A visual indicator which provides information on the operation of essential equipment. *See also **advisory signal**.*

advisory opinion
Such may be rendered by a court at the request of the government or an interested party indicating how the court would rule on a matter should adversary litigation develop. An advisory opinion is thus an interpretation of the law without binding effect.

advisory service
Federal Aviation Administration. Advice and information provided by a facility to assist

pilots in the safe conduct of flight and aircraft movement.

advisory signal
Any type of signal which indicates the condition of equipment or operations. *See also* ***advisory light***.

advocacy
The act of pleading for, supporting, or recommending active espousal.

advocate
One who assists, defends, or pleads for another. One who renders legal advice and aid and pleads the cause of another before a court or a tribunal; a counselor. A person learned in the law, and duly admitted to practice, who assists his/her client with advice, and pleads for him/her in open court.

AEC
U.S. Atomic Energy Commission (the former name of the present Nuclear Regulatory Commission).

aeolian deposit
Soil deposited by the wind.

aeration
A process which promotes biological degradation of organic water. The process may be passive (as when waste is exposed to air) or active (as when a mixing or bubbling device introduces the air).

aeration tank
A chamber used to inject air into water.

aerator
A device used to introduce air or oxygen into water or wastewater.

aeremia
The presence of air in the blood.

aerial application
Federal Aviation Administration. Any use of an aircraft for work purposes which concerns the production of foods, fibers, and health control in which the aircraft is used in lieu of farm implements or ground vehicles for the particular task accomplished. This includes fire fighting operations, the distribution of chemicals or seeds in agriculture, reforestation, or insect control.

aerial application flying
National Transportation Safety Board. The operation of aircraft for the purpose of dispensing any substance for plant nourishment, soil treatment, propagation of plant life, pest control, or fire control, including flying to and from the application site.

aerial observation
Federal Aviation Administration. Any use of an aircraft for aerial mapping and photography, survey, patrol, fish spotting, search and rescue, hunting, highway traffic advisory, or sightseeing; not included under FAR Part 135.

aerial refueling
Federal Aviation Administration. A procedure used by the military to transfer fuel from one aircraft to another during flight.

aerial tramway
(1) *American Public Transit Association.* An electric system of aerial cables with suspended unpowered passenger vehicles. The vehicles are propelled by separate cables attached to the vehicle suspension system and powered by engines or motors at a central location not on board the vehicle. (2) *Federal Transit Association.* Unpowered passenger vehicles suspended from a system of aerial cables and propelled by separate cables attached to the vehicle suspension system. The cable system is powered by engines or motors at a central location not on board the vehicle.

aeroallergen
Airborne material, such as particulates, pollen, dusts, and dander, that may precipitate an allergic response in susceptible persons.

aerobe
A microorganism whose existence requires the presence of air or free oxygen, as opposed to anaerobe.

aerobic
Describes an environment with molecular oxygen present; organisms that live or grow in the presence of molecular oxygen; reactions that occur in the presence of molecular oxygen.

aerobic bacteria
Bacteria that require free oxygen to sustain their life processes.

aerobic capacity
See ***maximal aerobic capacity.***

aerobic digestion
Sludge stabilization process in which aerobic biological reactions destroy biologically degraded organic components of sludge.

aerobic endurance capacity
*See **maximal aerobic capacity.***

aerobic energy
That energy which can be derived from foodstuffs by aerobic metabolism.

aerobic metabolism
The normally complete physiological oxidation of glucose or other bodily fuels in the presence of adequate oxygen to water and carbon dioxide.

aerobic treatment
Process by which microbes decompose complex organic compounds in the presence of oxygen and use the liberated energy for reproduction and growth. Types of aerobic processes include extended aeration, trickling filtration, and rotating biological contactors.

aerobic work capacity
*See **maximal aerobic capacity**.*

aerodrome
A defined area on land or water (including any buildings, installations, and equipment) intended to be used either wholly or in part for the arrival, departure, and movement of aircraft. Aerodromes may include airports, heliports, and other landing areas. *See also **airport** and **heliport**.*

aerodrome beacon
Aeronautical beacon used to indicate the location of an aerodrome from the air.

aerodrome control tower
A unit established to provide air traffic control service to aerodrome traffic.

aerodrome elevation
The elevation of the highest point of the landing area.

aerodrome traffic circuit
The specified path to be flown by aircraft operating in the vicinity of an aerodrome.

aerodynamic coefficient
Non-dimensional coefficients for aerodynamic forces and moments.

aerodynamic diameter
The diameter of a unit density sphere having the same settling velocity as the particle in question of whatever shape and density. It is also referred to as equivalent diameter.

aeromedicine
That branch of medicine concerned with disorders that result from or occur during flying.

aeronautical beacon
A visual Navigation Aid (NAVAID) displaying flashes of white and/or colored light to indicate the location of an airport, a heliport, a landmark, a certain point of a federal airway in mountainous terrain, or an obstruction.

aeronautical chart
A map used in air navigation containing all or part of the following topographic features: hazards and obstructions, navigation aids, navigation routes, designated airspace, and airports.

Aeronautical Information Manual (AIM)
A primary Federal Aviation Administration (FAA) publication whose purpose is to instruct airmen about operating in the National Airspace System of the U.S. It provides basic flight information, Air Traffic Control (ATC) procedures, and general instructional information concerning health, medical facts, factors affecting flight safety, accident and hazard reporting, and types of aeronautical charts and their use. Also referred to as *Airmen's Information Manual.*

aeronautics
Science that focuses on the operation of aircraft; also the art or science of operating aircraft.

aeroneurosis
A psychoneurotic condition occurring in airmen and aviators resulting from nervous tension, worry, or fatigue, characterized by mild depression, abdominal pain, insomnia, and nervous irritability.

aerosol
(1) A dispersion of solid or liquid particles of microscopic size in a gaseous medium. Smokes, fogs, fibers, dusts, and mists are examples of common aerosols. (2) Atomized particles ejected into the air from a pressurized can. (3) A solution of bactericidal substances that are atomized to sterilize the air of a room.

aerosol photometer
An instrument used for detecting aerosols (i.e., dusts, mists, fumes, etc.) by exposing them to a source of illumination, typically a beam of light, as they are drawn through an

enclosed volume and measuring the scattered light created by the aerosol as it passes through the light beam.

aerospace
Pertaining to equipment, vehicles, or activities in either or both a planetary atmosphere and space.

aerovane
A wind instrument that indicates or records both wind speed and wind direction.

Aesculapian
(1) Of or pertaining to Aesculapius, the Greco-Roman god of healing. (2) Referring to the art of healing.

aesthenic
See ***asthenic***.

AFARMRL Anthropometric Data Bank
A computerized database of several anthropometric surveys, consisting of both American and foreign subjects.

AFC
See ***automatic fare collection system***.

affect
To act upon; influence; change; enlarge or abridge; often used in the sense of acting injuriously upon persons and things.

afferent
Conducting a signal, information, or a substance toward a central point, usually referring to neural structures.

afferent nerve
A nerve which conducts sensory information from a receptor toward the spinal cord and/or the brain. Also referred to as a *sensory nerve*.

AFFF
See ***aqueous film forming foam***.

affirmative action program
Employment programs required by federal statutes and regulations designed to remedy discriminatory practices in hiring minority group members. Factors considered are race, color, sex, creed, and age.

affirmative defense
(1) *General Law*. In pleading, matter asserted by defendant which, assuming the complaint to be true, constitutes a defense to it. A response to a plaintiff's legal right to bring an action, as opposed to attacking the truth of claim. (2) *OSHA*. A category of defending against an OSHA citation which basically holds that the employer is not specifically arguing the fact that the cited condition(s) existed. The defense is really to the contrary. By not disputing the cite itself, the employer actually affirms the allegation of noncompliance but offers substantial proof to justify reasons for not complying with the cited standard(s).

affricate
The type of sound produced on complete closure of the vocal tract followed by a constriction.

AFO
Air fail open.

AFPA
American Forest and Paper Association.

afterburner
In incinerator technology, a burner located so that the combustion gases are made to pass through its flame in order to remove smoke and odors. It may be attached to or be separated from the incinerator proper.

aftercondenser
A condenser installed as the last stage of an evaporator venting system to minimize atmospheric steam discharge.

after-flame
The time a test specimen continues to flame after the flame source has been removed.

afterimage
An aftersensation in the visual system. *See also* ***aftersensation, positive afterimage*** *and* ***negative afterimage.***

aftersensation
A sensory impression in any modality which persists after cessation of the causing stimulus, but which may have different characteristics from the original stimulus.

aftershock
Any earthquake which occurs after a larger earthquake (a *mainshock*) within one rupture-length of the original fault rupture before the seismicity rate in that area has returned to the background (pre-mainshock) level is generally considered an aftershock. For some earthquakes, a specific *aftershock zone* may

be defined, in lieu of the one-rupture-length given above.

afterslip
A seismic slip, very similar to *creep,* that occurs along a fault ruptured by a large earthquake in the months following that event. *See also* ***creep***.

after the fact
Subsequent to an event from which time is reckoned, e.g., accessory after the fact is one who harbors, conceals, or aids in the concealment of the principal felon after the felony has been committed.

against public interest
An agreement or act which is or has been declared to be adverse to the general good or public welfare such that a judge may, on his/her own, declare it void.

agar
A gelatinous substance extracted from a red algae, commonly used as a medium for laboratory cultivation of bacteria.

agar plate
A circular glass plate, containing a nutrient, used to culture microorganisms.

agency
(1) A relationship between two persons, by agreement or otherwise, where one (the agent) may act on behalf of the other (the principal) and bind the principal by words and actions. (2) The relation created by express or implied contract or law whereby one party delegates the transaction of some lawful business with more or less discretionary power to another who undertakes to manage the affair and render an account thereof. (3) The location or place at which business of a company, organization, or individual is transacted by an agent. (4) A department, division, or administration within the federal government (e.g., OSHA, EPA)

agent
(1). *Science.* A biological, physical, or chemical entity capable of causing disease. (2) *Law.* A person authorized by another (the principal) to act for or in place of him/her. For example, an individual who, while not an employee, is authorized to act on behalf of the organization in activities or practices which, if abused, can expose the organization to criminal liability.

agent orange
A toxic herbicide and defoliant which was used in the Vietnam conflict. It contains 2,4,5-T) trichlorophenoxyacetic acid (2,4,5-T) and 2-4 dichlorophenoxyacetic acid (2,4-D) with trace amounts of dioxin.

agent-specific disease
A disease known to be caused only by one factor.

ageusia
An impairment in or loss of the sense of taste; taste blindness. Also referred to as *ageustia*.

AGGIE
A general purpose, 3-D nonlinear finite element structures numerical modeling program.

agglomeration
The process by which precipitation particles grow larger by collision or contact with cloud particles or other precipitation particles.

agglutination
The process of uniting solid particles coated with a thin layer of adhesive material or of arresting solid particles by impact on a surface coated with an adhesive.

aggravating factors
An increase in the amount of damages or penalties. Under the Federal Sentencing Guidelines (FSGs) for Organizations, factors considered in sentencing for an organization to receive increased penalties are (1) management involvement in or tolerance of criminal activities; (2) the organization's prior criminal, civil, and administrative history; (3) violations of a court order or probation; and (4) obstruction of justice. *See also* ***FSGs***.

aggregate
(1) *General.* The entire number, sum, mass, or quantity of something; total amount; complete whole. (2) *Ergonomics.* The combination of the tool being manipulated or mass being lifted/carried with those primary body parts affected or used in the operation.

aggregate ratio
See ***estimate ratio, mean and ratio estimate***.

aggressive sampling
A sampling procedure employed following asbestos removal activities to demonstrate

that the area is not contaminated with materials which contain asbestos fibers. It typically involves stirring up the air in the abated area to produce worst-case conditions, collecting air samples during this procedure, and analyzing the samples to determine the airborne level of asbestos fibers as structures per cubic centimeter.

aggrieved party
One whose legal right is invaded by an act complained of, or whose pecuniary interest is directly and adversely affected by a decree or judgment.

agitator body
Truck body designed and equipped to mix concrete in transit.

agnosia
Inability to comprehend a sensory perception, although the sensory sphere is intact. Agnosia results from disorders of the brain or nervous system.

agonist
See ***prime mover***.

agranulocyte
A leukocyte that does not have cytoplasmic granules.

agranulocytosis
A serious, destructive blood disease distinguished by a decrease of the leukocytes.

agravic illusion
See ***oculoagravic illusion***.

Agreement State
In ionizing radiation, a State that has signed an agreement with the U.S. Nuclear Regulatory Commission allowing the State to regulate certain activities for the use of radioactive materials not normally regulated by the State.

agrichemical
Any inorganic, artificial, or manufactured chemical substances used in agricultural processes, usually as fertilizers, herbicides, and pesticides.

agricultural
Pertaining to, or dealing with, agriculture; also, characterized by or engaged in farming as the leading pursuit.

agricultural commodity trailer
A trailer that is designed to transport bulk agricultural commodities from off-road harvesting sites to a processing plant or storage location, as evidenced by skeletal construction that accommodates harvest containers, a maximum length of 28 feet, and an arrangement of air control lines and reservoirs that minimizes damage in field operations.

agricultural labor
Services performed on a farm, for or on behalf of the owner or tenant.

agricultural pollution
The liquid and solid wastes from farming, including runoff and leaching of pesticides and fertilizers; erosion and dust from plowing; animal manure and carcasses; crop residues; and debris.

agricultural solid waste
Solid waste generated by the rearing of animals, and the producing and harvesting of crops or trees.

agriculture
The science or art of cultivating the soil, harvesting crops, and raising livestock; also as the science or art of the production of plants and animals useful to man and in varying degrees the preparation of such products for man's use and their disposal.

agronomy
Branch of agriculture that deals with the raising of crops and the care of the soil.

A-h
Ampere-hour.

AHERA
Asbestos Hazards and Emergency Response Act.

A-horizon
Topsoil, or the uppermost layer of soil containing the highest accumulation of mineral and organic matter.

AHP
Air horsepower.

AHS
Air handling system; aquatic humic substances.

AHU
See ***air-handling unit***.

AIA
Asbestos Information Association.

AICHE
American Institute of Chemical Engineers.

aid
To support, help, assist, or strengthen. To act in cooperation with; supplement the efforts of others.

aid and abet
To help, assist, or facilitate the commission of a crime, promote the accomplishment thereof, help in advancing or bringing it about, or encourage, counsel, or incite as to its commission. It includes all assistance rendered by words, acts, encouragement, support, or presence. actual or constructive, to render assistance if necessary.

AIDS
See ***acquired immune deficiency syndrome.***

AIG
*See **Airbus Industries Group**.*

AIHA
*See **American Industrial Hygiene Association**.*

AIHA accredited laboratory
A certification given by the AIHA to an analytical laboratory that has met specific requirements and successfully participated in the "proficiency Analytical Testing" program for quality control as established by the National Institute for Occupational Safety and Health.

AIHC
(1) American Industrial Health Council. (2) American Industrial Health Conference.

AIM
*See **Aeronautical Information Manual**.*

AIP
*See **airport improvement program**.*

AIPE
American Institute of Plant Engineers.

air
The mixture of gases that surrounds the earth. The major constituents of air are nitrogen (78.08%), oxygen (20.95%), argon (0.93%), and carbon dioxide (0.03%).

air-blower noise
*See **air-handler noise**.*

air bone gap
The decibel difference in the hearing ability level at a particular frequency as determined by air conduction and bone audiometric testing.

air-bound
Obstruction of water flow in a pipeline or pump due to the entrapment of air.

air brake
A brake in which the mechanism is actuated by manipulation of air pressure. The term is often used to describe brakes that employ air under pressure above atmospheric, in contrast to vacuum brakes, which employ pressure below atmospheric.

air cargo
Cargo is freight and mail, loaded and unloaded (arriving and departing) at the airport. It is the total volume of freight, mail, and express traffic transported by air and includes the following: Freight and Express commodities of all kinds, including small package counter services, express services, and priority reserved freight; U.S. Mail in all classes of mail transported for the U.S. Postal Service. *Freight* is any property carried on an aircraft other than mail, stores, and baggage. *Mail* comprises closed bags handed over by the postal service, whatever their contents may be and does not include passenger baggage and trucked freight.

air carrier
(1) A person who undertakes directly by lease, or other arrangement, to engage in air transportation. (2) Commercial system of air transportation, consisting of domestic and international scheduled and charter service. (3) Commercial system of air transportation consisting of certificated air carriers, air taxis (including commuters), supplemental air carriers, commercial operators of large aircraft, and air travel clubs. (4) As defined in the Federal Aviation Act of 1958, any citizen of the United States who undertakes, whether directly or indirectly, or by lease or any other arrangement, to engage in air transportation. (5) An air carrier holding a Certificate of Public Convenience and Necessity issued by the Department of Transportation to conduct scheduled services over specified routes and a limited amount of nonscheduled operations. (6) Any air operator operating under Federal Aviation Regulation (FAR) Parts 121, 127, or 135.

Air Carrier Activity Information System (ACAIS)
Database of revenue passenger enplanement and all-cargo landing data used in the distribution of AIP (Airport Improvement Program) entitlement funds.

air carrier operations
Arrivals and departures of air carriers certificated in accordance with Federal Aviation Regulations (FAR) Parts 121 and 127.

air cleaner
A device designed to remove airborne contaminants such as dusts, fumes, vapors, gases, etc. from the air.

air cleaning
An Indoor Air Quality (IAQ) control strategy to remove various airborne particulates and/or gases from the air. The three types of air cleaning most commonly used are particulate filtration, electrostatic precipitation, and gas sorption.

air commerce
Interstate, overseas, or foreign air commerce or the transportation of mail by aircraft or any operation or navigation of aircraft within the limits of any federal airway or any operation or navigation of aircraft which directly affects, or which may endanger safety in, interstate, overseas, or foreign air commerce.

air conditioning
A process of treating air to control factors such as temperature, humidity, and cleanliness, and to distribute the air throughout a space to meet the requirements of personal protective equipment.

air conduction
With regard to acoustics, the process by which sound is conducted through the air to the inner ear, with the outer ear canal serving as part of the pathway.

air contaminant
Any particulate matter, gas, or combination thereof, other than water vapor or natural air.

air contamination
Introduction of a foreign substance into the air to make the air impure. *See also* ***air contaminant.***

air courier service
Establishments primarily engaged in furnishing air delivery of individually addressed letters, parcels, and packages (generally under 100 pounds), except by the U.S. Postal service. While these establishments deliver letters, parcels, and packages by air, the initial pick-up and the final delivery are often made by other modes of transportation, such as by truck, bicycle, or motorcycle. Separate establishments of air courier companies engaged in providing pick-up and delivery only, drop-off points; or distribution centers are all classified in this industry.

air curtain
A method of containing oil spills. Air bubbling through a spill causes an upward water flow that slows the spread of oil. It can also be used to stop fish from entering polluted perforated pipe water.

Air Defense Identification Zone (ADIZ)
The area of airspace over land or water, extending upward from the surface, within which the ready identification, location, and control of aircraft are required in the interest of national security. Air Defense Identification Zone (ADIZ) locations and operating and flight plan requirements for civil aircraft operations are specified in Federal Aviation Regulation (FAR) Part 99.

air density
The weight of air in pounds per cubic foot. Dry standard air at T (temperature) = 70°F and BP (barometric pressure) = 29.92 inch Hg (mercury) has a weight density of 0.075 lbs./cu.ft (pounds per cubic foot). Also referred to as *weight density.*

air diffuser
A device designed to transfer atmospheric oxygen into a liquid.

air dose
In radiation, a dose of x-rays or gamma rays expressed in roentgens, delivered at a point in free air. In radiological practice, it consists of the radiation of the primary beam and that scattered from surrounding air.

air drying
A process to significantly reduce pathogens in solid waste by allowing liquid sludge to drain and/or dry on under-drained sand beds, or paved or unpaved basins in which the sludge is at a depth of nine inches. A minimum of three months is needed, two months of which

temperatures average on a daily basis above zero C.

air embolism
A form of decompression sickness in which an air bubble blocks blood flow in a blood vessel.

air exchange rate
(1) The number of times that the outdoor air replaces the volume of air in a building per unit time, typically expressed as air changes per hour. (2) The number of times that the ventilation system replaces the air within a room or area within the building.

air express (reserved priority air freight)
Freight shipments which are shipped on a guaranteed served flight basis at a premium rate, not to be confused with small package service which is restricted to shipments of 50 pounds or less.

air filter
A device for removing particulate matter from air.

Air Force pediscope
A system consisting of a pressure-transducing blanket and readout for measuring seat pressure on the ischial tuberosities.

air glow
A faint glow of light emitted by exited gases in the upper atmosphere. Air glow is much fainter than the aurora.

air-handler noise
That acoustic noise output from heating, ventilation, or air-conditioning fans and ducts. Also known as *air-blower noise.*

air-handling unit (AHU)
Refers to the ventilation equipment in heating, ventilation, and air-conditioning (HVAC) systems.

air horsepower
The theoretical horsepower required to drive a fan if there were no losses in the fan's efficiency (i.e., the fan is operating at 100% efficiency).

air infiltration
The uncontrolled leakage of air into a building through cracks, open windows, holes, etc. when the building is under negative pressure and/or as a result of the influence of wind or temperature differences.

air-line respirator
A respiratory protective device that is supplied breathing air through a hose line.

air lock
A system of enclosures or doors which prevent the transfer of air between one area and an adjacent one.

air, makeup
Air that replaces other air exhausted from a space. Insufficient makeup air is one possible cause of insufficient exhaust airflow.

air mass
A widespread body of air that gains certain meteorological or polluted characteristics, such as a heat inversion or smog while set in one location. The characteristics can change as it moves away.

air mass thunderstorm
A thunderstorm produced by local convection within an unstable air mass.

air mass weather
A persistent type of weather that may last for several days (up to a week or more). It occurs when an area comes under the influence of a particular air mass.

air mode
*See **aviation mode**.*

air monitoring
The sampling for, and measuring of, contaminants in the air. A form of environmental monitoring in which one or more quantities of environmental gases are taken and a determination of contents and proportions made. *See also **air sampling**.*

Air Movement and Control Association (AMCA)
An association which establishes performance classes for various types of fans.

air mover
Any type of device that is used to transfer air from one space/area to another.

air navigation facility
Any facility used in, available for use in, or designed for use in aid of air navigation, including landing areas, lights, any apparatus or equipment for disseminating weather information, for signaling, for radio-directional finding, or for radio or other electrical communication, and any other structure or mecha-

nism having a similar purpose for guiding or controlling flight in the air or the landing and takeoff of aircraft. *See also* ***navigational aid***.

Although a U.S. Landmark, the Arch in St. Louis is considered a hazard to air navigation

air navigation hazard

An object which, as a result of an aeronautical study, the Federal Aviation Administration (FAA) determines will have an adverse effect upon the safe and efficient use of navigable airspace by aircraft, operation of air navigation facilities, or existing or potential airport capacity.

air operator

A person or organization authorized to operate aircraft or aviation facilities under Federal Aviation Regulation (FAR) Parts 91, 121, 125, 127, 129, 133, 135, or 137.

air parcel

See ***parcel of air***.

air pollutant

(1) Any air pollution agent or combination of such agents, including any physical, chemical, biological, radioactive (including source material, special nuclear material, and byproduct material) substance or matter which is emitted into or otherwise enters the ambient air. Such term includes any precursors to the formation of any air pollutant, to the extent the EPA Administrator has identified such precursor or precursors for the particular purpose for which the term "air pollutant" is used. (2) Any substance in air which could, if in high enough concentration, harm man, other animals, vegetation, or material. Pollutants may include almost any natural or artificial composition of matter capable of being airborne. They may be in the form of solid particles, liquid droplets, gases, or in combinations of these forms. Generally, they fall into two main groups: those emitted directly from identifiable sources, and those produced in the air by interaction between two or more primary pollutants, or by reaction with normal atmospheric constituents, with or without photoactivation. Exclusive of pollen, fog, and dust, which are of natural origin, about 100 contaminants have been identified and fall into these categories: solids, sulfur compounds, volatile organic chemicals, nitrogen compounds, oxygen compounds, halogen compounds, radioactive compounds, and odors. (3) Any substance in air that could, in high concentration, harm people, animals, or vegetation, or damage non-living material. Such pollutants may be from solid particles, liquid droplets, gases, or any combination of these. (4) Dust, fume, mist, smoke or other aerosol, gas, odorous substance, or any combination of these which is emitted into the air or otherwise enters the ambient air.

air pollution

(1) Any undesirable substance mixed with open air. (2) The presence of contaminant or pollutant substances in the air that do not disperse properly and interfere with human health or welfare, or produce other harmful environmental effects (EPA). (4) The presence of an unwanted material in the air, such as dusts, vapors, smoke, etc., in sufficient concentration to affect the comfort, health, or welfare of residents or damage property exposed to the contaminated air. The deterioration of the quality of the air that results from the addition of impurities.

Air Pollution Control Agency

Under the Federal Clean Air Act (CAA). (1) A single state agency designated by the Governor of that state as the official state air pollution control agency for the purposes of the CAA. (2) An agency established by two or more States and having substantial powers or duties pertaining to the prevention and control of air pollution. (3) A city, county, or other local government health authority, or in the case of any city, county, or other local government in which there is an agency other than the health authority charged with responsibility for enforcing ordinances or laws relating to the prevention and control of air pollution. (4) An agency of two or more municipalities located in the same state or in different states and having substantial powers or duties pertaining to the prevention and control of air pollution. (5) An agency of an Indian tribe.

air pollution episode
A period of abnormally high concentration of air pollutants, often due to low winds and temperature inversion, that can cause illness and/or death.

air pressure
See ***atmospheric pressure****.*

air-purifying canister
An air-tight module containing absorptive and/or adsorptive substances for use in an air purifier.

Typical air-purifying respirator with cartridges installed

air-purifying respirator
A device worn by an individual that filters the air to be breathed and returns it to an acceptable state. Also known as *chemical cartridge respirators,* this type of respirator relies on the person's own breathing force to draw air through the filter medium, or it may utilize a powered blower to provide breathing air (i.e., a powered air-purifying respirator or PAPR). When external air is the only source of breathing air (i.e., no powered air supply is used), then this type of respiratory protection cannot be used in oxygen deficient atmospheres. It is a filter only, and subsequently, the air being filtered must have an oxygen content of at least 19.5%.

air quality control region (AQCR)
An area designated by the federal government in which communities share a common air pollution problem (sometimes several states are involved).

air quality criteria
The levels of pollution and lengths of exposure above which adverse health and welfare effects may occur.

air quality-related value (AQRV)
A value referring to the reduction in the visibility that may be caused by a new air emission.

air quality standards
The level of pollutants prescribed by regulations that may not be exceeded during a specified time in a defined area.

air rights
The right to use all or a portion of the air space above real estate. Such right is vested by grant (e.g., fee simple, lease, or other conveyance). While commercial airlines have a right to fly over one's land, if such "flight paths" interfere with the owner's use of such land, the owner is entitled to recover the extent of actual damage suffered by him/her. On the other hand, the owner of the land is precluded by state and federal laws from polluting the air.

air route surveillance radar (ARSR)
Air Route Traffic Control Center (ARTCC) radar used primarily to detect and display an aircraft's position while en route between terminal areas. The ARSR enables controllers to provide radar air traffic control service when aircraft are within the ARSR coverage. In some instances, ARSR may enable an Air Route Traffic Control Center (ARTCC) to

provide terminal radar services similar to but usually more limited than those provided by a radar approach control.

Air Route Traffic Control Center (ARTCC)
A facility established to provide air traffic control service to aircraft operating on an Instrument Flight Rules (IFR) flight plan within controlled airspace and principally during the en route phase of flight. When equipment capabilities and controller workload permit, certain advisory and assistance services may be provided to Visual Flight Rules (VFR) aircraft.

air sampling
The collection of samples of air to determine the presence of and the concentration of a contaminant, such as a chemical, aerosol, radioactive material, airborne microorganism, or other substance by analyzing the collected sample to determine the amount present and calculating the concentration based on the sample volume.

air scour
The agitation of granular filter media with air during the filter backwash cycle.

air, standard
Dry air at 70°F (21°C) and 29.92 inches of mercury barometric pressure. It is equivalent to 0.075 pounds per cubic foot.

air stripper
The process of removing volatile and semi-volatile contaminants from liquid; air and liquid are passed simultaneously through a packed tower.

air-supplied respirator
A respiratory protective device that provides a supply of breathable air from a source outside the contaminated work area. Includes airline respirators and self-contained respirators.

air-supply device
A hand- or motor-operated blower for a hose-mask type respirator, or a compressor or other source of respirable air (e.g., breathing air cylinder) for the air-line respirator.

air taxi
(1) A classification of air carriers, which transports, in accordance with Federal Aviation Regulations (FAR) Part 135, persons, property, and mail using small aircraft (under 30 seats or a maximum payload capacity of 7,500). (2) Used to describe a helicopter/Vertical Takeoff and Landing (VTOL) aircraft movement conducted above the surface but normally not above 100 feet above ground level. The aircraft may proceed either via hover taxi or fly at speeds of more than 20 knots. The pilot is solely responsible for selecting a safe airspeed/altitude for the operation being conducted. (3) A classification of air carriers which directly engages in the air transportation of persons, property, mail, or in any combination of such transportation and which do not directly or indirectly use large aircraft (over 30 seats or a maximum payload capacity of more than 7,500 pounds) and do not hold a Certificate of Public Convenience and Necessity or economic authority issued by the Department of Transportation. (4) An air taxi operator that a) performs at least five round trips per week between two or more points and publishes flight schedules which specify the times, days of the week and places between which such flights are performed; or b) transports mail by air pursuant to a current contract with the U.S. Postal Service. (5) An air carrier certificated in accordance with FAR Part 135 and authorized to provide, on demand, public transportation of persons and property by aircraft. Generally operates small aircraft "for hire" for specific trips.

air taxi/commercial operator (ATCO)
Commercial air carrier operating on-demand air taxi services on aircraft. Certificated in accordance with Federal Aviation Regulation Part 135.

air taxi survey
Federal Aviation Administration (FAA) form 1800-31, sent to carriers for reporting of air taxi/commercial operator (ATCO) activity.

air titration
A field analytical method involving the use of an impinger or bubble to draw air through a liquid reagent that changes color in direct proportion to the concentration of the contaminant in the air. Not as precise as laboratory methods.

air-to-cloth ratio
The ratio of the volumetric flow rate of a gas to be filtered to the fabric area of the filter.

air toxins
Chemical compounds that have been established as hazardous to human health. Also re-

ferred to as hazardous volatile organic compounds, including hydrocarbons such as benzene, halohydrocarbons such as carbon tetrachloride, nitrogen compounds such as amines, oxygen compounds such as ethylene oxide, and others.

air traffic
Aircraft operating in the air or on an airport surface, exclusive of loading ramps and parking areas. *Also referred to as* ***airport traffic****.*

air traffic clearance
An authorization by air traffic control, for the purpose of preventing collision between known aircraft, for an aircraft to proceed under specified traffic conditions within controlled airspace.

air traffic command and control center (ATCCC)
An Air Traffic Operations service facility consisting of four operational units: 1) *Central Flow Control Function (CFCF).* Responsible for coordination and approval of all major inter-center flow control restrictions on a system basis in order to obtain maximum utilization of the airspace. 2) *Altitude Reservation Concept.* Responsible for coordinating, planning, and approving special user requirements. 3) *Airport Reservation Office (ARO).* Responsible for approving Instrument Flight Rules (IFR) flights at designated high-density traffic airports (John F. Kennedy, LaGuardia, O'Hare and Washington National) during specified hours. 4) *Air Traffic Control (ATC) Contingency Command Post.* A facility that enables the Federal Aviation Administration (FAA) to manage the ATC system when a significant portion of the system's capabilities has been lost or threatened.

air traffic control (ATC)
(1) Service operated by an appropriate authority to promote the safe, orderly, and expeditious flow of air traffic. (2) The safety separation process to prevent collisions between aircraft and collisions with obstructions while expediting and maintaining an orderly flow of air traffic; an element of the air traffic management process.

air traffic control clearance
Authorization for an aircraft to proceed under conditions specified by an air traffic control unit.

air traffic control facility
A facility which provides air traffic control services located in the U.S., its possessions and territories, and in foreign countries especially established by international agreement.

air traffic control service
A service provided for the purpose of 1) Preventing collisions between aircraft and on the maneuvering area between aircraft and obstructions, and 2) Expediting and maintaining an orderly flow of air traffic.

air traffic controller
(1) A certified individual responsible for regulating aircraft traffic within a specified region. (2) A person authorized to provide air traffic control service. Also referred to as *air traffic control specialist, controller,* and *final controller.*

air traffic hub
Air traffic hubs are not airports; they are the cities and Standard Metropolitan Statistical Areas requiring aviation services. The hubs fall into four classes as determined by each community's percentage of the total enplaned passengers, all services, and all operations of U.S. certificated air carriers in the 50 states, the District of Columbia, and other U.S. areas designated by the Federal Aviation Administration. Large: a community enplaning 1.00% or more of the total enplaned passengers; Medium: 0.25% to 0.99%; Small: 0.05% to 0.24%; Nonhub: Less than 0.05%.

air traffic liabilities
The value of transportation sold, but not used or refunded (i.e., liabilities to passengers or liabilities to others. These include payables to other airlines for portions of interline passenger trip amounts the ticketing carrier owes the performing carrier. Also included are amounts the ticketing carrier owes to passengers prior to flights, which remain unearned revenue until air transportation is provided).

air traffic management (ATM)
The process used to ensure the safe, efficient, and expeditious movement of aircraft during all phases of operations. Air traffic management consists of air traffic control and traffic flow management.

air traffic service
A generic term meaning: 1) Flight Information Service; 2) Alerting Service; 3) Air Traf-

fic Advisory Service; 4) Air Traffic Control Service, Area Control Service, Approach Control Service, or Airport Control Service.

air transport movements
Landing and takeoff of an aircraft operating a scheduled or non-scheduled service.

air transportation
(1) Interstate, overseas, or foreign air transportation or the transportation of mail by aircraft. (2) Includes establishments that provide domestic and international passenger and freight services, and establishments that operate airports and provide terminal facilities. Also included are flying services such as crop dusting and aerial photography.

air travel club
An operator certificated in accordance with Federal Aviation Regulation (FAR) Part 123 to engage in the carriage of members who qualify for that carriage by payment of an assessment, dues, membership fees, or other similar remittance.

air travel insurance
A form of life insurance which may be purchased by air travelers according to the terms of which the face value of the policy is paid to the named beneficiary in the event of death resulting from a particular flight. *See also* ***insurance***.

air waybill
A bill of lading which covers both domestic and international flights transporting goods to a specified destination. Technically, it is a non-negotiable instrument of air transport which serves as a receipt for the shipper, indicating that the carrier has accepted the goods listed therein and obligates itself to carry the consignment to the airport of destination according to specified conditions.

airbag
A device which is pressurized to inflate on impact to protect the occupant in a vehicle.

airborne asbestos sample
A sample that has been collected in a prescribed manner for determining the concentration of asbestos fibers or structures in the air by a specific analytical method.

airborne dust
Airborne particulates, including the total dust and the respirable dust, present in the air.

airborne particulates
Total suspended particulate matter found in the atmosphere as solid particles or liquid droplets. The chemical composition of particulates varies widely, depending on location and time of year. Airborne particulates include windblown dust, emissions from industrial process, smoke from burning of wood and coal, and exhaust of motor vehicles.

airborne pathogen
A disease-causing microorganism which is transported through the ambient air or on particles present in the air.

airborne radioactive material
Radioactive material dispersed in the air in the form of a dust, fume, mist, vapor, gas, or other form.

airborne release
The release of any chemical into the air.

Airbus Industries Group (AIG)
A supernational management organization responsible for design, development, manufacture, marketing, sales and support of selected commercial aircraft.

aircraft
(1) All airborne vehicles supported either by buoyancy or by dynamic action. Used in a restricted sense to mean an airplane, any winged aircraft, including helicopters but excluding gliders and guided missiles. (2) Device(s) that are used or intended to be used for flight in the air. When used in air traffic control terminology may include the flight crew. *See also* ***airframe, airplane, and airship***.

aircraft accident
Occurrence incident to flight in which, as a result of the operation of an aircraft, any person (occupant or non-occupant) receives fatal or serious injury or any aircraft receives substantial damage. Substantial damage means: a) damage or failure which adversely affects the structural strength, performance, or flight characteristics of the aircraft, and would normally require major repair or replacement of the affected component; b) engine failure, damage limited to an engine, bent fairings or cowling, dented skin, small punctured holes in the skin or fabric, ground damage to rotor or propeller blades. Damage to landing gear, wheels, tires, flaps, engine accessories, brakes,

or wing tips is not considered "substantial damage." *See also* ***accident***.

aircraft accident incident rate (AIR)
A measure of the safety of flying, represented by the formula

$$AIR = \frac{number\ aircraft\ accidents \times 100{,}000}{number\ of\ flight\ hours}$$

aircraft agreement (Agreement On Trade In Civil Aircraft)
Negotiated in the Tokyo Round of the Multilateral Trade Negotiations, and implemented January 1, 1980, providing for elimination of tariff and non-tariff trade barriers in the civil aircraft sector.

aircraft and traffic servicing expenses
Compensation of ground personnel, in-flight expenses for handling and protecting all non-passenger traffic including passenger baggage, and other expenses incurred on the ground to a) protect and control the in-flight movement of aircraft, b) schedule and prepare aircraft operational crews for flight assignment, c) handle and service aircraft while in line operation, and d) service and handle traffic on the ground after issuance of documents establishing the air carrier's responsibility to provide air transportation.

aircraft approach category
A grouping of aircraft based on 1.3 times their stall speed in landing configuration at maximum certified landing weight, as follows:

Category A: Speed less than 91 knots.

Category B: Speed 91 knots or more but less than 121 knots.

Category C: Speed 121 knots or more but less than 141 knots.

Category D: Speed 141 knots or more but less than 166 knots.

Category E: Speed 166 knots or more.

aircraft contact
Aircraft with which the Flight Service Stations (FSS) have established radio communications contact. One count is made for each en route, landing, or departing aircraft contacted by an FSS regardless of the number of contacts made with an individual aircraft during the same flight. A flight contacting five FSSs would be counted as five aircraft contacted.

aircraft departure
An aircraft takeoff made at an airport.

aircraft engine
An engine that is used or intended to be used for propelling aircraft. It includes turbo-superchargers, appurtenances, and accessories necessary for its functioning, but does not include propellers.

aircraft facility
An area where aircraft can takeoff and land, usually equipped with associated buildings and facilities. *See also* ***airport and heliport***.

aircraft incident
An occurrence, other than an accident, associated with the operation of an aircraft that affects or could affect the safety of operations and that is investigated and reported on FAA Form 8020-5.

aircraft industry
Industry primarily engaged in the manufacture of aircraft, aircraft engines, and parts including propellers and auxiliary equipment.

aircraft miles
The distance flown by aircraft in terms of great circle airport-to-airport distances measured in statute miles.

aircraft miles scheduled
The sum of the airport-to-airport distances of all flights scheduled, excluding those operated only as extra sections to accommodate traffic overflow.

aircraft operations
The airborne movement of aircraft in controlled or non-controlled airport terminal areas, and counts at en route fixes or other points where counts can be made.

aircraft revenue departures performed
The number of aircraft takeoffs actually performed in scheduled passenger/cargo and all-cargo services.

aircraft revenue hours
The airborne hours in revenue service, computed from the moment an aircraft leaves the ground until it touches the ground again.

aircraft revenue mile
The miles (computed in airport-to-airport distances) for each inter-airport hop actually completed in revenue service, whether or not performed in accordance with the scheduled

pattern. For this purpose, operation to a flag stop is a hop completed even though a landing is not actually made. In cases where the inter-airport distances are inapplicable, aircraft miles flown are determined by multiplying the normal cruising speed for the aircraft type by the airborne hours.

aircraft type
(1) A term used in a number of Federal Aviation Administration (FAA) publications in grouping aircraft by basic configuration: fixed-wing, rotorcraft, glider, dirigible, and balloon. (2) A distinctive model of an aircraft as designated by the manufacturer.

airflow
The volumetric rate at which air flows through a space, usually measured in cubic feet per minute (CFM) or cubic meters per second (CMS). Also, the speed at which air moves through a space, usually measured in feet per minute or meters per second.

airfoil sill/jamb
Tapered openings on the bottom (sill) and sides (jamb) of laboratory type hoods.

airframe
(1) The fuselage, booms, nacelles, cowlings, fairings, airfoil surfaces (including rotors but excluding propellers and rotating airfoils of engines), and landing gear of an aircraft and their accessories and controls. (2) Structural components of an airplane, such as fuselage, empennage, wings, landing gear, and engine mounts, but excluding such items as engines, accessories, electronics, and other parts that may be replaced from time to time. *See also **aircraft, airplane,** and **airship**.*

airlift
A device for pumping liquid by injecting air at the bottom of a riser pipe submerged in the liquid to be pumped.

airline transport pilot
An individual who has been trained and has demonstrated proficiency in the operation of aircraft of a specific category and classification. An airline transport pilot may act as a pilot-in-command of an aircraft engaged in air carrier service.

airman
A pilot, mechanic or other licensed aviation technician.

airman certificate
A document issued by the Administrator of the Federal Aviation Administration (FAA) certifying that the holder complies with the regulations governing the capacity in which the certificate authorizes the holder to act as an airman in connection with aircraft. *See also **airman**.*

Airman's Information Manual
*See **Aeronautical Information Manual**.*

Airman's Meteorological Information (AIRMET)
In-flight weather advisories issued only to amend the area forecast concerning weather phenomena which are of operational interest to all aircraft and potentially hazardous to aircraft having limited capability because of lack of equipment, instrumentation, or pilot qualifications. AIRMETs concern weather of less severity than that covered by Significant Meteorological Convective Information's (SIGMET) or Convective SIGMETs. AIRMETs cover moderate icing, moderate turbulence, sustained winds of 30 knots or more at the surface, widespread areas of ceilings less than 1,000 feet and/or visibility less than 3 miles, and extensive mountain obstructions.

AIRMET
*See **Airman's Meteorological Information**.*

airplane
An engine-driven fixed-wing aircraft heavier than air, that is supported in flight by the dynamic reaction of the air against its wings. *See also **aircraft, airframe,** and **airship**.*

airplane design group
A grouping of airplanes based on wingspan as follows:

Design Group I: A wingspan up to but not including 49 feet (15m).

Design Group II: A wingspan 49 feet (15m) up to but not including 79 feet (24m).

Design Group III: A wingspan 79 feet (24m) up to but not including 118 feet (36m).

Design Group IV: A wingspan 118 feet (36m) up to but not including 171 feet (52m).

Design Group V: A wingspan 171 feet (52m) up to but not including 214 feet (65m).

Design Group VI: A wingspan 214 feet (65m) up to but not including 262 feet (80m).

airport
(1) An area of land or water that is used or intended to be used for the landing and take-off of aircraft, and includes its buildings and facilities, if any. (2) Facility used primarily by conventional, fixed-wing aircraft. (3) A facility, either on land or water, where aircraft can take off and land. Usually consists of hard-surfaced landing strips, a control tower, hangars and accommodations for passengers and cargo. (4) A landing area regularly used by aircraft for receiving discharging passengers or cargo. *See also **aerodrome, aircraft facility,** and **heliport**.*

airport acceptance rate (AAR)
A dynamic input parameter specifying the number of arriving aircraft which an airport or airspace can accept from the Air Route Traffic Control Center per hour. The AAR is used to calculate the desired interval between successive arrival aircraft.

airport advisory area
The area within ten miles of an airport without a control tower or where the tower is not in operation, and on which a Flight Service Station is located.

airport advisory service
A service provided by flight service stations at airports not served by a control tower. This service consists of providing information to arriving and departing aircraft concerning wind direction/speed-favored runway, altimeter setting, pertinent known traffic/field conditions, airport taxi routes/traffic patterns, and authorized instrument approach procedures. This information is advisory in nature and does not constitute an Air Traffic Control (ATC) clearance.

airport and airways trust fund
Mechanism for funding capital improvements for the nation's air traffic control system and airports. The fund is supported by taxes contributed by users of the aviation system, including an 10 percent tax of airline passengers and a tax on aviation fuel. Established by the Airport and Airway Revenue Act of 1970.

airport elevation
The highest point on an airport's usable runways, expressed in feet above mean sea level.

airport/facility directory
A publication designed primarily as a pilot's operational manual containing all airports, seaplane bases, and heliports open to the public including communications data, navigational facilities, and certain special notices and procedures. This publication is issued in seven volumes, according to geographical area.

airport improvement program (AIP)
Provides funding from the Airport and Airway Trust Fund for airport development, airport planning, noise compatibility planning, and to carry out noise compatibility programs. *See also **airport and airway trust fund**.*

airport information desk
An airport unmanned facility designed for pilot self-service briefing, flight planning, and filing of flight plans.

airport layout plan (ALP)
The plan of an airport showing the layout of existing and proposed airport facilities.

airport lighting
(1) *General.* Various lighting aids that may be installed on an airport. (2) *Runway Lights/Runway Edge Lights*. Lights having a prescribed angle of emission used to define the lateral limits of a runway. Runway lights are uniformly spaced at intervals of approximately 200 feet, and the intensity may be controlled or preset.

airport marking aids
Markings used on runway and taxiway surfaces to identify a specific runway, a runway threshold, a centerline, a hold line, etc. A runway should be marked in accordance with its present usage such as a) visual, b) non-precision instrument, or c) precision instrument.

airport noise
That environmental noise in the vicinity of an airport due primarily to engine noise from approaching and departing aircraft.

airport operations
The number of arrivals and departures from the airport at which the airport traffic control

tower is located. There are two types of operations: local and itinerant.

airport reference point (ARP)
The latitude and longitude of the approximate center of the airport.

airport reservation office (ARO)
Office responsible for monitoring the operation of the high density rule. Receives and processes requests for Instrument Flight Rules (IFR) operations at high density traffic airports.

airport runway centerline lighting
Flush centerline lights spaced at 60-foot intervals beginning 76 feet from the landing threshold and extending to within 75 feet of the opposite end of the runway.

airport runway end identifier lighting (REIL)
Two synchronized flashing lights, one on each side of the runway threshold, which provide rapid and positive identification of the approach end of a particular runway.

airport surface detection equipment (ASDE)
Radar equipment specifically designed to detect all principal features on the surface of an airport, including aircraft and vehicular traffic, and to present the entire image on a radar indicator console in the control tower. Used to augment visual observation by tower personnel of aircraft and/or vehicular movements on runways and taxiways.

airport surveillance radar (ASR)
Approach control radar used to detect and display an aircraft's position in the terminal area. ASR provides range and azimuth information but does not provide elevation data. Coverage of the ASR can extend up to 60 miles. *See also **ground controlled approach** and **precision approach radar**.*

airport taxi charts
Designed to expedite the efficient and safe flow of ground traffic at an airport. These are identified by the official airport name (e.g., Orlando International Airport).

airport threshold lighting
Fixed green lights arranged symmetrically left and right of the runway centerline, identifying the runway threshold.

airport touchdown zone lighting
Two rows of transverse light bars located symmetrically about the runway centerline normally at 100 foot intervals. The basic system extends 3,000 feet along the runway.

airport traffic
Aircraft operating in the air or on an airport surface exclusive of loading ramps and parking areas. *See also **air traffic**.*

airport traffic control service
Air traffic control service provided by an airport traffic control tower for aircraft operating on the movement area and in the vicinity of an airport.

airport traffic control tower (ATCT)
(1) A central operations facility in the terminal air traffic control system, which consists of a tower cab structure, including an associated IFR room if radar equipped, and uses air/ground communications, radar, visual signaling, and other services to provide safe and expeditious movement of terminal air traffic. (2) A terminal facility that uses air-ground radio communications, visual signaling, and other devices to provide air traffic control (ATC) services to aircraft operating in the vicinity of an airport or on the movement area. Authorizes aircraft to land or takeoff at the airport controlled by the tower or to transit the airport traffic area regardless of flight plan or weather conditions (instrument flight rules [IFR] or visual flight rules [VFR]). A tower may also provide approach control services.

airport visual approach slope indicator lighting (VASI)
An airport lighting facility providing vertical visual approach slope guidance to aircraft during approach to landing by radiating a directional pattern of high intensity red and white focused light beams which indicate to the pilot that he is "on path" if he sees red/white, "above path" if white/white, and "below path " if red/red. Some airports serving large aircraft have three-bar VASI which provide two visual glide paths to the same runway.

airports, flying fields, and airport terminal services
Establishments primarily engaged in operating and maintaining airports and flying fields; in servicing, repairing (except on a factory ba-

sis), maintaining and storing aircraft; and in furnishing coordinated handling services for airfreight or passengers at airports. This industry also includes private establishments primarily engaged in air traffic control operations.

Airports Grants In Aid Program
A grant of funds by the Secretary of Transportation under the Airport & Airway Improvement Act of 1982 to a sponsor for the accomplishment of one or more projects.

airports of entry
Aircraft may land at these airports without prior permission to land from U.S. Customs.

airship
An engine-driven lighter-than-air aircraft that can be steered. *See also* ***aircraft, airframe,*** *and* ***airplane***.

airspace hierarchy
Within the airspace classes, there is a hierarchy and, in the event of an overlap of airspace: Class A preempts Class B, Class B preempts Class C, Class C preempts Class D, Class D preempts Class E, and Class E preempts Class G.

airspeed
The speed of an aircraft relative to its surrounding air mass. The unqualified term "airspeed", means one of the following: (1) *Indicated airspeed.* The speed shown on the aircraft airspeed indicator. This is the speed used in pilot/controller communications under the general term "airspeed." (2) *True airspeed.* The airspeed of an aircraft relative to undisturbed air. Used primarily in flight planning and en route portion of flight. When used in pilot/controller communications, it is referred to as "true airspeed" and not shortened to "airspeed."

airspeed indicator
An aircraft display showing velocity relative to the surrounding air.

airspeed/mach indicator (AMI)
See ***airspeed indicator*** and ***mach indicator***.

airstart
The starting of an aircraft engine while the aircraft is airborne, preceded by engine shutdown during training flights or by actual engine failure.

airway
(1) *Anatomy.* The pathway through which air and other respiratory gases pass between the mouth or nostrils and the lung alveoli. (2) *Aviation.* A Class E airspace area established in the form of a corridor, the centerline of which is defined by radio navigational aids.

airway beacon
Aviation. Used to mark airway segments in remote mountain areas. The light flashes Morse Code to identify the beacon site.

airway resistance
That resistance which must be overcome for air to flow through the airway.

AISC
American Institute of Steel Construction.

AISI
American Iron and Steel Institute.

aitken nuclei
See ***condensation nuclei***.

akimbo span
See ***span akimbo***.

akinesia
A movement disorder in which the person executes no voluntary movements or exhibits a pause prior to initiation of a movement.

AL
Action level.

alabaster
A compact, fine-grained gypsum material.

Alachlor
A herbicide, marketed under the trade name Lasso, used mainly to control weeds in corn and soybean fields.

Alar
Trade name for daminozide, a pesticide that makes apples redder, firmer, and less likely to drop off trees before growers are ready to pick them. It is also used to a lesser extent on peanuts, tart cherries, concord grapes, and other fruits.

ALARA
See ***as low as reasonably achievable***.

alarm
An indicator that some condition exists which may or will require human action to correct in

order to prevent loss of life, property, or equipment.

alarm set point
The selected concentration at which an instrument is set to alarm.

albedo
The percent of radiation returning from a surface compared to that which strikes it.

albumin
A member of a class of water-soluble proteins that are found in the juices and tissues of animals, in the white or clear part of eggs, and in vegetables, and that contain sulfur, oxygen, hydrogen, carbon, and nitrogen.

albuminuria
Presence of serum albumin in the urine.

alcohol
An organic compound synthesized from petroleum or natural products or derived from a fermentation process. Widely used as a solvent and for chemical syntheses.

alcohol concentration (AC)
The concentration of alcohol in a person's blood or breath. When expressed as a percentage it means grams of alcohol per 100 milliliters of blood or grams of alcohol per 210 liters of breath.

alcohol involvement
A fatality or fatal crash as alcohol-related or alcohol involved if (1) Either a driver or a non-motorist (usually a pedestrian) had a measurable or estimated blood alcohol concentration (BAC) of 0.01 grams per deciliter (g/dl) or above. Probabilities of alcohol involvement are now calculated for each driver, pedestrian, or crash. (2) Coded by police when evidence of alcohol is present. This code does not necessarily mean that a driver, passenger or non-occupant was tested for alcohol.

alcoholic
Containing or pertaining to alcohol; a person addicted to alcohol.

alcoholism
Drunkenness, or long-continued, excessive consumption of alcohol. Generally refers to chronic alcoholism.

aldehyde
A class of organic compounds containing a CHO group, including formaldehyde and acetaldehyde.

alderman
Municipal officer; member of the legislative body of a municipality. Often called *councilman.*

Aldicarb
An insecticide sold under the trade name Temik. It is made from ethyl isocyanate.

alert notice
Aviation. A request originated by a Flight Service Station (FSS) or an air route traffic control center (ARTCC) for an extensive communication search for overdue, unreported, or missing aircraft. Also known as *alnot.*

alerting service
Aviation. A service provided to notify appropriate organizations regarding aircraft in need of search and rescue aid, and assist such organizations as required.

Aleutian low
The subpolar low-pressure area that is centered near the Aleutian Islands on charts that show mean sea-level pressure.

algae
Simple rootless plants that grow in sunlit waters in relative proportion to the amounts of nutrients available. They can affect water quality adversely by lowering the dissolved oxygen in the water. They are food for fish and small aquatic animals.

algae blooms
Sudden spurts of algae growth, which can affect water quality adversely and indicate potentially hazardous changes in local water chemistry.

algaecide
Any substance used to kill algae. Also spelled *algicide.*

algorithm
An accepted procedure that has been developed for the purpose of solving a specific problem.

alimentary canal
All the organs making up the route taken by food as it passes through the body from the

mouth to the anus. Also called the *digestive tract*.

aliphatic hydrocarbon
One of the major groups of organic compounds characterized by a straight- or branched-chain arrangement of carbon atoms. This group in composed of three subgroups: *alkanes* (paraffins), which are saturated and relatively unreactive; *alkenes,* which contain double bonds and are reactive; and, *alkynes* (acetylenes), which contain triple bonds and are highly reactive.

aliquot
A part which is a definite fraction of the whole, such as an aliquot of a sample for analysis.

alkali
A compound that has the ability to neutralize an acid and form a salt. Any substance which in water solution is bitter, more or less irritating or caustic to the skin, turns litmus blue, and has a pH value greater than 7. *See also* ***base***.

alkali-ash diet
A therapeutic diet prescribed to dissolve uric acid and cystine urinary calculi. This type of diet changes the urinary pH so that certain salts are kept in solution and excreted in the urine. Emphasis is placed on fruits, vegetables, and milk. Meat, eggs, bread, and cereals are restricted.

alkali metals
The elements lithium, sodium, potassium, rubidium, and cesium.

alkaline
Water containing sufficient amounts of alkalinity to raise the pH above 7.0

alkaline soil
Soil with a pH greater than 7.0.

alkalinity
The ability of a water to neutralize an acid due to the presence of carbonate, bicarbonate, and hydroxide ions.

alkaloid
One of a large group of organic, basic substances found in plants. They are usually bitter in taste and are characterized by powerful physiological activity. Examples are morphine, cocaine, atropine, quinine, nicotine, and caffeine. The term is also applied to synthetic substances that have structures similar to plant alkaloids, such as procaine.

alkalosis
A pathologic condition resulting from accumulation of base or loss of acid without comparable loss of base in the body, and characterized by a decrease in hydrogen ion concentration (increase in pH). Although the normal pH of the blood is slightly alkaline at 7.4, a drastic shift of the acid-base balance toward alkalinity can produce serious symptoms, including shallow or irregular respiration, prickling or burning sensation in the fingers, toes or lips, muscle cramps and, in severe cases, convulsions. The opposite of alkalosis is *acidosis*.

alkylating agent
A synthetic compound containing two or more end (alkyl) groups that combine readily with other molecules. Their action seems to be chiefly on the deoxyribonucleic acid (DNA) in the nucleus of the cell. They are used in chemotherapy of cancer although they do not damage malignant cells selectively, but also have a toxic action on normal cells. Locally, they cause blistering of the skin and damage to the eyes and respiratory tract. Systemic toxic effects are nausea and vomiting, reduction in both leukocytes and erythrocytes, and hemorrhagic tendencies.

all-cargo and mail aircraft
Movement by aircraft and helicopters operated for commercial transport operations involving freight and/or mail, but not passengers.

all-cargo carrier
(1) One of a class of a carriers holding an All-Cargo Air Service Certificate issued under Section 418 of the Federal Aviation Act and certificated in accordance with Federal Aviation Regulations Part 121 to provide domestic air transportation of cargo. (2) An air carrier certificated in accordance with Federal Aviation Regulations (FAR) Part 121 to provide scheduled air freight, express, and mail transportation over specified routes, as well as to conduct nonscheduled operations that may include passengers.

all clear
(1) *Emergency Response*. Term used to indicate the emergency area is secured and it is

safe to reenter the area. (2) *Marine Safety.* Term used in boating to mean a) that a tow boat is ready to leave barge, or b) is clear of an obstruction.

all-fire level
The minimum direct current or radio frequency energy that causes initiation of an electroexplosive initiator with a reliability of 0.999 at a confidence level of 95 percent, as determined by a Bruceton test. recommended operating level is all-fire current, as determined by test, at ambient temperature plus 150 percent of the minimum all-fire current. *See also* ***Bruceton test.***

all-or-none
Denoting either a complete response to a stimulus by a nerve or muscle, or none at all.

all-pass
Pertaining to a condition or piece of equipment in which all frequencies of a signal are processed equivalently without attenuation; also *all pass* and *allpass.*

all-risk insurance
Type of insurance policy which ordinarily covers every loss that may happen, except by fraudulent acts of the insured. A type of policy which protects against all risks and perils except those specifically enumerated.

all services
Transportation. The total of scheduled and nonscheduled transport services.

allergen
Any of a wide variety of substances or environmental conditions which may provoke an allergic reaction. Almost any substance in the environment can become an allergen. The list of known allergens (i.e., substances to which individuals have become sensitive) includes plant and tree pollens, spores of mold, animal hairs, dust, foods, feathers, dyes, soaps, detergents, cosmetics, plastics, some valuable medicines, including penicillin, and even sunlight. Allergens can enter the body by being inhaled, swallowed, touched, or injected. The allergen is not directly responsible for the allergic reaction, but sets off the chain of events that brings it about. When a foreign substance enters the body, the system reacts by producing antibodies that attack the substance and render it harmless. When their work is done, the antibodies attach themselves to tissue surfaces, where they remain in reserve, ready to be called into action if the same substance should enter the body again. Should the substance do so, the antibodies again enter into the immune reaction which is part of the body's valuable natural defense against invading disease germs. Also referred to as *sensitizers.*

allergic contact dermatitis
Initial exposure of an individual to a chemical may not cause a problem but will result in the formation of antigens. In some cases, subsequent exposure of the individual to the material results in an inflammatory response, with resulting erythema and edema, which is referred to as an allergic contact dermatitis.

allergic reaction
Abnormal response following exposure to a substance by an individual who is hypersensitive to that substance as a result of a previous exposure.

allergy
(1) An unusual or exaggerated response to a particular substance in a person sensitive to that substance. (2) The acquired hypersensitivity of an individual to a particular substance. (3) A hypersensitive or pathological reaction by a person to environmental factors or substances, such as pollens, foods, dust, or microorganisms, in amounts that do not affect most people. (4) An abnormal response of a hypersensitive person to a chemical or physical stimulus.

allesthesia
The perception of a given peripheral tactile stimulus as occurring at a point different from the actual point of stimulation.

allopathy
A method of treating disease by the use of agents producing effects different from those of the disease treated; opposite of *homeopathy.*

Allotment Management Plan
Under the Federal Land Policy and Management Act of 1976: A document prepared in consultation with the lessees or permittees involved, which applies to livestock operations on the public lands or on lands within National Forests in the eleven contiguous Western States and which prescribes the manner in, and extent to, which livestock operations will

be conducted in order to meet the multiple-use, sustained yield, economic and other needs and objectives as determined by the Secretary concerned. It also describes the type, location, ownership, and general specifications for the range improvements to be installed and maintained on the lands to meet the livestock grazing and other objectives of land management. It contains such other provisions relating to livestock grazing and other objectives found by the Secretary of the Interior to be consistent with the Federal Land Policy and Management Act of 1976 and applicable law.

allotropism
A condition in which an element is present in two or more distinct forms with unlike properties.

allowable level
As used in Rule 57 (pursuant to Act 245 of 1929), a discharge level which is acceptable once the discharge has mixed with a specified volume of receiving water (e.g., river or stream).

allowable load
*See **load limit**.*

allowance
(1) That specified minimal clearance between two parts which are to be assembled. (2) Some time value or factor by which the normal time required to complete a task is increased to allow for such things as delays, policy, fatigue, or personal needs. Also known as *allowed time* and *time allowance*.

allowance for depreciation
Asset valuation account which includes the balance of the offsetting credit to be a capitalized asset value to represent the loss and deterioration in the value of the asset over time, so that the asset is amortized over its useful life, and provisions are made for replacement of the asset at the end of its productive life.

allowance for uncollectable accounts
Accruals for estimated losses from uncollectable accounts.

allowed time
*See **allowance**.*

alloy
A combination of two or more metals to form an alloy in which the atoms of one metal replace or occupy interstitial positions between the atoms of the other metal.

alluvial soil
Soil formed of material that was carried by flowing water before being deposited. *See also **alluvial stream**.*

alluvial stream
Any stream whose banks are subject to attack, allowing channel meander. The stream has the property of depositing material such as soil, sand, or gravel and building up land in one area while washing it away in another. *See also **alluvial soil**.*

alluvium
Loose materials (clay, silt, sand, gravel, and larger rocks) washed down from hills and mountains and deposited in low areas.

alnot
*See **alert notice**.*

aloe
Medicinal plant that yields a purgative drug, aloin.

alongside
The side of the ship. Goods to be delivered "alongside" are to be placed on the dock or lighter within reach of the ship's tackle so that they can be loaded aboard the ship.

alopecia
Partial or total loss of hair from natural or abnormal causes.

ALP
*See **airport layout plan**.*

alpha
The probability of making a Type I error, represented by the symbol α. *See also **beta**.*

alpha emitter
A radioactive substance which gives off alpha particles during the decay process. Also referred to as *alpha decay*.

alpha factor
The ratio of oxygen transfer coefficients for water and wastewater at the same temperature and pressure; used in the sizing of aeration equipment.

alphanumeric
Any letter of the alphabet, numeral, punctuation mark, or other symbolic character.

alphanumeric display
Letters and numerals used to show identification (ID), altitude, beacon code, and other information concerning a target on a radar display.

alpha particle
A specific particle, consisting of two protons and two neutrons (a helium nucleus) ejected spontaneously from the nuclei of some radioactive elements. It has low penetrating power and short range. Even the most energetic alpha particles will generally fail to penetrate unbroken skin. The danger arises when matter containing alpha-emitting isotopes is introduced into the lungs or intestinal tract.

alpha radiation
A stream of alpha particles.

alpha ray
A strongly ionizing and weakly penetrating radiation stream of fast-moving helium nuclei.

alpha rhythm
A band of the EEG spectrum consisting of frequencies from about 8 Hz to 13 Hz.

alpha testing
The preliminary testing phase of a new software product outside the facility or company in which it was developed. *See also **beta testing**.*

alpine tundra
*See **tundra**.*

Alquist-Priolo Act
Zoning act passed in 1972 in response to the 1971 San Fernando earthquake to prevent building across the traces of active faults.

ALS
*See **approach light system**.*

alternate airport
An airport at which an aircraft may land if a landing at the intended airport becomes inadvisable.

alternate method
Any method of sampling and analyzing for an air pollutant which is not a reference or equivalent method but which has been demonstrated in specific cases to the EPA's satisfaction to produce results adequate for compliance.

alternating current (AC)
An electrical current flow which alternates in amplitude about a baseline.

alternative dispute resolution
Term refers to procedures for settling disputes by means other than litigation such as arbitration, mediation, or mini-trials. Such procedures, which are usually less costly and more expeditious, are increasingly being used in commercial and labor disputes, divorce actions, in resolving motor vehicle and medical malpractice tort claims, and in other disputes that would likely otherwise involve court litigation.

alternative energy
Energy obtained from sources other than traditional fossil fuels or nuclear energy, and which are usually renewable and nonpolluting. Alternative energy sources include solar energy, wave power, geothermal power, and biomass fuels.

alternative fuel
*See **alternative energy**.*

alternative fuel capacity
The on-site availability of apparatus to burn fuels other than natural gas.

alternative input device
*See **alternative pointing device**.*

alternative or innovative treatment technologies
Under the Comprehensive Environmental Response, Compensation, and Liability Act (CERCLA) of 1980: Those technologies, including proprietary or patented methods, which permanently alter the composition of hazardous waste through chemical, biological, or physical means so as to significantly reduce the toxicity, mobility, or volume (or any combination thereof) of the hazardous waste or contaminated materials being treated. The term also includes technologies that characterize or assess the extent of contamination, the chemical and physical character of the contaminants, and the stresses imposed by the contaminants on complex ecosystems at sites.

alternative pointing device
A device used to assist in a disabled individual's interaction with a computer. Also known as *alternative input device*.

alternative work schedule
Any work schedule other than standard work week.

altimeter
An instrument that indicates the altitude of an object above a fixed level (for example, an aircraft above sea level). Pressure altimeters use an aneroid barometer with a scale graduated in altitude instead of pressure.

altimeter setting
The barometric pressure reading used to adjust a pressure altimeter for variations in existing atmospheric pressure or to the standard altimeter setting (29.92).

altitude
The vertical distance of a level, a point, or an object considered as a point measured in feet above ground level (AGL) or from mean sea level (MSL). 1) *MSL altitude*. Altitude expressed in feet measured from mean sea level. 2) *AGL altitude*. Altitude expressed in feet measured above ground level. 3) *Indicated altitude*. The altitude as shown by an altimeter. On a pressure or barometric altimeter it is altitude as shown uncorrected for instrument error and uncompensated for variation from standard atmospheric conditions.

altitude encoding
An aircraft altitude transmitted via the Mode C transponder feature that is visually displayed in 100 feet increments on a ground radar scope having readout capability.

altitude engine
A reciprocating aircraft engine having a rated takeoff power that is producible from sea level to an established higher altitude.

altitude readout
An aircraft's altitude, transmitted via the Mode C transponder feature, that is visually displayed in 100-foot increments on a radar scope having readout capability.

altitude reservation (ALTRV)
Airspace utilization under prescribed conditions normally employed for the mass movement of aircraft or other special requirements which cannot otherwise be accomplished. ALTRVs are approved by the appropriate Federal Aviation Administration (FAA) facility. *See also* ***formation flight***.

altitude restriction
An altitude or altitudes, stated in the order flown, which are to be maintained until reaching a specific point or time. Altitude restrictions may be issued by Air Traffic Control (ATC) due to traffic, terrain, or other airspace considerations.

altitude restrictions are canceled
Adherence to previously imposed altitude restrictions is no longer required during a climb or descent.

altitude sickness
A syndrome caused by exposure to altitude high enough to cause significant hypoxia, or lack of oxygen. At high altitudes, the atmospheric pressure is decreased and consequently arterial oxygen content is also lowered. *Acute altitude sickness* may occur after a few hours' exposure to a high altitude. Mental function may be affected, and there may be lightheadedness and breathlessness. Eventually headache and prostration may occur. Older persons and those with pulmonary or cardiovascular disease are most likely to be affected. After a few hours or days of acclimation, the symptoms will subside. *Chronic altitude sickness* (sometimes referred to as *Monge's disease* or *Andes disease*) occurs in those in the high Andes above 15,000 feet. It resembles *polycythemia,* but is completely relieved if the patient is moved to sea level.

altocumulus
A middle cloud, usually white or gray. Often occurs in layers or patches with wavy, rounded masses or rolls.

altocumulus castellanus
An altocumulus showing vertical development. Individual cloud elements have tower-like tops, often in the shape of tiny castles (hence the name).

altostratus
A middle cloud composed of gray or bluish sheets or layers of uniform appearance. In the thinner regions, the sun or moon usually appears dimly visible.

ALTRV
See ***altitude reservation***.

alum
(1) A substance used, in the form of colorless crystals or white powder, as a styptic or he-

mostatic because of its astringent action. It also may be given by mouth to induce vomiting. Large doses may cause gastrointestinal disturbances. (2) Common name for aluminum sulfate [$Al_2(SO_4)_3 \cdot 14H_2O$], frequently used as a coagulant in water and wastewater treatment.

alum sludge
Sludge resulting from treatment processes where alum is used as a coagulant.

alumina
A form of aluminum oxide; chemical formula is AL_2O_3.

aluminosis
A pneumoconiosis that results from the inhalation of aluminum-bearing dusts.

aluminum
A lightweight, nonferrous metal with good corrosion resistance and electrical and thermal conductivity.

aluminum silicates
Compounds containing aluminum, silica, and oxygen as main constituents.

alveolar gas exchange
That gaseous exchange through the thin walls of the alveoli and the capillaries, normally such that oxygen is absorbed by the blood and carbon dioxide is released into the alveolus.

alveolar pressure
That combined air and water vapor pressure within an alveolus of the lung.

alveolar ventilation
The replenishment of alveolar gases by atmospheric air.

alveoli
Plural of alveolus. (1) Numerous small, terminal air sacs in the lungs where pulmonary capillary blood is in close juxtaposition to the alveolar gas, permitting the rapid exchange of carbon dioxide and oxygen in the lungs. There are approximately 300 million alveoli situated at the ends of small air passageways in the lungs. Alveoli are the main deposition site of respirable dust particles (1-10 microns in diameter) or respirable fibers (e.g., asbestos) that can result in various respiratory diseases such as silicosis and asbestosis. (2) The cavities or sockets of either jaw in which the roots of the teeth are embedded.

A/m
Amperes per meter.

A.M. peak period
Federal Transit Association. The period in the morning when additional services are provided to handle higher passenger volumes. The period begins when normal, scheduled headways are reduced and ends when headways return to normal. *See also* ***P.M. peak period.***

AMA
See ***American Medical Association.***

amalgam
Any mixture or alloy of mercury combined with other metals, such as zinc, gold, silver, or alloys.

amalgamation
(1) *Metallurgy.* The alloying of metals with mercury. (2) *Law.* Union of different races, or diverse elements, societies, unions, associations, or corporations, so as to form a homogeneous whole or new body; interfusion; intermarriage; consolidation; merger or coalescence.

amaurosis
Partial or complete loss of sight from loss of power in the optic nerve or retina, without any visible defect in the eye except an immovable septum.

ambassador
A public officer clothed with high diplomatic powers, commissioned by a government to transact the international business of his/her government with a foreign government.

ambidextrous
Having the faculty of using both hands with equal ease and facility.

ambient
The surroundings or the area encircled.

ambient air
(1) Any unconfined portion of the atmosphere (open air, surrounding air). (2) The surrounding air or atmosphere in a given area under normal conditions. (3) The part of the atmosphere that is external to structures and to which the public has access.

ambient air quality
A general term used to describe the quality of the open or ambient air.

Ambient Air Quality Standards
See ***Criteria Pollutants*** and ***National Ambient Air Quality Standards.***

ambient aquatic life advisory concentrations (AALACs)
The Environmental Protection Agency's advisory concentration limit for acute or chronic toxicity to aquatic organisms.

ambient noise
The noise associated with a given environment and composed of the sounds from many sources. It is the total noise energy, or the composite of sounds from many sources in an environment.

ambient temperature
The temperature of the medium which surrounds an object.

ambient water quality criteria (AWQS)
The Environmental Protection Agency's maximum acute or chronic toxicity concentrations for protection of aquatic life and its uses.

ambivert
A person possessing characteristics of both the introvert and the extrovert.

ambiyopia
Dullness or dimness of eyesight without any apparent defect in the organs; the first stage of amaurosis. *See also* ***amaurosis***.

ambulance or rescue service
Establishments primarily engaged in furnishing ambulances or rescue services, except by air, where such operations are primarily within a single municipality, contiguous municipalities, or a single municipality and its suburban areas.

ambulant
Able to move from place to place; not confined to a bed.

ambulatory
Pertaining to an illness or condition that can be treated while the patient is able to walk about and is not confined to bed.

ameba
A member of a genus of protozoa, a one-celled semi-fluid animal. Some species are parasitic in the human body and cause disease.

amebic dysentery
A form of dysentery caused by a protozoan parasite, usually resulting from poor sanitary conditions and transmitted by contaminated food or water. Also referred to as *amoebic dysentery*.

amelioration
Improvement of conditions immediately after an accident; the immediate treatment of injuries and conditions which endanger people and/or property.

amend
To improve. To change for the better by moving defects of faults. To change, correct, revise.

amended water
Water to which a wetting agent has been added to improve its ability to wet a material.

amendment
(1) *General.* Change for the better; a removal of faults, or a correction. (2) *Law.* The statement of a change, such as a law, bill, or motion. (3) *Environmental.* Organic material, such as wood chips or sawdust, added to sludge in a composting operation to promote uniform air flow.

amentia
Imbecility, idiocy, or dotage; deficiency of mental capacity.

American Academy of Industrial Hygiene (AAIH)
A professional society of board-certified industrial hygienists.

American Arbitration Association
A national organization of arbitrators from whose panel arbitrators are selected for labor and commercial disputes. The Association has produced a Code of Ethics and Procedural Standards for use and guidance of arbitrators.

American Association of Occupational Health Nurses (AAOHN)
An organization dedicated to promoting the field of occupational health nursing, formerly named the American Association of Industrial Nurses. It has numerous chapters at the state level throughout the United States. It provides board certification in the specialty of occupational health nurses.

American Bar Foundation
An outgrowth of the American Bar Association involved with sponsoring and funding projects in legal research, education, and social studies.

American Board of Industrial Hygiene (ABIH)
Specialty board whose objective is to improve the practice and educational standards of the profession of industrial hygiene, and that is authorized to certify qualified practitioners in the discipline of industrial hygiene.

American Conference of Governmental Industrial Hygienists (ACGIH)
A professional, non-governmental organization founded in 1938, composed of industrial hygienists employed in the government and academia. The ACGIH establishes threshold limit values (TLV) for certain chemicals, and co-sponsors (with the American Industrial Hygiene Association) the annual American Industrial Hygiene Conference. Their primary function is to encourage the exchange of experiences among governmental industrial hygienists, and to collect and make available information of value to their members.

American Industrial Hygiene Association (AIHA)
An association of professional industrial hygienists trained in the anticipation, recognition, evaluation, and control of health hazards, and the prevention of adverse health effects among personnel in the workplace.

American Medical Association (AMA)
Professional association of persons holding a medical degree or an unrestricted license to practice medicine with the purpose of promoting the science of medicine and the betterment of public health.

American National Standards Institute (ANSI)
A voluntary organization made up of members that coordinate, develop, and publish consensus standards for a wide variety of conditions, procedures, and devices.

American Occupational Medical Association (AOMA)
Professional society of medical directors and plant physicians, specializing in occupational medicine and surgery. The organization was established to encourage the study of problems peculiar to the practice of industrial medicine and to develop methods to conserve the health of workers and develop an understanding of medical care needs of workers.

American Society for Testing and Materials (ASTM)
Members are from business, the scientific community, government agencies, educational institutions, laboratories, etc., and establish voluntary consensus standards for materials, products, systems, and services.

American Society of Heating, Refrigerating and Air Conditioning Engineers (ASHRAE)
A professional society of heating, ventilating, refrigeration, and air conditioning engineers that carries out research programs and develops recommended practices/guidance in these areas. It is the primary association involved in filtration and comfort ventilation as well as indoor air quality (IAQ).

American Society of Safety Engineers (ASSE)
An international, multi-disciplinary, not-for-profit, professional organization with more than 34,000 members consisting primarily of individual safety professionals dedicated to the advancement of occupational safety, health, and environmental professions. Organized in 1911 as the United Association of Casualty Inspectors and incorporated in 1915, ASSE is one of the oldest sustaining professional membership societies based in the United States.

American Standard Code for Information Interchange (ASCII)
The most common convention for representing alphanumeric data for transmission or storage.

Americans with Disabilities Act of 1990 (ADA)
A law passed by Congress with the intent of aiding those with physical and mental disabilities by preventing employment discrimination, providing for public access to public transportation, and providing for the use of other facilities and services used by the public at large.

Ames test
A test used to determine the carcinogenicity of chemicals. It is often referred to as the *Salmonella test.* In the test, mutant strains of *Salmonella typhimurium* are cultured on a medium deficient in histidine while being exposed at the same time to a potential carcino-

gen and liver extracts. Mutagenic bacteria will back-mutate to contain a functional histidine gene, permitting bacterial growth. The level of mutagenicity can be determined by the number of colonies that develop.

ametropia
An abnormal condition of the eye with respect to refraction of light, as in myopia.

AMI
See ***Available Motions Inventory***.

amicus curiae
A Latin term meaning "friend of the court," referring to a party authorized by a court to submit a legal brief (but generally not oral argument or evidence) to assist the court in resolving the litigation. The term does not connote a full party in the litigation for purposes of making motions, conducting discovery, participating at trial, or appealing to a higher court. That would require *intervention*.

amino
The monovalent radical NH_2, when not united with an acid radical.

amino acid
An organic acid that is one of the building blocks in the formation of proteins. More than 20 different amino acids are commonly found in proteins. Some of them can be produced within the body, but there are eight that the human organism cannot manufacture; these essential amino acids are isoleucine, leucine, lysine, methionine, phenylalanine, threonine, tryptophan, and valine. Histidine and arginine, which may be manufactured in the body under certain circumstances, are sometimes considered essential. Protein foods that provide large amounts of essential amino acids are known as *complete proteins* and include proteins from animal sources such as meat, eggs, fish, and milk. Proteins that cannot supply the body with all essential amino acids are known as *incomplete proteins*; these are the vegetable proteins most abundantly found in peas, beans, and certain forms of wheat.

aminopyrine
A drug used as a fever preventive and pain reliever.

aminosis
Excessive formation of amino acids in the body.

aminuria
The presence of amines in the urine.

amitosis
The direct method of cell division, characterized by simple cleavage of the nucleus, without formation of chromosomes.

AML
Acute myelogenous leukemia.

ammeter
An instrument for measuring in amperes the strength of a current flowing in a circuit.

ammoaciduria
The presence of ammonia and amino acids in the urine.

ammonia
A colorless alkaline gas, NH_3, with a pungent odor and acrid taste, and soluble in water.

ammonia nitrogen
The quantity of elemental nitrogen present in the form of ammonia.

ammoniated mercury
A compound used as an antiseptic skin and ophthalmic ointment. It should be applied with caution as excessive use may irritate the skin and cause a dermatitis.

ammoniemia
The presence of ammonia or its compounds in the blood.

ammonification
Bacterial decomposition of organic nitrogen to ammonia.

ammonium
A hypothetical radical, NH_4, forming salts analogous to those of the alkaline metals.

ammoniuria
Excess of ammonia in the urine.

amnesia
Pathologic impairment of memory usually the result of physical damage to areas of the brain from injury, disease, or alcoholism. It may also be caused by a decreased supply of blood to the brain, a condition that may accompany senility. Another cause is psychological. A shocking or unacceptable situation may be too painful to remember, and the situation is then retained only in the subconscious mind. The technical term for this is repression. Rarely is the memory completely obliterated. Amnesia

takes different forms depending upon the area of the brain affected and how extensive the damage is. In *auditory amnesia,* or "word deafness," the patient is unable to interpret spoken language. In *visual amnesia,* or "word blindness," the written language is forgotten. *Tactile amnesia* is the inability to recognize once familiar objects by the sense of touch.

amoebic dysentery
See amebic dysentery.

amorphous
Non-crystalline and without definite shape or form.

amortization, capital leases
Charges applicable to assets recorded under capital leases.

amortization of developmental and preoperating expenses, etc.
Charges to expense for the 1) amortization of capitalized developmental and preoperating costs and other intangible assets; and 2) obsolescence and deterioration of flight equipment spare parts (included under depreciation and amortization).

amosite asbestos
An asbestiform mineral of the amphibole group made up of straight brittle fibers which are light gray to pale brown in color. Often referred to as brown asbestos.

amp
See ***ampere.***

ampacity
Current-carrying capacity of electric conductors expressed in amperes.

ampere (A)
A unit of electrical current; that amount of constant electrical current which, if maintained in two straight, infinitely long, parallel conductors having negligible cross-sectional area and separated by 1 meter in a vacuum, would produce a force of 2×10^{-7} newtons per meter of conductor.

amperometric titrator
Titration device containing an internal indicator or electrometric device to show when the reactions are complete.

amphetamine
A white crystalline powder used as a central nervous system stimulant. It is odorless and has a slightly bitter taste. Amphetamine has the temporary effect of increasing energy and apparent mental alertness. It is used in some cases of mental depression and alcoholism, in the chronic rigidity following encephalitis, in attacks of narcolepsy, and to control the appetite of obese people. It is also used to overcome the depressant effects of barbiturates. Caution is warranted when using amphetamine in persons hypersensitive to stimulants, those suffering from coronary or cardiovascular disease or hypertension, or women in the early stages of pregnancy.

Amphibia
A class of animals living both on land and in water.

amphibole
One of two major groups of minerals.

amphibole asbestos
Fibrous silicates of magnesium, iron, calcium, and sodium that are generally brittle. This form of asbestos is more resistant to heat than the serpentine (chrysotile) type.

ampholyte
An organic or inorganic substance capable of acting as either an acid or a base.

amphoteric
Material having the capacity of behaving as an acid or a base.

amphotericity
The power to unite with either positively or negatively charged ions, or with either basic or acid substances.

ampicillin
A broad-spectrum penicillin of synthetic origin, used in treatment of a number of infections, and available in oral preparations as well as ampules for intramuscular injections. It is active against many of the gram-negative pathogens, in addition to the usual gram-positive ones that are affected by penicillin.

amplification
(1) *General.* To make larger, as in the increase of an auditory or visual stimulus, as a means of improving its perception. (2) *Radiation.* As related to radiation detection instruments, the process (either gas, electronic, or both) by which ionization effects are magnified to a degree suitable for their measurement.

amplitude
(1) The instantaneous deviation or displacement from some baseline. (2) The peak-to-peak difference, maximum value, or averaged value of a signal.

amplitude modulation
The multiplication of an approximately constant higher frequency carrier signal by a second signal, usually of a much lower frequency.

ampoule
See ***ampule***.

ampule
A small, hermetically sealed glass flask which may contain medicine or some other chemical compound in a pre-measured dose for administration by a third party (e.g., parental administration of a medicine to a child). Also known as *ampoule*.

ampulla
The enlarged portion of a semicircular canal in the inner ear which contains the crista.

amputation
The removal of a limb or other appendage or outgrowth of the body. Amputation is sometimes necessary in cases of cancer, infection, and gangrene. It may be necessary after irreparable traumatic injury to a limb. Blood vessel disorders such as arteriosclerosis, often secondary to diabetes mellitus, account for the greatest percentage of non-injury-induced leg amputations.

AMSA
Association of Metropolitan Sewage Agencies.

amt.
Amount.

Amtrak
American Track operated by the National Railroad Passenger Corporation of Washington, DC. This rail system was created by President Nixon in 1970, and was given the responsibility for the operation of intercity, as distinct from suburban, passenger trains between points designated by the Secretary of Transportation.

amu
Atomic mass unit.

AMVER
See ***automated mutual assistance vessel rescue system***.

amyl nitrate
A vasodilator often used in the treatment of *angina pectoris* because of its quick relief of pain. Presumably it relaxes the smooth muscles of the coronary arteries, causing dilation of these blood vessels. The drugs are dispensed in pearls that are crushed and inhaled. It acts very quickly and its effects are brief. These effects include decreased blood pressure, irregular pulse, headache, and dizziness.

amyotrophic lateral sclerosis
A type of motor disorder of the nervous system in which there is destruction of the anterior horn cells and pyramidal tract. The cause is unknown. Early symptoms include weakness of the hands and arms, difficulty in swallowing and talking, and weakness and spasticity of the legs. As the disorder progresses there is increased spasticity and atrophy of the muscles, with loss of motor control and over-activity of the reflexes. There is no known specific or effective treatment. Although there may be periods of remission, the disease usually progresses rapidly with death occurring in 2 to 5 years in most cases.

an easy distance off
Marine Safety. A reasonably close (comfortable) distance off the bank, allowing ample room for maneuverability. This is a relative term depending upon size of stream and size of tow. The term "easy" pertains more to ease of mind than to closeness of distance and indicates that this portion of the river has no tight spots.

anabatic wind
A localized wind that flows up valley or mountainous slopes, usually in the afternoon, caused by the replacement of cool valley air with the warmer air above it.

anabolism
The synthesis of more complex living structures from simpler materials.

anadromous
Fish that spend their adult life in the sea but swim upriver to fresh water spawning grounds to reproduce.

anaerobe
Organisms unable to multiply in any environment that contains oxygen. Anaerobic microorganisms have oxygen-sensitive enzymes and cannot function in the presence of molecular oxygen. Some may be more air tolerant than others. Those severely affected by the presence of oxygen are called *strict anaerobes* or *obligate anaerobes.*

anaerobic
Meaning without oxygen. Also refers to cells or organisms that can live without oxygen or processes that occur in the absence of oxygen.

anaerobic bacteria
Bacteria that do not require free oxygen to live, or are not destroyed by its absence.

anaerobic digestion
Sludge stabilization process where the organic material in biological sludge is converted to methane and carbon dioxide in an airtight reactor. The process is conducted in the absence of air at residence times ranging from 60 days at 20°C to 15 days at 35 to 55°C, with a volatile solids reduction of at least 38 percent.

anaerobic energy
That energy derived from anaerobic metabolism.

anaerobic metabolism
The partial physiological oxidation of glucose or other bodily fuels in tissues without adequate oxygen, forming lactic acid with the release of energy. It can provide a brief reserve of energy under physical workloads.

anakusis
Complete deafness.

analog
A system, such as the output of a meter, where numerical data are represented by analogous physical magnitudes or electrical signals that vary continuously.

analog-to-digital conversion
The process of sampling the amplitude of a continuously varying signal at specified intervals and presenting a digital value to a resolution of some number of bits, typically carried out by an analog-to-digital converter.

analogue method of forecasting
A forecast made by comparison of past large-scale synoptic weather patterns that resemble a given (usually current) situation in its essential characteristics.

analysis
(1) *General.* A study or evaluation, usually performed to determine the current status of a given system or process. It will often utilize established standards or operating criteria as a baseline for comparison. (2) *Weather forecasting.* The drawing and interpretation of the patterns of various weather elements on a surface or upper air chart.

analysis of covariance (ANCOVA)
A modified analysis of variance involving compensation for covariates when random groups cannot be selected.

analysis of variance (ANOVA)
Any of a series of statistical tests in which variances are compared across two or more groups to make a determination as to whether the means of the groups are likely to be significantly different from one another.

analyte
The substance or contaminant being analyzed for in an analytical procedure.

analytical blank
Sampling media which has been set aside for analysis but which was not taken into the field.

analytical estimating
A technique in work measurement in which element times are estimated from previous experience and knowledge of the concerned elements.

analytical standard data
A set of time values represented in the form of or computed by a mathematical model. *See also* ***standard data****.*

analytical workplace design
The process of using established human factors' concepts to design a workplace suitable for human interaction.

analyzer
In acoustical science, a combination of filters and a system for indicating the relative energy that is passed through the filter system. The measurement is usually interpreted as giving

the distribution of energy of the applied signal as a function of frequency.

anaphylactic shock
A serious and profound state of shock brought about by hypersensitivity (*see **anaphylaxis***) to an allergen, such as a drug, foreign protein, or toxin. Insect bites and stings in hypersensitive persons may produce anaphylactic shock. Early symptoms are typical of an allergic reaction such as sneezing and edema or itching at the site of injection or sting. The symptoms increase in severity very rapidly and progress to dypsnea, cyanosis, and shock. The blood pressure drops rapidly, the pulse becomes weak and faint, and convulsions and loss of consciousness may occur. Severe anaphylactic shock can be fatal if immediate emergency measures are not taken.

anaphylaxis
An unusual or exaggerated allergic reaction of an organism to a foreign protein or other substance following previous contact with that material.

anaplasia
An irreversible alteration in adult cells toward more primitive or reversed development.

anatomic
(1) Of or pertaining to human anatomy or any of its various components. (2) Relating to the science of the morphology or structure of organisms.

anatomical position
A standard posture for defining certain aspects of the human body: the body is standing erect with the arms hanging at the sides and the wrists supinated such that the palms face forward/anterior.

anatomical reference point
*See **landmark**.*

anatomy
The study of the geometrical and topographical features of all body structures and of the body as a whole.

anchorage
An area where a vessel anchors or may anchor, either because of suitability or designation.

anchor it
Apply brakes for an emergency stop.

ancillary input
Under ISO 14000, material input that is used by the unit process producing the product, but is not used directly as a part of the product.

Andes disease
*See **altitude sickness**.*

androgen
Any substance that stimulates male characteristics. The two male androgens are androsterone and testosterone. The androgenic hormones are internal endocrine secretions circulating in the bloodstream and manufactured mainly by the testes under stimulation from the pituitary gland. To a lesser extent, androgens are produced in the adrenal glands in both sexes, as well as by the ovaries in women. Thus, women normally have a small percentage of male hormones, in the same way that men's bodies contain some female sex hormones, the *estrogens*. The androgens are responsible for the secondary sex characteristics, such as the beard and the deepening of the voice at puberty. They also stimulate the growth of muscle and bones throughout the body and thus account in the part for the greater strength and size of men as compared to women.

anechoic
The lack of significant reflected energy waves, usually with reference to sounds; having no echo.

anechoic chamber
*See **anechoic room**.*

anechoic room
A room whose boundaries (e.g., walls, ceiling, etc.) effectively absorb all the sound that is incident on their surface, thereby creating essentially a free-field condition. Also referred to as a *free-field room*.

aneisekonia
A condition in which different image sizes are experienced in the two eyes.

anemia
(1) A disorder of the blood as a whole; a deficiency in the number of red corpuscles or of hemoglobin. (2) A pathological deficiency of the oxygen-carrying material of the blood, measured in unit volume concentrations of hemoglobin, red blood cell volume, and red blood cell number.

anemometer
An instrument used to measure the motion of wind or air that employs a pitot tube directed by a vane or rotor, or a pressure plate deflected against a spring or gravity.

aneroid barometer
A barometer which measure atmospheric pressure using one or more aneroid capsules in series.

aneroid capsule
A thin metal disc partially evacuated of air and used to measure atmospheric pressure by measuring the expansion or contraction of the capsule as the pressure changes.

anesthesia
Loss of feeling or sensation. Artificial anesthesia may be produced by a number of agents capable of bringing about partial or complete loss of sensation. *See also **anesthetic**.*

anesthetic
(1) A chemical that has a depressant effect on the central nervous system, particularly the brain, and which induces insensibility to pain. (2) Lacking feeling or sensation.

anesthetic effect
A loss of the ability to perceive sensory stimulation that can be brought about by exposure to certain chemical substances either by inhalation, ingestion, injection, or dermal absorption.

aneurysm
A sac formed by dilation of the walls of a blood vessel, usually an artery, and filled with blood. There are two types of aneurysms: *true aneurysm,* in which the wall of the sac consists of one or more of the layers that make up the wall of the blood vessel, and *false aneurysm,* in which all the layers of the vessel are ruptured and the blood is retained by surrounding tissues. Aneurysms occur when the blood vessel wall becomes weakened by either physical injury to the vessel, a congenital defect, or a disease. They may occur in any vein or artery, but are most commonly found in the abdomen or chest. Certain infections may attack and weaken the tissues of the blood vessels; however, atherosclerosis is a common cause. A less common cause is syphilis. A person may have a small aneurysm for years without being aware of it; such aneurysms are often identified only accidentally, on x-ray examination for another purpose. An aneurysm may form a pulsating tumor which can be painful to the sufferer, especially if it is large enough to press against some other organ in the body. Aneurysms tend to increase in size, and there is a risk of rupture. If rupture occurs in the heart or brain or any other vital organ of the body, the results can be very serious.

angina
Any disease marked by spasmodic suffocative attacks, especially *angina pectoris*.

angina pectoris
Acute pain in the chest caused by interference with the supply of oxygen to the heart. Most sufferers from angina pectoris can readily distinguish it from other pains in the chest, such as might be caused by indigestion or coronary thrombosis, for the pain is usually of an unmistakable nature. It is generally described as a feeling of tightness, strangling, heaviness, or suffocation. The pain is usually just under the sternum and sometimes radiates down the neck, throat, lower jaw, left arm, and, more rarely, to the stomach, back, or across to the right side of the chest.

angiosarcoma
A malignant growth on the inner linings of blood vessels, typically found in areas of high blood vessel concentration, such as the liver. Vinyl chloride monomer is known to cause angiosarcoma of the liver.

angiospasm
The spasmodic contraction of blood vessels.

angle
The space or figure formed by two diverging lines, measured as the number of degrees one would have to be moved to coincide with the other.

angle collision
Collisions which are not head on, rear end, rear to rear, or sideswipe.

angle diagram
A graphical plot of the angular relationship over time between two joints as the joints move in some specified way.

angle of abduction
*See **abduction angle**.*

angle of incidence
That angle from the perpendicular to the surface of an object at which a light ray or other entity strikes the surface of that object.

angle of repose
The maximum angle that the inclined surface of a loosely divided material can make with the horizontal.

angle of resolution
*See **minimum resolution angle**.*

angle-torque curve
Any graphical relationship in which the maximum isometric force exerted at a given angle is plotted against that angle for the range of motion.

angstrom
A unit of length used chiefly in expressing short wavelengths. It is equal to 10^{-10} meter or 10^{-8} centimeter. It is typically represented in formula or by reference with the symbol Å.

angular acceleration (α)
The rate of change of angular velocity with time. Also referred to as *rotational acceleration*. Represented by the formula:

$$\alpha = \frac{d\omega}{dt}$$

angular deviation
That angle between the incident and transmitted light rays in a prism.

angular displacement
A vector representing the change in angle by rotation about some origin.

angular frequency
The oscillation frequency in an oscillating system multiplied by 2.

angular momentum (L)
A vector representing the rotational momentum of an object about an axis.

angular motion
The movement of a structure about its own local center of rotation.

angular velocity
The rate of change of angular displacement about some axis of rotation with time. Also referred to as *rotational velocity*.

anhydride
A chemical compound derived by the elimination of water.

anhydrous
A compound that does not contain water.

animal
All vertebrate and invertebrate species, including but not limited to man and other mammals, birds, fish, and shellfish.

animal feed
Under the Federal Food, Drug, and Cosmetic Act: An article which is intended for use as food for animals other than man and which is intended for use as a substantial source of nutrients in the diet of the animal, and is not limited to a mixture intended to be the sole ration of the animal.

animal starch
*See **glycogen**.*

anion
A negatively charged ion that migrates to the anode when an electrical potential is applied to a solution.

anionic polymer
A polyelectrolyte with a net negative electrical charge.

anisomelia
Inequality of length, as of a limb.

anistropic
Having physical properties which vary in different spatial directions.

ankle
The joint formed by the junction of the distal ends of the fibula and the tibia with the talus, including all the surrounding soft tissues.

ankle bone
*See **talus**.*

ankle breadth
*See **bimalleolar breadth**.*

ankle height
The vertical distance from the floor or other reference surface to the level of the ankle circumference measure. Measured with the individual standing erect.

annihilation radiation
Photons produced when an electron and a positron unit cease to exist. The annihilation of a positron-electron pair results in the pro-

duction of two photons, each of which has at least 0.511 MeV energy.

annoyance
A condition or stimulus which causes one to be disturbed, irritated, or troubled.

annual aggregate financial ability (accidental releases)
The amount of money that would be required to pay for accidental releases that may occur within 12 months.

annual committed effective dose
According to Chapter I (Environmental Protection Agency), Subchapter F (Radiation Protection Programs) of Title 40 (Protection of Environment) of the Code of Federal Regulations: The committed effective dose resulting from one-year intake of radionuclides released plus the annual effective dose subject to the Environmental Standards for Groundwater Protection under Subchapter F.

annual injury incidence
An OSHA formula used for determining the injury rate for comparison with other companies or industries, expressed as:

$$AII = \frac{\text{number of OSHA form 200 recordable injuries x 200,000}}{\text{number hours worked by company employee}}$$

annual operating factor
The annual fuel consumption divided by the product of design firing rate and hours of operation per year.

annual range of temperature
The difference between the warmest and coldest months at any given location.

annuity insurance
An insurance contract calling for periodic payments to the insured or annuitant for a stated period or for life. *See also* ***insurance***.

annulus
The fiber and cartilage structure surrounding the nucleus pulposus in an intervertebral disk. Also referred to as *annulus fibrosus*.

anode
Positive electrode. The electrode to which negative ions are attracted.

anorexia
The lack of, or loss of, appetite for food.

anosmia
The absence of the sense of smell.

anoxemia
The reduction of the oxygen content of the blood to below physiologic levels.

anoxia
The absence of, or a diminished amount of, oxygen in the blood, tissue, or a body of water. The deficiency of oxygen in organisms often results in an increased rate of breathing. Anoxia in humans is often accompanied by dizziness, rapid heartbeat, and headache. It can result in death.

anoxic
A condition characterized by the absence of free oxygen.

ANPRM
See ***Advanced Notice of Proposed Rule Making.***

ANSI
See ***American National Standards Institute.***

antagonism
(1) *General.* The competitive interaction or opposition of two or more agents to control or lessen the effect of an agent's individual effect(s). (2) *Chemistry.* The interaction of two chemicals having an opposing, or neutralizing, effect on each other, or given some specific biological effect, a chemical interaction that appears to have an opposing or neutralizing effect over what might otherwise be expected.

antagonist
An entity which opposes or competes with the action of another entity. It may be a person, group, muscle, or drug.

antagonistic
A substance that tends to nullify the action of, or acts against, another. Opposition in the action between similar things, as between medicines, chemicals, muscles, etc.

Antarctica mineral resource activity
Under the Federal Antarctic Protection Act of 1990: Prospecting, exploration, or development in Antarctica of mineral resources, but does not include scientific research within the meaning of Article III of the Antarctic treaty, done at Washington, DC on December 1, 1959.

Antarctic "ozone hole"
Refers to the seasonal depletion of ozone in a large area over Antarctica.

antenna
A metallic apparatus for sending and receiving electromagnetic waves.

antenna array
A group of directional antennas.

anterior
Pertaining to the front portion of the body or toward the front of the body.

anterior neck length
The surface distance from suprasternale to the junction of the posterior lower jaw and the neck in the midsagittal plane. It is measured with the individual standing erect and looking straight ahead.

anterior waist length
The surface distance from the most anterior point of the lower neck to the waist. Measured with the individual standing erect.

anthracosilicosis
A complex, chronic pneumoconiosis that is a combination of anthracosis and silicosis.

anthracosis
Also known as Collier's disease, Shaver's disease, miner's lung, and black lung; a usually asymptomatic pneumoconiosis resulting from the accumulation of carbon from inhaled smoke or coal dust in the lungs.

anthrax
An acute, highly infectious, bacterial disease usually affecting the skin. Also known as wool sorter's disease, rag picker's disease, or malignant edema, this disease is transmitted by contact with tissues of infected animals (cattle, sheep, goats, horses, and others) or contaminated hair, wool, or hides.

anthro-
Prefix; like or pertaining to man.

anthropogenic compound
A compound created by human beings, often relatively resistant to biodegradation.

anthropology
The science of man and mankind, including the study of the physical and mental constitution of man, his cultural development, and social conditions, as exhibited both in the present and in the past.

Anthropology Research Project (ARP)
A Department of Defense (DOD)-sponsored project to provide anthropometric surveys for USAF flying personnel.

anthropometer
A device for measuring linear dimensions of the body.

anthropometric
Relating to human body measurements and modes of action to determine their influence on the safe and efficient operation of equipment.

anthropometric evaluation
A study of body size and actions with the objective of improving the design of machines and tools to enable more effective use of them by humans.

anthropometric measurement
Any physical measurement derived from the body or its various parts.

anthropometrist
One who is qualified by education, training, and experience to practice anthropometry.

anthropometry
The measurement of the human body, including body dimensions, range of motion of body members, and strength (including both static and dynamic measurements). The branch of anthropology that deals with the comparative measurements of the human body. *See also* ***ergonomics***.

anthropomorphic
Having a form like a human or human parts.

anthropophobia
A pathological fear of human companionship or of society in general.

antibiosis
A relationship between two organisms that is harmful to one, as parasitism.

antibiotic
A chemical substance produced by living organisms that inhibits the growth of or kills other organisms.

antibody
A globulin found in tissue fluids and blood serum that is produced in response to the stimulus of a specific antigen, and is capable of combining with that antigen to neutralize

or destroy it. Also referred to as *immune substances.*

anticipated cost of removal
Under the Federal Forest and Range Land Renewable Resources Planning Act of 1974: The projected cost of removal of wood residues from timber sales areas to points of prospective use, as determined by the Secretary of Agriculture at the time of advertisement of the timber sales contract in accordance with appropriate appraisal and sale procedures.

anticipated value
Under the Federal Forest and Range Land Renewable Resources Planning Act of 1974: The projected value of wood residues as fuel or other merchantable wood products, as determined by the Secretary of Agriculture at the time of advertisement of the timber sales contract in accordance with appropriate appraisal and sale procedures.

anticipation error
An error produced due to an expectation of a change.

anticoagulant
Any substance that inhibits the blood clotting mechanism.

anticonvulsant
Inhibiting convulsions; an agent that suppresses convulsions.

anticyclone
An area of high pressure around which the wind blows clockwise in the Northern Hemisphere and counterclockwise in the Southern Hemisphere.

antidegradation clause
Part of federal air quality and water quality requirements prohibiting deterioration where pollution levels are above the legal limit.

antidote
A remedy to counteract the effects of a toxic substance.

anti-exposure suit
Any form of outer clothing to protect an individual from the elements, especially wind and cold temperatures.

antifoam agent
A surface active agent used to reduce or prevent foaming.

antifoulant
An additive or dispersant that prevents fouling and/or the formation of scale.

antifouling paint
Under the Federal Organotin Antifouling Paint Control Act of 1988: A coating, paint, or treatment that is applied to a vessel to control fresh water or marine fouling organisms.

anti-g straining maneuver (AGSM)
Any internally generated technique for temporarily increasing blood pressure in an attempt to withstand high positive g stresses in high performance aircraft and/or spacecraft. *See **M-1 maneuver*** and ***L-1 maneuver***.

anti-g suit
A special garment designed to apply counter pressure to the lower body during high positive g forces as an aid in preventing blackout of the wearer.

antigen
That portion or product of a biologic agent capable of stimulating the formation of specific antibodies.

anti-glare filter
A transparent device for reducing glare.

antihistamine
A drug that counteracts the effects of histamine, a normal body chemical that is believed to cause the symptoms of persons who are hypersensitive to various allergens. Antihistamines are used to relieve the symptoms of allergic reactions, especially hay fever and other allergic disorders of the nasal passages. Some antihistamines have an antinauseant action that is useful in the relief of motion sickness while others have a sedative and hypnotic action and may be used as tranquilizers.

anti-inflammatory
Counteracting or suppressing inflammation.

antiknock additive
A compound, usually tetraethyl lead, added to gasoline to minimize engine pre-ignition and its accompanying knocking and pinging. Pollution from the release of such compounds in auto emissions led to the introduction of unleaded gasoline.

antilogarithm
A number whose logarithm returns original number.

antimetabolite
A substance exerting its desired effect perhaps by replacing or interfering with the utilization of an essential metabolite.

antimicrobial
Agent that kills microbial growth. *See also* ***disinfectant, sanitizer,*** *and* ***sterilizer.***

antimony
A chemical element, atomic number 51, atomic weight 121.75, symbol Sb. Antimony compounds are used in medicine as anti-infective agents in the treatment of tropical diseases, especially those of protozoan origin. All antimony compounds are potentially poisonous and must be used with caution.

antimorphic
In genetics, antagonizing or inhibiting normal activity (as in an antimorphic mutant gene).

antioxidant
A chemical compound added to a substance to reduce deterioration from oxidation; a preservative.

antipole
A point on the skull opposite to the point of impact in an accident.

antiscalant
An additive that prevents the formation of inorganic scale.

antiseptic
Chemical compounds that are capable of reducing the number of microorganisms on body surfaces. Used primarily on humans and animals, in contrast to *disinfectants,* which are used primarily on non-living surfaces as a form of infection control. *See also* ***disinfectant.***

antitoxin
An antibody to the toxin of a microorganism, usually a bacterial exotoxin. Antitoxins combine with a specific toxin, in vivo or in vitro, with the consequent neutralization of toxicity.

antitrope
One of two structures that are similar but reverse oriented, like a right and left glove.

antivenin
A material used to neutralize the venom of a poisonous animal.

anuria
The absence of the excretion of urine from the body.

anxiety
A feeling of uneasiness, apprehension, or dread.

AOA
American Optometric Association.

AOC
See ***assimilable organic carbon.***

AOMA
See ***American Occupational Medical Association.***

AOPs
See ***advanced oxidation processes.***

aorta
The great artery arising from the left ventricle. *See also* ***circulatory system.***

aortoclasia
Rupture of the aorta.

apathy
Reactive absence of emotions.

APCA
Air Pollution Control Association.

apepsia
Cessation or failure of digestive function.

APF
See ***assigned protection factor.***

APHA
American Public Health Association.

aphagia
Loss of the power of swallowing and subsequent failure to eat, which can result in illness or death.

aphakic
Pertaining to an individual with the lens removed from one or more eyes.

API
American Petroleum Institute.

API gravity
An index inversely related to specific gravity used to identify liquid hydrocarbons.

API separator
Rectangular basin in which wastewater flows horizontally while free oil rises and is skimmed from the surface.

aplasia
Absence of an organ due to failure of development of the embryonic primordium.

aplastic
Pertaining to or characterized by aplasia; having no tendency to develop into new tissue.

aplastic anemia
A condition in which the bone marrow fails to produce an adequate supply of red blood cells.

apnea
The temporary cessation of breathing.

apneumia
A developmental anomaly with the absence of the lungs.

apocrine
Denoting that type of glandular secretion in which the secretory products become concentrated at the free end of the secreting cell and are thrown off, along with the portion of the cell where they have accumulated, as in the mammary gland. *See also* ***eccrine gland***.

aponeurosis
An expansion of a muscle tendon which serves to attach a muscle to bone at an origin or insertion, or to enclose a group of muscles.

apophysis
Any outgrowth or swelling, especially a bony outgrowth that has never been entirely separated from the bone of which it forms a part, such as a process, tubercle, or tuberosity.

apoplexy
Copious extravasation of blood into an organ; often used alone to designate such extravasations into the brain (cerebral apoplexy) after rupture of an intracranial blood vessel; stroke.

apparent color
The color in water caused by the presence of suspended solids.

apparent motion
An illusion of motion, regardless of the cause, whether by certain patterns of non-moving stimuli, by certain conditions under which non-moving stimuli are observed, or by stimulation of sensory receptors or the nervous system.

appeal
A legal term referring to carrying a matter to a higher tribunal, as from agency staff to a hearing officer, or a hearing officer to reviewing board, or reviewing board to court, or trial court to appellate court. The Supreme Court is the highest appeal court of a state. The United States Supreme Court is the highest appeal court in the federal judiciary.

appellate
Pertaining to or having cognizance of appeals and other proceedings for the judicial review of adjudications. The term has a general meaning, and it has a specific meaning indicating the distinction between original and jurisdiction and appellate jurisdiction.

appellate court
A court having jurisdiction of appeal and review; a court to which causes are removable by appeal, certiorari, error, or report.

appendicitis
Inflammation of the vermiform appendix. Appendicitis is a serious disease, usually requiring surgery.

appetite
The desire for food. It is stimulated by the sight, smell, or thought of food and accompanied by the flow of saliva in the mouth and gastric juice in the stomach. The stomach wall also receives an extra blood supply in preparation for digestive activity. Appetite is psychological, depending on memory and associations, as compared with hunger, which is physiologically aroused by the body's need for food. Appetite can be discouraged by unattractive food, surroundings, or company, and by emotional states such as anxiety, irritation, anger, and fear. Certain drugs may also affect appetite. Chronic loss of appetite is known as anorexia.

appliance
(1) *General.* Utilization equipment, generally other than industrial, normally built to perform one or more functions such as clothes washing, air conditioning, food mixing, deep frying, etc. (2) *Aviation.* Any instrument, mechanism, equipment, part, apparatus, appurtenance or accessory, including communications equipment, that is used or intended to be used in operating or controlling an aircraft in flight, is installed in or attached to the aircraft, and is not part of an airframe, engine, or propeller.

applicant
Transportation. A governmental entity, a nonprofit public-purpose organization, or any

responsible person having the legal, financial, and technical capacity to implement an intermodal passenger terminal project under 49 CFR 256. The applicant must have legal authority to receive and expend federal funds.

application
A software package for performing a specific type of task other than direct system support or system utilities.

applied load
The actual load (or stress) imposed on a structure in the service environment

applied sciences
Those disciplines involved in the use of information gathered by the basic sciences.

apportionment
(1) *Government*. The process by which legislative seats are distributed among units entitled to representation. (2) *Aviation*. Distribution of Airport Improvement Plan (AIP) funds from the Airport & Airways Trust Fund to airport sponsors based on enplanements or cargo landed weights. Also referred to as *entitlement*. *See also* ***passenger facility charge***.

approach clearance
Authorization by Air Traffic Control (ATC) for a pilot to conduct an instrument approach. The type of instrument approach for which clearance and other pertinent information are provided in the approach clearance when required.

approach control facility
A terminal air traffic control facility providing approach control service.

approach control service
Air traffic control service provided by an approach control facility for arriving and departing Visual Flight Rules (VFR)/Instrument Flight Rule (IFR) aircraft and, on occasion, en route aircraft. At some airports not served by an approach control facility, the Air Route Traffic Control Center (ARTCC) provides limited approach control service.

approach gate
An imaginary point used within Air Traffic Control (ATC) as a basis for vectoring aircraft to the final approach course. The gate will be established along the final approach course 1 mile from the outer marker (or the fix used in lieu of the outer marker) on the side away from the airport for precision approaches and 1 mile from the final approach fix on the side away from the airport for non-precision approaches. In either case, when measured along the final approach course, the gate will be no closer than 6 miles from the landing threshold.

approach light system (ALS)
An airport lighting facility which provides visual guidance to landing aircraft by radiating light beams in a directional pattern by which the pilot aligns the aircraft with the extended centerline of the runway on his final approach for landing. Condenser-Discharge Sequential Flashing Lights/Sequenced Flashing Lights may be installed in conjunction with the Approach Lighting System (ALS) at some airports. Types of approach light systems are 1) ALSF-1: Approach Light System with Sequenced Flashing Lights in Instrument Landing System (ILS) Cat-I configuration. 2) ALSF-2: Approach Light System with Sequenced Flashing Lights in ILS Cat-II configuration. The ALSF-2 may operate as a Simplified Short Approach Light System with Runway Alignment Indicator Lights (SSALR) when weather conditions permit. 3) SSALF: Simplified Short Approach Light System with Sequenced Flashing Lights. 4) SSALR: Simplified Short Approach Light System with Runway Alignment Indicator Lights. 5) MALSF: Medium Intensity Approach Light System with Sequenced Flashing Lights. 6) MALSR: Medium Intensity Approach Light System with Runway Alignment Indicator Lights. 7) LDIN: Lead-in-light system: Consists of one or more series of flashing lights installed at or near ground level that provide positive visual guidance along an approach path, either curving or straight, where special problems exist with hazardous terrain, obstructions, or noise abatement procedures. 8) RAIL: Runway Alignment Indicator Lights (Sequenced Flashing Lights which are installed only in combination with other light systems). 9) ODALS: Omnidirectional Approach Lighting System consists of seven omnidirectional flashing lights located in the approach area of a nonprecision runway. Five lights are located on the runway centerline with the first light located 300 feet from the threshold and extending at equal intervals up to 1,500 feet from the threshold. The other

two lights are located, one on each side of the runway threshold, at a lateral distance of 40 feet from the runway edge, or 75 feet from the runway edge when installed on a runway equipped with a Visual Approach Slope Indicator (VASI).

approach locking
Electric locking effective while a train is approaching, within a specified distance. A signal displaying an aspect to proceed, and which prevents, until after the expiration of a predetermined time interval after such signal has been caused to display its most restrictive aspect, the movement of any interlocked or electrically locked switch, movable-point frog, or derail in the route governed by the signal, and which prevents an aspect to proceed from being displayed for any conflicting route.

approach sequence
The order in which aircraft are positioned while on approach or awaiting approach clearance.

approach signal
A roadway signal used to govern the approach to another signal and, if operative, so controlled that its indication furnishes advance information of the indication of the next signal.

approach speed
The recommended speed contained in aircraft manuals used by pilots when making an approach to landing. This speed will vary for different segments of an approach as well as for aircraft weight and configuration.

approach velocity
The average water velocity of fluid in a channel upstream of a screen or other obstruction.

approachway
The airspace through which aircraft approach or leave a landing area.

appropriate air traffic service (ATS) authority
The relevant authority designated by the state responsible for providing air traffic services (ATS) in the airspace concerned. In the United States, the "appropriate ATS authority" is the Director, Office of Air Traffic System Management, ATM-1.

appropriate authority
(1) Regarding flight over the high seas, the relevant authority is the State of Registry. (2) Regarding flight over other than the high seas, the relevant authority is the state having sovereignty over the territory being overflown.

approved
(1) *Product Safety*. Item that has been tested and found to be acceptable by a recognized authority and approved for use under specified conditions. Testing agencies include the U.S. Bureau of Mines, Factory Mutual (FM), Underwriters Laboratories (UL), the U.S. Department of Agriculture, and others. (2) *Marine Safety*. A term used to indicate Coast Guard approval of a specific item among the limited number that the Coast Guard has been directed by law to test and "approve." Some of these items are personal flotation devices, fire extinguishers, carburetor backfire flame arresters, distress signals, and certain types of life rafts. The standards program has not required "approval" of any boat or item of associated equipment.

approved equipment
Equipment that has been designed, tested, found to be acceptable, and approved by an appropriate authority as safe for use in a specified hazardous location or atmosphere.

approved for the purpose
Approved for a specific use, purpose, environment, or application as described in a particular standard requirement. Suitability of equipment, or materials for a specific purpose, environment or application may be determined by a nationally recognized testing laboratory, inspection agency, or other organization concerned with product evaluation as part of its listing and labeling program.

approved landfill
Site that has been approved by a government environmental protection authority (federal or state) for the disposal of hazardous wastes.

approved refrigerant recycling equipment
Under the Clean Air Act (CAA): Equipment certified by the EPA Administrator (or an independent standards testing organization approved by the Administrator) to meet the standards established by the Administrator and applicable to equipment for the extraction and reclamation of refrigerant from motor vehicle air conditioners. Such standards shall, at a minimum, be at least as stringent as the standards of the Society of Automotive Engi-

neers (SAE) in effect as of November 15, 1990 and applicable to such equipment (SAE Standard J-1990).

apocrine
Sweat glands that open into hair follicles. Apocrine sweat glands are limited to a few regions of the body, notably the underarm and genital areas.

apraxia
Impairment of the ability to use objects correctly.

apron
(1) A floor or lining of resistant material at the toe of a dam or bottom of a spillway to prevent erosion from turbulent water flow. (2) That portion of a pier and wharf measured between the outer edges of the water-facing side and the transit shed or other inshore structure. (3) That portion of an airport runway area used for final aircraft checklist completion prior to departure. It is not intended for use in any actual aircraft landing or take-off activity.

aptitude
An innate ability for acquiring a particular skill or knowledge.

aptitude test
Any system or device for determining whether an individual is likely to be successful in an activity for which he/she has not yet been trained.

APWA
American Public Works Association.

AQCR
See ***Air Quality Control Region.***

aquatic chronic value (ACV)
As used in Rule 57 (pursuant to Act 245 of 1929), a value used to represent the higher concentration of a substance which does not cause an adverse effect to important aquatic species when exposure occurs continuously over the lifetime of the organism.

aquatic nuisance species
Under the Federal Non-indigenous Aquatic Nuisance Prevention and Control Act of 1990: A non-indigenous species that threatens the diversity or abundance of native species or the ecological stability of infested waters, or commercial, agricultural, or recreational activities dependent on such waters.

aqueduct
A structure designed to transport domestic or industrial water from a supply source to a distribution point, often by gravity. *See also* ***canal/ditch***.

aqueous
Of, relating to, or resembling water; made from, with or by water.

aqueous film forming foam (AFFF)
A fluorinated surfactant with a foam stabilizer which is diluted with water to act as a temporary barrier to exclude air from mixing with the fuel vapor by developing an aqueous film on the fuel surface of some hydrocarbons which is capable of suppressing the generation of fuel vapors.

aqueous humor
The fluid in the anterior (front) chamber of the eye.

aquifer
An underground geological formation, or group of formations, containing usable amounts of groundwater that can supply wells and springs.

aquifuge
An underground layer of impermeable rock that will not allow the free passage of groundwater.

arable
Land capable of being farmed.

arachnoid layer
A nonvascular membrane between the dura mater and pia mater surrounding the brain and spinal cord.

ARAR
(1). *System Safety*. Acronym for "accident risk assessment report." A detailed analysis of a facility and its operating system(s) to determine hazardous conditions and risk abatement measures. (2) *Environmental.* Acronym for "applicable or relevant and appropriate requirements." Cleanup standards, control standards, and other substantive environmental protection requirements, criteria, and limitations promulgated under federal, state, and local laws.

arbitrary delay
An unscheduled interruption of work which is unrelated to the task or job being performed.

arbitration

A process for the resolution of disputes. Decisions are made by an impartial arbitrator selected by the parties involved. These decisions are usually legally binding. *See also* ***mediation***.

arc

An anthropometric measurement following an open curved path, where the curve makes up the majority of the measurement value. Also referred to as *curvature*. *See also* ***circumference***.

arc lamp

An illumination source which operates using the principles of discharge of low cathode voltages and high currents.

arc-welder's disease

A pneumoconiosis resulting from the inhalation of iron particles. May also be referred to as *siderosis*.

arch

(1) *General*. A curved structure that supports the weight of material over an open space. (2) *Anatomy*. The curvature on the inferior surface of the foot.

arch height

The maximum vertical distance from the floor or other reference surface on which a person stands to the bottom of the foot tissue between the anterior and posterior support points. Measured with the individual standing erect, with body weight equally divided between both feet.

Archimedes' principle

The principle of buoyancy that states the force on a submerged body acts vertically upward through the center of gravity of the displaced fluid and is equal to the weight of the fluid displaced.

Archimedes' screw

See ***screw pump***.

arctic tundra

See ***tundra***.

arcus cloud

See ***roll cloud***.

area

The measure of the size or extent of a surface. Its dimensions are

$$\text{AREA} = (\text{LENGTH})^2$$

In the MKS System, area is measured in *meters*^{2}. In the CGS System, area is measured in *centimeters*^{2}. In the English System, area is measured in *feet*^{2} or, frequently, *inches*^{2}.

Area, A

The cross-sectional area through which air moves. The area could be the cross-sectional area of a duct, a lab fume hood, a door, or any space through which air moves. Measured in square feet.

area affected by Outer Continental Shelf activities

Under the Federal Outer Continental Shelf Lands Act Amendments of 1978: Any geographic area which is under oil or gas lease on the Outer Continental Shelf; where Outer Continental Shelf exploration, development, or production activities have been permitted, except geophysical activities; where pipeline rights-of-way have been granted; or otherwise impacted by such activities including but not limited to expired lease areas, relinquished rights-of-way and easements, Outer Continental Shelf supply vessel routes, or other areas as determined by the Secretary of the Commerce.

area control center

An ICAO (International Civil Aviation Organization) term for an air traffic control facility primarily responsible for Air Traffic Control (ATC) services being provided to Instrument Flight Rules (IFR) aircraft during the en route phase of flight. The U.S. equivalent facility is an air route traffic control center.

area control computer complex (ACCC)

The common automation system equipment and software that support control of aircraft in a specific area, and which are located within each area control facility. The ACCC is one portion of the AAS (Advanced Automation System).

area control facility (ACF)

As of 1992, the planned 23 facilities that result from consolidation of existing Air Route Traffic Control Center (ARTCC) and Terminal Radar Approach Control (TRACON)/Terminal Radar Approach Control in Tower Cab (TRACAB) facilities. An Area Control Facility (ACF) may be formed from an existing ARTCC or may be created in a new building. The number, location, and implementation dates of ACFs are in accordance with the National Airspace System Plan. There will be 20 Continental U.S. (CONUS) ACFs converted from ARTCCs plus Honolulu, Anchorage, and the New York TRACON. Each can accomplish either an en route or an approach/departure control.

Area Director

The employee or officer regularly or temporarily in charge of an Area Office of the Occupational Safety and Health Administration, U.S. Department of Labor, or any other person or persons who are authorized to act for such employee or officer. The term also includes any employee or officer exercising supervisory responsibilities over an Area Director. A supervisory employee or officer is considered to exercise concurrent authority with the Area Director.

area navigation (RNAV)

A method of navigation that permits aircraft operation on any desired course within the coverage of station-referenced navigation signals or within the limits of a self-contained system capability. Random area navigation routes are direct routes, based on area navigation capability, between waypoints defined in terms of latitude/longitude coordinates, degree/distance fixes, or offsets from published or established routes/airways at a specified distance and direction. The major types of equipment are 1) Combined VOR and TACAN navigational facility (VORTAC) referenced or Course Line Computer (CLC) systems, which account for the greatest number of Radio Navigation (RNAV) units in use. To function, the CLC must be within the service range of a VORTAC. 2) OMEGA/VLF. Although two separate systems, can be considered as one operationally. A long-range navigation system based upon Very Low Frequency (VLF) radio signals transmitted from a total of 17 stations worldwide. 3) Inertial navigation systems (INS), which are totally self-contained and require no information from external references. They provide aircraft position and navigation information in response to signals resulting from inertial effects on components within the system. 4) Microwave Landing System (MLS) Area Navigation (MLS/RNAV), which provides area navigation with reference to an MLS ground facility. 5) LORAN-C is a long-range radio navigation system that uses ground waves transmitted at low frequency to provide user position information at ranges of up to 600 to 1,200 nautical miles at both en route and approach altitudes. The usable signal coverage areas are determined by the signal-to-noise ratio, the envelope-to-cycle difference, and the geometric relationship between the positions of the user and the transmitting stations.

area of critical environmental concern

Under the Federal Land Policy and Management Act of 1976: Areas within the public lands where special management attention is required (when such areas are developed or used or where no development is required) to protect and prevent irreparable damage to important historic, cultural, or scenic values, fish and wildlife resources or other natural systems or processes, or to protect life and safety from natural hazards.

area of review

The area surrounding an injection well that is reviewed during the permitting process to determine whether the injection operation will induce flow between aquifers.

area sample

An environmental sample obtained at a fixed point in the workplace. Used to measure properties of the workplace itself, which may or may not correlate with personal results of individual worker samples.

area source

Any small source of non-natural air pollution that is released over a relatively small area but which cannot be classified as a point source. Such sources include vehicles and other small fuel combustion engines.

area to be submerged

The known extent of the intended lake that will be created behind a dam under construction.

area wide template

Transportation. A computerized format (spreadsheet) for data entry of system length, vehicle travel, population, net land area, fatal and injury accidents, and percent of travel by vehicle type.

argyria
Poisoning by silver or a silver salt. A prominent symptom is a permanent gray discoloration of the skin, conjunctiva, and internal organs.

arising from the operation of a railroad
Includes all activities of a railroad that are related to the performance of its rail transportation business.

arithmetic mean
The sum of values divided by the number of values.

arm
One of the pair of upper extremities, consisting of the humerus, radial, and ulnar bones, and other associated soft tissues.

arm circumference
See ***forearm circumference, upper arm circumference,*** and ***axillary arm circumference***.

arm/disarm device
An electrically or mechanically actuated switch that can make or break one or more electroexplosive firing circuits. It operates in a manner similar to a safe and arm (S&A) device, except they do not physically interrupt the explosive train. *See also* ***safe and arm device***.

arm-hand
Involving both the arm and the hand, generally referring to internally generated or motor activities. *See also* ***hand-arm***.

arm-hand steadiness
A measure of the ability to keep both the hand and arm steady, whether stationary or moving. *See also* ***hand steadiness***.

arm-hand-tool aggregate
The combination of the hand/arm and tool acting as a biomechanical unit.

arm reach from wall
The horizontal distance from the wall to the tip of the longest finger. Measured with the rear of both the individual's shoulders against the wall, and with both hands and arms extended forward parallel to the floor for symmetry.

arm work
That physical work which uses the arm(s), with essentially no or minimal trunk or leg involvement.

Armed Forces
The Army, Navy, Air Force, Marine Corps, and Coast Guard, including their regular and reserve components and members serving without component status.

arming plug
A removable device that provides electrical continuity when inserted in a firing circuit.

armored cable
Type AC armored cable is a fabricated assembly of insulated conductors in a flexible metallic enclosure.

armpit
See ***axilla***.

Armstrong starter
Old-fashioned hand crank (usually for automobiles, but generally used for any type of engine).

Army Aviation Flight Information Bulletin
A bulletin that provides air operation data covering Army, National Guard, and Army Reserve aviation activities.

ARO
See ***airport reservation office***.

aromatic hydrocarbon
A major group of unsaturated cyclic hydrocarbons containing one or more rings made up of 6 carbon atoms. This group, most notably benzene, is chiefly derived from petroleum and coal tar. The name is due to the strong and often pleasant odor characteristic of substances within this group.

aromatic process oils
See ***high boiling aromatic oils***.

arousal
The degree of awareness of the environment.

ARP
See ***airport reference point***.

arraignment
The stage of a criminal prosecution where the defendant pleads guilty, not guilty, or *nolo contendere* (no contest).

arrangement of passenger transportation
Includes establishments engaged in providing travel information and acting as agents in arranging tours, transportation, car rentals, and lodging for travelers.

arrangement of passenger transportation not elsewhere classified
Establishments primarily engaged in arranging passenger transportation (other than travel agencies and tour operators), such as ticket offices (not operated by transportation companies) for railroads, buses, ships, and airlines.

arrangement of workplace principals
See ***workplace layout principals***.

array
A group of solar collection devices arranged in a suitable pattern to efficiently collect solar energy.

arrestance
Refers to the ability of a filter to remove coarse particulate matter from air passed through it.

arresting system
Aviation. A safety device consisting of two major components, namely, engaging or catching devices and energy absorption devices for the purpose of arresting both tailhook and/or nontailhook-equipped aircraft. It is used to prevent aircraft from overrunning runways when the aircraft cannot be stopped after landing or during aborted takeoff. Arresting systems have various names (e.g., arresting gear, hook device, wire barrier cable).

arrhythmia
See ***cardiac arrhythmia***.

arrival aircraft interval (AAI)
An internally generated program in hundredths of minutes based upon the Airport Acceptance Rate. *Arrival airport interval* is the desired optimum interval between successive arrival aircraft over the vertex.

arrival center
The air route traffic control center having jurisdiction for the impacted airport.

arrival delay
A parameter which specifies a period of time in which no aircraft will be metered for arrival at the specified airport.

arrival post
A signboard placed approximately 1/2 mile below the lock on the upstream and the downstream side to inform the pilot of the towboat that he has arrived at the lock and his preference is rated upon his first arrival either below or above. This term is falling into disuse since the advent of radio communications between towboats and the lock.

arrival program sequencing
The automated program designed to assist in sequencing aircraft destined for the same airport.

arrival sector
An operational control sector containing one or more meter fixes.

arrival time
The time an aircraft touches down on arrival.

arroyo
A stream or watercourse that is often dry.

arsenic
A chemical element, atomic number 33, atomic weight 74.92, symbol As. Arsenic compounds have been widely used in medicine; however, they have been replaced for the most part by antibiotics, which are less toxic and equally effective. Some arsenic compounds are used for infectious disease, especially those caused by protozoa. Since arsenic is highly toxic, it must be administered with caution. The antidote for arsenic poisoning is *dimercaprol*.

arson
In common law, the malicious burning of the premises of another. More broadly, any act by a person with the specific intent of using fire to destroy a building or occupied structure of another, or causing the destruction or damage of any property, regardless of the owner, for the purpose of collecting insurance for the loss.

arson clause
A clause in insurance policies voiding coverage if a fire is set under the direction of or by the insured.

arsphenamine
A light yellow powder containing 30 to 32 percent of arsenic; used intravenously in syphilis, yaws, and other protozoan infections.

ARSR
See ***air route surveillance radar***.

ARTCC
See ***Air Route Traffic Control Center.***

arterenol
See ***norepinephrine***.

arterial highway
Arterial highways serve major traffic movements or major traffic corridors. While they may provide access to abutting land, their primary function is to serve traffic moving through the area. *See also* ***freeway, minor arterial,*** *and* ***principal arterial****.*

arterial street
A major thoroughfare, used primarily for through traffic rather than for access to adjacent land, that is characterized by high vehicular capacity and continuity of movement.

arteriosclerosis
Thickening and loss of the elasticity of the coats of the arteries, with inflammatory changes; popularly known as hardening of the arteries. There are two main types: *arteriosclerosis proper,* in which the hardening is the result of fibrous and mineral deposits in the middle layer of the artery wall; and *atherosclerosis*, in which fatty and other substances collect in the inner lining of the arteries to form what are known as atheromatous plaques. These plaques encroach upon the passageway and gradually obstruct the flow of blood. Of the two types, atherosclerosis is by far the more common and more serious condition. Atherosclerosis is one of the major killers in the United States today.

arteriovenous oxygen difference
The difference in oxygen content between the blood entering and leaving the pulmonary capillaries.

artery
A blood vessel that conveys blood from the heart to any part of the body.

artesian water
Bottled water from a well that taps a confined aquifer located above the normal water table.

artesian well
A well with sufficient pressure to produce water without pumping.

arthralgia
Pain in a joint.

arthritis
Inflammation of a joint. The term covers more than 100 different types of joint diseases, the most common types being *rheumatoid arthritis* and *osteoarthritis*. Arthritis may also arise as a side effect of a number of diseases, including tuberculosis, syphilis, gonorrhea, and viral diseases such as measles and influenza. Rheumatism is a general term for arthritis and is often applied to almost any pain in the joints or muscles. The symptoms of rheumatoid arthritis are usually mild with a gradual onset. Osteoarthritis is most likely to occur in the large and weight-bearing joints. It is a degenerative disorder that is commonly secondary to other joint diseases. Another common form of osteoarthritis affects the joints of the fingers; this form usually occurs in women. Osteoarthritis is much less crippling than severe rheumatoid arthritis because it does not cause the two bone surfaces to fuse and immobilize the joint.

arthroscope
An instrument having a small diameter tube used for visualizing the interior of some body part.

article
Under the Toxic Substances Control Act (TSCA), a manufactured item which is formed to a specific size and shape during manufacture whose end use function is dependent in whole or in part on that specific size and shape.

articular
Pertaining to one or more joints.

articulate
Produce speech sounds easily recognizable by another individual fluent in a given language.

articulated bus
A bus usually 55 feet or more in length with two connected passenger compartments that bend at the connecting point when the bus turns a corner.

articulated motor buses
Extra-long (54 ft. to 60 ft.) motor buses with the rear body section connected to the main body by a joint mechanism. The joint mechanism allows the vehicles to bend when in operation for sharp turns and curves, and yet have a continuous interior.

articulated total body model (ATB)
A computerized model developed for examining the biodynamic effects of ejection from high-performance aircraft on the various body segments.

articulation index (AI)
See ***speech articulation index****.*

artificial
(1) A replacement for a natural limb. (2) A mechanical device that can substitute temporarily or permanently for a body organ.

artificial gravity
That relative downward acceleration experienced by an individual or object on the interior of a larger, rotating object as a result of centrifugal force.

artificial horizon
A graphic or pictorial flight instrument display for providing the pilot with information about the orientation of the aircraft with respect to the ground.

artificial pupil
A small aperture in a manufactured or cultured disk or diaphragm used to restrict the amount of light entering the eye.

artificial radioactivity
(1) Radioactivity produced by the bombardment of a target element with nuclear particles. (2) The output from radioactive substances or from high energy electromagnetic wave production in instrumentation.

artificial reality
See ***virtual environment***.

ARTS
See ***automated radar terminal system***.

arytenoid
A skeletal muscle of the larynx which is involved in controlling pitch by regulating the length of the vibrating segment of the vocal cord.

as-built plan
A drawing which covers property boundaries, streets bordering the site and building layout, and provides accurate scale and a north arrow.

asbestiform mineral
(1) Minerals which, due to their crystalline structure and chemical composition, tend to be separated into fibers and can be classed as a form of asbestos. (2) The EPA defines asbestiform as a specific type of mineral fibrosity in which the fibers and fibrils possess high tensile strength and flexibility.

asbestos
A generic term used to describe a number of naturally occurring fibrous, hydrated mineral silicates differing in chemical composition. They are white, gray, green, or brown. Asbestos fibers are characterized by high tensile strength, flexibility, heat and chemical resistance, and good frictional properties. Chrysotile, crocidolite, amosite, anthophyllite, and actinolite are all forms of asbestos. Exposure to asbestos fibers is known to cause a variety of diseases, including *asbestosis* (a diffuse, interstitial non-malignant scarring of the lung tissues), *bronchogenic carcinoma* (a lung cancer), *mesothelioma* (a tumor of the lining of the chest cavity or lining of the abdomen), and *cancer* of the stomach, colon, and rectum.

asbestos abatement
Procedures to control the release of asbestos fibers from asbestos-containing materials.

asbestos bodies
Dumbbell-shaped bodies that may appear in the lungs and sputum of persons who have been exposed to asbestos. These are also called ferruginous bodies.

asbestos cement pipe
Pipe manufactured of a mixture of asbestos fiber and Portland cement.

asbestos-containing materials (ACM)
Any material which contains more than 1 percent asbestos by weight which can be released upon destruction or disturbance of the structural integrity of the material.

asbestos fiber
An asbestos fiber that is greater than 5 micrometers in length, with a length to width ratio equal to or greater than 3 to 1.

asbestos standard
Regulations promulgated by the federal Occupation Safety and Health Administration in 1986 that require major reductions in the level of airborne asbestos fibers in workplaces and that also prescribe a system of engineering controls and work practices related to asbestos. It is actually two standards, one for general industry and one for the construction industry, with somewhat different requirements.

asbestosis
A disease associated with chronic exposure to and inhalation of asbestos fibers. The disease makes breathing progressively more difficult and can lead to death.

ASCE
American Society of Civil Engineers.

ASCII
See ***American Standard Code for Information Interchange***.

ascorbic acid
Vitamin C, also called cevitamic acid; a substance found in fruits and vegetables, especially citrus fruits (oranges and lemons) and tomatoes. Ascorbic acid is an essential element of the diet. Lack of vitamin C can lead to scurvy or to less severe conditions, such as delayed healing of wounds. Solutions of vitamin C deteriorate very rapidly and the vitamin is not stored in the body to any extent. Large doses of commercial preparations of ascorbic acid may cause gastrointestinal irritation. There is no general agreement as to the normal and therapeutic daily requirements for vitamin C intake.

ASDE
See ***airport surface detection equipment***.

asemasia
Inability to make or comprehend signs or tokens of communication.

asepsis
Clean and free of microorganisms.

aseptic
Free from infection; sterile.

aseptic bone necrosis
See ***dysbaric osteonecrosis***.

aseptic technique
Procedures designed to exclude infectious agents; laboratory or clinical techniques that do not result in the transfer of disease-producing microorganisms from one surface to another.

ash
The mineral content of a product remaining after complete combustion.

ASHARA
Asbestos School Hazard Abatement Reauthorization Act.

ashing
The decomposition, prior to analysis, of the organic matrix constituents of a sampling media.

ASHRAE
See ***American Society of Heating, Refrigerating and Air Conditioning Engineers***.

asiderosis
Deficiency in the iron reserve of the body.

askarel
Generic term for a group of nonflammable synthetic chlorinated hydrocarbons that have been used as electrical insulating material. These are also referred to as *polychlorinated biphenyls*.

as low as reasonably achievable (ALARA)
A basic concept of radiation protection that specifies that radioactive discharges from nuclear plants and exposure of personnel to ionizing radiation be kept as far below regulatory limits as is reasonably achievable.

ASM
See ***available seat mile***.

ASME
American Society of Mechanical Engineers.

Asmussen dynamometer
A device using a piston drive for measuring strength, in either a push or pull mode.

aspect
The appearance of a roadway signal conveying an indication as viewed from the direction of an approaching train; the appearance of a cab signal conveying an indication as viewed by an observer in the cab.

aspect ratio
(1) With regard to asbestos fibers, the ratio of fiber length to fiber width. (2) *EPA*. The ratio of the length to width of a particle. The aspect ratio for counting structures, as defined in the transmission electron microscope (TEM) method of asbestos sample assessment, is equal to or greater than 5 to 1. (3) *OSHA*. To be counted as a fiber by the phase contrast microscopy (PCM) method of analysis, the fiber must be at least 5 micrometers in length and have a length to width ratio of at least 3 to 1.

aspergillosis
An infectious disease of the skin, lungs, and other parts of the body caused by certain fungi of the genus *Aspergillus*.

Aspergillus fumigatus
Airborne fungi that may result from composting operations and may cause human ear, lung, and sinus infections.

asphalt
A dark brown-to-black cement-like material containing bitumens as the predominant constituents obtained by petroleum processing. The definition includes crude asphalt as well as the following finished products: cements, fluxes, the asphalt content of emulsions (exclusive of water), and petroleum distillates blended with asphalt to make cutback asphalts.

asphalt-rubber
A mixture of ground rubber and bituminous concrete used as a pavement interlayer to reduce stress and prevent cracking.

asphyxia
(1) The state of respiratory distress or suffocation due to the lack of respirable oxygen. (2) A condition due to lack of oxygen in inspired air, resulting in loss of consciousness or actual cessation of life.

asphyxiant
Any commodity capable of reducing or depleting the oxygen content of a space to the point of asphyxiation. Asphyxiants may be simple or chemical. *Simple asphyxiants* are materials that can displace oxygen in the air (e.g., nitrogen). *Chemical asphyxiants* render the body incapable of utilizing an adequate supply of oxygen (e.g., carbon monoxide). Both types of anoxia can potentially result in insufficient oxygen to sustain life.

asphyxiation
(1). The depravation of oxygen caused by chemical or physical means. Chemical asphyxiants prevent oxygen transfer from the blood to the body cells. Physical asphyxiants prevent oxygen from reaching the blood. (2) That point where oxygen content is no longer at a level capable of supporting life; a cause of death resulting from a lack of sufficient oxygen; suffocation.

aspirate
(1) To remove (by suction) a gas or body fluid from a body cavity, from an unusual accumulation, or from a container. (2) The accidental passage of a liquid or solid substance into the lungs following attempted ingestion or during a vomiting sequence.

aspirating aerator
Aeration device that uses a motor-driven propeller to draw atmospheric air into the turbulence caused by the propeller to form small bubbles.

aspiration
A hazard to the lungs following the ingestion (accidental or on purpose) of a material, such as a solvent or solvent-containing product, when a small amount of the material is taken into or is aspirated into the lungs in liquid form. Aspiration can occur during ingestion, or if and when the material is later vomited.

aspirator
A hydraulic device that creates a negative pressure by forcing liquid through a restriction and increasing the velocity head.

aspirin
Acetylsalicylic acid, a common drug generally used to relieve pain and reduce fever, and specifically prescribed for rheumatic and arthritic disorders. Indiscriminate use of the drug may lead to toxic symptoms such as gastrointestinal disorders, ringing in the ears, headache, and, in severe toxicity, depression of heart rate.

ASQC
American Society for Quality Control.

ASR
See ***airport surveillance radar***.

ASSE
See ***American Society of Safety Engineers.***

assembly
(1) *General*. The concourse or meeting together of a considerable number of persons at the same place; also, the persons so gathered. The process of putting together individual pieces or components to make a whole or complete item. (2) *System Safety*. A combination of multiple components or parts grouped together to perform a single function or a specific set of functions within a system or subsystem.

assembly line
A work arrangement in which the product being assembled is delivered to each person who then performs a somewhat specialized task or job at a specific work site.

assessment
An evaluation or examination of a specific area of concern, such as a program, policy, or procedural assessment.

assessor's map
A map which covers property boundaries and gives the location of properties bordering the subject site.

assignable cause
Any identifiable source of deviation from the normal in some process or system.

assigned protection factor (APF)
A numerical indicator of how well a respirator can protect its wearer under optimal conditions of use. The numerical value, or assigned protection factor, is the ratio of the air contamination concentration outside a respirator to that inside the respirator. For example, an assigned protection factor of 10 means that $1/10^{th}$ the workspace exposure concentration is that which is inhaled by the wearer.

assigned vehicle
A vehicle provided to an organizational element of a government agency or contractor by General Services Administration's (GSA) Interagency Fleet Management System for a period of more than 30 days.

assimilable organic carbon (AOC)
The portion of dissolved organic carbon that is easily used by microbes as a carbon source.

assimilation
The ability of a body of water to purify itself of pollutants. *See also **color assimilation**.*

assimilative capacity
The ability of a water body to receive wastewater and toxic materials without deleterious effects on aquatic life or the humans who consume the water.

Assistant Regional Director
The employee or officer regularly or temporarily in charge of a Region of the Occupational Safety and Health Administration, U.S. Department of Labor, or any other person or persons who are specifically designated to act for such employee or officer in his/her absence. The term also includes any employee or officer in the Occupational Safety and Health Administration exercising supervisory responsibilities over the Assistant Regional Director.

assistive device
Any tool which either enables or enhances human-machine interaction for an individual with a physical handicap.

associated corpuscular emission
The full complement of secondary charged particles (usually limited to electrons) associated with an x-ray or gamma-ray beam in its passage through air.

associated equipment
Marine Safety. Any system, part, or component of a boat as originally manufactured or any similar part or component manufactured or sold for replacement, repair, or improvement of such system, part, or component; any accessory or equipment for, or appurtenance to, a boat; and any marine safety article, accessory, or equipment intended for use by a person on board a boat; but excluding radio equipment, as designated by the Secretary of Transportation under 46 U.S.C. 2101.

assumpsit
Law. (Latin) A promise or engagement by which one person assumes or undertakes to do some act or pay something to another. It may be either oral or in writing, but is not under seal. It is *express* if the promisor puts his engagement in the distinct and definite language; it is *implied* where the law infers a promise (though no formal one has passed) from the conduct of the party or the circumstances of the case.

assumption
The act of conceding or taking for granted. Laying claim to or taking possession of.

assumption of liability
As applied to the waste business, assumption of liability occurs when a licensed transporter facility automatically assumes responsibility and, accordingly, risk and liability for a generator's waste when the waste is accepted for transportation, storage, treatment, or other handling. This assumption does not, however, reduce or remove the generator's liability for responsibility.

assumption of risk
Law. The doctrine of assumption of risk, also known as *volenti non fit injuria,* means legally that a plaintiff may not recover for an injury to which he/she assents, i.e., that a person may not recover for an injury received when he/she voluntarily exposes himself to a known and appreciated danger. The requirements for the defense of assumption of risk are that a) a condition of risk must exist, b) he/she knows

the condition is dangerous, c) he/she appreciates the nature or extent of the danger, and d) he/she voluntarily exposes himself/herself to the danger.

AST
Abovcground storage tank.

astern
(1) Behind a vessel. (2) Move in a reverse direction.

asthenia
Lack or loss of strength or energy.

asthenic
A Kretschmer somatotype characterized by a slender, feeble build. Also referred to as *aesthenic*. *See also* ***Kretschmer somatotype***.

asthenopia
Impairment of vision, with pain in the eyes, back of the head, and the neck.

asthma
A disease of the bronchi, technically know as bronchial asthma. Constriction of the bronchial tube muscles, in response to irritation, allergy, or other stimulus. Symptoms include dyspnea, wheezing, and a sense of constriction in the chest.

astigmatism
An error of refraction in which parallel light rays fail to come to focus on the retina, owing to differences in curvature in various meridians of the refractive surfaces (cornea and lens) of the eye. The exact cause of astigmatism is not known.

ASTM
See American Society for Testing and Materials.

astringent
(1) Causing contraction and arresting discharges. (2) An agent that arrests discharges. Astringents act as protein precipitants; they arrest discharge by causing shrinkage of tissue.

astronautics
The art and science of designing, building, and operating manned or unmanned space objects.

astrosphere
A structure made up of a group of radiating fibrils that converge toward the centrosome and continue in the centrosphere of a cell.

asymbolia
Loss of ability to understand symbols, as words, figures, gestures, and signs.

asymmetric lift
A manual lifting task in which the load is not equally shared by paired limbs.

asymmetric membrane
Membranes that are not reversible and can only desalinate efficiently in one direction.

asymmetry
A lack of structural correspondence between two sides of a normally or especially symmetric structure, especially pertaining to paired members.

asymptomatic
The lack of identifiable signs or symptoms; without symptoms.

asymptote
That value represented by approximately a horizontal straight line which a curve approaches as the axis approaches infinity.

at anchor
Held in place in the water by an anchor. Includes "moored" to a buoy or anchored vessel and "dragging anchor."

at grade
See ***grade crossings*** *and* ***highway-rail crossing***.

at grade, exclusive right-of-way
Railway right-of-way from which all other traffic, mixed and cross, is excluded. Median strip right-of-way is included provided all crossings of the right-of-way pass over or under the median.

at grade, mixed and cross traffic
Railway right-of-way over which other traffic moving in the same direction or the cross directions may pass. City street right-of-way is included.

at grade, with cross traffic
Railway right-of-way over which no other traffic may pass, except to cross at grade-level crossings. A median strip right-of-way with grade-level crossings at intersecting streets is included.

ATAD
See ***autothermal thermophilic aerobic digestion*** *process*.

ataxia
A failure, or lack of muscular coordination.

ATB
See ***articulated total body model***.

ATC
See ***air traffic control***.

ATCCC
See ***air traffic command and control center***.

ATCO
See ***air taxi/commercial operator***.

ATCT
See ***airport traffic control tower***.

atelectasis
A collapsed or airless state of the lung, which may be acute or chronic, and may involve all or part of the lung. The primary cause of atelectasis is obstruction of the bronchus serving the affected area. Symptoms include sudden obstruction of the bronchus, pain in the affected side, dyspnea and cyanosis, elevation of temperature, and a drop in blood pressure or shock.

atherosclerosis
A condition characterized by degeneration and hardening of the walls of the arteries and sometimes the valves of the heart, related especially to thickening of the intimal layer. See also ***arteriosclerosis***.

athetosis
A movement disorder characterized by almost continuous involuntary slow, sinuous movements.

athlete's foot
A fungus infection of the skin of the foot; also called tinea pedis. Athlete's foot causes itching and often blisters and cracks, usually between the toes. Causative agents are *Candida albicans, Epidermophyton floccosum* and species of *Trichophyton*, which thrive on warmth and dampness. If not arrested, athlete's foot can cause a rash and itching in other parts of the body as well. It is likely to be recurrent, since the fungus survives under the toenails and reappears when conditions are favorable. Although athlete's foot is usually little more than an uncomfortable nuisance, the open sores it causes provide excellent sites for more serious infections. See ***dermatophytoses***.

athletic
A Kretschmer somatotype having a stocky, muscular build with little body fat. *See also* ***Kretschmer somatotype***.

atlanto-occipital joint
The junction of the atlas with the occipital bone of the skull.

atlas
Anatomy. The first cervical vertebra.

atm
Atmosphere.

ATM
See ***air traffic management***.

atmometer
An instrument used to measure the evaporative capacity of the air.

atmosphere
(1) A standard unit of pressure representing the pressure exerted by a 29.92-inch column of mercury at sea level at 45° latitude and equal to 1000 grams per square centimeter. (2) The whole mass of air surrounding the earth, composed largely of oxygen and nitrogen and extending to an altitude of 10 miles.

atmosphere-supplying respirator
A respiratory protective device which is designed to supply breathing air to the wearer. This type of respirator does not rely on the use of air from the work environment. The air is obtained from an independent source. Respirators of this type are classified as a supplied-air respirator or self-contained breathing apparatus.

atmospheric corrosion
Corrosion that results from exposure to the atmosphere.

atmospheric effect
See ***greenhouse effect***.

atmospheric pressure
The pressure exerted by the weight of the atmosphere, equivalent to 14.7 pounds per square inch at sea level. Also equivalent to the pressure exerted by a column of mercury 760 mm high or a column of water 406.9 inches high.

atmospheric stagnation
A condition of light winds and poor vertical mixing that can lead to a high concentration of pollutants. Air stagnation is most often as-

sociated with fair weather, an inversion, and the sinking air of a high-pressure area.

atmospheric window
The wavelength range between 8 and 11 m in which little absorption of infrared radiation takes place.

at. no.
Atomic number.

atom
The smallest particle of an element which can not be divided or broken by chemical means. It consists of a central core called the *nucleus,* which contains *protons* and *neutrons. Electrons* move in orbital fashion in the region surrounding the nucleus.

atomic absorption spectrophotometry (AAS)
A method commonly used for the analysis of heavy metals in water.

atomic energy
Under the Federal Atomic Energy Act of 1954: All forms of energy released in the course of nuclear fission or nucleus transformation.

atomic energy defense activity
Under the Federal Nuclear Waste Policy Act of 1982: Any activity of the Secretary of Energy performed in whole or in part in carrying out any of the following functions: naval reactors development; weapons activities including defense inertial confinement fusion; verification and control technology; defense nuclear materials production; defense nuclear waste and materials byproducts s management; defense nuclear materials security and safeguards and security investigations; and defense research and development.

atomic fission
See ***fission***.

atomic mass
The mass of a neutral atom of a nuclide, usually expressed in terms of atomic mass units.

atomic mass unit
One-twelfth the mass of one neutral carbon-12 atom equivalent to 1.6604 E-24 grams.

atomic number
The number of protons in the nucleus of an atom.

atomic power
The production of electricity through the use of a nuclear reactor.

atomic weapon
Under the Federal Atomic Energy Act of 1954: Any device utilizing atomic energy, exclusive of the means for transporting or propelling the device (where such means is a separable and divisible part of the device), the principal purpose of which is for use as, or for development of, a weapon, a weapon prototype, or a weapon test device.

atomic weight
Approximately the sum of the number of protons and the number of neutrons found in the nucleus of an atom, also known as the *mass number.*

atomize
To divide a liquid into extremely minute particles, either by impact with a jet of steam or compressed air, or by passage through some mechanical device.

atomizer
An instrument through which a liquid is sprayed to produce a fine mist.

atrial fibrillation
A cardiac arrhythmia characterized by extremely rapid, irregular atrial impulses, ineffectual atrial contractions, and irregular, rapid ventricular beats. Also known as *auricular fibrillation. See also* ***fibrillation***.

atrophy
Wasting away or diminution in the size of a cell, tissue, organ, or part, from defect, failure of nutrition, or lack of use.

atropine
A poisonous parasympatholytic alkaloid of belladonna, used in a variety of conditions. Actions include decrease of secretions, increased heart rate and rate of respiration, and relaxation of smooth muscle tissue. It may be used to dilate pupils, for general cerebral stimulation, for relief of gastrointestinal cramps and hypermotility, and locally to relieve pain. In various combinations with other drugs, atropine may be administered orally or intramuscularly, or applied topically. *Atropine methylnitrate* and *atropine sulfate* are soluble compounds of atropine, with similar uses.

ATSDR
Agency for Toxic Substances and Disease Registry.

attached growth process
See ***fixed film process.***

attachment plug
A device which, by insertion in a receptacle, establishes connection between the conductors of the attached flexible cord and the conductors connected permanently to the receptacle.

attainment area
An area considered to have air quality as good as or better than the National Ambient Air Quality Standards (NAAQS) as defined in the Clean Air Act. An area may be an attainment area for one pollutant and a non-attainment area for others.

attendant
(1) *Safety.* With regard to confined space entry, a trained individual who remains outside a confined space and acts as an observer of the authorized entrants within the space, keeping in constant, though not necessarily continuous communication with them, so the attendant can immediately call rescue services if needed. (2) *Law.* One who owes a duty or service to another, or in some way depends upon him or her. One who follows and waits upon another.

attended operation
An operation which is attended at all times by a person who is sufficiently knowledgeable to act should the need arise.

attention
The general, but not highly directed, allocation of sensory-perceptual functions, possibly involving motor functions as well, to a subset of the possible inputs. *See also* ***selective attention*** and ***divided attention***.

attention span
That length of time or number of items or tasks to which an individual can respond before performance deteriorates.

attenuate
To reduce in amount, concentration, intensity, strength, force, or amplitude some entity over a period of time or space.

attenuation
(1) *General.* The process by which a compound is reduced in concentration over time, through adsorption, degradation, dilution, and/or transformation. (2) *Acoustics.* The reduction, expressed in decibels, of the sound intensity at a designated position as compared to the sound intensity at a second position acoustically further from the source, or as a result of an intervening material. (3) *Ionizing Radiation.* The process by which a beam of ionizing radiation is reduced in intensity when passing through a material.

attitude
(1) *See* ***posture***. (2) The forward orientation of a vehicle which is capable of motion in all three spatial dimensions, especially an aircraft or spacecraft. (3) An individual's feeling or opinion about some issue or expected event which will shape his/her response.

atto
Prefix designating 1 E-18, (a).

attorney
In the most general sense, this term denotes an agent or substitute, or one who is appointed and authorized to act in the place or stead of another. An agent, or one acting on behalf of another. In its most common usage, however, unless a contrary meaning is clearly intended, this term means "attorney at law, "lawyer," or "counselor at law."

attorney at law
Person admitted to practice law in his/her respective state and authorized to perform both civil and criminal legal functions for clients, including drafting of legal documents, giving of legal advice, and representing such before courts, administrative agencies, boards, etc.

attorney-client privilege
In law of evidence, the client's privilege to refuse to disclose and to prevent any other person from disclosing confidential communications between the client and his/her attorney. Such privilege protects communications between the attorney and the client made for the purpose of furnishing or obtaining professional legal advice or assistance. That privilege also permits an attorney to refuse to testify as to communications between the client and the attorney. It is important to note that the privilege belongs to the client, not to the

attorney, and hence the client may waive it. Unless waived, the privilege protects a communication between privileged persons in confidence for the purpose of obtaining or providing legal assistance for the client.

Attorney General
Head of the U.S. Department of Justice and chief law officer of the Federal Government, represents the United States in legal matters generally and gives advice and opinions to the President and to the heads of the executive departments of the Government when so requested. In each State, there is also an Attorney General, who is the chief law officer of the state.

attractant
A chemical or agent that lures insects or other pests by stimulating their sense of smell.

attraction
The force of influence by which one object is drawn toward another.

attractive dirt
Any form of dirt which tends to remain attached to a surface through electrostatic forces.

attractive nuisance
An insurance term applicable to any existing or visible condition that may attract attention and the desire to inspect the condition or circumstance. An example would be a construction site containing interesting or unusual machinery or devices that might attract children or others for closer investigation, thereby risking harm or damage to person or property. Fencing and/or other security may be ways to control an attractive nuisance, although they may not be sufficient.

attributable risk
A measure of the occurrence of a specific disease or injury in those exposed to a particular situation or causal agent.

attribute
Some characteristic of an element or condition.

attrition
Wearing or grinding down of a substance by friction. A contributing factor in air pollution, as with dust.

at-will employee
An employee who works for a company under an at-will employment agreement.

at-will employment
An agreement or understanding, either written or verbal, between an employer and employee that the employer may terminate an employee at any time, with or without good reason or notice, and with no legal liability of the employer. Also referred to as *employment-at-will*.

at. wt.
Atomic weight.

audible frequency range
See ***audible sound***.

audible sound
Sound containing frequency components between approximately 16-20 and 20,000 Hz. Also referred to as *audible frequency range, audible range,* and *audio frequency range*.

audible range
See ***audible sound***.

audio frequency range
See ***audible sound***.

audio frequency spectrum
See ***spectra (1)***.

audiogenic
Resulting from sound.

audiogenic seizure
A reflex convulsion caused by exposure to an intense or sudden noise.

audiogram
A graphic or tabular record of hearing level measured at different sound frequencies produced by an audiometer in a control setting. Audiograms are used in the diagnosis and treatment of hearing loss. *See also* ***audiometer***.

audiologist
A professional, specializing in the study and rehabilitation of hearing, who is certified by the American Speech-Language-Hearing Association or licensed by a state board of examiners.

audiometer
A frequency-controlled audio signal generator that produces pure tones at various frequencies and intensities and that is used to measure

hearing sensitivity or acuity. Measurement of hearing threshold results in an audiogram, measured in decibels at selected frequencies. *See also* ***audiogram***.

audiometric reference level
The sound pressure level and specification to which an audiometer has been calibrated.

audiometric technician
An individual who is trained to perform audiometry.

audiometric testing room
A specialized chamber insulated for sound and equipped for hearing acuity measurement.

audiometric zero
The threshold of hearing, which is equivalent to a sound pressure of 2 E-4 microbars.

audiometry
The testing of the sense of hearing.

audiovisual
The simultaneous stimulation of both the sense of hearing and the sense of sight.

audit
A detailed and systematic inspection or review of an occupational health and safety program, environmental program, financial operating program, or some other program, to determine compliance with company policies, practices, and procedures, as well as the regulations that are applicable to the operations and work being performed.

audit conclusion
Under ISO 14000, professional judgment of opinion expressed by an auditor about the subject matter of the audit, based on and limited to reasoning the auditor has applied to audit findings.

audit criteria
Under ISO 14000, policies, practices, procedures, or requirements against which the auditor compares collected audit evidence about the subject matter.

audit evidence
Under ISO 14000, verifiable information, records, or statements of fact.

audit findings
Under ISO 14000, results of the evaluation of the collected audit evidence compared against the agreed-upon criteria.

audit team
Group of auditors, or a single auditor, designated to perform a given audit; the audit team may also include technical experts and auditors-in-training.

auditing procedures
Compliance procedures, which include employee training in the company's mandatory standards and procedures as well as regular and unannounced audits, conducted under the direction of counsel and the corporate compliance officer to ensure compliance. *See also* ***monitoring procedures.***

auditor
One who checks the accuracy, fairness, and general acceptability of business operations in general, and operating plans and procedures in specific. One who conducts an audit. *See also* ***audit, independent audit,*** *and* ***internal audit***.

auditory
Pertaining to the sense of hearing.

auditory absolute threshold
See ***threshold of audibility***.

auditory aftereffect
A phenomenon is which familiar sounds appear modulated for a period of time after listening to rapid, high-intensity impulses.

auditory attention
The ability to focus on a single auditory source in the presence of distracting auditory stimuli.

auditory canal
See ***Eustachian tube.***

auditory fatigue
A temporary increase in auditory threshold due to prolonged intense noise or a previous auditory stimulus.

auditory lateralization
The determination by a person that apparent direction of a sound is either to the left or right of the midsagittal plane of the head when wearing earphones.

auditory localization
The process of determining the apparent direction and/or distance of an external sound source.

auditory nerve
That portion of the vestibulocochlear nerve which carries auditory information from the inner ear to the brain. May be referred to as *acoustic nerve*.

auditory ossicle
Any of the three small bones in the middle ear used for hearing: the malleus (hammer), incus (anvil), and stapes (stirrup).

auditory sound
*See **background noise**.*

auditory stimulus
Any stimulus which excites the cochlea to convey signals indicating sound perception to the brain.

auditory system
The combined structures of the external, middle, and inner ear which are involved in the function of hearing, and the acoustic nerve.

aural
Pertaining to the ear or hearing.

aural insert protectors
A form of hearing protector commonly known as earplugs. They are available in numerous configurations as foam, plastic, fine glass fiber, and wax-impregnated cotton. The three types are formable, custom-molded, and premolded.

auricle
(1) The part of the ear that projects from the head. Also called the *pinna*. (2) One of the two upper chambers of the heart.

auricular fibrillation
*See **atrial fibrillation**.*

auricular point
That location on the longitudinal axis of the external auditory canal at which it passes to the exterior.

aurora
Glowing light display in the night time sky caused by excited gases in the upper atmosphere giving off light. In the Northern Hemisphere, it is called the *aurora borealis* (northern lights; in the Southern Hemisphere, the *aurora australis* (southern lights).

authentication
In the law of evidence, the act or mode of giving authority or legal authenticity to a statute, record, or other written instrument, or a certified copy thereof, so as to render it legally admissible in evidence.

authority
Permission. Right to exercise powers; to implement and enforce laws; to exact obedience; to command; to judge. Control over; jurisdiction. Often synonymous with power. The delegated power by one person to another. The lawful delegation of power by one person to another. Power of agent to affect legal relations of principal by acts done in accordance with principal's manifestations of consent to agent.

authority having jurisdiction
The organization, office, agency, or individual responsible for approving equipment and installation, or a procedure.

authorized altitude
A published altitude representing the maximum usable altitude or flight level for an airspace structure or route segment. It is the highest altitude on a federal airway, jet route, area navigation low or high route, or other direct route for which a Minimum En Route Instrument Flight Rules (IFR) Altitude (MEA) is designated in Part 95 at which adequate reception of navigation aid signals is assured.

authorized entrant
(1) As pertains to confined spaces, an employee who is authorized by the employer to enter a permit space. (2) In confined spaces, an employee who is authorized by the employer or its designee to enter a confined space.

auto restricted zone
An area in which normal automobile traffic is prohibited or limited to certain times, and vehicular traffic is restricted to public transit, emergency vehicles, taxicabs and, in some cases, delivery of goods.

AutoBill
*See **automotive billing module**.*

autoclave
A self-locking apparatus used for the sterilization of materials by steam under pressure. The autoclave allows steam to flow around each article placed in the chamber. The vapor penetrates cloth or paper used to package the articles being sterilized. Autoclaving is one of the most effective and proven methods for destruction of all types of microorganisms. The amount of time and degree of temperature

necessary for sterilization depend on the articles to be sterilized and whether they are wrapped or left directly exposed to the steam.

autoexec.bat
Computing. A text file that contains a list of commands that a computer executes every time it is started (booted) or restarted (rebooted).

autogenous combustion
Burning that occurs when the heat of combustion of a wet organic material or sludge is sufficient to vaporize the water and maintain combustion without auxiliary fuel.

autogenous temperature
Equilibrium temperature in sludge combustion where heat input from the fuel equals the heat loss and combustion is self-supporting.

autoignition
The ignition of a combustible material without initiation by a spark or flame; when the temperature of a material has been raised to a level at which self-sustaining combustion occurs.

autoignition temperature
The lowest temperature at which a flammable gas-air or vapor-air mixture ignites from its own heat source or a contacted hot surface but without the presence of a spark or flame.

autokinesis
Voluntary movement.

autokinetic illusion
An effect in which a stationary point light source in a dark background or with no visual reference frame appears to move. Also referred to as *autokinetic phenomenon.*

autokinetic phenomenon
See ***autokinetic illusion***.

autoland approach
Aviation. A precision instrument approach to touchdown and, in some cases, through the landing rollout. An autoland approach is performed by the aircraft autopilot which is receiving position information and/or steering commands from onboard navigation equipment. *See also* ***coupled approach***.

automated control
The use of feedback in a continuously monitored, computerized system to self-correct any output deviations.

automated flight service station
A station that provides interactive alphanumeric and graphic workstations for the flight service specialist.

automated guideway
(1) An electric railway operating without vehicle operators or other crew on board the vehicle. (2) One or more automatically controlled vehicles operating over an exclusive guideway.

automated guideway transit
Guided transit vehicles operating singly or in multi-car trains with a fully automated system (no crew on transit units). Service may be on a fixed schedule or in response to a passenger-activated call button. Automated guideway transit includes personal rapid transit, group rapid transit, and people mover systems.

automated guideway transit system
Fixed guideway transit system which operates with automated (driverless) individual vehicles or multi-car trains. Service may be on a fixed schedule or in response to a passenger-activated call button.

automated guideway vehicles
Guided transit passenger vehicles operating under a fully automated system (no crew on transit units).

automated mutual assistance vessel rescue system (AMVER)
A facility which can deliver, in a matter of minutes, a Surface Picture (SURPIC) of vessels in the area of a potential or actual search and rescue incident, including their predicted positions and their characteristics.

automated radar terminal system (ARTS)
The generic term for the ultimate in functional capability afforded by several automation systems. Each differs in functional capabilities and equipment. Automated Radar Terminal System (ARTS) plus a suffix roman numeral denotes a specific system. A following letter indicates a major modification to that system. In general, an ARTS displays for the terminal controller aircraft identification, flight plan data, and other flight associated information, e.g., altitude, speed, and aircraft position symbols in conjunction with the radar presentation. Normal radar co-exists with the alphanumeric display. In addition to enhancing visualization of the air traffic situation,

ARTS facilitates intra/inter-facility transfer and coordination of flight information. These capabilities are enabled by specially designed computers and subsystems tailored to the radar and communications equipment and operational requirements of each automated facility. Modular design permits adoption of improvements in computer software and electronic technologies as they become available while retaining the characteristics unique to each system.

automated transfer information
Aviation. A pre-coordinated process, specifically defined in facility directives, during which a transfer of altitude control and/or radar identification is accomplished without verbal coordination between controllers using information communicated in a full data block.

automatic
Self-acting, operating by its own mechanism when actuated by some impersonal influence, as, for example, a change in current strength, pressure, temperature, or mechanical configuration.

automatic altitude reporting
That function of a transponder which responds to Mode C interrogations by transmitting the aircraft's altitude in 100-foot increments.

automatic block sign system
Transportation. A block signal system wherein the use of each block is governed by an automatic block signal, cab signal, or both.

automatic carrier landing system
U.S. Navy final approach equipment consisting of precision tracking radar coupled to a computer data link to provide continuous information to the aircraft, monitoring capability to the pilot, and a backup approach system.

automatic direction finder (ADF)
An aircraft radio navigation system which senses and indicates the direction to a Low/Medium Frequency (L/MF) nondirectional radio beacon (NDB) ground transmitter. Direction is indicated to the pilot as a magnetic bearing or as a relative bearing to the longitudinal axis of the aircraft depending on the type of indicator installed in the aircraft. In certain applications, such as military, ADF operations may be based on airborne and ground transmitters in the VHF/UHF frequency spectrum.

automatic fare collection (AFC) system
The controls and equipment that automatically admit passengers on insertion of the correct fare in an acceptable form, which may be coins, tokens, tickets, or farecards (stored value farecards must be inserted again on exit, at which point an additional fare may be required).

automatic fire detection device
A device designed to automatically detect the presence of fire by heat, flame, light, smoke, or other products of combustion.

automatic interlocking
Transportation-Rail. An arrangement of signals, with or without other signal appliances, which functions through the exercise of inherent powers as distinguished from those whose functions are controlled manually, and which are so interconnected by means of electric circuits that their movements must succeed each other in proper sequence; train movements over all routes being governed by signal indication.

automatic pilot
Aviation. The roll, pitch, and yaw axis of an aircraft can be controlled by use of an automatic pilot. Information from very high frequency omni-directional radio range (VOR), instrument landing systems (ILS), microwave landing systems (MLS), and other navigation aids can be coupled to the automatic pilot for en route and approach flights.

automatic restraint system
Any restraint system that requires no action on the part of the driver or passengers to be effective in providing occupant crash protection (e.g., air bags or passive belts).

automatic terminal information service
The continuous broadcast of recorded noncontrol information in selected terminal areas. Its purpose is to improve controller effectiveness and to relieve frequency congestion by automating the repetitive transmission of essential but routine information.

automatic train control system
A system so arranged that its operation will automatically result in the following: 1) A full service application of the brakes which

will continue either until the train is brought to a stop, or, under control of the engineman, its speed is reduced to a predetermined rate; 2) When operating under a speed restriction, an application of the brakes when the speed of the train exceeds the predetermined rate and which will continue until the speed is reduced to that rate.

automatic train stop system
A system so arranged that its operation will automatically result in the application of the brakes until the train has been brought to a stop.

automatic vehicle location system
A system that senses, at intervals, the location of vehicles carrying special electronic equipment that communicates a signal back to a central control facility.

automatic vehicle monitoring system
A system in which electronic equipment on a vehicle sends signals back to a central control facility, locating the vehicle and providing other information about its operations or about its mechanical condition.

automation
The increased use of mechanization and/or computerization.

automatism
A movement disorder in which non-reflex motor actions occur during abnormal states of consciousness.

automobile
(1) Any 4-wheeled vehicle propelled by fuel which is manufactured primarily for use on public streets, roads, and highways (except any vehicle operated exclusively on a rail or rails), and that either a) is rated at 6,000 pounds gross vehicle weight or less; or b) which i) is rated more than 6,000 pound gross vehicle weight, but less than 10,000 pounds gross vehicle weight; ii) is a type of vehicle for which the National Highway Traffic Safety Administration (NHTSA) Administrator determines, under paragraph b) of 49 CFR 523, average fuel economy standards are feasible; and iii) is a type of vehicle for which the Administrator determines, under paragraph b) of 49 CFR 523, average fuel economy standards will result in significant energy conservation, or is a type of vehicle which the Administrator determines, under paragraph b) of 49 CFR 523, is substantially used for the same purposes as vehicles described in a) above. (2) A privately owned and/or operated licensed motorized vehicle including cars, jeeps, and station wagons. Leased and rented cars are included if they are privately operated and not used for picking up passengers in return for fare. (3) Passenger cars, up to and including station wagons in size. *See also* ***bus, car, minivan, motor vehicle, taxi,*** *and* ***vehicle.***

automobile insurance
(1) *General.* Insurance against the loss of or damage to a motor vehicle caused by fire, windstorm, theft, collision, or other insurable hazards, and also against legal liability from the operation of the vehicle. (2) A policy of indemnity to protect the operator and owner of a vehicle from liability to third persons as a result of the operation of the automobile. *See also* ***insurance, collision insurance,*** *and* ***no-fault insurance.***

automobile size classification
Automobile size classifications as established by the Environmental Protection Agency (EPA), as follows: Minicompact – less than 85 cubic feet of passenger and luggage volume; Subcompact – between 85 to 100 cubic feet of passenger and luggage volume; Compact-between 100 to 110 cubic feet of passenger and luggage volume; Midsize – between 110 to 120 cubic feet of passenger and luggage volume; Large – more than 120 cubic feet of passenger and luggage volume; Two seat – automobiles designed primarily to seat only two adults. Station wagons are included with the size class for the sedan of the same name.

automobile transporter body
Truck body designed for the transportation of other vehicles.

automotive billing module (AutoBill)
This module creates non-GSA customer billing tapes and General Services Administration (GSA) interfund transactions from billing records generated in the Transportation Interface and Reporting System (TIRES) and generates monthly accounting transaction information to send to the NEAR (National Electronic Accounting and Reporting) system.

automotive payment module (AutoPay)
This module processes all maintenance and extended warranty vendor invoices entered into

the Fleet Service Station (FSS) Fleet Management System by the Maintenance Control Centers and processes the rental authorization records for commercial rent-a-car rentals from the Fleet Management Center.

autonomic
Not subject to voluntary control, as the *autonomic nervous system.*

autonomic nervous system (ANS)
A generally efferent subdivision of the peripheral nervous system which is distributed to and directs the function of smooth muscle and glands throughout the body, normally at a subconscious level. *See also* ***parasympathetic*** *and* ***sympathetic***.

auto-oxidation
A self-induced oxidation process.

AutoPay
See ***automotive payment module.***

autophony
The sensation of abnormal loudness of one's own voice.

autopsy
The detailed examination of the body following death to determine the actual cause of death. An autopsy is ordered by the coroner or medical examiner when the cause of death is unknown or the death occurred under suspicious circumstances.

autorotation
Aviation. A rotorcraft flight condition in which the lifting rotor is driven entirely by action of the air when the rotorcraft is in motion. 1) Autorotative landing/touchdown autorotation. Used by a pilot to indicate that he will be landing without applying power to the rotor. 2) Low level autorotation. Commences at an altitude well below the traffic pattern, usually below 100 feet AGL and is used primarily for tactical military training. 3) 180 degrees autorotation. Initiated from a downwind heading and is commenced well inside the normal traffic pattern. "Go around" may not be possible during the latter part of this maneuver.

autospectral density
See ***power spectral density.***

autothermal thermophilic aerobic digestion
A biological digestion system that converts soluble organics to lower-energy forms through anaerobic, fermentative, and aerobic processes at thermophilic temperatures.

autothermic combustion
See ***autogenous combustion.***

autotrophic
An organism that produces food from inorganic substances.

autism
Morbid self-absorption with extreme withdrawal and failure to relate to other persons.

autumnal equinox
The equinox at which the sun approaches the Southern Hemisphere and passes directly over the equator. Occurs around September 23.

auxiliary lock
With respect to rail operations, a smaller secondary lock adjacent to the main lock.

auxiliary rotor
A rotor that serves either to counteract the effect of the main rotor torque on a rotorcraft or to maneuver the rotorcraft about one or more of its three principal axes.

availability
A measure of the likelihood of having a system in working order at any given time. Also referred to as *measure of availability.* Expressed in the formula:

$$Availability = \frac{uptime}{uptime + maintenance\ time}$$

available flight stage length
The average distance covered per aircraft hop in revenue services, from takeoff to landing. Derived by dividing the total aircraft miles flown in revenue service by the number of aircraft revenue departures performed.

available for work
To be considered "available" for purposes of eligibility for unemployment compensation, claimant must be ready, willing, and able to accept either temporary or permanent suitable employment at any time by another employer and be actually and currently attached to the labor force.

available machine time
That portion of the time during a task cycle in which a machine could be producing useful work but is not.

Available Motions Inventory (AMI)
A series of tests using equipment developed at Wichita State University with the Cerebral Palsy Research Foundation of Kansas to make objective determinations of the physical capabilities of handicapped persons.

available process time
That portion of the time during a processing cycle in which a worker or system could be performing useful work, but is not.

available seat mile (ASM)
(1) One seat transported one mile. (2) The aircraft miles flown in each inter-airport hop multiplied by the number of seats available on that hop for revenue passenger use.

available ton mile
(1) One ton of capacity (passengers and cargo) transported one mile. (2) The aircraft miles flown in each inter-airport hop multiplied by the capacity available (in tons) for that hop.

available tons per aircraft mile
The average total passenger/cargo carrying capacity (tons) offered for sale per aircraft per aircraft mile, derived by dividing the overall available ton miles by the total aircraft miles flown in revenue service.

avascular necrosis
*See **osteonecrosis**.*

average
An arithmetic term indicating the value arrived at by finding the sum of a number of values and dividing the sum by the number of values.

average acceleration
The results of the total change in velocity in a period of time divided by that time.

average fare per unlinked passenger trip
Passenger revenue divided by unlinked passenger trips.

average fleet age
The cumulative years active revenue vehicles are in service divided by the sum of all active revenue vehicles.

average flight stage length
The average distance covered per aircraft hop in revenue services, from takeoff to landing. Derived by dividing the total aircraft miles flown in revenue service by the number of aircraft revenue departures performed.

average length of haul
The average distance in miles one ton is carried. Computed by dividing total ton miles by tons of freight originated. *See also **ton mile**.*

average lifetime mileage per active vehicle
The cumulative mileage for each active vehicle from the date of manufacture through the end of the fiscal year divided by the number of active vehicles.

average man concept
The idea that using the average measurement on a human dimension is adequate for describing a population. It is not generally considered valid, but can be useful as a guideline.

average number of available seats per aircraft
Available seat-miles divided by the number of aircraft revenue miles in passenger service.

average power
The total amount of physical work done, involving moving objects, divided by the period of time during which it is accomplished.

average road width
The average width of the travelway.

average vehicle fuel consumption
A ratio estimate defined as total gallons of fuel consumed by all vehicles, divided by the total number of vehicles (for average fuel consumption per vehicle) or the total number of households (for average fuel consumption per household).

average vehicle miles traveled
A ratio estimate defined as total miles traveled by all vehicles, divided by the total number of vehicles (for average miles traveled per vehicle), or the total number of households (for average miles traveled per household).

average velocity
The total distance traveled in a period of time divided by that time value.

average weekday
Transportation. A representative weekday in the operation of the transit system computed as the mathematical average of several typical weekdays selected at random throughout the year. A typical weekday is one where there are no anomalies such as high ridership due to extra service added for a convention, or low

ridership due to a snowstorm. Average Saturday and Sunday data, including holiday service, are determined the same way.

averaging time
The time period over which a function is measured, yielding a time-weighted average.

avg.
Average.

Aviation Act
Federal law that created the Federal Aviation Administration (FAA) which is responsible for the regulation of aviation including aircraft safety, aircraft marking, etc. *See also **Federal Aviation Administration**.*

aviation gasoline
All special grades of gasoline for use in aviation reciprocating engines, as given in the American Society of Testing and Materials (ASTM) Specification D910. Includes all refinery products within the gasoline range that are to be marketed straight or in blends as aviation gasoline without further processing (any refinery operation except mechanical blending). Also included are finished components in the gasoline range which will be used for blending or compounding into aviation gasoline.

aviation gasoline blending components
Naphthas that are used for blending or compounding into finished aviation gasoline (e.g., straight-run gasoline, alkylate, and reformate). Excludes oxygenates (alcohols and ethers), butane, and pentanes plus.

aviation mode
Consists of airways and airports; airplanes, helicopters, and other flying craft for carrying passengers and cargo.

aviation weather service
A service provided by the National Weather Service (NWS) and Federal Aviation Administration (FAA) which collects and disseminates pertinent weather information for pilots, aircraft operators, and Air Traffic Control (ATC). Available aviation weather reports and forecasts are displayed at each NWS office and FAA Flight Service Station (FSS).

aviator's breathing oxygen (ABO)
A grade of commercial oxygen for high-altitude flying which has no water content.

aviator's vertigo
A disturbance in the pilot's orientation with respect to the earth caused by a conflict between gravitational and visual cues.

avionics
Communications, navigation, flight controls, and displays.

Avogadro's law
Equal volumes of perfect gases at the same temperature and pressure contain the same number of molecules. *See also **Avogadro's number**.*

Avogadro's number
One of the fundamental physical constants. It is expressed as 6.022×10^{23} atoms per gram-atomic weight (or molecules per gram-atomic weight). It is an expression of the number of atoms in a gram-atomic weight of any element. *See also **Avogadro's law** and **mole**.*

avoidable accident
Any accident which can be or could have been prevented by the implementation of appropriate controls/hardware, environmental conditions, or behaviors.

avoidable consequences doctrine
Law. Imposes duty on a person injured to minimize damages. The general rule relating to duty of a party who has been wronged by breach of contract to mitigate damages (i.e., not sit idly by and allow damages to accumulate). This doctrine basically provides that one injured by the tort of another is not entitled to recover damages for any harm that he/she could have avoided by the use of reasonable effort.

avoidable delay
A work element involving a pause or interruption which is unnecessary, due to factors under worker control, and which is not calculated for in standard time figures.

avoirdupois
A common system of weight used in English-speaking countries for all commodities except drugs, precious stones, and precious metals.

avulsion
The tearing away of a body part.

A-weighted sound level
The sound level determined by employing the A scale of a sound level meter, or other noise

survey meter equipped with this weighting network, and expressed as dBA.

AWQC
See ***ambient water quality criteria.***

AWS
American Welding Society.

axial flow
The flow of fluid in the same direction as the axis of symmetry of a tank or basin.

axial flow pump
A type of centrifugal pump in which fluid flow remains parallel to the flow path and develops most of its head by the lifting action of the vanes.

axilla
The somewhat hollow region beneath the junction of the shoulder and trunk. Also referred to as the *armpit.*

axillary arm circumference
The surface distance around the arm at level of the axillary fold. Measured with the individual standing erect and the arms hanging naturally at the sides.

axillary fold
The junction of the torso skin and the arm skin beneath the shoulder at the axilla.

axis
(1) The second cervical vertebra. (2) A graphical or imaginary line representing one of the dimensions in a coordinate system, at which the value of all other dimensions is zero.

axon
The long outgrowth of the body of a nerve cell which conducts impulses from the body toward the next neuron; sometimes spelled axone. *See also* **neuron.**

Ayerza's disease
A form of erythremia marked by chronic cyanosis, chronic dyspnea, chronic bronchitis, bronchiectasis, hepatosplenomegaly and hyperplasia of bone marrow, and associated with sclerosis of the pulmonary artery.

azeotrope
A liquid mixture that has a constant boiling point different from that of its constituents and that distills without change of composition.

azimuth
A magnetic bearing extending from a microwave landing system navigation facility.

azotemia
An excess of urea or other nitrogenous bodies in the blood.

B

B-10 life
The rated life defining the number of revolutions that 90 percent of a group of identical bearings will complete before first evidence of failure develops.

"B" basis allowables
The mechanical strength values specified by material producers and/or suppliers such that at least 90 percent of the materials they produce or supply will meet or exceed the specified values with a 95 percent confidence level.

B display
A sound display in which the data are presented on a rectangular coordinate system with range and azimuth comprising the axes. Also referred to as *range-bearing display*.

B-horizon
The intermediate soil layer, usually having a high clay content, where minerals and other particles washed down from the A-horizon accumulate.

B scale
A sound weighing system which approximates the response characteristics of the human ear in the 40- to 70-phon equal loudness contour range.

B shift
See ***second shift***.

B-weighted sound pressure level
That sound pressure level measured using the B scale. Represented as dB(B). *See also* ***A-weighted sound pressure level***.

Babes-Ernst granules
Metachromatic granules present in many bacterial cells.

Babinski reflex
A reflex action of the toes, indicative of abnormalities in the motor control pathways leading from the cerebral cortex and widely used as a diagnostic aid in disorders of the central nervous system. It is elicited by a firm stimulus (usually scraping) on the sole of the foot, which results in dorsiflexion of the great toe and fanning of the smaller toes. Normally such a stimulus causes all the toes to bend downward. Also called *Babinski's sign*.

BAC
Biologically active carbon. *See also* ***Blood Alcohol Concentration***.

bacillary
Pertaining to bacilli or to rod-like structures.

bacillemia
The presence of bacilli in the blood.

bacillicide
An agent that destroys bacilli.

bacilluria
The presence of bacilli in the urine.

bacillus
The shape of a bacteria cell, also commonly referred to as a *rod*. Bacilli (plural) are generally shaped like a cylinder. They may also be curved, spiral, or helical shaped. Bacilli that are curved are designated as *vibrios*, spiral rods are designated as *spirilla*, and helical rods are designated *spirochetes*.

back
(1) *Anatomy*. The posterior aspects of the ribs, muscles, and all other tissues associated with the posterior trunk/torso from the thoracic vertebrae to the base of the spine. (2) *Law*. To endorse; to sign on the back; to sign generally by way of acceptance or approval; to substantiate; to countersign; to assume financial responsibility for.

back chute
An old channel no longer used that may be located behind an island adjacent to the present navigable channel. It is sometimes used during high water stages to navigate without having to buck the strong currents in the main channel.

back curvature
The surface distance across the back as measured from the right midaxillary line at the posterior axillary fold level to the corresponding point on the left. Measured with the individual standing erect and the arms hanging naturally at the sides.

back door cold front
A cold front moving south or southwest along the Atlantic seaboard of the United States.

back 'er down
To stop headway of a tow.

back injury
Any injury involving the spine, the spinal cord, the nerves exiting from the spinal cord through the spine, a rib-vertebral junction, and/or the muscles of the back.

back pressure
Pressure due to a force operating in a direction opposite to that required.

back siphonage
A backflow of water of questionable quality that results from a negative pressure within the water distribution system.

back to work agreement
An agreement between union and employer covering terms and conditions upon which employees will return to work following settlement of a strike.

back up
To kill headway; to flank or twist tow at foot of crossing or head of bend.

backache
Any pain in the back, usually the lower part. The pain is often dull and continuous, but sometimes sharp and throbbing. Backache, or lumbago, is one of the most commonly encountered ailments and can be caused by a wide variety of disorders, some serious and some not. Occasionally, backache is a symptom of spinal arthritis, peptic ulcer, enlargement of the pancreas, sciatica, diseases of the kidney, or other serious disorders, but usually backache is caused simply by strain of the back in such a way that the bones, ligaments, nerves, and muscles of the spine are compressed or stretched. A sudden action, using muscles that are already fatigued or out of condition, is particularly likely to cause acute strain. A sharp and persistent pain, following the use of unusual force against something could indicate a slipped disk or sacroiliac strain.

backbone
See ***spine***.

backdating
Predating a document or instrument prior to the date it was actually drawn. The negotiability of an instrument is not affected by the fact that has been backdated.

backfill
The material used to refill a ditch or excavation, or the process of refilling.

backflow
(1) Flow reversal in a water distribution system that may result in contamination due to a cross-connection. (2) Abnormal backward flow of fluids; regurgitation.

backflow prevention device
A device used to prevent cross-connection or backflow of non-potable water into a potable water system. *See also* ***backflow***.

background concentration
See ***background level.***

background contamination
(1) Contamination introduced accidentally into dilution waters, reagents, rinse water, or solvents that can be confused with constituents in the sample being analyzed. (2) Substance in the air that is typically present from sources other than those from which one is trying to assess an exposure.

background level
(1) In air pollution control, the concentration of air pollutants in a definite area during a fixed time prior to the starting up or on the stoppage of a source of emission under control. Also referred to as *background concentration*. (2) In toxic substances monitoring, the average presence in the environment, originally referring to naturally occurring phenomena.

background luminance
The luminous intensity of a region within which a target is to be viewed or detected.

background noise
Noise which is coming from sources other than that which is being measured. The total of all sources of interference in a system, apart from the signal. Simply referred to as *noise*. Also known as *auditory ground*.

background processing
The data processing or transmission which is performed secondary to a primary operation or higher priority operations.

background radiation
That which arises from radioactive material, other than that which is being measured. Background radiation due to cosmic rays and natural radioactivity is always present.

background soil pH
The pH of the soil prior to the addition of substances that alter the hydrogen ion concentration.

backhaul
In freight transportation, to carry a shipment back over a segment of a route already covered.

backing line
A line used on a tow to keep barges from running ahead.

backing rudders
*See **flanking rudder**.*

backing wind
A wind that changes direction in a counterclockwise sense (e.g., north to northwest to west).

backlands
A term of no very definite import, but generally signifying lands lying back from (not contiguous to) a highway or water course.

backlash
A control system response in which the direction of movement is momentarily reversed when the movement of a control is stopped.

backlight
The use of a lighting source behind an object to separate that object or region from the background.

backrest
Any structure which is capable of supporting the back.

backrest reference plane
The plane established by a backrest.

backrest-to-seat angle
*See **seatback angle**.*

backup
A copy of a file on a computer hard drive, diskette, or Zip disk, that is kept in case the original is lost or damaged. It is a good practice to keep backup copies of all important files.

backward chaining
A reasoning or control strategy in which the beginning point is the final or desired state with the process extending backward to a known point. Synonymous with *goal-oriented problem solving*.

backward masking
A type of masking in which the masking stimulus occurs following the test stimulus.

backwash
A high-rate reversal of flow for the purpose of cleaning or removing solids from a filter bed or screening medium.

backwash rate
The flow rate used during filter backwash, when the direction of flow through the filter is reversed for cleaning.

backwater
(1) Water in a stream which, in consequence of some dam or obstruction below, is detained or checked in its course, or flows back. (2) Water backed up by a tributary stream. (3) An area of calm water unaffected by the current of a stream.

backwater curve
The term applied to the profile of the water surface above a dam or other obstruction in a channel. This may also be stated as the effect on the natural water surface profile of either of two confluent streams upstream from their confluence due to flow conditions in the other stream.

backyard boatbuilder
Person who builds a boat for his own use and not for the purposes of sale. A backyard boatbuilder may subcontract all work.

Bacon-Davis Act
Federal law (1931) granting the Secretary of Labor power to set wage rates on public construction work to meet wages in the private sector (40 U.S.C.A. § 276a).

BACT
*See **best available control technology**.*

bacteria
(1) Nonspore-forming or non-motile microorganism. Term is applied loosely to any microorganism of the order *Eubacteriales* and popularly called *germ*. Bacteria are one-celled organisms visible only through a microscope. There are many varieties, only some of which cause disease. Most are non-pathogenic, and many are useful. Bacteria are forms of plant life, and are found almost everywhere. They reproduce about every 20 minutes. Their population remains in check, however, since many bacteria feed on each other. Bacteria are classified into three basic groups according to their shape. The rod-shaped bacteria are called *bacilli,* the spiral-

shaped bacteria are named *spirilla*, and dot-shaped bacteria are referred to as *cocci*. The latter may appear in pairs (*diplococci*), in chains like strings of beads (*streptococci*) or in clusters that resemble a bunch of grapes (*staphylococci*). Helpful bacteria existing in the human intestine feed on other microscopic organisms that might be harmful. They also produce some vitamins, including the vitamin B complex and vitamins C and K. Most pathogenic bacteria that invade the body produce toxins. The body's defenses fight back against the invader by rushing leukocytes (white blood cells) and antitoxins to the area of infection; some of the leukocytes engulf the bacteria while the antitoxins neutralize the poisons. The extra blood supply contributes to the inflammatory process. The resulting fever and pain also help by enforcing rest and thus conserving the body's energies to fight off the invader. (2) Microscopic living organisms which can aid in pollution control by consuming or breaking down organic matter in sewage, or by similarly acting on oil spills or other water pollutants. Bacteria in soil, water, or air can also cause human, animal, and plant health problems. The singular form of bacteria is *bacterium*.

bactericidal effectiveness
A measure of the ability of various regions of the ultraviolet spectrum to kill bacteria. May also be referred to as *germicidal effectiveness*.

bactericidal lamp
A light source outputting a high level of ultraviolet-C radiation. Also known as *germicidal lamp*.

bactericide
Any substance that kills bacteria.

bacterid
A skin condition due to hypersensitivity to a bacterial infection.

bacteriocidin
A bactericidal substance present in the blood.

bacterioclasis
The breaking up of bacteria into fragments.

bacteriologist
An expert in the study of bacteria.

bacteriolysin
A substance formed in the blood as a result of infection, and capable of destroying the bacteria causing the infection.

bacteriophage
A virus that destroys bacteria. Several varieties exist, and usually each attacks only one kind of bacteria. Certain types of bacteriophages attach themselves to the cell membrane of the bacterium and instill a charge of DNA into the cytoplasm. DNA carries the genetic code of the virus, so that rapid multiplication of the virus can and does take place inside the bacterium. The growing viruses act as parasites, using the metabolism of the bacterial cell for growth and development. Eventually the bacterial cell bursts, releasing many more viruses capable of destroying similar bacteria.

bacteriosis
Pertains to any disease or abnormal condition caused by a bacterium.

bacteriostat
A substance that inhibits or retards the growth of bacteria but does not necessarily kill them. Also known as *bacteriostatic agent*.

bacteriostatic agent
*See **bacteriostat**.*

bacterium
*See **bacteria**.*

bad character
Absence of moral virtue; the predominance of evil habits in a person. In law of evidence, such character may be shown to affect credibility of a witness by introduction of records of convictions for crimes or by reputation.

bad faith
(1) *General Law*. The opposite of "good faith," generally implying or involving actual or constructive fraud, or a design to mislead or deceive another, or neglect or refusal to fulfill some duty or some contractual obligation, not prompted by an honest mistake as to one's rights or duties, but by some interested or sinister motive. The term "bad faith" is not simply bad judgment or negligence, but rather it implies the conscious doing of a wrong because of dishonest purpose or moral obliquity. It is different from the negative idea of negligence in that it contemplates a state of mind affirmatively operating with furtive design or ill will. (2) *Insurance*. Bad faith on the part

of the insurer is any frivolous or unfounded refusal to pay proceeds of a policy; it is not necessary that such refusal be fraudulent. For purposes of an action against an insurer for failure to pay a claim, such conduct imports a dishonest purpose and means a breach of a known duty (i.e., good faith and fair dealing), through some motive of self-interest or ill will; mere negligence or bad judgment is not bad faith.

bad motive
Intentionally doing a wrongful act knowing at the time that it is wrongful. *See also* ***bad faith***.

bad title
One which conveys no property to the purchaser of the estate. One which is so radically defective that it is not marketable, and hence such that a purchaser cannot be legally compelled to accept it. *See also* ***marketable title***.

badger
To harass, pester, or bedevil persistently, especially in a manner likely or designed to confuse, annoy, or wear down.

baffle
(1) A plate used to provide even distribution, or to prevent short-circuiting or vortexing of flow entering a tank or vessel. (2) A partition used to shield/absorb sound or light energy transmission. (3) To confound or confuse.

baffle chamber
In incinerator design, a chamber designed to promote the settling of fly ash and coarse particulate matter by changing the direction and/or reducing the velocity of the gases produced by the combustion of the refuse or sludge.

bagasse
Crushed sugar cane or sugar beet refuse from sugar making.

bagassosis
A lung disease, or pneumoconiosis, produced as a result of the inhalation of the dust of bagasse, the waste of sugar cane after the sugar has been extracted. Bagasse itself, which is moist and recently ground, is not believed to cause this disease. It is similar to farmer's lung disease. *See also* ***farmer's lung disease*** *and* ***bagasse***.

baghouse filter
Large fabric bag, usually made of glass fibers, used to eliminate intermediate and large (greater than 20 microns in diameter) particles. This device operates in a way similar to the bag of an electric vacuum cleaner, passing the air and smaller particulate matter, while entrapping the larger particulates.

bakeout
The procedure of overheating a new building or space for several days before occupancy and then flushing it out with 100% outside air to remove contaminants that may contribute to poor indoor air quality.

balance
(1) A condition in which working times, tasks, activities, and output are coordinated between the hands of an individual worker, between workers, or between groups so that an operation proceeds smoothly without building excessive inventory or wasting time. (2) A condition of stable posture in which muscle forces exactly counteract the gravitational or other forces imposed on the body. (3) A condition in which the outputs from all speakers in an audio system provide the same output intensity. (4) The difference between the sum of debit entries minus the sum of credit entries in an account. If positive, the difference is called debit balance; if negative, a credit balance.

balance of trade
The difference between a country's total imports and exports; if the exports exceed the imports, a "favorable" balance of trade exists.

balanced motion pattern
A sequence or succession of movements using both the right and left hands/arms which enables the worker to establish and maintain coordination and an efficient rhythm.

balanced transportation
See ***intermodalism***.

balancing by dampers
Method for designing a local exhaust system and its ductwork using adjustable dampers to distribute airflow after system installation.

balancing by static pressure
Method for designing a local exhaust system and its ductwork by selecting the duct diameters that generate the static pressure to distribute the desired airflow throughout the system without the use of dampers.

balancing delay (BD)
The waiting or non-productive time of one hand/arm of a single worker or of another worker or group in an operation due to a lack of balance.

balancing of equities
See ***equity***.

balancing test
A constitutional doctrine in which the court weighs the right of an individual to certain rights guaranteed by the Constitution with the rights of a state to protect its citizens from the invasion of their rights; used in cases involving freedom of speech and equal protection.

Balantidium
A genus of ciliated protozoa, including many species found in the intestine in vertebrates and invertebrates.

Baldrige Award
See ***Malcolm Baldrige National Quality Award***.

balefill
A land disposal site where solid waste material is compacted and baled prior to disposal.

baler
A machine used to compress and bind solid recyclable materials such as cardboard or paper.

baling
Compacting solid waste into blocks to reduce volume and simplify handling.

balk
That customer or user behavior of not participating in an activity or entering a line because of some real or perceived objectionable quality associated with that activity or line, such as its length or waiting period.

Balkan frame
An apparatus for continuous extension in treatment of fractures of the femur, consisting of an overhead bar, with pulleys attached, by which the leg is supported in a sling.

ball-of-foot circumference
The maximum surface distance measured around the distal ends of the protuberances of the metatarsal bones of the foot. Measured with the individual standing erect, with both feet on the floor, and the body weight equally distributed on both feet.

ball valve
A valve utilizing a rotating ball with a hole through it that allows straight-through flow in the open position.

ballast
(1) An electrical transformer for producing the required current, voltage, and wave form to operate certain types of luminaries. (2) Any heavy substance, as sand, stone, etc., laid in the hold of a vessel to steady it. (3) Gravel or broken stone laid down as a stabilizer for a rail bed. (4) Material place on a track bed to hold the track in line and elevation and to distribute its load. Suitable material consists of hard particles (e.g., crushed rock, slag, gravel) that are stable, easily tamped, permeable, and resistant to plant growth.

ballast water
Any water and associated sediments used to manipulate the trim and stability of a vessel. It usually must be treated as an oily wastewater.

ballism
A movement disorder characterized by flinging movements of the limbs. Also called *ballismus*.

ballismus
See ***ballism***.

ballistics
The science of gun examination frequently used in criminal cases, especially cases of homicide, to determine the firing capacity of a weapon, its ability to fire, and whether a given bullet was fired from a particular gun.

ballistic lift
A ballistic movement in which an object is being lifted, with the momentum resulting from the initial motion moving the object through much of the terminal portion of the trajectory.

ballistic motion
See ***ballistic movement***.

ballistic movement
A rapid, gross, relatively smooth change in position of a bodily extremity which is initiated by one or more protagonist muscles which are active only during the initial phase of the motion. May also be referred to as *ballistic motion* and/or *preprogrammed movement*.

ballistic separator
A machine that sorts organic from inorganic matter for composting.

ballistocardiography
Graphic recording of forces imparted to the body by cardiac ejection of blood.

ballistophobia
A morbid dread or fear of missiles.

balloon
A lighter-than-air aircraft that is not engine driven.

balloon freight
Lightweight freight.

balm
(1) *See* ***balsam***. (2) A soothing or healing medicine.

balsam
A semi-fluid, fragrant, resinous vegetable juice. It is used in various preparations to treat irritated or denuded areas of the skin and mucous membranes.

Ban-Lon
A fabric made from a combination of nylon, polyester, and other fiber blends.

band
(1) *General.* A strip that constricts or binds a part. (2) *Acoustics.* A segment of the frequency spectrum of noise.

band application
In pesticides, the spreading of chemicals over, or next to, each row of plants in a field. *See also* ***emissions trading***.

band-pass
Pertaining to a limited range of frequencies which are transmitted or allowed beyond a certain point within a system.

band-pass filter
A filter with a single transmission band extending from lower to upper cutoff frequencies.

band pressure level
The sound pressure level within a specified frequency bandwidth.

band rejection
See ***bandstop***.

bandage
A strip or piece of gauze or other fabric for wrapping or covering any part of the body. Bandages may be used to stop the flow of blood, to provide a safeguard against contamination, or to hold a medicated dressing in place. They may also be used to hold a splint in position or otherwise immobilize an injured part of the body to prevent further injury and to facilitate healing.

bandscreen
See ***traveling water screen***.

bandstop
Pertaining to a limited range of frequencies which are not allowed to pass through a system, or which pass at a much lower intensity than the higher and lower frequencies. Synonymous with *band rejection*.

bandwidth
(1) That range of continuous frequencies capable of being processed or output by a system. (2) That maximum rate at which information can be transferred over a channel, typically with units in some multiple of bits per second.

bang-bang control
A type of discrete system control using relays to control input and in which the operator moves a control from essentially maximum deflection in one direction to essentially maximum deflection in another direction.

bank
(1) That excess amount or material or numbers of product which are allowed to accumulate at some point in a production line or operation without being currently worked to provide for reasonable fluctuations in flow. (2) The elevation of the outer margin relative to the inner margin of the curves on a roadway. (3) To roll an aircraft about its longitudinal axis. (4) A place of storage for such materials as blood (blood bank), or for other human tissue (bone bank, eye bank, skin bank, etc.) to be used in reparative surgery. (5) An institution, usually incorporated, whose business it is to receive money on deposit, cash checks or drafts, discount commercial paper, make loans, and issue promissory notes payable to bearer, known as bank notes.

bank sand
Sand excavated from a natural deposit, usually not suitable for use in filter processing or grading.

banking
A system of recording qualified air emission reductions for later use in bubble, offset, or netting transactions.

bar
(1) Shortened for *barometer*. (2) A unit or pressure equal to 0.9869 atmospheres, 10^6 dyn/cm^2, and 14.5 lb/in^2. (3) A structure that hinders or impedes. (4) The court, in its strictest sense, sitting in full term; the presence, actual or constructive, of the court. (5) A submerged or emerged mound, ridge, or succession of ridges of sand or other material extending across the bottom and which may obstruct navigation.

bar association
An association of members of the legal profession.

bar chart
See ***bar graph, Gantt chart***.

bar code
A set of parallel lines of differing widths which contain coded information.

bar code reader
See ***bar code scanner***.

bar code scanner
A laser-based device which views a bar code and transmits the information to a computer, Also referred to as *bar code reader*.

bar graph
A graphical representation of the frequency of occurrence within a set of discrete groupings or values in which the length of the bar is proportional to frequency. *See also* ***Gantt chart***.

bar screen
(1) In wastewater treatment, a device used to remove large solids. (2) A screening device that utilizes mechanically operated rakes to remove solids retained on a stationary bar rack.

Bárány chair
A rotating chair used for vestibular or nystagmus experimentation.

barbituism
A toxic condition produced by use of barbital and its derivatives.

barbiturate
One of a group of organic compounds derived from barbituric acid, and commonly described as *sleeping pills*.

barbituric acid
A compound $C_4H_4N_2O_3$, the parent substance of barbiturates.

bare conductor
See ***conductor***.

bareback
Tractor without its semitrailer.

barf bag
Common slang for a plastic disposable bag used for collecting or capturing expelled vomitus during motion sickness.

bargaining unit
Labor union or group of jobs authorized to carry on collective bargaining on behalf of employees. A particular group of employees with a similar community of interest appropriate for bargaining.

barge
(1) A non-powered, flat bottom, shallow draft vessel including river barges, scows, car floats, and lighters. The term does not typically include ship-shaped or deep draft barges. (2) Shallow, non-self-propelled vessels used to carry bulk commodities on rivers and the Great Lakes. (3) A non-self-propelled vessel. *See also* ***boat*** *and* ***tanker***.

barge carriers
Ships designed to carry barges. Some are fitted to act as full container ships and can carry a varying number of barges and containers at the same time.

baritosis
A form of pneumoconiosis resulting from the inhalation of barium sulfate or other barium compounds.

barium
A chemical element, atomic number 56, atomic weight 137.34, symbol Ba.

barium sulfate
A fine, white, bulky powder, used as an opaque medium for x-ray examination of the digestive tract.

barium test
X-ray examination using a barium mixture to help locate disorders in the esophagus, stom-

ach, duodenum, and the small and large intestines.

bark pocket
See ***pitch and bark pocket.***

barnacles
A marine crustacean with a calcareous shell that attaches itself to submerged objects.

barn door
One of a set of adjustable light shields which may be used in conjunction with a luminaire to partially direct and control the luminance emitted from the luminaire.

baroclinic
The state of the atmosphere where surfaces of constant pressure intersect surfaces of constant density. On an isobaric chart, isotherms cross the contour lines and temperature advection exists.

baroclinic instability
A type of instability arising from a meridional (north to south) temperature gradient, a strong vertical wind speed shear, temperature advection, and divergence in the flow aloft. Many mid-latitude cyclones develop as a result of this instability.

barodontalgia
A form of decompression sickness resulting in tooth pain from the expansion of trapped air within a tooth or between a filling and the tooth material.

barograph
A continuous recording barometer.

barometer
An instrument used to measure atmospheric pressure.

barometric condenser
A condenser in which vapor is condensed by direct contact with water.

barometric damper
A pivoting plate used to regulate the amount of air entering a duct or flue to maintain a constant draft within an incinerator.

barometric leg
(1) A condensate discharge line submerged below the liquid level of an atmospheric tank. (2) A gravity tailpipe from a vacuum barometric condenser.

barometric pressure
Ambient or local pressure surrounding a gauge, evaporator shell, vent pipe, etc.

barometric tendency
See ***pressure tendency.***

barosinusitis
Pain in the sinus due to rapid changes in atmospheric air pressure.

barotalgia
That sensation of pressure or ear pain due to an inequality of air pressure between the middle ear and the environment. May be referred to as *ear squeeze*.

barotitis media
Barotrauma to the middle ear.

barotrauma
Any injury resulting from expansion or contraction of gases in closed spaces within certain structures of the body due to pressure changes in the ambient environment.

barotropic
A condition in the atmosphere where surfaces of constant density parallel surfaces of constant pressure.

barrel (bbl)
According to the Federal Oil Pollution Act of 1990, a measure equal to forty-two United States gallons at 60 degrees Fahrenheit.

barrier
(1) A control (device, mechanism, structure, sign, etc.) intended to prevent the transfer of energy from one element of a system to another. (2) Any object, individual, or structure which impeded progress toward a goal or which prevents entry to a region for safety reasons.

barrier landscape water renovation system (BLWRS)
A wastewater treatment and denitrification system where wastewater is applied to the top of a mound of soil overlaying a water barrier and microbes oxidize soluble organics as the water percolates through the soil.

barrier coating(s)
A layer of a material that acts to obstruct or prevent passage of something through a surface that is to be protected, e.g., grout, caulk, or various sealing compounds. Sometimes used with polyurethane membranes to prevent

corrosion or oxidation of metal surfaces, chemical impacts on various materials, or, for example, to prevent soil-gas-borne radon from moving through walls, cracks, or joints in a house.

barrier cream
A protective viscous substance that may be used for preventing skin contact with harmful agents and/or percutaneous absorption of toxic materials. Used as a supplement (not a replacement) to personal protective equipment.

barrier equivalent velocity (BEV)
The effective velocity at which a vehicle impacts a barrier in crash testing.

barrier free
An ideal condition in which handicapped individuals have full and equal access to all facilities accessible to able-bodied individuals.

barrier guard
Any protective device designed to prevent access to hazardous areas, or to prevent inadvertent operation of controls or equipment.

barrier remediation
Exhaust radon from the enclosure.

BARS
*See **behavior-anchored rating scale**.*

Barthel Index (BI)
A numerical score based on 10 items of a physically disabled individual's ability to care for himself or herself by performing some of the activities of daily living.

barylalia
Indistinct, thick speech, resulting from a lesion of the central nervous system.

basal
(1) Pertaining to a base; fundamental. (2) In physiology, pertaining to the lowest possible level, resting level.

basal application
In pesticides, the application of a chemical on plant stems or tree trunks just above the soil line.

basal cell carcinoma
A cutaneous cancer of relatively low-grade malignancy, arising from the basal layers of the epidermis. *See also **carcinoma**.*

basal conditions
Those conditions under which the basal metabolic rate measures are taken: (a) fast for at least 12 hours, (b) following a night of restful sleep, (c) no strenuous exercise since sleep, and (d) comfortable, relaxed conditions with the air temperature about 70-75°F, depending on clothing.

basal metabolic rate (BMR)
(1) Rate of heat production by the human body under neutral conditions. (2) That energy expenditure per unit time by the body while an individual is awake, but at rest under basal conditions.

basal metabolism
That minimal metabolism required to maintain cellular function.

basal temperature
The normal body temperature of a healthy individual following sleep in the morning.

base
(1) The lower part of an object, or the broadest part of the conical or pyramidal structure. (2) The main ingredient of a compound. (3) A compound that reacts with an acid to form a salt; another term for *alkali*. It turns litmus paper blue.

base period
(1) *General*. The reference period of time (year, month, etc.) against which some current period is judged. (2) *Transit*. The time of day during which vehicle requirements and schedules are not influenced by peak-period passenger volume demands (e.g., between morning and afternoon peak periods). At this time, transit riding is fairly constant and usually low to moderate in volume when compared with peak-period travel. Also referred to as *off-peak period*.

base time
*See **normal time**.*

base wage rate
The hourly monetary compensation paid to a normal operator working at a standard pace on a specified task.

baseline
(1) A sample used as a comparative reference point when conducting further tests or calculations. (2) According to the Clean Air Act of 1990: The annual quantity of fossil fuel

consumed by an affected unit, measured in millions of British Thermal Units (mmBtus), calculated as follows:

a) For each utility unit that was in commercial operation prior to January 1, 1985, the baseline shall be the annual average quantity of mmBtus consumed in fuel during calendar years 1985, 1986, and 1987, as recorded by the Department of Energy pursuant to Form 767. For any utility unit for which such form was not filed, the baseline shall be the level specified for such unit in the 1985 National Acid Precipitation Assessment Program (NAPAP) Emissions Inventory, Version 2, National Utility Reference File (NURF) or in a corrected data base as established by the EPA Administrator pursuant to paragraph (3). For nonutility units, the baseline is the NAPAP Emissions Inventory, Version 2. The Administrator, in the Administrator's sole discretion, may exclude periods during which a unit is shutdown for a continuous period of four calendar months or longer, and make appropriate adjustments under this paragraph. Under petition of the owner or operator of any unit, the Administrator may make appropriate baseline adjustments for accidents that caused prolonged outages.

b) For any other nonutility unit that is not included in the NAPAP Emissions Inventory, Version 2, or a corrected data base as established by the Administrator pursuant to paragraph (3), the baseline shall be the annual average quantity, in mmBtu consumed in fuel by that unit, as calculated pursuant to a method which the Administrator shall prescribe by regulation to be promulgated not later than eighteen months after enactment of the CAA Amendments of 1990.

c) The Administrator shall, upon application or on his own motion, by December 31, 1991, supplement data needed in support of this subchapter and correct any factual errors in data from which affected Phase II units' baselines or actual 1985 emission rates have been calculated. Corrected data shall be used for purposes of issuing allowances under the subchapter. Such corrections shall not be subject to judicial review, nor shall the failure of the Administrator to correct an alleged factual error in such reports be subject to judicial review.

baseline audiogram
The audiogram against which all future audiograms are compared. *See also* ***audiogram***.

baseline concentration
Under the Clean Air Act of 1990: With respect to a pollutant, the ambient concentration levels which exist at the time of the first application for a permit in an area subject to the CAA, based on air quality data available in the EPA or a state air pollution control agency on such monitoring data as the permit applicant is required to submit. Such ambient concentration levels shall take into account all projected emissions in, or which may affect, such area from any major emitting facility on which construction commenced prior to January 6, 1975, but which has not begun operation by the date of the baseline air quality concentration determination. Emissions of sulfur oxides and particulate matter from any major emitting facility on which construction commenced after January 6, 1975, shall not be included in the baseline and shall be counted against the maximum allowable increases in pollutant concentrations established under the CAA.

baseline data
Data that describe the magnitude and range of exposures for a homogeneous exposure group and stressor (e.g., an airborne contaminant, physical agent, etc.).

basic division of accomplishment
See ***therblig***.

basic division of work
See ***therblig***.

basic element
See ***therblig***.

basic elements of performance
A quantitative technique for measuring the residual capabilities of disabled individuals.

basic event
As pertains to fault tree analysis (FTA) and/or the management oversight and risk tree (MORT), a root fault event, or the first in the process to have occurred that requires no further development or analysis. Represented graphically as a circle.

basic grant
The funds available to a state for carrying out an approved State Enforcement Plan (SEP),

which include, but are not limited to: 1) recruiting and training of personnel, payment of salaries and fringe benefits, the acquisition and maintenance of equipment except those at fixed weigh scales for the purposes of weight enforcement, and reasonable overhead costs needed to operate the program; 2) commencement and conduct of expanded systems of enforcement; 3) establishment of an effective out-of-service and compliance enforcement system; and 4) retraining and replacing staff and equipment.

basic input output system (BIOS)
A process that allows one computer to "speak the same language" as every other computer.

basic measurement cycle
Any of four basic scales used for classifying data in statistical analyses. *See also* ***nominal scale, ordinal scale, equal-interval scale,*** and ***ratio scale***.

basic motion
Any fundamental, complete motion using the primary physiological and/or biomechanical performance capabilities of the body or its member parts, as determined by motion analysis studies. *See* ***therblig***.

Basic Motion Time Study (BMTS)
A predetermined motion time system.

basic research
That fundamental research performed to acquire scientific knowledge, without concern for immediate practical application. Also referred to as *pure research*.

basic sciences
Those disciplines involving the study of mathematics, chemistry, physics, biology, and psychology.

Basic T
A reference to the arrangement of four basic flight instruments in a standard pattern, with the airspeed and attitude indicators and altimeter in a horizontal line across the top, and the heading indicator centered below.

basic time
That time allowed or required for performing a work element at a standard rate. Represented in the formula:

$$BT = \frac{\textit{observed time x observed rating}}{\textit{standard rating}}$$

basic units
There are seven basic or fundamental units of measure in use throughout the world. The most widely recognized set of these units, known as the *International System of Units (SI)*, was initially adopted in 1960 and is reviewed and amended, as required, at one of the General Conferences on Weights and Measures, an international meeting that convenes periodically. In addition, there are two common "metric" systems, often referred to as the *MKS System* (meters, kilometers, and seconds) and the *CGS System* (centimeters, grams, and seconds). Also, there is the *English System* (obsolete almost every on earth except for the Untied States).

basic utility stage I airport
This type of airport serves 75 percent of the single-engine and small twin-engine airplanes used for personal and business purposes. Precision approach operations are not usually anticipated. This airport is designed for small airplanes in Airport Reference Code B-I.

basic utility stage II airport
This type of airport serves all the airplanes of stage I plus some small business and air taxi-type twin-engine airplanes. Precision approach operations are not usually anticipated. This airport is also designed for small airplanes in Aircraft Reference Code B-I.

basicity factor
Factor used to determine neutralization capabilities of alkaline reagents used to treat acidic wastes.

basilar membrane
That membrane in the cochlea to which the organ of Corti is attached.

basin
(1) When speaking of a large river, ordinarily means or includes the entire area drained by the main stream and its tributaries. (2) Any bowl-shaped depression in the surface of the land or ocean floor. *See also* ***inundation area***.

basin and range
An area of the southwestern United States characterized by roughly parallel mountain ranges and valleys, formed by a series of tilted fault blocks, and brought about by

tectonic extension of the region. As is true of any region experiencing crustal extension, normal faulting is common here. This name can apply generally to any zone of similar landforms and tectonics.

basket centrifuge
Batch-type centrifuge where sludge is introduced into a vertically mounted spinning basket and separation occurs as centrifugal force drives the solids to the wall of the basket.

basket guard
*See **cage**.*

bass
Those sound frequencies in the lower portion of the audio range, generally below about 250 Hz.

bastard tow
A tow made up of uneven or dissimilar barges.

BAT
*See **best available technology**.*

batch process
A non-continuous treatment process in which a discrete quantity or batch of liquid is treated or produced at one time.

batch reactor
A reactor where the contents are completely mixed and flow neither enters nor leaves the reactor vessel.

batching
The process of scheduling work in small increments. Synonymous with *short-interval scheduling*.

BATEA
Acronym for best available technology economically available.

bathophobia
Morbid dread of depths or looking down from high places, with fear of falling.

bathtub curve
A graphic representation of the life cycle of products, systems, or individual components in terms of frequency of failures relative to periods of usefulness. In system safety, it is also known as a *reliability curve*. *See also **life cycle characteristic curve**.*

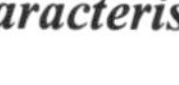

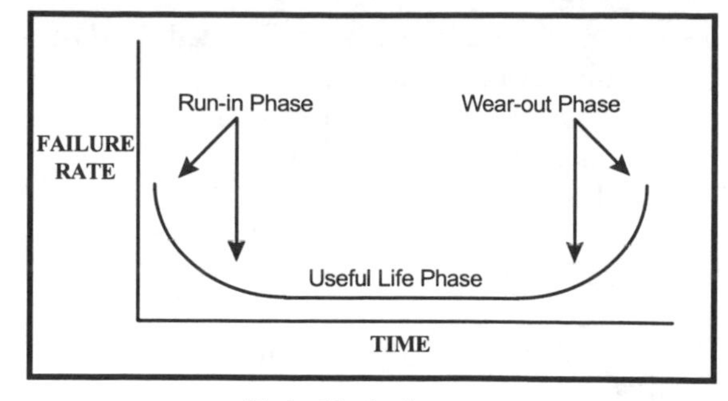

Typical bathtub curve

bathycardia
A condition in which the heart is positioned abnormally low in the thorax.

batt
A section of insulting material.

battery
(1) *General*. Any unit, apparatus, or grouping in which a series or set of parts or components is assembled to serve a common end. (2) *Law*. Intentional and wrongful physical contact with a person without his or her consent that entails some injury or offensive touching. (3) *Electrical*. One or more cells operating together as a single source of direct current. (4) *Military*. An artillery unit equivalent to an infantry company; a group of guns, rockets, or related equipment forming an artillery unit.

battery life
With instrumentation, the time period over which the battery of an instrument will provide sufficient power for uninterrupted operation of the device.

battery limit
The boundary limits of equipment, or a process unit that defines interconnecting points for electrical piping or wiring.

battle of the forms
In commercial law, term used to describe the effect of a multitude of forms used by buyers and sellers to accept and to confirm terms expressed in other forms.

batture
A marine term, used to denote a bottom of sand, stone, or rock, mixed together, and rising towards the surface of the water; as a technical word and also in common parlance, an elevation of the bed of a river, under the surface of the water. The term is, however, sometimes used to denote the same elevation of the bank, when it has risen above the sur-

face of the water, or is as high as the land on the outside of the bank.

baud
A unit of serial transmission speed, usually equivalent to bits per second. If each signal event is represented by one bit, baud is equivalent to bits per second.

bauxite
Ore containing alumina monohydrate or alumina trihydrate, which is the principal raw material for alumina production.

bauxite fume pneumoconiosis
A pulmonary fibrosis due to the inhalation of aluminum ore or processing dust fumes. Also referred to as *Shaver's disease*.

bay/inlet
A water area that is an opening of the sea/ocean into the land, or of an estuary, lake, or river into its shore.

baygall
A low-lying wet land matter with vegetable fibers and often with gallberry and other thick-growing bushes.

Bayle's disease
Progressive general paralysis.

Bazin's disease
Tuberculosis indurativa, chronic tuberculosis of the skin, characterized by indurated nodules.

bbl
See ***barrel***.

BBL
Burst before leak.

bcf
Billion cubic feet.

BCF
See ***bioconcentration factor***.

BCS
See ***border cargo selectivity***.

BCSP
See ***Board of Certified Safety Professionals***.

BDAT
See ***best demonstrated available technology***.

BDOC
See ***biodegradable dissolved organic carbon***.

beachwell
A shallow intake well making use of beach sand and structure as a filter medium.

beacon
A fixed signal, mark, or light and associated facilities erected for the guidance of mariners or airplane pilots.

beam
(1) *Illumination*. A concentrated emission of light energy along a definite projection in a single direction. (2) *Material Handling*. The horizontal support member of a lifting mechanism. (3) *Longshoring*. A portable transverse or longitudinal beam which is placed across a hatchway and acts as a bearer to support the hatch covers. Also referred to as *strongback*. (4) *Boating*. The width of a ship.

beam angle
A measure of light beam spread; the angle between diametrically opposed edges of a projected light beam at which the luminous intensity is some stated percentage of the maximum along the beam axis, with all measures taken in a wavefront equidistant from the source. *See also* ***beam axis***.

beam axis
(1) An imaginary straight line representing the direction along which a light beam is projected. (2) A line from the source through the centers of the x-ray fields.

beam element
A modeling structure which is capable of bending, torsion, and axial stiffness.

beam spread
The lateral distribution of a projected light beam from the beam axis. *See also* ***beam axis***.

bean hauler
A driver who transports fruits and vegetables.

bear trap
A section of movable dam with concrete piers in either side (generally about 100 feet wide) and provided with a gate which may be raised or lowered by compressed air. The bear trap serves as a type of safety valve. When the pool level maintained at the dam becomes too high, the bear trap is lowered to permit the excess water to run out. This pool control feature of movable wicket-type dams is found on the Ohio and Illinois rivers. The bear trap will always be located on the opposite side of the river from the lock and is very dangerous to approach when open.

bearer
In scaffolding construction, a horizontal member of a scaffold upon which the platform rests and which may be supported by ledgers.

bearing
Navigation. The horizontal direction to or from any point, usually measured clockwise from true north, magnetic north, or some other reference point, through 360 degrees. *See also course and flight path.*

beat
A periodic variation in the intensity of sound generated from the combination of two simple tones of slightly different frequencies having approximately the same orders of magnitude.

beat elbow
Bursitis of the elbow joint that can result from the use of heavy vibrating tools.

beat frequency
The number of occasions per unit time at which a beat occurs.

beat knee
A bursitis in the knee joint due to prolonged vibration, pressure, or repeated friction.

Beaufort scale
A numerical scale of wind force on which a Beaufort force 0 wind is calm and a force 12 wind indicates hurricane force with winds in excess of 120 km/hr (75 mph).

becquerel
A unit expressing the rate of radioactive disintegration. One becquerel is equal to one radioactive disintegration per second. There are 3.7 E+10 becquerels per curie of radioactivity, (Bq).

bed depth
The depth of filter media or ion exchange resin contained in a vessel.

bed-load movement
Solids which are transported along the riverbed as a semi-suspended sediment.

bed rest
The maintaining of an individual in bed, either for therapy or experimental purposes.

Bedaux plan
A wage incentive plan in which performance above the rated standard 60 B units per hour would result in merit bonuses.

bedbug
A bug of the genus *Cimex.* A flattened, oval, reddish insect that inhabits houses, furniture, and neglected beds, and feeds on man, usually at night.

bedrock
Solid rock encountered below the mantle of loose rock and soil on the earth's surface.

Beer-Lambert law
Holds that the absorptive characteristics of a substance is a constant with respect to changes in concentration.

beggiatoa
Filamentous microbe, commonly associated with sludge bulking, that results from low dissolved oxygen levels and/or high sulfide levels.

beginning milepost
The continuous milepost notation to the nearest 0.01 mile that marks the beginning of any road or trail segment.

beginning spurt
A briefly higher than normal level of activity at the beginning of the work period or by an employee new to the job.

behavior
Any and all responses of an individual, group, or system.

behavior-anchored rating scale (BARS)
Any rating scale developed to evaluate individual behavior patterns.

behavioral competence
Having the ability to integrate those psycho- and sensorimotor patterns required to complete one or more specified tasks.

behavioral dynamics
The behavioral operating characteristics of individuals and groups as they are conditioned by the external working environment and/or individual and group interactions.

behavioral resistance
An opposition to carrying out an order, directive, or someone's expressed desire.

behavioral rigidity
An inability to effectively deal with new situations.

behavioral toxicology
The study and assessment of neural impairment due to toxic chemical exposure through the use of psychological methods.

behaviorism
A theory of psychology based upon a purely objective observation and analysis of human and animal behavior without reference to the complexities and nuances of psychoanalytic depth psychology.

Behring's law
Blood and serum of an immunized person, when transformed to another subject, will render the latter immune.

BEI
See ***biological exposure index****.*

BEIR Committee
Biological Effects of Radiation Committee of the National Academy of Sciences. Reports on the health effects of ionizing radiation.

Békésy audiometry
An auditory threshold determination procedure involving observer control in which an individual alternately, over several cycles, presses a switch to reduce signal level when the sound is heard and releases it when the sound becomes inaudible. Also referred to as *Békésy tracking procedure*.

Békésy tracking procedure
See *Békésy audiometry*.

bel (B)
A non-dimensional measure of the intensity of some energy, corresponding to the ratio of two intensity or power levels. *See also* ***decibel****.*

Belding-Hatch heat stress index
An estimate for body heat stress based on the ratio of actual evaporative heat loss to the maximum possible evaporative heat loss for the given environment, which may be determined by temperature, humidity, air velocity, workload, clothing, and their interactions. Commonly referred to simply as *heat stress index*.

belief-action distinction
The distinction noted in analysis of cases under the First Amendment, U.S. Constitution (freedom of speech and religion) to the effect that one is guaranteed the right to any belief he/she chooses, but when that belief is translated into action, the state also has rights under its police power to protect others from such actions.

Bell's palsy
Neuropathy of the facial nerve, resulting in paralysis of the muscles of the face, usually on one side. Victims usually are unable to close their mouths, so that they drool and cannot whistle. If they are unable to close the eye on the affected side, it may become tearful and inflamed. Facial palsy is often no more than a temporary condition lasting a few days of weeks. Occasionally, the paralysis results from a tumor pressing on the nerve, or from physical trauma to the nerve. More often, however, the cause is unknown. In many cases, the deformity may be reduced by plastic surgery.

belly
(1) The fleshy central portion of a muscle along its longitudinal axis. (2) Slang term for the stomach-abdominal region of the frontal portion of the torso.

belly button rule
A task design guideline stating that the hands should remain close to the abdominal region when lifting or handling items.

below knee (BK)
Pertaining to a prosthesis for or an amputee/amputation for which some part of the lower leg, and all of the ankle and foot, are missing/taken.

below minimums
Weather conditions below the minimums prescribed by regulation for the particular action involved (e.g., landing minimums, takeoff minimums).

belt conveyor
A device used to transport material, consisting of an endless belt that revolves around head and tail pulleys.

belt filter press
See ***belt press****.*

belt press
A sludge dehydration device utilizing two fabric belts revolving over a series of rollers to squeeze water from the sludge. May be referred to as a *belt filter press*.

belt thickener
Mechanical sludge processing device that uses a revolving horizontal filter belt to prethicken sludge prior to dehydration and/or disposal.

Benadryl

Trademark for diphenhydramine, an antihistamine. *See also* ***antihistamine***.

bench

A seat of judgment or tribunal for the administration of justice. The seat occupied by judges in courts. Also, the court itself, or the aggregate of the judges composing a court itself, as in the phrase "before the full bench." *See also* ***bar***.

bench conference

A meeting at the judge's bench prior to, during, or after a trial, or hearing between counsel and the judge to discuss a matter pertaining to such proceeding. Commonly called to discuss questions of evidence out of hearing of jury; it may or may not be made part of the written record of the proceeding.

bench test

A small-scale test or study used to determine whether a technology is suitable for a particular application.

bench warrant

Process issued by the court itself, or "from the bench," for the attachment or arrest of a person in case of contempt, or where an indictment has been found, or to bring a witness who fails to obey a subpoena.

benchmark

A thoroughly documented reference value or standard or measurement against which performance, response, or other characteristics may be compared with confidence.

benchmark job

A job or task having enough common characteristics with one or more other jobs such that it may be used as a predictor for those jobs in such aspects as evaluations of worker output and time standards.

bend

Curve in the river, analogous to a curve in a highway.

bends

A form of decompression sickness in which pain occurs in joints, muscles, and/or bones. The condition results from a too rapid decrease in atmospheric pressure (as with deep sea divers). The term is derived from the bodily contortions its victims undergo when atmospheric pressure is abruptly changed from a high pressure to a relatively lower pressure. A form of altitude sickness suffered by aviators who ascend too rapidly to high altitudes is similar to the bends. Bends may also be a complication in a type of oxygen therapy called *hyperbaric oxygenation* in which the person is placed in a high-pressure chamber to increase the oxygen content of the blood. The phenomenon of bends is explained in terms of a law of physics: The greater the atmospheric pressure, the greater the amount of gas that can be dissolved in a liquid. The gas involved in bends is the air we breathe, composed chiefly of nitrogen and oxygen. Under normal atmospheric pressure (about 15 lb. per square inch), nitrogen is present in the blood in dissolved form. If the atmospheric pressure is substantially increased, a proportionately greater amount of nitrogen will be dissolved in the blood. The same is true of oxygen, and this is the basis for hyperbaric oxygenation in the treatment of oxygen deficiency. The increase in pressure causes no ill effects. Nor will there be any ill effects if the pressure is gradually brought back to normal. When the decrease in pressure is slow, the nitrogen escapes safely from the blood as it passes through the lungs to be exhaled. If the pressure drops abruptly back to normal, the nitrogen is suddenly released from its state of solution in the blood and forms bubbles. Although the body is now under normal air pressure, expanding bubbles of nitrogen are present in the circulation and force their way into the capillaries, blocking the normal passage of the blood. This blockage (or *embolism*) starves cells dependent on a constant supply of oxygen and other blood nutrients. Some of these cells may be nerve cells located in the limbs or in the spinal cord. When they are deprived of blood, an attack of the bends occurs. The oxygen in the blood reacts similarly when abnormal pressure is abruptly relieved. But because oxygen is dissolved more easily than nitrogen, and because some of the oxygen combines chemically with hemoglobin, the oxygen released in decompression forms fewer bubbles, and is therefore less troublesome.

beneficial impact

A purposeful impact, as in performing some task. Also referred to as *functional impact*.

beneficial organism
A pollinating insect, pest predator, parasite, pathogen, or other biological control agent that functions naturally or as part of an integrated pest management program to control another pest.

benefit
Advantage; profit; fruit; privilege; gain; interest. The receiving, as the exchange for promise, some performance or forbearance which the promisor was not previously entitled to receive.

benefit-of-the bargain damages
The difference between the value received and the value of the fraudulent party's performance as represented.

benign
Pertaining to the mild character of an illness or the non-malignant character of a neoplasm.

bent
Deformed in a shape deviation from the original line or plane; creased, kinked, or folded.

bent torso breadth
The horizontal linear distance across the shoulders with the individual in the bent torso position. *See also* ***bent torso position***.

bent torso height
The vertical linear distance from the floor or other reference surface to the highest point on the head in the bent torso position. *See also* ***bent torso position***.

bent torso position
A posture with the individual standing with feet separated by 18 inches, leaning forward with hands on knees, and looking straight ahead.

benthal oxygen demand
The oxygen demand exerted by the organic mud and sludge deposits on the bottom of a river or stream.

benthic
Relating to the bottom environment of a water body.

benthic organism
Benthos. A form of aquatic plant or animal life that is found on or near the bottom of a stream, lake, or ocean.

benthic region
The bottom layer of a body of water.

bentonite
Colloidal clay-like mineral that can be used as a coagulant aid in water treatment systems. Also used as a landfill liner because of its limited permeability.

benzene
An aromatic hydrocarbon used as a solvent; it has carcinogenic properties and is often characterized by its ring structure. Chemical formula is C_6H_6, represented by the structure:

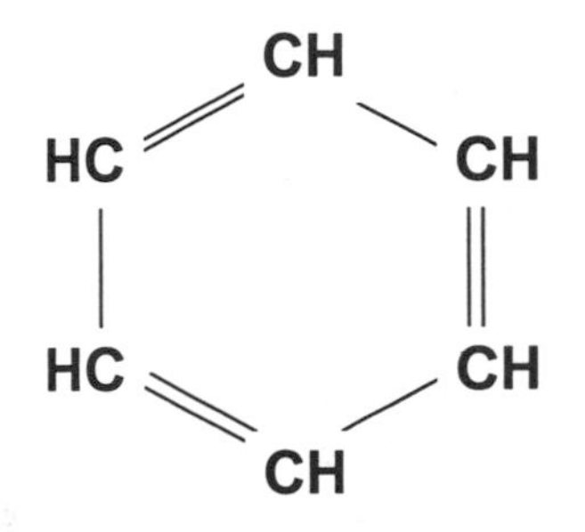

beriberi
A disease caused by vitamin B (thiamin) deficiency common among populations that survive on polished rice. It is characterized by loss of muscle power, emaciation, and exhaustion.

berm
A horizontal, earthen ridge or bank. The sharp definitive edge of a dredged channel such as in a rock cut.

Bermuda high
See ***subtropical high***.

Bernoulli's equation
Energy equation commonly used to calculate head pressure; it considers velocity head, static head, and elevation.

berth
A specific segment of wharfage where a ship ties up alongside at a pier, quay, wharf, or other structure that provides a breasting surface for the vessel. Typically, this structure is a stationary extension of an improved shore and intended to facilitate the transfer of cargo or passengers. *See also* ***wharfage***.

beryl
A silicate of beryllium and aluminum that is considered a carcinogen.

berylliosis
Chronic poisoning caused by exposure to the dust or fume of beryllium metal, beryllium oxide, or soluble beryllium compounds.

Symptoms include a loss of appetite and weight, weakness, cough, extreme difficulty in breathing, cyanosis, and cardiac failure. The disease may appear 5-20 years after exposure has ceased. It is commonly progressive in severity, will cause fibrous growth in the lungs, can create kidney stones, and can be accompanied by enlargement of the heart, liver, and spleen.

beryllium
(1) A chemical element, atomic number 4, atomic weight 9.012, symbol Be. (2) An airborne metal that can be hazardous to human health when inhaled. It is discharged by machine shops, ceramic and propellant plants, and foundries.

best available control technology (BACT)
An emission limitation based on the maximum degree of emission reduction (considering energy, environmental, and economic impacts and other costs) achievable through application of production processes and available methods, systems, and techniques. In no event does BACT permit emissions in excess of those allowed under any applicable Clean Air Act provisions. Use of the BACT concept is allowable on a case by case basis for major new or modified emissions sources in Attainment Areas and applies to each regulated pollutant.

best available technology (BAT)
The best technology, treatment techniques, or other means available after considering field, rather than solely laboratory, conditions.

best demonstrated available technology (BDAT)
A technology demonstrated in full-scale commercial operation to have statistically better performance than other technologies.

best evidence rule
Rule which requires that the best evidence available be presented in lieu of less satisfactory evidence.

best management practice (BMP)
The schedules of activities, methods, measures, and other management practices to prevent pollution of waters and facilitate compliance with applicable regulations.

beta
(1) The probability of making a Type II error. Represented by the symbol β. (2). A measure of a system's response bias in signal detection theory, represented by the ratio at the criterion level of the height of the signal + noise distribution to the height of the noise distribution alone.

beta coefficient
The weighting factor preceding a variable in a regression equation.

beta decay
Radioactive change by emission of a beta particle. In beta decay, a neutron decays into a proton, with the emission of an *electron* (or beta particle); or, a proton transforms into a neutron and emits a *positron*. In both cases, the charge of the nucleus is changed without changing the number of nucleons.

beta particle
A small particle ejected spontaneously from the nucleus of a radioactive element. It has the mass of the electron, has a charge of either -1 or +1, and has a mass of 1/1840 that of a proton or neutron. It has low penetrating power and short range. The most energetic of beta particles can penetrate the skin (causing a skin burn effect) and other tissues.

beta radiation
See ***beta particle***.

beta ray
A stream of beta particles of nuclear origin more penetrating but less ionizing than alpha rays per unit length of travel; a stream of beta particles emitted in certain radioactive disintegrations.

beta ray irradiation
A process to reduce pathogens in solid waste by irradiating sludge with beta rays from an accelerator at dosages of a least 1.0 megarad at room temperature.

beta rhythm
An EEG frequency band consisting of frequencies greater than 13 Hz.

beta testing
The second release phase of software evaluation, just prior to release in the commercial market. *See also* ***alpha testing***.

Betadine
Trademark for preparations of providone-iodine, which have a longer antiseptic action than most iodine solutions.

betatron
A circular electron accelerator providing a pulsed beam of high-energy electrons or x-rays by magnetic induction.

BeV
Billion electron volts, 1 E+9 eV.

BEV
See ***barrier equivalent velocity***.

beverage semitrailer
A van-type, drop-frame semitrailer designed and used specifically for the transport and delivery of bottled or canned beverages that has side-only access for loading and unloading this commodity.

beyond a reasonable doubt
In evidence, means fully satisfied, entirely convinced, satisfied to a moral certainty; the phrase is the equivalent of the words clear, precise, and indubitable.

beyond compliance
A regulatory trend becoming increasingly more prevalent as a voluntary alternative to existing statutes. Companies performing in this mode would perform beyond what is expected by existing laws or regulations, established performance limitations and technical requirements. The trend may promote increased operational flexibility and more open community reporting, and is used in many EPA incentive programs for industry.

bezel
A rim for holding a piece of transparent glass or plastic for a display on a meter or other indicator.

Bezold spreading effect
See ***color assimilation***. Also known as *assimilation* and *similitude effects*.

Bezold-Brücke effect
See ***Bezold-Brücke phenomenon***.

Bezold-Brücke hue shift
See ***Bezold-Brücke phenomenon***.

Bezold-Brücke phenomenon
A change in the apparent color of a visual stimulus with a change in stimulus intensity or illumination. Also referred to as *Bezold-Brücke effect* and *Bezold-Brücke hue shift*.

BGT
Black globe temperature. *See also* ***black globe thermometer***.

bhp
See ***brake horsepower***.

bi-
A prefix denoting a relationship to two symmetrical or approximately symmetrical parts.

BIA
See ***bioelectrical impedance analysis***.

biacromial breadth
The horizontal distance across the shoulders from right to left acromion. Measured with the individual standing erect, and the shoulders straight. Also referred to as *shoulder breadth*.

bias
(1) An individual preference or prejudgment on an issue. (2) A systematic error represented by the difference between the mean of repeated measurements and the true value; a tendency to over- or underestimate the true/actual value. (3) A relatively constant voltage offset from zero.

biauricular breadth
The horizontal linear distance from the most lateral point of the right ear to the same point of the left ear. Measured without auricular compression.

bicanthic diameter
See ***ectocanthic breadth***.

bicarbonate
Any chemical compound containing a HCO_3 group.

bicarbonate alkalinity
Alkalinity caused by bicarbonate ions.

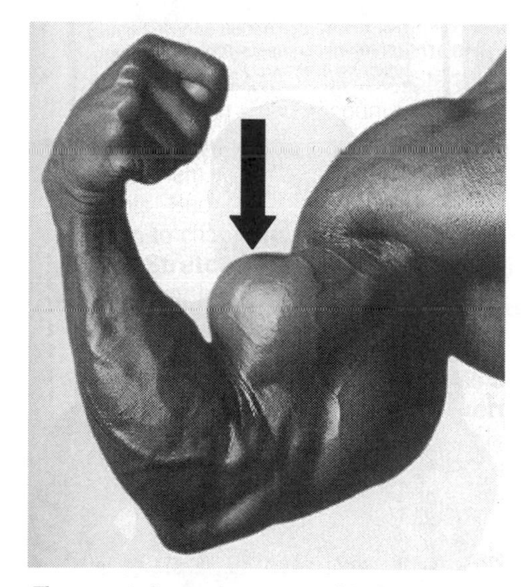

The arrow indicates the position of the biceps brachii

biceps brachii
The large, two-headed muscle in the anterior upper arm.

biceps circumference, flexed
The maximum surface distance around the biceps brachii. Measured with both the shoulder and elbow flexed 90 degrees, such that the upper arm is horizontal, and the hand clenched into a fist.

biceps circumference, relaxed
The maximum surface around the upper arm at the level of the biceps brachii belly. Measured with the arm hanging relaxed at the individual's side.

biceps femoris
A large, two-headed muscle in the posterior thigh; one of the hamstring muscles.

biceps muscle
*See **biceps brachii,** and **biceps femoris**.*

biceps skin fold
The thickness of a vertical skin fold on the anterior midline of the upper arm over the belly of the biceps brachii muscle at the level of the upper arm circumference measure. Measured with the individual standing erect and the arms hanging naturally at the sides.

bicipital
Pertaining to a muscle having two heads, often specifically to the biceps brachii and biceps femoris muscles.

bicristale breadth
*See **biiliocristale breadth**.*

bicycle
A vehicle having two tandem wheels, propelled solely by human power, upon which any person or persons may ride.

bicycle ergometer
A stationary cycle (typically with only one wheel) used to measure or work against a fixed or adjustable force.

bicycle lane
A portion of a roadway which has been designated by striping, signing and pavement markings for the preferential or exclusive use of bicyclists. Also referred to as *bike lane*.

bicycle path
A bikeway physically separated from motorized vehicular traffic by an open space or barrier and either within the highway right-of-way or within an independent right-of-way. Also referred to as *bike path*.

bicycle route
A segment of a system of bikeways designated by the jurisdiction having authority with appropriate directional and information markers, with or without a specific bicycle route number. Also referred to as *bike route*.

bicycles
Includes bicycles of all speeds and sizes that do not have a motor.

bideltoid breadth
The horizontal linear distance across the maximum lateral protrusions of the right and left deltoid muscles. Measured with the individual standing erect with the arms hanging naturally at the sides. See also ***biachromial breadth***.

biflow filter
Granular media filter characterized by water flow from both top and bottom to a collector located in the center of the filter bed.

bifocals
A pair of lenses in glasses having two correction portions, one for distance vision, the other for near vision.

big hat
(slang). State Trooper.

big rigger
Trucking (slang). An arrogant driver, or one who will drive only long trailers.

bight of a bend
Deepest portion of a bend (not in depth of water); sharpest part of a curve.

bigonial breadth
The horizontal linear distance across the gonial angles of the jaw. Measured with the jaw muscles relaxed and the individual sitting or standing erect.

biiliac breadth
*See **biiliocristale breadth**.*

biiliocristale breadth
The horizontal linear distance across the torso measured between the superior points of the iliac crests. Measured with the individual standing erect, with weight equally balanced on both feet. Also known as *biiliac breadth, transverse pelvic breadth, pelvic breadth,* and *bicristale breadth*.

bike lane
See ***bicycle lane***.

bike path
See ***bicycle path***.

bike route
See ***bicycle route***.

bikeway
Any road, path, or way which in some manner is specifically designated as being open to bicycle travel, regardless of whether such facilities are designated for the exclusive use of bicycles or are to be shared with other transportation modes.

bilabial height
The vertical linear distance between the most superior point on the upper lip and the most inferior point on the lower lip.

bilateral
Pertaining to similar structures present on both sides of a symmetric or approximately symmetric body.

bilateral contract
A term, used originally in civil law but now generally adopted, denoting a contract in which both the contracting parties are bound to fulfill obligations reciprocally toward each other; as a contract of sale where one becomes bound to deliver the thing sold, and the other to pay the price of it.

bilateral teleoperator
A teleoperator system in which force and motion can be transmitted in both directions (from the operator to teleoperator and vice versa).

bilateral trade agreement
Commerce between two countries based on a reciprocal trade agreement which specifies the quantity of goods to be traded, the time limit of the agreement, and that the balances due be remitted directly between the countries.

bile
The yellowish-brown or green fluid secreted by the liver and discharged into the small intestine where it aids in the emulsification of fats, increases peristalsis, and retards putrefaction.

bilharzia
Waterborne disease also known as schistosomiasis. *See* ***schistosomiasis***.

bilharziasis
See ***schistosomiasis.***

bilingual
Capable of speaking, writing, and understanding two languages.

bilirubin
An orange bile pigment produced by the breakdown of hemoglobin and excreted by the liver cells. Failure of the liver cells to excrete bile, or obstruction of the bile ducts, can cause an increased amount of bilirubin in the body fluids and thus lead to obstructive jaundice. Normally, the body produces a total of approximately 260 mg of bilirubin per day. Almost 99% is excreted in the feces; the remaining 1% is excreted in the urine (as urobilinogen). A test for bilirubin in the blood is called the *van den Bergh test.* Normal range for this test is 0.0 to 0.1 mg/100 mL of serum for direct bilirubin, and 0.2 to 1.4 mg/100 mL of serum for total bilirubin.

Bill of Lading
A document that establishes the terms of a contract between a shipper and a transportation company. It serves as a document of title, a contract of carriage, and a receipt for goods.

Bill of Rights
The first ten Amendments to the U.S. Constitution providing for individual rights, freedoms, and protections.

billing
A transaction conducted by a carrier involving the determination of the proper rate and total charges for a shipment and the issuance of a freight bill.

billow clouds
Broad, nearly parallel lines of clouds oriented at right angles to the wind.

bimalleolar breadth
The horizontal linear distance across the protrusions of the medial and lateral ankle bones. Measured with the individual standing erect and the body weight distributed evenly on both feet. Also referred to as *ankle breadth*.

bimanual
Performed with both hands.

bimetallic thermometer
A thermometer which consists of two different metal strips that are brazed together and

the differences of expansion of the metal strips, due to a temperature change, are used to provide an indication of temperature.

bimodal
(1) Pertaining to or affecting two sensory modalities simultaneously. (2) A statistical distribution having two modal values.

binary digit
See ***bit***.

binary fission
Asexual reproduction in some microbes where the parent organism splits into two independent organisms.

binaural
Having input to both ears simultaneously.

binaural hearing
The perception of sound by both ears.

bind
To obligate; to bring or place under definite duties or legal obligations, particularly by a bond or covenant. To affect one in a constraining or compulsory manner with a contract or a judgment.

binders
Brakes.

binding
Restricted in movement or a tightening or sticking condition resulting from high or low temperatures, foreign materials in the mechanism, surface friction, etc.

binding agreement
A contract which is enforceable such as an offer to buy or sell when a person to whom it is made accepts it and communicates his/her acceptance.

binding energy
The energy represented by the difference in mass between the sum of the component parts of a nucleus and the actual mass of the nucleus. It is the energy that holds the neutrons and protons together and, subsequently, it is the amount of energy required to separate the individual nucleons.

binocular
Pertaining to the use of or input to both eyes simultaneously.

binocular accommodation
The process of both eyes accommodating simultaneously.

binocular disparity
The difference in visual images on the right and left retinas resulting from the lateral separation of the eyes. May also be referred to as *lateral retinal image disparity* or *binocular parallax*.

binocular fusion
The merging of images from the two eyes into a single perception. May sometimes be referred to more simply as *fusion*.

binocular parallax
See ***binocular disparity***.

binocular portion of the visual field
See ***binocular visual field***.

binocular rivalry
A phenomenon in which an alternation of partial or entire images is perceived when the two eyes are stimulated simultaneously with different images. Also known as *retinal rivalry*.

binocular suppression
A loss of all or some portion of one eye's visual field resulting from conflicting information being presented to the fusion region of the outer eye's visual field.

binocular vision
The quality of vision existing by virtue of having two eyes in which the visual fields of the two eyes overlap. More technically referred to as *stereopsis*.

binocular visual field
The portion of the visual field where the monocular visual fields of the two eyes overlap. *See also* ***monocular visual field***.

binomial distribution
A distribution of data or results describing probabilities of the outcome of trials that can have one or two mutually exclusive results (e.g., exposure above or below a PEL). This theoretical discrete probability distribution for a binomial random variable is represented as:

$$P = n/r \quad p^n (1 - p)^{n-r}$$

where:
n = total number of outcomes
r = number of successful outcomes

(n/r) = number of combinations of n outcomes, taken r at a time.

Used to approximate the normal distribution for large sample sizes.

bio-
Prefix. Pertaining to living systems or those components which may be or have been a part of a living system.

bioaccumulative
Substances that increase in concentration in living organisms (that are very slowly metabolized or excreted) as they breathe contaminated air, drink contaminated water, or eat contaminated food.

bioacoustic
Pertaining to the effect of sound on the body.

bioacoustics
The study or use of the relationships between sound and living organisms.

bioaerosol
Any aerosol consisting primarily of biological entities such as microbes; the presence of biological entities in aerosol form.

bioassay
Use of living organisms to measure the effect of a substance, factor, or condition by comparing pre- and post-exposure data.

bioastronautics
The study of medicine, biology, and physiology of one or more substances.

bioavailability
The amount of a chemical that becomes available to the target organ/tissue after the material has entered the body.

biobrick
A building brick made of kiln-dried municipal wastewater sludge.

biochemical
Describes the event or action involving chemistry of living organisms and the chemical changes occurring therein.

biochemical oxidation
Oxidative reactions caused by biological activity which result in chemical combination of oxygen with organic matter. *See also* ***biological oxidation***.

biochemical oxygen demand
See ***biological oxygen demand (BOD)***.

biochemistry
The study of chemical reactions occurring in living organisms.

biocidal
Causing the death of living organisms.

biocide
A chemical used to inhibit or control the population of troublesome microbes.

bioclimatology
Scientific study of effects on living organisms of conditions of natural environment (rainfall, daylight, temperature, etc.) prevailing in specific regions of the earth.

bioconcentration
The net increase in concentration of a chemical and its metabolites in an organism relative to the concentration of the chemical of concern in the ambient water or air.

bioconcentration factor (BCF)
The accumulation of chemicals that live in contaminated environments equal to the quotient of the concentration of a substance in aquatic organisms divided by the concentration in the water during the same time period.

biocontactor
A unit process such as an aeration basin, trickling filter, rotating biological contactor, or digester where microbes degrade/transform organic matter.

biocontainment
Any technique used to achieve bioisolation of one or more substances.

bioconversion
The conversion of organic waste products into an energy resource through the action of microbes.

biocular breadth
See ***ectocanthic breadth***.

biocycle
The sequence of certain rhythmically repeated phenomena observed in living organisms.

biodegradable
The ability to break down or decompose rapidly under natural conditions and processes.

biodegradable dissolved organic carbon (BDOC)
The portion of total organic carbon that is easily degraded by microbes.

biodegradable material
Organic waste material that can be broken down into basic elements by the action of microorganisms.

biodegradation
Chemical reaction of a substance induced by enzymatic activity of microorganisms.

biodiversity
An environment where multiple organisms coexist.

biodynamicist
One who works in the field of biodynamics.

biodynamics
That field concerned with the effects of external forces or dynamic conditions on biological systems. Also known as *impact biodynamics.*

bioelectrical impedance analysis (BIA)
A technique for estimating/measuring total body fat/lean body mass by observing the impedance of electricity passed through a part of the body.

bioelectricity
Electrical phenomena apparent in living cells.

bioengineering
The integration and application of knowledge in the fields of human biology, medicine, and engineering.

biofeasibility
A bioremediation feasibility study done to determine the applicability and potential success of a bioremediation technique or procedure for a given site.

biofeedback
The use of instrumentation to provide information to an organism which enables that organism to alter its behavior accordingly.

biofilm
An accumulation of microbial growth.

biofilter
See ***biological filter***.

bioflavonoid
A generic term for a group of compounds widely distributed in plants and concerned with maintenance of a normal state of the walls of small blood vessels.

biofoul
Presence and growth of organic matter in a water system.

biogas
The gases produced by the anaerobic decomposition of organic matter.

biogenesis
The theory that living organisms arise only from other living organisms.

biogeography
The scientific study of the geographic distribution of living organisms.

biogravics
A branch of science, developed since the start of space flight, that studies the effects of weightlessness and excessive gravitational force on living organisms.

biohazard
(1) Term applied to organisms or products of organisms that present a health risk to humans. (2) It is also used to identify human blood, bloodborne products, or other forms of human wastes that may contain human bodily fluids which may present a health hazard to anyone who may come into unprotected contact with these materials. Derived from a combination of the words *biological* and *hazard.*

biohazard area
Any area in which work has been or is being performed with biohazardous agents or materials. The area is usually identified using the biohazard sign and/or symbol.

Biohazard warning sign and symbol

bioisolation
A condition in which biological systems are effectively separated from each other using any one or more of physical, chemical, or biological methods.

biokinetics
The branch of science that pertains to the study of living organisms.

biologic agents
Biologic organisms which cause infections or a disease, such as the spores that cause anthrax, which can be considered an occupational disease under certain circumstances.

biologic test
A measurement taken from biological media to determine the presence of a specific material or metabolite, or some other measurable effect on a worker which is a result of an exposure to a specific substance.

biological clock
Any hypothesized internal mechanism responsible for maintaining one or more biological rhythms. Sometimes referred to as *body clock.*

biological contaminants
Agents derived from or that are living organisms (e.g., viruses, bacteria, fungi, mammal and bird antigens) that can be inhaled and can cause many types of health effects including allergic reactions, respiratory disorders, hypersensitivity diseases, and infectious diseases. Also referred to as *microbiologicals* or *microbials.*

biological control
In pest control, the use of animals and organisms that eat or otherwise kill or out-compete pests.

biological electricity
The electricity created by living beings and cells.

biological exposure index (BEI)
Set of reference values established by the American Conference of Government Industrial Hygienists as guidelines for the evaluation of potential health hazards in biological specimens collected from a healthy worker who has been exposed to chemicals to the same extent as a worker with inhalation exposure to the threshold limit value. The values apply to 8-hour exposures, 5 days per week.

biological filter
A bed of sand, stone, or other media through which wastewater flows that depends on biological action for its effectiveness. Sometimes referred to as *biotower.*

biological half-life
The time required for the body to eliminate one-half of an administered dose of any substance by regular process of elimination. This time is approximately the same for both stable and radioactive isotopes of a particular element.

biological hazard
See ***biohazard.***

biological magnification
Refers to the process whereby certain substances such as pesticides or heavy metals move up the food chain, work their way into a river or lake, and are eaten by aquatic organisms such as fish, which in turn are eaten by large birds, animals, or humans. The substances become concentrated in tissues or internal organs as they move up the chain. *See also* ***bioaccumulative.***

biological monitoring
(1) Periodic examination of blood, urine, or any other body substance to determine the extent of body absorption and retention of toxic materials. (2) The determination of the effects on life, including the accumulation of pollutants in tissue, due to the discharge of pollutants by techniques and procedures. This includes sampling of organisms representative of appropriate levels of the food chain and the volume and the physical, chemical, and biological characteristics of the effluent. Appropriate frequencies and locations are also considered.

biological needs
The basic physiological needs for a living entity to function, including air (oxygen), water, and food.

biological oxidation
The way bacteria and microorganisms feed on and decompose complex organic materials. Used in self-purification of water bodies and in activated sludge wastewater treatment.

biological oxygen demand (BOD)
A measure of the amount of oxygen consumed in the biological processes that break

down organic matter in water. The greater the BOD, the greater the degree of pollution.

biological rhythm
A self-maintained, cyclic variation within a relatively fixed period in a living organism. *See also* ***circadian rhythm, infradian rhythm, ultradian rhythm,*** *and* ***circannual rhythm***.

biological treatment
The treatment technology that utilizes bacteria to consume waste. This treatment breaks down organic materials.

biolysis
Decomposition of organic matter by living organisms.

biomarker
A measurable biologic characteristic which has a definable relation to prior exposure to a substance.

biomass
All of the living material in a given area; often refers to vegetation. Also called *biota*.

biome
A biological community or ecosystem characterized by a specific habitat and climate such as a tropical rain forest or a desert.

biomechanical profile
Any combined set of biomechanical, electromyographic, motion, and other data recorded simultaneously during some activity.

biomechanics
The study of the human body acting as a system under the laws of Newtonian mechanics and the biological laws of life.

biomedicine
A field of medical science concerned with the ability of a human being to live and function in abnormal environments.

biometeorology
Scientific study of effects on living organisms of the extra-organic aspects (temperature, humidity, barometric pressure, rate of air flow, and air ionization) of the physical environment, whether natural or artificially created, and also their effects in closed ecological systems, as in satellites or submarines.

biometer
An instrument for measuring carbon dioxide given off by living tissue.

biometrics
See ***biometry***.

biometry
The measurement of biological parameters and the use of simple descriptive statistics for the data obtained. Also referred to as *biometrics*. *See also* ***biostatistics***.

biomonitoring
The use of living organisms to test water quality at a site further downstream.

bion
An individual living organisms.

bionics
The study of biological systems to derive knowledge for use in the design, modeling, development, and/or implementation of artificial systems.

bio-oxidation
See ***biochemical oxidation***.

biophysics
Science dealing with the application of physical methods and theories to biological problems/effects, such as the interaction of radio frequency energies with living systems.

biopsy
The removal and examination of tissue from living mammals.

biopure water
Water that is sterile, pyrogen free, and has a total solids content of less than 1 mg/L.

bioremediation
Application of the natural ability of microbes to use waste materials in their metabolic processes and convert them into harmless end products.

BIOS
See ***basic input output system***.

biosafety cabinet
A hood to control the dissemination of viable or non-viable particles of biological origin, microorganisms, and their decay products.

biosampling
The collection of samples (e.g., air, surface wipes, settling plates, etc.) to identify and quantify the presence of bioaerosols in the work environment.

biosensor
Any sensor in biotechnology which is com-

posed of living tissue, biological materials, or fabricated from basic biological materials.

biosolids
Primarily organic sludge or byproducts of wastewater treatment that can be beneficially recycled.

biosorption process
*See **contact stabilization process**.*

biospheres
The mass of living organisms found in a thin belt at the Earth's surface.

biostabilizer
A machine that converts solid waste into compost by grinding and aeration.

biostatistics
The use of statistical methodology to describe biological data or draw inferences from those data.

biosynthesis
The synthesis of a chemical within and by a living organism.

biota
All living organisms within a system. *See also **biomass**.*

biotechnology
Techniques that use living organisms or parts of organisms to produce a variety of products (from medicines to industrial enzymes) to improve plants or animals or to develop microorganisms for specific uses such as removing toxics from bodies of water, or as pesticides.

biotherapy
Treatment by means of living organisms and their products, including vaccines, immune serum, and blood transfusions.

biotic
Of or pertaining to life.

biotic agent
Microorganisms and parasites which act on the skin and body to produce disease.

biotic community
A naturally occurring assemblage of plants and animals that live in the same environment and are mutually sustaining and interdependent.

biotower
*See **biological filter**.*

biotoxicology
The scientific study of toxins produced by organisms, their effects, and the treatment of conditions they produce.

bioturbation
The net effect of the activity of benthic organisms at wastewater treatment plant discharges, which may aid in the dispersion of contaminants and increase the exchange of oxygen and nutrients between the sediment and water.

bipolar
(1) Having two poles. (2) Pertaining to both poles.

bipolar disorder
A psychosis characterized by abrupt or subtle behavioral manifestations of opposite extremes, such as manic-depressive disorder.

bird fancier's disease
Extrinsic allergic alveolitis observed in some individuals who have been exposed to birds. The condition is accompanied by breathlessness or tightness of the chest, coughing, and wheezing. Extensive fibrosis can be seen in the chronic form.

birdyback
Intermodal transportation system using highway freight containers carried by aircraft.

birth rate
The ratio of live births per some unit of existing population within a given period of time.

birth-death process
A queuing system in which units to be served or worked on arrive and depart in a random fashion.

biserial correlation
That correlation existing between two continuously distributed variables, but in which one of the variables has been scored as a dichotomous variable.

bispinous breadth
The transverse distance between the centers of the anterior superior iliac spines. Measured with the individual standing erect with weight evenly distributed on both feet.

BIT
*See **Built-in Test**.*

bit

(1) A numerical value in the binary scale, either zero (0) or one (1). The basic unit in a digital electronic system; contraction of binary digit. (2) That amount of information obtained when one of two equally likely alternatives is given or specified; the basic unit of information. *See also* ***information theory***.

BITE

See ***Built-in Test Equipment***.

bitmap

Computing. A graphic made up of a collection of colored dots. The computer stores the graphic as one or more *bits* of information for each dot (hence the name bitmap). Some filename extensions for graphic files that are bitmaps include .PCX, .TIFF, .BMP, and .GIF.

bitragion breadth

The transverse width of the head as measured from right to left tragion. Measured with the individual sitting or standing erect with the scalp muscles relaxed.

bitragion-crinion arc

The surface distance from right tragion, over the anterior hairline, to the left tragion. Measured with the individual sitting or standing erect with the scalp muscles relaxed.

bitragion-inion arc

The surface distance from right tragion, over inion (including the hair), to left tragion. Measured with the individual sitting or standing erect with the scalp muscles relaxed.

bitragion-menton arc

The surface distance from right tragion, under the anterior/inferior tip of the chin, to the left tragion. Measured with the individual sitting or standing erect with the jaws closed and the facial muscles relaxed.

bitragion-minimum frontal arc

The surface distance from right tragion, over the forehead just above the brow, to left tragion. Measured with the scalp and facial muscles relaxed.

bitragion-posterior arc

The surface distance from right tragion, across the base of the skull, to left tragion. Measured with the individual sitting or standing erect with the scalp muscles relaxed.

bitragion-submandibular arc

The surface distance from right tragion, under the gonial angles of the jaw, to left tragion. Measured with the individual sitting or standing erect with the jaws closed and the scalp and facial muscles relaxed.

bitragion-subnasale arc

The surface from right tragion, across subnasale to left tragion. Measured with the individual sitting or standing erect and the scalp and facial muscles relaxed.

bitrochanteric breadth

The horizontal linear distance between the most lateral projections of the right and left greater trochanters. Measured with the flesh compressed and the individual standing erect with weight distributed equally on both feet. Also referred to as *bitrochanteric width*.

bituminous coal

A coal high in carbonaceous matter that yields a considerable amount of volatile waste matter when burned.

bivalent

Having a valence of two.

bivariate regression

A special case of multiple regression in which the number of predictor variables is two.

bizygomatic breadth

The transverse width of the face across the most lateral protrusions of the zygomatic arches. Also referred to as *face breadth*.

black

Having the property of absorbing all or most of the incident visible light.

black body

A hypothetical ideal body which absorbs all incident radiation, independent of wavelength and direction.

Black Death

The bubonic plague, which first occurred in Europe in epidemic form during the fourteenth century; characterized by black spots on the skin.

black eye

A bruise of the tissue around the eye marked by discoloration, swelling, and pain.

black globe thermometer

Typically a 6-inch hollow, thin-wall, copper sphere painted flat black with an ordinary thermometer placed into the globe at the center.

black light
The region of the electromagnetic spectrum between 300 nm and 400 nm in the ultraviolet (UV) region. It is the region responsible for the added pigmentation of the skin (burning and tanning) following exposure to UV light.

black liquor
Strong organic waste generated during kraft pulping process.

Black Lung Benefits Act
Federal statute benefiting coal miners who are stricken with pneumoconiosis, a chronic dust disease of the lung. Benefits under the Act are administered by the Department of Labor. *See also **black lung disease**.*

black lung disease
A disease contracted by coal miners and marked by varying degrees of pulmonary impairment, including x-ray abnormalities, cough, breathlessness, massive progressive fibrosis, formation of nodules and scar tissue in the lungs. Also known as *coal miner's pneumoconiosis, Collier's disease,* and *Shaver's disease.*

black sand
Discoloration of filter sand resulting from manganese deposits.

blackbody
A surface which, ideally, absorbs all incident visible light energy and emits radiant energy with a spectral distribution varying according to the absolute temperature of the surface. Synonyms include *ideal blackbody, blackbody source, Plancklan radiator, blackbody radiator, full radiator, standard radiator, ideal radiator,* and *complete radiator.*

blackbody locus
A set of points representing the chromaticities of a potential set of blackbodies with various color temperatures on a chromaticity diagram. Also referred to as *Planckian locus.*

blackbody radiator
*See **blackbody**.*

blackbody source
*See **blackbody**.*

blackmail
The unlawful demand of money or property under threat to do bodily harm, to injure property, to accuse of crime, or to expose disgraceful defects. This crime is commonly included under extortion or criminal coercion statutes. *See also **extortion**.*

blackout
A temporary loss of vision, regardless of the cause. *See also **grayout** and **gravity-induced loss of consciousness**.*

blackwater
Water that contains animal, human, or food wastes.

blackwater fever
An acute form of malaria occurring in tropical and semitropical regions, characterized by febrile paroxysms and bloody urine.

bland
Not having a stimulating taste characteristic.

blank QA (spiked) sample
Sampling media spiked by quality assurance personnel with selected compounds at known amounts for submitting to the analytical laboratory along with regular samples to determine analyte recovery effectiveness, possible effects of sample storage/shipment/etc.

blank sample
A non-contaminated or otherwise clean sample medium sent along to a testing laboratory along with actual sample results that is used to help determine sample inaccuracies or other compromising conditions. *See also **field blank sample**.*

blanket certificate (authority)
Permission granted by the Federal Energy Regulatory Commission (FERC) for a certificate holder to engage in an activity (such as transportation service or sales) on a self implementing or prior notice basis, as appropriate, without case-by-case approval from FERC.

blanking
The absolute closure of a pipe, line, or duct by the fastening of a solid plate (such as a spectacle blind or skillet blind) that completely covers the bore and that is capable of withstanding the maximum pressure of the pipe, line, or duct with no leakage beyond the plate. Also called *blinding.*

blast furnace
Furnace used in the iron-making process in which hot blast air flows upward through the raw materials and exits at the furnace top.

blast gate
A devicc that regulates airflow in duct work, similar to a damper, but usually operated by positioning a sliding metal plate across a duct.

blasting agent
Any material or mixture consisting of fuel and oxidizer intended for blasting, not otherwise classified as an explosive and in which none of the ingredients is classified as an explosive, provided that the finished product as mixed and packaged for use or shipment cannot be detonated by means of a number 8 blasting cap when confined.

blastomycosis
Term for any infection caused by a yeast-like organism.

bleach
Oxidizing compound, usually containing chlorine combined with calcium or sodium.

bleed
To draw accumulated liquid or gas from a line or container.

bleeder
(1) The popular term for a person who bleeds freely, especially one suffering from a condition in which the blood fails to clot properly. *See **hemophilia**.* (2) A large blood vessel divided during surgery. (3) In pressure systems, a faulty valve or pressure relief device that is releasing gaseous materials unexpectedly or at an inappropriate time.

bleeding
(1) The escape of blood, as from an injured vessel. (2) Purposeful withdrawal of blood from a vessel of the body. (3) Intentional release of gaseous of liquid commodities from a pressure system for the purpose of reducing system pressures.

BLEVE
*See **boiling liquid expanding vapor explosion**.*

blind
(1) Not having certain information regarding ongoing activities in an experiment. *See also **double blind**.* (2) Having no visual capability, or having a Snellen visual acuity less than 20/200 even using corrective lenses. *See also **Snellen acquity** and **Snellen test**.* (3) Pertaining to a ship or other military vehicle which has lost its radar or other sensing capabilities. (4) Typically, a metal plate that serves as an absolute means to seal off a pipe, line, or duct from another section of the process. It completely covers the bore of the pipe, line, or duct and is capable of withstanding the maximum pressure present with no leakage beyond the plate.

blind flange
A pipe flange with a blind end used to close the end of a pipeline.

blind positioning
A movement which requires the placement of one or more objects at some orientation or point in space without visual cues.

blind sample
A sample medium sent along to a testing laboratory that has been pre-conditioned at the sample site with known contaminant levels that are not reported to the laboratory and is used to determine the accuracy of a laboratory analysis.

blind side
Right side of truck and trailer.

blind speed
The rate of departure or closing of a target relative to the radar antenna at which cancellation of the primary radar target by moving target indicator (MTI) circuits in the radar equipment causes a reduction or complete loss of signal.

blind spot
(1) *Physiology*. Normal defect of vision caused by the position of the optic nerve at the point where it enters the eye; that region of the posterior eyeball where no photoreceptors are located due to the optic neural fibers exiting the eyeball. (2) *Transit*. Any region on or around a vehicle at which another object may not be readily seen due to lack of mirror coverage or inability to view directly. (3) *Communication*. An area from which radio transmissions and/or radar echoes cannot be received. The term is also used to describe portions of the airport not visible from the control tower. Also referred to as *blind zone*.

blind thrust fault
Seismology. A shallow-dipping reverse fault which terminates before it reaches the surface.

When it breaks, therefore, it may produce uplift, but never any clear surface rupture. Many still-unknown blind thrust faults may exist in southern California. Two examples of known blind thrust faults are the Elysian Park Thrust, which runs underneath downtown Los Angeles and the Northridge Thrust Fault, which ruptured in the 1994 Northridge quake.

blind velocity
The radial velocity of a moving target such that the target is not seen on primary radar fitted with certain forms of fixed echo suppression.

blind zone
*See **blind spot**.*

blinding
The reduction or cessation of flow through a filter resulting from solids restricting the filter openings. *See also **blanking**.*

blinding glare
Any extremely intense glare which interferes with vision for a significant period of time after removal of the glare source.

blink
(1) A unit of time equal to 0.864 seconds or 10-5 day. (2) Turn quickly on and then off at approximately regular intervals. (3) *See **eye blink**.*

blink coding
The use of a blinking stimulus as a highlighting or attention-getting technique.

blink rate
(1) That number of occasions which a light or segment of a display turns on and off within a specified interval. (2) *See **eye blink rate**.*

blip
A brief visual signal of higher intensity or different quality from the background, which may enable or enhance detection.

blizzard
A severe weather condition characterized by low temperatures and strong winds (greater than 32 mph) bearing a great amount of snow. When these conditions continue after falling snow has ended, it is termed a *ground blizzard*.

BLM
*See **Bureau of Land Management**.*

block
Rail. A length of track of defined limits, the use of which by trains is governed by block signals, cab signals, or both.

block and tackle
A combination of a rope or other line material and an independent pulley. Used to increase mechanical efficiency.

block signal
A roadway signal operated either automatically or manually at the entrance to a block.

block signal system
A method of governing the movement of trains into or within one or more blocks by block signals or cab signals.

block to block time
*See **flight time**.*

blocked
Communication. Phraseology used to indicate that a radio transmission has been distorted or interrupted due to multiple simultaneous radio transmissions.

blood
The viscous red bodily fluid, consisting of plasma and the formed elements, which carries nutrients, waste products, and body defensive mechanisms through the cardiovascular system.

Blood Alcohol Concentration (BAC)
Measured as a percentage by weight of alcohol in the blood (grams/deciliter). A positive BAC level (0.01 g/dl and higher) indicates that alcohol was consumed by the person tested. A BAC level of 0.10 g/dl or more indicates that the person was intoxicated.

blood alcohol count
Refers to the standard measure for legal intoxication under state DWI laws. In most states, a person can be charged with "driving while intoxicated" with a blood alcohol level of .10 percent or higher. See also ***driving while intoxicated, blood alcohol concentration,*** *and* ***breathalyzer test***.

blood count
The number of erythrocytes or white blood cells in a cubic millimeter of blood.

blood dyscrasia
Any persistent change from normal of one or more components of blood.

blood-forming organs (BFO)
The red bone marrow tissue and the spleen.

blood level
The concentration of a material, such as lead, in the blood. Typically reported as micrograms per 100 grams of blood or micrograms per 100 mL (i.e., deciliter) of blood).

blood plasma
The clear, almost colorless fluid of the blood when separated from blood corpuscles by centrifuging; used in blood transfusions, since it clots as easily as whole blood.

blood platelet
A minute circular or oval body found in blood, necessary for blood clotting.

blood poisoning
*See **toxemia**.*

blood pressure
That force exerted on the internal heart and vessel walls of the circulatory system by the blood. *See also **systolic blood pressure** and **diastolic blood pressure**.*

blood priority
Figurative reference to management's approach to accident investigation, in the early years during and following the industrial revolution (United States). Very simply, if there was no blood spilled, then there was no real priority for any action (or budget), and even less management interest.

blood products
Any product derived from blood, including but not limited to blood plasma, platelets, red and white blood corpuscles, and other derived licensed products such as interferon.

blood serum
The yellowish, clear liquid remaining after all solid constituents of the blood have been removed.

blood sugar
Glucose, supplied by the liver, circulating in the blood.

blood test
The test of a blood sample to determine such qualities as blood type, or such quantities as sugar content. Also used to determine if a person has ingested quantities of substances beyond a legally established limit (e.g., alcohol, drugs).

blood test evidence
Blood may be extracted against the will of a person without offense to Fifth Amendment rights (U.S. Constitution) when arrested for driving while intoxicated. *See also **blood alcohol count** and **DNA identification**.*

blood type
The phenotype of erythrocytes defined by one or more antigenic determinants. Under the usual system of blood typing, there are four main blood types of blood groups: A, B, O, and AB. The ABO blood typing system was first introduced in 1900 by Karl Landersteiner and is still generally used today as the basis for transfusing whole blood. It is now known, however, that many different antigens exist in the red blood cells, and that as many as 11 or more different antigenic systems of grouping blood can be recognized. Even within the ABO system, numerous subgroups of the main groups exist.

blood worm
The larval stage of the midge fly.

bloodborne pathogens
Pathogenic microorganisms that are present in human blood and can cause disease in humans. These pathogens include, but are not limited to, hepatitis B virus (HBV) and the human immune deficiency virus (HIV).

blow down
A discharge from a recirculating system designed to prevent a buildup of some material.

blow him down
Boating Safety. To sound the danger signal in case of misunderstood passing signals, when the pilot on the other boat refuses to obey signals, or when just desiring to pass information.

blower
Air-conveying equipment that generates pressures up to 103 kPa (15 pounds per square inch), commonly used for wastewater aeration systems.

blowout
An uncontrolled flow of gas, oil, or other well fluids into the atmosphere.

BLS
Bureau of Labor Statistics.

blue
A primary color, corresponding to that hue

apparent to the normal eye when stimulated only with electromagnetic radiation approximately between wavelengths from 455 nm to 490 nm.

blue asbestos
See ***crocidolite asbestos***

blue baby syndrome
See ***methemoglobinemia****.*

blue blindness
See ***tritanopia****.*

blue collar
Pertaining to those workers typically doing production work, as opposed to management personnel strictly doing administrative work.

blue signal
Railroad Safety. A clearly distinguishable blue flag or blue light by day and a blue light at night. When attached to the operating controls of a locomotive, it need not be lighted if the inside of the cab area of the locomotive is sufficiently lighted so as to make the blue signal clearly distinguishable.

Typical blue signals. Arrows indicate a blue flag (on right) and a blue light (on left, not activated during daylight hours)

blue vitrol
Common name for copper sulfate, used to control algae. *See* ***copper sulfate****.*

blue-yellow blindness
A rare form of color blindness in which the individual cannot differentiate between blue and yellow.

bluff bar
A sandbar having a sharp drop-off into deep water. Also called a *bluff reef.*

blur
A condition in which an image is not well focused.

BLWRS
See ***barrier landscape water renovation system****.*

BMI
See ***body mass index****.*

BMP
See ***best management practices****.*

BMR
See ***basal metabolic rate****.*

BNR
Biological nutrient removal.

Board
(1) An official or representative body organized to perform a trust or to execute official or representative functions of having the management of a public office or department exercising administrative or governmental functions. (2) Lodging, food, and entertainment, furnished to a guest at an inn or boarding house.

Board of Adjustment
Public and quasijudicial agency charged with the duty to hear and determine zoning appeals. Also called *Board of Zoning Appeals* in certain cities.

Board of Aldermen
The governing body of a municipal corporation. *See also* ***alderman****.*

Board of Appeals
A non-judicial, administrative tribunal which reviews the decision made by the hearing officer or by the head of the agency.

Board of Certified Safety Professionals (BCSP)
The BCSP was originally organized as a peer certification board in 1969 with the purpose of certifying those who practice in the safety profession. The specific functions of the BCSP, as outlined in its charter, are to evaluate the academic and professional experience qualifications of safety professionals, to administer examinations, and issue certifications to those professionals with demonstrated qualifications who have met the BCSP criteria and successfully passed its examinations.

Board of Commissioners
A legal body of 3 to 9 individuals having broad administrative authority over a river port's operation. This board is primarily con-

cerned with the development and determination of policies of the port authority.

Board of Directors
The governing body of a corporation elected by the stockholders; usually made up of officers of the corporation and outside (non-company) directors. The board is empowered to elect and appoint officers and agents to act on behalf of the corporation, declare dividends, and act on other major matters affecting the corporation.

Board of Fire Underwriters
An unincorporated voluntary association composed exclusively of persons engaged in the business of fire insurance, for consolidation and cooperation in matters affecting the business.

Board of Governors of the Federal Reserve System
Seven-member board, with fourteen-year terms, which governs the twelve Federal Reserve Banks and branches. The Board of Governors determines general monetary, credit, and operating policies for the System as a whole and formulates the rules and regulations necessary to carry out the purposes of the Federal Reserve Act. The Board's principal duties consist of exerting an influence over credit conditions and supervising the Federal Reserve Banks and member banks.

Board of Health
A municipal or state board or commission with certain powers and duties relative to the preservation and improvement of the public health.

Board of Pardons
State board, of which the governor is usually a member, authorized to review and grant pardons and clemency to convicted prisoners.

Board of Review
Board authorized to review administrative agency decisions and rulings. Body authorized to review alleged improper valuation and assessment of property. In some cities, a board charged with responsibility to review alleged police brutality or excessive force.

Board of Zoning Appeals
See ***Board of Adjustment.***

boat
See ***barge, general cargo ship, motorboat, towboat, tugboat,*** *and* ***vessel.***

boat trailer
A trailer designed with cradle-type mountings to transport a boat and configured to permit launching of the boat from the rear of the trailer.

boat transporters
Any vehicle combination designed and used specifically to transport assembled boats and boat hulls. Boats may be partially disassembled to facilitate transporting.

Boating Safety Circular (BSC)
Published by COMDT (G-NAB) for free distribution to boat and equipment manufacturers, dealers, marinas, yacht clubs, OCMI personnel, and other boating organizations. Information in the BSC concerns boating standards and boating safety in general.

boatswain
A seaman who superintends the work of the crew. The foreman of sailors.

boatswain's chair
A seat, suspended from a higher level by slings attached to a rope, that allows the occupant to safely perform work at heights above the ground level but below the chair's suspension point. It has provisions for proper occupant securing and protection.

Worker seated in boatswain's chair

BOCA
Building Officials and Code Administrators

BOD
See ***biological oxygen demand.***

BOD_5
Five-day carbonaceous or nitrification-inhibited biological oxygen demand. *See also* ***biological oxygen demand***.

BOD_u
See ***ultimate BOD***.

bodily injury
(1) Any injury to an individual from mechanical or physical processes. (2) Injury to the body, sickness, or disease including death resulting from any of these.

body
(1) *Anatomy*. The human frame, including all its organs, tissues, and other normal materials. (2) *Automotive*. Semitrailer. *See also* ***chassis***.

body breadth, maximum
The maximum linear horizontal distance across the body, including the arms. Measured with the individual standing erect and the arms hanging naturally at the sides; for accuracy and future reference, specify the level at which the measure is taken.

body burden, maximum permissible
(1) *Radiation*. An amount of radioactive material in a critical organ such that the whole-body dose is 0.3 rem per week or less; in case of an alpha or beta emitter that is deposited in the bone, body burden is derived from the long-established maximum permissible body burden of radium (0.1 microcurie) adjusted for possible less uniform deposition. (2) *Biological*. The total amount of a substance stored in the body following exposure. The body burden of a particular substance is a function of its biological half-life and its biochemical uptake and elimination rate.

body clock
See ***biological clock***.

body composition
The proportions of tissue makeup in the body, generally classified by two primary categories as a function of body mass: lean body mass and body fat.

body depth, maximum
The maximum horizontal distance between two vertical planes which represent the most anterior and posterior aspects of the torso. Measured with the individual standing erect and arms hanging naturally at the sides; for accuracy and future reference, specify exactly where the measurement is taken.

body envelope
That volume which includes the body and any protective clothing or other items required during performance of a specific task.

body fat
That portion of body composition which is composed of adipose tissue

body feed
Coating material added to the influent of precoat filters during filtration cycle.

body fluids
Liquid emanating or derived from humans including blood, dialysis, amniotic, cerebrospinal, synovial, pleural, peritoneal, and pericardial fluids, as well as semen and vaginal secretions.

body heat content (H_b)
The mathematical product of the body's heat capacity and the mean temperature of body tissues.

body height
See ***stature***.

body mass
The total mass of the body.

body mass index (BMI)
A guideline for estimating the percentage of body fat and nutritional status of the body. Represented as

$$BMI = \frac{weight}{(stature)^2}$$

body motion
The movement of one or more body parts which involves a mass redistribution within some coordinate system.

body of the crime
See ***body of the offense***.

body of the offense
When applied to any particular offense, means that the particular crime charged has actually been committed by someone. Also referred to as *body of the crime*.

body position
See ***posture***.

body proportionality
The distribution of an individual's circumference measurements.

body segment
Any portion of the body located between two joints, or the terminal portion of a body part from a joint, which has a relatively constant geometry when moved.

body surface
Any part or all of the total surface area of the body.

body surface area
The total surface area of the body.

body temperature and pressure, saturated conditions (BTPS)
The air mixture saturated with water vapor at ambient body temperature, as found in the lung alveoli or exhaled air.

body type
Automotive. Refers to the individual classifications of motor vehicles by their design structure based on definitions developed by the Society of Automotive Engineers, such as 1) the appearance of the vehicle, and 2) detailed type of motor vehicle within a vehicle type.

body typology
Any of various attempts to ascribe behaviors and personality characteristics to the shape or composition of an individual's body. *See also* ***somatotype***.

body versus machine rule
A task design guideline that the machine should not be capable of injuring the worker during any phase of a task.

body volume
The total volume occupied by the body.

body weight
The nude weight of an individual. Measured under standard conditions.

body-load aggregate
The combined effect of the weight being manipulated and the weight of those parts of the body involved in a materials handling or lifting task.

BOEMAN
A computerized, human modeling package for aiding design and evaluation of reach capabilities in cockpits and other aircraft workstations.

bog
Poorly drained land filled with decayed organic matter that is wet and spongy and unable to support any appreciable weight.

bogey
Automotive. An assembly of two or more axles.

bogie
(1) *Automotive.* A set of wheels built specifically as rear wheels under the container. (2) *Aviation (slang).* Term used to describe an unidentified "target" sited by the pilot of an aircraft. Usually a military term.

BOHS
British Occupational Health Society.

boil
(1) *Physiology.* A local infection of the skin containing pus and showing on the surface as a reddened, tender swelling; a type of skin abscess. Also called *furuncle.* (2) *Hydrology.* Turbulence in the water caused by deep holes, ends of dikes, channel changes, or other such submerged obstructions. Indicates a changing channel condition. A boil is easily detected by electronic depth sounders by rapidly changing depths appearing as waves on the tracing paper.

boil out
An evaporator-cleaning process where wash water is boiled in an evaporator to remove scale deposits.

boiler
A pressure vessel in which water is continuously vaporized into steam by the application of heat.

boiler deck
See ***cabin deck***.

boiler feed water
Water that, in the best practice, is softened and/or demineralized and heated to nearly boiling temperature and deaerated before being pumped into a steam boiler.

boilermaker's deafness
A form of hearing impairment in which an individual hears better under noisy conditions than in quiet. Caused by working for long periods around loud noises.

boilerplate
(1) Language which is used commonly in documents having a definite meaning in the same context without variation. (2) Used to describe standard language in a legal document that is identical in instruments of a like nature.

boiling liquid expanding vapor explosion (BLEVE)
A violent rupture of a pressure vessel containing saturated liquid/vapor at a temperature well above its normal boiling point. A BLEVE often occurs when a fire adjacent to a tank holding a volatile flammable commodity causes the commodity's temperature to increase thereby causing a subsequent increase in the pressure inside the tank. As the liquid reaches its boiling point, the pressure becomes too great for the tank to contain. The resulting explosion is violent and, once the contained liquid ignites (either as a result of the explosion or upon contact with the fire outside), the resulting deflagration can be devastating.

boiling point
The temperature at which a liquid's vapor pressure equals the pressure acting on the liquid.

boiling point elevation (BPE)
The difference between the boiling point of a solution and the boiling point of pure water at the same pressure.

bold reef
A bluff reef which acts like a weir and is plainly visible for quite some distance.

bold right-hand reef
A sandbar or group of rocks which can be seen or detected by water turbulence, located on the right bank of the channel.

boll weevil
(1) *Agriculture*. A pest (beetle) that typically infests and destroys cotton bolls. (2) *Transit (trucking slang)*. A novice truck driver.

bolometer
An instrument which measures radiant heat by correlating the radiation-induced change in electrical resistance of a blackened metal foil with the amount of radiation absorbed.

bolt lock
With respect to rail operations, a mechanical lock so arranged that if a switch, derail or movable-point frog is not in the proper position for a train movement, the signal governing that movement cannot display an aspect to proceed; and that will prevent a movement of the switch, derail or movable-point frog unless the signal displays its most restrictive aspect.

bolus
A cohesive mass, either of food material for swallowing or of fecal material following defecation.

BOM
*See **Bureau of Mines**.*

bomb calorimeter
An instrument used to determine the heat content of sludge or other material.

bona fide
In or with good faith; honestly, openly, and sincerely; without deceit or fraud.

bond
(1) A form of monetary security given to secure the performance of some act or to provide funds if some problem arises. (2) An equalization of electrical potential between objects. (3) The linkage between atoms or radicals of a chemical compound, or the symbol representing this linkage and indicating the valance of the atoms or radicals.

bonded petroleum imports
Petroleum imported and entered into Customs bonded storage. These imports are not included in the import statistics until they are 1) withdrawn from storage free of duty for use as fuel for vessels and aircraft engaged in international trade, or 2) withdrawn from storage with duty paid for domestic use.

bonding
(1) An electrical conductor, or the act of attaching such conductor, to eliminate a difference in electrical or electrostatic potential that would cause a spark to occur between objects. (2) The permanent joining of metallic parts to form an electrically conductive path which will assure electrical continuity and the capacity to safely conduct any current likely to be imposed.

bonding jumper
A reliable conductor to assure the required electrical conductivity between metal parts required to be electrically connected.

bone
The skeletal tissue of vertebrates consisting of cells arranged in a matrix of collagen fibers and cells containing calcium and phosphate.

bone conduction
The passage of sound waves to the inner ear via the bones of the skull.

bone conduction test
A hearing test in which the audiometer oscillator or tuning fork is placed against the mastoid process of the temporal bone.

bone marrow
Soft material that fills the cavity in most bones. It manufactures most of the formed elements of the blood.

bone seeker
Any compound or ion in the body that migrates preferentially to the bone.

BOO
Acronym for build, own, operate.

booking
Arrangements with steamship companies for the acceptance and carriage of freight.

boom
(1) A floating barrier used to contain oil on a body of water. (2). In rigging, a boom is a timber or metal section or strut, pivoted or hinged at the heel (lower end) at a location fixed in height on a frame or mast or vertical member, and with its point (upper end) supported by chains, ropes, or rods to the upper end of the frame, mast, or vertical member.

boom harness
The block and sheave arrangement on the boom point to which the topping lift cable is reeved for lowering and raising the boom. *See also* ***boom (2)***.

boom it down
Tighten chains around freight.

boom point
The outward of the top section of the boom.

boomers
Binder devices used to tighten chains around cargo on flatbed trailers.

boosted fire
A fire wherein some inflammable substance, other than that which the building was constructed or which it contained, contributed to it burning. *See also* ***arson***.

booster pump
A pump used to raise the pressure of the fluid on its discharge side.

boot
(1) A covering for the foot. (2) The act of restarting computer hardware, usually so that newly installed software and/or peripherals can be properly sequenced into the system's startup configuration process. As opposed to a *warm boot,* this process usually requires turning the computer off and then back on again. *See also* ***warm boot***.

BOOT
Acronym for build, own, operate, transfer.

borborygmus
The involuntary rumbling sound caused by the movement of gas or fluid in the large intestine.

border cargo selectivity (BCS)
An automated cargo selectivity system based on historical and other information. The system is designed to facilitate cargo processing and to improve Customs enforcement capabilities by providing targeting information to border locations. The system is used for the land-border environment.

borderline between comfort and discomfort
See ***comfort-discomfort boundary***.

bore hole
A manmade hole in a geological formation.

boreal forest
See ***taiga***.

boredom
A form of mental fatigue generally due to lack of stimulation, lack of interest in the ongoing activity, isolation, performance of a monotonous task, other similar situations, or some combination of these situations.

Borg scale
See ***rating of perceived effort scale***.

boric acid
A crystalline powder, formerly used as a household antiseptic for treating minor irritations of the skin and eyes. Because the pow-

der is highly poisonous when taken internally, and since other antiseptics are more effective, boric acid is no longer recommended. Boric acid ointment (for external use only) is occasionally helpful in cases of mild skin irritations or in keeping a gauze dressing from sticking to a wound.

boron
A chemical element, atomic number 5, atomic weight 10.811, symbol B.

borrowed employee
An employee of one employer who provides services to another, under an agreement between the two employers. Before a person may be considered an borrowed employee, his/her services must be loaned with his acquiescence or consent and he/she must become wholly subject to control and direction of the second employer, and free during the temporary period from the control of the original employer. Under the borrowed employee doctrine, if one to whom an employee is lent is the "master of the servant" at the very time a negligent act occurs, it is upon the master, as a special employer, that liability rests. But if the one lending the employee is considered the master at the very time of injury, then he/she, as general employer, incurs liability.

botanical insecticide
A pesticide whose active ingredient is a plant-produced chemical such as nicotine or strychnine.

Botsball
A small copper sphere, painted black and covered with a sized black mesh wetted fabric, which contains a thermometer for estimating heat stress. *See also* ***wet globe temperature***.

Bottle Bill
Term applied to proposed or enacted legislation which requires a returnable deposit on beer or soda containers and provides for retail store or other redemption centers. Such legislation is designed to discourage the use of throwaway containers.

bottlers body
Truck body designed for hauling cased, bottled beverages.

bottom
The portion of the ground surface which lies below water.

bottom ash
The noncombustible particles that fall to the bottom of a boiler furnace.

bottom dumps
Trailer that unloads through bottom gates.

bottom land hardwoods
Forested freshwater wetlands adjacent to rivers in the southeastern United States. They are especially valuable for wildlife breeding and nesting and habitat areas.

bottom shell
That portion of a tank car tank surface, excluding the head ends of the tank car tank, that lies within two feet, measured circumferentially, of the bottom longitudinal centerline of the tank car tank.

bottom time
That length of time a diver has been at depth in an underwater dive or at maximum pressure in a hyperbaric chamber for treatment of decompression sickness.

botulin
A toxin sometimes found in imperfectly preserved or canned meats and vegetables.

botulism
A severe illness resulting from ingestion of the toxin from the strictly anaerobic bacillus *Clostridium botulinium.* The illness may cause blurred vision, sore throat, or other symptoms of a nervous system disorder. Since these toxins generally disrupt nerve impulse transmission, they are referred to as neurotoxins. Vomiting, diarrhea, or constipation may also occur. If death occurs, it is usually the result of respiratory paralysis.

Boulder winds
Fast-flowing, local downslope winds that may attain speeds of 100 knots or more. They are especially strong along the eastern foothills of the Rocky Mountains near Boulder, Colorado (hence the name).

bound water
Water held on the surface or interior of colloidal particles.

boundary
(1) Every separation, natural or artificial,

which marks the confines or line of division of two contiguous properties. (2) A non-physical line indicating the limit or extent of an area or territory.

boundary representation
A technique used in solid computer modeling where the geometry is defined in terms of its edges and surfaces.

bounty
A gratuity, or an unusual or additional benefit conferred upon, or compensation paid to, a class of persons.

bounty hunter provision
Under the Clean Air Act of 1990, a provision which authorizes EPA to pay a bounty of up to $10,000.00 to anyone who provides information that leads to a civil penalty or criminal conviction. This provision applies to current as well as past employees.

Bourdon tube
A closed, curved, flexible tube of elliptical cross-section which responds to changes in barometric pressure and provides a measurement of that parameter.

bovine
Pertaining to, characteristic of, or derived from the ox (cattle).

bow
(1) *Structural dynamics*. The deflection of a portion of structure caused by a pressure differential on the two sides. (2) *Boating*. The front of a vessel.

Bowen's disease
A pre-cancerous condition characterized by scaly skin lesions resembling psoriasis and showing microscopic changes in the epidermal cells.

bowleg
A deformity in which the space between the knees is abnormally large.

box
Transit. (1) A semitrailer. (2) The transmission part of the tractor.

boxcar
A closed rail freight car.

boycott
Concerted refusal to do business with a particular person or business to obtain concessions or to express displeasure with certain acts or practices of the person or business.

Boyle's law
The volume of a mass of gas is inversely proportional to the pressure, provided the temperature remains the same.

bp
See ***boiling point***.

BPE
See ***boiling point elevation***.

BPR
(1) Biological phosphorus removal. (2) Boiling point rise.

Bq
See ***becquerel***.

brace
In scaffolding construction, a tie that holds one scaffold member in a fixed position with respect to another member.

brachium
The upper arm.

brackish water
Water containing a low concentration of soluble salts, usually between 1000 and 10,000 mg/L.

bradyarthria
See ***bradylalia***.

bradycardia
A lower than normal heart rate, usually less than 60 beats per minute. The condition may occur following an infectious or febrile disease or it may be a symptom of a disorder of the conduction system of the heart. It sometimes occurs with increased intracranial pressure, obstructive jaundice, and myxedema. It should be noted that a heart rate and pulse of less than 60 beats per minute can occur in normal persons, particularly during sleep. Trained athletes usually have a slow heart and pulse rate; opposite of *tachycardia*.

bradykinesia
Any movement disorder in which body movements are slowed.

bradylalia
A very slow articulation in speaking due to central nervous system lesion.

bradylexia
An abnormal slowness in reading.

Braille
A communication system for the blind which uses tactile characters.

brain
The mass of soft, spongy, pink-gray nerve tissue occupying the cranial cavity, consisting of the cerebrum, cerebellum, pons, and medulla oblongata, and connecting at its base with the spinal cord. The human brain weighs about 3 pounds. The brain consists of billions of nerve cells, intricately connected with each other. It contains centers (groups of neurons and their connections) which control many involuntary functions, such as circulation, temperature regulation, and respiration, and interpret sensory impressions received from the eyes, ears, and other sense organs. Consciousness, emotion, thought, and reasoning are functions of the brain. It also contains centers or areas for associative memory which allow for recording, recalling, and making use of past experiences.

brain potential
Any recordable electrical difference between two or more locations on the scalp or brain. *See also* ***electroencephalogram*** *and* ***evoked potential****.*

brain stem
That portion of the brain which is continuous with the spinal cord and lies beneath the cerebellum and cerebral hemispheres, containing neurons governing many of the body's vital functions.

brain wave
The recorded or observed varying electrical potentials from the brain. *See also* ***electroencephalogram*** *and* ***evoked potential****.*

brainstorm
Propose and discuss ideas, freely and without criticism, in an attempt to discover all possible approaches to a situation.

brake
An energy conversion mechanism used to stop, or hold a vehicle stationary.

brake horsepower (bhp)
The power developed by an engine as measured by a dynamometer applied to the shaft or flywheel.

brake pipe
A pipe running from the engineman's brake valve through the train, used for the transmission of air under pressure to charge and actuate the automatic brake equipment and charge the reservoirs of the electro-pneumatic brake equipment on each vehicle of the train.

brake shoe
The non-rotating portion of a tread or disc brake assembly. The shoe is pressed against the tread, disc, or drum when the brake is applied.

brake tubing/hose
Metallic brake tubing, nonmetallic brake tubing and brake hose are conduits or lines used in a brake system to transmit or contain the medium (fluid or vacuum) used to apply the motor vehicle's brakes.

braking action
Aviation. A report of conditions on the airport movement area providing a pilot with a degree/quality of braking that he/she might expect. Braking action is reported in terms of good, fair, poor, or nil.

Braking Action Advisories
Aviation. When tower controllers have received runway braking action reports which include the terms "poor" or "nil," or whenever weather conditions are conducive to deteriorating or rapidly changing runway braking conditions, the tower will include on the Automated Terminal Information Service (ATIS) broadcast the statement, "BRAKING ACTION ADVISORIES ARE IN EFFECT." During the time Braking Action Advisories are in effect, Air Traffic Control (ATC) will issue the latest braking action report for the runway in use to each arriving and departing aircraft. Pilots should be prepared for deteriorating braking conditions and should request current runway condition information if not volunteered by controllers. Pilots should also be prepared to provide a descriptive runway condition report to controllers after landing.

braking distance
Total distance required to stop a motor vehicle from the time the driver recognizes the need to stop until the vehicle is standing still. Influencing factors include the speed of the vehicle, the weather, the road conditions, the vehicle's tires and condition of its brakes, etc. Sometimes referred to as *stopping distance*.

branch
In ventilation, a duct or pipe connecting an exhaust hood to a main or sub-main.

branch circuit
The circuit conductors between the final over-current device protecting the circuit and the outlet(s).

branch duct entry
The point in a ventilation system where a branch or secondary duct joins a main duct.

branch of greater resistance
The path from a hood or duct opening to the fan and exhaust stack in a ventilation system which causes the most pressure loss.

branch railroad
A lateral extension of a main line; a road connected with or issuing from a main line.

branch sewer
A sewer that receives wastewater from a small area and discharges into a main sewer serving more than one area.

brass
A copper alloy containing up to 40% zinc.

brass-founders ague
Metal fume fever that may occur in workers in brass foundries.

brattice
Partitions that are placed throughout underground mines to control the flow of ventilation. These are often made of heavy cloth such as canvas, or of plywood.

Brayfield-Rothe Scale of Job Satisfaction
A commercially available standardized questionnaire for surveying job satisfaction among employees.

breach
(1) The breaking or violating of a law, right, obligation, engagement, or duty, either by commission or omission. Exists where one party to a contract fails to carry out term, promise, or condition of the contract. (2) Bypass, avoid, or dismantle a safety or security mechanism.

breach of contract
(1) Failure, without legal excuse, to perform any promise which forms the whole or part of a contract. (2) Unequivocal, distinct, and absolute refusal to perform an agreement.

breach of duty
In a general sense, any violation or omission of a legal or moral duty. More particularly, the neglect or failure to fulfill, in a just and proper manner, the duties of an office or fiduciary employment.

breach of warranty
(1) In real property law and the law of insurance, the failure or falsehood of an affirmative promise or statement, or the nonperformance of an executory stipulation. (2) As used in the law of sales, breach of warranty, unlike fraud, does not involve guilty knowledge, and rests on contract. (3) Under the Uniform Commercial Code, consists of a violation of either an express or implied warranty relating to title, quality, content, or condition of goods sold for which an action in contract will lie.

breadth
Width; a straight-line horizontal measurement having only lateral extent, from one side of the body or a body segment to the other.

break
(1) *General-Structure Mechanics*. A fracture resulting in complete separation into parts. (2) *Hydrology*. A surface disturbance of the water similar to a boil, caused by an underwater obstruction. *See also* ***boil***.

break-bulk
Packages of hazardous materials that are handled individually, palletized, or unitized for purposes of transportation as opposed to bulk and containerized freight.

break-even analysis
An quantitative technique used to determine the sales necessary to achieve the break-even point. *See also* ***break-even point***.

break-even chart
A graphical representation of the relationships between income and costs, usually based on different levels of volume for production and sales.

break-even point
That economic level at which total operating costs equal total income, and the company neither makes a profit nor has a loss.

break the unit
To uncouple the tractor from the trailer.

break time
See ***rest period***.

break up tow
To disassemble the tow either at the end of the voyage or inadvertently on a sandbar.

breakbone fever
See ***dengue***.

breakbulk cargo
Packaged products that can be palletized into larger parcels and assembled together, for example, on pallet boards bound by wire, or gathered up in rope cargo slings as a means of lifting on and off a vessel.

breakdown
(1) A decomposition of some process or activity into its component parts. (2) The ceasing of operation of a system, subsystem, or component due to some fault or failure.

breakdown bar
A length of pipe used to increase the leverage in setting up ratchets when connecting tow rigging. Also called *cheater bar*.

breakdown maintenance
See ***corrective maintenance***.

breaking strength
That stress level at which a material fails.

breakout tank
A tank used to 1) relieve surges in an oil or hazardous liquid pipeline system, or 2) receive and store oil or hazardous liquid transported by a pipeline for reinjection and continued transportation by pipeline.

breakpoint
That readily distinguishable point in time which represents a boundary between two task elements, at which one element is completed and the other is begun. Also referred to as *reading point* and *endpoint*.

breakpoint chlorination
Addition of chlorine until the chlorine demand has been satisfied. Further addition will result in a chlorine residual so that disinfection can be assured.

breakthrough
(1) *NIOSH*. The presence of 25% or more of a contaminant in the rear portion of a sorbent tube. (2) *Water Treatment*. That point in the granular media filter cycle when the filtrate turbidity begins to increase because the filter bed is full and no longer able to retain solids.

breakwater
An offshore barrier, often connected to shore, that breaks the force of waves and provides shelter from wave action.

breast
(1) The anterior thorax, especially in the region of the nipple. (2) The human female mammary gland.

breast line
Any line that leads straight in or square. Keeps a barge from moving out from its mooring facilities.

breastbone
See ***sternum***.

breathalyzer test
Test to determine content of alcohol in a person arrested for operating a motor vehicle while under the influence of liquor. The results of such test, if properly administered, are admissible evidence.

breathe
To alternately inhale and exhale air from the lungs.

breathing air
Air that equals or exceeds Grade D specifications for gaseous air in accordance with ANSI/CGA G-7.1-73, and that does not present a health hazard to anyone breathing the air.

breathing zone
(1). Usually the air within a 12 to 24 inch radius surrounding a person's head. (2) Area of a room in which occupants breathe as they stand, sit or lie down.

breathing zone sample
An air sample collected in the breathing area of a worker to assess exposure to an airborne contaminant.

bremsstrahlung
Radiation. A German word meaning "braking radiation," it is the secondary x-radiation (ionizing photon radiation) that is produced when a beta particle is slowed down or stopped by a high-density surface.

BRI
Building-related illness.

bricklayer's square scaffold
A scaffold composed of framed wood squares which support a platform limited to light and medium duty.

bridge
A structure including supports erected over a depression or an obstruction, such as water, highway, or railway, and having a track or passageway for carrying traffic or other moving loads, and having an opening measured along the center of the roadway of more than 20 feet between undercopings of abutments or spring lines of arches, or extreme ends of openings for multiple boxes; it may also include multiple pipes, where the clear distance between openings is less than half of the smaller contiguous opening.

bridge foundation bearing material
The type of material supporting the substructure of a bridge. Code as follows: GW, well-graded gravel; GP, poorly graded gravel; GM, silty gravel; GC, clay gravel; SW, well-graded sand; SP, poorly graded sand; SM, silty sand; SC, clay sand; RK, bedrock; UK, unknown; O, other.

bridge number
The number of the installation, consisting of the full route number (including segment and spur) plus the milepost location of the bridge to the nearest one hundredth of a mile.

bridge posted load restrictions
Load restrictions posted at a bridge structure. Entry order: single axle, dual axle, load type 3, load type 3S2, load type 3-3, and Special.

bridge posted speed restrictions
A speed limit posted at a bridge structure, in miles per hour.

bridge structure
A two-character code for recording the type of bridge structure. Code as follows: SS, simple span; CS, continuous span; SC, combination simple and cantilever; CC, combination continuous and cantilever; O, other.

bridge superstructure
Those elements of the bridge structure which are above the uppermost deck.

bridging encapsulant
A material, generally in a liquid form, that is employed to seal the surface of an asbestos-containing material or other product, to prevent the release of fibers.

bridle line
The wire cable used to connect a barge in trailing fashion behind the towboat.

bright
A highlighting technique in which one or more portions of a display appear brighter than the remainder.

brightener
Any colorless, fluorescent dye which causes washed clothing to appear brighter under certain lighting conditions by converting ultraviolet light into visible light, normally at the blue end of the spectrum. Also referred to as *whitener, whitening agent,* and *optical brightener.*

brightness
A subjective judgment of the relative amount of light projected or reflected from a surface or object, ranging from brilliant to dark. *See also* ***luminance***.

brightness contrast
The subjective difference between the brightness of an object and the background against which that object is located. *See also* ***luminance contrast***.

brightness control
A potentiometer or other adjustment device for varying the luminance on a display. Also known as *brilliance control.*

brightness enhancement
The use of a flashing light within a certain flashing frequency range (about 2-20 Hz) to make a light appear brighter than if the same average light intensity were used from a steady light.

bril
A subjective scale for judging brightness.

brilliance
See ***brightness***.

brilliance control
See ***brightness control***.

brine
Water saturated with, or containing a high concentration of, salts, usually in excess of 36,000 mg/L.

brine concentrator
A vertical tube falling film evaporator employing special scale control techniques to maximize concentration of dissolve solids.

brine heater
The heat input section of a multistage flash evaporator where feed water is heated to the process' top temperature.

brine staging
*See **reject staging**.*

brinelled
Defaced or distorted surfaces typically caused by shock of impact between surfaces.

British Thermal Unit (BTU)
The amount of energy required to raise the temperature of 1 pound of water 1 degree Fahrenheit (F) at or near 39.2 degrees F and 1 atmosphere of pressure. One British Thermal Unit (BTU) is about equal to the heat given off by a blue-tip match.

brittle fracture
(1) A type of failure mode in structural materials that usually occurs without prior plastic deformation and at extremely high speed. (2) A type of failure mode such that burst of the vessel is possible during cycling. Normally this mode of failure is a concern when cycling to the maximum expected operating pressure (MEOP) or when the vessel is under sustained load at MEOP. (3) A type of fracture that is characterized by a flat fracture surface with little or no shear lips (slant fracture surface) and at average stress levels below those of general yielding.

Brl
*See **building restriction line**.*

broad-crested weir
A weir having a substantial crest width in the direction parallel to the direction of water flowing over it.

broadband
Containing many frequencies.

broadband noise
Noise with components extending over a wide frequency range. *See also **white noise**.*

broadcast
(1) A message sent to all stations connected to a computer network. (2) Transmission of information for which an acknowledgment is not expected.

broadcast application
In pesticide application, the spreading of chemicals over an entire area.

broke
Paper waste generated prior to completion of the paper-making process.

broken shift
*See **split shift**.*

broken train collision
A collision in which a moving train breaks into parts and an impact occurs between these parts, or when a portion of the broken train collides with another consist.

broker
A person who arranges for transportation of loads for a percentage of the revenue from the load. *See also **customs house broker** and **freight forwarder**.*

brokerage
Freight forwarder/broker compensation as specified by ocean tariff.

bromine
(1) A chemical element, atomic number 35, atomic weight 79.909, symbol Br. (2) A halogen used as a water disinfectant in combination with chlorine and as a chlorine-bromide mixture.

bronchial tubes
Branches or subdivisions of the trachea (windpipe). A *bronchiole* (the narrowest of the tubes which carry air into and out of the lungs) is a branch of the *bronchus* which is a branch of the trachea. Also referred to simply as *bronchi*.

bronchiectasis
Chronic dilation of the bronchi with spasmodic coughing and production of phlegm.

bronchioles
*See **bronchial tubes**.*

bronchitis
An inflammation of the bronchi or bronchial tubes. It can be either acute or chronic; an acute case occasionally develops into a chronic one. If the inflammation reaches the bronchioles and the alveoli, the condition is bronchopneumonia.

bronchogenic carcinoma
A carcinoma of the lung.

bronchopneumonia
Term indicating inflammation of the lungs, usually beginning in the terminal bronchioles, followed by their becoming clogged.

bronze
A copper-tin alloy, or any other copper alloy that does not contain zinc or nickel as the principal alloying element.

brow
*See **forehead** and **eyebrow**.*

brow ridges
The bony ridges of the forehead that lie above the orbits of the eyes.

brown asbestos
*See **amosite asbestos**.*

brown coal
A common term for lignite.

brown lung
*See **byssinosis**.*

Brownfield
(1) A contaminated property, either abandoned or underutilized because the perceived cost of remediation exceeds the perceived value. Often located in urban and economically distressed areas. (2) Former industrial sites that, either because of actual or perceived contamination, lie idle or underutilized because of fear of hazardous waste liability attached to their ownership or operation.

Brownfield Initiative
An EPA program begun in 1985 to fund pilot projects with investors, businesses, and developers to redevelop selected Brownfield sites. The initiative eliminated 25,000 +/- sites from the National Priority List.

Brownian motion
Erratic movement of colloidal particles that results from the impact of molecules and ions dissolved in the solution.

brownie
Automotive. An auxiliary transmission.

brucellosis
An illness caused by the bacterium of the Genus *Brucella.* Symptoms include fever, chills, headache, muscle aches, malaise, weakness, loss of appetite and subsequent loss of weight. Mortality is possible but rare. Contact with infectious materials such as animal blood is an important mode of infection for livestock growers, veterinarians, and processing plant workers. Intact skin is an effective barrier, but cuts and abrasions provide a direct route of exposure. Inhalation and ingestion are also potential routes of infection. *See also **undulant fever**.*

Bruceton test method
A statistical method for determining the all-fire and no-fire characteristics of an electro-explosive device using a small sample size, but with high reliability.

bruise
An injury characterized by capillary or venous hemorrhaging beneath an unbroken skin. *See also **hematoma**.*

brush aerator
Mechanical aeration device most frequently used in oxidation ditch wastewater treatment plants, consisting of a horizontal shaft with protruding paddles that are rapidly rotated at the water surface. *Also called a **rotor**.*

brush out
To clear out the brush or vegetation around a light or day mark so that the structure is visible to navigation in all necessary directions. An aid should be cleared or brushed out so as to be completely visible to navigation from the beginning of its use in a set of marks until it is no longer being used in that or another set of marks. *See also **landscaping**.*

bruxism
Grinding of the teeth.

BRYNTRN
A computer model for determining the effects of nucleons on target materials.

BSC
*See **Boating Safety Circular**.*

BS&W
Bottom sediments & water.

BTPS conditions
*See **body temperature and pressure, saturated conditions**.*

BTS
*See **Bureau of Transportation Statistics**.*

BTU
*See **British Thermal Unit**.*

bubble
(1) A system under which existing emissions sources can propose alternate means to comply with a set of emissions limitations. Under the bubble concept, sources can implement more than the required controls at one emission point where control costs are relatively low in return for a comparable relaxation of controls at a second emission point where costs are higher. (2) A trapped volume of air or other gas(es) within a more viscous fluid or solid.

bubble meter
A burette, or other similar volumetric device, that can be used with a soap solution to form a bubble for calibrating a sampling device, such as a pump, by timing the period it takes for the bubble to traverse a specific volume and using this data to calculate its flow rate. This method is considered a primary calibration method. Also referred to as a *soap-film* or *soap-bubble flow meter.*

bubble policy
*See **emissions trading**.*

bubble tube
A simple device used to calibrate air-sampling pumps.

bubbler
A device used to collect air contaminants by bubbling sampled air through a liquid medium (e.g., absorbent) contained in the bubbler. The sampling tube of the bubbler typically has a glass frit at the end which is immersed in the collecting solution or sampling medium.

bubbler system
Common terminology for a pneumatic-type differential level controller.

bubonic plague
An acute infectious disease usually transmitted from infected animals to humans by the bite of a rat flea.

buchner funnel
A laboratory funnel with a perforated bottom that utilizes a disposable filter paper to evaluate wastewater and sludge dehydration.

bucket elevator
A conveying device consisting of a head and foot assembly that supports and drives an endless chain or belt to which buckets are attached.

buddy-breathing device
An accessory to self-contained breathing apparatus which permits a second person to share the same air supply as that of the wearer of the apparatus.

buddy system
A system organizing employees into work groups in such a manner that each employee of a work group is designated to be observed by another person in the work group.

Buerger's disease
A disease affecting the medium-sized blood vessels, particularly the arteries of the legs, which can cause severe pain and in serious cases, lead to gangrene. Also called *thromboangitis obliterans,* a term that refers to the clotting, pain, and inflammation occurring in this disease and to the fact that it can obliterate, or destroy, blood vessels. The cause of this violent reaction has been thought to be excessive use of tobacco over a long period of time. The number of cases has diminished strikingly in recent years. The intense pain that is a symptom of the disease is caused by the formation of blood clots, or thrombosis, in the lining of the arterial blood vessels. When the clots grow larger, the blood slows and may stop entirely. Since every part of the body depends on the continuous flow of blood, affected areas such as fingers and toes, soon begin to atrophy or develop ulcers. If the causes of the disease are not completely arrested, amputation may be necessary.

buffer
(1) A substance that stabilizes the pH value of solutions. (2) A region separating one area from another for safety, habitability, or other reasons. (3) A temporary computer storage location in which data may be kept while awaiting transfer to another, more permanent location.

buffer strips
Strips of grass or other erosion-resisting vegetation between or below cultivated strips or fields.

buffering capacity
The capacity of a solution to resist a change in composition, especially changes in pH.

bug it
Transit. To carry freight from the front to the back of a truck.

builder
(1) Any chemical used in the laundry process which acts to soften water for improved detergent activity. (2) One whose occupation is the building or erection of structures, the controlling and directing of construction, or the planning, constructing, remodeling and adapting to particular uses buildings and other structures.

building block
One of a fixed group of elements or modules which may be joined to form a system or complete some activity.

building code
(1) A set of regulations that provides standards to which structures must be built. These may be issued by local, county, state, regional, or national agencies. (2) Laws, ordinances, or government regulations concerning fitness for habitation setting forth standards and requirements for the construction, maintenance, operation, occupancy, uses or appearance of buildings, premises, and dwelling units. *See also* ***code***.

building envelope
Elements of the building, including all external building materials, windows, and walls, that enclose the internal space.

building-related illness
A diagnosable illness whose symptoms can be identified and whose cause can be directly attributed to airborne building pollutants (e.g., *Legionnaire's Disease, hypersensitivity pneumonitis*).

building restriction line (Brl)
A line which identifies suitable building area locations on airports.

Built-in Test (BIT)
A circuit or other equipment located within a system which automatically or on direction by an operator verifies system function.

Built-in Test Equipment (BITE)
That circuitry or other hardware incorporated into a system for monitoring that system's function and analyzing faults when they occur.

bulb
The primary source of light in an electrically powered lamp.

bulk arrival
The arrival of several customers or users at a location at one time or as part of a single event.

bulk cargo
(1) Cargo not packaged or broken into smaller units. Bulk cargo is either dry (grain) or liquid (petroleum) and cannot be counted. (2) The tonnes of bulk cargo assessed at the Bulk rate of tolls as defined in the St. Lawrence Seaway Tariff of Tolls. (3) Cargo that is unbound as loaded and carried aboard ship; it is without mark or count, in a loose unpackaged form, and has homogeneous characteristics.

bulk cargo spout
A spout, which may or may not be telescopic and may or may not have removable sections, but is suspended over the vessel from some overhead structure by wire rope or other means. Such a spout is often used with a thrower or trimming machine. A grain loading spout is an example of spouts covered by this definition.

bulk cargo sucker
A pneumatic conveyor which utilizes a spout-like device, which may be adjustable vertically and/or laterally, and which is suspended over a vessel from some overhead structure by wire rope or other means. An example of an installation of this nature is the grain sucker used to discharge grain from barges.

bulk carriers
All vessels designed to carry bulk cargo such as grain, fertilizers, ore, and oil.

bulk density
The density/volume ratio for a solid including the voids contained in the bulk material.

bulk materials
Any powdery, granular, or lumpy substance in loose form.

bulk packaging
A packaging, other than a vessel or a barge, including a transport vehicle or freight container, in which hazardous materials are loaded with no intermediate form of containment and which has: 1) a maximum capacity greater than 450 L (119 gallons) as a recepta-

cle for a liquid; 2) a maximum net mass greater than 400 kg (882 pounds) and a maximum capacity greater than 450 L (119 gallons) as a receptacle for a solid; or 3) a water capacity greater than 454 kg (1000 pounds) as a receptacle for a gas as defined in 49 CFR 173.115.

bulk sample
(1) As related to asbestos, a small portion of suspect building materials that are collected and sent to a laboratory for analysis by polarized light microscopy coupled with dispersion staining, or by electron microscopy for verification.

bulk terminal
(1) A facility used primarily for the storage and/or marketing of petroleum products, which has a total bulk storage capacity of 50,000 barrels or more and/or receives petroleum products by tanker, barge, or pipeline. (2) A purpose-designed berth or mooring for handling liquid or dry commodities, in unpackaged bulk form, such as oil, grain, ore, and coal. Bulk terminals typically are installed with specialized cargo-handling equipment such as pipelines, conveyors, pneumatic evacuators, cranes with clamshell grabs, and rail lines to accommodate cargo-handling operations with ships or barges. Commodity-specific storage facilities such as grain silos, petroleum storage tanks, and coal stock yards are also located at these terminals.

bulkhead
(1) A partition of wood, rock, concrete, or steel used for protection from water, or to segregate sections of tanks or vessels. (2) A partition separating one part of a ship, freight car, aircraft, or truck from another part.

bulking sludge
A poorly settling activated sludge that results from the predominance of filamentous organisms.

bulky waste
Large items of solid waste such as household appliances, furniture, large auto parts, trees, branches, stumps, or other oversize wastes whose large size precludes or complicates their handling by normal solid wastes collection, processing, or disposal methods.

bull hauler
One who hauls livestock.

bullae
Bladder or sac containing liquid, such as occurs when lungs become emphysematous.

bulletin board
(1) *General*. A posting board usually located in a common area of access used to post information of general or specific interest. (2) *Marine Safety*. A board located at each dam upon which is displayed information concerning the navigability of the dam, such as indicating when movable dams are down and open river conditions exist. Also located elsewhere such as at gauges to publish gauge readings and river level trend.

bulling
The horizontal dragging of cargo across a surface with none of the weight of the cargo supported by the fall.

bullnose
A slanted riverward end of the intermediate lock wall.

bump
(1) *General*. a) A rise (or dip) or slight elevation (or depression) above (or below) normal grade. b) A minor (negligible) collision or contact between two or more bodies. (2) *Marine Navigation*. Usually used in the phrase "watch the bump," a term used on board tows when one or more barges are likely to make contact. May also mean a momentary grounding, usually due to excess speed in shallow water.

bumpers
(1) Fenders. (2) Pads made out of Styrofoam, old ropes, old tires, or similar material, which are hung over the side of a water vessel to prevent damage to the vessel when berthing or locking through dams.

bundle
Asbestos (EPA). A structure composed of three or more fibers in a parallel arrangement with each fiber closer than one fiber diameter.

bunker
A storage tank.

bunker C/Number 6 fuel oil
A high viscosity oil used mostly by ships, industry, and large-scale heating installations. This heavy fuel requires preheating in the storage tank to permit pumping and additional preheating to permit atomizing at the burners.

bunkering fuels
Fuels stored in ship bunkers.

bunkers
Fuels supplied to ships and aircraft in international transportation, irrespective of the flag of the carrier, consisting primarily of residual, distillate, and jet fuel oils.

buoy
A float moored or anchored in water.

buoy line
A line formed by two or more buoys marking a contour edge of a channel.

buoy range markers
Painted stakes set up on shore so placed as to form a range through the exact location of a buoy. Used only on the Tennessee River to mark buoys in dredged cuts.

buoyancy
The tendency of a body to rise or float in a liquid.

burden
Capacity for carrying cargo. Something that is carried. Something oppressive or worrisome.

burden of persuasion
The onus on the party with the burden of proof to convince the trier of fact of all elements of his/her case.

burden of producing evidence
The obligation of a party to introduce evidence sufficient to avoid a ruling against him/her on the issue.

burden of proof
In the law of evidence, the necessity or duty of affirmatively proving a fact or facts in dispute on an issue raised between the parties in a cause. The obligation of a party to establish by evidence a requisite degree of belief concerning a fact in the mind of the trier of fact or the court.

bureau
(1) An office for the transaction of business. (2) A name given to the several departments of the executive or administrative branch of government, or their divisions. (3) A specialized administrative unit. (4) Business establishment for exchanging information, making contacts, coordinating activities, etc.

Bureau of Land Management
Established July 16, 1946 by the consolidation of the General Land Office (created in 1812) and the Grazing Service (formed in 1934). The Bureau manages the national resource lands (some 450 million acres) and their resources. It also administers the mineral resources connected with acquired lands and the submerged lands of the Outer Continental Shelf (OCS). It is within the U.S. Department of the Interior.

Bureau of Mines (BOM)
A research and fact-finding agency in the U.S. Department of the Interior with the goal of stimulating private industry to produce the country's mineral needs in ways that protect workers and the public interest.

Bureau of Transportation Statistics (BTS)
The Bureau was organized pursuant to section 6006 of the Intermodal Surface Transportation Efficiency Act (ISTEA) of 1991 (49 U.S.C. 111), and was formally established by the Secretary of Transportation on December 16, 1992. BTS has an intermodal transportation focus whose missions are to compile, analyze, and make accessible information on the nation's transportation systems; to collect information on intermodal transportation and other areas; and to enhance the quality and effectiveness of DOT's statistical programs through research, the development of guidelines, and the promotion of improvements in data acquisition and use. The programs of BTS are organized in six functional areas and are mandated by ISTEA to: 1) compile, analyze, and publish statistics; 2) develop a long-term data collection program; 3) develop guidelines to improve the credibility and effectiveness of the Department's statistics; 4) represent transportation interests in the statistical community; 5) make statistics accessible and understandable; and 6) identify data needs.

BuRec
U.S. Bureau of Reclamation.

burette
A glass tube with fine gradations and bottom stopcock used to accurately dispense fluids.

burial ground
A disposal site for radioactive waste materials that uses earth or water as a shield. Also referred to as *graveyard*.

burn
Injury caused by contact with dry heat (fire), moist heat (steam or liquid), chemicals, electricity, lightning, or ultraviolet rays of the sun. Burns are classified according to degree.

- A *first degree* burn involves a reddening of the skin area;
- A *second degree* burn causes the skin to blister;
- A *third degree* burn is the most serious, involving damage to the deeper layers of the skin. In some cases, the growth cells of the tissues in the affected area may be destroyed.

burn-in test
A period of time in which a completed system or set of subsystems is observed under expected operating or more extreme conditions to determine if any of the components will fail prematurely. Synonymous with *debug*.

burning rate
The rate at which solid waste is incinerated or heat is released during incineration.

burnishing
A surface-finishing process in which surface irregularities are displaced rather than removed.

burr
A ragged edge or sharp point on a surface, possibly as a result of some faulty machining process or as a natural characteristic of the material.

bursa
A fluid-filled, sac-like structure having a slippery surface and located at joints or other tissues to reduce friction in movement.

bursitis
An inflammation of the joints of the body, occasionally with calcium deposit development. *See also* ***cumulative trauma disorder***.

burst
A rapid decrease in pressure within a container of specified volume as it ruptures under pressure and the contents spread rapidly to the external environment.

burst factor
A multiplying factor applied to the maximum expected operating pressure (MEOP) to obtain the design burst pressure. Synonymous with *ultimate pressure factor*.

burst lung
See ***pulmonary hyperinflation syndrome***.

bus
(1) Any of several types of self-propelled vehicles, generally rubber-tired, intended for use on city streets, highways, and busways, including but not limited to minibuses, forty- and thirty-foot buses, articulated buses, double-deck buses, and electrically powered trolley buses, used by public entities to provide designated public transportation service and by private entities to provide transportation service including, but not limited to, specified public transportation services. Self-propelled, rubber-tired vehicles designed to look like antique or vintage trolleys are considered buses. (2) Any motor vehicle designed, constructed, and or used for the transportation of passengers, including taxicabs. (3) A vehicle designed to carry more than 15 passengers, including the driver. (4) Large motor vehicles used to carry more than ten passengers, including school buses, intercity buses, and transit buses. (5) Includes intercity buses, mass transit systems, and shuttle buses that are available to the general public. Also includes Dial-A-Bus and Senior Citizen buses that are available to the public. *See also* ***automobile, minivan, motor vehicle,*** *and* ***vehicle***.

bus charter service
(Except Local) Establishments primarily engaged in furnishing passenger transportation charter service where such operations are principally outside a single municipality, outside one group of contiguous municipalities, or outside a single municipality and its suburban areas.

bus lane
A street or highway lane intended primarily for buses, either all day or during specified periods, but sometimes also used by carpools meeting requirements set out in traffic laws.

bushing
(1) A short threaded tube that screws into a pipe fitting to reduce its size. (2) The bearing surface for pin rotation when a chain revolves around a sprocket.

business
(1) Employment, occupation, profession, or commercial activity engaged in for gain or livelihood. (2) Activity or enterprise for gain, benefit, advantage, or livelihood. (3) An enterprise in which a person engaged shows willingness to invest time and capital on future outcome.

business district
The territory contiguous to and including a highway when within any 600 feet along such highway there are buildings in use for business or industrial purposes, including but not limited to hotels, banks, or office buildings which occupy at least 300 feet of frontage on one side or 300 feet collectively on both sides of the highway.

business flying
The use of aircraft by pilots (not receiving direct salary or compensation for piloting) in connection with their occupation or in the furtherance of a private business.

business insurance
A type of insurance which protects a business on the disability or death of a key employee. *See also* ***insurance***.

business interruption insurance
A type of insurance which protects a business from losses due to an inability to operate because of fire or other hazards. *See also* ***insurance***.

business tort
A noncontractual breach of a legal duty by a business directly resulting in damages or injury to another.

business transportation
Use of an aircraft not for compensation or hire by individuals for the purpose of transportation required by businesses in which they are engaged.

bust depth
The horizontal linear distance from the most posterior protrusion at the bra tip level of an individual's back to the bust point. Measured with the individual standing erect; for females only.

bust point
The most anterior external protrusion of the bra pocket (for females only).

bust point - bust point breadth
The horizontal distance between bust points. Measured with the individual standing erect; for females only.

bust point height
The vertical distance from the floor or other reference surface to the bust point. Measured with the individual standing erect and the weight distributed evenly on both feet; for females only. Also known as *chest height*.

busway
(1) Exclusive freeway lane for buses and carpools. (2) A roadway reserved for buses only. It may be a grade-separated or controlled access roadway.

but for rule
See ***cause in fact***.

butterfly valve
A valve equipped with a stem-operated disk that is rotated parallel to the liquid flow when opened and perpendicular to the flow when closed.

buttock
The mass of fleshy tissue posterior to the hip, consisting largely of the gluteus maximus and other muscles.

buttock circumference
The surface distance around the body without tissue compression at the level of the maximum posterior protuberance of the buttocks. Measured with the individual standing erect and the weight balanced evenly on both feet. Also referred to as *hip circumference, standing*.

buttock circumference, sitting
The surface distance around the buttocks and diagonally across the lap. Measured with the individual sitting erect. Also referred to as *hip circumference, sitting*.

buttock depth
The horizontal linear distance from the maximum posterior protrusion of the buttocks to the most anterior portion of the torso at that level. Measured with the individual standing erect with the hip and thigh muscles relaxed.

buttock - heel length
See ***buttock - leg length***.

buttock height
The vertical distance from the floor to the maximum posterior protrusion of the buttock.

Measured with the individual standing erect and the weight balanced evenly on both feet.

buttock - knee length
The horizontal distance from the rearmost point of the buttocks to the front of the knee-caps. Measured with the individual sitting erect, the knees flexed 90°, feet flat on the floor, and the upper leg parallel to the floor.

buttock - leg length
The horizontal distance from the wall or the most posterior point of the buttocks to the underside of the heel. Measured with the individual sitting erect on the floor or other flat surface (possibly against a wall but with no tissue compression), the knee fully extended, and the long longitudinal axis of the foot perpendicular to the leg. Also referred to as *buttock - heel length.*

buttock - popliteal length
The horizontal distance from the rearmost surface of the buttock to the back of the lower leg. Measured with the individual sitting erect, knees flexed 90°, the feet flat on the floor, and the upper leg parallel to the floor.

buttock protrusion
The point of maximum posterior protrusion of the buttock.

button
(1) A fastening device, usually used in garments. (2) A heavy steel casting found mostly on lock walls, designed to hold the eye of a line or wire. It is also used as deck fittings on towboats and on barges. (3) A short mushroom-shaped bit or a short timberhead.

button her up
Tie down the load on a truck or trailer.

Buys-Ballot's Law
Describes the relationship between the wind direction and the pressure distribution. In the Northern Hemisphere, if you stand with your back to the wind, lower pressure will be to your left. In the Southern Hemisphere, it is reversed.

BWI
British drinking Water Inspectorate.

bylaws
Regulations, ordinances, rules, or laws adopted by an association or corporation or the like for its internal governance. Bylaws define the rights and obligations of various officers, persons, or groups within the corporate structure and provide rules for routine matters such as calling meetings and the like.

bypass
(1) A channel or pipe arranged to divert flow around a tank, treatment process, or control device. (2) A surgical procedure where a critical vessel that has been blocked (due to plaque) or otherwise damaged and is not able to properly function, is clamped and re-routed or replaced by an vessel segment obtained from another part of the body (e.g., femoral artery) to ensure proper and continuous flow of blood to a critical area (such as the heart).

bypass fume hood
A laboratory fume hood constructed such that, as the sash is closed, air bypasses the hood face via an opening that is typically located above the sash, thereby providing a reasonably constant velocity of air entering the hood face.

byproduct
(1) Under the Federal Geothermal Energy Research, Development, and Demonstration Act of 1974: Any mineral or minerals which are found in solution or in association with geothermal resources and which have a value of less than 75 percent of the value of the geothermal steam and associated geothermal resources or are not, because of quantity, quality, or technical difficulties in extraction and production, of sufficient value to warrant extraction and production by themselves. (2) A material or substance that is not a primary product of a process and is not separately produced.

byproduct material
Under the Federal Atomic Energy Act of 1954: Any radioactive material (except special nuclear material) produced in or made radioactive by exposure to the radiation incident to the process of producing or utilizing special nuclear material. Also the tailings or wastes produced by the extraction or concentration of uranium or thorium from any ore processed primarily for its source material content.

byssinosis
A disease of the lungs caused by chronic exposure to cotton and/or linen fibers and dusts.

byte
A group of bits which may be treated as a single unit in a digital computer. The number depends on the type of hardware, but there are typically 8 bits to a byte.

C

C-scale
A sound weighting system having flat response characteristics for high sound pressure levels up to about 8 kHz.

C-shift
See ***third shift***.

C-weighted sound level (dBC)
The sound level as determined on the C-scale of a sound level meter or other noise survey meter with the weighting network. *See also C-scale.*

C&W
See ***caution and warning***.

CA
See ***cab-to-axle dimension***.

CAA
See ***Clean Air Act***.

CAAA
See ***Clean Air Act Amendments***.

CAB
See ***Civil Aeronautics Board***.

cab
(1) That portion of the superstructure designed to be occupied by the crew operating a locomotive. (2) The compartment of a locomotive from which the propelling power and power brakes of the train are manually controlled. (3) Portion of truck where the driver sits; tractor. The passenger compartment of a vehicle.

cab beside engine
The cab is located to left or right side of the engine.

cab forward of the engine
The engine is directly behind the cab.

cab over
A vehicle with a substantial part of its engine located under the cab. Also known as *snubnose*.

cab-over-engine (COE)
A truck or truck-tractor, having all, or the front portion, of the engine under the cab.

cab-over-engine (COE) high profile
A COE having the door sill step above the height of the front tires.

cab signal
Rail Transportation. A signal located in the engineman's compartment or cab, indicating a condition affecting the movement of a train and used in conjunction with interlocking signals and in conjunction with or in lieu of block signals.

cab-to-axle dimension (CA)
The distance from the back of a truck cab to the centerline of the rear axle. For trucks with tandem rear axles, the CA dimension is given midway between the two rear axles.

cabin
The occupied portion of the interior of a passage vehicle.

cabin deck
The second deck on most river steamboats. It was lined with staterooms surrounding the main cabin. It was also called the *boiler deck* even though the boilers were on the cargo deck below.

cabin motorboat
Motorboats with a cabin which can be completely closed by means of doors or hatches. Large motorboats with cabins, even though referred to as yachts, are considered to be cabin motorboats.

cabin pressure
The atmospheric pressure within a cabin.

cabin temperature
The dry-bulb temperature within a cabin.

cabinet
(1) An independent structure containing drawers and/or shelves. (2) An enclosure designed either for surface or flush mounting, and provided with a frame, mat, or trim in which a swinging door or doors are or may be hung.

cable car
An electric railway operating in mixed street traffic with unpowered, individually controlled transit vehicles propelled by moving cables located below the street surface and powered by engines or motors at a central location not on board the vehicle.

cable cars
Streetcar type of passenger vehicles operating by means of an attachment to a moving cable located below the street surface and powered

by engines or motors at a central location not on board the vehicles.

cable tray system
A unit or assembly of units or sections, and associated fittings, made of metal or other noncombustible materials forming a rigid structural system used to support cables. Cable tray systems include ladders, troughs, channels, solid bottom trays, and other similar structures.

cablebus
An approved assembly of insulated conductors with fittings and conductor termination in a completely enclosed, ventilated, protective metal housing.

cableway
A conveyor system in which carrier units run on wire cables strung between supports.

caboose
A car in a freight train intended to provide transportation for crew members.

cabotage
A law which requires coastal and intercoastal traffic to be carried by vessels belonging to the country owning the coast.

cacesthesia
Disordered sensibility.

cache
Computing. A pool of memory set aside to store items from a slower device, such as a hard disk. By using a cache, the computer can retrieve often-used information much faster than if it were required to find it on the hard disk each time.

cachexia
A state of malnutrition, emaciation, and debility, usually in the course of a chronic illness.

cackle crate
Truck that hauls live poultry.

CAD
Computer-aided design.

cadaver
The body of a deceased human.

cadmium
A chemical element, atomic number 48, atomic weight 112.40, symbol Cd.

cadmium sulfide
A light yellow or possibly orange powder used, in a 1 per cent suspension, in treatment of seborrheic dermatitis of the scalp (dandruff).

caduceus
The wand of Hermes or Mercury; used as a symbol of the medical profession and as the emblem of the Medical Corps of the U.S. Army.

Caduceus

CAE
See ***cost of accidents per employee***.

CAFÉ
See ***corporate average fuel economy standard***. *See also* ***fuel economy standard***.

cafeteria benefit plan
See ***cafeteria plan***.

cafeteria plan
A means of handling fringe benefits in which the employer allocates a certain amount of money to each employee for such benefits, and the employee is able to select the distribution of those benefits to his/her own best advantage. Also referred to as *cafeteria benefit plan*.

caffeine
A white powder, slightly soluble in water and having a bitter taste, found in coffee and tea.

It is an alkaloid and acts as a central nervous system stimulant and a mild diuretic.

cage
An enclosure that is fastened to the side rails of a fixed ladder or to the ladder's supporting structure to encircle the climbing space of the ladder for the safety of the person who must climb the ladder. May also be referred to as a *basket guard* or *cage guard*.

cage guard
See ***cage***.

CAI
See ***computer-aided instruction***.

caisson
Watertight structure used for underwater work.

caisson disease
Decompression sickness, a condition suffered by underwater workers and caused by a too rapid decrease in atmospheric pressure. The condition is named after the pressurized, watertight compartments (caissons) in which underwater construction personnel work. The main symptoms are dizziness, staggering, muscle spasms, difficulty in breathing, abdominal pain, and partial paralysis. Caisson disease is a form of the *bends*.

cake
Dehydrated sludge with a solids concentration sufficient to allow handling as a solid material.

cake filtration
Filtration classification for filters where solids are removed on the entering face of the granular media.

cal
See ***calorie***.

Cal
See ***Calorie***.

CAL-3D crash victim simulator (CAL-3D-CVS)
A computer modeling program for simulating the biomechanical responses of an individual in a vehicular crash.

calamine
A mixture of zinc and ferric oxides, used topically in lotions and ointments.

calamity
A state of extreme distress or misfortune, produced by some adverse circumstance or event. Any great misfortune or cause of loss or misery, often caused by natural forces (e.g., hurricane, flood, or the like). *See also* ***act of God***.

calandria
The heating element in an evaporator consisting of vertical tubes that act as the heating surface.

calcaneus
The heel bone.

calcareous
Composed of or containing calcium compounds, particularly calcium carbonate.

calcicosis
A lung disease due to the inhalation of marble dusts.

calcification
The deposit of calcium salts in a tissue. The normal absorption of calcium is facilitated by parathyroid hormone and by vitamin D. When there are increased amounts of parathyroid hormone in the blood (as in hyperparathyroidism), there is deposition of calcium in the alveoli of the lungs, the renal tubules, the thyroid gland, the gastric mucosa, and the arterial walls. Normally calcium is deposited in the bone matrix to ensure stability and strength of the bone. In osteomalcia, there is decalcification of bone because of a failure of calcium and phosphorus to be deposited in the bone matrix.

calcify
To become stone-like or chalky due to deposition of calcium salts.

calciner
A device in which the moisture and organic matter in phosphate rock is reduced in a combustion chamber.

calcining
Exposure of an inorganic compound to a high temperature to alter its form and drive off a substance that was originally part of the compound.

calciokinesis
Mobilization of calcium stored in the body.

calcium
A chemical element, atomic number 20, atomic weight 40.08, symbol Ca. Calcium is the most abundant mineral in the body. In combination with phosphorus, it forms calcium phosphate, the dense, hard material of the bones and teeth.

calcium carbonate
A white, chalky substance that is the principal hardness- and scale-causing compound in water. Chemical formula is $CaCO_3$.

calcium carbonate equivalent (mg/L as CaCO3)
A convenient unit of exchange for expressing all ions in water by comparing them to calcium carbonate, which has a molecular weight of 100 and an equivalent weight of 50.

calcium hypochlorite
A chlorine compound frequently used as a water or wastewater disinfectant. The chemical formula is Ca(OCl).

calcium sulfate
A white solid known as the mineral "anhydrite" with the chemical formula $CaSO_4$ and gypsum, which has the chemical formula $CaSO_4 \cdot H_2O$.

calculated
Adopted by calculation, forethought, or contrivance to accomplish a purpose; likely to produce a certain effect.

calculated landing time
Aviation. A term that may be used in place of tentative or actual calculated landing time, whichever applies.

Caldwell regimen
A procedure for static strength assessment, involving providing to the subject the details of the experiment and the necessary instructions, noting the posture and muscles involved, and having the subject maintain a four-second hold on the measuring device.

calendar year
The period of time between January 1 and December 31 of any given year.

calender
A machine which passes some pliable material between rollers or plates to make a relatively smooth, continuous or long sheet. Normally, this machine presents a clear nip point safety hazard.

calf
The fleshy part of the posterior lower leg, consisting largely of the gastrocnemius muscle.

calf circumference
The surface distance around the lower leg in a horizontal plane at the vertical level which gives the greatest value. Measured with the individual standing erect, with weight equally distributed on both feet.

calf circumference, recumbent
The calf circumference of a reclining individual. Measured with the individual supine, the knee and hip both flexed 90°, and the longitudinal axis of the foot perpendicular to that of the leg.

calf depth
The linear horizontal distance from the posterior surface to the anterior surface on the lower leg at the level of the calf circumference. Measured with the individual standing erect and weight equally distributed on both feet.

calf length
The linear distance parallel to the longitudinal axis of the lower leg between the knee joint level and the medial malleolus.

calibrate
Instrument. The adjustment or standardization of a measuring instrument. To adjust the span or gain of an instrument so that it indicates the actual concentration of a specific substance or mixture which is present at the sensor.

calibrated airspeed
The indicated airspeed of an aircraft, corrected for position and instrument error. Calibrated airspeed is equal to true airspeed in standard atmosphere at sea level.

calibration
Determination of variation from standard, or accuracy, of measuring instruments to ascertain necessary correction factors.

calibration gas
A gas of accurately known concentration which is used as a comparative standard in determining instrument performance and to adjust the instrument to indicate the true concentration.

California current
The ocean current that flows southward along the west coast of the United States from about Washington to Baja California.

California norther
A strong, dry, northerly wind that blows in late spring, summer, and early fall in northern and central California. Its warmth and dryness are due to downslope compression heating.

caliper
A device for obtaining accurate measurements of relatively short linear measures.

calisthenics
A form of exercise performed to improve strength, endurance, and/or grace.

call for release
Wherein the overlying Air Route Traffic Control Center (ARTCC) requires a terminal facility to initiate verbal coordination to secure Air Route Traffic Control Center (ARTCC) approval for release of a departure into the en route environment.

call out
A vocal method for presenting information to be heard by an individual. *See also* ***read-out***.

call up
Initial voice contact between a facility and an aircraft, using the identification of the unit being called and the unit initiating the call.

calorie (cal)
The amount of heat required to raise the temperature of one gram of water by one degree Celsius. The calorie used in nutrition and metabolism is spelled with a capital "c" as Calorie. *See also* ***Calorie***.

Calorie (Cal)
The unit for heat (energy) production in body nutrition and metabolism; equal to 1 Kcal. Also may be referred to as *kilocalorie* or *large calorie*. *See also* ***calorie***.

CALSPAN
A computer modeling program for simulating crash victim dynamics.

calumniator
In the civil law, one who accused another of a crime without cause; one who brought a false accusation.

calumny
Defamation; slander; false accusation of a crime or offense.

CAM
See ***continuous air monitor***.

camel back body
Truck body with floor curving downward at the rear.

camera ready
The detailed preparation of data (e.g., writing, drawings, figures, photographs, etc.) in a manner and format that is ready for immediate reproduction in printed form. This means that the materials to be printed must be absolutely error-free since there will be no opportunity for corrective action once they are printed.

camera study
See ***memomotion study***.

CAMP
Continuous air monitoring program.

camp car
Any on-track vehicle, including outfit, camp, or bunk cars or modular homes mounted on flat cars used to house rail employees. It does not include wreck trains.

campaign
Maritime. A Defect/Noncompliance Campaign Program carried out by the manufacturer and initiated under 46 U.S.C. 4310. Starts as a case. A campaign may involve only one boat.

Canadian Minimum Navigation Performance Specification Airspace
That portion of Canadian domestic airspace within which Minimum Performance Specifications Airspace (MNPSA) separation may be applied.

canal caps
A type of personal hearing protection which blocks noise from entering the external ear canal by placing a tight fitting cap over them.

canal/ditch
Artificial waterway used for navigation, drainage, or irrigation of land. An artificial open waterway constructed to transport water, to irrigate or drain land, to connect two or more bodies of water, or to serve as a waterway for watercraft. *See also* ***aqueduct***.

canard
The forward wing of a canard configuration that may be a fixed, movable, or variable geometry surface, with or without control surfaces.

canard configuration
A configuration in which the span of the forward wing is substantially less than that of the main wing.

cancellation
(1) *General.* To destroy the force, effectiveness, or validity of. To annul, abrogate, or terminate. Words of revocation written across an instrument. (2) *FIFRA.* Under the Federal Insecticide, Fungicide, and Rodenticide Act: The authorization to cancel a pesticide registration if unreasonable adverse effects to the environment and public health develop when a product is used according to widespread and commonly recognized practice, or if its labeling or other material required to be submitted does not comply with FIFRA provisions.

cancellation clause
A provision in a contract or lease which permits the parties to cancel or discharge their obligations thereunder.

cancellation of insurance
The withdrawal of insurance coverage by either the insurer or the insured.

cancellation test
A clerical aptitude test for speed and accuracy in crossing letters or numbers in a sequence.

cancellous bone
That interior portion of some bones which contains a criss-crossed matrix of calcified bone tissue with the remaining volume filled with marrow. Also called *spongy bone.*

cancer
A malignant neoplasm. Cancer is a neoplastic disease in which there is new growth of abnormal cells. Normally, the cells that compose body tissues grow in response to a normal stimulus. Worn-out body cells are regularly replaced by new cell growth which stops when the cells are replaced; new cells form to repair tissue damage and stop forming when healing is complete. Why they stop is unknown, but clearly the body in its normal processes regulates cell growth. In cancer, cell growth is unregulated. The cells continue to reproduce until they form a mass of tissue known as a tumor. Not all tumors are malignant; those which are non-cancerous are referred to as benign tumors. *Benign* tumors vary in size, and may grow so large that they obstruct organs or cause ulceration and bleeding. They are encapsulated, do not metastasize, and usually can be removed by surgery without difficulty. *Malignant* tumors grow in a disorganized fashion, interrupting body functions and robbing normal cells of their food and blood supply. The malignant cells may spread to other parts of the body by a) direct extension into adjacent tissue, b) permeation along lymphatic vessels, c) traveling in the lymph stream to the lymph nodes, d) entering the blood circulation, and/or e) invasion of a body cavity by diffusion.

cancer potency factor
See ***cancer slope factor.***

cancer slope factors (CSF)
Used in assessing toxicity in the risk assessment process. These factors are estimates of risk of developing cancer per unit of exposure and have units of 1/(mg/kg/d).

candela (cd)
A unit of luminous intensity equivalent to one lumen per square foot. Formerly called the *candle.* It is equal to the intensity of 555 nm (5.40×10^{14} Hz) point source radiating 1.464×10^{-3} watts per steradian. Also called *new candle.*

candle
An outdated term. A unit of luminous intensity equal to the intensity of light from a 7/8-inch sperm candle burning at the rate of 120 grains an hour.

candlepower
A luminous intensity expressed in *candelas;* a candle one inch in diameter produces one candela in a horizontal direction.

canister
A container filled with a sorbent and possibly catalysts, for removing contaminants (gases or vapors) from air being inspired through the device.

cannabis
Commonly referred to as marihuana, *cannabis sativa L* embraces all marihuana-producing

cannabis. All parts of the plant *cannabis sativa L,* whether growing or not; the seeds thereof; the resin extracted from any part of such plant; and every compound, manufacture, salt, derivative, mixture, or preparation of such plant, its seeds or resin, are included in the term "marihuana." *See also **controlled substance** and **marihuana**.*

canopy
(1) The large fabric part of a parachute which fills with air to slow the fall of an individual or object. (2) The transparent cover for the cockpit of an aircraft.

canopy hood
A one- or two-sided exhaust hood designed to capture contaminants or heated air rising from an open tank, placed some distance from the tank.

cant
In the civil law, a method of dividing property held in common by two or more joint owners. It may be avoided by the consent of all of those who are interested, in the same manner that any other contract or agreement may be avoided.

canthus
The corner or angle formed by the junction of the eyelids. *See also **endocanthus** and **ectocanthus**.*

cap
(1) A soft type of head wear which is preformed and sized. (2) A covering for a jar, bottle, or other rigid structure to contain the enclosed items or to prevent access by moisture, children, mold, or other entities. (3) A layer of clay, or other highly impermeable material, installed over the top of a closed landfill to prevent entry of rainwater and minimize production of leachate.

CAP
Chemical accident prevention. *See also **Control Assessment Protocol**.*

cap cloud
*See **pileus cloud**.*

cap lamp
A lamp worn by miners or others working in dark areas which is attached to a safety cap or helmet.

capable
Susceptible, competent; qualified; fitting; possessing legal power or capacity.

capacitance (C)
The value of the ratio between the absolute charge of two equal but opposite charged conductors to the potential difference between them.

capacitive touchscreen
A video display having a thin layer of material over its front which uses a change in capacitance to indicate a touch location.

capacity
(1) The upper limit of an individual's ability to learn, understand, or perform through inherent ability, training, practice, and any other means. (2) *See **endurance**.*

capacity defense
Generic term to describe lack of fundamental ability to be accountable for actions, as one under duress lacks the capacity to contract, and hence when sued on such contract he/she interposes defense of lack of capacity.

capacity factor
The ratio between the actual electric output from a unit and the potential electric output from that unit.

capacity per aircraft mile
The average total passenger/cargo carrying capacity (tons) offered for sale per aircraft in revenue services, derived by dividing the overall available ton miles by the total aircraft miles flown in revenue services.

CAPE
*See **Computerized Accommodated Percentage Evaluation**.*

cape
A relatively extensive land area jutting into a water body, which prominently marks a change in or notably interrupts the coastal trend of that water body.

capillarity
The ability of a soil to retain a film of water around soil particles and in pores through the action of surface tension.

capillary
A small, thin-walled blood vessel connecting an artery with a vein.

capillary action
The action or movement of surface water, or water in very small interstices, due to the relative attraction of molecules of a liquid for each other and for those of a solid.

capillary fringe
The zone of porous material above the zone of saturation that may contain water due to capillarity.

capital case (or crime)
One in or for which the death penalty may, but need not necessarily, be imposed.

capital employee
Transportation. An employee involved with construction or capital procurement and who has no involvement with operation of the transit system.

capital gains or losses, operating property
Gains or losses on retirements of operating property and equipment, equipment expendable parts or miscellaneous materials and supplies when sold or otherwise retired in connection with a general retirement program as opposed to incidental sales performed as a service to others.

capital gains or losses, other
Gains or losses on no operating assets, investments in other than marketable equity securities, and troubled debt restructuring.

capital program funds
Transportation. Financial assistance from the Capital Program of 49 U.S.C. (formerly Section 3). This program enables the Secretary of Transportation to make discretionary capital grants and loans to finance public transportation projects divided among fixed guideway (rail) modernization; construction of new fixed guideway systems and extensions to fixed guideway systems; and replacement, rehabilitation, and purchase of buses and rented equipment, and construction of bus-related facilities.

capitalized interest
Adjustment to income for interest capitalized on funds actually committed as equipment purchase deposits or actually used to finance the construction or acquisition of operating property.

capitate bone
One bone of the distal group of bones in the wrist.

capitulum
A smooth hemispherical protuberance at the anterior distal end of the humerus which forms part of the joint with the radius head.

capricious disregard
A willful and deliberate disregard of competent testimony and relevant evidence which one of ordinary intelligence could not possibly have avoided in reaching the result.

capsizing
Overturning of a vessel. The bottom must be uppermost, except on the case of a sailboat, which lies on its side.

captain of the port (COTP)
The officer of the Coast Guard, under the command of a District Commander, designated by the Commandant for the purpose of giving immediate direction to Coast Guard law enforcement activities within an assigned area. The term *captain of the port* includes an authorized representative of the captain of the port.

captive imports
Products produced overseas specifically for domestic manufacturers.

capture efficiency
The fraction of all organic vapors generated by a process that are directed to an abatement or recovery device.

capture gamma ray
A high energy gamma ray that is emitted when the nucleus of an atom captures a neutron and becomes intensely excited.

capture velocity
The air velocity at any point in front of a hood or at the hood opening necessary to overcome opposing or ambient air currents and to capture air contaminants at that point by causing them to flow into the hood. Also referred to as *control velocity*.

capturing hood
A hood with sufficient airflow to reach outside the hood and draw in contaminants.

CAR
See ***Computer Assessment of Reach***.

car
(1) Common term for an automobile. (2) Any unit of on-track equipment designed to be hauled by locomotives. (3) Any unit of on-track work equipment such as a track motor-car, highway-rail vehicle, push car, crane, ballast tamping machine, etc. (4) A railway car designed to carry freight, railroad personnel, or passengers. This includes boxcars, covered hopper cars, flatcars, refrigerator cars, gondola cars, hopper cars, tank cars, cabooses, stock cars, ventilation cars, and special cars. It also includes on-track maintenance equipment. *See also* ***automobile, minivan, motor vehicle, taxi,*** *and* ***vehicle***.

car capacity
Load limitation of a freight car in terms of volume or weight.

car-mile
The movement of a car a distance of one mile.

car shop repair track area
One or more tracks within an area in which the testing, servicing, repair, inspection, or rebuilding of railroad rolling equipment is under the exclusive control of mechanical department personnel.

car sickness
That motion sickness due to travel in a road vehicle. *See also* ***motion sickness***.

CARB
California Air Resources Board.

carbohydrate
A compound of carbon, hydrogen, and oxygen, the latter two in the proportions of water, synthesized by green plants. Carbohydrates in food are an important and immediate source of energy for the body. One gram of carbohydrate yields 4 calories. They are present, at least in small quantities, in most foods, but the chief sources are the sugars and starches. Carbohydrates may be stored in the body as glycogen for future use. If they are eaten in excessive amounts, however, the body changes them to fats and stores them in that form.

carbohydrate loading
The purposeful intake of large amounts of carbohydrates prior to a long-duration, physically fatiguing event in an attempt to generate additional glycogen reserves.

carbon
A chemical element, atomic number 6, atomic weight 12.011, symbol C. An element present in all materials of biological origin.

carbon-14
A naturally occurring radioactive isotope of carbon that emits beta particles when it undergoes radioactive decay.

carbon adsorber
An add-on device that uses activated carbon to adsorb volatile organic compounds from a gas stream. These compounds can later be recovered from the carbon for analysis.

carbon arc lamp
An arc lamp using carbon rods.

carbon black
An additive that prevents degradation of thermoplastics by ultraviolet light.

carbon cycle
A graphical presentation of the movement among living and nonliving matter.

carbon dioxide (CO_2)
(1) A minor component of air representing about 0.4% of the atmosphere that is released by respiration and removed from the atmosphere by photosynthesis. It is a non-combustible gas. (2) A fluid consisting of more than 90 percent carbon dioxide molecules compressed to a supercritical state.

carbon dioxide production rate ($\dot{V}co_2$)
That rate at which carbon dioxide is exhaled from the lungs.

carbon fixation
A process occurring in photosynthesis where atmospheric carbon dioxide gas is combined with hydrogen obtained from water molecules.

carbon monoxide (CO)
A colorless, odorless, and tasteless gas formed as a product of the incomplete combustion of organic materials. This gas has an affinity for red blood cells approximately 220 times that of oxygen which causes a decrease in pulmonary and cardiac function upon exposure. It is lethal to humans at concentrations exceeding 5000mg/L. An EPA-listed criterion pollutant.

carbon steel
A general-purpose steel whose major properties depend on its 0.1 to 2% carbon content

without substantial amounts of other alloying elements.

carbon tetrachloride
A clear, colorless, mobile liquid (formula CCl_4) once used widely as a solvent and in insecticides. It is highly toxic and even carcinogenic to the liver, kidney, and heart when persons are exposed to high concentrations. It is represented by the following structure:

```
        Cl
        |
  Cl—   C   —Cl
        |
        Cl
```

carbonaceous biochemical oxygen demand (CBOD)
The portion of biochemical oxygen demand where oxygen consumption is due to oxidation of carbon, usually measured after a sample has been incubated for 5 days. Also called *first stage BOD*.

carbonate
A compound containing the anion radical of carbonic acid CO_3.

carbonate alkalinity
Alkalinity caused by carbonate ions.

carbonate hardness
The hardness in water caused by bicarbonates and the carbonates of calcium and magnesium.

carbonation
The diffusion of carbon dioxide gas through a liquid.

carbonator
A device used to carbonate or re-carbonate water.

carbonemia
An excess of carbon dioxide in the blood.

carboxyhemoglobin (HbCO)
Hemoglobin in which the iron is associated with carbon monoxide (CO).

carboy
A large container used to store or transport liquid chemicals or water samples.

carburetor
A fuel delivery device for producing a proper mixture of gasoline vapor and air, and delivering it to the intake manifold of an internal combustion engine. Gasoline is gravity fed from a reservoir bowl into a throttle bore, where it is allowed to evaporate into the stream of air being inducted by the engine. The fuel efficiency of carburetors is more temperature dependent than fuel injection systems. *See also **diesel fuel system** and **fuel injection**.*

carcinogen
A substance known to cause cancer in humans and animals representing a broad range of organic and inorganic chemicals, hormones, immuno-suppresants, and solid-state materials.

carcinogen risk assessment verification endeavor (CRAVE) work group
An EPA work group formed to validate Agency carcinogen risk assessments and resolve conflicting potency values among various program offices.

carcinogenesis
The beginning or origin of a cancer.

carcinogenic
Describes agents known to induce cancers.

carcinoma
Malignant neoplasm composed of epithelial cells, regardless of the derivation. A form of cancer, carcinoma makes up the majority of the cases of malignancy of the breast, uterus, intestinal tract, skin, and tongue.

carcinosis
Widespread dissemination of cancer throughout the body.

cardholder
Refers to a member of a group such as a union wherein the card is the symbol and identification of membership.

cardia
That region of the superior stomach which contains the esophageal sphincter.

cardiac
Relating to the heart.

cardiac arrhythmia
An abnormality of the heart rhythm. Also commonly referred to as *arrhythmia* and *heart arrhythmia*.

cardiac index
The cardiac output per square meter of body surface area.

cardiac muscle
That branched, somewhat striated muscle comprising the wall of the heart which is involved in heart contractions.

cardiac output
The volume of blood pumped by the left ventricle of the heart in a given period of time. Usually expressed in liters per minute.

cardiac pacemaker
An electronic device which may be implanted in the body to provide regular stimulation to the heart.

cardiac reserve
That ability of the heart to increase its output above normal to meet an increased workload.

cardinal altitude
"Odd" or "even" thousand-foot altitudes or flight levels, e.g., 5,000, 6,000, 7,000, FL 250, FL 260, FL 270. *See also* ***flight level***.

cardinal planes
The three standard planes used for describing the human body in the anatomical position: sagittal, coronal/frontal, and transverse/horizontal.

cardiogram
A record produced by cardiography.

cardiograph
An instrument for recording the movements of the heart.

cardiography
The graphic recording of a physical or functional aspect of the heart, e.g., electrocardiography, kinetocardiography, phoncardiography, vibrocardiography.

cardiomyopathy
A sub-acute or chronic disorder of the heart muscle, often with involvement of the endocardium and sometimes of the pericardium.

cardiopulmonary
Pertaining to the heart and lungs.

cardiovascular
Pertaining to the heart, blood, or blood-carrying vessels.

cardiovascular shock
Any condition exemplified by a sudden fall in blood pressure following an injury, operation, loss of blood, or administration of anesthesia. Often referred to simply as shock.

cardiovascular system
The system of the human body, including the heart, vessels, and veins, associated with blood distribution and transmission of cellular nutrients.

care
Law. Watchful attention; concern; custody; diligence; discretion; caution; prudence; regard; preservation; security; support; vigilance. Opposite of negligence or carelessness. To be concerned with, and to attend to, the needs of oneself or another. In the law of negligence, the amount of care demanded by the standard of reasonable conduct must be in proportion to the apparent risk. As the danger becomes greater, the actor is required to exercise caution (care) commensurate with it. There are three degrees of care which are frequently recognized, corresponding (inversely) to the three degrees of negligence: slight care, ordinary care, and great care. There is also the concept of reasonable care. *See also* ***slight care, ordinary care, great care,*** *and* ***reasonable care***.

careless
Absence of care; negligence; reckless.

carelessness
That behavior or mental functioning which does not exhibit adequate attention or concern for the task being performed.

carfloat
(1) A vessel that operates on a short run on an irregular basis and serves one or more points in a port area as an extension of a rail line or highway over water, and does not operate in ocean, coastwise, or ferry service. (2) A barge equipped with tracks on which up to about 12 railroad cars are moved in harbors or inland waterways.

cargo
(1) The load (i.e., freight) of a vessel, train, truck, airplane, or other carrier. (2) Freight carried by a barge. (3) Property, mail and express; other than passengers transported. *See also* ***commodity, freight, goods,*** *and* ***product***.

cargo aircraft
An aircraft that is used to transport cargo and is not engaged in carrying passengers. The terms cargo aircraft only, cargo-only aircraft, and cargo aircraft have the same meaning.

cargo-carrying unit
Any portion of a commercial motor vehicle (CMV) combination (other than a truck tractor) used for the carrying of cargo, including a trailer, semitrailer, or the cargo-carrying section of a single-unit truck.

cargo crane
A crane especially adapted to the transferring of cargo between a vessel's hold and a wharf.

cargo insurance
Insures risk that cargo will not be delivered in the same condition in which it was initially shipped. *See also* ***insurance***.

cargo insurance and freight (CIF)
Refers to car-go for which the seller pays for the transportation and insurance up to the port of destination.

cargo tank
A bulk packaging which a) is a tank intended primarily for the carriage of liquids or gases and includes appurtenances, reinforcements, fittings, and closures; b) is permanently attached to or forms a part of a motor vehicle, or is not permanently attached to a motor vehicle but which, by reason of its size, construction, or attachment to a motor vehicle is loaded or unloaded without being removed from the motor vehicle; and c) is not fabricated under a specification for cylinders, portable tanks, tank cars, or multi-unit tank car tanks.

cargo tank motor vehicle
A motor vehicle with one or more cargo tanks permanently attached to or forming an integral part of the motor vehicle.

cargo ton miles
One ton of cargo transported one mile.

cargo tonnes
The tonnes of cargo carried by a vessel on each or any transit.

cargo transfer system
A component, or system of components functioning as a unit, used exclusively for transferring hazardous fluids in bulk between a tank car, tank truck, or marine vessel and a storage tank.

cargo vessel
(1) Any vessel other than a passenger vessel. (2) Any ferry being operated under authority of a change of character certificate issued by a Coast Guard Officer-in-Charge, Marine Inspection.

carload
(1) The quantity usually contained in an ordinary freight car used for transporting the particular commodity involved. A commercial unit which, by commercial usage, is a single whole for purposes of sale and division. (2) Shipment of freight required to fill a rail car. (3) A shipment of not less than 10,000 pounds of one commodity from one consignor to one consignee.

Carmack Act
Amendment to the Interstate Commerce Act prescribing liability of carrier for loss, damage, or injury to property carried in interstate commerce.

carpal
Pertaining to the carpus, or wrist.

carpal tunnel
An internal passage in the wrist between the extensor retinaculum and the carpal bones through which the median nerve, finger flexor tendons, and blood vessels pass from the arm to the hand.

carpal tunnel syndrome (CTS)
A cumulative trauma disorder (CTD) often associated with activities involving flexing or extending the wrists, or repeated force on the base of the palm and wrist. The *carpal tunnel* is an opening in the wrist under the carpal ligament on the palmar side of the carpal bones in the wrist. The median nerve, the finger flexor tendons, and blood vessels all pass through this tunnel. Overuse of the tendons can cause them to become inflamed and swollen, creating pressure against the adjacent median nerve and resulting in CST. Symptoms include tingling, pain, or numbness in the thumb and first three fingers. *See also* ***repetitive motion injury***.

carpenter's bracket scaffold
A scaffold consisting of wood or metal brackets supporting a platform.

carpenter's elbow
A type of cumulative trauma disorder (CTD) associated with repeatedly pushing the palm downward in such a way that a deviation of

the ulnar nerve occurs. Symptoms include pain in the elbow, foreman, and hand.

carpool
An arrangement where two or more people share the use and cost of privately owned automobiles in traveling to and from prearranged destinations together.

carpus
The eight bones composing the articulation between the hand and the forearm; the wrist.

carriage
Transportation of goods, freight, or passengers.

Carriage of Goods by Sea Act
Federal act governing the most important of the rights, responsibilities, liabilities, and immunities arising out of the relation of the issuer to the holder of the ocean bill of lading, with respect to loss or damage of goods.

carried carload
Any carload which travels on a particular railroad.

carrier
(1) An agent by which something is carried, especially an individual harboring pathogenic microorganisms and capable of transmitting them to others. (2) Individual or organization engaged in transporting passengers of goods for hire. (3) Any person engaged in the transportation of passengers or property by land, as a common, contract, or private carrier, or a freight forwarder, as those terms are used in the Interstate Commerce Act (as amended), and officers, agents, and employees of such carriers. *See also* ***common carrier, contract carrier***, *and* ***private carrier***.

carrier gas
Gases such as nitrogen, helium, argon, and hydrogen that are used in gas chromatography or other laboratory procedures to sweep (or "carry") another gas or vapor through a system.

carrier group
A grouping of certificated air carriers determined by annual operating revenues as follows: majors, >$1 billion; nationals, $100 million to $1 billion; large regionals, $20 million to $99.9 million; medium regionals, $0 to $19.9 million or that operate aircraft with 60 or less seats or maximum payload capacity of 18,000 lbs.

carrier liability
A common carrier is liable for all loss, damage, and delay with the exception of an act of God, act of a public enemy, act of a public authority, act of the shipper, and the inherent nature of the goods. Carrier liability is specified in the terms of the bill of lading.

carrier's lien
The right to hold the consignee's cargo until payment is made for the work of transporting it.

Carroll doctrine
Rule of law to the effect that an existing licensee has standing to contest the grant of a competitive licensee because of economic injury to an existing station becomes important when, on the facts, it spells diminution or destruction of service.

carrot and stick approach
The Federal Sentencing Guidelines (FSGs) for organizations address business conduct through this approach. The "stick" is a mandatory schedule of stiff fines, which can increase with the severity of the violation and can reach hundreds of millions of dollars. The "carrot" lies in the prospect of mitigation credit if, before an offense occurs, a company has instituted an effective program to prevent and detect violations of law. *See also* ***positive incentives, interactive corporate compliance, effective compliance program***.

carry costs
A verdict is said to carry costs when the party for whom the verdict is given becomes entitled to the payment of his/her costs incurred as a result of or as incident to such verdict.

carrying capacity
(1) *Recreation Management*. The amount of use a recreation area can sustain without deterioration of its quality. (2) *Wildlife Management*. The maximum number of animals an area can support during a given period of the year.

carryout collection
The collection of solid waste from a storage area proximate to the dwelling unit(s) or establishment.

cartage
Usually refers to intra-city hauling on drays or trucks.

Cartesian coordinate system
See ***rectangular coordinate system***.

cartilage
A tough, fibrous, non-vascular connective tissue frequently found at the articulating ends of bones or as a forming material in tubular structures in the body. Also referred to as *gristle*.

cartridge filter
(1) Filter unit with cylindrical replaceable elements or cartridges. (2) A small canister that is employed to remove contaminants from inspired air.

Carver-Greenfield process
A multiple effect evaporation process to extract water from sludge.

CAS
See ***Chemical Abstracts System***.

CAS number
Chemical Abstracts Service Registry number which is a unique identification number that is assigned to each chemical. The Chemical Abstract Service is a division of the American Chemical Society.

cascade impactor
A device used to measure the size range of airborne particles based on the principle that a high velocity air stream striking a flat surface at a 90° angle will cause a sudden change in air direction and momentum. This will also cause the dust in the air to be deposited on a plate and to be separated from the air stream. A series of plates are used to capture different-sized particles, which can then be analyzed for total weight, particle count, and chemical composition.

cascade method
An experimental technique for determining visual stimuli relationships in which an observer sequentially adjusts the wavelength of one of a pair of visual stimuli until a minimal difference exists.

cascading failure
Any secondary or other failure which results from the failure of another system or component.

case
(1) *Law.* A general term for an action, cause, suit, or controversy, at law or in equity; a question contested before a court of justice; an aggregate of facts which furnishes occasion for the exercise of the jurisdiction of a court of justice. A judicial proceeding for the determination of a controversy between parties wherein rights are enforced or protected, or wrongs are prevented or redressed; any proceeding judicial in its nature. (2) *Medical.* A particular instance of disease, as a case of leukemia; sometimes used incorrectly to designate the patient with the disease. (3) *Maritime.* a) An investigation of a particular boating problem or incident to determine if there is a substantial risk to the public or violation of the regulations. A case may become a campaign. b) A single incident of distress to which one or more Coast Guard units respond.

case control study
An epidemiology study which starts with the identification of individuals with a disease or adverse health effect of interest and a suitable control group without the disease.

case in chief
That part of a trial in which the party with the initial burden of proof presents his/her evidence, after which he/she rests.

case law
The aggregate of reported cases as forming a body of jurisprudence, or the law of a particular subject as evidenced or formed by the adjudged cases, in distinction to statutes and other sources of law. It includes the aggregate of reported cases that interpret statutes, regulations, and constitutional provisions.

casing
A pipe or tube placed in a bore hole to support the sides of the hole and to prevent other fluids from entering or leaving the hole.

cask
A thick-walled container (usually lead) used to transport radioactive material. Also called a *coffin*.

cassette
A light-proof housing for x-ray film, containing front and back intensifying screens, between which the film is placed.

cassette loop analysis
The selection of some videotaped task or operation with cutting and splicing or copying to form a continuous loop for repeated viewing. *See also* ***film loop analysis***.

cast iron
A general description for a group of iron-carbon-silicon metallic products obtained by reducing iron ore with carbon at temperatures high enough to render the metal fluid and cast it in a mold.

caster
A small, either fixed or swiveling, wheel attached to the base of an object for ease of movement across a surface.

casual employment
Employment at uncertain or irregular times. Employment for a short time and limited, temporary purpose. Occasional, irregular or incidental employment. Such employee does not normally receive seniority rights nor, if hours worked are below a certain number each week, fringe benefits. By statute in many states, such employment may or may not be subject to workers' compensation at the election of the employer. The test is the nature of the work or the scope of the contract of employment or the continuity of employment.

casualty
(1) *General.* A serious or fatal accident. A person or thing injured, lost, or destroyed. A disastrous occurrence due to sudden, unexpected, or unusual cause. Accident, misfortune or mishap; that which comes by chance or without design. A loss from such an event or cause; as by fire, shipwreck, lightning, etc. (2) *Federal Railroad Administration.* a) A fatality, a nonfatal injury, or an occupational illness resulting from railroad operations. b) A reportable death, injury, or illness arising from the operation of a railroad. Casualties may be classified as either fatal or nonfatal. *See also* ***accident, loss,*** *and* ***unavoidable casualty***.

casualty insurance
That type of insurance that is primarily concerned with losses caused by injuries to persons and legal liability imposed upon the insured for such injury or for damages to the property of others. *See also* ***insurance***.

casualty loss
A casualty loss is defined for tax purposes as the complete or partial destruction of property resulting from an identifiable event of a sudden, unexpected, or unusual nature (e.g., floods, storms, fires, auto accidents). Individuals may deduct business casualty losses in full. Losses include those in a trade or business or incurred in a transaction entered into for a profit. Personal or non-business casualty losses are deductible by individuals as itemized deductions.

casus fortuitus
Latin. An inevitable accident, a chance occurrence, or fortuitous event. A loss happening in spite of all human effort and sagacity.

CAT
See ***computerized axial tomography***. *See also* ***clear air turbulence***.

cat-scratch disease
An infection most frequently acquired through the scratch of a cat. It is actually sterile regional lymphadenitis. The disease is probably caused by a virus that is found between the claws of cats and kittens. Here, the virus usually does no harm to the cat, and the animal appears healthy. But a scratch may transfer the virus to a human being. In approximately half the cases, after several days there is a persistent sore at the site of the scratch, and fever and other symptoms of infection may develop. There is also swelling of the lymph nodes draining the infected part. In milder cases, the symptoms soon disappear, with no aftereffects. Sometimes the attack is more serious and the glands may require surgical incision and drainage. This disease is generally mild and lasts for about 2 weeks. In rare cases, it may persist for a period of up to 2 years.

catabolism
Destructive metabolism; the process by which an organism reconverts living, organized substances into simpler compounds, with release of energy for its use.

catalepsy
A movement disorder in which the body experiences a loss of voluntary motion and a rigidity of passively moved parts for prolonged periods of time.

catalyst
A substance which can alter the speed of a chemical reaction without being chemically altered itself by the reaction.

catalytic converter
An air pollution abatement device that removes pollutants from motor vehicle exhaust, either by oxidizing them into carbon dioxide and water or reducing them to nitrogen and oxygen.

catalytic incinerator
A control device which oxidizes volatile organic compounds by using a catalyst to promote the combustion process. Catalytic incinerators require lower temperatures than conventional thermal incinerators, with resultant fuel and cost savings.

catalytic sensor
Instruments. A sensor with heated active and reference elements (i.e., each a platinum wire). The heat of combustion of the contaminant on the active element produces an imbalance in a bridge circuit such that the amount of imbalance is proportional to the concentration of the contaminant in the sampled air. This type detector can detect and measure the concentration of combustible gases or vapors well below their lower flammable/combustible limit.

catanadramous
Fish that swim downstream to spawn.

cataplexy
A movement disorder characterized by rapid onset of partial or complete loss of muscle tone as a result of extremely intense emotion.

cataract
A clouding of the crystalline lens of the eye. The clouding obstructs the passage of light. Cataracts are caused by free-radical damage that clouds the eye's lens (which inhibits its ability to focus light) and reduces or scatters light entering the eye. Depending on their size and location, cataracts can reduce vision slightly or cause blindness.

catastrophe
A notable disaster; a more serious calamity than might ordinarily be understood from the term *casualty*. An utter or complete failure. An event resulting in injury, death, and damage or destruction of relatively great proportion. Often also considered relative to the scope of activities (i.e., at an individual or system level). *See also* ***casualty***.

catastrophic
A loss of extraordinary magnitude in physical harm to people or damage and destruction of property.

catastrophic event
System Safety. An occurrence, subsequent to the introduction of a hazard or set of hazards into a system, that results in a level of injury, damage, or loss of such severe magnitude that quick or total recovery would be highly improbable (e.g., death, crippling injuries, total system loss, irreplaceable property or equipment loss or damage, etc.). The parameters for this categorization are usually established by management in the System Safety Program Plan, or other policy-making documentation.

catastrophic release
According to OSHA, a major uncontrolled emission, fire, or explosion, involving one or more highly hazardous chemicals, that presents serious danger to employees in the workplace.

catatonia
A form of schizophrenia marked by conspicuous motor disturbances (retardation and stupor, or excessive activity and excitement).

catch basin
A well or reservoir for storm water runoff which can be located in paved areas, unpaved locations such as near roof drains, or in draining areas.

catchment
A barrel, cistern, or other container used to catch water.

catchment area
The area of land bounded by watersheds draining into a river, lake, or reservoir.

catch-up growth
A period of rapid growth following a growth retarding event, such as a severe illness or malnutrition.

catecholamine
Any of a group of chemical substances consisting of a benzene ring with adjacent hydroxyl groups and an amine group on a carbon chain which may serve as a

neurotransmitter and/or a hormone. *See also* ***epinephrine*** *and* ***norepinephrine***.

categorical exclusion
A class of actions which either individually or cumulatively would not have a significant effect on the human environment and therefore would not require preparation of an environmental assessment or environmental impact statement under the National Environmental Policy Act (NEPA).

categorical pretreatment standard
A technology-based effluent limitation for an industrial facility which discharges into a municipal sewer system. Analogous in stringency to Best Available Technology (BAT) for direct discharges.

category
Aviation. (1) As used with respect to the certification, ratings, privileges, and limitations of airmen, means a broad classification of aircraft. Examples include airplane, rotorcraft glider, and lighter-than-air. (2) As used with respect to the certification of aircraft, means a grouping of aircraft based upon intended use or operating limitations. Examples include transport, normal, utility, acrobatic, limited, restricted, and provisional.

category I contaminant
U.S. EPA contaminant category indicating that sufficient evidence of carcinogenicity via ingestion in humans or animals exists to warrant classification as "known or probable human carcinogens via ingestion."

category II contaminant
U.S. EPA contaminant category for which limited evidence of carcinogenicity via ingestion exists to warrant classification as "possible human carcinogens via ingestion."

category II operations
Aviation. With respect to the operation of aircraft, means a straight-in Instrument Landing System (ILS) approach to the runway of an airport under a Category II ILS instrument approach procedure issued by the Federal Aviation Administration (FAA) Administrator or other appropriate authority.

category III contaminant
U.S. EPA contaminant category of substances for which insufficient or no evidence of carcinogenicity via ingestion exists.

category III operations
Aviation. With respect to the operation of aircraft, means an Instrument Landing System (ILS) approach to, and landing on, the runway of an airport using a Category III ILS instrument approach procedure issued by the Federal Aviation Administration (FAA) Administrator or other appropriate authority.

category A
Aviation. With respect to transport category rotorcraft, means multi-engine rotorcraft designed with engine and system isolation features specified in 14 CFR Part 29 and utilizing scheduled takeoff and landing operations under a critical engine failure concept which assures adequate designated surface area and adequate performance capability for continued safe flight in the event of engine failure.

category A EED/ordnance
An electroexplosive device (EED) or other such ordnance that, by the expenditure of their own energy or because they initiate a chain of events, may cause serious injury or death to personnel and/or damage to property.

category B
Aviation. With respect to transport category rotorcraft, means single-engine or multi-engine rotorcraft which do not fully meet all Category A standards. Category B rotorcraft have no guaranteed stay-up ability in the event of engine failure and unscheduled landing is assumed.

category B EED/ordnance
An electroexplosive device or other such ordnance that, by the expenditure of their own energy or because they initiate a chain of events, will not cause serious injury or death to personnel and/or damage to property.

catenary bar screen
Mechanical screening device using revolving chain-mounted rakes to clean a stationary bar rack.

catheter
A tubular instrument of rubber, plastic, metal, or other material, used for draining or injecting fluids through a body passage.

catheterization
Passage of a catheter into a body channel or cavity. The most common usage of the term

is in reference to the introduction of a catheter via the urethra into the urinary bladder.

cathode
The negative electrode where the current leaves an electrolytic solution.

cathode ray tube (CRT)
A vacuum tube in which an electron beam is directed at a phosphor-coated screen. The component of a video display terminal that generates the display.

cathodic protection
A technique to prevent corrosion of a metal surface by making that surface the cathode of an electrochemical cell.

cation
A positively charged ion that migrates to the cathode when an electrical potential is applied to a solution.

cationic detergent
Any of a group of detergents having a quaternary ammonium salt cation with a hydrocarbon chain.

cationic polymer
A polyelectrolyte with a net positive electrical charge.

cauliflower ear
A thickened and deformed ear caused by the accumulation of fluid and blood clots in the tissue following repeated injury. It is most commonly seen in boxers, for whom it is almost considered an occupational hazard. A cauliflower ear will not recover its normal shape but it can be restored to normal by means of plastic surgery.

causal association
Having a demonstrable connection between the occurrence of some factor and an incident, where the presence of that factor will increase the probability and the absence of that factor will decrease the probability of that incident.

causal factors
A combination of simultaneous or sequential circumstances which contribute directly or indirectly to an accident, occupational disease, or other effect.

causalgia
Persistent, diffuse, and burning pain associated with tropic skin changes in the hand or foot following injury of the part. The syndrome may be aggravated by the slightest stimulus or it may be intensified by emotions. Causalgia usually begins several weeks after the initial injury and the pain is described as intense.

causation
The fact of being the cause of something produced or of happening. The act by which an effect is produced. An important doctrine in the fields of negligence and criminal law. Sometimes referred to as *proximate cause.*

causator
A litigant; one who takes the part of the plaintiff or defendant in an action.

cause
(1) *General (verb).* To be the cause or occasion of; to effect as an agent; to bring about; to bring into existence; to make to induce; to compel. (2) *Safety (noun).* An event, situation, or condition which results, or could result *(potential cause),* directly or indirectly, in an accident or incident. Each separate antecedent of an event. Something that proceeds and brings about an effect or result. (3) *Law (noun).* A suit, litigation, or action. Any question, civil or criminal, litigated or contested before a court of justice.

cause-effect diagram
A graphic display of the causes linked to an effect.

cause in fact
That particular cause which produces an event and without which the event would not have occurred. Courts express this form of a rule commonly referred to as the "but for" rule: the injury to an individual would not have happened but for the conduct of the wrongdoer. *See also* ***proximate cause.***

cause of action
The claim or theory invoked by a plaintiff in a court case.

cause of injury
That which actually produces it.

causeway
A raised roadbed through low lands or across wet ground or water.

caustic
(1) Any substance that strongly irritates, burns, corrodes, or destroys living tissue. (2)

A class of substances, also known as bases or alkalis, with high pH. Strong caustics are corrosive.

caustic soda
Common term for sodium hydroxide. Chemical formula is NaOH.

caution
To warn, exhort, to take heed, or give notice of danger. *See also* ***caution signal***.

caution and warning (C&W)
A system of classification for providing information to the operator or crew of a vehicle that some life- or vehicle-threatening hazardous situation exists.

caution signal
A signal provided for or presented to the operator or crew of a vehicle that some hazardous condition exists or will soon exist, and that action will be required to correct the situation. *See also* ***caution***.

cautionary instruction
Law. That part of a judge's charge to a jury in which he/she instructs them to consider certain evidence only for a specific purpose, e.g., evidence that a criminal defendant committed crimes other than the crime for which he/she is on trial may be admitted to prove a scheme or to show intent as to this crime, but not to prove that he/she committed this particular crime and such evidence requires cautionary instructions. Also, instructions by a judge to a jury to not be influenced by extraneous matters on outside forces, or to talk about cases to anyone outside of trial.

caval
See ***kevel***.

caveat
Latin. Meaning "let him beware." A warning to one to be careful. A formal notice or warning given by an interested party to a court, judge, or ministerial officer against the performance of certain acts within his/her power and jurisdiction.

caveat emptor
Meaning "let the buyer beware," without a warranty the buyer takes the risk of quality upon himself/herself.

caveat venditor
Meaning "let the seller beware."

caving bank
A bank which is eroding because of swift running currents along the shore or because of eddies below bends or along the shoreline whether on right- or left-handed drafts. More prevalent on rivers with unstable channels and during periods of high water; very common occurrence on the Lower Mississippi River.

cavitation
(1) A selective corrosion that results from the collapse of air or vapor bubbles with sufficient force to cause metal loss or pitting. (2) The action of a pump attempting to discharge more water than suction can provide. Vibration, noise, and/or physical damage to equipment can result.

CBA
See ***cost-benefit analysis***.

CBC
See ***complete blood count***.

CBD
See ***Central Business District***.

CBOD
See ***carbonaceous biochemical oxygen demand***.

CCB
Coal combustion byproducts.

CCL
See ***convective condensation level***.

CDC
See ***Centers for Disease Control***.

CDL
See ***commercial driver's license***.

CDLIS
See ***commercial driver's license information system***.

cd/m^2
Candela per square meter.

CDT programs
See ***controlled departure time programs***.

Ce
See ***coefficient of entry***.

cease
To stop; to become extinct; to pass away; to suspend, or to forfeit.

cease and desist order
An order issued by an administrative agency or court prohibiting a person or business firm from continuing a particular course of conduct.

CEEL
Community emergency exposure limit.

CEF
Cellulose ester filter.

ceiling
(1) The upper limit of performance measured by a test. (2) The upper interior surface of a large enclosed volume, such as a room. (3) The height of the lowest layer of clouds when the weather reports describe the sky as broken or overcast, or, the height above the earth's surface of the lowest layer of clouds or obscuring phenomena that is reported as "broken", "overcast", or "obscured", and not classified as "thin" or "partial." The height above the ground or water of the base of the lowest layer of the cloud below 6,000 meters (20,000 feet) covering more than half the sky.

ceiling area lighting
A form of general illumination in which the ceiling area comprises essentially one large luminaire.

ceiling balloon
A small balloon used to determine the height of the cloud base. The height is computed from the balloon's ascent rate and the time required for its disappearance into the cloud.

ceiling exposure limit (CEL)
(1) The absolute concentration of a chemical to which workers should never be exposed, even instantaneously, during any part of a working day. (2) An OSHA standard setting the maximum concentration of a contaminant to which a worker may be exposed. (3) The ACGIH has established ceiling limits for some substances as part of its threshold limit value table (TLV-C).

ceiling plenum
Space below the flooring and above a suspended ceiling that accommodates the mechanical and electrical equipment and that is used as part of the air distribution system. The space is kept under negative pressure.

ceilometer
An instrument that automatically records cloud height.

CEL
See ***ceiling exposure limit***.

cell
(1) The fundamental unit of structure and function in organisms. Although cells may be highly differentiated and highly specialized in their function, they all have the same basic structure; that is, they have an outer covering called the membrane, a main substance called the cytoplasm, and a control center called the nucleus. The cytoplasm and the substance of the nucleus (nucleoplasm or karyoplasm) are collectively referred to as protoplasm. Cell membranes are capable of selection in the passage of substances into and out of the cell. Cells in the body are organized into tissue and tissues into organs. The fluid within the cell (60 to 90 percent of the protoplasm is water) is called intracellular fluid. The fluid surrounding the cell and within the tissues is called interstitial fluid or tissue fluid. The molecules and ions in these fluids are essential to the life of the cell. (2) In solid waste disposal, cells are holes where waste is dumped, compacted, and covered with layers of dirt on a daily basis. (3) A place of confinement, as a prison cell.

cell life
Instrument. The period over which an instrument detector can reasonably be expected to meet the performance specifications for the device.

cellular refractory period
That time following an action potential in a neuron or muscle cell during which the cell has reduced excitability or is incapable of normal excitation. Also referred to simply as the refractory period.

cellulitis
An inflammation of tissues that produces pain, edema, swelling, and functional difficulties. It may be caused by streptococcal, staphylococcal, or other organisms. It usually occurs in the loose tissues beneath the skin, but may also occur in tissues beneath mucous membranes or around muscle bundles or surrounding organs.

cellulose
The structural form of polysaccharides in plants, acting as a support for plant tissues.

Celsius degree (°C)
A division of the Celsius temperature scale which divides the range between the freezing and boiling points of water into 100 equal intervals. Synonymous with *Centigrade degree.*

Celsius thermometer
Centigrade thermometer on which the ice point is at 0 and the normal boiling point of water is at 100 degrees (100°C), with the interval between these two established points divided into 100 equal units. The abbreviation 100°C should be read "one hundred degrees Celsius." It relates to the absolute (Kelvin) scale by $T_C = T_K - 273.15$.

CEMA
Conveyor Equipment Manufacturers Association.

cement
A powder that, mixed with water, binds a stone and sand mixture into strong concrete when dry.

cement dermatitis
An inflammation of the skin's surface resulting from an allergic reaction on exposure to cement or cement mixing products. *See also* ***industrial dermatitis***.

cement kiln dust
Alkaline material produced during the manufacture of cement that may be used to stabilize sludge.

cement mixer
Transportation (slang). A truck with a noisy engine or transmission.

cementing
The process of pumping a cement slurry into a drilled hole and/or forced behind the casing.

cementitious material
Asbestos-containing materials that are densely packed and are non-friable.

CEMS
See ***continuous emission monitoring system***.

censored data
Monitoring results that are non-quantified because they are less than the limit of detection.

censure
The formal resolution of a legislative, administrative, or other body reprimanding a person, normally one of its own members, for specified conduct. An official reprimand or condemnation.

census
The complete enumeration of a population or groups at a point in time with respect to well-defined characteristics. For example, population, production, and/or traffic on particular roads. In some connection the term is associated with the data collected rather than the extent of the collection so that the term *sample census* has a distinct meaning. The partial enumeration resulting from a failure to cover the whole population, as distinct from a designed sample inquiry, may be referred to as an "incomplete census."

census division
A geographic area consisting of several states defined by the U.S. Department of Commerce, Bureau of the Census. The states are grouped into nine divisions and four regions.

center area
Aviation. The specified airspace within which an Air Route Traffic Control Center (ARTCC) provides air traffic control and advisory service.

center frequency
The geometric mean of a frequency band.

center of gravity
A point representing a body or system at which the force due to a uniform gravitational attraction acts. With regard to human factors, the *center of mass* and center of gravity can normally be assumed to be the same point. *See also* ***center of mass***.

center of mass
That point of an object or system which may be treated as if the entire mass of the object or system were concentrated at that point, and any external translational forces appear to act through that point. With regard to human factors, the center of mass and the *center of gravity* can normally be assumed to be the same point. *See also* ***center of gravity***.

center of rotation
That point about which a rotational movement occurs.

center pivot irrigation machine
A center pivot irrigation machine is a multi-motored irrigation machine which revolves around a center pivot point and employs alignment switches or similar devices to control individual motors.

Center Radar Approach Control (CERAP)
A combined Air Route Traffic Control Center (ARTCC) and a Terminal Radar Approach Control facility (TRACON).

center weather advisory (CWA)
An unscheduled weather advisory issued by Center Weather Service Unit meteorologists for Air Traffic Control (ATC) use to alert pilots of existing or anticipated adverse weather conditions within the next 2 hours. A CWA may modify or redefine a SIGMET.

Centers for Disease Control (CDC)
A U.S. Department of Health agency responsible for surveillance of disease patterns, developing disease control and prevention procedures, and public health education.

centi-
A prefix; one one-hundredth or 10^{-2} of a base unit.

centigrade
Having 100 gradations (steps or degrees), as in the Celsius temperature scale (thermometer). *See also* ***Celsius thermometer***.

centigrade degree
See ***Celsius degree***.

centile point
A point within a centile scale. *See also* ***centile scale***.

centile rank
That position or score based on a centile scale. *See also* ***percentile***.

centile scale
A dispersion scale having a range of 100 in which each point represents one percent of the population along some dimension.

centimeter
One-hundredth of a meter, abbreviated cm.

centimeter-gram-second (CGS) system
A coherent system of units for mechanics, electricity, and magnetism, in which the basic units of length, mass, and time are the centimeter, gram, and second. Sometimes referred to as the *CGS System*. *See also* **basic units**.

centipoise (cp)
(1) One one-hundredth of a poise. The poise is the metric system unit of viscosity. *See also* ***poise***. (2) A unit of the dynamic viscosity of a liquid. The dynamic viscosity of water at 20°C is 1 centipoise.

centistoke (cSt)
One one-hundredth of a stoke, the kinematic unit of viscosity. It is equal to the viscosity in poise divided by the density of the fluid in grams per cubic centimeter, both measured at the same temperature.

central blindness
The lack of visual function due to optic nerve or visual cortex damage. *See also* ***foveal blindness***.

Central Business District (CBD)
The downtown retail trade and commercial area of a city or an area of very high land valuation, traffic flow, and concentration of retail business offices, theaters, hotels, and services.

central city
Usually one or more legally incorporated cities within the metropolitan statistical area (MSA) that is significantly large by itself or large relative to the largest city in the MSA. Additional criteria for being classified as a "central city" include having at least 75 jobs for each 100 employed residents and having at least 40 percent of the resident workers employed within the city limits. Every MSA has at least one central city, usually the largest city. Central cities are commonly regarded as relatively large communities with a denser population and a higher concentration of economic activities than the outlying or suburban areas of the MSA. "Outside central city" are those parts of the MSA not designated as central city. *See also* ***metropolitan statistical area (MSA),*** *and* ***standard metropolitan statistical area***.

central deafness
See ***central hearing loss***.

Central East Pacific
Aviation. An organized route system between the U.S. West Coast and Hawaii.

central hearing loss
A hearing impairment of deafness due to auditory nerve or auditory cortex damage. Synonymous with *central deafness*.

Central Intelligence Agency (CIA)
An agency of the federal government charged with the responsibility of coordinating all information relating to security of the country. All such intelligence information, recommendations, etc. are reported to the National Security Council, to whom the CIA is responsible and under the direction of.

central nervous system (CNS)
The portion of the human control and sensory feedback system consisting of the brain, spinal cord, and tributary nerve endings.

central tendency
Having a typical, average, or expected value within a frequency distribution. A finer characterization of data beyond the distribution.

central vision
See ***foveal vision***.

central visual field
That portion of the visual field which falls on the foveal or macula lutea portion of the retina. Opposite of *peripheral visual field*.

central visual field blindness
See ***foveal blindness***.

centralization
Concentration of power and authority in a central organization or government. For example, power and authority over national and international matters are centralized in the federal government.

centrate
Dilute stream remaining in a centrifuge after solids are removed.

centrifugal
Moving away from a center.

centrifugal collector
A mechanical system using centrifugal force to remove aerosols from a gas stream.

centrifugal force
That outwardly directed radial force in a rotating reference frame. Opposite of *centripetal force*.

centrifugal pump
A pump with a high-speed impeller that relies on centrifugal force to throw incoming liquid to the periphery of the impeller housing where velocity is converted to head pressure.

centrifugation
The use of centrifugal force to separate solids from liquids based on density differences.

centrifuge
A laboratory device used to subject substances in solution to centrifugal forces 20,00-25,000 times gravity.

centripetal acceleration
The inward-directed acceleration on a particle moving in a curved path.

centripetal force
That radial force directed toward the center of rotation of an object which keeps an object moving in a circular path. Opposite of *centrifugal force*.

CEO
Abbreviation for Chief Executive Officer of a corporation.

CEQ
See ***Council on Environmental Quality***.

CEQ regulations
Under the National Environmental Policy Act (NEPA), the regulations called for federal agencies to integrate NEPA requirements to ensure that plans and decisions reflect environmental values, avoid delays later in the process, and head off potential conflicts. *See also* ***Council on Environmental Quality***.

CERAP
See ***Center Radar Approach Control***.

CERCLA
See ***Comprehensive Environmental Response, Compensation, and Liabilities Act***.

CERCLIS
A database maintained by the EPA and the states which lists sites where releases of contamination either have been addressed or need to be addressed.

cerebellum
That part of the hind-brain lying dorsal to the pons and medulla oblongata, comprising a median portion (the vermis) and a cerebellar hemisphere on each side; the cerebellum is concerned with coordination of movements.

cerebral
Pertaining to the cerebrum.

cerebral cortex
The outer layer of gray matter of the brain, which governs thought, reasoning, memory, sensation, and voluntary movement.

cerebral palsy
Any impairment of motor, perceptual, or behavioral function dating from birth or infancy without worsening of symptoms.

cerebrospinal fluid (CSF)
The fluid in the subarachnoid spaces surrounding the brain and spinal cord, and in the ventricles of the brain. The fluid is formed continuously by the choroid plexus in the ventricles, and, so that there will not be an abnormal increase in the amount and pressure, it is reabsorbed into the blood by the arachnoid villi at approximately the same rate at which it is produced. The fluid aids in the protection of the brain, spinal cord, and meninges by acting as a watery cushion surrounding them to absorb the shocks to which they are exposed. There is a blood-cerebrospinal fluid barrier that prevents harmful substances, such as metal poisons, some pathogenic organisms, and certain drugs from passing from the capillaries into the cerebrospinal fluid.

cerebrum
The main portion of the brain occupying the upper part of the cranium; the two cerebral hemispheres, united by the corpus callosum, form the largest part of the central nervous system in humans.

CERES
See ***Coalition for Environmentally Responsible Economies***.

cerium
A chemical element, atomic number 58, atomic weight 140.12, symbol Ce.

CERMS
Continuous emissions rate monitoring system.

certainty
Absence of doubt; accuracy; precision; definite. The quality of being specific, accurate, and distinct.

certificate
A written assurance, or official representation, that some act has or has not been done, or some event occurred, or with which some legal formality has been compiled. A written assurance made or issuing from some court, and designed as a notice of things done therein, or as a warrant or authority, to some other court, judge, or officer. A statement of some fact in a writing signed by the party certifying.

certificate of destruction
See ***certificate of disposal***.

certificate of disposal (COD)
A document that verifies destruction, the receipt of destruction, or successful delivery of waste to an ultimate or intermediary location prior to destruction. While these documents are considered by some transporters and generators as legal proof of the end of their liability for contamination, they may only serve to prove responsibility for contribution to a site that is later identified by the EPA as a Superfund site. Also referred to as *certificate of destruction*.

certificate of inspection
A document certifying that merchandise (such as perishable goods) was in good condition immediately prior to shipment. Pre-shipment inspection is a requirement for importation of goods into many developing countries.

certificate of origin
A form showing the country of production of export shipment, frequently required by customs officials of an importing country. The certificate enables customs officials to determine which goods being imported are entitled to preferential tariff treatment. It is usually endorsed by a consular official of the country of destination at the port of shipment.

Certificate of Public Convenience and Necessity
(1) *Aviation.* A certificate issued to an air carrier under Section 401 of the Federal Aviation Act by the U.S. Department of Transportation (DOT) authorizing the carrier to engage in air transportation. (2) *Maritime.* The grant of operating authority (issued by the Interstate Commerce Commission (ICC) and the Federal Maritime Commission (FMC) that is given to common carriers. A carrier must prove that a public need exists and that the carrier is fit, willing, and able to provide the needed service. The certificate may specify the commodities to be hauled, the area to be served, and the routes to be used.

certificated
Holding a currently valid Certificate of Public Convenience and Necessity.

certificated air carrier
An air carrier holding a Certificate of Public Convenience and Necessity issued by DOT to conduct scheduled services interstate. Non-scheduled or charter operations may also be conducted by these carriers. These carriers operate large aircraft (30 seats or more or a maximum payload capacity of 7,500 pounds or more) in accordance with Federal Aviation Regulations (FAR) Part 121.

certificated airport
An airport operating under Federal Aviation Regulations (FAR) Part 139. The Federal Aviation Administration (FAA) issues airport operating certificates to all airports serving scheduled air carrier aircraft designed for more than 30 passenger seats. Certificated airports must meet minimum safety standards in accordance with Federal Aviation Regulations (FAR) Part 139.

certification
(1) Granted by some states to certain laboratories; ensures that laboratories meet certain minimum standards. (2) A manufacturer's statement that the boat they manufacture is subject to the Federal regulations indicated in the certification statement and has been designed and constructed to comply with those regulations.

certification of labor union
Declaration by a labor board (such as the National Labor Relations Board) that a union is the bargaining agent for a group of employees.

certification of public road mileage
An annual document (certification) that must be furnished by each state to the Federal Highway Administration (FHWA) certifying the total public road mileage (kilometers) in the state as of December 31 of the preceding year.

certification to federal court
Method of taking a case from the U.S. Court of Appeals to the Supreme Court in which the former court may certify any question of law in any civil or criminal case as to which instructions are requested.

certified
(1) *Equipment.* Equipment is considered "certified" if it a) has been tested and found by a nationally recognized testing laboratory to meet nationally recognized standards or to be safe for use in a specified manner, or b) is of a kind whose production is periodically inspected by a nationally recognized testing laboratory, and c) it bears a label, tag, or other record of certification. (2) *Personnel.* Persons who have met a professional evaluation process or standard (e.g., by examination, experience, education, etc.) of a specific certification agency or authority to receive the designation of "certified."

certified capacity
The capability of a pipeline project to move gas volumes on a given day, based on a specific set of flowing parameters (operating pressures, temperature, efficiency, and fluid properties) for the pipeline system as stated in the dockets filed (and subsequently certified) in the application for the Certificate of Public Convenience and Necessity at the Federal Energy Regulatory Commission. Generally, the certificated capacity represents a level of service that can be maintained over an extended period of time and may not represent the maximum throughput capability of the system on any given day. *See also* ***design capacity***.

certified carriers
Carriers using highways of a state to whom certificates of public convenience and necessity have been issued.

certified gas-free
When a tank, compartment, or container on a vessel is certified gas-free it means that it has been tested using an approved testing instrument, and proved to be sufficiently free, at the time of the test, of toxic or explosive gases for a specified purpose, such as hot work, by an authorized person and that a certificate to this effect has been issued.

certified glazing
Railroad. A glazing material applied to railroad passenger car windows that has been certified by the manufacturer as having met the testing requirements set forth in Appendix A of 49 CFR 223 and that has been installed in such a manner that it will perform its intended function.

Certified Health Physicist (CHP)
An individual who has been certified in this discipline by the American Board of Health Physicists.

Certified Industrial Hygienist (CIH)
An industrial hygienist who has met the education, experience, and examination requirements of the American Board of Industrial Hygiene and possess current ABIH certification as an industrial hygienist (i.e., has been certified as competent in one or more aspects of this discipline by the American Board of Industrial Hygiene).

Certified Safety Professional (CSP)
A professional safety practitioner who has been certified in one or more aspects of this discipline by the Board of Certified Safety Professionals.

cerumen
A waxy secretion of the glands of the external acoustic meatus; ear wax.

cervical
Pertaining to the neck or to the cervix.

cervical spine
That portion of the spinal column consisting of the seven cervical vertebrae in the neck.

cervicale
The protruding tip of the 7th cervical vertebrae at the base of the neck. Also known as the nuchale tubercle.

cervicale height
The vertical linear distance from the upper sitting surface to cervicale. Measured with the individual sitting erect. *See also* ***cervicale***.

cervix
A constricted structure in the body, typically referring to the neck or the narrow part of the uterus.

cesium
A chemical element, atomic number 55, atomic weight 132.905, symbol Cs.

cesspool
A covered tank with open joints constructed in permeable soil to receive raw domestic wastewater and allow partially treated effluent to seep into the surrounding soil, while solids are contained and undergo digestion.

CET
Certified Environmental Trainer. *See also* ***corrected effective temperature***.

CFB
Circulating fluidized bed.

CFC
See ***chlorofluorocarbon***.

cfm
See ***cubic feet per minute***.

cfm/sq ft
Cubic feet per minute per square foot.

CFR
See ***Code of Federal Regulations***.

cfs
Cubic feet per second.

CFU
See ***colony-forming units***.

CFU/m^3
Colony-forming units per cubic meter.

CGA
See ***Compressed Gas Association***.

CGI
See ***combustible gas indicator***.

CGMP
Current good manufacturing practice.

CGS System
See ***centimeter-gram-second system***. *See also* ***basic units***.

chafe
To irritate the skin through friction.

chafed
Wear damage resulting from friction between two parts rubbed together with limited and usually repeated motion.

chaff
Aviation. Thin, narrow metallic reflectors of various lengths and frequency responses, used to reflect radar energy. These reflectors, when dropped from aircraft and allowed to drift downward, result in large targets on the radar display.

chafing
Irritation of the skin by friction, usually from clothing or the rubbing together of body surface, such as the thighs, when they are damp with perspiration.

chain and flight collector
A sludge collector mechanism utilized in rectangular sedimentation basins or clarifiers.

chain of custody
In legal terms, regulatory agencies as well as employers must be able to verify the chain of possession and custody of any physical samples (air, water, soil, biological, etc.) that may be used to support litigation. Procedures to ensure this chain-of-custody include written records that can be used to trace possession and handling of the sample from its point of origin through analysis and its introduction as evidence. Without a continuous record of chain-of-custody, the validity of any sample or the results of any tests/analyses may be questioned.

chain of custody form
A form used for tracking samples from the time the samples are obtained, through their transportation, receipt at the laboratory, and analysis.

chain of infection
A series of related factors or events that must occur before an infection will occur. These factors can be identified as host, agent, source, and transmission factors.

chain of title
Record of successive conveyances, or other forms of documentation, affecting a particular parcel of land, arranged consecutively, from the government or original source of title down to the present holder.

chain reaction
(1) In chemical or nuclear processes, the energy or byproducts released cause a continuation of the process. (2) In the analysis of accident cause, the sequence of events that resulted in the accident. *See also* ***domino effect***.

chains and links
As used in real estate measurement, chain is equal to 66 feet long or 100 links.

chair
(1) *General.* A place for sitting. (2) *Administration.* The person designated as "in charge" or responsible for the actions and output of a committee or group. (3) *Transportation. See* ***sidecar***.

chalazion
A cyst or tumor on the eyelid caused by an infection of a sebaceous (oil) gland. A chalazion can sometimes be treated at home with the application of hot compresses, but while this method is usually successful with a sty, a similar infection that has not yet formed a cyst, chalazion often requires incision and drainage be performed by a physician. Also called *meibomian cysts*.

challenge
To object or take exception to; to proffer objections to a person, right, or instrument; to question formerly the legality or legal qualifications of; to invite into competition; to formerly call into question the capability of a person for a particular function, or the existence of a right claimed, or the sufficiency of validity of an instrument; to call or put in question; to put into dispute; to render doubtful.

chamfer
The bevel at the end of an object.

chance
The absence of explainable or controllable causation; accident; fortuity; hazard; result or issue of uncertain and unknown conditions or forces; risk; unexpected, unforeseen, or unintended consequences of an act.

chance variable
See ***random variable***.

change of grade
Usually understood as an elevation or depression of the surface of a street, or a change of the natural contour of its face so as to facilitate travel over it.

change of venue
The removal of a suit begun in one county or district to another county or district for trial, though the term is also sometimes applied to the removal of a suit from one court to another court of the same county or district. In criminal cases, a change of venue will be permitted if, for example, the court feels that the defendant cannot receive a fair trial in a given venue because of prejudice.

changeover
The process of modifying or replacing an existing workstation, workplace, or other facility, including the setup and tear-down.

changeover allowance
A special time allowance given a worker to compensate for the changeover time. *See also* ***setup allowance*** *and* ***tear-down allowance.***

changeover time
That temporal period required to affect a changeover.

channel
The bed in which the main stream of a river flows, rather than the deep water of the stream as followed in navigation. The deeper part of a river, harbor, or strait. It may also be used as a generic term applicable to any water course, whether a river, creek, or canal. The channel of a river is to be distinguished from a branch.

channel bottom
Project depth or grade elevation.

channel capacity
The maximum rate at which information can be received, transmitted, or processed at a given point, for either the human or instrumentation.

channel gradient
The slope of the water surface of a stream channel through the bridge site to the nearest one tenth of a percent.

channel meander
An unstable river channel that changes its location after high water periods.

channel report
A report of channel conditions, soundings, etc. found by an aids-to-navigation tender on routine patrol; includes report of courses steered. The channel report is issued only for open rivers and is written in pilots' jargon.

channel width
The upstream channel width (bank to bank dimensions of the defined channel, not the flood plain) at a bridge site; to the nearest foot.

channeling
A condition that occurs in a filter or other packed bed when water finds furrows or channels through which it can flow without effective contact with the bed.

channelization
The process of straightening and deepening streams so water will move faster. A flood-reduction or marsh-drainage tactic that can interfere with waste assimilation capacity and disturb fish and wildlife habitats.

channoine weir
A section of a dam, built in the form of a spillway, lying between the anchor weir proper on the land side of a lock wall and the bear trap which is adjacent to the navigable path.

character
(1) The aggregate of the moral qualities which belong to and distinguish an individual person; the general result of one's distinguishing attributes. (2) An image of a letter or symbol appearing on a video display terminal or in printed media.

character evidence
Evidence of an individual's moral standing in a community based upon reputation.

character height
The vertical distance assigned to or occupied by a character on a display.

character of vessel
The type of service in which the vessel is engaged at the time of carriage of a hazardous material.

character width
The horizontal distance on a line of text from one point of one character to a corresponding point of the next character.

characteristic hazardous waste
Any one of four categories used in defining hazardous waste: ignitability, corrosivity, reactivity, and toxicity.

characteristics of easy movement
See ***motion efficiency principals.***

charcoal tube (CT)
A glass tube of specified dimensions and assembly, containing 100 mg of 20/40 mesh activated coconut shell charcoal in a front section and 50 mg in a backup section. Larger tubes are available.

charge density
In a polyelectrolyte, the mole ratio of the charged monomers to non-charged monomers.

charge it
Transportation (slang). To let brake air flow into semitrailer lines.

Charles's law
The volume of a mass of gas is directly proportional to the absolute temperature, provided the pressure remains the same.

charley horse
A minor muscle disorder resulting from the violent use of a muscle or group of muscles in strenuous work or play. It usually occurs when muscles that have not been conditioned for hard use are put under a strain, with the result that some of the muscle fibers are strained or may actually tear. It is characterized by soreness, stiffness, and pain which often comes on very suddenly.

chart
Any form of graphical or tabular data which provides information about one or more variables or activities.

chart recorder
See ***oscillograph, kymograph,*** *and* ***polygraph***.

charted visual flight procedure approach
An approach wherein a radar-controlled aircraft on an Instrument Flight Rules (IFR) Flight plan, operating in Visual Flight Rules (VFR) conditions and having an Air Traffic Control (ATC) authorization, may proceed to the airport of intended landing via visual landmarks and altitudes depicted on a charted visual flight procedure.

charted visual flight rules (VFR) flyways
Flight paths recommended for use to bypass areas heavily traversed by large turbine-powered aircraft. Pilot compliance with recommended flyways and associated altitudes is strictly voluntary. Visual Flight Rules (VFR) flyway planning charts are published on the back of existing VFR terminal area charts.

charter bus
A bus transporting a group of persons who pursuant to a common purpose, and under a single contract at a fixed price, have acquired the exclusive use of a bus to travel together under an itinerary.

charter party
Contract between the owner of a vessel and a shipper for letting of the vessel or a part thereof.

charter service
(1) A commercial passenger vehicle trip not scheduled, but specially arranged. The charter contract normally commits the carrier to furnish the agreed to transportation service at a specified time between designated locations. (2) A vehicle hired for exclusive use that does not operate over a regular route, on a regular schedule and is not available to the general public.

charter service hours
The total hours traveled/operated by a revenue vehicle while in charter service. Charter service hours include hours traveled/operated while carrying passengers for hire, plus associated deadhead hours.

charter transportation of passengers
Transportation, using a bus, of a group of persons who pursuant to a common purpose, under a single contract, at a fixed charge for the vehicle, have acquired the exclusive use of the vehicle to travel together under an itinerary either specified in advance or modified after having left the place of origin.

chase
Aviation. An aircraft flown in proximity to another aircraft normally to observe its performance during training or testing.

chassis
(1) The load-supporting frame in a truck or trailer, exclusive of any appurtenances which might be added to accommodate cargo. (2) A frame with wheels and container locking devices in order to secure the container for movement. *See also* ***body***.

chassis cab
An incomplete vehicle consisting of a cab on a bare frame rail chassis, needing a body or load platform in order to become complete.

cheater bar
See ***breakdown bar***.

check
(1) A mental skill involving the comparison of a finished product with what was planned to verify if the goals were met or standards achieved. (2) A written promise from one party to another ensuring financial reimbursement to the second party.

check line
Maritime. A line used to help check a boat's headway when landing or entering a lock.

check post
Maritime. A mooring bit on a lock wall.

check study
A timing review of a job to evaluate the appropriateness of the standard time for that job.

check time
The time period between the start time of a time study and the beginning of the first work element observed or between the completion of the last element and the stop time of the study.

check valve
A valve that opens in the direction of normal flow and closes with flow reversal.

checked baggage
Baggage accepted by the air carrier for transportation in the hold of the aircraft.

checkoff
The withholding of union dues from a worker's paycheck by agreement.

cheek
(1) The tissue comprising the side of the face from the zygomatic bone to the mandible. (2) Slang term commonly used for a buttock.

cheekbone
*See **zygomatic bone**.*

cheilion
The lateral corner of the mouth opening formed by the junction of the lips.

chelate
A chemical compound in which a metallic ion is combined with a molecule with multiple chemical bonds.

chelating agent
A type of organic sequestering agent that reduces water hardness and inactivates certain metal ions in water. Sometimes used in detergent formulations to reduce the effects of metals in water.

chelation
A treatment that removes harmful substances from the body. The chelating agent bonds to the contaminant which, due to the resulting poor absorption, is excreted from the body.

Chemical Abstract Service (CAS)
A numerical index listing chemical compounds and substances, each with its own distinct CAS identification number.

chemical agent
A hazardous substance, chemical compound, or mixture of these.

chemical analysis
Any form of examination through the use of chemicals, as in blood tests to determine a person's sobriety, the presence of drugs, etc.

chemical asphyxiant
A substance that chemically interferes with the respiratory process. There may be sufficient atmospheric oxygen present, but the body is unable to utilize it because the physiological mechanism for use and transport of oxygen is blocked (such is the case with carbon monoxide).

chemical burn
The tissue damage or destruction occurring as a direct result of chemical exposure.

chemical cartridge respirator
An air-purifying respirator capable of filtering out chemical contaminants from the air that is breathed. It usually acts through the use of chemical sorbant pads encased in cartridges that are attached to the respirator facepiece.

chemical compound
A substance composed of two or more elements combined in a fixed and definite proportion by weight.

chemical dosimeter
A self-indicating device for determining total (or accumulated) radiation exposure dose based on color change accompanying chemical reactions induced by the radiation.

chemical element
The smallest substance into which some physical or chemical entity can be chemically divided and still retain its chemical properties.

chemical emergency
An occurrence, such as a transportation accident, equipment failure, container rupture, or control equipment failure, that results in an uncontrollable release of a hazardous chemical into the environment or work place.

chemical feeder
A device used to dispense chemicals at a predetermined rate.

chemical fixation
The transformation of a chemical compound to a new, nontoxic form.

Chemical Hygiene Plan (CHP)
Part of a laboratory safety plan which must be established by laboratories handling hazardous chemicals due to a set of requirements mandated by OSHA.

chemical manufacturer
A person or business who imports, produces, or manufacturers a chemical substance.

Chemical Manufacturers Association (CMA)
An Association of chemical product manufacturers that disseminates information on the safe handling, transportation, and use of chemicals. In addition, it develops labeling guidelines and provides medical advice on the prevention and treatment of chemical injuries.

chemical oxygen demand (COD)
A measure of the oxygen required to oxidize all compounds in water, both organic and inorganic.

chemical pneumonitis
Pneumonitis or inflammation of the lung parenchyma as a result of the aspiration of a hydrocarbon solvent which spreads rapidly as a film over the lung's surfaces. The inhalation of beryllium of cadmium fumes or dust can cause an acute pneumonitis.

chemical protective clothing
Clothes made from various materials that exhibit chemical-resistant properties to an offending agent.

chemical reaction
A change in the arrangement of atoms or molecules to yield substances of different composition and properties.

chemical sludge
Sludge resulting from chemical treatment processes of inorganic wastes that are not biologically active.

chemical substance
Any organic or inorganic substance of a particular molecular identity, including any combination of such substances occurring in whole or in part as a result of a chemical reaction or occurring in nature; and any element or uncombined radical.

Chemical Transportation Emergency Center (CHEMTREC)
A section of the Chemical Manufacturers Association that provides emergency response information upon request to control an emergency.

chemical treatment
Any one of a variety of technologies that use chemicals or a variety of chemical processes to treat waste.

chemical waste
The waste generated by chemical, petrochemical, plastic, pharmaceutical, biochemical, or microbiological manufacturing processes.

chemicals of potential concern
Chemicals that are potentially site related and whose data are of sufficient quality for use in the quantitative risk assessment of that site.

chemiluminescence
The emission of absorbed energy as light, due to a chemical reaction of the compounds of the system. This principle is employed in some instruments for determining the airborne concentration of some substances (e.g., ozone).

chemiluminescence detector
A detector that is designed to detect light produced in chemical reactions, such as that between ozone and ethylene or nitric oxide. This phenomenon is employed in determining ambient levels of ozone and oxides of nitrogen.

chemisorption
The formation of an irreversible chemical bond between the sorbate molecule and the surface of the adsorbent.

chemistry
The area of science that deals with the elements and atomic structure of matter and the compounds of the elements.

chemoreceptor
A portion of a large protein or other cellular molecule which has the three-dimensional capacity for accepting and/or binding to a specific chemical substance.

chemosterilant
A chemical that controls pests by preventing reproduction.

chemosurgery
The destruction of tissue by chemical agents; originally applied to chemical fixation of malignant, gangrenous, or infected tissue, with

use of frozen sections to facilitate systematic microscopic control of its excision.

chemotaxis
The response of an individual toward a chemical stimulus.

chemotherapy
The treatment of illness by chemical means; that is, by medication. The term was first applied to the treatment of infectious diseases, but it now is used to include treatment of mental illness and cancer with drugs.

chemotrophs
Organisms that extract energy from organic and inorganic oxidation/reduction reactions.

CHEMTREC
See ***Chemical Transportation Emergency Center***.

chest
The thorax.

chest breadth
The horizontal linear width of the torso without tissue compression at the nipple level (males) and at the level where the fourth rib meets the sternum (females). Measured with the individual standing erect with the arms hanging naturally at the sides, and breathing normally.

chest breadth to bone
The horizontal linear width of the torso at the nipple level with tissue compression. Measured with the individual standing erect, and breathing normally.

chest/bust circumference
See ***chest circumference***.

chest/bust depth
See ***chest depth***.

chest circumference
The surface distance around the torso at the nipple level. Measured with the individual standing erect, breathing normally, and with arms slightly abducted. Also referred to as *chest/bust circumference*.

chest circumference at scye
The surface distance around the torso at the level of the axillary folds. Measured with the individual standing erect.

chest circumference below bust
Term applies to females only. The surface distance around the chest just below the cups of the bra. Measured with the individual standing erect and breathing normally.

chest depth
The anterior-posterior horizontal linear depth of the torso measured at the nipple level (males) and above the breasts at the level where the fourth rib joins the sternum (females). Measured with the individual standing erect, the arms hanging naturally at the sides, and breathing normally. Also know as the *chest/bust depth*.

chest depth at scye
The anterior-to-posterior horizontal linear depth of the torso measured at the scye level. Measured with the individual standing erect and breathing normally.

chest depth below bust
The transverse depth of the chest at the level of the inferior margin of the xiphoid process. Measured with the individual standing erect and breathing normally.

chest height
The vertical distance from the floor to the center of the nipples (males) or point of the bra (females). Measured with the individual standing erect and his/her weight balanced on both feet. *See also* ***bust point height***.

chest skinfold
See ***pectoral skinfold***.

Cheyne-Stokes respiration
A form of respiration in which the individual appears to have stopped breathing for 10 to 50 seconds, then breathing starts again with increasing intensity, then stops as before, and then repeats the previous breathing rhythm.

CHI
See ***Comfort-Health Index*** *and* ***computer-human interface***.

chi square (X^2)
A statistical test using differences in frequency data, especially for small samples, based on obtained vs. theoretical/expected frequency counts, to determine significance. Also referred to as *chi square test*. Represented by the formula:

$$X^2 = \sum_{i=1}^{n} \frac{(f_o - f_t)^2}{f_t}$$

where:
f_o = observed frequency
f_t = theoretical or expected frequency
n = sample size

chi square distribution
A mathematical or graphical function for chi square, having the probability distribution function

$$f(X^2) = G_v (X^2)^{\frac{(v-2)}{2}} e^{-\frac{X^2}{2}}$$

where:
v = degrees of freedom
G_v = a constant for a given v
Note: The shape of the distribution varies with degrees of freedom, approaches the normal distribution as degrees of freedom increase.

chi square test
See ***chi square***.

Chicago grips
A colloquial term used for a wire come-along used in hoisting wire rope.

chicane
A plow or other obstacle used on a belt thickener or belt press to mix or turn sludge to facilitate sludge dehydration.

chicken board
See ***crawling board***.

chief hood lifter
Transit (slang). Garage superintendent.

chigger
The so-called red-bug or larva of the mite family *Trombiculidae* whose bite produces a welt with itching and dermatitis.

chilblain
A localized painful erythema of the fingers, toes, or ears produced by excessive exposure to cold. The basic cause of chilblain is sensitivity to cold, sometimes resulting from circulatory disturbances, which may be corrected in part by exercise and proper diet; severe cases require medical attention. Extreme heat or cold applications should not be applied directly to chilblains. This condition should not be confused with frostbite, another type of skin damage caused by exposure to cold.

child
An individual younger than the age of puberty.

chill
A feeling of cold, with convulsive shaking of the body. A true chill, or rigor, results from an increase in chemical activity within the body and usually ushers in a considerable rise in body temperature. The pallor and coldness of a chill, and the goose flesh that often accompanies it, are caused by constriction of the peripheral blood vessels. Chills are symptomatic of a wide variety of diseases. They usually do not accompany well-localized infections.

chilling effect
The lowering of the Earth's temperature because of an increased level of particles in the air blocking the sun's rays. *See also* ***greenhouse effect***.

chime
The act of turning a cylindrical container on the edge of its base to assist in moving it from one location to another.

chimney effect
The tendency of air or gas in a vertical passage to rise when it is heated because its density is lower than the surrounding air or gas.

chin
The anterior lower part of the jaw, including the anterior lower portion of the mandible and all surrounding tissues.

chin prominence to wall
The horizontal distance from the wall to the most anterior protrusion of the chin. Measured with the individual standing erect with his back and head against the wall, facing straight ahead.

chin strap
Any thin, flexible, strong material or device which is attached to headgear and can be passed underneath the chin for aiding in headgear retention.

Chinese restaurant disease
Often called the *Chinese restaurant illness* or syndrome, this condition is due to the ingestion of large amounts of food containing *monosodium glutamate (MSG),* a flavoring

additive. Symptoms include headaches, tightness in the face, and lightheadedness.

Chinese wall
A fictional device used as a screening procedure which permits an attorney involved in an earlier adverse role to be screened from other attorneys in the firm so as to prevent disqualification of the entire law firm simply because one member of the firm previously represented a client who is now an adversary of the client currently represented by the firm.

Chinook
A warm, dry wind on the eastern side of the Rocky Mountains. In the Alps, this wind is called a *Foehn*.

Chinook wall cloud
A bank of clouds over the Rocky Mountains that signifies the approach of a Chinook.

chiropractic
A method of detecting and correcting by manual or mechanical means structural imbalance, distortion, or subluxations in the human body to remove nerve interference where such is the result of or related to distortion, misalignment, or subluxations of or in the vertebral column. A system of therapeutic treatment, through adjusting of articulations of the human body, particularly those of the spine. The specific science that removes pressure on the nerves by the adjustment of the spinal vertebrae.

chloracne
A disfiguring skin condition noted among workers who have had significant contact with certain chemicals such as chlorinated diphenyls, chlorinated dioxins, and chlornaphthalenes.

chloramines
Disinfecting compounds of organic or inorganic nitrogen and chlorine.

chloride
A salt of hydrochloric acid; any binary compound of chlorine in which the latter carries a negative charge of electricity.

chlorinated
(1) The condition of water or wastewater that has been treated with chlorine. (2) A description of an organic compound to which chlorine atoms have been added.

chlorinated hydrocarbons
These include a class of persistent, broad-spectrum insecticides that linger in the environment and accumulate in the food chain. Among them are DDT, aldrin, dieldrin, heptachlor, chlordane, lindane, endrin, mirex, hexchloride, and toxaphene.

chlorinated solvent
An organic solvent containing chlorine atoms such as methylene chloride and trichloromethane.

chlorination
The application of chlorine to drinking water, sewage, or industrial waste to disinfect or to oxidize undesirable compounds. Liquid chlorine has been found to be the most effective water disinfectant, and is almost invariably used in the United States for the purification of both public water supplies and swimming pools. This addition of chlorine is harmless, since enough chlorine to affect the health of those using the chlorinated water would also make the water too unpalatable to drink.

chlorinator
A metering device used to add chlorine to water or wastewater.

chlorine
A gaseous chemical element, atomic number 17, atomic weight 35.453, symbol Cl. It is a disinfectant, bleaching agent, and irritant poison. It is used for disinfecting, fumigating, and bleaching, either in an aqueous solution or in the form of chlorinated lime.

chlorine contact chamber
A detention chamber to diffuse chlorine through water or wastewater while providing adequate contact time for disinfection.

chlorine demand
The difference in the amount of chlorine added to a water or wastewater and the amount of residual chlorine remaining after a specific contact duration, usually 15 minutes.

chlorine residual
The amount of chlorine remaining in water after application at some prior time. *See also* ***free chlorine residual***.

chlorine tablets
Common term for pellets of solidified chlorine compounds such as calcium hypochlorite used for water disinfection.

chlorofluorocarbons (CFC)
A family of inert, nontoxic, and easily liquefied chemicals used in refrigeration, air conditioning, packaging, insulation, or as solvents and aerosol propellants. Because CFCs are not destroyed in the lower atmosphere, they drift into the upper atmosphere where their chlorine components destroy the ozone.

chloroform
A colorless, mobile, highly reactive, volatile liquid with a characteristic sweet odor and taste. It is used in industry as a solvent, as a cleansing agent, in the manufacture of refrigerant, and in fire extinguishers. It is also used in the manufacture of fluorocarbon plastics, in analytical chemistry, as a fumigant, and an insecticide, In the past, it was used extensively as an anesthetic. However, due to its toxic effects, this use has been abandoned. It is represented by the formula $CHCl_3$ and the structure:

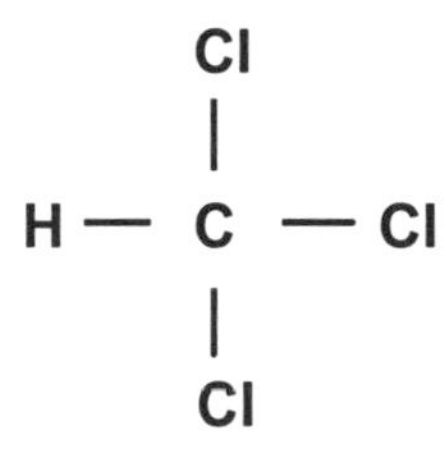

chlorophyll
The green photosynthetic pigment contained in many vegetable organisms.

chloropsia
A defect of vision in which objects appear green.

chlorosis
Discoloration of normally green plant parts that can be caused by disease, lack of nutrients, or various air pollutants.

choice reaction time (CRT)
That temporal interval measured for an individual or group after the presentation onset of one or a group of stimuli to decide which of more than one possible responses is appropriate and initiate that response. Generally represents an average time over several trials.

chokes
A form of decompression sickness in which a choking sensation, difficult breathing, and/or substernal pain are experienced due to air bubbles in the lungs.

cholalic acid
An acid formed in the liver from cholesterol that plays, with other bile acids, an important role in digestion.

cholecystic
Pertaining to the gallbladder.

cholera
Highly infectious disease of the gastrointestinal tract caused by waterborne bacteria.

cholesterol
The principal animal sterol, occurring in faintly yellow, pearly leaflets or granules in all animal tissues. Research has suggested the possibility that eating foods high in cholesterol may be a contributing factor in heart and circulatory disease, particularly in the formation of fatty deposits in the arteries (atherosclerosis).

cholinesterase
(1) An enzyme that hydrolyzes acetylcholine within the central nervous system. This enzyme can be depressed following exposure to organophosphate pesticide compounds. Workers using organophosphate pesticides should be routinely monitored for cholinesterase levels. (2) An enzyme that splits acetylcholine into acetic acid and choline. This enzyme is present throughout the body, but is particularly important at the myoneural junction where the nerve fibers terminate and become embedded in muscle fibers. Acetylcholine, which is formed when a nerve impulse reaches a myoneural junction, acts as a stimulant to the muscle fibers, causing them to contract. Immediately after acetylcholine has sparked a contraction it must be removed so that the muscle fiber will repolarize, or recharge itself; otherwise, it would not be ready to contract the next time it is stimulated. Cholinesterase performs this service by splitting acetylcholine into its components, thus rendering it ineffective.

cholinesterase inhibition
The loss or decrease of enzymatic activity of cholinesterase caused by binding of the enzyme with another chemical.

chorea
A movement disorder in which a series of complex, involuntary writhing movements are made, generally involving distal extremities and/or the face, tongue, and swallowing mus-

cles. *See also* ***Huntington's chorea, Sydenham's chorea.***

choreologist
One who has been trained and is competent to record human movement in some system of notation.

CHP
See ***Certified Health Physicist*** *and/or* ***Chemical Hygiene Plan.***

CHRIS
Chemical Hazard Response Information System.

chroma
That apparent degree to which a color compares to a similarly illuminated white or achromatic reference. *See also* ***Munsell chroma.***

chromate
A salt of chromium trioxide (chromic acid).

chromatic
Having a hue; colored; pertaining to any color except white, black, or gray.

chromatic aberration
An image containing colored fringes around the border, resulting from unequal refraction of light of different wavelengths causing focusing at different points in an optical lens system.

chromatic adaptation
That modification of the color sensory properties of the visual system by observing colored stimuli.

chromatic audition
See ***chromatism.***

chromatic contrast
That apparent contrast due to the presence of differing adjacent hues or colors. Synonymous with *color contrast, hue contrast,* and *simultaneous color contrast. See also* ***luminance contrast.***

chromatic contrast threshold
That minimal difference in the combined aspects of luminance and chromaticity which is detectable for a given pair of adjacent stimuli. Also referred to as *color contrast threshold.*

chromatic diagram
See ***chromaticity diagram.***

chromatic vision
See ***photopic vision.***

chromaticity
A measure of the quality of colored light, defined either by its chromaticity coordinates or by its dominant wavelength and excitation purity. Luminance or brightness is not involved.

chromaticity coordinate
Any of a set of numbers representing the proportions of two of the three normalized primary colors, usually x and y, required to produce a given color, with the brightness variable eliminated. Synonymous with *trichromatic coefficients* and *CIE chromaticity coordinates. See also* ***CIE color system.*** Represented by the formula (represented by x, y, z in the CIE system):

$$x = \frac{X}{X+Y+Z} \quad y = \frac{Y}{X+Y+Z} \quad z = \frac{Z}{X+Y+Z}$$

chromaticity diagram
A planar diagram based on the CIE color system and produced by using two of the chromaticity coordinates as axes in a rectangular coordinate system. Also referred to as *chromatic diagram.*

chromaticness
A visual attribute in which a perceived color appears more or less chromatic.

chromatin
The more readily stainable portion of a cell nucleus.

chromatism
Sensing an image of color when stimulated by a sensory modality other than vision. Synonymous with *chromatic audition. See also* ***synesthesia.***

chromatogram
For the differentiating type detector, which is the most common type in a gas chromatograph instrument, the chromatogram is a graphical presentation corresponding to the components present in the sample introduced into the instrument. The elapsed time from sample injection to each peak is a means to identify the components in the sample. The area under each peak is proportional to the total mass of that component in the sample. The chromatogram for the integrating type detector is a series of plateaus with each pla-

teau proportional to the total mass of the component in the eluted zone.

chromatographic detector
There are two types of chromatographic detectors: the differentiating type and the integrating type. The *integrating type* detector gives a response proportional to the total mass of the component in the eluted zone. The *differentiating type* gives a response proportional to the concentration or mass flow rate of the eluted component.

chromatography
A practical analytical methodology involving the separation of complex mixtures and the detection of each component of the mixture.

chrominance
The coloring power of a stimulus.

chromium
A chemical element, atomic number 24, atomic weight 51.996, symbol Cr.

chromosome
One of the thread-like bodies (normally 46 in humans) of chromatin that are found in the nucleus and that are the bearers of genes.

chromosphere
The sun's atmosphere just above the photosphere.

chromostereopsis
See ***color pseudo-stereopsis.***

chronic
With reference to disease, of long duration, or characterized by slowly progressive symptoms. Deep-seated or obstinate, or threatening a long continuance.

chronic alcoholism
A medically diagnosable disease characterized by chronic, habitual, or periodic consumption of alcoholic beverages resulting in substantial interference with an individual's social or economic functions in the community, and/or the loss of self-control with respect to the use of such beverages.

chronic carrier
A person who continues to harbor an infectious agent without showing symptoms of the disease. Chronic carriers are possible in many illnesses. Salmonellosis is an example.

chronic disease
Long-lasting, persistent, prolonged, repeated, or frequently recurring over a long period.

chronic effect
An effect which is the result of exposure to a toxic substance over a long period. The daily dose is insufficient to elicit an acute response, but it may have a cumulative effect over a period of time. Oftentimes, the rate of absorption of the toxic agent exceeds the rate of elimination, thereby resulting in a buildup of the substance in the body.

chronic exposure
(1) *Chemical.* Continual exposure to low levels of a chemical over a long period of time (usually 3 years or more), which can produce symptoms and disease. (2) *Radiation.* Exposure to radiation for long duration by fractionation or protraction. Generally, any dosage absorbed over a 24-hour period or longer.

chronic oxygen poisoning
See ***chronic oxygen toxicity.***

chronic oxygen toxicity
A lung disorder due to breathing higher than normal oxygen partial pressures at normal barometric pressure for 24 hours or more and characterized by chest pain, pulmonary edema, and possibly damage to the alveoli and bronchi. Also known as chronic oxygen poisoning. *See also* ***acute oxygen toxicity.***

chronic RfDs
An estimate (with uncertainty spanning perhaps an order of magnitude or greater) of a lifetime daily exposure level for the human population, including sensitive subpopulations, that is likely to be without an appreciable risk of deleterious effects. Chronic RfDs are specifically developed to be protective for long-term exposure to a compound (7 years to lifetime).

chronic toxicity
The capacity of a substance to cause long-term poisonous human health effects.

chronic toxicity test
Test method used to determine the concentration of a substance that produces an adverse effect on a test organism over an extended period of time.

chronobiology
The study of the effects of time on varying biological systems, including psychobiological rhythms.

chronocyclegram
See ***chronocyclegraph***. May also be known as *chronocyclogram*.

chronocyclegraph
The single negative or photograph from a chronocyclegraphic measurement; also *chronocyclograph*.

chronocyclegraphy
The use of a motion tracking system comprised of (a) one or more small electric light bulbs which flash at known, regular intervals and are attached to the fingers or other body part and (b) a still camera, ideally using a stereoscopic camera to obtain three-dimensional data, for recording motions on a single negative or print to determine velocitics and accelerations of the body parts. Typically the subject is in a darkened area; the exposure time is greater than or equal to one motion cycle. Also referred to as *chronocyclography*. *See also* ***cyclegraphy***.

chronograph
A constant-speed recording device which marks a paper or tape at known intervals so that timing during an ongoing process can be determined. Also referred to as *marstochron* and *marstograph*.

chronological age
The age as of the previous birthday or the age as of the previous birthday plus 0.5 years. *See also* ***developmental age*** *and* ***mental age***.

chronological study
The observation and recording of events or data in the order in which they occur over time.

chrysotile asbestos
Asbestiform mineral in the serpentine group that has been used as an insulation material in buildings. It is referred to as white asbestos and is the type that has been the most widely used in the U.S.

chunk
Differential housing on powered axles.

chute
(1) Section of a river that is narrower than ordinary and through which the river current increases, often navigable from bank to bank. (2) A narrow sloping passage by which water falls or flows to a lower level (between an island and a bank).

chyle
The product of intestinal digestion absorbed into the lymphatic system through the lacteals and conveyed through the thoracic duct to empty into the venous system at the root of the neck.

chyme
The semi-liquid mass into which food is converted by the action of gastric secretions during the digestive process.

CIA
See ***Central Intelligence Agency***.

cicatrix
The mark left in the flesh or skin after healing of a wound and having the appearance of a seam or of a ridge of flesh.

CID
See ***cubic inch displacement***.

CIE
Commission Internationale de l'Eclairage. *See* ***CIE color system***.

CIE color rendering index (CRI)
A measure of the amount of color shift which an object appears to present when illuminated by one source compared to that of a reference source having similar color temperature.

CIE color system
A standard color reference system established in 1931 by the Commission Internationale de l'Eclairage (CIE) based on the technique of flicker photometry and using a chromaticity diagram to specify color coordinates. Generally considered the world standard.

CIE Standard Observer
A table representing an observer having normal color vision which is developed from experimental data in color-matching using the primary colors with a 2° field of view. Also referred to as *standard observer* and *2° observer*. *See also* ***CIE Supplemental Standard Observer***.

CIE Standard Observer response curve
See ***spectral luminous efficiency function***.

CIE Supplementary Standard Observer
A variant of the CIE Standard Observer adopted in 1964 which accommodates a 10° field of view and permits better judgment of color matching in the shorter wavelengths (blue, violet). Also referred to as *10° observer*.

CIF
See ***cargo insurance and freight***. *See also* ***Cost, Insurance, Freight***.

CIH
See ***Certified Industrial Hygienist***.

CIIT
Chemical Industry Institute of Toxicology.

cilia
Short, tiny, hair-like processes on the surface of protozoan or certain metazoan cells that, by their constant motion, accomplish locomotion or produce a water current. Examples are found in the bronchi and respiratory tract where they aid in the removal of dusts.

ciliary muscle
An intrinsic smooth muscle of the eye, which is involved in lens accommodation.

CIM
See ***computer-integrated manufacturing***.

cinema verité
The use of only naturally available, not additional photographic, lighting for photography or videography.

cinematography
Motion picture photography.

CIR
See ***Crash Injury Research project***.

circadian
Having a period of approximately 24 hours.

circadian pacemaker
An internal timing mechanism which maintains circadian rhythms. *See also* ***internal clock*** *and* ***biological rhythm.***

circadian rhythm
A biological activity that recurs in periods of 24 hours under natural environmental conditions. Sleep patterns in mammals and leaf movements in some plants are examples of circadian rhythms.

circannual rhythm
A biological rhythm with a period of about one year.

circle to land maneuver
Aviation. A maneuver initiated by the pilot to align the aircraft with a runway for landing when a straight-in landing from an instrument approach is not possible or is not desirable. This maneuver is made only after Air Traffic Control (ATC) authorization has been obtained and the pilot has established required visual reference to the airport.

circle to runway (runway number) maneuver
Aviation. Used by Air Traffic Control (ATC) to inform the pilot that he must circle to land because the runway in use is other than the runway aligned with the instrument approach procedure. When the direction of the circling maneuver in relation to the airport/runway is required, the controller will state the direction (eight cardinal compass points) and specify a left or right downwind or base leg as appropriate (e.g., "Cleared Very High Frequency Omni-Directional Radio Range (VOR) Runway Three Six Approach circle to Runway Two," or "Circle northwest of the airport for a right downwind to Runway Two.").

circling approach
See ***circle to land maneuver***.

circuit
A conductor or a system of conductors through which electric current flows.

circuit breaker
(1) *600 Volts Nominal, or Less*. A device designed to open and close a circuit by non-automatic means and to open the circuit automatically on a predetermined overcurrent without damage to itself when properly applied within its rating. (2) *Over 600 volts, nominal.* A switching device capable of making, carrying, and breaking currents under normal circuit conditions, and also making, carrying for a specified time, and breaking currents under specified abnormal circuit conditions, such as those of short circuit.

circuit controller
A device for opening and closing electric circuits.

circuit courts
Courts whose jurisdiction extends over several counties or districts, and of which terms are held in the various counties or districts to which their jurisdiction extends.

circuit courts of appeals
The former name for federal intermediate appellate courts, changed in 1948 to the present designation of United States Courts of Appeals.

circuit-mile
The total length in miles of separate circuits regardless of the number of conductors used per circuit.

circulation
(1) The movement of blood through the circulatory system. (2) The movement of people, information, supplies, equipment, or other items within a building or other structure where work is being accomplished.

circulator bus
A bus serving an area confined to a specific locale, such as a downtown area or suburban neighborhood with connections to major traffic corridors.

circulatory system
The major system concerned with the movement of blood and lymph; it consists of the heart, blood vessels, and lymph vessels. The circulatory system transports to the tissues and organs of the body the oxygen, nutritive substances, immune substances, hormones, and chemicals necessary for normal function and activities of the organs; it also carries away waste products and carbon dioxide. It equalizes body temperature and helps maintain normal water and electrolyte balance. An adult male has an average of 5 quarts of blood in his body; the circulatory system carries this entire quantity on one complete circuit through the body every minute. In the course of 24 hours, 7200 quarts of blood pass through the heart. The rate of blood flow through the vessels depends upon several factors: force of the heartbeat, rate of the heartbeat, venous return, and control of the arterioles and capillaries by chemical, neural and thermal stimuli.

circumaural protector
A form of hearing protector commonly known as the earmuff, consisting of two cup-shaped devices that fit over the entire external ear and are sealed against the side of the head.

circumduction
A basic type of joint motion occurring in those joints capable of three-dimensional movement in which the proximal end of a bone in its socket provides the apex of a cone and the distal end of that bone moves in a circular pattern, sweeping out a conical volume.

circumference
(1) A curved, closed, anthropometric measurement that follows a body contour. It need not be circular. *See also* ***arc***. (2) The length comprising the perimeter of a circle.

circumoral paresthesia
A burning sensation around or near the mouth.

circumstances
Attendant or accompanying facts, events, or conditions. Subordinate or accessory facts (e.g., evidence that indicates the probability or improbability of an event).

circumstantial evidence
Testimony not based on actual personal knowledge or observation of the facts in controversy, but of other facts from which deductions are drawn, showing indirectly the facts sought to be proved.

circus wagon
Transportation (slang). Low-sided trailer with high bow tarp.

cirrhosis
Interstitial inflammation of an organ, particularly the liver. Cirrhosis is marked by degeneration of the liver cells and thickening of the surrounding tissue.

cirrocumulus
A high cloud that appears as a white patch of cloud without shadows. It consists of very small elements in the form of grains or ripples.

cirrostratus
A high cloud appearing as a whitish veil that may totally cover the sky. Often produces halo phenomena.

cirrus
A high cloud composed of ice crystals in the form of thin, white, feather-like clouds in patches, filaments, or narrow bands.

cistern
A small covered tank for storing water, usually placed underground.

citation
(1) A writ issued out of a court of competent jurisdiction, commanding a person therein

named to appear on a day named and do something therein mentioned, or show cause why he/she should not. (2) A written notice from a regulatory agency alleging an employer's non-compliance with a specific standard or regulation, or group of standards or regulations, or the General Duty Clause of the OSHAct of 1970.

citizen suit

(1) *General.* A type of legal action in court brought by persons (or organizations on behalf of members) to enforce laws against violators, usually invoking a statutory right to sue without showing traditional standing. (2) *CERCLA.* A provision under CERCLA which permits any person to initiate a civil action against any other person, including the United States, for violations of any standard, regulation, condition, requirement, or order effective under CERCLA, and against any officer of the United States for failure to perform a non-discretionary act under CERCLA.

citric acid

A crystalline acid present in citrus fruits. Chemical formula is $C_6H_8O_7 \cdot H_2O$.

city flyer

Transportation (slang). Short, low trailer with high bow tarp.

city gate

A point or measuring station at which a distribution gas utility receives gas from a natural gas pipeline company or transmission system.

City Solicitor

See ***Town Counsel.***

city trip

A commercial vehicle trip within a single city, town, county, or other geographic jurisdiction.

civil

An area of the law where matters are decided with no criminal consequences, as in contracts, torts, eminent domain, licensing, grants, Civil Penalties, and most administrative enforcement. Contrasted with investigations and criminal prosecutions carried out following Criminal Procedure rather than Civil Procedure.

civil action

Action brought to enforce, redress, or protect private rights. In general, all types of actions other than criminal proceedings. The term includes all actions, or, in other phraseology, both suits in equity and actions, at law.

Civil Aeronautics Board (CAB)

A defunct organization. Originally, an independent regulatory commission that was established under the Civil Aeronautics Act of 1938. Its functions were terminated or transferred to other agencies beginning in 1966, with all remaining functions transferred to the U.S. Secretary of Transportation by 1985.

civil aircraft

Aircraft other than public aircraft.

civil death

The state of a person who, though possessing natural life, has lost all civil rights and as to them is considered civilly dead.

civil jury trial

Trial of civil action before a jury rather than before a judge. In suits at common law in federal court where the value in controversy exceeds $20.00, there is a constitutional right to a jury trial.

civil law

That body of law which every particular nation, commonwealth, or city has established peculiarly for itself; more properly called "municipal" law, to distinguish it from the "law of nature," and from international law. These laws are concerned with civil or private rights and remedies, as contrasted with criminal laws.

civil liberties

Personal, natural rights guaranteed and protected by the Constitution. Examples are freedom of speech, press, freedom from discrimination, etc. The body of law dealing with natural liberties, shorn of excesses which invade equal rights of others. Constitutionally, they are restraints on government.

civil nuisance

At common law, anything done to hurt or annoyance of lands, tenements, or hereditaments of another.

civil obligation

One which binds in law, and may be enforced in a court of justice.

civil offense

Term used to describe violations of statutes making the act a public nuisance.

civil penalties
Represents punishment for specific activities, e.g., violation of antitrust or securities laws, usually in the form of fines or money damages.

civil procedure
Body of law concerned with methods, procedures, and practices used in civil litigation.

Civil Rights Act
An act designed to prohibit public accommodations and employment discrimination due to a person's color, race, religion, sex, or national origin.

claim
(1) To demand as one's own or as one's own right; to assert; to urge; to insist. A cause of action. Means by or through which a claimant obtains possession or enjoyment of a privilege or thing. Demand for money or property as a right (e.g., an insurance claim).

claim adjuster
An independent agent or employee of an insurance company who negotiates and settles claims against the insurer.

claimant
One who claims or asserts a right, demand, or claim.

Claims Court, U.S.
This federal court was established in 1982 and succeeds to all the original jurisdiction formerly exercised by the Court of Claims. The court has jurisdiction to render money judgments upon any claim against the United States founded under either (a) upon the Constitution, or (b) any act of Congress or any regulation of an executive department, or (c) upon any express or implied in fact contract with the United States, or (d) for liquidated or unliquidated damages in cases not sounding in tort. Judgments of the Court are final and conclusive on both the claimant and the United States subject to an appeal as of right to the U.S. Court of Appeals for the Federal Circuit. Authority also rests with the court to furnish reports on any bill that may be referred by either House of Congress. Jurisdiction of the Court is nationwide, and jurisdiction over the parties is obtained when suit is filed and process is served on the United States through the Attorney General.

clarification
Clearing action that occurs during wastewater treatment when solids settle out. This is often aided by centrifugal action and chemically induced coagulation in wastewater.

clarifier
A quiescent tank used to remove suspended solids by gravity settling. Also called sedimentation or settling basins, they are usually equipped with a motor-driven rake mechanism to collect settled sludge and move it to a central discharge point.

clarifying agent
Any substance used to remove turbidity from drinks.

clarifying lotion
A substance for removing oil and grease from the face.

clash point
A point at which the human body or its reach envelope, whether physically or in computer modeling, intersects some equipment, instrumentation, or workspace boundaries in a workplace.

class
(1) With respect to the certification, ratings, privileges, and limitations of airmen, means a classification of aircraft within a category having similar operating characteristics. Examples include single engine, multiengine, land, water, gyroplane, helicopter, airship, and free balloon. (2) With respect to the certification of aircraft, means a broad grouping of aircraft having similar characteristics of propulsion, flight, or landing. Examples include airplane, rotorcraft, glider, balloon, land plane, and seaplane.

Class I biological safety cabinet
An open-front, negative pressure, ventilated cabinet with a minimum inward face velocity at the work opening of at least 75 feet per minute with the exhaust air filtered through a HEPA filter.

Class II biological safety cabinet-laminar flow
An open-front, ventilated cabinet with an average inward face velocity at the work opening of at least 75 feet per minute and providing HEPA-filtered recirculated airflow in the cabinet workspace and exhaust air passed through a HEPA filter.

Class III biological safety cabinet
A totally enclosed cabinet of gas-tight construction, such as a glove-box. The exhaust fan for this cabinet is a dedicated unit with exhaust air discharged directly to the outdoors. Air entering the cabinet is passed through a HEPA filter, with operations conducted in the enclosure using glove ports. In use, the cabinet is maintained at 0.5 inches water gauge negative pressure.

Class 100 Clean Room
An area or room in which the particle count in the air does not exceed 100 particles per cubic foot in the size range of 0.5 micrometers and larger.

Class 10,000 Clean Room
An area or room in which the particle count in the air does not exceed 10,000 particles per cubic foot larger than 0.5 micrometers or 65 particles per cubic foot larger than 5 micrometers in size.

Class 100,000 Clean Room
An area or room in which the particle count in the air does not exceed 100,000 particles per cubic foot larger than 0.5 micrometers or 700 particles per cubic foot larger than 5 micrometers.

Class I Freight Railroad
Defined by the Interstate Commerce Commission each year based on annual operating revenue. For 1988, the threshold for Class I railroads was $87.9 million. A railroad is dropped from the Class I list if it fails to meet the annual earnings threshold for three consecutive years.

Class I laser
Referred to as an exempt laser. Under normal conditions, these do not emit a hazardous level of optical radiation.

Class I Motor Carrier
Motor carrier with annual revenues greater than $10 million. Prior to January 1, 1994, the revenue classification level was $5 million.

Class I Railroad
A railroad with an annual gross operating revenue in excess of $250 million based on 1991 dollars.

Class II laser
A low-power laser which may cause retinal injury if viewed for long periods of time.

Class II Motor Carrier
Motor carrier with annual revenues between $3 and $10 million. Prior to January 1, 1994, the revenue classification level was between $1 and $5 million.

Class III-A laser
A visible laser which can cause injury to the eyes. Class III laser devices are classed as medium-power laser devices.

Class III-B laser
Can cause injury to the eye as a result of viewing the direct or reflected beam. Class III laser devices are classed as medium-power laser devices.

Class III Motor Carrier
Motor carrier with annual revenues over $1 million and less than $3 million. Prior to January 1, 1994, the revenue classification level was under $1 million.

Class IV laser
These are high-powered laser systems that require extensive exposure controls for preventing eye and skin exposure to both the direct and reflected laser beam.

Class V laser
Includes any Class II, III, or IV laser device which, by virtue of appropriate design or engineering controls, cannot directly irradiate the eye at levels in excess of established exposure limits.

Class I Location
Those in which flammable gases or vapors are or may be present in the air in quantities sufficient to produce explosive or ignitable mixtures. Class I locations include

Class I, Division 1. A location in which a) hazardous concentrations of flammable gases or vapors may exist under normal operating conditions, or b) hazardous concentrations of such gases or vapors may exist frequently because of repair or maintenance operations or because of leakage, or c) breakdown or faulty operation of equipment or processes might release hazardous concentrations of flammable gases or vapors, and might also cause simultaneous failure of electric equipment. This classification usually includes locations where volatile flammable liquids or liquefied flammable gases are transferred from one container to another; interiors of spray booths and areas in the vicinity of spraying and painting operations where volatile flammable solvents are used; locations containing open tanks or vats of volatile flammable liquids; drying rooms or compartments for the evaporation of flammable solvents; locations containing fat and oil extraction equipment using volatile flammable solvents; portions of cleaning and dying plants where flammable liquids are used; gas generator rooms and other portions of gas manufacturing plants where flammable gas may escape; inadequately ventilated pump rooms for flammable gas or for volatile flammable liquids; the interiors of refrigerators and freezers in which volatile flammable materials are stored in open, lightly stopped, or easily ruptured containers; and all other locations where ignitable concentrations of flammable vapors or gases are likely to occur in the course of normal operations.

Class I, Division 2. A location in which a) volatile flammable liquids or flammable gases are handled, processed, or used, but in which the hazardous liquids, vapors, or gases will normally be confined within closed containers or closed systems from which they can escape only in the case of accidental rupture or breakdown of such containers or systems, or in the case of abnormal operation of equipment, or b) hazardous concentrations of gases or vapors are normally prevented by positive mechanical ventilation, and which might become hazardous through failure or abnormal operations of the ventilating equipment, or c) that is adjacent to a Class I, Division 1 location, and to which hazardous concentrations of gases or vapors might occasionally be communicated unless such communication is prevented by adequate positive-pressure ventilation from a source of clean air, and effective safeguards against ventilation failure are provided. This classification usually includes locations where volatile flammable liquids or flammable gases or vapors are used, but which would become hazardous only in case of an accident or of some unusual operating condition. The quantity of flammable material that might escape in case of accident, the adequacy of ventilating equipment, the total area involved, and the record of the industry or business with respect to explosions fires are all factors that merit consideration in determining the classification and extent of each location. Piping without valves, checks, meters, and similar devices would not ordinarily introduce a hazardous condition even though used for flammable liquids or gases. Locations used for the storage of flammable liquids or liquefied or compressed gases in sealed containers would not normally be considered hazardous unless also subject to other hazardous conditions. Electrical conduits and their associated enclosures separated from process fluids by a single seal or barrier are classed as a Division 2 location if the outside of the conduit and enclosure is a nonhazardous location.

Class II Location

Those that are hazardous because of the presence of combustible dust. Class II locations include the following:

Class II, Division 1. A location in which a) combustible dust is or may be in suspension in the air under normal operating conditions, in quantities sufficient to produce explosive or ignitable mixtures, or b) where mechanical failure or abnormal operation of machinery or equipment might cause such explosive or ignitable mixtures to be produced, and might also provide a source of ignition through simultaneous failure of electric equipment, operation of protection devices, or from other causes, or c) in which combustible dusts of an electrically conductive nature may be present. This classification may include areas of grain handling and processing plants, starch plants, sugar-pulverizing plants, malting plants, hay-grinding plants, coal pulverizing plants, areas where metal dusts and powders are produced or processed, and other similar locations which contain dust producing machinery and equipment (except where the equipment is dust-tight or vented to the outside). These areas would have combustible dust in the air, under normal operating conditions, in quantities sufficient to produce explosive or ignitable mixtures. Combustible dusts which are electrically nonconductive include dusts produced in the handling and processing of grain products, pulverized sugar and coca, dried egg and milk powders, pulverized spices, starch and pastes, potato and wood flour, oil meal and beans and seed, dried hay, and other organic materials which may produce combustible dusts when processed or handled. Dusts containing magnesium or aluminum are particularly hazardous and the use of extreme caution is necessary to avoid ignition and explosion.

Class II, Division 2. A location in which a) combustible dust will not normally be in suspension in the air in quantities sufficient to produce explosive or ignitable mixtures, and dust accumulations are normally insufficient to interfere with the normal operation of electrical equipment or other apparatus, or b) dust may be in suspension in the air as a result of infrequent malfunctioning of handling or processing equipment, and dust accumulations resulting therefrom may be ignitable by abnormal operation or failure of electrical equipment or other apparatus. This classification includes locations where dangerous concentrations of suspended dust would not be likely but where dust accumulations might form on or in the vicinity of electric equipment. These areas may contain equipment from which appreciable quantities of dust would escape under abnormal operating conditions or be adjacent to a Class II, Division 1 location, as described above, into which an explosive or ignitable concentration of dust may be put into suspension under abnormal operating conditions.

Class III Location

Those locations that are hazardous because of the presence of easily ignitable fibers or flying but in which such fibers or flyings are not likely to be in suspension in the air in quantities sufficient to produce ignitable mixtures. Class III locations include the following:

Class III, Division 1. A location is a location in which easily ignitable fibers or materials producing combustible flyings are handled, manufactured, or used. Such locations usually include some parts of rayon, cotton, and other textile mills; combustible fiber manufacturing and processing plants; cotton gins and cotton-seed mills; flax-processing plants; clothing manufacturing plants; woodworking plants, and establishment; and industries involving similar hazardous processes or conditions. Easily ignitable fibers and flyings include rayon, cotton (including cotton linters and cotton waste), sisal or henequen, istle, jute, hemp, tow, cocoa fiber, oakum, baled waste kapok, Spanish moss, excelsior, and other materials of similar nature.

Class III, Division 2. A location in which easily ignitable fibers are stored or handled, except in thc process of manufacturing.

Class 1 Road

Hard surface highways including interstate and U.S. numbered highways (including alternates), primary state routes, and all controlled access highways.

Class 2 Road

Hard surface highways including secondary state routes, primary county routes, and other highways that connect principal cities and towns, and link these places with the primary highway system.

Class 3 Road

Hard surface roads not included in a higher class and improved, loose surface roads passable in all kinds of weather. These roads are adjuncts to the primary and secondary highway systems. Also included are important private roads such as main logging or industrial roads which serve as connecting links to the regular road network.

Class 4 Road

Unimproved roads which are generally passable only in fair weather and used mostly for local traffic. Also included are driveways, regardless of construction.

Class 5 Road

Unimproved roads passable only with 4-wheel drive vehicles.

Class A Explosive

Possessing detonating or otherwise maximum hazard, such as dynamite, nitroglycerin, picric acid, lead azide, fulminate of mercury, black powder, blasting caps, and detonating primers.

Class A and B Explosives In Bulk

The transportation, as cargo, of any Class A or B explosive(s) in any quantity.

Class A by Inland and Coastal Waterways Carrier

A class A carrier by water is one with an average annual operation revenue that exceeds $500,000.

Class A Fire

A fire involving ordinary combustible materials such as paper, wood, cloth, and some rubber and plastic materials. *See also* ***fire classification***.

Class B by Inland and Coastal Waterways Carrier
A class B carrier by water is one with an average annual operating revenue greater than $100,000 but less than $500,000.

Class B Explosive
Possessing flammable hazard, such as propellant explosives (including some smokeless propellants), photographic flash powders, and some special fireworks.

Class B Fire
A fire involving flammable or combustible liquids, flammable gases, greases, and similar materials, and some rubber and plastic materials. *See also* ***fire classification***.

Class C Explosive
Includes certain types of manufactured articles which contain Class A or Class B explosives, or both, as components but in restricted quantities.

Class C Fire
A fire involving energized electrical equipment where the safety to firefighter requires the use of electrically non-conductive extinguishing media. *See also* ***fire classification***.

Class D Fire
A fire involving combustible metals such as magnesium, titanium, zirconium, sodium, lithium, and potassium. *See also* ***fire classification***.

classical anthropometry
The measurement of various static body girths and lengths with measurement devices such as a simple tape measure, anthropometer, and calipers. Synonymous with *conventional anthropometry* and *traditional anthropometry*.

classical conditioning
A type of learning in which an initially neutral stimulus is paired with a natural stimulus and response such that after some number of trials the neutral stimulus will elicit the natural response.

classification
Arrangement into groups or categories on the basis of established criteria. The word may have two meanings, one primarily signifying a division required by statutes, fundamental and substantial, and the other secondary, signifying an arrangement or enumeration adopted for convenience only.

classification of risks
Term used in fire insurance to designate the nature and situation of the articles insured, and in accident insurance to the occupation of the applicant.

classified waste
Waste material that has been given security classification in accordance with the U.S. Code and Executive Order.

classifier
A device used to separate constituents according to relative sizes or densities.

clastogenic
Substance that damages chromosomes.

clathrate
A compound formed by the inclusion of molecules in cavities formed by crystal lattices.

claused bill of lading
A bill of lading which has exceptions to the receipt of merchandise in "apparent good order" noted.

clavicle
The bone which connects the sternum and the scapula. Synonymous with *collarbone*.

clay
A fine-grained earthy material that is plastic when wet, rigid when dried, and vitrified when fired to high temperatures.

clay liner
A layer of clay soil added to the bottom and sides of an earthen basin for use as a disposal site of potentially hazardous wastes.

clean
(1) To remove dirt, impurities, or other undesired entities. (2) Pertaining to a condition in which specified or implied standards are met for cleanliness.

clean air
Air that is free of any substance that will adversely affect the operation or cause a response of an instrument.

Clean Air Act (CAA)
1970 U.S. federal law requiring air pollutant emission standards; reauthorized in 1977 and again in 1990.

Clean Air Act Amendments (CAAA)
Amendments issued in 1990 to expand the EPA's enforcement powers and place restrictions on air emissions.

clean alternative fuel
Under the Clean Air Act (CAA), any fuel, including methanol, ethanol, or other alcohols (including any mixture thereof containing 85 percent or more by volume of such alcohol with gasoline or other fuels, reformulated gasoline, diesel, natural gas, liquefied petroleum gas, and hydrogen) or power source (including electricity) used in a clean-fuel vehicle that complies with the standards and requirements applicable to such vehicle under CAA when using such fuel or power source. In the case of any flexible fuel vehicle or dual-fuel vehicle, the term "clean alternative fuel" means only a fuel with respect to which such vehicle was certified as a clean-fuel vehicle meeting the standards applicable to clean-fuel vehicles under the CAA when operating on clean alternative fuel (or any California Air Research Board standards which replace such standards pursuant to the CAA).

clean area
In asbestos abatement, a controlled environment which is maintained and monitored to assure a low probability of asbestos contamination in that space.

clean bill of lading
A bill of lading which covers goods received in apparent good order and condition and without qualification.

clean bore
A single tank without compartments inside.

clean coal technology
Under the Clean Air Act (CAA), any technology, including technologies applied at the pre-combustion, combustion, or post-combustion stage, at a new or existing facility which will achieve a significant reduction in air emissions of sulfur dioxide or oxides of nitrogen associated with the utilization of coal in the generation of electricity, process steam, or industrial products, which was not in widespread use as of November 15, 1990.

clean-fuel vehicle
Under the Clean Air Act (CAA), a vehicle in a class or category of vehicles which has been certified to meet, for any model year, the clean-fuel vehicle standards applicable under the CAA for that model year to clean-fuel vehicles in that class or category.

clean room
A specially constructed area or space that is carefully controlled for airborne aerosols, temperature, humidity, air flow, and, in some cases, air pressure. Personnel who enter or work in such a space must follow very strict protocols with regard to clothing and other coverings (e.g., hair, face, etc.) to ensure the integrity of the clean room environment. Periodic measurements are taken inside the room to determine the level of contaminants present. The level of these protocols is usually dependent upon the class of clean room. *See also* ***Class 100/10,000/100,000 Clean Room***.

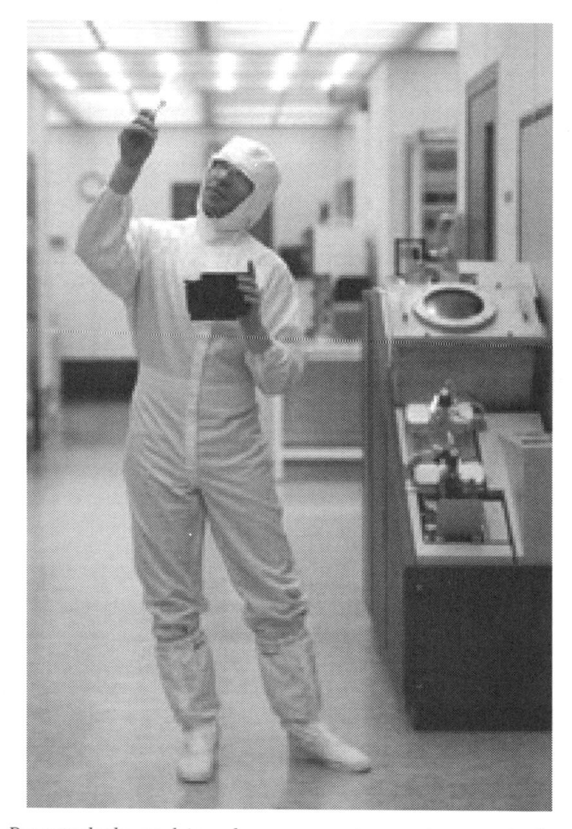

Personnel who work in a clean room environment are required to take specific precautions and wear special clothing that ensure the preservation of clean room integrity.

Clean Water Act (CWA)
1972 U.S. federal law regulating surface water discharges; updated in 1987.

cleaning allowance
That paid time given an employee for personal hygiene required due to the working en-

vironment and for workspace and tool cleaning. Also referred to as *cleanup time*.

cleanup
Actions taken to deal with the release or threat of release of a hazardous substance that could affect humans and/or the environment. The term "cleanup" is sometimes used interchangeably with the terms *remedial action, removal action, response action,* or *corrective action*.

cleanup time
See ***cleaning allowance***.

clear
(1) A function which removes the current selection from the display. (2) To remove any turbidity from a fluid. (3) Without conflict or confusion in understanding or action.

clear air turbulence (CAT)
Turbulence encountered by aircraft flying through cloudless skies. Thermals, wind shear, and jet streams can each be a factor in producing this phenomenon.

clear and convincing proof
That proof which results in reasonable certainty of the truth of the ultimate fact in a controversy. This is proof which requires more than a preponderance of evidence, but less than proof beyond a reasonable doubt.

clear and present danger
Doctrine in constitutional law providing that governmental restrictions on First Amendment freedoms of speech and press will be upheld if necessary to prevent grave and immediate danger to those interests which the government may lawfully protect. Speech which incites to unlawful action falls outside the protection of the First Amendment where there is a direct connection between the speech and violation of the law (this is the "clear and present danger test").

clear cutting
The practice of completely felling a stand of trees, usually followed by the replanting of a single species.

clear evidence or proof
Evidence which is positive, precise, and explicit, which tends directly to establish the point which is adduced and is sufficient to establish a prima facie case. It necessarily means a clear preponderance of proof.

clear ice
A layer of ice that appears transparent because of its homogeneous structure and small number and size of air pockets.

clear of the runway
(1) A taxiing aircraft, which is approaching a runway, is clear of the runway when all parts of the aircraft are held short of the applicable holding position marking. (2) A pilot or controller may consider an aircraft, which is exiting or crossing a runway, to be clear of the runway when all parts of the aircraft are beyond the runway edge and there is no Air Traffic Control (ATC) Federal Aviation Administration (FAA) Glossary C 2 restriction to its continued movement beyond the applicable holding position marking. (3) Pilots and controllers shall exercise good judgment to ensure that adequate separation exists between all aircraft on runways and taxiways at airports with inadequate runway edge lines or holding position markings.

clearance lamp
A lamp used on the front and the rear of a motor vehicle to indicate its overall width and height.

clearance limit
The fix, point, or location to which an aircraft is cleared when issued an air traffic clearance.

clearance sampling
A sampling procedure carried out at the end of an asbestos abatement activity to determine whether the asbestos abatement has been effective and the fiber concentration is acceptable. Typically, the acceptable concentration is the background level, or that which has been specified in the abatement contract.

clearance traffic control
Authorization for an aircraft to proceed under conditions specified by an air traffic control unit.

clearance void if not off by time
Used by Air Traffic Control (ATC) to advise an aircraft that the departure clearance is automatically canceled if takeoff is not made prior to a specified time. The pilot must obtain a new clearance or cancel his Instrument Flight Rules (IFR) flight plan if not off by the specified time.

clearance void time
A time specified by an air traffic control unit at which a clearance ceases to be valid unless the aircraft concerned has already taken action to comply therewith.

cleared approach
Air Traffic Controller (ATC) authorization for an aircraft to execute any standard or special instrument approach procedure for that airport. Normally, an aircraft will be cleared for a specific instrument approach procedure.

cleared as filed
Means the aircraft is cleared to proceed in accordance with the route of flight filed in the flight plan. This clearance does not include the altitude, standard instrument departure (SID), or SID transition.

cleared for takeoff
Air Traffic Control (ATC) authorization for an aircraft to depart. It is predicated on known traffic and known physical airport conditions.

cleared for the option
Air Traffic Control (ATC) authorization for an aircraft to make a touch-and-go, low approach, missed approach, stop-and-go, or full-stop landing at the discretion of the pilot. It is normally used in training so that an instructor can evaluate a student's performance under changing situations. Also known as *option approach*.

cleared through
Air Traffic Control (ATC) authorization for an aircraft to make intermediate stops at specified airports without refiling a flight plan while enroute to the clearance limit.

cleared to land
Air Traffic Control (ATC) authorization for an aircraft to land. It is predicated on known traffic and known physical airport conditions.

clearway
Aviation. (1) For turbine engine powered airplanes certificated after August 29, 1959, an area beyond the runway, not less than 500 feet wide, centrally located about the extended centerline of the runway, and under the control of the airport authorities. The clearway is expressed in terms of a clearway plane, extending from the end of the runway with an upward slope not exceeding 1.25 percent, above which no object nor any terrain protrudes. However, threshold lights may protrude above the plane if their height above the end of the runway is 26 inches or less and if they are located to each side of the runway. (2) For turbine engine powered airplanes certificated after September 30, 1958, but before August 30, 1959, an area beyond the takeoff runway extending no less than 300 feet on either side of the extended centerline of the runway, at an elevation no higher than the elevation of the end of the runway, clear of all fixed obstacles, and under the control of the airport authorities.

clearwell
A tank or reservoir of filtered water used to backwash a filter.

cleats
With regard to ladders, cross-pieces of rectangular cross-section placed on edge which a person may step in, ascending or descending.

cleavage line
Any line of tension in the skin along which a tear will tend to occur from a penetrating object, producing a slit rather than a rounded opening. Also referred to as *Langer's line*.

Clerical Task Inventory (CTI)
A compilation of over 100 clerical or office-type tasks for job evaluation or wage determination purpose.

click
Press and release a button on an input device such as a mouse or track ball to provide a command or other input to a computer.

client
An individual, corporation, trust, or estate that employs a professional to advise or assist it in the professional's line of work. Professionals include but are not limited to safety and health professionals, industrial hygienists, environmental personnel, attorneys, accountants, architects, etc.

client's privilege
The right of a client to require an attorney not to disclose confidential communications made to him/her in the attorney-client relationship, including disclosure on the witness stand.

climate
The accumulation of daily and seasonal weather events over a long period of time.

climatic optimum
A period in geological history (approximately 7000 to 5000 years ago) when temperatures were warmer than at present.

climatological forecast
A weather forecast, usually a month or more in the future, which is based upon the climate of a region rather than upon current weather conditions.

climb to VFR
Air Traffic Control (ATC) authorization for an aircraft to climb to Visual Flight Rules (VFR) conditions within clause B, C, D, and E surface areas when the only weather limitation is restricted visibility. The aircraft must remain clear of clouds while climbing to Visual Flight Rules (VFR).

climbout
That portion of flight operation between takeoff and the initial cruising altitude.

climbout speed
With respect to rotorcraft, means a referenced airspeed which results in a flight path clear of the height-velocity envelope during initial climbout.

clinical laboratory
A workplace where diagnostic or other screening procedures are performed on blood or other potentially infectious materials.

clinical tests
Tests involving direct observation of the patient, including laboratory and diagnostic examinations.

clinker
A fused byproduct of the combustion of coal or other solid fuels.

clino
*See **clinoptilolite**.*

clinoptilolite
A naturally occurring clay that can be used in an ion exchange process for ammonia removal.

clipboard
A temporary computer-editing buffer which is independent of, but able to interface with, other system applications. Also referred to as *temporary editing buffer*.

clo
A unit for the thermal insulation provided by clothing, not counting the approximately 25% for heat loss via the respiratory system and passive diffusion through the skin; that amount of insulation needed in a sitting and resting average individual to be thermally comfortable in a normally ventilated room (approximately 10 cm/sec air velocity, 21°C temperature, and 50% relative humidity).

clockwise rotating shift
Pertaining to a rotating shift work schedule in which the shift worked is periodically delayed by increments (i.e., from the first shift to the second, or from the second to the third).

cloning
In biotechnology, obtaining a group of genetically identical cells from a single cell. This term has assumed a more general meaning that includes making copies of a gene.

close-coupled pump
A pump coupled directly to a motor without gearing or belting.

close the gates
Transportation (slang). To close the rear doors of a trailer.

Close View
A feature which permits enlargement of the display characters for easier reading by visually impaired individuals.

closed-circuit SCBA
A self-contained respiratory protective device in which the breathing air is recirculated and rebreathed after carbon dioxide has been removed to maintain the quality of the breathing air.

closed-cycle cooling system
A cooling water system in which heat is transferred by recirculating water contained within the system, producing a relatively small blow downstream of concentrated solids.

closed insurance policy
Insurance contract, the terms and rates of which cannot be changed.

closed-loop recycling
Reclaiming or reusing wastewater for non-potable purposes in an enclosed process.

closed-loop system
Any type of system in which the output or some derivative of the output from the system is directed back into the system itself. Synonymous with *feedback control loop*.

closed respiration system
A breathing gas system that is self-contained and provides a continuing and proper oxygen/nitrogen supply ratio and pressure for its personnel, with removal of carbon dioxide and excess water vapor.

closed runway
A runway that is unusable for aircraft operations. Only the airport management/military operations office can close a runway.

closed shop
Exists where workers must be members of a union as a condition of their employment. This practice was made unlawful by the Taft-Hartley Act.

closed shop contract
A contract requiring an employer to hire only union members and to discharge nonunion members and requiring that employees, as a condition of employment, remain union members. A "closed shop" provision in a collective bargaining agreement requires membership in the contracting union before a job applicant can be employed and for the duration of his/her employment.

closed traffic
Successive operations involving takeoffs and landings or low approaches where the aircraft does not exit the traffic pattern.

closed union
A labor union whose membership rolls have closed. *See also* ***closed shop***.

closed window
A display window not accessible to the user without taking some specific action to gain access.

closing argument
The final statements by an attorney to the jury or the court summarizing the evidence that they think they have established and the evidence that they think the other side has failed to establish. The argument *does not* constitute evidence and may be limited in time by rule of the court.

closing dam
An earthen, sand, rock, or rock and brush structure built across sloughs or back channels to stop current flow at water stages below the crest elevation of the structure. Low flows are thus diverted to the main channel.

closure
Occurs when a series of pattern elements are perceived as a single unit, rather than unrelated parts.

closure plan
Written plan to decommission and secure a hazardous waste management facility.

clothes changing allowance
Any work time for which an employee is paid due to a requirement for removing one clothing assembly and donning another. Also called *clothes changing time*.

clothes changing time
See ***clothes changing allowance***.

clothing
Any tailored or processed material or combination of materials which may be used to cover the body or its parts, for whatever purposes.

clothing area factor (f_{cl})
That portion of increased surface area over the nude body which is added by clothing. Represented by the formula:

$$f_{cl} = \frac{\textit{clothed body surface area}}{\textit{nude body surface area}}$$

clothing assembly
See ***clothing system***.

clothing ensemble
See ***clothing system***.

clothing fastener
Any device, mechanism, or system for attaching different articles of clothing or portions of a single piece of clothing together.

clothing insulation value
See ***thermal insulation value of clothing***.

clothing system
The combination of garments and their arrangement being worn on the body at any one time. Also referred to as *clothing assembly* and *clothing ensemble*.

clotting
The formation of a jelly-like substance over the ends or within the walls of a blood vessel, with resultant stoppage of the blood flow. A natural defense mechanism of the body. A clot usually forms within 5 minutes after a blood vessel wall has been damaged.

cloud
A mass of small water droplets in the atmosphere that are not of sufficient size to fall to the earth.

cloud chamber
A device for observing the paths of ionizing particles, based on the principle that supersaturated vapor condenses more readily on ions than on neutral molecules.

cloud seeding
The introduction of artificial substances such as silver iodide or dry ice into clouds to induce rain.

cloud streets
Term used to describe lines or rows of cumuliform clouds.

cloud-to-ground lightning
See ***lightning***.

cloudburst
Any sudden and heavy rain shower.

CLP
See ***Contract Laboratory Program***.

cluster
(1) *Epidemiology*. An increased incidence or suspected excess occurrence of a disease in time, location, area, occupation, etc. (2) *Fibers*. A structure with fibers in a random arrangement such that all fibers are intermixed and no single fiber is isolated from the group, with the groupings having more than two intersections.

cluster workstation
A multi-person workstation built around a central core to provide some separation from co-workers.

cluster zoning
Such zoning modifies lot size and frontage requirements on certain conditions involving setting aside of land by the developer for parks, schools, or other public needs.

clutter
In radar operations, clutter refers to the reception and visual display of radar returns caused by precipitation, chaff, terrain, numerous aircraft targets, or other phenomena. Such returns may limit or preclude Air Traffic Control (ATC) from providing services based on radar.

cm
Centimeter.

CMA
See ***Chemical Manufacturers Association***.

CMOS
Complimentary metal oxide semiconductor sensor.

CMSA
See ***consolidated metropolitan statistical area***.

CMV
See ***commercial motor vehicle***.

CNG
See ***compressed natural gas***.

CNS
See ***central nervous system***.

CNS effect
An effect which occurs to the central nervous system, including drowsiness, dizziness, loss of coherence, and reasoning, as well as other effects.

coach passenger revenue
Revenues from the air transportation of passengers moving at fares reduced from the first class or premium fares which are predicated upon both the operation of specifically designated aircraft space and a reduction in the quality of service regularly and ordinarily provided.

coach service
Transport service established for the carriage of passengers at special reduced passenger fares that are predicated on both the operation of specifically designed aircraft space and a reduction in the quality of service regularly and ordinarily provided.

co-administrator
One who is a joint administrator with one or more others.

coagulation
The destabilization and initial aggregation of finely divided suspended solids by the addition of a polyelectrolyte or a biological process.

coagulin
An antibody (precipitin) that coagulates its antigen.

coal

A black or brownish-black solid, combustible substance formed by the partial decomposition of vegetable matter without access to air. The rank of coal, which includes anthracite, bituminous coal, sub-bituminous coal, and lignite, is based on fixed carbon, volatile matter, and heating value. Coal rank indicates the progressive alteration, or coalification, from lignite to anthracite. Lignite contains approximately 9 to 17 million British Thermal Unit (BTU) per ton. The heat contents of sub-bituminous and bituminous coal range from 16 to 24 million BTU per ton, and from 19 to 30 million BTU per ton, respectively. Anthracite contains approximately 22 to 28 million BTU per ton.

coal gasification

The conversion of solid coal to a gas mixture to be used as a fuel.

coal miner's pneumoconiosis

A pneumoconiosis resulting from the deposition of coal dust in the lungs. Characterized by emphysema. Also referred to as *black lung disease* and *coal worker's pneumoconiosis (CWP).*

coal pile runoff

Rainfall runoff from or through a coal storage pile.

coal slurry

Finely crushed coal mixed with sufficient water to form a fluid.

coal tar

A black viscous liquid with a naphthalene-like odor that is obtained by the destructive distillation of bituminous coal and used as a raw material for dyes, solvents, and many other products. Coal tar is known to contain many carcinogens and, thus, its use has been extremely curtailed.

coal worker's pneumoconiosis

See ***coal miner's pneumoconiosis***.

coalesce

The merging of two droplets to form a single, larger droplet.

coalescence

The merging of cloud droplets into a single larger droplet.

Coalition for Environmentally Responsible Economies (CERES)

Ten environmentally responsible principles originally known as the "Valdez Principles" with which a company voluntarily complies initially and in subsequent years. Some of the benefits to the company are positive publicity, lowering costs, and increasing revenues by recycling activities, and strengthening its environmental standards to avoid disasters.

Coanda effect

The tendency of a liquid coming out of a nozzle or orifice to travel close to the wall contour even if the wall curves away from the jet's axis.

coarse-bubble aeration

An aeration system that utilizes submerged diffusers which release relatively large bubbles.

coarse sand

Sand particles, usually larger than 0.5 mm.

coarse screen

A screening device usually having openings greater than 25 mm (1").

coast

(1) *General.* To continue moving without the additional application of mechanical or physical power, as in coasting in a motor vehicle, on a bicycle, or other type of vehicle. (2) *Ecology.* The edge or margin of a country bounding on the sea. The term includes small islands and reefs naturally connected with the adjacent land, and rising above the surface of the water. This word is particularly appropriate to the edge of the sea, while "shore" may be used to describe the margins of inland waters. In precise modern usage, the term "shore" denotes a line of low-water mark along a mainland, while the term "coast" denotes a line of shore plus the line where inland waters meet the open sea.

Coast Guard

The Coast Guard is responsible for enforcing federal laws on the high seas and navigable waters of the United States and its possessions. Navigation and vessel inspection laws are specific responsibilities. Under the provisions of the Federal Boating Act of 1958, Coast Guard boarding teams inspect small boats to insure compliance with required safety measures. The Coast Guard cooperates with other agencies in their law enforcement responsibilities, including enforcement of drug, conservation, and marine environmental laws.

coast waters
Tide waters navigable from the ocean by sea-going craft, the term embracing all waters opening directly or indirectly into the ocean and navigable by ships coming in from the ocean as great as that of the larger ships which traverse the open seas.

coastal
Means transits to or from the Maritimes and U.S. Atlantic Ports.

Coastal Air Defense Identification Zone
An Air Defense Identification Zone (ADIZ) over the coastal waters of the United States.

coastal ecosystem
A system of interacting biological, chemical, and physical components throughout the water column, water surface, and benthic environment of coastal waters.

coastal fix
A navigation aid or intersection where an aircraft transitions between the domestic route structure and the oceanic route structure.

coastal reclamation
Reclaiming land from shallow coastal areas of the sea by dumping rubble and refuse, or by constructing breakwaters and sea walls and drainage of the enclosed area.

coastal state
A state of the United States in, or bordering on, the Atlantic, Pacific, or Arctic Oceans, the Gulf of Mexico, Long Island Sound, or one or more of the Great Lakes. Under the Federal Coastal Wetlands Planning, Protection and Restoration Act, the term may also include Puerto Rico, the Virgin Islands, Guam, the Commonwealth of the Northern Mariana Islands, and the Trust Territories of the Pacific Islands and American Samoa.

coastal water quality (national coastal monitoring)
The physical, chemical and biological parameters that relate to the health and integrity of coastal ecosystems.

coastal water quality monitoring (national coastal monitoring)
A continuing program of measurement, analysis, and synthesis to identify and quantify coastal water quality conditions and trends to provide a technical basis for decision making.

coastal waters (United States)
Waters of the Great Lakes, including their connecting waters and those portions of rivers, streams, and other bodies of water having unimpaired connection with the open sea up to the head of tidal influence, including wetlands, intertidal areas, bays, harbors, and lagoons, including waters of the territorial sea of the United States and the contiguous zone.

Coastal Wetlands Conservation Project
Under the Federal Coastal Wetlands Planning, Protection and Restoration Act, the obtaining of a real property interest in coastal lands or waters, if the obtaining of such interest is subject to terms and conditions that will ensure that the real property will be administered for the long-term conservation of such lands and waters and the hydrology, water quality, and fish and wildlife dependent thereon; and the restoration, management, or enhancement of coastal wetlands ecosystems if such restoration, arrangement, or enhancement is conducted on coastal lands and waters that are administered for the long-term conservation of such lands and waters and the hydrology, water quality, and fish and wildlife dependent thereon.

Coastal Wetlands Restoration Project
Under the Federal Coastal Wetlands Planning, Protection and Restoration Act, any technically feasible activity to create, restore, protect, or enhance coastal wetlands through sediment and freshwater diversion, water management, or other measures that the Task Force (Louisiana Coastal Wetlands Conservation and Restoration Task Force) consisting of the Secretary of the Army, who shall serve as chairman, the EPA Administrator, the Governor of Louisiana, the Secretary of the Interior, the Secretary of Agriculture and the Secretary of Commerce, finds will significantly contribute to the long-term restoration or protection of the physical, chemical and biological integrity of coastal wetlands in the State of Louisiana, and includes any such activity authorized

under the Coastal Wetlands Planning, Protection and Restoration Act or under any other provision of law, including, but not limited to, new projects, completion or expansion of existing or on-going projects, individual phases, portions, or components of projects and operation, maintenance and rehabilitation of completed projects. The primary purpose of a "coastal wetlands restoration project" shall not be to provide navigation, irrigation or flood control benefits.

coastal zone

(1) Lands and waters adjacent to the coast that exert an influence on the uses of the sea and its ecology, or, inversely, whose uses and ecology are affected by the sea. (2) In the United States the coastal waters (including the lands therein and thereunder) and the adjacent shore lands (including the waters therein and thereunder), strongly influenced by each other and in proximity to the shorelines of the several coastal states, and include islands, transitional and intertidal areas, salt marshes, wetlands, and beaches. The zone extends, in Great Lakes waters, to the international boundary between the United States and Canada and, in other areas, seaward to the outer limit of State title and ownership under the Federal Submerged Lands Act, the Act of March 2, 1917, the Covenant to Establish a Commonwealth of the Northern Mariana Islands in Political Union with the United States of America, as approved by the Act of March 24, 1976, or the Act of November 20, 1963, as applicable. The zone extends inland from the shorelines only to the extent necessary to control shore lands, the uses of which have a direct and significant impact on the coastal waters, and to control those geographical areas which are likely to be affected by or vulnerable to sea level rise. Excluded from the coastal zone are lands the use of which is by law subject solely to the discretion of or which is held in trust by the Federal Government, its officers, or agents. The term "coastal zone" refers to all United States waters subject to the tide, waters of the Great Lakes and Lake Champlain, specified ports and harbors on inland rivers, waters of the contiguous zone, other waters of the high seas subject to the National Contingency Plan, and the land surface or land substrate, groundwaters, and ambient air proximal to those waters. The term "coastal zone" delineates an area of federal responsibility for response action. Precise boundaries are determined by agreements between the Environmental Protection Agency (EPA) and the U.S. Coast Guard (USCG), and are identified in Federal Regional Contingency Plans and Area Contingency Plans.

Coastal Zone Management Act (CZMA)

An Act requiring all federal agencies and permittees who conduct activities affecting a state's coastal zone to comply with an approved state coastal zone management program.

coastwise traffic

Domestic traffic which moves over the ocean, or the Gulf of Mexico, e.g., between New Orleans and Baltimore, New York and Puerto Rico, San Francisco and Hawaii, Puerto Rico and Hawaii. Traffic between Great Lakes ports and seacoast ports, when having a carriage over the ocean, is also deemed to be coastwise. The Chesapeake Bay and Puget Sound are considered internal bodies of water rather than arms of the ocean; traffic confined to these areas is deemed to be "internal" rather than coastwise.

cobalt

A chemical element, atomic number 27, atomic weight 58.933, symbol Co. Radioisotopes of cobalt are used for implantation in the treatment of various forms of malignancy and also serve as the radioactive source in teletherapy machines.

COC

See ***cycles of concentration***.

cocaine

A white crystalline narcotic alkaloid extracted from coca leaves. Used as a local anesthetic. A "controlled substance" as included in narcotic laws.

co-carcinogen

A substance that works symbiotically with a carcinogen in the development of cancer.

coccidioidomycosis

See ***desert fever***.

coccus
Spherical bacteria cells. Cocci (plural) may appear singly, in pairs called diplococci, in chains, or in grape-like clusters.

coccyx
A triangular-shaped bone at the base of the spine formed by the fusion of the lowest four (sometimes five or three) vertebrae, and forming the caudal extremity of the vertebral column.

cochlea
A spiral cavity of the inner ear, shaped like a snail shell, that contains the organ of hearing. The cochlea is filled with fluid and is connected with the middle ear by two membrane-covered openings, the oval window (fenestra vestibuli) and the round window (fenestra cochleae). Inside the cochlea is the organ of Corti, a structure of highly specialized cells that translate sound vibrations into nerve impulses. The cells of this organ have tiny hair-like strands (cilia) that protrude into the fluid of the cochlea. Sound vibrations are relayed from the tympanic membrane (eardrum) to the bones of hearing in the middle ear to the oval window of the cochlea, where they set up corresponding vibrations in the fluid of the cochlea. These vibrations move the cilia of the organ of Corti, which then sends nerve impulses to the brain.

cochlear duct
A tube-shaped structure within the cochlea which is filled with endolymph and contains the organ of Corti and the tectorial membrane.

cochleitis
Inflammation of the cochlea.

cochleovestibular
Pertaining to the cochlea and vestibule of the ear.

cockpit
The location within a vehicle from which control of the vehicle and observation of the external environment and events may occur. Usually refers to the operating/control area of an aircraft.

The cockpit (or cabin) of an aircraft with the pilot in command seated on the left and the co-pilot seated on the right

COD
See ***certificate of disposal***. Also, acronym for chemical oxygen demand.

code
(1) To translate information or data from one form or symbol to another form or symbol which has a meaning in its own context. (2) A set of mandatory standards or regulations adopted by a local, regional, national, or international governmental agency which has the force and/or effect of law; a set of recommended rules or guidelines within an industry. (3) A sequence of steps in some process, such as a computer program or task. (4) A system of symbols which can be used to organize and/or communicate information about conditions, processes, or entities. (5) The number assigned to a particular multiple pulse reply signal transmitted by a transponder.

Code of Federal Regulations
The Code of Federal Regulations (CFR) is the annual accumulation of executive agency regulations published in the daily Federal Register (FR), combined with regulations is-

sued previously that are still in effect. Divided into 50 titles, each representing a broad subject area, individual volumes of the Code of Federal Regulations are revised at least once each calendar year and issued on a staggered quarterly basis. The CFR contains the general body of regulatory laws governing practice and procedure before federal administrative agencies. The CFR title for Labor, which includes occupational safety and health regulations, is number 29. Environmental regulatory requirements can be found in Title 40 and Transportation regulations are contained in Title 49.

Code of Hammurabi
Prescribed indemnification for major injuries or death in circa 1792-1750 B.C. Required the punishment for causing the injury or death of another person to match the loss incurred, literally *an eye for an eye.* The Code's primary purpose was to assess blame and provide indemnification (or revenge), rather than to determine the cause of the accident itself. These set of laws were once considered the oldest promulgation of laws in human history prepared by a Babylonian king.

Code of Military Justice
This code, which is uniformly applicable in all its parts to the Army, the Navy (including the Marines), the Air Force, and the Coast Guard, covers both the substantive and the procedural law governing military justice and its administration in all of the armed forces of the United States. The Code established a system of military courts, defines offenses, authorizes punishment, provides broad procedural guidance, and statutory protection which conform to the due process safeguards preserved and established by the Constitution. As an additional safeguard for an accused person, the Code also provides for a system of automatic appellate review. A Court of Military Review is established within each service to review all court-martial cases where the sentence includes death, punitive discharge, or confinement for one year or more. Appellate review in this court is automatic. No approved sentence of a court-martial may be executed unless such findings and sentence are affirmed by a Court of Military Appeals, which was established to review certain cases from all the Armed Forces. The latter Court consists of three civilian judges. Automatic review before the Court is provided for all cases in which the sentence, as affirmed by a Court of Military Review, affects a general or flag officer or extends to death. In addition, the Judge Advocate General of each service may direct that a case be reviewed by the Court. An accused may petition the Court for review.

coded track circuit
A track circuit in which the energy is varied or interrupted periodically.

codeine
An alkaloid obtained from opium or prepared from morphine by methylation.

codification
The process of collecting and arranging systematically, usually by subject, the laws of a state or country, or the rules and regulations covering a particular area or subject of law or practice. Examples include the United States Code, the Code of Military Justice, the Code of Federal Regulations, and the California Evidence Code. The end product may be called a code, revised code, or revised statutes.

codisposal
A method of sludge disposal where the digested sludge is mixed with sorted refuse and incinerated, composted, or treated by pyrolysis prior to final disposal.

COE
See ***cab-over-engine***.

coefficient
A number by which one value is to be multiplied in order to give another value, or a number that indicates the range of an effect produced under certain conditions.

coefficient alpha
A measure of the internal consistency and/or reliability of a scale.

coefficient of alienation (k)
A measure of the lack of relationship between two variables

$$k = \sqrt{1 - r^2}$$

where:
r = correlation coefficient

coefficient of concordance
See ***Kendall's coefficient of concordance***.

coefficient of correlation
*See **correlation coefficient**.*

coefficient of determination (r^2, d)
The proportion of the variance accounted for by the Pearson product-moment correlation coefficient; equal to the Pearson product-moment correlation coefficient squared. Synonymous with *generality*. *See also **Pearson product-moment correlation coefficient**.*

coefficient of entry (Ce)
The actual rate of flow caused by a given hood static pressure compared to the theoretical flow which would result if the static pressure could be converted to velocity pressure with 100% efficiency.

coefficient of evaporative heat transfer
*See **evaporative heat transfer coefficient**.*

coefficient of friction (μ)
*See **coefficient of rolling friction, coefficient of sliding friction, coefficient of static friction.*** Synonymous with *friction coefficient*.

coefficient of kinetic friction
*See **coefficient of rolling friction, coefficient of sliding friction.***

coefficient of haze (COH)
A measure of air visibility determined by the darkness of the stain remaining on white paper after it has been used to filter air.

coefficient of multiple correlation
*See **multiple correlation coefficient**.*

coefficient of non-determination (k^2)
That proportion of the variance between two variables not accounted for by the coefficient of determination.

coefficient of reflection
*See **reflection coefficient**.*

coefficient of reliability
*See **reliability coefficient**.*

coefficient of rolling friction
The ratio of the magnitude of the rolling force to the magnitude of the perpendicular force between two objects/surfaces at the point where their surfaces are parallel.

coefficient of sliding friction (μ)
The ratio of the magnitude of the sliding force to the magnitude of the perpendicular force between two objects/surfaces.

coefficient of static friction (μ)
The ratio of the magnitude of the static force to the magnitude of the perpendicular force between two objects/surfaces.

coefficient of utilization (CU)
The value of the ratio of the luminous flux reaching the workplace to the total luminous flux emitted from a lighting source, as in the formula:

$$CU = \frac{\textit{lumens reaching work surface}}{\textit{lumens emitted by lamp}}$$

coefficient of variation (Cv)
A statistical parameter equal to the standard deviation of the sample data divided by the mean of the data. It is often expressed as the percent coefficient of variation. Another term for it is the *relative standard deviation*.

coerce
Compelled to compliance; constrained to obedience, or submission in a vigorous or forcible manner.

COFC
*See **container on flatcar**.*

coffee break
*See **rest period**.*

cofferdam
A temporary dam, usually of sheet piling, built to provide access to an area that is normally submerged.

coffin
(1) A box or container used to contain the remains of a deceased animal, most typically humans, in perpetuity. (2) See ***cask***.

coffin-box
Transportation (slang). Sleeper compartment independent of truck cab.

cogeneration
A power system that simultaneously produces both electrical and thermal energy from the same source.

cognition
Those higher mental activities or intellectual functions.

cognitive disability
Any disability involving literacy, mental capacity, learning, non-motor speech processes, or perceptual processes.

cognitive dissonance
A discrepancy which exists between a person's attitudes or statements and behaviors.

cognitive reaction time
That temporal interval between the receipt of a stimulus and the initiation of a response in a task which requires some type of choice and which is presumed to involve cognitive processing. Also referred to as *decision time.*

cognitive restructuring
A mental exercise in which attempts are made by an individual to change certain personal beliefs.

cognizance
Jurisdiction, or the exercise of jurisdiction, or power to try and determine causes. Judicial examination of a matter, or power and authority to make it.

Cognizant Officer in Charge of Marine Inspection (OCMI)
The Officer in Charge Marine Inspection in which the manufacturer responsible for defect notification (or other corrective action) is located. Commandant (G-MVI) directs specific actions for OCMIs to take in cases where a single cognizant OCMI cannot be identified.

COH
See ***coefficient of haze.***

COHb
See ***carboxyhemoglobin.***

coherence
A measure of the correlation at each frequency, or within each frequency band, between two time series signals.

coherent
A light beam is coherent when its waves have a continuous relationship among phases.

cohort
A group of individuals selected for scientific study in toxicology, epidemiology, or some other study focus.

cohort analysis
A method used in employment discrimination suits to test for race discrimination whereby all employees who start together at the same level are surveyed over the course of an observation period and their comparative progress in salary and promotion is evaluated.

cohort study
An epidemiological study where population subgroups with a common exposure to a suspected disease-causing agent are studied over time to determine the risk of developing disease.

coincidental peak-day flow
The volume of gas that moves through a pipeline or section thereof or is delivered to a customer on the day of the year when the pipeline system handles the largest volume of gas.

coinsurance
A relative division of risk between the insurer and the insured, dependent upon the relative amount of the policy and the actual value of the property insured, and taking effect only when the actual loss is partial and less than the amount of the policy; the insurer being liable to the extent of the policy for a loss equal to or in excess of that amount. Insurance policies that protect against hazards such as fire or water damage often specify that the owner of the property may not collect the full amount of insurance for a loss unless the insurance policy covers at least some specified percentage, usually 80 percent, of the replacement cost of the property. *See also* ***insurance.***

coinsurance clause
Provision in insurance policy requiring a property owner to carry insurance up to an amount determined in accordance with the provisions of the policy.

coke
The solid carbon residue resulting from the distillation of coal or petroleum.

coke tray aerator
An aerator where water is sprayed or flows over coke-filled trays.

cold
(1) An acute and highly contagious virus infection of the upper respiratory tract. *See also* ***common cold.*** (2) A relatively low temperature as compared with the normal temperature; the lack of heat. A total absence of heat is absolute zero, at which all molecular motion ceases. A body temperature below 94°F results in impairment of the heat-regulating center in the hypothalamus. As the temperature drops, sleepiness and coma develop, and

as a result the central nervous system heat-control mechanism is depressed and shivering (a means of heat production) is prevented.

cold clouds
*See **supercooled cloud**.*

cold fog
*See **supercooled cloud**.*

cold front
A transition zone where a cold air mass advances and replaces a warm air mass.

cold lime-soda softening
Lime-soda softening process of water treatment at ambient temperatures.

cold occlusion
*See **occluded front**.*

cold start fluorescent lamp
*See **instant start fluorescent lamp**.*

cold stress
A form of environmental/thermal stress in which too much body heat is lost to a cold environment.

cold wave
A rapid fall in temperature within 24 hours that often requires increased protection for agriculture, industry, commerce, and human activities.

cold work
Mechanical or other type of work of a non-sparking nature that presents no risk of fire or explosion.

colic
A severe cramping in the abdomen.

coliform bacteria
Rod-shaped bacteria living in the intestines of humans and other warm-blooded animals.

coliform index
A rating of the purity of water based on the count of fecal bacteria present in it.

coliform organism
Microorganisms found in the intestinal tract of humans and animals. Their presence in water indicates potentially dangerous bacterial contamination.

collagen
A scleroprotein present in connective tissue of the body.

collagen diseases
A group of poorly understood diseases that cause deterioration of the connective tissues. Their cause is unknown, and the relationships among them are unclear. Apparently they are not infectious. Widely varying symptoms often make early diagnosis difficult. The four major forms of collagen diseases include systemic lupus erythematosus, periarteritis nodosa, scleroderma, and dermatomyositis. In addition to these four diseases, rheumatoid arthritis and rheumatic fever are frequently held to belong to the group.

collapse
(1) To break down or flatten. (2) A state of extreme prostration.

collar cloud
*See **wall cloud**.*

collarbone
*See **clavicle**.*

collateral source rule
Under this rule, if an injured person receives compensation for his/her injuries from a source wholly independent of the tort-feasor, the payment should not be deducted from the damages which he/she would otherwise collect from the tort-feasor. In other words, a defendant tort-feasor may not benefit from the fact that the plaintiff has received money from other sources as a result of the defendant's tort (e.g., sickness and health insurance).

collection
In solid waste management, the act of removing solid waste (or materials which have been separated for the purpose of recycling) from a central storage point.

collection efficiency
A measure of sampler performance as determined from the ratio of the material collected to the amount present in the sampled air. Typically expressed as a percentage.

collection main
The public sewer to which a building service or individual system is connected.

collective bargaining
(1) As contemplated by National Labor Relations Act (N.L.R.A), a procedure looking toward the making of collective agreements between an employer and accredited representatives of union employees concerning

wages, hours, and other conditions of employment, and requires that both parties deal with each other with open and fair minds and sincerely endeavor to overcome obstacles existing between them to the end that employment relations may be stabilized and obstruction to the free flow of commerce prevented. (2) Negotiation between an employer and organized employees as distinguished from individuals, for the purpose of determining by joint agreement the conditions of employment.

collective bargaining agreement
An agreement between an employer and a labor union which regulates terms and conditions of employment. The joint and several contract of members of a union made by officers of the union as their agents establishing, in a general way, the reciprocal rights and responsibilities of employer, employees collectively, and the union. Such is enforceable by and against the union in matters which affect all members, particularly those who are employees of the other party to contract.

collective bargaining unit
All of the employees of a single employer, unless the employees of a particular department or division have voted otherwise.

collective labor agreement
Also called "trade agreement." Bargaining agreement as to wages and conditions of work entered into by groups of employees, usually organized into a brotherhood or union on one side and groups of employers or corporations on the other side.

collector chain
Chain used to convey sludge scraper in a rectangular sludge collector.

collector highway
Collector highways are those highways which link local highways to arterial highways.

collector ring
An assembly of slip rings for transferring electrical energy from a stationary to a rotating member.

collectors
Transportation. In rural areas, routes serving intra-county, rather than statewide travel. In urban areas, streets providing direct access to neighborhoods as well as direct access to arterials.

Collier's disease
*See **black lung disease**.*

collimate
Make parallel to a certain path.

collimated beam
A beam of light or electromagnetic radiation with parallel waves.

collimating optics
Those optical components, such as lenses, used to produce parallel rays of light.

collimation
The confining of a beam of particles or rays to a defined cross-section.

collimator
A device for confining a beam, such as radiation, within a solid angle.

collision
(1) *General.* Striking together of two objects, one of which may be stationary. Act or instance of colliding; state of having collided. The term implies an impact or sudden contact of a moving body with an obstruction in its line of motion, whether both bodies are in motion or one stationary and the other, no matter which, is in motion. (2) *Railroad.* An impact between on-track equipment consists while both are on rails and where one of the consists is operating under train movement rules or is subject to the protection afforded to trains. This includes instances where a portion of a consist occupying a siding is fouling the main line and is struck by an approaching train. It does not include impacts occurring while switching within yards, as in making up or breaking up trains, shifting or setting out cars, etc.

collision accident
An accident involving a collision between a commercial motor vehicle and another object. Collision objects include trains, other motor vehicles, pedestrians, bicyclists, animals, and fixed objects.

collision between aircraft
Is so classified only when both aircraft are occupied. This includes collisions wherein both aircraft are airborne (midair); both on the ground or where one is airborne and the other

on the ground. A collision with a parked unoccupied aircraft is classified under the broad category of collision with objects.

collision insurance
A form of automobile insurance that covers loss to the insured vehicle from its collision with another vehicle or object, but not covering bodily injury or liability also arising out of the collision. A type of coverage which protects the insured for damages to his/her own property in an accident as contrasted with liability insurance which protects him/her in an action or claim for loss to another person's property. *See also* ***insurance*** *and* ***convertible collision insurance***.

collision with another vessel
Any striking together of two or more vessels, regardless of operation at time of the accident, is a collision. (Also includes colliding with the tow of another vessel, regardless of the nature of the tow, i.e., surfboard, ski ropes, tow line, etc.)

collision with fixed object
The striking of any fixed object, above or below the surface of the water.

collision with floating object
Collision with any waterborne object above or below the surface that is free to move with the tide, current, or wind, except another vessel.

collision with object
(1) *Transportation.* An incident in which a transit vehicle strikes an obstacle other than a vehicle or person (e.g., building, utility pole). Reports are made if the accident results in a death, injury, or property damage over $1,000. (2) *Aviation.* Where an occupied aircraft collides with a parked unoccupied aircraft or some other object.

collision with other vehicles
An incident involving one or more transit agency vehicles and any other vehicle. Report collisions between rail cars from coupling operations. Report fatalities or injuries that occur inside the transit vehicle as well as fatalities or injuries that occur inside other involved vehicles.

collision with people
An incident in which a transit vehicle strikes a person. Except where specifically indicated, collisions with people do not include suicide attempts. Reports are made if the incident results in death, injury, or property damage over $1,000.

collision with vehicle
An incident in which a transit vehicle strikes or is struck by another vehicle. Reports are made if the accident results in death, injury, or property damage over $1,000.

colloid
Suspended solid with a diameter less than one micron that cannot be removed by sedimentation alone.

collusion
The behavior in which one person acts on behalf or to the benefit of himself/herself plus one or more others.

colony-forming units (CFU)
The number of bacteria present in a sample as determined in a laboratory plate count test where the number of visible bacteria colony units present is counted.

colophony
Rosin, such as that used in rosin core solder.

color
(1) A property of a surface or substance due to absorption of certain light rays and reflection of others within the range of wavelengths (roughly 370 to 760 mμ) adequate to excite the retinal receptors. (2) Radiant energy within the range of adequate chromatic stimuli of the retina, i.e., between the infrared and ultraviolet. (3) A sensory impression of one of the rainbow hues. (4) Water condition resulting from presence of colloidal material (see ***apparent color***) or organic matter (see ***true color***), measured by visual comparison with lab prepared standards. (5) That aspect of visual perception due solely to stimulation of the retinal cones by different wavelengths of electromagnetic radiation within the visible spectrum, and neglecting such aspects of a stimulus such as structure, size, and pattern.

color additive
Under the Federal Food, Drug and Cosmetic Act, a material which (a) is a dye, pigment, or other substance made by a process of synthesis or similar artifice, or extracted, isolated, or otherwise derived, with or without intermediate or final change of identity, from a vegetable, animal, mineral, or other source;

and (b) when added or applied to a food, drug, or cosmetic, or to the human body or any part thereof, is capable (alone or through reaction with other substance) of imparting color thereto.

color assimilation
A type of chromatic induction in which the difference or contrast between adjacent, differently colored fields diminishes. Also known as *Bezold spreading effect.*

color blindness
Inability to distinguish between certain colors. Genuine color blindness, a complete inability to see colors, is quite rare, affecting only one person in 300,000. Generally, the term describes some form of deficiency of color vision. The most common form is red-green confusion, which affects approximately 8 million people in the United States. There is no known cure for color deficiency. Color vision is a function of the cones in the retina of the eye, which are stimulated by light and transmit impulses to the brain. It is now thought that there are three types of cones, each type stimulated by one of the primary colors in light (red, green, and violet). Most cases of color deficiency affect either the red or green receptors, so that the two colors do not appear distinct from each other. *See also* ***monochromasia, protanopia, deuteranopia, trianopia, monochromat,*** *and* ***color vision deficiency***.

color coding
The use of multiple colors for easier, more rapid visual identification, access, and/or processing of groups of organized materials.

color constancy
The phenomenon in which an object appears to have approximately the same color under different lighting conditions.

color contrast
See ***chromatic contrast***.

color contrast threshold
See ***chromatic contrast threshold***.

color correction
An adjustment made for the presentation of color image, usually to make perceived colors appear more natural.

color deficiency
See ***color vision deficiency***.

color discrimination
The ability to perceive visual matches or note differences between hue, saturation, and brightness of two or more colored stimuli. Synonymous with *visual color discrimination.*

color formulation
The use of any one or combination of methods for making a desired color, including mathematical models, materials, colorants, particle sizes, absorption coefficients, and scattering coefficients.

color grade
A measure of the color appearance of a product, which may be used to determine price or quality.

Color index
A publication by the American Association of Textile Chemists and Colorists which provides a large number of reference dyes and pigments.

color match
(1) A condition in which two colored stimuli appear identical. (2) Make one variable color stimulus appear the same as a reference stimulus.

color matching function
See ***tristimulus value***.

color mixing
The bending of colored lights or materials to alter an existing color. *See also* ***additive color mixing*** *and* ***subtractive color mixing***.

color ordering system
Any method for the unambiguous interpolation between closely related colors within a large set of colors. Also referred to as *CIE color system, Munsell color system, Federal Standard 595a, coloroid color system, Inter-Society Color Council - National Bureau Color System.*

color pseudo-stereopsis
The visual perception of depth or of structure being out of the background plane from objects emitting or reflecting different dominant frequencies or wavelengths (especially blues and reds) within a dark background. Also referred to as *chromostereopsis*.

color rendering
That effect which a light source other than a standard illuminant has on the apparent color of an object.

color rendering index
See ***CIE color rendering index.***

color saturation
A perceptual attribute pertaining to the strength or vividness of a particular hue.

color system
See ***color ordering system.***

color temperature (T_c)
The temperature (in Kelvin) of a radiating blackbody having the same chromaticity or spectral distribution as a given color light source.

color temperature scale
A scale by which the color of light emitted from an incandescent source is related to temperature, normally corresponding to the Kelvin scale. Also referred to as *temperature color scale.*

color vision
See ***photopic vision.***

color vision deficiency
Having some form of reduced color sensation ability. *See also* ***protanomaly, deuteranomaly, tritanomaly, color blindness.***

color wheel
A disk consisting of multiple colored and appropriately interleaved radial segments, each segment being a single color, for providing a desired perceptual color mixture when spun rapidly.

colorant
Any substance added to a product to provide a different color. Also referred to as *coloring agent.*

Colorcurve®
A color ordering system based on the physical brightness of gray levels.

colorfulness
That attribute of a visual sensation which appears to exhibit more or less of its hue.

colorimeter
A photoelectric instrument used to measure the amount of light of a specific wavelength absorbed by a solution.

colorimetric tube
See ***detector tube.***

colorimetry
An analytical method in which color is developed in a reaction between the sorbent and a contaminant with the resulting color intensity measured photometrically for determining contaminant concentration. *See also* ***photometry.***

coloring agent
See ***colorant.***

colorless
Having no chromatic color; achromatic.

coloroid color system
A color ordering system which attempts to provide equal aesthetic spacing between colors and is based on specifications of hue (A), chromatic content (T), and lightness (V). *See also* ***color ordering system.***

column
A vertical arrangement of numbers, text, or other information in a matrix or table.

coma
A state of unconsciousness from which the person cannot be aroused by physical stimulation. *See also* ***unconsciousness.***

COMBIMAN
See ***Computer Biomechanical Man.***

combination chain
Chain used in conveyor applications, having cast block links with steel pins and connecting bars.

combination export manager
A firm which acts as an export sales agent for two or more U.S. manufacturers, all of which are noncompetitive with the others.

combination packaging
A combination of packaging, for transport purposes, consisting of one or more inner packagings secured in a non-bulk outer packaging. It does not include a composite packaging.

combination passenger and cargo ships
Ships with a capacity for 13 or more passengers.

combination tone
An apparent secondary tone heard when two pure primary tones having widely separated frequencies are presented simultaneously.

combination trailer
A trailer used to handle freight in the transportation of goods for others; excludes house trailers, light farm trailers, and car trailers.

combination truck
(1) Consists of a power unit (a truck tractor) and one or more trailing units (a semitrailer or trailer). The most frequently used combination is popularly referred to as a "tractor-semitrailer" or "tractor trailer." (2) A tractor not pulling a trailer; a tractor pulling at least one full or semitrailer, or a single-unit truck pulling at least one trailer.

combined available chlorine
The concentration of chlorine combined with ammonia as chloramine and still available to oxidize organic matter.

combined center
An air traffic facility which combines the functions of an air route traffic control center and a radar approach control facility.

combined household energy expenditures
(1) The total amount of funds spent for energy consumed in, or delivered to, a housing unit during a given period of time and for fuel used to operate the motor vehicles that are owned or used on a regular basis by the household. (2) The total dollar amount for energy consumed in a housing unit includes state and local taxes but excludes merchandise repairs or special service charges. Electricity and natural gas expenditures are for the amount of those energy sources consumed. Fuel oil, kerosene, and LPG expenditures are for the amount of fuel purchased, which may differ from the amount of fuel consumed. (3) The total dollar amount of fuel spent for vehicles is the product of fuel consumption and price.

combined motions
Two or more parallel elemental movements performed by a given body segment.

combined sewer overflow (CSO)
Wastewater flow that consists of storm water and sanitary sewage.

combined sewers
A sewer system that carriers both sewage and stormwater runoff. Normally, its entire flow goes to a waste treatment plant, but during a heavy storm, the stormwater volume may be so great as to cause overflows. When this occurs, untreated mixtures of storm water and sewage may flow into receiving waters. Stormwater runoff may also carry toxic chemicals from industrial areas or streets into the sewer system.

combined work
A job or task involving any combination of two or more workers or workers and multiple machines.

combustible
Capable of being ignited with resultant burning or explosion.

combustible dust
A dust that is capable of undergoing combustion or of burning when subjected to a source of ignition.

combustible gas indicator (CGI)
An instrument for determining the presence and concentration of a combustible and/or flammable hydrocarbon vapor/gas-air mixture relative to the lower explosive limit of the substance. Essentially all combustible or flammable vapors or gases can be detected with this type of device, but their concentration cannot be determined accurately unless the instrument has been calibrated for the specific substance or mixture. It is essential that adequate oxygen be present (i.e., above approximately 12%) for the proper operation of this type of detector.

combustible gases
The mixture of gases and vapors produced by burning.

combustible liquid
Any liquid that has a flash point at or above 100 degrees Fahrenheit and below 200 degrees Fahrenheit.

combustibles
Refers to materials that can be ignited at a specific temperature in the presence of air to release heat.

combustion
A chemical process that involves oxidation sufficient to produce light or heat. Also referred to as *fire*.

combustion product
Substance produced during burning or oxidation of a material.

comfort
(1) A state of subjective well-being in relation to one's external environment. (2) The absence of significant or excessive physical and/or mental stressors.

comfort-discomfort boundary
That threshold luminance in a glare condition at which visual discomfort becomes apparent. Synonymous with *borderline between comfort and discomfort*.

Comfort-Health Index (CHI)
A table based on a computed effective temperature, assuming a 50% relative humidity, for determining expected thermal sensations.

comfort rating
An expressed measure of the level of satisfaction with one or more aspects of an individual's current environment.

comfort rating scale
Any of a number of ranking techniques for rating comfort.

comfort ventilation
Airflow intended to remove heat, odors, smoke, etc. from an inside location and provide a comfortable environment for occupants.

comfort zone
The range of effective temperatures, as identified by ASHRAE, over which the majority (50% or more) of adults feel comfortable. ASHRAE has identified combinations of dry- and wet-bulb temperatures and air movement for summer and winter conditions that provide comfort for room occupants.

comfortable reach
That range through which an individual can reach without straining excessively against gravity or a restraint.

comma cloud
A band of organized cumuliform clouds that looks like a comma on a satellite photograph.

command
Any statement which may potentially be input to a computer system and which calls for one or more specific actions.

command area
A region within a display in which user input commands are presented for viewing.

command error
An inappropriate or incorrect command entered into a computer.

command input
The entering of a command to a system.

command language
A clearly defined specific set of terms for directing control of a computer.

command line
A command area composed of a single line on a text display which is reserved for user-entered commands.

command post
Facility located at a safe distance upwind from an accident site, where the on-scene coordinator, responders, and technical representatives can make response decisions, deploy manpower and equipment, maintain liaison with the news media, and handle communications.

comment period
Time provided for the public to review and comment on a proposed EPA action or rulemaking after it is published in the Federal Register.

commerce
(1) The exchange of goods, productions, or property of any kind; the buying, selling, and exchanging of articles. Also, the transportation of persons and property by land, water, and/or air. (2) Under the Toxic Substances Control Act (TSCA), the term means trade, traffic, transportation, or other commerce between a place in a state and any other place outside a state or actions which might affect such trade, traffic, or commerce. (3) Any trade, traffic or transportation within the jurisdiction of the United States between a place in a state and a place outside of such state, including a place outside of the United States and trade, traffic, and transportation in the United States which affects any trade, traffic, and transportation described in the first part of this definition.

commerce clause
A provision in the U.S. Constitution (Art. I, § 8, cl. 3) granting Congress the power to authorize administrative agencies such as OSHA to act. Specifically, the commerce

clause grants Congress the power *"to regulate commerce ...among the States."*

Commerce Department
Part of the executive branch of the federal government headed by a cabinet member (Secretary of Commerce) which is concerned with the promotion of domestic and international business and commerce. It may also be a department of a state government with similar functions.

commerce power
*See **police power**.*

commercial
Relates to or is connected with trade and traffic or commerce in general; is occupied with business and commerce.

commercial activity
Term includes any type of business or activity which is carried on for a profit.

commercial air carrier
An air carrier certificated in accordance with Federal Aviation Regulations (FAR) Part 121 or 127 to conduct scheduled services on specified routes. These air carriers may also provide non-scheduled or charter services as a secondary operation. Four carrier groupings have been designated for statistical and financial data aggregation and analysis: *Majors* (annual operating revenues greater than $1 billion), *Nationals* (annual operating revenues between $100 million and $1 billion), *Large Regionals* (annual operating revenues between $10 million and $99,999,999), *Medium Regionals* (annual operating revenues less than $10 million).

commercial airport
A public airport which is determined to enplane annually 2,500 or more passengers and receive scheduled passenger service of aircraft. *See also **commercial service airport**.*

commercial body
Transportation. A body type not normally furnished by the original equipment manufacturer as a standard option but available from other manufacturers (e.g., dump, compactor tank, and utility).

commercial driver's license (CDL)
A license issued by a state or other jurisdiction, in accordance with the standards contained in 49 CFR 383 to an individual, which authorizes the individual to operate a class of a commercial motor vehicle.

commercial driver's license information system (CDLIS)
A database containing information on CDLs issued in the United States. Established by Federal Highway Administration (FHWA) pursuant to section 12007 of the Commercial Motor Vehicle Safety Act of 1986.

commercial insurance
Indemnity agreements, in the form of insurance bonds and policies, whereby parties to commercial contracts are, to a designated extent, guaranteed against loss by reason of a breach of contractual obligations on the part of the other contracting party. To this class belong policies of contract and title insurance. *See also **insurance**.*

commercial invoice
A document of the transaction between a buyer and a seller.

commercial law
A phrase used to designate the whole body of substantive jurisprudence (e.g., Uniform Commercial Code; Truth in Lending Act) applicable to the rights, intercourse, and relations of persons engaged in commerce, trade, or mercantile pursuits.

commercial motor vehicle (CMV)
(1) *Under 49 CFR 383.* A motor vehicle or combination of motor vehicles used in commerce to transport passengers or property if the motor vehicle: a) has a gross combination weight rating of 26,001 or more pounds inclusive of a towed unit with a gross vehicle weight rating of more than 10,000 pounds; or b) has a gross vehicle weight rating of 26,001 or more pounds; or c) is designed to transport 16 or more passengers, including the driver; or d) is of any size and is used in the transportation of materials found to be hazardous for the purposes of the Hazardous Materials Transportation Act and which require the motor vehicle to be placarded under the Hazardous Materials Regulations (49 CFR 172, subpart F). (2) *Under 49 CFR 350 and 49 CFR 390.* Any self-propelled or towed vehicle used on public highways in interstate commerce to transport passengers or property when: a) the vehicle has a gross vehicle weight rating or gross combination weight

rating of 10,001 or more pounds; or b) the vehicle is designed to transport more than 15 passengers, including the driver; or c) the vehicle is used in the transportation of hazardous materials in a quantity requiring placarding under regulations issued by the Secretary of Transportation under the Hazardous Materials Transportation Act (49 U.S.C. App. 1801-1813).

commercial motor vehicle traffic violation
See ***serious traffic violation*****.**

commercial operator
A person who, for compensation or hire, engages in the carriage by aircraft in air commerce of persons or property, other than as an air carrier or foreign air carrier or under the authority of Part 375 of this title. Where it is doubtful that an operation is for *compensation or hire,* the test applied is whether the carriage by air is merely incidental to the person's other business or is, in itself, a major enterprise for profit.

commercial passengers
Number of revenue and nonrevenue passengers arriving or departing via commercial aircraft and helicopters on a scheduled or nonscheduled flight.

commercial pilot
A commercial pilot may act as pilot-in-command of an aircraft carrying passengers for compensation or hire and act as pilot-in-command in an aircraft that is being operated for compensation or hire.

commercial rental motor vehicle
A motor vehicle obtained from a commercial source such as the Military Traffic Management Command (MTMC) rental car agreements for a period of 60 days or less.

commercial sector
As defined economically, consists of business establishments that are not engaged in transportation or in manufacturing or other types of industrial activity (agriculture, mining, or construction). Commercial establishments include hotels, motels, restaurants, wholesale businesses, retail stores, laundries, and other service enterprises; religious and nonprofit organizations (health, social, and educational institutions); and federal, state, and local governments. Street lights, pumps, bridges, and public services are also included if the establishment operating them is considered commercial.

commercial service airport
A public airport that is determined by the Secretary [of Transportation] to enplane annually 2,500 or more passengers and receive scheduled passenger service of aircraft. The commercial service airports are further categorized as *primary* and *non-primary*. *See also* ***commercial airport*****.**

commercial service non-primary airport
A commercial service airport which is determined by the Secretary [of Transportation] to enplane between 2,500 and 10,000 passengers annually.

commercial service primary airport
A commercial service airport which is determined by the Secretary [of Transportation] to have more than 10,000 passengers enplaned annually. A *primary airport* can be further classified as a large, medium, small, or non-hub.

commercial use
Use in commercial enterprise providing salable goods or services.

commercial waste
Solid waste from nonmanufacturing establishments such as office buildings, markets, restaurants, and stores.

comminuted
Broken into small pieces, as a type of bone fracture. *See also* ***comminutor*****.**

comminution
Mechanical shredding or pulverizing of waste. Used in both solid waste management and wastewater treatment.

comminutor
A circular screen with cutters that grinds large sewage solids into smaller, settling particles.

commission
(1) A warrant or authority, issuing from the government, or one of its departments, or a court, empowering a person or persons named to do certain acts, or to exercise the authority of an office. An example would be the *Occupational Safety and Health Review Commission (OSHRC).* (2) An incentive plan which represents an award to the employee of some

specified portion of the selling price of some service or product.

commission to examine witnesses
A commission issued out of the court in which an action is pending, to direct the taking of the depositions of witnesses who are beyond the territorial jurisdiction of the court.

commissioned agent
An agent who wholesales or retails a refined petroleum product under a commission arrangement. The agent does not take title to the product or establish the selling price, but receives a percentage of fixed fee for serving as an agent.

commissioner
A person to whom a commission is directed by the government or a court. A person with a commission. An officer who is charged with the administration of laws relating to some particular subject matter, or the management of some bureau or agency of the government. Member of a commission or board. A specially appointed officer of court.

commissioning
The initial acceptance process in which the performance of equipment/system is evaluated, verified, and documented to assure its proper operation in accordance with codes, standards, design, specifications, etc.

committee
(1) A person, or an assembly or board of persons, to whom the consideration, determination, or management of any matter is committed or referred, as by a court or legislature. (2) An individual or body to whom others have delegated or committed a particular duty, or who have taken on themselves to perform it in the expectation of their act being confirmed by the body they profess to represent or act for.

commodity
(1) *General.* Something bought and sold. Anything of use or profit. (2) *Transportation.* The classification of commodities is based on that prescribed by the Canadian Transport Commission and the Interstate Commerce Commission of the U.S.

commodity classes
The primary commodities involved in waterways transportation are fuels, chemicals, grains, and metals.

common authority
Refers to that entity at a multi-employer worksite (usually the general contractor or owner) who has the authority to permit entry of an OSHA representative to conduct an inspection thereby waiving any right or expectation of privacy that other contractors working on the property may have.

common carrier
Those that hold themselves out or undertake to carry persons or goods of all persons indifferently, or of all who choose to employ it. A for-hire carrier that holds itself out to serve the general public at reasonable rates and without discrimination. The carrier must secure (from ICC and FMC) a certificate of public convenience and necessity to operate.

common cause failure analysis
A system safety analytical technique (also known as *common cause analysis*) used primarily in the evaluation of multiple failures that have the occurrence of a single event as a common causal factor.

common cold
An acute and highly contagious virus infection of the upper respiratory tract, also called acute rhinitis. At least 20 identifiable viruses have been found to cause colds, and they may attack anyone with lowered resistance. Cold viruses are resistant to present antibiotics, and there is no really effective preventive vaccine as yet that will work against them in all situations for all people.

common instrument flight rules room
A highly automated terminal radar control facility. It provides terminal radar service in an area encompassing more than one major airport that accommodates instrument flight rule (IFR) operations.

common knowledge
Refers to what the court may declare applicable to action without necessity of proof. It is knowledge that every intelligent person has, and includes matters of learning, experience, history, and facts of which judicial notice may be taken.

common laboratory contaminants
Certain organic chemicals (considered by EPA to be acetone, 2-butanone, methylene chloride, toluene, and phthalate esters) that are commonly used in the laboratory and thus may be

introduced into the sample from a laboratory cross-contamination, not from the site.

common law
As distinguished from *statutory law* created by the enactment of legislatures, the common law comprises the body of law that develops and derives through judicial decisions, as distinguished from legislative enactment. It consists of those principles, usage, and rules of action applicable to government and security of persons and property which do not rest for their authority upon any express and positive declaration of the will of the legislature.

common nuisance
A nuisance is a common nuisance or a *public nuisance,* the terms being synonymous, where it affects the rights enjoyed by citizens as part of the public, that is, the rights to which every citizen is entitled.

common point
Aviation. A significant point over which two or more aircraft will report passing or have reported passing before proceeding on the same or diverging tracks. To establish/maintain longitudinal separation, a controller may determine a common point not originally in the aircraft's flight plan and then clear the aircraft to fly over the point.

common return circuit
A term applied where one wire is used for the return of more than one electric circuit.

common route
Aviation. That segment of a North American route between the inland navigation facility and the coastal fix.

common sense initiative
An EPA incentive program initiated in 1994 to bring together all levels of government officials, environmentalists, and industry leaders to create strategies that work cleaner, cheaper, and smarter to protect the health of the U.S. population and its natural resources.

common traffic advisory frequency (CTAF)
A frequency designed for the purpose of carrying out airport advisory practices while operating to or from an airport without an operating control tower. The CTAF may be a UNICOM, Multicom, F99, or I tower frequency and is identified in appropriate aeronautical publications.

communicable
As applied to disease, it is one that results from the spread or transmission of an infectious agent. The causative agent of the disease can be transmitted from one infected individual to another. Some diseases of animals are transmissible to man and are thus considered *communicable diseases.*

communicable disease
A disease spread by direct contact with the infectious agents causing it. Modes of transmission include a) direct contact with body excreta or discharges from an ulcer, open sore, etc.; b) indirect contact with inanimate objects such as drinking glasses, toys, tools, clothing, etc.; c) by vectors (flies, mosquitoes, or other insects capable of spreading the disease).

communication
The meaningful interchange using some form of language or other set of signals between individuals, groups, or instrumentation.

community
Aviation. A city, group of cities, or a Standard Metropolitan Statistical Area receiving scheduled air service by a certificated route air carrier.

community relations
The EPA effort to establish two-way communication with the public to create understanding of EPA programs and related actions, to assure public input into decision-making processes related to affected communities, and to make certain that the Agency is aware of and responsive to public concerns. Specific community relations activities are required in relation to Superfund remedial actions.

Community Right-to-Know
Shortened name for the *Emergency Planning and Community Right-to-Know Act of 1986,* also known as Title III of the Superfund Amendments and Reauthorization Act of 1986 (SARA). Covers facilities that keep specified quantities of extremely hazardous chemicals and contains reporting and emergency planning requirements.

community water system
A public water system which serves at least fifteen service connections used by year-round residents or regularly serves at least twenty-five year-round residents.

commutation ticket

In rail systems, a ticket sold at a reduced rate for a fixed or unlimited number of trips in a designated area during a specified time period.

commute

Regular travel between home and a fixed location (e.g., work, school).

commuter

General. A person who travels regularly between home and work or school.

commuter air carrier

(1) *FAA1 and FAA14.* A Federal Aviation Regulation (FAR) Part 135 operator who carries passengers in an aircraft with a maximum of 60 seats, on at least five round trips per week or at least one route between two more points, or that carries mail according to its published flight schedule that specifies the times, days of the week, and places between which those flights are performed. (2) *FAA11.* An air carrier that operates aircraft with 30 seats or less and a maximum payload capacity of 7,500 pounds or less and performs at least five round trips per week between two or more points and publishes a flight schedule. (3) *FAA2 and FAA9.* An air taxi that performs at least five scheduled round trips per week between two or more points or carries mail. (4) *FAA6.* An air taxi operator which performs at least five round trips per week between two or more points and publishes flight schedules which specify the times, days of the weeks, and plans between which such flights are performed.

commuter authority

Railroad. Any state, local, regional authority, corporation, or other entity established for purposes of providing commuter rail transportation (including, but not necessarily limited to, the New York Metropolitan Transportation Authority, the Connecticut Department of Transportation, the Maryland Department of Transportation, the Southeastern Pennsylvania Transportation Authority, the New Jersey Transit Corporation, the Massachusetts Bay Transportation Authority, the Port Authority Trans-Hudson Corporation, and any successor agencies) and any entity created by one or more such agencies for the purpose of operating, or contracting for the operation of, commuter rail transportation.

commuter bus service

Fixed route bus service, characterized by service predominantly in one direction during peak periods, limited stops, use of multi-ride tickets, and routes of extended length, usually between the central business district and outlying suburbs. Commuter bus service may also include other service, characterized by a limited route structure, limited stops, and a coordinated relationship to another mode of transportation.

commuter lane

See ***high-occupancy vehicle lane.***

commuter rail

(1) Long-haul passenger service operating between metropolitan and suburban areas, whether within or across the geographical boundaries of a state, usually characterized by reduced fares for multiple rides, and commutation tickets for regular, recurring riders. (2) Urban passenger train service for short distance travel between a central city and adjacent suburbs. Does not include heavy rail or light rail service. (3) Railroad local and regional passenger train operations between a central city, its suburbs, and/or another central city. It may be either locomotive-hauled or self-propelled, and is characterized by multi-trip tickets, specific station-to-station fares, railroad employment practices, and usually only one or two stations in the central business district. Also known as *suburban rail.*

commuter rail car

A rail passenger car obtained by a commuter authority for use in commuter rail transportation.

commuter rail locomotives

Commuter rail vehicles used to pull or push commuter rail passenger coaches. Locomotives do not carry passengers themselves.

commuter rail passenger coaches

Commuter rail passenger vehicles not independently propelled and requiring one or more locomotives for propulsion.

commuter rail self-propelled passenger cars

Commuter rail passenger vehicles not requiring a separate locomotive for propulsion.

commuter rail transportation

Short-haul rail passenger service operating in metropolitan and suburban areas, whether within or across the geographical boundaries of a state, usually characterized by reduced fare, multiple ride, and commutation tickets and by morning and evening peak period operations. This term does not include light or rapid rail transportation.

commuter railroad
Those portions of mainline railroad (not electric railway) transportation operations which encompass urban passenger train service for local travel between a central city and adjacent suburbs. Commuter railroad service using both locomotive hauled and self-propelled railroad passenger cars is characterized by multi-trip tickets, specific station-to-station fares, and usually only one or two stations in the central business district. Also known as *suburban railroad*.

commuter train
A short-haul passenger train operating on track which is part of the general railroad system of transportation, within an urban, suburban, or metropolitan area. It includes a passenger train provided by an instrumentality of a state or political subdivision thereof. Includes commuter trains and passenger trains other than elevated trains and subways. Includes local and commuter train service. Does not include intercity service by Amtrak.

compact bone
The dense outer tissue portion of a bone.

compaction
Reduction of the bulk of solid waste by rolling and tamping.

compactor collection vehicle
A vehicle with an enclosed body containing mechanical devices that convey solid waste into the main compartment of the body and compress it into a smaller volume of greater density.

company automotive outlet
Any retail outlet selling motor fuel under a reporting company brand name. 1) *Company operated.* A company retail outlet which is operated by salaried or commissioned personnel paid by the reporting company. 2) *Lessee.* An independent marketer who leases the station and land and has use of tanks, signs, etc. A lessee dealer typically has a supply agreement with a refiner or a distributor and purchases products at dealer tank wagon prices. The term "lessee dealer" is limited to those dealers who are supplied directly by a refiner or any affiliate or subsidiary company of a refiner. "Direct supply" includes use of commission agent common carrier delivery. 3) *Open.* An independent marketer who owns or leases (from a third party who is not a refiner) the station or land of a retail outlet and has use of tanks, pumps, signs, etc. An open dealer typically has a supply agreement with a refiner or a distributor and purchases products at or below dealer tank wagon prices.

company outlet
Any retail outlet (e.g., service station) selling gasoline or diesel fuel that has the ability to set the retail product price and directly collect all or part of the retail margin. This category includes retail outlets being operated by salaried employees of the company and/or its subsidiaries and affiliates, and/or involving personnel services contracted by the company.

comparative negligence
Under comparative negligence statutes or doctrines, negligence is measured in terms of percentage, and any damages allowed shall be diminished in proportion to the amount of negligence attributable to the person for whose injury, damage, or death recovery is sought.

comparison group
*See **control group**.*

comparison stimulus
Any variable stimulus which is presented in addition to a reference stimulus in certain experimental designs for determining differences in thresholds.

compass
(1) A sliding caliper. (2) A magnetic sensing device used in navigation for determining one's heading relative to magnetic north.

compass calibration pad
An airport facility used for calibrating an aircraft compass.

compass locator
Aviation. A low power, low or medium frequency (L/MF) radio beacon installed at the site of the outer or middle marker of an instrument landing system (ILS). It can be used for navigation at distances of approximately 16 miles or as authorized in the approach pro-

cedure. *See also* ***middle marker*** *and* ***outer marker***.

compass rose

Aviation. A circle, graduated in degrees, printed on some charts or marked on the ground at an airport. It is used as a reference to either true or magnetic direction.

compatibility

(1) A measure of how well spatial movements of controls, display behavior, or conceptual relationships meet human expectations. (2) That combination of characteristics and attributes which permit two or more individuals, groups, or systems to work together without significant interference of conflict. (3) The ability of two or more materials or substances to come in contact without altering their structure or causing an unwanted reaction in terms such as permeability, flammability, ignition, combustion, functional or material degradation, contamination, toxicity, pressure, temperature, shock, oxidation, or corrosion.

compatibility group letter

A designated alphabetical letter used to categorize different types of explosive substances and articles for purposes of stowage and segregation.

compatible or compatibility

In relation to state laws and regulations pertaining to commercial motor vehicle safety, having the same effect as the Federal Motor Carrier Safety Regulations (FMCSR) or Federal Hazardous Materials Regulations (FHMR) in that those state rules are either identical or fall within the tolerance guidelines in appendix C, 49 CFR 350.

compensable death

Within workers' compensation acts, is one which happens to an employee from injury by an accident arising out of and in the course of employment.

compensable injury

Such an injury, within workers' compensation acts, is one caused by an accident arising out of and in the course of the employment and for which the injured employee is entitled to receive compensation under such law.

compensation

(1) Indemnification; payment of damages; making amends; making whole; giving an equivalent or substitute of equal value. That which is necessary to restore an injured party to his/her former position. (2) Remuneration for services rendered, whether in salary, fees, or commissions. (3) A movement of a part of the body to restore or maintain equilibrium as another body part moves. (4) Any behavior which attempts to minimize the effect of a weakness in one process by relying on another, stronger process or improving another process.

compensation plan

That rule or set of rules which an organization follows in setting payment rates for jobs or type of work done.

compensatory

Pertaining to the use of error information only in generating control inputs.

compensatory damages

That monetary value awarded a victim by a court to pay for his/her injuries or losses. Damages awarded a person as compensation, indemnity, or restitution for harm sustained by him/her. Compensatory or actual damages consist of both general and special damages. General damages are the natural, necessary, and usual result of the wrongful act or occurrence in question. Special damages are those which are the natural, but not the necessary and inevitable result of the wrongful act. *See also* ***damages, general damages, special damages,*** *and* ***punitive damages.***

competency

In the law of evidence, the presence of those characteristics, or the absence of those disabilities, which render a witness legally fit and qualified to give testimony in a court of justice; applied, in the same sense, to documents or other written evidence. Competency differs from credibility. The former denotes the personal qualification of the witness; the latter his/her veracity. A witness may be competent, and yet give incredible testimony; he/she may be incompetent, and yet his/her evidence, if received, may be perfectly credible. Competency is for the court, credibility is for the jury. *See also* ***credibility***.

competent

Duly qualified; answering all requirements; having sufficient capacity, ability, or authority; possessing the requisite physical, mental, natural, or legal qualifications; able; adequate; suitable; sufficient; capable; legally fit.

competent authority
Transportation. A national agency responsible under its national law for the control or regulation of a particular aspect of the transportation of hazardous materials (dangerous goods). The term *appropriate authority,* as used in the International Civil Aviation Organization (ICAO) Technical Instructions, has the same meaning as Competent Authority. The Associate Administrator [of the Research and Special Programs Administration (RSPA)] for Hazardous Materials Safety is the competent authority for the United States.

competent witness
One who is legally qualified to be heard to testify in a case. A witness may not testify to a matter unless evidence is introduced sufficient to support a finding that the witness has personal knowledge of the matter.

competition
That condition in which more than one person or group vies against each other for a limited number of prizes, positions, market share, or other reward.

compilation
The development of higher order skills from lower level processes.

complainant
One who applies to the courts for legal redress by filing a complaint (i.e., the plaintiff). Also, one who instigates prosecution or who proffers accusation against a suspected person.

complaint
The original or initial pleading by which an action is commenced under codes or the Rules of Civil Procedure. The pleading which sets forth a claim for relief.

complementary color
That perceptual color on the opposite side of the achromatic point in the chromaticity diagram from a given color, which, when mixed in proper proportions, will produce a gray or white. *See also* ***complementary wavelength***.

complementary wavelength ($_c$)
The wavelength designated on the spectrum locus of a chromaticity diagram by an extension of the line determining the dominant wavelength in the opposite direction from the achromatic point. *See also* ***complementary color***.

complete blood count (CBC)
A measure of the hemoglobin concentration, and the number of red blood cells, white blood cells, and platelets in one cubic millimeter of blood. In addition, the proportion of various white blood cells is determined and the appearance of red and white cells is noted.

complete carcinogen
A carcinogen which produces cancer without subsequent exposure to any other substance.

complete diffusion
A condition in which a diffusing medium so scatters the incident flux that no image can be formed from the transmitted flux.

complete menu hierarchy
A menu hierarchy having the same number of menus along each branch from top to bottom.

complete radiator
See ***blackbody***.

completion
Refining. The installation of permanent equipment for the production of oil or gas. If a well is equipped to produce only oil or gas from one zone or reservoir, the definition of a well (classified as an oil well or gas well) and the definition of a completion are identical. However, if a well is equipped to produce oil and/or gas separately from more than one reservoir, a well is not synonymous with a completion.

complex carbohydrates
Carbohydrates with a relatively complex molecular structure. Known also as polysaccharides. Bread, rice, and pasta are primarily complex carbohydrates.

complex projects
Those projects requiring the coordination of numerous tasks and personnel or those that involve unusual logistics to complete the work.

complex reaction time
The temporal interval required to react to a stimulus situation when a choice or discrimination needs to be made before responding.

complex sound
Any sound composed of a large number of multiple sinusoidal components and their harmonics/overtones.

complex spectrum
Those coefficients resulting from a Fourier or other transform of a time series signal which contain both real and complex values.

complex tone
An auditory signal composed of multiple simple sinusoidal components with different frequencies.

complexion
The color and overall appearance of facial skin.

compliance
(1) To comply with both laws and general organizational requirements. *See also* ***compliance program***. (2) A measure of the softness of a system or structure, represented by the reciprocal of the stiffness.

compliance coating
A coating whose volatile organic compound content does not exceed that allowed by regulation.

compliance monitoring
A strategy or technique to determine compliance with a government standard. One compliance monitoring method is to identify the maximally exposed worker and, if that exposure is less than the standard, then all worker exposures are assumed to be below the exposure limit.

compliance plan
(1) *OSHA*. A documented approach to OSHA compliance required by some specific standards, such as Hazard Communication, which essentially establishes the employer's intended methods of achieving compliance. (2) *EPA-CAA*. Either a statement that the source will comply with all applicable requirements under CAA; or where applicable, a schedule and description of the method or methods for compliance and certification by the owner or operator that the source is in compliance with the requirements of CAA.

compliance procedures manual
A manual used by the corporate compliance officer and other compliance managers to monitor and audit compliance with the corporate code and supplemental compliance materials. Often includes a management policy stating the organization's intentions to comply with the law, areas of the law related to the organization's operations, and a statement of employees' and agents' requirements to follow this policy. May also include detailed operating procedures to instruct the employees or agents within particular areas of the organization.

compliance program
A tool for achieving an objective which is the responsibility of management: the proper control of the organization, its employees, and its assets. The creation of a system of internal controls to safeguard the interests of the organization, its investors, and the public's image of the corporation. An effectively administered program will deter illegal activity by employees, provide an effective mechanism and establish procedures for dealing with misconduct, generally resulting in more favorable treatment for the organization by enforcement authorities if illegal conduct is discovered and reported, and make it more likely that the disclosure of illegal activity need not result in a finding that the corporation is "presently responsible." *See also* ***effective compliance program, compliance plan,*** *and* ***compliance program inventory***.

compliance program inventory
A liability and risk-assessment inventory of an organization which includes the legal requirements and areas which a company may want to consider for its compliance program. Legal risk areas include, among many, OSHA, RCRA, and CAA. *See also* ***compliance program***.

Compliance Safety and Health Officer (CSHO)
A person authorized by the Occupational Safety and Health Administration, U.S. Department of Labor, to conduct inspections.

compliance schedule
A negotiated agreement between a pollution source and a government agency that specifies dates and procedures by which a source will reduce emissions and, thereby, comply with a regulation.

compliance strategy
Method an employer will develop and implement to achieve and maintain compliance with a regulation. It may include engineering and administrative controls, adherence to established procedures and work practices, the use of personal protective equipment, as well

as training of personnel regarding hazards, and making available hazard information (i.e., signs, materials safety data sheets, hazard communication training, etc.) to personnel.

compliance training
Training to educate senior management regarding the organization's mandatory standards and procedures and overall compliance program, as well as employees or agents on the Code of Conduct and other issues or procedures. Also includes detailed training tailored to each employee's job.

complimentary metal oxide semiconductor
A type of detector used in the detection of gases or vapors.

component
(1) *Systems Engineering.* A functional part of a subsystem or equipment which is essential to operational completeness of the subsystem or equipment and which may consist of a combination of parts, assemblies, accessories, and attachments. (2) *Hazardous Materials Management.* Any part, or system of parts functioning as a unit, including, but not limited to, piping, processing equipment, containers, control devices, impounding systems, lighting, security devices, fire control equipment, and communication equipment, whose integrity or reliability is necessary to maintain safety in controlling, processing, or containing a hazardous fluid. (3) *Hydraulics.* Any part of a pipeline which may be subjected to pump pressure including, but not limited to, pipe, valves, elbows, tees, flanges, and closures.

composite flight plan
A flight plan which specifies Visual Flight Rules (VFR) operation for one portion of flight and Instrument Flight Rules (IFR) for another portion. It is used primarily in military operations.

composite maintenance
The integration or simultaneous use of several types of maintenance.

composite material
The combinations of materials differing in composition or form on a macroscale. The constituents retain their identities in the composite. Normally, the constituents can be physically identified, and there is an interface between them.

composite packaging
A packaging consisting of an outer packaging and an inner receptacle, so constructed that the inner receptacle and the outer packaging form an integral packaging. Once assembled it remains thereafter an integrated single unit; it is filled, stored, shipped, and emptied as such.

composite route system
Aviation. An organized oceanic route structure, incorporating reduced lateral spacing between routes, in which composite separation is authorized. *See also* ***composite separation***.

composite sample
A water or wastewater sample made up of a number of samples taken at regular intervals over a 24-hour period.

composite separation
Aviation. A method of separating aircraft in a composite route system where, by management of route and altitude assignments, a combination of half the lateral minimum specified for the area concerned and half the vertical minimum is applied. *See also* ***composite route system***.

compost
The end product of composting; a mixture of garbage and degradable trash with soil in which certain bacteria in the soil break down the garbage and trash into organic fertilizer.

composting
(1) Sludge stabilization process relying on aerobic decomposition of organic matter in sludge by bacteria and fungi. (2) The natural biological decomposition of organic material in the presence of air to form a humus-like material. Controlled methods of composting include mechanical mixing and aerating, ventilating the materials by dropping them through a vertical series of aerated chambers, or placing the compost in piles out in the open air and mixing or turning it periodically. *See also* ***compost***.

compound
(1) Chemical combination of two or more elements in a fixed and definite proportion by weight. (2) A substance composed of atoms or ions of two or more elements in chemical combination. The constituents are bound by bonds or valence forces.

compound fracture
A broken bone in which at least one of the ends protrudes through the skin surface. Also referred to as *open fracture*.

Comprehensive Environmental Response Compensation and Liability Act (CERCLA) of 1980
Generally referred to as Superfund, CERCLA establishes a tax on certain chemical feedstocks used to fund the clean up of abandoned hazardous waste sites. Superfund was designated to provide immediate remedial action for highly contaminated areas. The federal government can then seek to recover these costs through negotiation or legal action against the contributor or sources of pollution.

comprehensive insurance
See ***all-risk insurance***.

comprehensive transportation
See ***intermodalism*** *(3)*.

comprehensive zoning plan
A general plan to control and direct the use and development of property in a municipality or in a large part thereof by dividing it into districts according to the present and potential use of the properties.

compress
Reduce the volume of a substance or material, or the duration of some event.

compressed drive
Computing. A drive has had its data compressed to take up less space. Special software must be running in the system to read from and write to a drive that is set up this way.

compressed gas
Any material or mixture having in the container an absolute pressure exceeding 40 psi at 70 degrees Fahrenheit or having an absolute pressure exceeding 104 psi at 13 degrees Fahrenheit or any liquid flammable material having a vapor pressure exceeding 40 psi absolute at 100 degrees Fahrenheit, according to the laws of chemical combination. Each compound has it own characteristic properties different from those of its constituent elements.

Compressed Gas Association (CGA)
A professional association of gas producers, suppliers, equipment manufacturers, and representatives from related industries that develops safety standards, makes recommendations to improve methods of handling, transporting, storing of compressed gases, and advising regulatory agencies concerned with the safe handling of compressed gases.

compressed natural gas (CNG)
Natural gas which is comprised primarily of methane, compressed to a pressure at or above 2,400 pounds per square inch (psi), and stored in special high-pressure containers. It is used as a fuel for natural gas-powered vehicles. *See also* ***compressed gas***.

compressed seat height
The height of a cushioned chair seat pan from the floor or other reference surface when an individual is seated in it.

compressed spectral array (CSA)
A three-dimensional display or hard copy of a sequential series of spectra as a function of time, with time being the depth axis.

compressed workweek
A work schedule in which employees provide approximately 40 hours of work in less than five days.

compressibility
That property of a tissue or other soft material to be locally depressed or of a gas to be reduced in volume when external pressure is applied.

compression
Internal stress created in a material by forces acting inward, in opposite directions, in a manner which decreases the size of the material by closing or tightening its molecular structure.

compression failure
In wood, a deformation (buckling) of the fibers due to excessive compression along the grain.

compression ratio
The volume of the combustion chamber and cylinder when the piston is at the bottom of its stroke, divided by the volume of the combustion chamber when the piston is at the top of its stroke.

compression settling
Phenomenon referring to sedimentation of particles in a concentrated suspension where further settling can occur only by compression of the existing structure of settled particles.

compression wood
An aberrant (abnormal) and highly variable type of wood structure occurring in softwood species. The wood commonly has density somewhat higher than does normal wood, but somewhat lower stiffness and tensile strength for its weight in addition to high longitudinal shrinkage.

compressor
A mechanical device used to increase the pressure of a gas or vapor.

Compton effect
The glancing collision of a gamma-ray with an electron wherein the gamma-ray gives up part of its energy to the electron. Also referred to as *Compton scattering*.

compulsory reporting points
Aviation. Reporting points which must be reported to Air Traffic Control (ATC). They are designated on aeronautical charts by solid triangles or filed in a flight plan as fixes selected to define direct routes. These points are geographical locations which are defined by navigation aids/fixes. Pilots should discontinue position reporting over compulsory reporting points when informed by (ATC) that their aircraft is in "radar contact."

computer-aided instruction (CAI)
The use of computers and displays for presenting information to be learned. Also referred to as *computer-assisted instruction*.

computer-aided manufacturing
See ***computer-integrated manufacturing***.

computer anxiety
A state of apprehension or fear when required to interact with a computer, which is out of proportion to any reasonable danger posed by the computer.

computer anxiety scale
A survey consisting of 10 test items dealing with feelings about computers, on which an individual judges a rank for each item according to a Likert scale. Synonymous with *Raub scale*.

Computer Assessment of Reach (CAR)
A crew station modeling program which attempts to determine what percentage of the air crew will be able to function in a given design.

computer-assisted instruction
See ***computer-aided instruction***.

computer-assisted tomography (CAT)
See ***computerized axial tomography***.

Computer Biomechanical Man (COMBIMAN)
A three-dimensional, interactive computer graphics modeling software package which can be used in the physical evaluation of pilots and other air crew members for crew station design, including sizing, reach, strength, and visual field.

computer enhancement
(1) The use of computer technology (hardware/software) to enhance or make better use of a particular task or process, such as computer-enhanced photography. (2) In weather forecasting, a process in which the temperatures of radiating surfaces are assigned different shades of gray on an infrared picture. This allows special features to be more clearly delineated.

computer graphics
The input, processing, or output of any pictorial or graphical data displayed on a computer monitor or hard copy.

computer-human interface (CHI)
See ***human-computer interface***.

computer input device
Any type of hardware tool which can be used by an individual to get text, graphics, commands, or data into a computer.

computer-integrated manufacturing (CIM)
The use of computers in the actual manufacturing process.

computer model
Any numerical or graphical representation of objects, systems, or processes using a computer.

computer vision
The integration of one or more video cameras with appropriate software into a computer processing system for any purpose, such as electronic scene comparison, to simulate human vision for mobile robots, or other uses.

Computerized Accommodated Percentage Evaluation (CAPE)
A modeling tool for determining what percentage of the air crew population could

function satisfactorily in a given crew station design. (An old model, no longer used.)

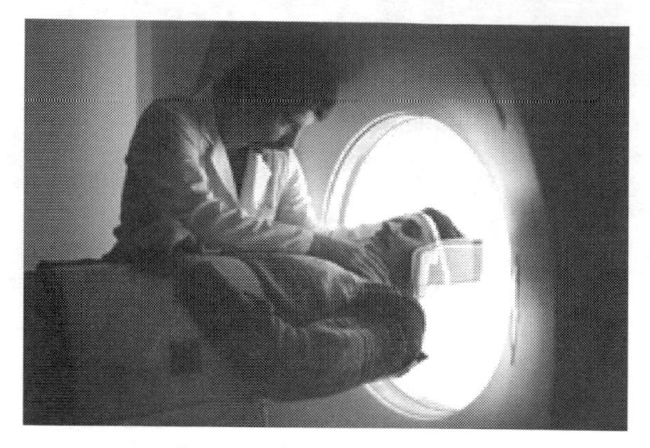

Patient undergoing a CAT Scan

computerized axial tomography (CAT)
The use of computers for control, acquisition, storage, processing, and display of a series of single planes of x-ray images along the longitudinal axis of the body or other x-ray transparent objects. Also referred to as *computer-assisted tomography*.

Computerized Relationship Layout Planning (CORELAP)
A computer model for developing a plant layout based on relationships when large numbers of groups are involved.

Computerized Relative Allocation of Facilities Technique (CRAFT)
A computer model for improving a plant layout, with the priority of minimizing transportation costs.

conative
Pertaining to the basic strivings of a person, as expressed in his/her behavior and actions.

concave
Rounded and somewhat depressed or hollowed out.

concave function
A mathematical relationship or graph which has a negative second derivative during the interval of interest, resulting in an inverted U-shaped curve.

CONCAWE
Conservation of Clean Air and Water in Europe.

concealed
Rendered inaccessible by the structure or finish of a building. Wires in concealed raceways are considered concealed, even though they may become accessible by withdrawing them. *See also **accessible** (2).*

concentration
(1) The amount of a substance dissolved or suspended in a unit volume of solution. (2) The process of increasing the amount of a substance per unit volume of solution. (3) Increase in strength by evaporation. (4) Medicine that has been strengthened by evaporation of its nonactive parts. (5) The relative content of a contained or dissolved substance in a solution.

concentration polarization
A phenomenon in which solutes form a dense, polarized layer next to a membrane surface, which eventually restricts flow through the membrane.

concentration ratio
The ratio of the concentration of solids in a water system to those of the dilute makeup water added to the system.

concentric action
A dynamic muscle action which involves active muscle shortening against a resistance. Also referred to as *concentric contraction* and *concentric muscle contraction*.

concentric contraction
*See **concentric action**.*

concentric muscle contraction
*See **concentric action**.*

concept
A abstract idea or notion which enables an individual to generalize from known specific examples not previously encountered.

concept hierarchy
An organization in which the most general aspects of a concept are located at the top, with subsidiary aspects branching beneath.

concept hierarchy analysis
The examination of a concept hierarchy to determine if a better arrangement can be made or to compare with related structures.

concept phase
That portion of a system's, product's, or other yet to be developed program's life cycle during which ideas are first conceptualized; precedes the design phase.

concept trainer
A training aid used when the principles to be learned are too complex to be easily understood from verbal descriptions or when simulation with actual physical objects appears to be the optimum method.

concordance coefficient
*See **Kendall's coefficient of concordance**.*

concrete
A mixture of water, sand, stone, and a binder that hardens to a stone-like mass.

concurrent causes
Causes acting contemporaneously and together causing injury, which would not have resulted in the absence of either. Two distinct causes operating at the same time to produce a given result, which might produced by either, are considered concurrent causes. However, two distinct causes, successive and unrelated in an operation, cannot be concurring, and one will be regarded as the proximate and efficient and responsible cause, and the other will be regarded as the remote cause. *See also **cause, proximate cause,** and **efficient cause**.*

concurrent insurance
Insurance coverage under two or more similar policies of varying dates and amounts. *See also **insurance**.*

concurrent loading
A test or working condition in which an individual is required to perform both a fatiguing exercise and a criterion task simultaneously.

concurrent negligence
Consists of the negligence of two or more persons concurring, not necessarily in point of time, but not in point of consequence, in producing a single indivisible injury.

concurrent validity
Having a high correlation between job incumbent test scores and performance on the job.

concussion
A violent jar or shock, or the condition that results from such an injury. A clinical condition caused by a sudden, strong, mechanical force applied to the head and characterized by temporary impairment of neural function such as an alteration in consciousness or disturbances of vision, equilibrium, and/or reflexes.

condensate
Water obtained by evaporation and subsequent condensation.

condensate polishing
Treatment of condensate water to achieve required purity.

condensation
(1) The change in state from vapor to liquid; opposite that of evaporation. (2) The act of rendering, or the process of becoming, more compact. (3) Pathologic hardening of a part. (4) The unconscious union of concepts to produce a new idea or mental picture.

condensation nuclei
Small particles on which water vapor condenses. In the development of precipitation, if enough vapors are present in the atmosphere, the nuclei eventually become large enough and heavy enough to fall to the earth as precipitation (rain, snow, hail, etc.) or remain suspended (fog, clouds, etc.). Small nuclei less than 0.2 m radius are called *aitken nuclei.* Those with radii between 0.2 and 1 m are *large nuclei,* while *giant nuclei* have radii larger than 1 m.

condensation trail
*See **contrail**.*

condenser
(1) A vessel or apparatus for condensing gases or vapors. (2) A device for illuminating microscopic objects. (3) An apparatus for concentrating energy or matter.

condiment
Any flavoring added to food to improve taste or increase stimulation of the taste buds (such as spice, salt, etc.), or an item having such effect (such as gum or mint).

conditional cues
Any displayed information which provides the user with a brief indicator of the current operating rules or conditions.

conditional event
As pertains to fault tree analysis (FTA) and/or the Management Oversight and Risk Tree (MORT), an occurrence that, based upon its own unique characteristics, imposes conditions or exclusions on the occurrence of other events in the fault path. Represented graphically as an oval. *See also **exclusive event**.*

conditional reflex

A learned response to a stimulus which did not originally cause that response.

conditional registration

Under special circumstances, the Federal Insecticide, Fungicide, and Rodenticide Act (FIFRA) permits registration of pesticide products that is "conditional" upon submission of additional data. These special circumstances include a finding by the EPA Administrator that a new product or use of an existing pesticide will not significantly increase the risk of unreasonable adverse effects. A product containing a new (previously unregistered) active ingredient may be conditionally registered only if the Administrator finds that such conditional registration is in the public interest, that a reasonable time for conducting the additional studies has not elapsed, and the use of the pesticide for the period of conditional registration will not present an unreasonable risk.

conditionally exempt

An exemption applied to those *small quantity generators* that generate less than 100 kilograms of hazardous waste during any month.

conditionally unstable air

An atmospheric condition that exists when the environmental lapse rate is less than the dry adiabatic rate but greater than the moist adiabatic rate. Also called *conditional instability.*

conditioned air

Air that has been heated, cooled, humidified, or dehumidified to maintain an interior space within the comfort zone. Sometimes referred to as *tempered air. See also* ***comfort zone.***

conditioned reflex

A reflex that does not occur naturally in the animal but that may be developed by regular association of some physiologic function with an unrelated outside event, such as ringing a bell or flashing of a light.

conditioning

(1) *Environmental.* Pretreatment of a wastewater or sludge, usually by means of chemicals, to facilitate removal of water in a subsequent thickening or dehydration process. (2) *Physiology.* Any physical (or mental) activity or training which prepares an individual for a given task.

conductance

The ability to conduct or transmit, as electricity or other energy or materials. A measure of a solution's electrical conductivity that is equal to the reciprocal of the resistance.

conduction

(1) The transfer of heat by direct contact from one body to another. (2) The transfer of heat by molecular activity from one substance to another, or through a substance. This transfer is always from warmer to colder regions.

conduction deafness

See ***conductive hearing loss.***

conductive hearing loss

(1) A type of hearing loss that is not caused by noise, but is due to any disorder that prevents sound from reaching the inner ear. It is a hearing loss that is due to poor transmission of sound from the outer ear to the cochlea. (2) A physical defect or condition of the outer or middle ear that interferes with the passage of sound. This can be the end result of physical obstruction within the outer or middle ear, a birth defect, the aging process, or disease, all of which can affect conversion of sound energy into mechanical energy. This type of hearing loss involves a reduction in the perception of loudness and not in clarity.

conductive heat loss

That amount of heat eliminated from the body via heat conduction, indicated by an equation of the form:

$$H = kA \frac{\Delta T}{\Delta x}$$

where:

H	= heat loss
k	= thermal conductivity coefficient
A	= body surface area in contact with another object
$\Delta T/\Delta x$	= temperature gradient

conductivity

The ability of a substance to conduct electricity; directly related to the mineral content of water.

conductivity detector

A detection method based on the absorption of a gas by an aqueous solution with the formation of electrolytes, thereby producing a change in the electrical conductivity of the solution, which can be measured by this type detector and equated to gas concentration.

conductor

(1) *Bare Conductor.* A conductor having no covering or electrical insulation whatsoever. (2) *Covered Conductor.* A conductor encased within material of a composition and thickness that is not recognized as electrical insulation. (3) *Insulated Conductor.* A conductor encased within material of a composition and thickness that is recognized as electrical insulation.

conduit body

A separate portion of a conduit or tubing system that provides access through a removable cover(s) to the interior of the system at a junction of two or more sections of the system or at a terminal point of the system. Boxes such as FS and FD or larger cast or sheet metal boxes are not classified as conduit bodies.

condyle

A rounded projection on a bone surface, often associated with a joint.

cone

A solid figure or body having a circular base and tapering to a point, especially one of the structures of the retina, which, with the rods, form the light-sensitive elements of the retina. The cones make possible the perception of color. *See also* ***color blindness***.

cone monochromatism

A condition in which an individual has only a single type of retinal cone, thus seeing only one color, while having normal color brightness discrimination.

conference

(1) *General.* A discussion or consultation on some important matter; also, a formal meeting for this. A league or association, as of athletic teams, schools, churches, specific professionals, etc. (2) *Maritime.* An association of ship owners operating in the same trade route who operate under collective conditions and agree on tariff rates.

confidence

A measure of a material's ability to conduct electrons.

confidence interval

A range, or interval, that has a specified probability of including the true value of a parameter of a distribution.

confidence limits

Confidence limits are mathematically determined intervals, defined as upper and lower limits, for which one is confident (e.g., 90%, 95%, etc.) that the true value is greater than, less than, or between.

config.sys

Computing. A text file that contains a list of commands that MS-DOS executes every time a computer is started (booted) or restarted (rebooted). The config.sys file contains MS-DOS settings and drives that run operations such as those pertaining to a network and/or devices such as a CD-ROM drive.

configuration control

A design or procedure for the controlled development, operation, and maintenance of a system or process.

confined aquifer

An aquifer in which groundwater is confined under pressure that is significantly greater than atmospheric pressure.

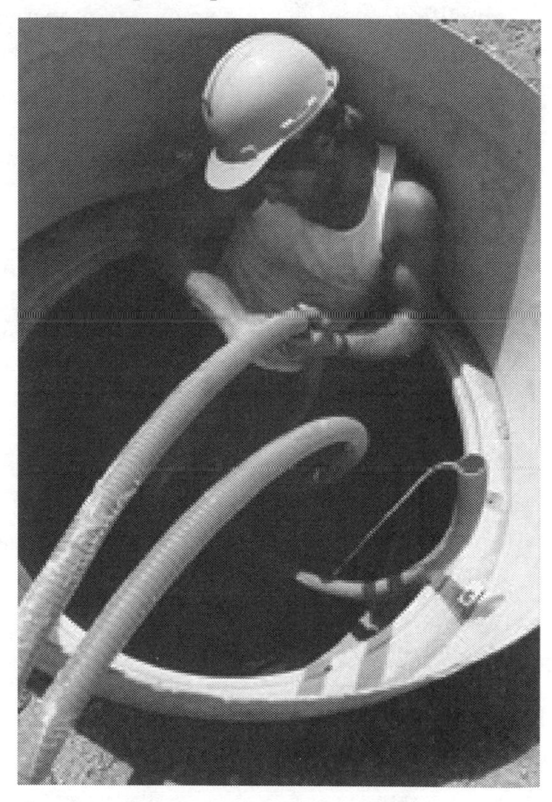

A worker prepares to enter a permit-required confined space

confined space
Any space not designed or intended for continuous occupancy that has a limited or restricted means of entry or exit, and that is subjcct to thc accumulation of toxic or flammable contaminants or has an oxygen-deficient atmosphere. Confined spaces must be large enough for an employee to enter and perform assigned work. Where confined spaces are categorized, there are two levels of classification: permit-required confined space and low-hazard permit space.

confined space entry
The entry of personnel (one or more) into a confined space. *See also* ***confined space*** *and* ***confined space, permit required***.

confined space, permit required
According to OSHA, a confined space that contains or has the potential to contain a hazardous atmosphere, a material with the potential to engulf an entrant, is configured such that an entrant could be trapped or asphyxiated, or contains any other recognized health or safety hazard.

conflagration
A fire extending over a considerable area and engulfing a considerable amount of property.

conflict
A state resulting from an individual having incompatible desires or two or more individuals or groups having different goals or means to achieve a goal.

conflict alert
Aviation. A function of certain air traffic control automated systems designed to alert radar controllers to existing or pending situations between tracked targets that require immediate attention/action. *See also* ***mode C intruder alert***.

conflict of interest
Term used in connection with public officials and fiduciaries and their relationship to matters of private interest or gain to them.

conflict resolution
Aviation. The resolution of potential conflicts between aircraft that are radar identified and in communication with Air Traffic Control (ATC) by ensuring that radar targets do not touch. Pertinent traffic advisories shall be issued when this procedure is applied. Note: This procedure shall not be provided utilizing mosaic radar systems.

conflicting evidence
Evidence offered by the plaintiff and defendant, or prosecutor and defendant which is inconsistent and cannot be reconciled.

conflicting movement
Aviation. Movements over conflicting routes. *See also* ***conflicting routes***.

conflicting routes
Aviation. Two or more routes, opposing, converging or intersecting, over which movements cannot be made simultaneously without possibility of collision.

confluence
The point where the flows of streams or rivers meet.

confounding variable
A variable which is uncontrolled and which has, or is likely to have, some effect in an experiment.

congeal
To thicken, jell, or solidify, usually by cooling or freezing.

congenital
Refers to certain mental or physical traits, abnormalities, malformations, or diseases that may be either inherited or due to an influence that occurred between conception and birth.

congenital abnormality
Any defect in the structure or function of an individual existing before or at birth. More commonly referred to as *congenital defect*.

congenital defect
See ***congenital abnormality***.

congestive hypoxia
A form of hypokinetic hypoxia in which venous blood flow is reduced.

Congress
Formal meeting of delegates or representatives. The Congress of the United States was created by Article I, Section 1, of the Constitution, adopted by the Constitutional Convention on September 17, 1787, providing that "All legislative Powers herein granted shall be vested in a Congress of the United States, which shall consist of a Senate and House of Representatives." The first Congress under the Constitution met on March 4, 1789, in the Federal Hall

in New York City. The membership then consisted of 20 Senators and 59 Representatives.

Congressional committee
A committee of the House of Representatives or of the Senate or a joint committee formed for some particular purpose.

Congressional district
A geographical unit of a state from which one member of the House of Representatives is elected.

Congressional immunity
*See **legislative immunity**.*

Congressional powers
The authority vested in the Senate and House of Representatives to enact laws, etc. as provided in the U.S. Constitution.

conjugate
Seismology. Describes a pair of intersecting (or nearly intersection) faults, the slip motions of which are opposite (e.g., right-lateral and left-lateral), so as to accommodate the rotation of the block they bound. Conjugate faults will sometimes slip roughly simultaneously (within hours of each other), causing pairs of earthquakes.

conjunctiva
The delicate membrane lining the eyelids and covering the eyeball.

conjunctivitis
Inflammation of the conjunctiva, the thin membrane that covers the eyeball and lines the eyelid. This disorder may be caused by bacteria or a virus, or by allergic, chemical, or physical factors. Its infectious form (of bacterial or viral origin) is highly contagious. The type of conjunctivitis known as *pinkeye* is an example of a highly contagious conjunctivitis and must be handled with extreme care to prevent its spread.

connate water
Water trapped in sedimentary rocks during their formation. Also known as *fossil water*.

connected word recognition
A capability in which a phrase or a sequence of a few meaningfully connected words may be understood by an artificial system.

connecting carrier
One of several common carriers whose united lines or parts constitute the route over which a shipment is to pass, and which participates in transportation of such shipment as a common carrier furnishing a necessary link in transportation. *See also **common carrier**.*

connective tissue
A fibrous type of body tissue with varied functions. The connective tissue system supports and connects internal organs, forms bones and walls of blood vessels, attaches muscles to bones, and replaces tissues of other types following injury. Connective tissue consists mainly of long fibers embedded in non-cellular matter, the ground substance. The density of these fibers and the presence or absence of certain chemicals make some connective tissues soft and rubbery and others hard and rigid. Compared to most other kinds of tissue, connective tissue has few cells. The fibers contain a protein called collagen, and the tissue for that reason is often called collagen tissue. Collagen tissue can develop in any part of the body, and the body uses the ability to help repair or replace damaged areas. Scar tissue is the most common form of this substitute.

connector
Computing. An electrical connection that allows the computer to send and receive data to and from other devices and/or computers. Standard connectors on a computer include the serial and parallel connectors, which allow users to send information from their computer to a printer, for example.

consciousness
An awareness of one's external environment.

consensual standard
*See **consensus standard**.*

consensus standard
A standard of approach developed through a consensus process of agreement among representatives of interested or affected industries, organizations, or individual members of a nationally recognized standards producing organization. A standard of conduct that has been developed by a nationally recognized organization having understood expertise in a given field (such as the American National Standards Institute and the National Fire Protection Association). Such standards carry no force of law unless adopted and implemented

by a regulatory agency, such as OSHA. Also referred to as *consensual standard*.

consent decree
(1) *General.* Under the Federal Sentencing Guidelines (FSGs), an agreement between a party charged with a regulatory violation and a regulatory agency whereby the party charged agrees to accept a penalty specified by the agency without admitting a violation. Consent decrees under the FSGs may impose a comprehensive compliance program or enhancements to the organization's existing program. (2) *EPA-Specific.* A legal document, approved by a judge, that formalizes an agreement reached between the EPA and potentially responsible parties (PRPs) through which the PRPs will conduct all or part of a cleanup action at a Superfund site; cease or correct actions or processes that are polluting the environment; or otherwise comply with regulations where the PRP's failure to comply caused the EPA to initiate regulatory enforcement actions. The consent decree describes the actions PRPs will take and may be subject to a public comment period. *See also **potentially responsible party**.*

consequation
Any aspect of the environment which changes the behavior of an individual encountering it.

consequential damages
Such damage, loss, or injury as does not flow directly and immediately from the act of the party, but only from some of the consequences or results of such act. Damages which arise from intervention of special circumstances not ordinarily predictable. Those losses or injuries which are a result of an act but are not direct and immediate.

conservation
(1) The act of conserving, or the actions taken in the interests of conserving, environmental resources and/or ecosystems, wildlife, or any other interests of focus or concern. The protection, improvement, and use of natural resources according to principles that will assure their highest economic or social benefits. (2) Avoiding waste of, and renewing when possible, human and natural resources. *See also **conserve**.*

conservation group
Any concerted gathering of two or more persons dedicated in principle and practice to the protection and preservation of certain resources or attributes normally associated with the natural environment, including wildlife, or other such areas of common interest.

conservation management
Under the Federal Marine Mammal Protection Act of 1972, the collection and application of biological information for the purpose of increasing and maintaining the number of animals within species and populations of marine mammals at their optimum sustainable population. Such terms include the entire scope of activities that constitute a modern scientific resource program, including, but not limited to research, census, law enforcement, and habitat acquisition and improvement. Also included within these terms, when and where appropriate, is the periodic or total protection of species or populations as well as the regulated taking of such.

conservation of angular momentum
The principle that the angular momentum of an object will remain unchanged unless the object is acted upon by a net force.

conservation of linear momentum
The principle that the linear momentum of an object will remain unchanged unless the object is acted on by a net force.

conservation of momentum
*See **conservation of angular momentum** and **conservation of linear momentum**.*

conserve
Under the Federal Endangered Species Act of 1973, to use, and the use of, all methods and procedures which are necessary to bring any endangered species or threatened species to the point at which the measures provided pursuant to the Act are no longer necessary. Such methods and procedures include, but are not limited to, all activities associated with scientific resources management such as research, census, law enforcement, habitat acquisition and maintenance, propagation, live trapping, and transplantation, and in the extraordinary case where population pressures within a given ecosystem cannot be otherwise relieved, may include regulated taking.

consideration
The inducement to a contract. The cause, motive, price, or impelling influence which induces a contracting party to enter into a contract. The reason or material cause of a contract. Some right, interest, profit or benefit accruing to one party, or some forbearance, detriment, loss, or responsibility, given, suffered, or undertaken by another. It is a basic, necessary element for the existence of a valid contract that is legally binding on the parties.

consignee
A person or company to whom commodities are shipped. Officially, the legal owner of the cargo.

consignee mark
A symbol placed on packages for export identification purposes; generally consisting of a triangle, square, circle, diamond, or cross, with letters and/or numbers as well as port of discharge.

consignment
The physical transfer of goods from a seller (the consignor), who retains title, to the consignee, who acts as selling agent by selling the goods for commission, remitting the net proceeds to the consignor.

consignor
A person or company shown on the bill of lading as the shipper.

consist
Railroad. On-track railroad equipment such as a train, locomotive, group of railcars, or a single railcar not coupled to another car or to a locomotive.

consist responsibility
The railroad employing the crew members operating the consist at the time of the accident determines the consist owner for reporting purposes only.

consistency
A level of performance which repeatedly falls within certain specified limits.

consolan
Aviation. A low frequency, long-distance navigational aid (NAVAID) used principally for transoceanic navigation.

consolidated metropolitan statistical area (CMSA)
A metropolitan complex of 1 million or more population, containing two or more component parts designated as primary metropolitan statistical areas (PMSAs).

consolidated vehicle
A vehicle transferred, with or without reimbursement, to General Services Administration (GSA) by another government agency for participating in the Introductory Fleet Management System (IFMS).

constant
(1) Persistent and continuous; unrelenting. (2) A fixed numerical value, or a symbol representing such a value.

constant air volume system
An air handling system that provides a constant air flow while varying the temperature to meet heating and cooling needs.

constant dollars
(1) A dollar value adjusted for changes in the average price level. A constant dollar is derived by dividing a current dollar amount by a price index. The resulting constant dollar value is that which would exist if prices had remained at the same average level as in the base period. (2) A series of figures is expressed in constant dollars when the effect of change in the purchasing power of the dollar has been removed. Usually the data are expressed in terms of dollars of a selected year or the average of a set of years.

constant element
A job or task in which a worker exhibits consistency of performance time, even if minor changes in processing or product dimensions are made.

constant error
The difference between the point of subjective equality and the known standard value in psychophysical testing.

constant-height chart
In meteorology, a chart (also referred to as a *constant-level chart*) showing variables, such as pressure, temperature, and wind, at a specific altitude above sea level. Variation in horizontal pressure is depicted by isobars. The most common constant-height chart is the

surface chart, which is also called the *sea level chart*.

constant pressure chart

In meteorology, a chart (also referred to as an *isobaric chart*) showing variables such as temperature and wind, on a constant pressure surface. Variations in height are usually shown by lines of equal height (see *contour lines*).

constant-rate filtration

Filter operation where flow through the filter is maintained at a constant rate by an adjustable effluent control valve.

constipation

A condition in which the waste matter in the bowels is too hard to pass easily, or in which bowel movements are so infrequent that discomfort or uncomfortable symptoms result.

Constitutional Record

Proceedings of Congress are published in the *Congressional Record,* which is issued daily when Congress is in session. Publication of the *Record* began March 4, 1873; it was the first series officially reported, printed, and published directly by the Federal Government. The Daily Digest of the *Congressional Record,* printed in the back of each issue of the *Record,* summarizes the proceedings of that day in each House and before each of their committees and subcommittees, respectively. The Digest also presents the legislative program for each day, and at the end of the week, gives the program for the following week. Its publication began on March 17, 1947. Members of Congress are allowed to edit their speeches before printing and may insert material never actually spoken by securing from their respective houses permission to print or to extend their remarks.

constraint

A restriction affecting the degree of freedom to act or move; a boundary or condition which may dictate performance in other than the desired or intended manner.

constrictor

A muscle which contracts to close or reduce the cross-section of an opening.

construct

(1) To build or erect. (2) A postulated attribute of an individual assumed to be reflected in observable behaviors.

construct validity

The extent of the relationship between what a test measures and how test scores are reflected in behavior or performance.

constructed solid geometry (CSG)

A technique in solid modeling where primitive solids are generated and combined to produce more complex forms.

constructed wetlands

A wastewater treatment system using the aquatic root system of cattails, reeds, and similar plants to treat wastewater applied either above or below the soil level.

construction and demolition waste

Waste consisting of building materials, packaging, and rubble resulting from construction, remodeling, repair, and demolition operations on pavements, houses, commercial buildings, and other structures.

construction/maintenance zone

An area, usually marked by signs, barricades, or other devices indicating that highway construction or highway maintenance activities are ongoing.

construction pipeline

Nuclear. The various stages involved in the acquisition of a nuclear reactor by a utility. The events that define these stages are the ordering of a reactor, the licensing process, and the physical construction of the nuclear generating unit. A reactor is said to be "in the pipeline" when the reactor is ordered and "out of the pipeline" when it completes low-power testing and begins operation toward full power.

consular invoice

An invoice covering a shipment of export goods certified by a consular official of the country of destination. The invoice shows the value of the shipment in the currency of the country of export. It is used by customs officials of the country of entry to verify the value, quantity, and nature of the shipment.

consultant

An individual or group who is uniquely qualified or has claimed expertise in a particular field and may be called upon to perform some specialized technical function on a one-time or an occasional basis.

consumer
One who purchases goods or services for final use, not having the intent to reprocess or repackage for resale.

consumer commodity
A material that is packaged and distributed in a form intended or suitable for sale through retail sales agencies or instrumentalities for consumption by individuals for purposes of personal care or household use. This term also includes drugs and medicines.

consumer complaint
Oral or written communication from a consumer indicating a possible problem with a product.

Consumer Price Index (CPI)
An index issued by the U.S. Department of Labor, Bureau of Labor Statistics. The CPI is designed to measure changes in the prices of goods and services bought by wage earners and clerical workers in urban areas. It represents the cost of a typical consumption bundle at current prices as a ratio to its cost at a base year.

consumer product
A product intended for final use primarily by the general public, as opposed to industrial use.

Consumer Product Safety Act
Established the Consumer Product Safety Commission. Definition of consumer product does not include boats which are covered under the statutes.

Consumer Product Safety Commission
An independent federal regulatory agency established by act of 27 October 1972 (86 Stat. 1207) to administer and implement the Consumer Product Safety Act. The Commission has primary responsibility for establishing mandatory product safety standards, where appropriate, to reduce the unreasonable risk of injury to consumers from consumer products. In addition, it has authority to ban hazardous consumer products. The Consumer Product Safety Act also authorizes the Commission to conduct extensive research on consumer and industry information and education programs, and establish a comprehensive Injury Information Clearinghouse.

consumption
(1) *General.* The act or process of consuming; waste, decay, destruction. Using up of anything, as food, natural resources, heat, or time. (2) *Environmental.* With respect to any substance, the amount of that substance produced in the United States, plus the amount imported, minus the amount exported to Parties to the Montreal Protocol. Such term shall be construed in a manner consistent with the Montreal Protocol. (3) *Medical.* A wasting away of the body, applied especially to pulmonary tuberculosis.

consumption unit value
Total price per specified unit, including all taxes, at the point of consumption.

consumptive waste
Water that returns to the atmosphere without beneficial use.

contact
(1) To establish communication with (followed by the name of the facility and, if appropriate, the frequency to be used). (2) A flight condition wherein the pilot ascertains the attitude of his/her aircraft and navigates by visual reference to the surface.

contact approach
Aviation. An approach wherein an aircraft on an Instrument Flight Rules (IFR) flight plan, having an air traffic control authorization, operating clear of clouds with at least 1 mile flight visibility and a reasonable expectation of continuing to the destination airport in those conditions, may deviate from the instrument approach procedure and proceed to the destination airport by visual reference to the surface. This approach will only be authorized when requested by the pilot and the reported ground visibility at the destination airport is at least 1 statute mile.

contact condenser
A device in which steam is condensed through direct contact with a cooling liquid.

contact dermatitis
Inflammation of dermal tissue that is caused by contact with a primary irritant. A delayed type of induced sensitivity of the skin

resulting from cutaneous contact with a specific allergen.

contact irritants
Chemicals that produce visible signs of skin and eye irritation upon contact. Rubber, plastics, resins, glues, cement, oil, and organic solvents are examples of contact irritants.

contact lens
A thin, curved shell of glass or plastic that is applied directly to the cornea to correct refractive errors. They do not actually touch the surface of the eye, but float on a thin layer of the fluid that naturally moistens the eyeball. *See **also lens**.*

A "soft" contact lens

contact pesticide
A chemical that kills pests when it touches them, rather than by being eaten (stomach poison). Also, soil that contains the minute skeletons of certain algae that scratch and dehydrate waxy-coated insects.

contact process
Wastewater treatment process where diffused air is bubbled over fixed media surfaces.

contact rate
Amount of medium (e.g., groundwater, soil) contacted per unit time or event (e.g., liters of water ingested per day).

contact stabilization process
Modification of the activated sludge process where raw wastewater is aerated with activated sludge for a short time prior to solids removal and continued aeration in a stabilization tank. Also called *biosorption process*.

contagion
Literally the transmission of infection by direct contact.

contained-in principal
Under RCRA, the EPA view that soil, groundwater, surface water, and debris that are contaminated with a listed hazardous waste must be regulated.

container
(1) *General.* Something that contains, such as a box, can, drum, etc. (2) *Shipping.* A large standard size metal box into which cargo is packed for shipment aboard specially configured oceangoing containerships and designed to be moved with common handling equipment enabling high-speed intermodal transfers in economically large units between ships, railcars, truck chassis, and barges using a minimum of labor. The container, therefore, serves as the transfer unit rather than the cargo contained therein. (3) *Hazardous Materials Management.* A component other than piping that contains a hazardous fluid.

container cargo
The tonnes of containerized cargo assessed at the *container rate* of tools as defined in the St. Lawrence Seaway Tariff of Tolls.

container chassis
Transportation. A semitrailer of skeleton construction limited to a bottom frame, one or more axles, specially built and fitted with locking devices for the transport of cargo containers, so that when the chassis and container are assembled, the units serve the same function as an over-the-road trailer.

container load
A load sufficient in size to fill a container either by cubic measurement or by weight.

container on flatcar (COFC)
This is accomplished with containers resting on railway flatcars.

container terminal
An area designated for the stowage of cargoes in container; usually accessible by truck, railroad, and marine transportation. Here containers are picked up, dropped off, maintained, and housed.

containerization
Stowage of general or special cargoes in a container for transport in the various modes.

containerize
To place a material into a container.

containerized cargo
(1) Cargo shipped or stored in containers. (2) Cargo which is practical to transport in a container, and results in a more economical shipment than could be affected by shipping the cargo in some other form of unitization.

containership
A cargo vessel designed and constructed to transport, within specifically designed cells, portable tanks and freight containers which are lifted on and off with their contents intact. There are two types of containerships: *full* and *partial*. Full containerships are equipped with permanent container cells with little or no space for other types of cargo. Partial containerships are considered multi-purpose container vessels, where one or more but not all compartments are fitted with permanent container cells, and the remaining compartments are used for other types of cargo. This category also includes container/car carriers, container/rail car carriers, and container/roll-on/roll-off vessels.

containment
(1) A process, structure, or system within a specified area or volume for preventing an entity from spreading and/or interacting with other materials or another environment. (2) Control of the expansion or propagation of accidental loss. Commonly used in fire control and hazardous chemical spills.

containment level
That degree of independence or separation in containment provided by a specified system.

containment system
The system, including the structure, ventilation method, entry and/or egress routes, contaminant collection equipment, etc., that will be utilized to prevent the spread of contamination from a work site into the surroundings.

contaminant
Any foreign material not normally found in a substance. Also, any physical, chemical, biological, or radiological substance or matter that has a diverse effect on air, soil, or water.

contaminate
The placement of one or more contaminants in a location where they may degrade the environment.

contaminated sharps
Any contaminated objects that can penetrate the skin including, but not limited to, needles, scalpels, broken glass, broken capillary tubes, and exposed ends of dental wire. In this context, the "contaminant" is normally considered to be blood, blood byproducts, or other infectious materials.

contamination
The degradation of natural water, air, or soil quality resulting from human activity.

contempt
(1) A willful or intentional disregard or disobedience of a public authority. (2) The stage in civil litigation where it is alleged that a party has violated an injunction, with the consequences that the court can order an appropriate remedy to cure the contempt. Also, in any court proceeding, civil or criminal, where a person before the court engages in disrespectful or disruptive behavior.

contempt of Congress
Deliberate interference with the duties and the powers of Congress. Both houses of Congress may cite an individual for such contempt.

contempt of court
Any act which is calculated to embarrass, hinder, or obstruct the court in the administration of justice, or which is calculated to lessen its authority or its dignity. Committed by a person who does any act in willful contravention of the authority or dignity of the court, or tending to impede or frustrate the administration of justice, or by one who, being under the court's authority as a party to a proceeding therein, willfully disobeys its lawful orders or fails to comply with an undertaking which he/she has given.

content validity
The extent to which a test samples a domain of important job behaviors.

conterminous U.S.
The 48 adjoining States and the District of Columbia.

contiguous zone
The entire zone established or to be established by the United States under the Convention of the Territorial Sea and the Contiguous Zone.

continent
One of the large, unbroken masses of land into which the Earth's surface is divided.

continental drift
See ***plate tectonics***.

contingency
A possible situation or event, usually referring to an undesirable or abnormal situation or occurrence.

contingency allowance
A small time allowance included within the standard time to cover for legitimate, expected additional work and delays. Usually not measured precisely because of its infrequent occurrence.

contingency analysis
An analysis performed to identify what abnormal situations, errors, or malfunctions a system may develop or encounter to improve system performance or establish what special human responses may be required under those circumstances.

contingency plan
(1) A written plan describing in detail the actions that will be taken in the event certain defined events should occur during the normal course of business operations. It usually refers to unwanted events or occurrences, such as disasters (floods, hurricanes, earthquakes, etc.) or other emergency situations (bomb threats, hazardous materials spill, fire, etc.) and generally includes the assignment of specific responsibilities to persons within the organization and may also include information on recovery or post-event activities that will be taken to return to normal operation once the situation has been corrected. (2) A document setting out an organized, planned, and coordinated course of action to be followed in case of a fire, explosion, or other accident that releases toxic chemicals, hazardous wastes, or radioactive materials which threaten human health or the environment.

contingent valuation survey (CVM)
A survey technique for assigning value to injured natural resources based on respondents' willingness to support various resources in monetary terms.

continual improvement
Under ISO 14000, the process of enhancing the environmental management system to achieve improvements in overall environmental performance in line with the organization's environmental policy.

continuance
The adjournment or postponement of a session, hearing, trial, or other proceeding to a subsequent day or time; usually on the request or motion of one of the parties.

continuing damages
Those that accrue from the same injury, or from the repetition of similar acts, between two specified periods of time.

continuous air monitor (CAM)
An instrument which is typically located in a potentially contaminated location to detect a specific contaminant, such as flammable or toxic gas or vapor, and which will alarm if a preset concentration is exceeded. It can be a passive type sampler or an active type sample.

continuous discharge capacity
Measured as percent of rated energy capacity. Energy delivered in a constant power discharge required by an electric vehicle for hill climbing and/or high-speed cruise, specified as the percent of its rated energy capacity delivered in a one-hour constant-power discharge.

continuous emissions monitoring
The continuous measurement of pollutants emitted into the atmosphere from combustion or industrial processes.

continuous emission monitoring system (CEMS)
The equipment as required by the Clean Air Act (CAA), used to sample, analyze, measure, and provide on a continuous basis a permanent record of emissions and flow (expressed in pounds per million British thermal units (lbs./m Btu), pounds per hour (lbs./hr) or such other form as the EPA Administrator may prescribe by regulations under the CAA.

continuous exposure
Exposure to a health hazard throughout the workday.

continuous forms
Having each individual form or sheet attached to the next, usually with guides for a printer, and which necessitates separation after printing.

continuous function
Any mathematical function which has no breaks or gaps in its extent.

continuous monitoring
Usually refers to air sampling or radiation monitoring conducted at locations where leaks may occur, or where hazardous materials are handled in high quantities.

continuous noise
(1) That noise which is persistent over long periods of time. (2) According to OSHA, variations in noise level involving maxima at intervals of 1 second or less.

continuous passive motion machine
A device which repeatedly cycles automatically to passively flex and extend one or more joints through their ranges of motion.

continuous reading method
See ***cumulative timing***.

continuous spectrum
(1) *General.* A range of frequencies within which all frequencies are present. (2) *Acoustics.* A spectrum which is continuous in the frequency domain.

continuous speech recognition
See ***speech recognition***.

continuous timing
See ***cumulative timing***.

continuous timing method
See ***cumulative timing***.

continuous variable
A variable which may take any value within a specified range of values.

continuous wave (CW)
A laser system which provides a constant, steady-state delivery of laser power.

continuous work
A sustained workload without any rest period.

contour line
A line that connects points of equal elevation above a reference level, most often sea level.

contour plowing
Farming methods that break ground following the shape of the land in a way that discourages erosion.

contract
An enforceable agreement, written or oral, between two or more persons which creates an obligation to do or not to do a particular thing.

contract carrier
(1) A carrier which furnishes transportation service to meet the special needs of shippers who cannot be adequately served by common carriers. A transportation company that carriers, for pay, the goods of certain customers only as contrasted to a common carrier that carries the goods of the public in general. *See also* ***carrier*** *and* ***common carrier***. (2) For-hire interstate operators [which] offer transportation services to certain shippers under contracts.

contract demand
Refining. The level of service in terms of the maximum daily and/or annual volumes of natural gas sold and/or moved by the pipeline company to the customer holding the contract. Failure of a pipeline company to provide service at the level of the contract demand specified in the contract can result in a liability for the pipeline company.

contract laboratory program (CLP)
Analytical program developed for Superfund waste site samples to fill the need for legally defensible analytical results supported by a high level of quality assurance and documentation.

contract labs
Laboratories under contract to EPA, which analyze samples taken from wastes, soil, air, and water, or carry out research.

contract operations
Private operation of municipal facilities, such as water and wastewater treatment plants.

contract-required detection limit (CRDL)
A term that is equivalent to contract-required quantitation limit, but used primarily for inorganic substances. *See also* ***contract-required quantitation limit***.

contract-required quantitation limit (CRQL)
Substance-specific level that a contract laboratory program (CLP) laboratory must be able to routinely and reliably detect in specific sample matrices. It is not the lowest detectable level achievable, but rather the level that a CLP laboratory should reasonably quantify. The CRQL may or may not be equal to the quantitation limit of a given substance in a given sample. For hazardous Ranking System purposes, the term CRQL refers to both the contract-required quantitation limit and the contract-required detection limit. *See also **contract laboratory program** and **hazardous ranking system**.*

contracted gas
Any gas for which Interstate Pipeline has a contract to purchase from any domestic or foreign source that cannot be identified to a specific field or group. This includes tailgate plant purchases, single meter point purchases, pipeline purchases, natural gas imports, SNG purchases, and LNG purchases.

contraction
A shortening or reduction in some dimension of a structure.

contractor
One who contracts to do work for another. This term is strictly applicable to any person who enters into a contract, but is commonly reserved to designate one who, for a fixed price, undertakes to procure the performance of works or services on a large scale, or the furnishing of goods in large quantities, whether for the public or a company or individual. Such are typically classified as *general contractors* (responsible for entire job) and *subcontractors* (responsible for only a portion of the job). A contractor is a person who, in pursuit of any independent business, undertakes to do a specific piece of work for another or other persons, using his/her own means and methods without submitting to their control in respect to all its details, and who renders service in the course of an independent occupation representing the will of his/her employer only as to the result of the work and not as to the means by which it is accomplished.

contractor employee
Railroad. A person employed by a contractor hired by a railroad to perform normal maintenance work to railroad rolling stock, track structure, bridges, buildings, etc.

contractory tissue
Any tissue which is capable of shortening in response to stimulation.

contractual relationship
Under the Comprehensive Environmental Response, Compensation, and Liabilities Act (CERCLA), includes, but is not limited to, land contracts, deeds or other instruments transferring title or possession, unless the real property on which the facility concerned is located was acquired by the defendant after the disposal or placement of the hazardous substance on, in, or at the facility, and one or more of the circumstances described in the list below is also established by the defendant by a preponderance of the evidence:

1. At the time the defendant acquired the facility the defendant did not know and had no reason to know that any hazardous substance which is the subject of the release or threatened release was disposed of on, in, or at the facility.
2. The defendant is a government entity which acquired the facility by escheat, or through any other involuntary transfer or acquisition or through the exercise of eminent domain authority by purchase of condemnation.
3. The defendant acquired the facility by inheritance or bequest.

contracture
Abnormal shortening of muscle tissue, rendering the muscle highly resistant to stretching. A contracture can lead to permanent disability. It can be caused by fibrosis of the tissues supporting the muscle or the joint, or by disorders of the muscle fibers themselves.

contraflow lane
Reserved lane for buses on which the direction of bus traffic is opposite to the flow of traffic on the other lanes.

contrail
Long, narrow clouds caused when high-flying jet aircraft disturb the atmosphere. Best formed in clear, cold, humid air. Also referred to more accurately but less commonly as *condensation trail.*

Contrails created by a high-performance jet aircraft

contralateral
Located on or pertaining to the opposite side of the body.

contrast
See ***chromatic contrast*** *and* ***luminous contrast***.

contrast attenuation
A decrease in the amount of contrast over a space or time.

contrast detection
A basic visual task in which the visual system perceives a difference in luminance, creating an object and a background.

contrast ratio
A mathematical relationship involving some form of a ratio between figure luminance or reflectance and background luminance or reflectance. *See also* ***luminance contrast***.

contrast sensitivity
A measure of the ability to perceive a visual contrast between two regions; the reciprocal of the contrast threshold.

contrast threshold
The smallest difference between two visual stimuli which is perceptible to the human eye under specified conditions of adaptation, luminance, and visual angle on a certain proportion of a set of trials. Also referred to as *liminal contrast, liminal contrast threshold,* and *threshold contrast*.

contribution
An area of the law dealing with seeking reimbursement from other responsible parties for an appropriate share of damages or expenses which must be paid.

contributory event
As pertains to fault tree analysis (FTA) and/or the management oversight and risk tree (MORT), an event that significantly influences the outcome of the top or primary event. Represented graphically as a rectangle and may also be referred to as a *main event* or *secondary event.*

contributory negligence
The act or omission amounting to want of ordinary care on the part of the complaining party which, occurring with the defendant's negligence, is a proximate cause of the injury. The proving of contributory negligence can significantly decrease or possibly eliminate any subsequent recovery on the part of the plaintiff(s).

control
(1) *Epidemiology/Toxicology.* The nature, number, and reproducibility of the controls (unexposed or unaffected) to determine the accuracy and significance of the conclusions from the experimental (exposed) cohort results. A most important factor in any study of humans, animals, or biological organisms. *See also* ***cohort study***. (2) *Industrial Hygiene.* Measures, including engineering and administrative means, as well as the use of personal protective equipment, that are implemented to reduce, minimize, or otherwise reduce exposure to a health hazard. (3) *Radiation Protection.* Any action to stabilize, inhibit future misuse of, or reduce emissions or effluents from uranium byproduct materials.

control area
(1) *Interstate Highway System.* A metropolitan area, city or industrial center, a topographic feature such as a major mountain pass, a favorable location for a major river crossing, a road hub which would result in material traffic increments on the interstate route, a place on the boundary between two states agreed to by the states concerned, or other similar point of significance. (2) *Aviation.* A controlled airspace extending upward from a specified limit above the earth.

control arrangement
See ***control layout***.

Control Assessment Protocol (CAP)
A systematic procedure for clinicians to follow in the evaluation of disabled individuals for assistive devices.

control cab locomotive
A locomotive without propelling motors but with one or more control stands.

control circuit
An electrical circuit between a source of electric energy and a device which it operates.

control coding
The use of any of a variety of coding methods for labeling a control. *See* ***color coding, shape coding, size coding, label coding,*** *and* ***location coding***.

control device
See ***direct manipulation device***.

control display layout
See ***display-control layout***.

control-display ratio
The ratio of movement of a control to the movement or change of an indicator on a display. Synonymous *with control-response ratio*.

control efficiency
The ratio of the amount of pollutant removed from a source of release or emission by a control device, to the total amount of pollutant before control, and expressed as a percentage.

control force
That amount of force required to operate a control. *See also* ***control torque*** *and* ***actuation force***.

control group
A group of individuals or items selected from what is believed to be the same population as an experimental group, but which is not exposed to the experimental treatment(s) under consideration. Synonymous with *comparison group*.

control layout
The grouping of manual controls within a location at a workplace. Also referred to as control arrangement. *See also* ***control location*** *and* ***display-control layout***.

control limit
That boundary value which a measurement on some aspect or dimension of a product or system must not exceed.

control location
The general placement of controls for use by an operator. *See also* ***control layout***.

control machine
An assemblage of manually operated devices for controlling the functions of a traffic control system; it may include a track diagram with indication lights.

control operator
An employee assigned to operate the control machine of a traffic control system.

control placement
See ***control location***.

control precision
A psychomotor ability involving the positioning of larger muscle groups to make rapid, repeated adjustments to one or more controls.

control-response ratio
See ***control-display ratio***.

control rod
A rod used to control the nuclear power of a nuclear reactor. The reactor functions through the fission of nuclear fuel by neutrons. The control rod absorbs neutrons that would normally produce fission in the atoms of the fuel. Pushing the rod into the reactor reduces the release of nuclear power and pulling the rod out increases the rate.

control sector
Aviation. An airspace area of defined horizontal and vertical dimensions for which a controller or group of controllers has air traffic control responsibility, normally within an air route traffic control center or an approach control facility. Sectors are established based on predominant traffic flows, altitude strata, and controller workload. Pilot communications during operations within a sector are normally maintained on discrete frequencies assigned to the sector.

control sensitivity
The ratio between the amount of movement or change on a display and the control movement.

control slash
Aviation. A radar beacon slash representing the actual position of the associated aircraft. Normally, the control slash is the one closest to the interrogating radar beacon site. When

Air Route Traffic Control Center (ARTCC) radar is operating in narrow band (digitized) mode, the control slash is converted to a target symbol.

control spacing
That distance between the human-operated mechanism for two or more control devices.

control station
The place where the control machine of a traffic control system is located. *See also* ***control machine***.

control stick
The primary control device on many types of aircraft, generally consisting of a rod-shaped structure extending from the floor in front of the pilot's seat with aircraft handling and other controls.

control system
(1) A system whose primary function is the monitoring of outputs from a given set of functions and using that data or information to regulate that set in some specified manner or to propose new regulations. (2) A component, or system of components functioning as a unit, including control valves and sensing, warning, relief, shutdown, and other control devices, which is activated either manually or automatically to establish or maintain the performance of another component.

control technique guidelines (CTG)
A series of EPA documents designed to assist states in defining reasonable available control technologies (RACT) for major sources of volatile organic compounds (VOCs).

control torque
That amount of torque required to operate a rotary control. *See also* ***control force*** *and* ***actuation force***.

control velocity
See ***capture velocity***.

controllable emergency
An emergency where reasonable and prudent action can prevent harm to people or property.

controlled access rights-of-way
Lanes restricted for at least a portion of the day for use by transit vehicles and other high occupancy vehicles (HOVs). Use of controlled access lanes may also be permitted for vehicles preparing to turn. The restriction must be sufficiently enforced so that 95 percent of the vehicles using the lanes during the restricted period are authorized to use them. *See also* ***right-of-way***.

controlled airspace
An airspace of defined dimensions within which air traffic control service is provided to Instrument Flight Rules (IFR) flights and to Visual Flight Rules (VFR) flights in classification.

controlled area
(1) *General*. Specific area designated for the performance of certain work, usually work that is hazardous or sensitive in nature, where control measures have been implemented to prevent the unauthorized access of personnel into the area while the work is being performed. Control measures may include physical barriers, warning signs and signals, personnel monitoring, or any combination of these and/or other measures. (2) *Radiation*. A defined area in which the occupational exposure of personnel to radiation or radioactive material is under the supervision of an individual responsible for radiation protection.

controlled departure time programs (CDT programs)
These programs are the flow control process whereby aircraft are held on the ground at the departure airport when delays are projected to occur in either the en route system or the terminal of intended landing. The purpose of these programs is to reduce congestion in the air traffic system or to limit the duration of airborne holding in the arrival center or terminal area. A CDT is a specific departure slot shown on the flight plan as an expected departure clearance time (EDCT).

controlled experiment
An experimental investigation in which the relevant independent variables are directly and systematically manipulated and/or controlled and the effects of such manipulation are measured. Also referred to as a *controlled study*.

controlled motion
See ***controlled movement***.

controlled movement
Any controlled bodily movement in which prime mover and antagonist muscles are integrated using muscle contraction throughout

the range of the motion to generate a desired force and/or velocity. Also referred to as *non-ballistic movement* and *tension movement*.

controlled point
Aviation. A location where signals and/or other functions of a traffic control system are controlled from the control machine. *See also* ***control machine*** *and* ***control operator***.

controlled study
See ***controlled experiment***.

controlled substance
Any drug so designated by law whose availability is restricted, i.e., so designated by federal or state Controlled Substances Acts. Included in such classification are narcotics, stimulants, depressants, hallucinogens, and marihuana.

Controlled Substances Acts
Federal and state acts (the latter modeled on the Uniform Controlled Substances Act) the purpose of which is to control the distribution, classification, sale, and use of drugs. The majority of the states have such acts.

controlled time
That elemental time which is governed solely by some external process.

controlled time of arrival
The original estimated time of arrival adjusted by the ATCSCC ground delay factor.

controller
(1) Any device used for operating and/or regulating a system. (2) A device or group of devices that serves to govern, in some predetermined manner, the electric power delivered to the apparatus to which it is connected. (3) A person authorized to provide air traffic control services. *See* ***air traffic controller***.

controlling depth
The least available water in a navigable channel which limits the amount of cargo that may be carried by the vessel.

controlling locomotive
A locomotive arranged as having the only controls over all electrical, mechanical and pneumatic functions for one or more locomotives, including controls transmitted by radio signals if so equipped. It does not include two or more locomotives coupled in multiple which can be moved from more than one set of locomotive controls.

contusion
Injury to tissues without breakage of the skin; a bruise. In a contusion, blood from the broken vessels accumulates in surrounding tissues, producing pain, swelling, and tenderness. A discoloration appears as a result of blood seepage under the surface of the skin. Serious complications may develop in some cases of contusion. Normally, blood is drawn off from the bruised area in a few days. But there is a possibility that blood clotted in the area will form a cyst or calcify and require surgical treatment. The contusion may also be complicated by infection.

convection
The transfer of heat from one place to another by moving fluid (a gas or a liquid). Natural convection results from differences in temperature. The rising of heated surface air and the sinking of cooler air aloft is often called *free convection* (as opposed to *forced convection*).

convection heat load
The amount of heat energy transferred between the skin and the air. Human skin is normally 95°F (35°C). Air in excess of that temperature will warm the body, whereas air below that temperature will cool the body.

convective condensation level (CCL)
The level above the surface marking the base of a cumuliform cloud that is forming due to surface heating and rising thermals.

convective heat loss
That amount of heat eliminated from the body via convection, indicated by an equation of the form:

$$H = h_c A\,(T_s - T_a)$$

where:
H = convective heat loss
h_c = convective heat transfer coefficient
A = body surface area
T_s = weighted mean skin temperature
T_a = air temperature

convective heat transfer coefficient (h_c)
A number which includes factors for clothing thermal characteristics and environmental conditions.

convective instability
Instability arising in the atmosphere when a column of air exhibits warm, moist, nearly saturated air near the surface and cold, dry air aloft. When the lower part of the layer is lifted and saturation occurs, it becomes unstable.

conventional anthropometry
See ***classical anthropometry***.

conventional cab
A cab design in which the engine is located ahead, or mostly ahead, of the cowl.

conventional memory
Computing. The base RAM on a computer, typically the first 640 kilobytes. Conventional memory is the only kind of RAM that MS-DOS-based applications can use, unless an expanded memory manager (EMM) is used. *See also* ***extended memory*** *and* ***expanded memory***.

conventional pollutants
Statutory listed pollutants the nature of which is understood well by the scientific community. These may be in the form of organic waste, sediment, acid, bacteria and viruses, oil and grease, or heat.

conventional systems
Systems that have been traditionally used to collect municipal wastewater in gravity sewers and convey it to a central primary or secondary treatment plant prior to discharge to surface waters.

convergence
(1) *General*. A coordinated inward rotation of the eyes about their vertical axis to fixate on a point near the observer to obtain fusion. (2) *Meteorology*. An atmospheric condition that exists when the winds cause a horizontal net inflow of air into a specified region.

convergence angle
That angle formed by the intersection of the line of sight of each eye when both eyes are fixated at a single point.

convergence point
That location on a curve at which a worker's learning curve achieves standard performance.

convergent phoria
A tendency for an observer to fixate in front of a stationary target.

conversion
(1) The act of changing into something of different form or properties. (2) The transformation of emotions into physical manifestations.

conversion factor
A number that translates units of one system into corresponding values of another system. Conversion factors can be used to translate physical units of measure for various fuels into British Thermal Unit (BTU) equivalents. *See also British Thermal Unit*.

converter dolly
A motor vehicle consisting of a chassis equipped with one or more axles, a fifth wheel and/or equivalent mechanism, and drawbar, the attachment of which converts a semitrailer to a full trailer.

convertible
A truck or trailer that can be used either as a flatbed or open-top by removing side panels.

convertible collision insurance
Type of collision coverage generally carrying lower premium but requiring higher premium after the first loss or claim (an alternative form of deductible collision coverage). *See also* ***insurance***.

convertible insurance
A policy that may be changed to another form by contractual provision and without evidence of insurability. Usually used to refer to term life insurance convertible to permanent insurance. *See also* ***insurance***.

convertible life insurance
Generally a form of term life insurance which gives the insured the right to change the policy to permanent life insurance without requiring a medical examination of the insured party. *See also* ***insurance***.

convex function
A mathematical relationship or graph having positive second derivative over a specified interval of interest, resulting in a U-shaped curve.

conveyance
In its most common usage, the transfer of title to land from one person, or class of persons, to another by deed.

conviction
The end of a criminal legal proceeding by a determination of guilt.

convulsion
Involuntary spasm or contraction of muscles. In general, there are three types of convulsions: *clonic,* in which opposing muscles contract and relax alternately producing rhythmic movements; *tonic,* in which all the muscles tighten until the victim becomes rigid; and those that occur in *Jacksonian epilepsy,* in which the muscular twitching begins in one area and spread to another.

cooking unit, counter-mounted
A cooking appliance designed for mounting in or on a counter and consisting of one or more heating elements, internal wiring, and built-in or separately mountable controls.

cool color
A blue or green color, or a color which appears less bright than another for a given intensity.

coolant
A liquid or gas used to reduce the heat generated by power production in nuclear reactors, electric generators, various industrial and mechanical processes, and automobile engines.

cooling degree day
A form of degree-day used in estimating the amount of energy necessary to reduce the effective temperature of warm air. A cooling degree-day is a day on which the average temperature is one degree above a desired base temperature.

cooling pond
A pond where water is cooled by contact with air prior to reuse or discharge.

cooling tower
An open water recirculating device that uses fans or natural draft to draw or force ambient air through the device to cool warm water by direct contact.

cooling tower blowdown
A side-stream of water discharged from a cooling tower recirculation system to prevent scaling or precipitation of saturated salts or minerals.

cooling water
Water used, usually in a condenser, to reduce the temperature of liquids or gases.

Cooper Scale
A no longer used rating scale with a range of 1 (excellent) through 10 (fatal) which was developed in an attempt to have pilots provide more objective evaluations of aircraft handling qualities. Also referred to as the *Cooper Rating Scale.*

Cooper-Harper scale
An ordinal rating procedure using a decision tree on a scale of 1 (excellent) through 10 (major deficiencies) for task difficulty. Designed originally for use by test pilots for evaluating aircraft handling, but has been used in other physical workload situations as well. Also referred to as *Cooper-Harper aircraft handling characteristics scale.*

Cooper-Harper Scale, modified
An ordinal rating procedure using a decision tree on a scale of task difficulty ranging from 1 (very easy) through 10 (impossible) for mental workload determinations.

cooperation clause
That provision in insurance policies which requires the insured to cooperate with the insurer in defense of a claim.

coordinate
(1) A position in space, time, amplitude, or some other dimension. (2) To cause separate entities to act together harmoniously toward a final goal.

coordinate system
A spatial reference system with a defined origin and rules for defining locations within that system. *See also* ***rectangular coordinate system***.

coordinate transformation
Any mathematical or graphical process for modifying or shifting a coordinate system.

coordinates
The intersection of lines of reference, usually expressed in degrees/minutes/seconds of latitude and longitude, used to determine position or location.

coordination fix
The fix in relation to which facilities will handoff, transfer control of an aircraft, or coordinate flight progress data. For terminal facilities, it may also serve as a clearance for arriving aircraft.

co-partnership incentive plan
An incentive plan in which workers have the opportunity to own a share of the business enterprise, thus obtaining some portion of the profits resulting from that ownership.

coping
The top or covering of an exterior masonry wall.

copolymer
A long-chain molecule resulting from the reaction of more than one monomer species with another.

copper
A chemical element, atomic number 29, atomic weight 63.54, symbol Cu. It is necessary for bone formation and for the formation of blood because it acts as a catalyst in the transformation of inorganic iron into hemoglobin. There is little danger of deficiency in ordinary diets because of relatively abundant supply and minute daily requirements.

copper-nickel
A copper alloy containing 10-30% nickel to increase resistance to corrosion and stress corrosion cracking. Also called *cupronickel*.

copper sulfate
Chemical used for algae control, also called *blue vitriol*. Chemical formula is $CuSO_4$.

copperas
Common name for ferrous sulfate heptahydrate, a common coagulant. Chemical formula is $FeSO_4{\cdot}7H_2O$.

coproporphyrin
A porphyrin that is formed in the blood-forming organs and found in the urine and feces.

copter
See ***helicopter***.

copy
A computer operation system function which duplicates a file or segment in another location while leaving the original file or segment intact.

copyright
The right of literary property as recognized and sanctioned by positive law. An intangible, incorporeal right granted by statute to the author or originator of certain literary or artistic productions, whereby he/she is invested, for a specified period, with the sole and exclusive privilege of multiplying copies of the same and publishing and selling them.

cordelle
A hawser; a towline such as those used to pull keelboats in the French-speaking parts of North America.

core
The heart of the nuclear reactor where the nuclei of the fuel undergo fission (spilt) and release energy. The core is usually surrounded by a reflecting material that bounces stray neutrons back to the fuel.

core temperature
The temperature in the central part of the body. Rectal temperature is considered a measure of core temperature.

CORELAP
See ***Computerized Relationship Layout Planning***.

core-shell model
A simple thermodynamic concept in which the human is treated as having a heat-producing core and a surrounding shell, with heat exchange occurring through the shell to the environment.

Coriolis acceleration
That acceleration generated by the simultaneous exposure to rotational motion about two axes in an internal reference frame.

Coriolis effect
The misperception of body orientation, commonly accompanied by nausea and vertigo on exposure to Coriolis acceleration.

Coriolis force
An apparent force observed on any free-moving object in a rotating system. On the earth, this deflective force results from the earth's rotation and causes moving particles (including the wind) to deflect to the right in the Northern Hemisphere and to the left in the Southern Hemisphere.

corium
The fibrous inner layer of the skin, derived from the embryonic mesoderm, varying from 1/50 to 1/8 inch in thickness, well supplied with nerves and blood vessels and containing hair roots and sebaceous and sweat glands; on the palms and soles it bears ridges whose arrangement in whorls and loops is peculiar to the individual (fingerprints and footprints).

cornea
The clear, transparent anterior covering of the eye. The cornea is subject to injury by foreign bodies in the eye, bacterial infection, and viral infection, especially by the herpes simplex virus. The herpes zoster virus, which causes "shingles," can also infect the cornea. Prompt treatment of any corneal injury or infection is essential to avoid ulceration and loss of vision.

corneal reflex
A reflex action of the eye resulting in automatic closing of the eyelid when the cornea is stimulated. The corneal reflex can be elicited in a normal person by gently touching the cornea with a wisp of cotton. Absence of the corneal reflex indicates deep coma or injury of one of the nerves carrying the reflex action.

corneo-retinal potential (CRP)
The bioelectric potential between the anterior and posterior eyeball.

corner sweep
Scraper used to remove sludge from the corner of a square clarifier.

corona
(1) *General.* A crown-like structure or part, as the top of the head or the upper part of a tooth. (2) *Astronomy.* A luminous circle around one of the heavenly bodies, as when seen through cloud or mist. (3) *Astrophysics.* The luminous envelope of ionized gases visible during a total eclipse of the sun. (4) *Electricity.* The luminous discharge appearing at the surface or between the terminals of an electrical conductor under high voltage.

coronal plane
See ***frontal plane***. (Note: The term is often used instead of *frontal plane* in conjunction with the brain.)

coronary
Pertaining to blood vessels or nerves which encircle an organ or other structure, especially the heart.

coronary occlusion
The blockage of an artery supplying blood to the muscle tissue of the heart.

coronoid process
A projection from the proximal end of the ulna which fits into the coronoid fossa on flexion of the elbow.

corporate average fuel economy (CAFE) standards
CAFE standards were originally established by Congress for new automobiles, and later for light trucks, in Title V of the Motor Vehicle Information and Cost Savings Act (15 U.S.C. 1901, et seq.) with subsequent amendments. Under CAFE, automobile manufacturers are required by law to produce vehicle fleets with a composite sales-weighted fuel economy which cannot be lower than the CAFE standards in a given year, or for every vehicle which does not meet the standard, a fine of $5.00 is paid for every one-tenth of a mpg below the standard.

corporate code of conduct
A statement of the corporation's ethical standards and special goals, which is clearly stated and widely publicized. Includes standards of business conduct related to both legal compliance and general organizational requirements and legal responsibilities to not engage in conduct that would otherwise result in legal liability.

corporate compliance officer
A high-level employee with a substantial role in the making of policy within an organization to create and monitor the compliance program oversight and enforcement. The individual should be of the highest integrity and be knowledgeable of every function of the organization's business. The individual should also have sufficient command authority so his or her decisions will not be easily countermanded.

corporate/executive flying
The use of aircraft owned or leased, and operated by a corporate or business firm for the transportation of personnel or cargo in furtherance of the corporation's or firm's business, and which are flown by pilots working for compensation.

corporate probation
Under the Federal Sentencing Guidelines (FSGs), an appropriate sentence for an organization under the FSGs when needed to ensure that another sanction will be fully implemented, or to ensure that steps will be taken within the organization to reduce the likelihood of future criminal conduct. Often mandated under the FSGs for an organization with 50 or more employees without an effective compliance program. Conditions of probation can include requiring the organization to publicize the nature of the offense or to develop a compliance program satisfactory to the court. *See also* ***effective compliance program***.

corpuscle
A blood cell.

corrected effective temperature (CET)
A measure of environmental heat stress which includes average radiant temperature and globe temperature effects.

correction
Aviation (communication) An error has been made in the transmission and the correct version follows.

corrective action
See ***cleanup***.

corrective lens
An eyeglass lens that has been ground to the wearer's individual prescription to enable normal visual acuity.

corrective maintenance
A form of maintenance which is intended to return a system or piece of equipment to proper operating status after it has failed. Also referred to as *breakdown maintenance, unscheduled maintenance,* and *remedial maintenance*.

correlated color temperature
That temperature of a Planckian radiator whose perceived color most closely resembles that of a given stimulus source when viewed at the same brightness and under specified viewing conditions.

correlated work crew
A group of workers who interact with each other or work together on a task, such that each individual's work is not independent.

correlation
The degree of association between variables. The simultaneous increase or decrease in the value of two random variables (positive correlation), or the simultaneous increase in the value of one and decrease in the value of the other (negative correlation).

correlation coefficient
A number between 1.0 and -1.0 which represents the degree and direction of correlation between two variables.

correlative kinesiology
See ***electromyographic kinesiology***.

corridor
Transportation. A broad geographical band that follows a general directional flow connecting major sources of trips that may contain a number of streets, highways, and transit route alignments.

corrode
The gradual breaking down, wearing away, or alteration of a structure due to the action of air, moisture, or a chemical.

corrosion
Physical damage, usually in the form of deterioration or destruction caused by chemical or electrochemical action as contrasted with erosion caused by mechanical reaction.

corrosive
A chemical agent that reacts with the surface of a material (including skin) causing it to deteriorate or wear away.

corrosive waste
Waste having the ability to corrode standard containers or to dissolve toxic components or other waste.

corrosivity
The ability of a substance to produce corrosion.

corrugated plate interceptor (CPI)
Oil separation device utilizing inclined corrugated plates to separate free non-emulsified oil and water based on their density difference.

cortex
The outer portion of an organ or structure, usually referring to the brain, adrenal gland, or bone.

cortical bone
The compact bone tissue next to the surface of a bone.

corticospinal system
*See **pyramidal system**.*

corticosteroid
Any of the hormones elaborated by the cortex of the adrenal gland.

corundum
Natural aluminum oxide material that may contain traces of iron, magnesium, and silica.

coryza
An acute inflammation of the nasal mucous membrane, with profuse discharge.

cosine
A trigonometric function; the value of the ratio of the adjacent side of an acute angle to the hypotenuse in a right triangle.

cosine law of illumination
A rule that the illumination on any surface changes according to the cosine of the incident light angle from perpendicular to the surface. Represented as:

$$E = \frac{I \cos \Theta}{d^2}$$

where:

E = illumination level
I = intensity of light source
Θ = the angle of incidence of the light from perpendicular
d = the distance from the light source

cosmetic
Under the Federal Food, Drug, and Cosmetic Act, articles intended to be rubbed, poured, sprinkled, or sprayed on, introduced into, or otherwise applied to the human body or any part thereof or cleansing, beautifying, promoting attractiveness, or altering the appearance; and articles intended for use as a component of any such articles; except that such term shall not include soap.

cosmic radiation
Penetrating ionizing radiation, both particulate and electromagnetic, originating in outer space. Secondary cosmic rays, formed by interactions in the earth's atmosphere, add to the general background radiation.

cosmic ray
*See **cosmic radiation**.*

cost
Those expenses incurred in producing a product, delivering goods, or providing a service, whether financial, human, or metabolic.

cost-benefit analysis
(1) *General.* The determination or estimation and evaluation of the weighted relative financial, social, and/or other costs to the same or other categories of rewards or compensation. It should be performed prior to undertaking the endeavor being considered. (2) *System Safety.* A system safety analytical technique used to evaluate various possible courses of action with respect to the costs that are incurred compared to the benefit of the results.

cost-effective alternative
An alternative control or corrective regulatory or compliance method identified by the EPA after analysis as being the best available in terms of reliability, performance, and economic considerations (i.e., when selecting a method for cleaning up a site on the Superfund National Priorities List, the EPA balances costs with the long-term effectiveness of the various methods proposed).

cost-effectiveness
The relative financial or other benefits obtained compared to the cost of alternatives.

Cost, Insurance, Freight (CIF)
A type of sale in which the buyer of the product agrees to pay a unit price that includes the f.o.b. value of the product at the point of origin plus all costs of insurance and transportation. This type of transaction differs from a delivered purchase in that the buyer accepts the quantity as determined at the loading port (as certified by the Bill of Lading and Quality Report) rather than pay on the basis of the quantity and quality ascertained at the unloading port. It is similar to the terms of an f.o.b. sale, except that the seller, as a service for which he/she is compensated, arranges for transportation and insurance.

cost of accidents per employee (CAE)
The cost of the accidents incurred per year spread across the average number of employees, as in the following formula:

$$CAE = \frac{\textit{total accident costs}}{\textit{average number of employees}}$$

cost recovery
A legal process by which potentially responsible parties (PRPs) who contributed to the contamination at a Superfund site can be required to reimburse the Trust Fund for money spent during any cleanup actions by the federal government.

costal cartilage
That segment of cartilage which attaches a rib to the sternum or, in some cases, to adjacent ribs.

COTP
See ***captain of the port***.

coulomb
A quantity of electric charge equal to one ampere second.

coulomb friction
That friction from movement between dry surfaces.

Council on Environmental Quality (CEQ)
An agency created under Section 202 of the National Environmental Policy Act (NEPA) in January of 1970. Before its abolishment by President Clinton on 8 February 1993, the CEQ consisted of three members, appointed by the President, with advice and consent from the Senate. Under NEPA, the CEQ provided advice to the President on federal programs and policies affecting the environment and also prepared an annual Environmental Quality Report (EQR) for Congress that described the state of the environment in the United States and reported the status of specific initiatives during the previous year. Prior to 1978, the CEQ only issued guidelines, which were merely advisory and had no force of law behind them. Federal agencies could either follow these recommendations or pursue an alternate course of action. But in 1978, as a result of an Executive Order, CEQ issued regulations that required all federal agencies to implement NEPA. Upon its abolishment, President Clinton created a new office called the Office of Environmental Policy (OEP). *See also* ***Office of Environmental Policy***.

councilman
See ***alderman***.

count
The external indication of a device designed to enumerate ionizing events. It may refer to a single detected event or the total measured in a given time period.

counter
(1) The top flat surface of a work space. (2) A device for counting nuclear disintegrations used to measure radioactivity. (3) Any device or system which keeps track of incrementing or decrementing numbers of objects or events.

counterfeit drug
Under the Federal Food, Drug and Cosmetic Act, a drug which, or the container or labeling of which, without authorization, bears the trademark, trade name, or other identifying mark, imprint, or device, or any likeness thereof, of a drug manufacturer, processor, packer, or distributor other than the person or persons who in fact manufactured, processed, packed, or distributed such drug and which thereby falsely purports or is represented to be the product of, or to have been packed or distributed by, such other drug manufacturer, processor, packer, or distributor.

countermeasure
An action taken in opposition to another.

country breeze
A light breeze that blows into a city from surrounding countryside. It is best observed on clear nights when the urban heat island is most pronounced. *See also* ***urban heat island***.

county attorney
See ***district attorney***.

coupled approach
Aviation. A coupled approach is an instrument approach performed by the aircraft autopilot which is receiving position information and/or steering commands from onboard navigation equipment. In general, coupled non-precision approaches must be discontinued and flown manually at altitudes lower than 50 feet below the minimum descent altitude, and coupled precision approaches must be flown manually below 50 feet ALG. Note: Coupled and autoland approaches are flown in Visual Flight Rules (VFR) and Instrument Flight Rules (IFR). It is common for carriers to require their crews to fly coupled approaches and autoland approaches (if certified) when

the weather conditions are less than approximately 4,000 Runway Visual Range (RVR). *See also* ***autoland approach***.

coupler

A device for locking together the component parts of a tubular metal scaffold. The material used for the couplers shall be of a structure type, such as a drop-forged steel, malleable iron, or structural grade aluminum. The use of gray cast iron is prohibited.

coupling

Any of a variety of possible interfaces between the hand or robotic grapple fixture and another objects for purposes of gripping or touching.

coupon test

A method of determining the rate of corrosion or scale formation by placing metal strips, or coupons, of a known weight in a tank or pipe.

courier services (except by air)

Establishments primarily engaged in the delivery of individually addressed letters, parcels, and packages (generally under 100 pounds), except by means of air transportation or by the U.S. Postal Service.

course

(1) The intended direction of flight in the horizontal plane measured in degrees from north. (2) The Instrument Landing System (ILS) localizer signal pattern usually specified as the front course or the back course. (3) The intended track along a straight, curved, or segmented Microwave Landing System (MLS) path. *See also* ***bearing*** *and* ***flight path***.

course of employment

These words, as applied to compensation for injuries within the purview of workers' compensation acts, refer to the time, place, and circumstances under which the accident takes place. A worker is in the course of employment when, within the time covered by employment, he/she is doing something which he/she might reasonably do while so employed at a proper place.

courseware

That application or system software and the programmed/coded information base which are used to provide the information and interactions in a computer-based instruction system.

Court of Appeals

In those states with courts of appeals, such courts are usually intermediate appellate courts (with the highest appellate court being the state Supreme Court).

covariate

A variable which is related to and varies as the predictor and outcome variables do.

covenant

One of a number of enforceable promises which can govern the use of land. Others are called *restriction, easement,* and *equitable servitude*.

cover

Vegetation or other material providing protection as ground cover.

cover material

Soil or other suitable material used to cover compacted solid waste in a sanitary or secure landfill.

coverage

The number of jobs or the number of personnel whose jobs have been assigned standards during a particular period.

coverage ratios

Transportation. The ratio used to measure the degree to which expenditures are funded or "covered" by the various types of revenues. This ratio indicates the percent of expenditures that is funded by identifiable transportation-related tax receipts, fees, etc.

covered fleet

Under the provisions of the Clean Air Act (CAA), 10 or more motor vehicles that are owned or operated by a single person. In determining the number of vehicles owned or operated by a single person for purposes of the CAA, all motor vehicles owned or operated, leased, or otherwise controlled by such person, by any person who controls such person, by any person controlled by such person, and by any person under common control with such person shall be treated as owned by such person.

covered fleet vehicle

Under the provisions of the Clean Air Act (CAA), only a motor vehicle which is in a vehicle class for which standards are applicable under the CAA and in a covered fleet which is centrally fueled (or capable of being centrally fueled). No vehicle which under normal operations is garaged at a personal residence at night shall be considered to be a vehicle which is capable of being centrally fueled within the meaning of the CAA.

covert behavior
Any behavior consisting of actions not directly viewable by an external observer.

covert lifting task
An operation in which body parts are moved, thus involving biomechanical aspects of the body, but which doesn't involve the handling of a load other than the body parts themselves.

cowboy
Transportation (slang). Reckless driver.

cowl
The front part of a cab or body directly below the base of the windshield, between fire wall and instrument panel, and usually including the hood.

coxal bone
A bone consisting of the fused ileum, pubis, and ischium making up part of the pelvic girdle. Commonly referred to as *hip bone, pelvic bone,* and also known as *innominate bone.*

CPI
See ***corrugated plate interceptor****. See also* ***Consumer Price Index****.* Also, an acronym for *chemical process industry.*

cpm
Counts per minute.

CPM
See ***Critical Path Method****.*

cps
Cycles per second.

CPSC
Consumer Product Safety Commission (United States).

CPVC
Chlorinated polyvinyl chloride. A chlorinated form of PVC that provides increased heat resistance.

crack
(1) *General* An illicit drug of abuse derived primarily from cocaine but also may contain a variety of fillers and/or other materials that may or may not be considered "drugs." (2) *Structural Dynamics.* A fracture without complete separation into parts, except that castings with shrinkage cracks or hot tears that do not significantly diminish the strength of the member are not considered to be cracked.

cradle-to-grave
Under RCRA, the common term used to emphasize the extent of hazardous waste management responsibilities. Basically, hazardous wastes must be properly managed and those who participate in that management are held responsible from the moment it is generated (i.e., its creation) up to the time it is either neutralized, destroyed, or otherwise disposed of properly.

CRAFT
See ***Computerized Relative Allocation of Facilities Technique****.*

cramps
Common term for sudden, involuntary muscular contractions which cause severe pain. Painful muscle spasms in the extremities, back, or abdomen, as a result of, or due in part to excessive loss of salt during sweating.

crane
A mechanical device intended for lifting or lowering a load and moving it horizontally, in which the hoisting mechanism is an integral part of the machine. A crane may be a fixed or mobile machine.

cranial
Pertaining to the cranium.

cranial length
The linear distance from glabella to opisthocranion.

cranial nerves
Nerves that are attached to the brain and pass through the openings of the skull. There are 12 pairs of cranial nerves, symmetrically arranged so that they are distributed mainly to the structures of the head and neck. The one exception, the vagus nerve, extends beyond the head and carries among its fibers the motor fibers that go to the bronchi, stomach, gallbladder, small intestine, and part of the

large intestine. It also carries the fibers that control the release of secretions of the gastric glands and the pancreas, and inhibitory fibers to the heart.

cranial suture
A suture between two bones enclosing the brain.

craniosacral
See ***parasympathetic***.

craniostat
A device for measuring the facial angle.

cranium
The skeleton of the head, exclusive of the mandible and facial bones.

crash
An event that produces injury and/or damage, involves a motor vehicle in transport, and occurs on a traffic way or while the vehicle is still in motion after running off the traffic way. *See also* ***accident, casualty, collision, derailment, fatality, event, hit and run, incident, injury, vehicle maneuver,*** *and* ***vehicle role***.

Crash Injury Research (CIR) project
A U.S. government-sponsored project intended to determine the causes of aircraft accidents and record the injuries sustained in each accident. (Note: An older program; now referred to as Aviation Safety Engineering and Research.)

crash safety
A measure of a vehicle's ability for the occupant(s) to survive an impact and evacuate the vehicle following impact.

crash severity
The most severe injury sustained in the crash as recorded on the police accident report: Property Damage Only (no injuries), Minor or Moderate (Evident, but not incapacitating; complaint of injury; or injured, severity unknown), Severe or Fatal (killed or incapacitating).

Crash Survival Design Guide (CSDG)
A multiple volume document providing information on various aspects of aircraft design criteria which enhance crew and passenger survival during and following a crash.

crash type
Single vehicle or multiple vehicle crash.

crash worthiness
A measure of the capability of a vehicle to act as a protective container and energy absorber during impact.

CRAVE
See ***carcinogen risk assessment verification endeavor work group***.

crawl
A type of locomotion which involves moving in approximately a prone position, using the hands/elbows and knees for support and movement.

crawl space
A region of low height, generally under a large structure of some type, through which a worker may access certain utilities connections or other equipment.

crawling board
A plank with cleats spaced and secured at equal intervals, for use by a worker on roofs, not designed to carry any material. Also referred to as a *chicken ladder*.

creativity
The ability to generate ideas for novel approaches, devices, or artistic works through imagination, thinking, or considering a situation from a different perspective.

credibility
Worthiness of belief; that quality in a witness which renders his/her evidence worthy of belief. After the competence of a witness is allowed, the consideration of his/her credibility arises, and not before. As to the distinction between competency and credibility, see competency.

credible failure
Any failure that can physically occur without violating any scientific law.

creep
(1) *Work Mechanics*. A change in a work method within a task by a worker occurring over an extended period of time. (2) *Seismology*. Relatively slow, quiet movement along a fault. It is sometimes called *seismic creep* to distinguish it from the slumping of rock or soil on slopes (which is also known as creep), and sometimes called *aseismic creep* since it does not trigger events greater than *microearthquakes*.

creeper gear
Lowest gear or combination of gears used for extra power. Also known as *grandma*.

crenothrix
*See **iron bacteria**.*

crepuscular rays
Alternating light and dark bands of light that appear to fan out from the sun's position, usually at twilight.

crest factor
The ratio of the peak value of a vibratory motion to the root mean square value of that motion over a specified time interval.

crest of flood
Rise in river has reached its peak.

crevasse
A deep fissure in snow or ice.

crevasse field
An area of deep fissures in the surface of an ice mass caused by breaking or parting.

crevice corrosion
Localized corrosion in narrow crevices filled with liquid.

CREW CHIEF
A computerized, 3-dimensional human modeling program for simulating an aircraft maintenance person with respect to accessibility of components for maintenance and ultimately to the incorporation of such data into aircraft design.

crew-induced load
The reaction forces exerted by an individual on a structure as a result of that individual exerting effort with or reacting to external forces caused by another object on another portion of the body.

crew load
The number of personnel used to perform work on a certain product or component.

crew member
(1) *Aviation.* A person assigned to perform duty in an aircraft during flight time. (2) *Railroad.* A person, other than a passenger, who is assigned to perform either a) on-board functions connected with the movement of the train (i.e., an employee of a railroad, who is assigned to perform service subject to the federal hours of service laws during a tour of duty) or, b) on-board functions in a sleeping car or coach assigned to intercity service, other than food, beverage, or security service.

crew station
Any workstation or work site within a vehicle intended for use during vehicular operation by one or more members of the crew of that vehicle.

CRI
*See **CIE color rendering index**.*

crib
Maritime. A crate-like construction of logs or beams, usually filled with stones, placed in water as a free-standing mooring device or as the foundation of a pier or wharf.

crib area
Containing one or more cribs, (frames of logs or beams filled with heavy material that are sunk and used as foundations or retaining walls for docks, piers or similar structures, or as supports for pipelines).

cricoarytenoid
*See **posterior cricoarytenoid** and **lateral cricoarytenoid**.*

cricoid cartilage
A ring-shape piece of cartilaginous tissue encircling the airway passage in the larynx.

cricothyroid
A skeletal muscle in the larynx involved in producing tension and elongation of the vocal cords.

criminal
The term applied to any proceeding the resolution of which can result in incarceration and/or monetary fines. Most alleged crimes are either misdemeanors or felonies.

criminal damage
(1) Willfully injuring, damaging, mutilating, defacing, destroying, or substantially impairing the use of any property in which another has an interest without the consent of such other person. (2) Injuring, damaging, mutilating, defacing, destroying, or substantially impairing the use of any property with intent to injure or defraud an insurer or lien holder. Note: Criminal damage to property is by means other than fire or explosive.

crinion
The point in the midsagittal plane where the hairline meets the forehead. In a balding or

hairless individual, estimate where the hair growth line would be if he had normal hair.

crisis management team
Corporate executives or key officers and advisors of an organization who would form the organization's nerve center in a crisis situation. This group, usually headed by the organization's chief executive officer, would receive information about a crisis directly and would formulate the organization's responses. The team should be trained and warned of the types of crises to be expected. They should monitor relevant legal trends and conduct mock exercises. The team may also decide upon whether to self-report a violation of the law to a regulatory agency.

crista
The sensory structure within the ampulla of a semicircular canal, which detects motion of the head; composed primarily of the cupula and sensory hair cells.

cristobalite
A crystalline form of silica.

critale
*See **iliac crest**.*

criteria
(1) Exposure values or concentrations based on scientific information and used in the regulation of substances. (2) Descriptive factors taken into account by EPA in setting standards for various pollutants. These factors are used to determine limits on allowable concentration levels and to limit the number of violations per year. When issued by EPA, the criteria provide guidance to the states on how to establish their standards.

criteria pollutants
The 1970 amendments to the Clean Air Act required the EPA to set National Ambient Air Quality Standards (NAAQS) for certain pollutants known to be hazardous to human health. The term "criteria pollutants" derives from the requirements that the EPA must describe the characteristics and potential health and welfare effects of these pollutants. It is on the basis of these criteria that standards are set or revised. These pollutants include carbon monoxide, hydrocarbons, lead, nitrogen dioxide, sulfur dioxide, ozone, and suspended particulates.

criterion
A standard, rule, or test on which a judgment or decision can be based.

criterion-related validity
The usefulness of some test as a predictor in job performance.

criterion sound level
A sound level of 90 decibels (OSHA).

criterion variable
The variable consisting of the observed result in a correlation or regression study. Analogous to the dependent variable in experiments.

critical
Pertaining to an aspect of such importance that an operation cannot proceed without it or a situation may become life-threatening.

critical altitude
The maximum altitude at which, in standard atmosphere, it is possible to maintain, a specified power or a specified manifold pressure at a specified rotational speed. Unless otherwise stated, the critical altitude is the maximum altitude at which it is possible to maintain, at the maximum continuous rotational speed, one of the following: a) the maximum continuous power, in the case of engines for which this power rating is the same at sea level and at the rated altitude; b) the maximum continuous rated manifold pressure, in the case of engines, the maximum continuous power of which is governed by a constant manifold pressure.

critical condition
The most severe environmental condition in terms of loads, pressures, and temperatures, or combinations thereof. Imposed on structures, systems, subsystems, and components during service life.

critical damping
The minimum viscous damping that will allow a displaced system to return to its initial position without oscillation.

critical engine
The engine whose failure would most adversely affect the performance or handling qualities of an aircraft.

critical equipment
Equipment that is likely to result in a major problem or loss if damaged, operates improperly, or ceases to operate for whatever cause, and is therefore considered vital to the continued effective safe operation of the system or process.

critical event
System Safety. An occurrence, subsequent to the introduction of a hazard or set of hazards into a system, that results in a level of injury, damage, or loss of a magnitude for which quick or total recovery would be possible, although extremely difficult (e.g., personnel injuries, partial system loss, property or equipment damage, etc.). The parameters for this categorization are usually established by management in the System Safety Program Plan, or other policy-making documentation.

critical flaw
A specific shape of flaw with sufficient size that unstable growth will occur under the specific operating load and environment.

critical flicker frequency (cff)
*See **flicker fusion frequency**.*

critical flow
The rate of flow of a fluid equal to the speed of sound in that fluid.

critical flow orifice
A device used for determining volumetric flow rate with an accuracy of plus or minus 10% if made to standardized dimensions.

critical function
An activity or operation which can have a major impact on system performance or can endanger workers or the project if it fails.

critical fusion frequency (cff)
*See **flicker fusion frequency**.*

critical habitat
Under the Federal Threatened Species Act of 1973, the specific areas occupied by the species, at the time it is listed in accordance with the provisions of the Act on which are found those physical or biological features essential to the conservation of the species and which may require special management considerations or protection; and specific areas outside the geographical area occupied by the species at the time it is listed in accordance with the provisions of the Act, upon a determination of the Secretary of the Interior that such areas are essential for the conservation of the species.

critical incident method
A performance appraisal technique for either a system or employee. For a system: the process of gathering data by asking the users of that system to describe significant incidents, according to some established criteria. For an employee: the maintenance of a log documenting both favorable and unfavorable behaviors exhibited during an evaluation period. Synonymous with *critical incident technique*.

critical job
A job task within an occupation that has been associated with major loss more frequently than others. It could also be a job where an error has the potential for resulting in a major loss.

critical load
A load consisting of critical hardware and/or any load that includes personnel.

critical organ
(1) *Ionizing Radiation.* The body organ receiving the radionuclides that results in the greatest overall damage to the body. Usually, but not necessarily, it is the organ receiving the greatest concentration. (2) *Toxicology.* The organ in the body which receives the greatest damage as a result of exposure to a health hazard.

critical path analysis
*See **Critical Path Method**.*

Critical Path Method (CPM)
The development and use of a networked model containing the times required for different phases of a job, from which the critical path is determined and a decision made as to how the job will be carried out.

critical pitting temperature
A value used to compare a material's resistance to pitting corrosion.

critical point
The combination of pressure and temperature at which point a gas and liquid become indistinguishable.

critical pressure
The pressure required to liquefy a gas at the critical temperature. *See also **critical temperature**.*

critical ratio
The value of the ratio of a deviation from a mean to the standard deviation for that distribution.

critical score
That score which appears to separate those most likely to be successful from the most likely to fail.

critical speed
Any rotating speed which is associated with high vibration amplitude.

critical stress intensity factor
The stress intensity factor at which an unstable fracture occurs.

critical temperature
The temperature above which a gas cannot be liquefied by pressure alone. *See also* ***critical pressure***.

critical value
That value which lies on a boundary for rejection or acceptance of a hypothesis.

criticality
A scale or ranking of the possible types of failures in a system as to the importance of continued functioning of that system.

crocidolite asbestos
An amphibole variety of asbestos containing approximately 50% combined silica and nearly 40% combined iron (valence 2/3). This type of asbestos fiber has been considered the most toxic form of asbestos by some health professionals and regulatory agencies. Often referred to as *blue asbestos*.

Crohn's disease
Inflammation of the terminal portion of the ileum; also called regional enteritis and regional ileitis.

cross-auditing
Audits conducted within a facility but for a different work unit or facility.

cross-boundary interaction analysis
A study of the work-related interactions between workers on different tasks to determine the interdependence between tasks.

cross-collector
A mechanical sludge collector mechanism, extending the width of one or more longitudinal sedimentation basins, used to consolidate and convey accumulated sludge to a final removal point.

cross-connection
A physical connection in a plumbing system through which a potable water supply could be contaminated.

cross-coupling
A situation in which an event occurring in one aspect affects or causes an event to occur in another aspect.

cross (fix) at (altitude)
Used by Air Traffic Control (ATC) when a specific altitude restriction at a specified fix is required.

cross (fix) at or above (altitude)
Used by Air Traffic Control (ATC) when an altitude restriction at a specified fix is required. It does not prohibit the aircraft from crossing the fix at a higher altitude than specified; however, the higher altitude may not be one that will violate a succeeding altitude restriction or altitude assignment.

cross (fix) at or below (altitude)
Used by Air Traffic Control (ATC) when a maximum crossing altitude at a specific fix is required. It does not prohibit the aircraft from crossing the fix at a lower altitude; however, it must be at or above the minimum Instrument Flight Rules (IFR) altitude.

cross grain
A deviation of the fiber direction in a piece of wood from a line parallel to the sides of the piece. Also referred to as *slope of grain*.

cross light
Provide equivalent illumination on a subject using a pair of luminaries arranged at equal angles from the plan generated by the subject and the viewing axis of the viewer or camera.

cross-modality matching
A research technique. *See* ***cross-sensory matching***.

cross protection
An arrangement to prevent the improper operation of a signal, switch, movable-point frog, or derail as the result of a cross in electrical circuits.

cross-sectional area
Exposed area when an object or image is cut perpendicular on its longitudinal axis and viewed along the longitudinal axis.

cross-sectional design
A research methodology in which all samples are taken at approximately the same point in time.

cross-sectional study
A study using a cross-sectional design.

cross-sensory matching
A research technique in which the intensity of stimulation on one modality is compared or matched to the intensity of stimulation in another modality. Synonymous with *cross-modality matching.*

cross-sequential design
A research methodology in which independent groups of individuals from the same birth cohort are measured at different times or ages.

cross-servicing
Petroleum products, repairs, supplies, and services provided to General Services Administration (GSA) by other federal agencies, or vice versa. Cross-servicing may also refer to commercial firms where GSA or other federal agencies have agreements with these firms to supply services, repairs, or fuel.

cross training
A technique in which a worker may be trained on the job of one or more co-workers, usually with the co-workers being likewise trained.

crossboard
Maritime Navigation. A simple type of daymark in the shape of a "X" formerly used extensively on the Missouri River only.

crosscurrent
Stream flow across navigable portion of river.

crossed eyes
An eye condition in which both eyes cannot be focused on the same object at the same time; the result is that one eye focuses on the object, while the other eye is turned away from it. *Also called* ***strabismus***.

crossflow filtration
Method of filtration where the feed water flows parallel to the surface of the filter medium.

crossing
A place where a channel moves from along one bank of the river over to the other bank of the river.

crossing daymark
A diamond-shaped daymark erected at the head and foot of crossings and used by pilots to steer.

crossing light
A light located at the head and foot of crossings used by pilots to steer; may be equipped with an additional high intensity directional light.

crosslinkage
The degree of bonding of a monomer or set of monomers to form an insoluble, three-dimensional resin matrix.

crossmatching
A procedure vital in blood transfusions, testing for agglutination of donor erythrocytes by recipient's serum, and of recipient's red cells by donor serum.

crossover analysis
An evaluation for costing purposes of what alternative work methodologies should be used for different production levels.

crosstalk
A signal which is communicated to another channel in a system where it is not desired.

crosstie
Railroad. The transverse member of the track structure to which the rails are fastened. Its function is to provide proper gauge and to cushion, distribute, and transmit the stresses of traffic through the ballast to the roadbed.

crosstown
Non-radial bus or rail service which does not enter the Central Business District (CBD).

crosswind
Aviation. (1) When used concerning the traffic pattern, the word means *crosswind leg.* (2) When used concerning wind conditions, the word means a wind not parallel to the runway or the path of an aircraft.

crosswind component
The wind component measured in knots at 90 degrees to the longitudinal axis of the runway.

crotch
A location between two structures which emanate from adjacent points and are interconnected by some tissue or other material.

croup
A condition resulting from acute obstruction of the larynx caused by allergy, foreign body, infection, or new growth, occurring chiefly in infants and children.

crown
See ***vertex***.

crown-rump length, reclining
The linear horizontal distance from the top of the head to the bottom of the buttocks. Measured with the individual supine on a recumbent length table, the hips flexed 90°, and the head oriented so the Frankfort plane is perpendicular to the board surface.

CRP
See ***corneo-retinal potential***.

CRT
See ***cathode ray tube*** *or* ***choice reaction time***.

crude oil
(1) Unrefined petroleum as produced from underground formations. (2) A mixture of hydrocarbons that exists in the liquid phase in natural underground reservoirs and remains liquid at atmospheric pressure after passing through surface-separating facilities.

crude oil imports
The volume of crude oil imported into the 50 states and the District of Columbia, including imports from U.S. territories, but excluding imports of crude oil into the Hawaiian Foreign Trade Zone.

crude oil petroleum
A naturally occurring, oily, flammable liquid composed principally of hydrocarbons. Crude oil is occasionally found in springs or pools but usually is drilled from wells beneath the earth's surface.

crude oil production
The volume of crude oil produced from oil reservoirs during given periods of time. The amount of such production for a given period is measured as volumes delivered from lease storage tanks (i.e., the point of custody transfer) to pipelines, trucks, or other media for transport to refineries or terminals with adjustments for a) net differences between opening and closing lease inventories, and b) basic sediment and water (BS&W).

cruise
Used in an Air Traffic Control (ATC) clearance to authorize a pilot to conduct flight at any altitude from the minimum Instrument Flight Rulcs (IFR) altitude up to and including the altitude specified in the clearance. The pilot may level off at any intermediate altitude within this block of airspace. Climb/descent within the block is to be made at the discretion of the pilot. However, once the pilot starts descent and verbally reports leaving an altitude in the block, he may not return to that altitude without additional ATC clearance. Further, it is approval for the pilot to proceed to and make an approach at destination airport and can be used in conjunction with a) an airport clearance limit at locations with a standard special instrument approach procedure. The Federal Aviation Regulations (FAR) require that if an instrument letdown to an airport is necessary, the pilot shall make the letdown in accordance with a standard/special instrument approach procedure for that airport, or b) an airport clearance limit at locations that are within/below/outside controlled airspace and without a standard or special instrument approach procedure. Such a clcarance is NOT AUTHORIZATION for the pilot to descend under Instrument Flight Rules (IFR) conditions below the applicable minimum IFR altitude nor does it imply that ATC is exercising control over aircraft in Class G airspace; however, it provides a means for the aircraft to proceed to destination airport, descend, and land in accordance with applicable FAR's governing Visual Flight Rules (VFR) flight operations. Also, this provides search and rescue protection until such time as the IFR flight plan is closed. *See also* ***instrument approach procedure***.

cruise climb
A climb technique employed by aircraft, usually at a constant power setting, resulting in an increase of altitude as the aircraft weight decreases.

cruising
Proceeding normally, unrestricted, with an absence of drastic rudder or engine changes.

cruising altitude
An altitude or right level maintained during en route level flight. This is a constant alti-

tude and should not be confused with a cruise clearance.

cruising level
A level maintained during a significant portion of a flight.

crumb rubber
Ground or shredded rubber produced by shredding used automobile tires. It can be recycled in asphalt-rubber or other products.

crush injury
Any injury in which bodily tissues are severely compressed and possibly torn due to mechanical forces.

cryesthesia
Abnormal sensitiveness to cold.

cryogenic
Producing low temperatures.

cryogenic gas
A liquefied gas that exists in its containers at temperatures far below normal atmospheric temperatures.

cryogenic liquid
A refrigerated liquid gas with a boiling point below -130°F (-90°C).

cryogenics
The field of science dealing with the behavior of matter at very low temperatures.

cryosurgery
The destruction of tissue by application of extreme cold, as in the destruction of lesions in the thalamus for the treatment of Parkinson's disease and the treatment of certain malignant lesions of the skin and mucous membranes. The method has also been used successfully in some types of surgery of the eye, for example, in the removal of cataracts and the repair of retinal detachment.

cryptosporidiosis
Gastrointestinal disease caused by the ingestion of waterborne *Cryptosporidium parvum,* often resulting from drinking water contaminated by runoff from pastures or farmland.

cryptosporidium
A protozoan parasite that can live in the intestines of humans and animals.

Cryptosporidium parvum
A species of *Cryptosporidium* known to be infective in humans.

cryptotoxic
Having hidden toxic properties.

crystal
A homogenous chemical substance that has a definite geometric shape, with fixed angles between its faces and distinct edges of faces.

crystalline
Having a regular molecular structure evidenced by crystals.

crystallization
The process of forming crystals.

crystallizer
Common term for a forced circulation evaporator.

CSA
Canadian Standards Association. *See also* ***compressed spectral array***.

CSDG
See ***Crash Survival Design Guide***.

CSF
See ***cerebrospinal fluid***. *See also* ***cancer slope factors***.

CSG
See ***constructed solid geometry***.

CSHO
See ***Compliance Safety and Health Officer***.

CSO
See ***combined sewer overflow***.

CSP
See ***Certified Safety Professional***.

CSTR
Completely stirred tank reactor.

CT
See ***charcoal tube***.

CTAF
See ***common traffic advisory frequency***.

CTD
See ***cumulative trauma disorder***.

CTG
See ***control technique guidelines***.

CTI
See ***Clerical Task Inventory***.

CTS
See ***carpal tunnel syndrome***.

cubic feet per minute (cfm)
A measure of the volume of a substance flowing within a fixed period of time.

cu. ft
Cubic foot or cubic feet, ft^3. Conversion equivalents: 1,728 cubic inches, 60 pints, 8/10 bushel, 0.028 cubic meter, 28.32 liters.

cubic inch displacement (CID)
A measure of the physical size of the engine.

cubic meter (m^3)
A volume measurement equal to 1000 L or 264.2 gallons. One cubic meter of water weighs approximately 1 metric ton.

cuboid bone
One of the foot tarsus bones, lying between the calcaneus and the lateral two metatarsals. Also referred to as *os cuboideum*.

cue
A stimulus which is a signal to respond.

cul-de-sac
The round or circular section of the end of a dead-end street.

culm
Coal dust or anthracite tailings.

cultivation
The propagation of living organisms, applied especially to the growth of microorganisms or other cells in artificial media.

cultural eutrophication
Increasing rate at which water bodies "die" by pollution from human activities.

culture
(1) The propagation of microorganisms or of living tissue cells in special media conducive to their growth. (2) A growth of microorganisms propagated on or in the medium. (3) The social heritage. The totality of behaviors, values, attitudes, and customs shared by a group.

culvert
An enclosed channel serving as a continuation of an open stream where a stream meets a roadway or other barrier.

cu. m.
Cubic meter, m^3.

cumulative distribution function
See ***cumulative probability distribution***.

cumulative dose
Radiation. The total dose resulting from repeated exposures to radiation of the same region or of the whole body.

cumulative error
An error whose sum dose not converge to zero as the number of samples increases.

cumulative exposure
A weighted sum intended to represent an individual's effective exposure to some environmental condition over a period of time when the levels or intensity of that condition vary throughout the period of interest. Represented by the formula

$$E_c = \sum_{i=1}^{n} L_i T_i$$

where:

L_i = level of exposure (intensity, concentration, etc.)
T_i = length of time at exposure level i
n = number of exposure intervals used

cumulative frequency distribution
A graphical or tabular representation of an ever increasing curve corresponding to the summation of all scores of a dataset such that for each point on the distribution, the ordinate value represents the sum of all scores less than the corresponding point on the abscissa. *See also* ***cumulative probability distribution***.

cumulative pathogenesis
The development of some type of trauma through continued stress on one or more parts of the body.

cumulative probability distribution
A graphical, mathematical, or tabular representation of the integration or summation of some probability distribution function, yielding the cumulative probability of all events occurring in that set. *See also* ***cumulative frequency distribution***.

cumulative sum chart
A statistical quality control chart where the sum of product deviations is plotted against time. Also referred to as *cusum chart*.

cumulative timing
A work timing technique in which the timing device is permitted to run continuously across all elements of the task being measured. Also referred to as *continuous timing, continuous timing method, cycle timing,* and *continuous reading method*.

cumulative trauma disorder (CTD)
A collective term used to describe syndromes characterized by discomfort, impairment, disability, or persistent pain in the joints, muscles, tendons, and other soft tissues, with or without physical manifestations. It is often caused, precipitated, or aggravated by repetitive or forced motions which may occur in many differed occupational activities. Also referred to as *repetitive motion disorder*.

cumulative working level months (CWLM)
The sum of lifetime exposure to radon working levels expressed in total working level months.

cumulonimbus
An exceptionally dense and vertically developed cloud, often with a top in the shape of an anvil. The cloud is frequently accompanied by heavy rain showers, lightning, thunder, and sometimes hail. It is also known as a *thunderstorm cloud* or a *thunder head.*

A developing cumulonimbus cloud

cumulus
A cloud in the form of individual, detached domes or towers that are usually dense and well defined. It has a flat base with a bulging upper part that often resembles cauliflower. Cumulus clouds of fair weather are called *cumulus humilis*. Those that exhibit much vertical growth are called *cumulus congestus* or *towering cumulus*.

cumulus stage
The initial stage in the development of an air mass thunderstorm in which rising, warm, humid air develops into a cumulus cloud.

cup screen
A single-entry, double-exit drum screen.

cupric
Of or containing copper.

cupronickel
See ***copper-nickel***.

cupula
A gelatinous mass enclosing the sensory hair cells of a crista for detecting motion within the semicircular ducts.

curb cut
A section of curb at which a ramp has been laid, usually at an intersection, from the sidewalk to the street for the passage of wheeled vehicles or handicapped individuals.

curb weight
The weight of a motor vehicle with standard equipment, maximum capacity of fuel, oil, and coolant: and, if so equipped, air conditioning and additional weight of optional engine. Curb weight does not include the driver.

curie
A unit of measure formally defined as a quantity of any radioactive nuclide producing 3.7×10^{10} disintegrations per second. Now the curie (Ci) is officially a unit of activity rather than a quantity (i.e., a unit of radioactivity which is a measure of the rate at which a radioactive material emits particles). The new definition is $1 \text{ Ci} = 3.7 \times 10^{10}$ disintegrations $\times s^{-1}$. The higher the rate of disintegration, the greater the hazard.

current
(1) The flow of electrons through a conductor.
(2) *See* ***drift***.

current assets
Cash and cash equivalents, as well as current receivables and short-term investments, deposits and inventories.

current dollars
(1) The dollar value of a good or service in terms of prices current at the time the good or

service is sold. This contrasts with the value of the good or service measured in constant dollars. (2) Represents dollars current at the time designated or at the time of the transaction. In most contexts, the same meaning would be conveyed by the use of the term "dollars."

current flight plan
The flight plan, including changes, if any, brought about by subsequent clearances.

current liabilities
Aviation. Current portion of long-term debt and of capital leases, air travel liabilities, and other short-term trade accounts payable.

current of traffic
The movement of trains on a specified track in a designated direction.

cursor
A movable symbol, icon, or other element on a display to indicate position or pointing. *See also* ***pointing cursor*** *and* ***place-holding cursor***.

cursor control device
See ***direct manipulation device***.

curtain wall
An external wall that is not load bearing. Usually refers to a wall that extends down below the surface of the water to prevent floating objects from entering a screened area.

curvature
See ***arc***.

curvature effect
In cloud physics, as cloud droplets decrease in size, they exhibit a greater surface curvature that causes a more rapid rate of evaporation.

curve
A line that is not straight, or that describes part of a circle, especially a line representing varying values in a graph.

curve fitting
The process of determining which particular curve/line or function best fits the known data points.

curvilinear
Pertaining to one or more lines which are not straight.

curvilinear correlation
See ***nonlinear correlation***.

curvilinear regression
See ***nonlinear regression***.

Cushing's syndrome
A group of serious symptoms caused by overactivity of the cortices of the adrenal gland. Symptoms include painful, fatty swellings on the body, moonlike fullness of the face, distention of the abdomen, impairment of sexual function, high blood pressure, and general weakness. There may also be unusual growth of body hair (hirsutism) and streaked purple markings on the body.

cushion
Any form of soft material which acts to increase body tactile comfort.

custom house
The government office where duties and/or tolls are placed on imports or exports and are paid on vehicles or vessels entered or cleared.

customs
Duties, tolls, or taxes imposed by a government on commodities imported into or exported from that country.

customs house broker
A person licensed by the Treasury Department to transact business at a custom house on behalf of other persons. *See also* ***broker*** *and* ***freight forwarder***.

customs tariff
A schedule of charges assessed by the federal government on imported and/or exported goods.

cusum chart
See ***cumulative sum chart***.

cut
(1) *Editing.* The removal of a selected block of text, data, or graphics from the display for storage in a temporary buffer, for possible recall and placement in another location. (2) *Medical.* A tissue injury of varying depth but with much greater length than width. (3) *Construction.* An excavation of the Earth's surface to provide passage for a road, railway, canal, etc.

cut her loose
Maritime (slang). To untie all lines.

cut-in circuit
Railroad. A roadway circuit at the entrance to automatic train stop, train control or cab sig-

nal territory by means of which locomotive equipment of the continuous inductive type is actuated so as to be in operative condition.

cut-section
Railroad. A location other than a signal location where two adjoining track circuits end within a block.

cut set
As pertains to fault tree analysis (FTA) and/or the Management Oversight and Risk Tree (MORT), a defined set of events, under the top event, that can be isolated from the remainder of the fault tree and examined as contributory to the occurrence of the top or primary event.

cutaneous
Of or relating to the skin, its sensory receptors, or to the sensations produced by those receptors.

cutaneous lip
The area between the upper lip and the nose.

cuticle
See ***eponychium***.

cutie pie
A portable instrument used to determine the level of ionizing radiation.

cutoff
A cut made by dredging that eliminates a bend in the river or curve. Usually refers to a new channel made by entering at the head of a bend, passing through the cut, and emerging at the end of the bend on the downstream side.

cutoff frequency
That frequency at which an electrical filter begins to attenuate a signal. The direction of the attenuation depends on the type of filter.

cutoff low
A cold upper-level low that has become displaced out of the basic westerly flow and lies to the south of this flow.

cutout
As pertains to systems of over 600 volts (nominal), an assembly of a fuse support with either a fuse holder, fuse carrier, or disconnecting blade. The fuse holder or fuse carrier may include a conducting element (fuse link), or may act as the disconnecting blade by the inclusion of a non-fusible member.

cutout box
An enclosure designed for surface mounting and having swinging doors or covers secured directly to, and telescoping with, the walls of the box proper.

cutting fluid
An oil-water emulsion that is used for cooling and lubricating the tool and the work in machining and grinding operations.

cutting oil
An oil that is used for cooling and lubricating the tool and the work in machining and grinding operations.

cutting plane
An imaginary surface along which a computer model is "sliced" to yield a cross-section.

Cv
See ***coefficient of variation***.

CVM
See ***contingent valuation survey***.

CW
See ***continuous wave***.

CW laser
Continuous wave laser as opposed to a pulsed type laser.

CWA
See ***Clean Water Act***. *See also* ***center weather advisory***.

CWLM
See ***cumulative working level months***.

CWP
See ***coal miners' pneumoconiosis***.

cyanate
A salt of cyanic acid that contains the radical CNO.

cyanazine
A common, and potentially carcinogenic, herbicide sometimes found in drinking water.

cyanhemoglobin
A compound formed by the action of hydrocyanic acid on hemoglobin, which gives the bright red color to blood.

cyanic acid
A highly irritant compound, HOCN.

cyanide
A binary compound of cyanogen. Some inorganic compounds, such cyanide salts, potas-

sium cyanide, and sodium cyanide, are important in industry for extracting gold and silver from their ores and in electroplating. Other cyanide compounds are used in the manufacture of synthetic rubber and textiles. Cyanides are also used in pesticides. Most cyanide compounds are deadly poisons.

cyanmethemoglobin
A crystalline, colored substance formed by the action of hydrocyanic acid on oxyhemoglobin at body temperature; used in measuring hemoglobin in the blood.

cyanoderma
Blueness of the skin.

cyanosis
A bluish discoloration of the skin and mucous membranes due to excessive concentration of reduced hemoglobin in the blood.

cybernation
The use of computers in automating industry.

cybernetics
The study of communication and automated feedback control functions between living organisms and machined systems with an emphasis on gaining an understanding of living organisms by using machine analogies.

cyberspace
An abstract version of a virtual environment which extends beyond three dimensions.

Cybex dynamometer
A commercial dynamometer which can measure static or dynamic isokinetic strength.

cycle
A succession or recurring series of events. A complete sequence of elements or events making up a unit process or activity in a repetitive, periodic operation.

cycle life
The number of times a material can be stressed at a given level before it fails or is expected to fail. *See also* ***fatigue life***.

cycle per second (cps)
See ***Hertz***. (Note: cps is an outdated term.)

cycle time
That time required or used, whether by man or machine, to perform all the elements in a complete work cycle.

cycle timing
See ***cumulative timing***.

cyclegram
See ***cyclegraph***.

cyclegraph
A photographic record of the motion obtained in cyclegraphy.

cyclegraph technique
See ***cyclegraphy***.

cyclegraphy
The process of making a single photograph using one or more small light bulbs which are on at all times during the process for tracking the body or its parts with an exposure time on the same negative of at least one cycle of a repetitive motion. Typically the subject is in a darkened area. Synonymous with *cyclegraph technique*. *See also* ***chronocyclegraphy***.

cycles of concentration (COC)
The ratio of the total dissolved solids concentration in a recirculating water system to the total dissolved solids concentration of the makeup water.

cyclic
Pertaining to a periodic event.

cyclic element
An element of some operation or process which occurs at least once in every period of that operation or process.

cyclic timing
See ***cumulative timing***.

cyclogenesis
The development or strengthening of extratropical cyclones.

cyclograph
See ***cyclegraph***.

cyclone
An area of low pressure around which the winds blow counterclockwise in the Northern Hemisphere and clockwise in the Southern Hemisphere. Often considered synonymous with tornado, however a tornado is a more compact and generally more destructive phenomenon.

cyclone collector
A size selective device which is designed to separate coarse particulates from finer

particles. In industrial hygiene sampling, a cyclone is used to separate the respirable fraction of particulates in the sampled air from the total particulates drawn into the cyclone. The respirable particles are collected on a filter positioned downstream from the cyclone.

cyclonic scale
See ***synoptic scale***.

cyclotron
A particle accelerator which uses a magnetic field to confine a positive ion beam while an alternating electric field accelerates the ions in a spiral path.

cylinder
(1) A pressure vessel designed for pressures higher than 40 psi and having a circular cross-section. It does not include a portable tank, multi-unit car tank, cargo tank, or tank car. (2) In a reciprocating engine, a cylinder is the chamber in which combustion of fuel occurs and the piston moves, ultimately delivering power to the wheels. Common engine configurations include 4, 6, and 8 cylinders. Generally, the more cylinders a vehicle has, the greater the amount of engine power it has. However, more cylinders often result in less fuel efficiency. *See also* ***engine displacement*** *and* ***engine size***.

cylindrical grip
A type of grip in which the flexed fingers and the palm are used as if to hold an object of constant diameter with an extended length, where the degree of flexion of each finger joint is similar for each finger.

cyst
(1) A sac or capsule containing a liquid or semisolid substance. Most cysts are harmless. Nevertheless they should be removed when possible because they occasionally may change into malignant growths, become infected, or obstruct a gland. There are four main types of cysts: retention cysts, exudation cysts, embryonic cysts, and parasitic cysts. Removal is usually performed by surgical incision, typically preceded by a biopsy to verify the nature and the composition of the cyst (e.g., cancerous, malignant, benign, etc.). (2) A resting stage formed by some bacteria and protozoa in which the whole cell is surrounded by a protective layer.

cystitis
Inflammation of the urinary bladder.

cytogenics
The branch of genetics devoted to the study of the cellular constituents concerned in heredity, that is, chromosomes. The scientific study of the relationship between chromosomal aberrations and pathological conditions.

cytologist
A medical professional who limits his/her practice to cytology.

cytology
Study of the function and structure of living cells.

cytopenia
A deficiency in the cellular elements of the blood.

cytoplasm
The protoplasm of a cell exclusive of the nucleus.

cytolysis
Disruption of cells, resulting in the destruction and breakdown of the cell membrane.

cytotoxin
(1) An toxic agent that brings about destructive action on certain cells. (2) A toxin or antibody that has a specific toxic action upon cells of special organs. For example, a nephrotoxin would be a toxin that has a specific destructive effect on kidney cells.

cyturia
Presence of cells in the urine.

D

d'

A statistical index of an individual's sensitivity in estimating the distance between the mean of a noise distribution alone to the mean of the signal plus noise distribution. The units are expressed in standard deviations. Also referred to as *d prime*. *See also* ***signal detection theory***.

D-weighted noise level

Weighting level on some sound level meters for determining the offensiveness of aircraft noise.

dacryagogue

(1) Causing a flow of tears. (2) An agent that provokes a flow of tears.

dacryoadenalgia

Pain in a lachrymal gland.

dacryoadenitis

Inflammation of a lachrymal gland.

dacryorrhea

Excessive flow of tears.

dactylion

(1) Webbing of the fingers or toes. (2) The most distal point of the fleshy part of the middle finger, excluding the nail.

dactylion height

The vertical distance from the floor to the tip of the middle finger. Measured with the individual standing erect and the arm, hand, and fingers extended downward at the side.

dactylography

The scientific study of fingerprints as a means of identification.

dactyloscopy

Examination of fingerprints for the purpose of identification.

DAF

See ***dissolved air flotation***.

DAFT

Dissolved air flotation thickener.

daily average flow

The volume of gas that moves through a section of pipe determined by dividing the total annual volume of gas that moves through a section of pipe by 365 days. Volumes are expressed in million cubic feet per day measured at a pressure of 14.73 psia and a temperature of 60 degrees Fahrenheit. For pipes that operate with bi-directional flow, the volume used in computing the average daily flow rate is the volume associated with the direction of flowing gas on the peak day.

daily cover

Cover material spread and compacted on the top and side slopes of compacted solid waste at the end of each day to control fire, moisture, and erosion, and to ensure an aesthetic appearance.

daily living tasks

Those necessary tasks for normal housekeeping, cleanliness around the home. *See also* ***activities of daily living*** *and* ***instrumental activities of daily living***.

daily range of temperature

The difference between the maximum and minimum temperatures for any given day.

daily vehicle travel

Is the amount of vehicle travel (in thousands) accumulated over a 24-hour day, midnight to midnight, traversed along a "public road" by motorized vehicles, excluding construction equipment or farm tractors. Vehicle travel not occurring on public roads, such as that occurring on private land roads (private roads in parking lots, shopping centers, etc.) must also be excluded.

DAIR

See ***direct altitude and identity readout system.***

Dakin's solution

An aqueous solution of chlorine compounds of sodium used primarily as a germicide.

DALR

Dry adiabatic lapse rate. *See* ***adiabatic lapse rate***.

dalton

A nominal unit of weight equal to that of a single hydrogen atom; 1×10^{-24} grams.

Dalton's law

The pressure exerted by a mixture of nonreacting gases is equal to the sum of the partial pressures of the separate compounds.

daltonism
Red-green color blindness.

dam
(1) *General.* A naturally occurring or, more typically, a manmade structure erected in the flow of a waterway to act as a barrier for the purpose of water-retention, flow control, flow restriction, or a combination of these factors (note: certain animals, such as the beaver, are also known to erect dams in the path of waterways). (2) *Medical/Dental.* A sheet of latex rubber used to isolate teeth from fluids of the mouth during dental treatments; used also in surgical procedures to isolate certain tissues or structures.

dam bulletin board
A bulletin board located at certain dams to give stage readings and indicate whether to use the lock or go over the dam. "N.P." means use the lock. "P" means go over the dam. Largely obsolete in use since the advent of radio communications.

dam open
The situation when the gates are open so as to pass water unimpeded.

dam warning buoys
Buoys placed above the face of a dam to warn traffic of danger. These buoys may be of peculiar shape and generally have the word "danger" posted on them.

dam/weir
A barrier constructed to control the flow or raise the level of water.

damage
(1) *General.* Any loss of material value or usefulness. (2) *Law.* Loss, injury, or deterioration, caused by the negligence, design, or accident of one person to another, with respect to the latter's person or property. The word is to be distinguished from its plural *damages,* which means a compensation in money for a loss or damage. (3) *System Safety.* The partial or total loss of hardware caused by component failure; exposure to heat, fire, or other environments; human errors; or other inadvertent events or conditions.

damage to person
The measure of injury (physical, mental, and emotional) resulting from another person's action or omission, whether such action or omission be intentional or negligent. "Damages" and "injury" are commonly used interchangeably, but they are different to the extent that injury is what is actually suffered while damage is the measure of compensation for such suffering.

damage to property
Injury to property and generally does not include conversion of such property or taking of such property by public authority (i.e., eminent domain).

damage tolerance
A measure of the ability of structures to retain their load-carrying capability after exposure to sudden loads (for example, ballistic impact).

damages
A pecuniary compensation or indemnity, which may be recovered in the courts by any person who has suffered loss, detriment, or injury, whether to his/her person, property, or rights, through the unlawful act or omission or negligence of another. A sum of money awarded to a person injured by the tort of another person. Money compensation sought or awarded as a remedy for a breach of contract or for tortious acts. Damages may be *compensatory* or *punitive* depending on whether they are awarded as the measure of actual loss suffered or as punishment for outrageous conduct and to deter future transgressions. Nominal damages are awarded for the vindication of a right where no real loss or injury can be proved. Generally, punitive or exemplary damages are awarded only if compensatory or actual damages have been sustained. *See also* ***actual damages, benefit-of the bargain damages, compensatory damages, consequential damages, continuing damages, criminal damage, damages ultra, direct damages, excessive damages, excess liability damages, exemplary (or punitive) damages, expectancy damages, fee damages, foreseeable damages, general damages, hedonic damages, inadequate damages, incidental damages, irreparable damages, land damages, limitation of damages, liquidated damages and penalties, mitigation of damages, necessary damages, nominal damages, pecuniary damages, presumptive damages, prospective damages, proximate damages, remote damages, rescissory damages, severance damages, special damages, speculative***

***damages, statutory damages, substantial damages, temporary damages, treble damages, unliquidated damages,** and **vindictive damages**.*

damages ultra

Additional damages claimed by a plaintiff not satisfied with those paid into the court by the defendant. *See also* ***damages***.

damnify

To cause damage or injurious loss to a person or put him/her in a position where he/she must sustain it. A surety is "damnified" when a judgment has been obtained against him/her,

damnum absque injuria

Latin. Loss, hurt, or harm without injury in the legal sense; that is, without such breach of duty as is redressible by a legal action. A loss or injury which does not give rise to an action for damages against the person causing it.

damp

A noxious gas in a mine. A gaseous mixture formed in a mine by the gradual absorption of the oxygen and the giving-off of carbon dioxide by coal.

damp location

See ***location***.

damper

A device used to regulate airflow in ducts, often used to balance airflow in branch ducts. A damper position may be immovable, manually adjustable, or part of an automated control system.

damping

Steady diminution of the amplitude of successive vibrations of electric wave or current or sound wave. The dissipation of energy within a dynamic system over time for whatever reason.

damping factor

The ratio of actual system damping to critical damping for a system.

dander

Scales, dust, and dirt from the fur or feathers of animals that may cause allergic reactions in susceptible persons.

dandruff

A scaly material from or on top of the scalp. The condition may spread unless checked and, in rare cases, may extend to the eyebrows, ears, nose, and neck, causing a reddening of the skin in those areas. Also known as *seborrheic dermatitis*.

danger

Term of warning applied to an condition, operation, or situation that has the potential for physical harm to personnel and/or damage to property.

danger area

(1) *General.* An area or space, either well defined or not, within the boundaries of which there exists some degree of danger to person, property, or both. The danger, or dangerous conditions, within the area may be due to the presence of known (e.g., recognized) or unknown (e.g., unrecognized) hazards. Danger areas are typically posted as such and entry is usually controlled or prohibited.

Sample DANGER sign posted to control area access

(2) *Aviation.* An airspace of defined dimensions within which activities dangerous to the flight of aircraft may exist at specified times. Note: The term "Danger Area" is not used in reference to areas within the United States or any of its possessions or territories.

danger invites rescue

Term used in law of torts and, in a limited manner, in law of crimes to describe where liability is borne by one who creates a dangerous condition for one person when another person comes to his/her rescue and is injured. The liability to the second person is founded on this maxim.

danger signal

Maritime Safety. Four or more short blasts of the boat's or lockmaster's whistle to indicate danger or the need for special caution.

danger zone
A physical location in which some type of hazard exists.

dangerous
Attended with risk; perilous; hazardous; unsafe.

dangerous condition
One in which there exists a substantial and probable risk of injury and/or property damage. The risk may be imminent or merely possible when such a condition exists.

dangerous criminal
One convicted of a particularly heinous crime or one who has escaped or tried to escape from penal confinement by use of force of an aggravated character. An armed criminal. Such criminals may be segregated from others in a prison.

dangerous instrumentality
Anything which has the inherent capacity to place people in peril, either in itself (e.g., dynamite), or by a careless use of it (e.g., a boat). Due care must be exercised in use to avoid injury to those reasonably expected to be in proximity. In certain cases, absolute liability may be imposed.

dangerous machine
A machine is considered "dangerous" in such sense that the employer is required to guard it if, in the ordinary course of human affairs, danger may be reasonably anticipated from its use unless proper protection is provided.

dangerous occupation
Term used to describe hazardous work for purposes of worker's compensation laws, as well as in wage and hour determinations, and child labor laws.

dangerous per se
A thing that may inflict injury without the immediate application of human aid or instrumentality.

dangerous place
One where there is considerable risk, or danger, or peril; one where accidents or injuries are very apt to occur.

dangerous-tendency test
Propensity of a person or animal to inflict injury; used in dog bite cases to describe the vicious habits of dogs.

dangerous weapon
One dangerous to life; one by the use of which a serious or fatal wound or injury may probably or possibly be inflicted. In the context of criminal possession of a weapon, it can be any article which in circumstances in which it is used, attempted to be used, or threatened to be used, is readily capable of causing death or other serious physical injury. What constitutes a "dangerous weapon" depends not on the nature of the object itself but on its capacity, given manner of its use, to endanger life or inflict great bodily harm. As the manner of use enters into the consideration as well as other circumstances, the question is often one of fact for the jury, but not infrequently one of law for the court.

dangers of navigation
*See **dangers of the river** and **dangers of the sea**.*

dangers of the river
This phrase, as used in bills of lading, means only the natural accidents that are incident to river navigation, and does not embrace such as may be avoided by the exercise of that skill, judgment, or foresight which are demanded from persons in a particular occupation. It includes dangers arising from unknown reefs which have suddenly formed in the channel, and are not discoverable by care and skill.

dangers of the sea
Refers to those accidents that are peculiar to navigation that are of an extraordinary nature, or arise from irresistible force or overwhelming power, which cannot be guarded against by the ordinarily exertions of human skill and prudence.

darcy
A unit of measure used to indicate permeability, standardized by the American Petroleum Institute.

dark
(1) Having little or less reflected light, as in a dark color. (2) A severely reduced light level in the visual environment, as in a dark room.

dark adaptation
The process of undergoing neurochemical changes in the eye after being placed in darkness or low light levels, during which the vis-

ual system becomes more sensitive to light. Also referred to as *scotopic adaptation*.

dashpot
A symbol for a viscous damper for mechanical modeling, representing a vane placed within a viscous fluid.

DAT
See ***Differential Aptitude Test.***

data
A formalized representation of numbers or characters which have meaning for communication, interpretation, or processing purposes.

data bank
Any location, but typically in a computer system, where large amounts of a specific type of data are stored for retrieval by users.

data call-in
A part of the Office of Pesticide Programs, (OPP) process of developing key required test data, especially on the long-term, chronic effects of existing pesticides, in advance of scheduled Registration Standard reviews. Data call-in is an adjunct of the Registration Standards program intended to expedite re-registration and involves the "calling in" of data from manufacturers.

data category
Under ISO 14000, classificatory division of the input and output flows from a unit process or product system.

data display code
A graphical symbol representing a data point on some type of graphic output.

data entry
The process of inputting data into a computer using a pre-established format, regardless of the technique used.

data inquiry
The process of requesting and retrieving information from a computer and viewing it on a display or on a hard copy.

data quality
Under ISO 14000, the nature or characteristic of collected or integrated data.

database
A file of records, or a collection of data containing comparable information on different items and which provides a means for organized information retrieval.

date of injury
Means the inception date of the injury and is regarded as coincident with the date of occurrence or happening of the accident which caused the injury.

datum
Reference point for elevations of structures and water level.

daughter products
Isotopes formed by the radioactive decay of some other isotope.

Davis-Bacon Act
Federal law which deals with rate of pay for laborers and mechanics on public buildings and public works (40 U.S.C.A § 256a).

day in court
The right and opportunity afforded a person to litigate his/her claims, seek relief, or defend his/her rights in a competent judicial tribunal.

day-night sound level
The 24-hour time of day weighted equivalent sound level, in decibels, for any continuous 24-hour period, obtained after addition of ten decibels to sound levels produced in the hours from 10:00 p.m. to 7:00 a.m. It is abbreviated as L_{dn}.

day shift
See ***first shift.***

day tank
A tank used to store chemicals or diluted polymer solution for 24 hours or less.

daylight
That light present during the daytime hours from the sun, or the corresponding artificial illumination in terms of spectrum and intensity.

daylight availability
That amount of sunlight received from the sun with reference to certain conditions, such as location, intervening substances, date, and time.

daylight lamp
Any artificial light source with an output having a spectrum similar to that of a certain type of daylight. *See also* ***standard illuminant.***

daymark
Maritime Safety. An unlighted shore aid to navigation, either diamond, square, or triangle shaped.

daywork
That work for which compensation is based on time present, not output.

daywork drilling contract
A drilling contract under which the drilling contractor is compensated on the basis of the amount of time spent in drilling operations. Essentially, the lease owner hires the drilling rig and its staff to work under his direction. Broad discretion is given to the contracting party to give instructions to the drilling contractor as to how to conduct drilling operations. The courts impose broad liability upon the contracting party as a result of his/her broad discretion.

dazzle
Experiencing a condition of extreme brightness due to reflected and scattered light from particles in the atmosphere, resulting in viewing difficulties.

dB
See ***decibel***.

dBA
Refers to decibels measured on the A scale which is a frequency weighting network that approximates the response of the human ear.

DBA
Abbreviation for Doing Business As.

dBB
Sound level in decibels as determined on the B scale of a sound-level meter.

dBC
Sound level in decibels as determined on the C scale of a sound-level meter. The dBC value approximates the overall noise level.

DBP
See ***disinfection byproducts***.

D/DBP
Disinfectants, disinfection byproducts.

D/DBP Rule
A U.S. EPA rule to limit the maximum contaminated level of trihalomethanes.

DBRITE
See ***digital brite radar indicator tower equipment.***

DBT
See ***dry-bulb temperature***.

DC
Abbreviation for direct current. *See* ***direct current***.

DCS
See ***distributed control system***.

DDT
The first chlorinated hydrocarbon insecticide (chemical name: dichloro-diphsdyl-trichloromethane). It has a half-life of fifteen years and can collect in the fatty tissue of certain animals. The EPA banned registration and the interstate sale of DDT for virtually all but emergency uses in the United States in 1972 because of its persistence in the environment and accumulation in the food chain.

dead axle
Non-powered rear axle on tandem truck or tractor.

dead band
See ***dead zone***.

dead freight
The amount paid by a charterer for that part of the vessel's capacity which he/she does not occupy although he/she has contracted for it.

dead front
Without live parts exposed to a person on the operating side of the equipment.

dead hands
See ***Raynaud's syndrome***.

dead heading
Running empty.

dead locomotive
A locomotive other than a control cab locomotive that does not have any traction device supplying traction power; or a control cab locomotive that has a locked and unoccupied cab.

dead man control
A device requiring a constant force of a minimum magnitude applied to the device for operating a piece of equipment, and having a default mode which turns off or stops the equipment if that force is not applied.

dead reckoning
As applied to flying, the navigation of an airplane solely by means of computations based on airspeed, course, heading, wind direction, speed, groundspeed, and elapsed time.

dead room
A room that is characterized by an unusually large amount of sound absorption.

dead section
Railroad. A section of track, either within a track circuit or between two track circuits, the rails of which are not part of a track circuit.

dead time
(1) *Instrumentation.* The interval of time between the instant of introducing a sample into the instrument to the first indication of response. Also referred to as *lag time.* (2) *Radiation.* The time during which a Geiger-Mueller detector is insensitive to incoming radiation.

dead zone
That region, usually around the neutral position of a knob, hand controller, or lever, where there is no output from a device, even though an input may be provided.

deadhead
(1) *Transportation.* Miles and hours that a vehicle travels when out of revenue service. This includes leaving and returning to the garage, changing routes, etc., and when there is no reasonable expectation of carrying revenue passengers. However, it does not include charter service, school bus service, operator training, maintenance training, etc. For non-scheduled, non-fixed-route service (demand responsive), deadhead mileage also includes the travel between the dispatching point and passenger pickup or drop-off. (2) *Maritime.* Any water-soaked wooden pile, tree, or log that is floating just awash in a nearly vertical position. A menace to small boats and to the propellers of vessels. Also, a tow returning from a trip without barges or with empty barges. (3) *Rail Operations.* Refers to a lone locomotive traveling back to a terminal or yard. *See also* ***deadhead transportation***.

deadhead transportation
Railroad. Occurs when an employee is traveling at the direction or authorization of the carrier to or from an assignment, or the employee is involved with a means of conveyance furnished by the carrier or compensated for by the carrier.

deadly force
The degree of force that may result in the death of the person against whom the force is applied. Force likely or intended to be used is known to be capable of producing death or great bodily harm.

deadly weapon
Any firearm, or other weapon, device, instrument, material, or substance, whether animate or inanimate, which in the manner it is used or is intended to be used is known to be capable of producing death or serious bodily injury. Such weapons or instruments are made and designed for offensive or defensive purpose, or for the destruction of life or the infliction of injury.

deadman control
Railroad. A pedal, handle, or other form of switch, or combination thereof, that the operator must keep in a depressed or twisted position while a rail vehicle (or train) is moving. If the control is released, the power is cut off and the brakes are applied.

deadweight tons
The lifting capacity of a ship expressed in long tons (2,240 lbs.), including cargo, commodities, and crew.

deaerator
A device used to remove dissolved gases from solution.

deaf
The inability to hear any airborne or bone-conducted sounds due to some defect or damage in the auditory system or the brain. (Note: Many terms that use "deafness" are in reality only hearing reductions.)

deaf-mute
A person unable to speak or hear.

deaf person
Any person whose hearing is totally impaired or whose hearing is so seriously impaired as to prohibit the person from understanding oral communications when spoken in a normal conversational tone.

deafened
Having a loss of hearing ability after normal speech and hearing patterns had been established.

deafness
Impairment of hearing. Total deafness is quite rare, but partial deafness is common; an estimated 15 million Americans suffer from

some degree of deafness, and of these, perhaps 2.5 million are children whose defective hearing either is congenital (from birth) or developed before the age of five. The two major types of deafness are *conductive deafness* and *sensorineural* (nerve) *deafness*. In some cases, both types may be present; this is called *mixed deafness*. In *conductive deafness,* sound vibrations are interrupted in the outer or middle ear before they reach the nerve endings of the inner ear. In the outer ear, a foreign body or an accumulation of cerumen (earwax) may block the external acoustic meatus. These cases generally can be cured by removal of the obstruction. In the middle ear, infections, often entering through a perforated tympanic membrane (eardrum) of the Eustachian tube, may fill the chamber with fluid, hampering the passage of vibrations. The small bones of the middle ear (ossicles) may be damaged by injury or fixed in place by otosclerosis. In sensorineural deafness, the outer and middle ear function normally, but damage to the nerve endings of the inner ear, the cochlear portion of the vestibulocochlear (eight cranial) nerve or the hearing center in the brain causes either interruption or confusion of the sound messages. This damage may be caused by disease, head injury, tumor, excessively loud and sudden noise, or a continuous loud noise. A great many cases of congenital deafness are caused by infectious diseases, especially viral infections, contracted by the mother during pregnancy. Of these, rubella (German measles) is the most common.

dealkalization
Any process that removes or reduces alkalinity of water.

dealkalizer
Ion exchange unit with a strong anion bed used to reduce bicarbonate alkalinity.

deamidization
The liberation of ammonia from an amide.

deaminase
A enzyme that promotes the removal of an amino group from a compound.

dearterialization
(1) Conversion of arterial blood into venous blood. (2) Interruption of the supply of arterial (oxygenated) blood to an organ or part.

deashing
*See **demineralizing**.*

death
The apparent extinction of life, as manifested by the absence of heartbeat and respiration.

death benefits
Amount paid under insurance policy on the death of the insured. A payment made by an employer to the beneficiary or beneficiaries of a deceased employee on account of the death of the employee. A death benefit is also provided for under the Social Security Act.

death rate
*See **mortality rate**.*

death trap
A structure or situation involving imminent risk of death or a place apparently safe but actually very dangerous to life.

debarment
To bar, exclude, or preclude from having or doing something. Exclusion from government contracting and subcontracting.

debug
*See **burn-in test**.*

debridement
Removal of all foreign material and aseptic excision of all contaminated and devitalized tissues.

debris
Includes any abandoned or dilapidated structure or any sunken vessel or other object that can reasonably be expected to collapse or otherwise enter the navigable waters as drift within a reasonable period.

decacurie
A unit of radioactivity, being 10 curies.

decalcification
(1) Removal of calcerous matter from tissues. (2) The loss of calcium salts from bone or teeth.

decant
Separation of a liquid from settled solids by pouring or drawing off the upper layer of liquid after the solids have settled.

decarbonator
A device used to remove alkalinity from solution by conversion to CO_2 prior to air stripping.

decarboxylation
The removal of the carboxyl group from a compound.

decay
(1) The gradual decomposition of dead organic matter. (2) The process or stage of decline; old age and its effects on the mind and body. (3) In radiation, the gradual degradation of a radioisotope. (4) The disintegration of wood substance due to action of wood-destroying fungi. It is also known as *dote* and *rot*.

decay constant
The fraction of the number of radioisotope atoms that decay in a unit of time. The decay constant is $0.693/T$, where T is the half-life.

decay curve
A graph showing the decreasing radioactivity of a radioactive source as time passes.

decay product
A nuclide resulting from the radioactive disintegration of radionuclide or series of radionuclides. A decay product may be stable or radioactive.

decay, radioactive
The decrease in activity of any radioactive material with the passage of time, due to the spontaneous emission from the atomic nuclei of either alpha or beta particles, sometimes accompanied by gamma radiation.

deceleration
Acceleration in the direction opposite to that of the velocity vector to affect a slowing of motion. Also referred to as *negative acceleration.*

deception
The act of deceiving; intentional misleading by falsehood spoken or acted. Synonymous with fraud. Knowingly and willfully making a false statement or representation, express or implied, pertaining to a present or past existing fact.

decertification
(1) Process through which a group of employees decides it no longer wants a union to be its bargaining unit. The process involves an election conducted by the National Labor Relations Board. (2) An action conducted by a professional certification board, such as the Board of Certified Safety Professionals, to intentionally remove a professional certification designation from one of its charges for a temporary or permanent period of time. Such action may be a form of disciplinary action resulting from the some breach of conduct or a delinquency in dues payment on the part of the professional who is subject to the decertification.

dechlorination
Removal of chlorine from a substance by chemically replacing it with hydrogen or hydroxide ions in order to detoxify the substances involved.

deci-
Prefix. One-tenth of a base unit.

decibel (dB)
(1) One tenth of a bel. A non-dimensional logarithmic ratio of the measured quantity and a reference quantity for expressing power, pressure, or amplitude. The most common unit of sound or other signal intensity. (2) A means for expressing the logarithmic level of sound intensity, sound power, or sound pressure above an arbitrary reference value of 20 micropascals in air. *See also* ***dBA***.

decide
To arrive at a determination. To "decide" includes the power and right to deliberate, to weigh the reasons for and against, to see which preponderate, and to be governed by that preponderance.

decigram
One-tenth of a gram; 1.54 grains.

deciliter
One-tenth of a liter; 3.38 fluid ounces.

decimeter
One-tenth of a meter; 3.9 inches.

decipol
A unit for judging the perceived quality of outdoor air.

decision
A determination arrived at after consideration of facts and, in a legal context, law. A popular rather than technical or legal word; a comprehensive term having no fixed, legal meaning. It may be used as referring to ministerial acts as well as to those that are judicial or of a judicial character.

decision delay
See ***cognitive reaction time***. Also referred to as *decision time*.

decision height (DH)
With respect to the operation of aircraft, means the height at which a decision must be made, during an Instrument Landing System (ILS) or Precision Approach Radar (PAR) instrument approach, to either continue the approach or to execute a missed approach.

decision making
The process of evaluating information which results in the selection of a course of action.

decision time
See ***cognitive reaction time***. Also referred to as *decision delay*.

deck plate
A horizontal surface designed to provide a person with stable footing for the performance of work such as the connection and disconnection of air and electrical lines, gaining access to permanently mounted equipment or machinery or for similar needs.

Decker test
Under the Decker test, an employee of a corporation, though not a member of its control group, is sufficiently identified with the corporation so that his/her communication to the corporation's lawyer is privileged if the employee made the communication at the direction of his/her supervisors and the subject matter pertained to the performance of the employee's normal employment duties.

declaration
In common-law pleading, the first of the pleadings on the part of the plaintiff in an action at law, being a formal and methodical specification of the facts and circumstances constituting his/her cause of action.

declaratory judgment
Statutory remedy for the determination of a justifiable controversy where the plaintiff is in doubt as to his/her legal rights. A binding adjudication of the rights and status of litigants even though no consequential relief is awarded. Such judgment is conclusive in any subsequent action between the parties.

declaratory statute
One enacted for the purpose of removing doubts or putting an end to conflicting decisions in regard to what the law is in relation to a particular matter.

declination
A document filed in court by a fiduciary who chooses not to serve in his/her named capacity. At common law, a plea to the courts' jurisdiction on the ground that the judge is personally interested in the suit.

declining-rate filtration
Filter operation where the rate of flow through the filter declines and the level of the liquid above the filter bed rises throughout the length of the filter run.

decoder
The device used to decipher signals received from Air Traffic Control Radar Beacon System (ATCRBS) transponders to affect their display as select codes. *See also* ***code (5)*** *and* ***discrete code***.

decollement
Seismology. A detachment fault; a fault where crustal deformation causes separation along a boundary of rock types, typically between so-called crystalline "basement" rock and overlying sedimentary rocks.

decompensation
The inability of the heart to maintain adequate circulation; it is marked by dyspnea, venous engorgement, cyanosis, and edema.

decomposition
The breakdown of dead organic material into smaller or simpler parts that are then recirculated. Bacteria, fungi, heterotrophic protists, and saprophagous insects are important in the process of decomposition.

decompression
The return to normal environmental pressure after exposure to greatly increased pressure.

decompression sickness
Illness or injury associated with exposure to high-pressure atmospheres followed by rapid exposure to normal pressure. Also known as *the bends* and *caisson disease*.

decongestant
(1) Tending to reduce congestion or swelling. (2) An agent that reduces congestion or swelling, usually of the nasal membranes. Decongestants may be inhaled, taken as a spray or nose drops, or used orally in liquid or tablet

form. The medication acts by reducing swelling of the nasal passages. Among the leading medications used as decongestants are epinephrine, ephedrine, and phenylephrine. Antihistamines alone or in combination with decongestants may also be effective. A decongestant must be used several times a day to be helpful; but excessive use may cause headaches, dizziness, or other disorders and sometimes the medicine itself may cause reactive nasal swelling.

decontaminate
To render safe or harmless by the removal or elimination of poisonous, noxious, or otherwise harmful agents.

decontamination
The freeing of a person or an object of some contaminating substance such as radioactive material, chemical compounds, etc.

decree
The judgment of a court of equity or chancery, answering for most purposes to the judgment of a court of law.

decrement
(1) A deterioration in some performance measure, *see also* ***performance decrement***. (2) A decrease of a counter value in computing.

decriminalization
An official act generally accomplished by legislation, in which an act or omission, formerly criminal, is made non-criminal and without punitive sanctions.

decussation
The position of one part crossing another, similar part. The point of crossing.

dedicated
With regard to systems design, a feature that serves a single function (such as a power source serving a single load).

dedicated funds
Transportation. Any funds raised specifically for transit purposes and which are dedicated at their source (e.g., sales taxes, gasoline taxes, and property taxes), rather than through an allocation from the pool of general funds.

dedicated tow
A single commodity moved from origin to destination by the same towboat without picking up or dropping off other barges. Often used in the movement of grain, coal, and bulk liquid.

dedifferentiation
Regression from a more specialized or complex form to a simpler state.

deductive reasoning
The ability to apply general rules to specific problems and arrive at a logical conclusion.

deep bed filter
Granular media filter with a sand or anthracite filter bed up to 1.8 meters (6 feet) deep.

deep pocket
A person or corporation of substantial wealth and resources from which a claim or judgment may be made.

deep sea domestic transportation of freight
Establishments primarily engaged in operating vessels for transportation of freight on the deep seas between ports of the United States, the Panama Canal Zone, Puerto Rico, and United States island possessions or protectorates.

deep sea transportation of passengers
Establishments primarily engaged in operating vessels for the transportation of passengers on the deep seas except by ferry.

deep well injection
Disposal technique where raw or treated wastes are discharged through a properly designed well into a geological stratum.

deerfly
A member of the genus Chrysops, an important vector of various organisms. Its bite can cause a large, inflamed, welt-like pump to appear on the surface of the skin. In most cases, this pump may not appear for several hours after the bite. Even in mildly sensitive individuals, there can be intense itching, swelling, and redness around the area of the bite. The bite mark can remain clearly definable on the surface of the skin for three or more weeks, long after the itching sensation has subsided.

defamation
An intentional false communication, either published or publicly spoken, that injures another person's reputation or good name. Subjecting a person to ridicule, scorn, or contempt in a respectable and considerable part of the

community; may be criminal as well as civil. Includes both libel and slander.

defat
To deprive of fat, as when a solvent contacts the surface of the skin and removes the natural protective oily barrier and renders the skin more susceptible to infection.

default
A value, condition, or state which is automatically selected by a computer or other system unless overridden by an operator or program.

default judgment
A judgment entered against a party who has failed to defend against a claim that has been brought by another party. Under the Rules of Civil Procedure, when a party against whom a judgment for affirmative relief is sought has failed to plead (i.e., answer) or otherwise defend, he/she is in default and a judgment by default may be entered either by the clerk or the court.

defecation
The elimination of wastes and undigested food, as feces, from the rectum.

defect
Substandard physical condition, either inherent in the material or created through another action or event.

defect notification system
A computerized system that enables the Coast Guard to monitor the efforts of boat and equipment manufacturers in complying with 46 U.S.C. 4310.

defective
(1) Imperfect, or lacking in some specified area or some prescribed attribute. (2) A person lacking some physical, mental, or moral quality.

defective condition
In product liability law, a product is in a defective condition and considered unreasonably dangerous to the user when it has a propensity for causing physical harm beyond that which would be contemplated by the ordinary user or consumer who purchases it, with the ordinary knowledge common to the foreseeable class of users as to its characteristics. A product is not defective or unreasonably dangerous merely because it is possible to be injured while using it.

defend
To prohibit or forbid. To deny. To contest and endeavor to defeat a claim or demand made against one in a court of justice. To oppose, repel, or resist.

defendant
The person defending or denying; the party against whom relief or recovery is sought in an action or suit or the accused in a criminal case.

defense
(1) *General.* Resistance to or protection from attack. (2) *Law.* That which is offered and alleged by the party proceeded against in an action or suit, as a reason in law or fact why the plaintiff should not recover or establish what he/she seeks. That which is put forward to diminish a plaintiff's cause of action or defeat recovery. Evidence offered by the accused to defeat a criminal charge.

defense attorney
A lawyer who files an appearance on behalf of a defendant and represents such in a civil or criminal case.

defense mechanism
A psychologic reaction or technique for protection against a stressful environmental situation or against anxiety.

defense visual flight rules (DVFR)
Rules applicable to flights within an Air Defense Identification Zone (ADIZ) conducted under the visual flight rules in Federal Aviation Regulation, Part 91.

defensive response
*See **startle response**.*

defer
To delay, put off, remand, or otherwise postpone to a future time. The term does not, however, have the same meaning as abolish or omit.

deferent
Conducting or progressing away from a center or specific site of reference.

deferred credits
Items for which additional information or events are required to determine their ultimate disposition and accounting classification, in-

cluding deferred taxes, deferred investment tax credits, and other suspense items.

deferred income taxes
Tax effects which are deferred for allocation to income tax expense of future periods.

deferred investment tax credits
Investment tax credits deferred for amortization over the service life of the related equipment.

defervescence
The decline of high temperature (fever) to normal.

defibrillation
The stoppage of fibrillation of the heart. *See also **fibrillation***.

defibrillator
An apparatus that counteracts fibrillation by applying electric impulses to the heart and is used successfully in many cases of cardiac resuscitation.

deficiency
A lack or shortage; a condition characterized by the presence of less than the normal or necessary supply or competence.

deficiency disease
Avitaminosis, or other condition produced by dietary or metabolic deficiency; the term includes beriberi, scurvy, pellagra, etc.

defined mandatory use service area
Transportation. That listed in the determination of each Fleet Management Center or Fleet Management Subcenter.

definite sentence
A sentence calling for imprisonment for a specified number of years, as contrasted with an indeterminate sentence which leaves the duration to the prison authorities (e.g., parole board) and allows for the consideration of the good behavior of the prisoner.

deflagration
The thermal decomposition that proceeds at less than sonic velocity and may or may not develop hazardous pressures.

deflection
A turning aside. In psychoanalysis, an unconscious diversion of ideas from conscious attention. In the electrocardiogram, a deviation of the curve from the isoelectric baseline, that is, any wave or complex.

defluxion
A copious discharge or loss of any kind.

defoliant
Under the Federal Insecticide, Fungicide, and Rodenticide Act: Any substance or mixture of substances intended for causing the leaves or foliage to drop from a plant, with or without causing abscission.

deforestation
The permanent clearing of forest land and its conversion to non-forest uses.

deformable element
Any structure, whether physical or modeled, which is no rigid.

deformity
A deformed or misshapen condition; an unnatural growth, or a distorted or misshapen part or member; disfigurement, as a bodily deformity.

defraud
To make a misrepresentation of an existing material fact, knowing it to be false or making it recklessly without regard to whether it is true or false. Such statements are made with the specific intent to mislead another person to the point that such misleading is to the advantage of the person making the false statements.

degasifier
A device used to remove dissolved gases from a solution, usually by means of an air stripping column.

degassing
The release of gases dissolved in hot, molten rock.

degenerate
(1) To change from a higher to a lower form. (2) Characterized by degeneration. (3) A person whose moral or physical state is below the normal.

degradation
(1) The process by which a chemical is reduced to a less complex form. Conversion of a chemical compound to one less complex by splitting off one or more groups of atoms. (2) A deprivation of dignity; dismissal from rank or office; act or process of degrading. Moral or intellectual decadence; degeneration; deterioration.

degreaser
A chemical agent, usually a solvent, that is used to remove grease and oil from machinery. Because these chemicals will also remove the protective layer of oil on human skin, their use without protection can result in dermatitis.

degree
(1) The extent, measure, or scope of an action, condition, or relation. (2) The legal extent of guilt or negligence. (3) A title conferred on graduates of school, college, or university. (4) The state of civil condition of a person. (5) A unit of angular displacement; 1° = 1/360 of a circle. (6) A unit of temperature, either Fahrenheit, Celsius, or Kelvin/Absolute.

degree day
A unit used to estimate heating and cooling costs. For example, on a day when the mean temperature is less than 65°F, there is the same number of degree days as if the mean temperature of the day is below 65°F.

degree of freedom
(1) The minimum number of independent generalized coordinates required to completely define the positions of all parts of a system at any given time. (2) The number of values which are free to vary within a sample, given specified sampling constraints and experimental design.

degree of hazard (critical)
Aviation. A situation in which collision avoidance was due to chance rather than an act on the part of the pilot. Less than 100 feet of aircraft separation would be considered "critical."

degree of hazard (no hazard)
Aviation. A situation in which direction and altitude would have made a midair collision improbable regardless of evasive action taken.

degree of hazard (potential)
Aviation. An incident which would have resulted in a collision if no action had been taken by either pilot. Closest proximity of less than 500 feet would usually be required in this case.

degree of proof
That measure of cogency required to prove a case depending upon the nature of the case. In a criminal case, such proof may be beyond a reasonable doubt, whereas in most civil cases, such proof is by a fair preponderance of evidence.

degrees of negligence
The different grades of negligence which govern the liability of persons (e.g., ordinary negligence as contrasted with gross negligence).

dehumidifier
A device for lowering the moisture content of air.

dehydration
Removal of water from the body, a tissue, or any material or compound that naturally contains some degree of water; or, the condition that results from undue loss of water. Severe dehydration is a serious condition that may lead to fatal shock, acidosis, and the accumulation of waste products in the body (a condition known as uremia). Water accounts for more than half the body weight. Under normal conditions, a certain amount of fluid is lost daily. About 1.5 liters is removed by urination, and another 90 ml is lost from the digestive tract in the feces. Through vaporization, another liter is given off through the skin and lungs. To make up for these losses, about 2.5 liters of fluid must be taken into the body in food and fluids, and the cells contribute to another 250 ml through chemical activities. When the fluid intake is insufficient or the output is excessive, dehydration occurs.

dehydrogenase
An enzyme that catalyzes the transference of hydrogen ions.

dehydrogenate
(1) To remove hydrogen from. (2) A compound from which hydrogen has been removed.

deinking
The process of removing ink from secondary fibers.

deionization
The removal of ions from a compound.

déjà vu
French. An illusion that a new situation is a repetition of a previous experience.

delacrimation
An excessive flow of tears.

delamination
The separation of one layer of a material from another.

delay
(1) *General.* One of a set of basic work elements which involves some pause or interruption in an ongoing process or activity. (2) *Aviation.* Delays are incurred when any action is taken by a controller that prevents an aircraft from proceeding normally to its destination for an interval of 15 minutes or more. This includes actions to delay departing, enroute, or arriving aircraft as well as actions taken to delay aircraft at departing airports due to conditions en route or at destination airports.

delay allowance
(1) A credit of time or money given the operator to compensate for incentive on a specific delay incident not covered by the piece rate or standard. (2) A period of time which is added to the normal time to compensate for contingencies and minor delays beyond the control of the operator. Also referred to as *unavoidable delay allowance.*

delay indefinite (reason if known) expect further clearance (time)
Used by Air Traffic Control (ATC) to inform a pilot when an accurate estimate of the delay time and the reason for the delay cannot immediately be determined (e.g., a disabled aircraft on the runway, terminal or center area saturation, weather below landing minimums, etc).

delay time
(1) *Industrial Operations.* Any temporal interval during which a worker is idle due to any cause beyond the worker's control, such as a equipment breakdown, a lack of tools or parts, or a shortage of materials. May also be referred to as *waiting time* or *inherent delay. See also* ***idle time****.* (2) *Railroad.* As applied to an automatic train stop or train control system, the time which elapses after the onboard apparatus detects a more restrictive indication until the brakes start to apply. (3) *Aviation.* The amount of time that the arrival must lose to cross the meter fix at the assigned meter fix time. This is the difference between Actual Calculated Landing Time (ACLT) and Vertex Time of Arrival (VTA).

delead
To remove lead from the tissues by use of a chelating agent which is then excreted in the urine. Also, the term applies to the removal of lead-based paint in dwelling units.

delegate
A person who is appointed, authorized, delegated, or commissioned to act in the stead of another. The transfer of authority from one to another. A person to whom affairs are committed by another.

delegated state
A state (or other governmental entity) which has applied for, and received, authority to administer, within its territory, its state regulatory program as the federal program required under a particular federal statute.

delegation doctrine
A principle of constitutional law based upon the classic understanding that Congress, as the duly elected representative of the people, is the repository of all legislative power. Only the people can grant this power to the Congress. According to the delegation doctrine, Congress cannot, in turn, delegate this legislative power to another party, such as an administrative agency, because the agency has not been elected by the people. Under strict application of this doctrine, Congress is required to provide reasonably clear and specific statutory standards to guide agency decision making.

delegation of powers
The transfer of authority by one branch of government in which such authority is vested to some other branch or administrative agency. The U.S. Constitution delegates different powers to the executive, legislative, and judicial branches of government. Exercise by the executive branch of the powers delegated to the executive branch offends this separation and the delegation of powers and hence is unconstitutional. Certain powers may not be delegated from one branch of the government to another, such as the judicial powers or such congressional powers as the power to declare war, impeach, or admit new states.

deleterious
Refers to an agent (physical, chemical, or microbial) that is injurious or capable of causing harm.

deliberate
To weigh, ponder, discuss, regard upon, consider. To examine and consult in order to form an opinion. To weigh in the mind; to consider the reasons for and against; to consider maturely; reflect upon, as to deliberate a question; to weigh the arguments for an against a proposed course of action.

delinquent
(1) Lacking in some respect; characterized by antisocial, illegal, or criminal behavior. (2) A person whose conduct is antisocial, illegal, or criminal; applied to a minor exhibiting such conduct (juvenile delinquent).

deliquescence
The process of becoming liquid by absorption of water from the air.

delirium
A disordered mental state with excitement and illusions. Almost any acute illness accompanied by very high fever can bring on delirium. Other causes are physical and mental shock, exhaustion, fear and anxiety, alcoholism, drug overdose, and insulin shock.

delirium tremens (DTs)
Delirium from the excessive, chronic use of alcoholic beverages. It may also occur in cases of addiction to narcotics. Delirium tremens is a serious mental illness that is characterized by illusions and vivid hallucinations, extreme restlessness, agitation, uncontrollable shacking and, in general, an increased body metabolism. The victim is usually extremely fearful and apprehensive because the illusions and hallucinations are very real in his/her mind.

delist
Use of the petition process to have a facility's toxic designation rescinded.

deliverability
Refining. Represents the number of future years during which a pipeline company can meet its annual requirements for its presently certificated delivery capacity from presently committed sources of supply. The availability of gas from these sources of supply shall be governed by the physical capabilities of these sources to deliver gas by the terms of existing gas purchase contracts, and by limitations imposed by state or federal regulatory agencies.

delivered
Refining. The physical transfer of natural, synthetic, and/or supplemental gas from facilities operated by the responding company to facilities operated by others or to consumers.

delivered energy
The amount of energy delivered to the site (building); no adjustment is made for the fuels consumed to produce electricity or district sources. This is also referred to as *net energy*.

DeLorme boot
A special boot used to exercise the quadriceps muscles.

DeLorme exercises
See ***progressive resistance exercises***.

Delphi method
See ***Delphi technique***.

Delphi technique
A process designed to obtain a consensus of experts by successive iterations of questioning interspersed with feedback on others' opinions and supporting reasons. Also referred to as *Delphi method*.

delta
(1) The flat alluvial area at the mouth of some rivers where an accumulation of river sediment is deposited in the sea or a lake. (2) Term used to describe the change, or differential, that may be inflicted upon a single variable following the action by some other influencing factor(s).

delta P
Differential pressure.

delta ray
The track of electrons recoiling from ionizing or atomic reactions in tissue.

delta rhythm
An EEG frequency band consisting of frequencies less than 4 Hz.

delta T
Differential temperature.

deltoid arc
The surface distance from acromiale to the point where the deltoid muscle disappears from view.

deltoid muscle
The large skeletal muscle extending over the superior and lateral part of the shoulder.

deluge shower
A shower unit which enables the user to have water cascading over the entire body. A minimum flow rate of water and time of use are recommended for effective contaminant removal.

delusion
A false belief inconsistent with an individual's own knowledge and experience.

demand
The assertion of a legal right; a legal obligation asserted in the courts. An imperative request preferred by one person to another, under a claim of right, requiring the latter to do or yield something or to abstain from some act.

demand air taxi
Use of an aircraft operating under Federal Aviation Regulations, Part 135, passenger and cargo operations, including charter and excluding commuter air carrier.

demand airline device
Respirator in which air enters the facepiece only when the wearer inhales.

demand respirator
See ***demand airline device.***

demand response
Transportation. (1) Non-fixed-route service utilizing vans or buses with passengers boarding and alighting at pre-arranged times at any location within the system's service area. Also called *dial-a-ride.* (2) Passenger cars, vans, or Class C motor buses operating in response to calls from passengers or their agents to the transit operator, who then dispatches a vehicle to pick the passengers up and transport them to their destinations. A demand response operation is characterized by the following: a) the vehicles do not operate over a fixed route or on a fixed schedule except, perhaps, on a temporary basis to satisfy a special need; and b) typically, the vehicle may be dispatched to pick up several passengers at different pick-up points before taking them to their respective destinations and may even be interrupted en route to these destinations to pick up other passengers. (3) Personal transit service operated on roadways to provide service on demand. Vehicles are normally dispatched and used exclusively for this service.

demand responsive system
Any system of transporting individuals, including the provision of designated public transportation service by public entities and the provision of transportation service by private entities, including but not limited to specified public transportation service, which is not a fixed route system.

demand variability
A change in the desire to purchase a product over time.

demanded motions inventory
The various motions which are required to perform a given task.

demeanor
With regard to a witness or other person, relates to physical appearance, outward bearing, or behavior. It embraces such facts as the tone of voice in which a witness's statement is made, the hesitation or readiness with which his/her answers are given, the look of the witness, his/her carriage, any evidence of surprise, gestures, zeal, bearing, expression, the use of his/her eyes, any yawning, the pitch of voice, any embarrassment, candor, or seeming levity.

dementia
Progressive mental deterioration due to organic disease of the brain. A disorder in which cognitive and intellectual functions of the mind are prominently affected. Impairment of memory is an early sign and total recovery is not possible since organic cerebral disease is involved.

demineralizing
The process of removing minerals from water, most commonly through an ion exchange process.

de minimis doctrine
See ***de minimis non curat lex.***

de minimis no curat lex
Latin. The law does not care for, or take notice of, very small or trifling matters. The law does not concern itself about trifles. Provision is made under certain criminal statutes for dismissing offenses which are de minimis.

de minimis settlements
Under CERCLA, a final settlement between EPA and a company which disposed of relatively small quantities of hazardous substances.

de minimis violation
As defined in the Occupational Safety and Health Act of 1970, a violation which has no direct or immediate relationship to safety or health.

demographics
The gathering, analysis, and/or use of information such as occupation, income, education, family size, and ethnic background from those populating a certain region.

demography
The science dealing with social statistics, including questions of health, disease, births, and mortality.

demonstrative evidence
That evidence addressed directly to the sense without intervention of testimony. Such evidence is concerned with real objects which illustrate some verbal testimony and has no probative value in itself.

demulsify
To resolve or break an emulsion, such as water and oil, into its components.

demurrage
In domestic U.S. transportation, a penalty charge against shippers or consignees for delaying the carrier's equipment beyond the allowed free time provision of the tariff at the rail ramp; in international transportation, a storage charge to shippers which starts accruing after a container is discharged from a vessel. The charge varies according to rules of the appropriate tariff.

de Musset's sign
Rhythmic oscillation of the head caused by pulsation of the carotid arteries; a sign of aortic insufficiency.

demyelination
The destruction of the myelin sheath of a nerve or nerves.

dendrite
A long, branching protoplasmic process, such as the branches conducting impulses toward the body of a nerve cell.

dendrochronology
The analysis of the annual growth rings of trees as a means of interpreting past climatic conditions.

dendron
One of the branching processes of a nerve cell or neuron that conveys impulse. *See* ***dendrite***.

dengue
A viral disease carried by the *Aedes* mosquito. Also known a *breakbone fever* because of the intense joint pain associated with it.

denial
A traverse in the pleading of one party of an allegation of fact asserted by the other; a defense. A response by the defendant to matter(s) alleged by the plaintiff in the complaint. Under the Rules of Civil Procedure, denials must be specific and directed at the particular allegations controverted. Denials may be made in part (a *specific denial*) or in whole (a *general denial*), but in the main should be specific and fairly meet the substance of the averments denied.

denitrification
The anaerobic biological reduction of nitrate nitrogen to nitrogen gas.

denitrogenation
Remove nitrogen from a system, in particular the body, by breathing a nitrogen-free gas mixture.

dense nonaqueous-phase liquids (DNAPL)
Liquids that are not miscible in and denser than water.

densimeter
An apparatus for determining density or specific gravity.

densitometry
Determination of variations in density by comparison with that of another material or with a certain standard.

density
The ratio of the mass of a material to its volume expressed as g/cm^3, lb/ft^3, etc.

density current
A flow of water through a larger body of water that retains its unmixed identity due to a difference in density.

densography
The measurement of the contrast densities in a roentgen negative.

dentifrice
A preparation for cleaning and polishing the teeth.

dentistry

(1) That branch of the healing arts concerned with the teeth and associated structures of the oral cavity. (2) The work done by dentists, e.g., the creation of restorations, crowns and bridges, and surgical procedures performed in and about the oral cavity. (3) The practice of the dental profession collectively.

denture

A complement of teeth, either natural or artificial; ordinarily used to designate an artificial replacement for the natural teeth and adjacent tissues.

denuder

A device used to remove a gaseous contaminant from sampled air when monitoring for a substance(s) with which the denuded material would interface.

deoxidation

The process by which oxygen is removed from a chemical compound.

deoxycholic acid

One of the bile acids, capable of forming soluble, diffusible complexes with fatty acids, and thereby allowing for their absorption in the small intestine.

A model of a strand of DNA

deoxyribonucleic acid (DNA)

The type of nucleic acid that contains *deoxyribose sugar* and is found mainly in the chromosomes of animal and vegetable cells. DNA is considered to be the repository of hereditary characteristics and the auto-reproducing constituent of chromosomes and many viruses.

Department of Defense Standard (MIL-STD)

A U.S. Department of Defense Standard which uses metric values. *See also* ***Military Standard***.

Department of Energy (DOE)

United States federal agency responsible for research and development of energy technology. The DOE provides the framework for a comprehensive and balanced national energy plan through the coordination and administration of the energy functions of the federal government. The Department is responsible for the research, development, and demonstration of energy technology; the marketing of federal power; energy conservation; the nuclear weapons program; regulation of energy production and use; pricing and allocation, and a central energy data collection and analysis program.

Department of Transportation (DOT)

Establishes the nation's overall transportation policy. Under its umbrella there are ten administrations whose jurisdictions include highway planning, development and construction; urban mass transit; railroads; aviation; and the safety of waterways, ports, highways, and oil and gas pipelines. The DOT was established by Act of October 15, 1966, as amended (49 U.S.C. 102 and 102 note) "to assure the coordinated, effective administration of the transportation programs of the Federal Government" and to develop "national transportation policies and programs conducive to the provision of fast, safe, efficient, and convenient transportation at the lowest cost consistent therewith."

departure angle

Transit. The smallest angle, in a plane side view of an automobile, formed by the level surface on which the automobile is standing and a line tangent to the rear tire static-loaded radius arc and touching the underside of the automobile rearward of the rear tire.

departure center

The air route traffic control center having jurisdiction for the airspace that generates a flight to the impacted airport.

departure control

A function of an approach control facility providing air traffic control service for departing Instrument Flight Rules (IFR) and, under certain conditions, Visual Flight Rules (VFR) aircraft.

departure time

The time an aircraft becomes airborne.

dependence
The total psychophysical state of an addict in which the usual or increasing doses of the drug are required to prevent the onset of abstinence symptoms.

dependent variable
A response variable whose value is determined, wholly or in part, by one or more independent variables within an experimental situation.

depersonalization
A feeling of unreality or strangeness, related to oneself or to the external environment.

depilatory
Chemical having the ability to remove or destroy hair.

depleted uranium
Uranium having a smaller percentage of urnaium-235 than that found in uranium as it occurs naturally.

depletion curve
In hydraulics, a graphical representation of water depletion from storage-stream channels, surface soil, and groundwater. A depletion curve can be drawn for base flow, direct runoff, or total flow.

deployment
The distribution of workers to specific work sites.

depolarization
The abolition or disappearance of a difference in electrical charge.

depolarize
Reduce the amount of electrical charge across some structure, usually with reference to a neuronal or muscle cell membrane.

depolymerization
The breakdown of an organic compound into two or more less complex molecules.

depose
To make a deposition; to give evidence in the shape of a deposition; to make statements which are written down and sworn to; to give testimony which is reduced to writing by a duly qualified officer and sworn to by the deponent.

deposit
(1) Sediment or dregs. (2) Extraneous inorganic matter collected in the tissues or in an organ of the body.

deposition
(1) *Law.* The testimony of a witness taken upon oral questions or written interrogatories. Depositions are not taken in open court, but in response to an order to take testimony issued by a court, or under a general law or court rule on the subject, and reduced to writing. Depositions must be duly authenticated and are intended to be used in preparation and upon the trial of a civil action or a criminal prosecution. (2) *Meteorology.* A process that occurs in subfreezing air when water vapor changes directly to ice without becoming a liquid first. Also referred to as *sublimation.*

deposition nuclei
Tiny particles (ice nuclei) upon which an ice crystal may grow by the process of deposition. Also called *sublimation nuclei.*

depraved mind
An inherent deficiency of moral sense and rectitude, equivalent to the statutory phrase "depravity of heart" defined as the highest grade of malice. A corrupt, perverted, or immoral state of mind.

depreciate
To spread the cost of a system, piece of equipment, structure, or facility over time, usually for tax or accounting purposes, to allow for its reduction in value.

depressant
(1) Depressing or retarding. (2) An agent that retards any function, especially a drug that slows a function of the body or calms and quiets nervous excitement; a sedative. Among the best-known depressants are barbiturates. Alcohol is also a depressant, although its first effect is sometimes stimulating.

depression
(1) A hollow or fossa. (2) Reduction of vital functional activity; in psychiatry, a morbid sadness or melancholy, distinguished from grief, which is realistic and proportionate to a personal loss. Depression may be symptomatic of a psychiatric disorder or it may constitute the principal manifestation of a neurosis or psychosis. (3) A period of economic stress; usually accompanied by poor business conditions and high unemployment.

depressive reaction
A mental or emotional condition, precipitated by some external factor and manifested by

guilt, self-depreciation, psychomotor retardation, defection, and/or a sense of inadequacy. It is generally considered to be a neurosis.

depressor
(1) Any muscle producing a downward movement. (2) Any device effecting a downward movement of some structure.

depressurization
A condition that occurs when the air pressure inside a structure is lower than the air pressure outside. Depressurization can occur when household appliances, such as fireplaces or furnaces, consume or exhaust house air and are not supplied with enough makeup air. Radon-containing soil gas may be drawn into a house more rapidly under depressurization conditions.

depth
A straight-line measurement with anterior to posterior extent in any sagittal plane and perpendicular to the frontal plane of the body.

depth cueing
The process of making a complicated image more readily understandable by distinguishing between elements in the foreground and background.

depth filtration
Filtration classification for filters where solids are removed within the granular media.

depth of field
The distance from the lens of a camera to the farthest subject (infinity) in which the image will appear sharp. "Shallow" depth of field is a result of larger apertures; greater depth of field is achieved when smaller aperture openings are used.

depth perception
The ability to distinguish relative distances of two or more objects or the distance of a single object from the observer.

deputy
A substitute; a person duly authorized by an officer to exercise some or all of the functions pertaining to the office, in the place and stead of the latter.

DeQuervain's disease
A type of tenosynovitis of the exterior tendons of the wrist or abductors of the thumb. The inflammation of the synovial lining is often pronounced under conditions of highly repetitive hand usage or is due to poor design of the workplace or tools. *See also* ***cumulative trauma disorder***.

derailment
A derailment occurs when one or more than one unit of rolling stock equipment leaves the rails during train operations for a cause other than collision, explosion, or fire.

derailment/bus going off road
A non-collision incident which occurs as a result of rolling equipment leaving the rail, or buses leaving the roadway, and for rollovers.

derailment/left roadway
A non-collision incident in which a transit vehicle leaves the rails or road on which it travels. This also includes rollovers.

deregulation
Revisions or complete elimination of economic regulations controlling transportation. For example, the Motor Carrier Act of 1980 and the Staggers Act of 1980 revised the economic controls over motor carriers and railroads, respectively.

derepression
In psychiatry, the coming back of ideas or impulses into conscious awareness that were earlier pushed from such awareness into the unconscious because they were personally intolerable.

derivative
A chemical substance derived from another substance either directly or by modification or partial substitution.

derived from rule
Under RCRA, this special rule stipulates that waste generated from the treatment, storage, or disposal of hazardous waste is itself a hazardous waste, unless it does not exhibit any of the hazardous characteristics or it is not a listed waste.

derived requirement
A requirement not imposed by some original or high-level document or management, but which is imposed by secondary documents or lower levels of management.

dermabrasion
Removal, by sandpaper or high-speed brush, of acne scars or nevi.

dermal
Relating to the skin.

dermal toxicity
The ability of a pesticide or toxic chemical to poison people or animals by contact with the skin. See also contact pesticide.

dermatitis
Any inflammation of the skin surface from any cause. A skin abnormality resulting from an occupational exposure. Dermatitis can result from various animal, vegetable, and chemical substances, from heat or cold, from mechanical irritation, from certain forms of malnutrition, or from infectious disease. In some cases, dermatitis may have a psychologic rather than a physical cause. The symptoms may include itching, redness, crustiness, blisters, watery discharges, fissures, or other changes in the normal condition of the skin. May also be referred to as *industrial dermatitis, occupational contact dermatitis, professional eczema, cement dermatitis, chrome ulcers, oil acne, rubber itch,* or *tar warts.*

dermatology
That branch of medicine dealing with diseases of the skin.

dermatome
That region of the skin innervated with sensory fibers from a single spinal nerve.

dermatome chart
A graphic or visual display of the regions of the body surface innervated by specific spinal nerves.

dermatomyositis
An acute, subacute, or chronic disease involving constant inflammation of the skin and muscles, leading to muscular decomposition and atrophy. It is included among the group of illnesses known as collagen diseases. Among the variety of symptoms that point to the onset of the disease are fever, loss of weight, skin lesions, and aching muscles. As the disease progresses, there may be loss of the use of the arms and legs. Complications such as hardening may occur, similar to the changes seen in *scleroderma*.

dermatophytoses
A group of diseases caused by fungi and often found among farmers, animals handlers, pet and hide handlers, wool sorters, cattle ranchers, athletes. lifeguards, gymnasium employees, and animal laboratory workers. Ringworm of the hands and feet is the most common form and is usually prevented by recognition of the disease in animals, sterilization and proper laundering of towels, general cleanliness of showering facilities, and proper personal hygiene.

dermatosis
Generic term for skin disorders, particularly those not involving inflammation.

dermis
See ***corium***.

dermographic pencil
An instrument for marking landmarks or point marks on skin for taking anthropometric measurements.

derrick
A mechanical device intended for lifting, with or without a boom supported at its head by a topping lift from a mast, fixed A frame, or similar structure. The mast or equivalent member may or may not be supported by guys or braces. The boom, where fitted, may or may not be controlled in the horizontal plane by guys (vangs). The term also includes shear legs.

DES
Diethylstilbestrol, a synthetic estrogen, is used as a growth stimulant in food animals. Residues in meat are thought to be carcinogenic.

desalinization
Removing salt from ocean or brackish water.

Descemet's membrane
The posterior lining membrane of the cornea.

descent speed adjustments
Aviation. Speed deceleration calculations made to determine an accurate vertex time of arrival (VTA). These calculations start at the transition point and use arrival speed segments to the vertex.

describing function
Any mathematical model or representation of a time-varying system involving humans,

generally consisting of some transfer function plus remnants.

descriptive statistics
The collection, tabulation, and analysis of data in such a manner as to yield measures, (i.e., mean, variance, standard deviation, etc.) that describe the population, group, or sample data.

desensitization
The abolition of sensitivity to a particular antigen.

desensitize
To render less sensitive.

desert
A region characterized by a climatic pattern where evaporation exceeds precipitation.

desert fever
A fungal disease usually affecting the respiratory tract and lungs, although it may involve any or all of the body's organs; also called *San Joaquin Valley fever, desert rheumatism* and *coccidioidomycosis*.

desert pavement
An arrangement of pebbles and large stones that remains behind as finer dust and sand particles are blown away by the wind.

desert rheumatism
*See **desert fever**.*

desertification
The process where the biological productivity of land is reduced, resulting in desert-like conditions.

desiccant
(1) *General.* Chemicals, such as silica gel, that absorb moisture and are typically used to promote or ensure dryness. (2) Under the Federal Insecticide, Fungicide, and Rodenticide Act: Any substance or mixture of substances intended for artificially accelerating the drying of plant tissue.

design
The process of developing the requirements, structure, dimensions, tolerances, and materials to be used for an entity.

design burst pressure
The calculated pressure (the analytical value that was calculated using an acceptable industry and/or government practice to determine its design pressure) that components must withstand without rupture and/or burst to demonstrate design adequacy in a qualification test. The actual burst pressure for a tested component must demonstrate, during qualification testing, that the design burst pressure is less than the actual burst pressure. Safety factors are based on design burst pressure not actual burst pressure of a particular component.

design capacity
Pipeline. The capacity associated with the direction of the flow observed on the peak day. *See also **certified capacity**.*

design driver
A requirement which causes a system to be designed in a specific way.

design eye point
A fixed point providing a line of sight within which all controls and displays at a workstation should be located. This point is only recommended for use when the head position of the operator is severely constrained during the task or job. *See also **design eye volume**.*

design eye position
*See **design eye volume**.*

design eye volume
That region within which an operator's head is free to move and provide appropriate lines of sight at a workstation.

design for maintenance
A design priority concept which emphasizes the future maintenance aspects of a product's structure. Also referred to as *design for maintainability*.

design for manufacturing
A design priority concept which emphasizes the assembly aspects of manufacturing in structural design.

design for reliability
A design priority which emphasizes minimizing the chances of failure and/or maximizing the mean time between failures.

design for use
A design priority which emphasizes the ease of use of a product.

design head
Hydroelectric Engineering. The achieved river, pond, or reservoir surface height (forebay elevation) that provides the water level to

produce the full flow at the gate of the turbine in order to attain the manufacturer's installed nameplate rating for generation capacity.

design live road
Transit. The live road that the structure was designed to carry (85 psf, H-10, H-15, and HS-20).

design load
The weight which can be safety supported by a structure, as specified in its design criteria.

design safety factor
A factor used to account for uncertainties in material properties and analysis procedures. It is often referred to as *design factor of safety,* or simply *safety factor*.

design solution
An engineering design which meets or exceeds a set of requirements.

design speed
(1) *General.* That rate at which a mechanically driven operation is intended to occur or at which a piece of equipment is intended to move or rotate. (2) *Transit.* Design speed determines the maximum degree of road curvature and minimum safe stopping, meeting, passing, or intersection sight distance.

designated associated equipment
Maritime. Inboard engine, outboard engine, and stern drive unit. Specific equipment, in addition to completed boats, which has been designated in 33 (CFR) 179.03 as being subject to the requirements of 46 (U.S.C.) 4310. Other items of associated equipment may be the cause for recall of boats, but the manufacturers of those items of associated equipment are not subject to the requirement for recall.

designated facility
A hazardous waste treatment, storage, or disposal facility that has been designated on the manifest by the generator.

designated pollutant
An air pollutant which is neither a criteria nor hazardous pollutant, as described in the Clean Air Act, but for which new source performance standards exist. The Clean Air Act does require states to control these pollutants, which include acid mists, total reduced sulfur (TRS), and fluorides.

designated public transportation
Transportation provided by a public entity (other than public school transportation) by bus, rail, or other conveyance (other than transportation by aircraft or intercity or commuter rail transportation) that provides the general public with general or special service, including charter service, on a regular and continuing basis.

designated representative
Any person or party acting on behalf of an employee with the full permission of that employee. Can include labor unions, relatives, and attorneys. Under certain circumstances, designated representatives may have access to employee exposure and medical records.

designated seating capacity
Transportation. The number of designated seating positions provided.

designated seating position
Any plan view location capable of accommodating a person at least as large as a 5th percentile adult female, if the overall seat configuration and design and vehicle design are such that the position is likely to be used as a seating position while the vehicle is in motion, except for auxiliary seating accommodations such as temporary or folding jump seats. Any bench or split-bench seat in a passenger car, truck, or multipurpose passenger vehicle with a Gross Vehicle Weight Rating (GVWR) less than 10,000 pounds, having greater than 50 inches of hip room (measured in accordance with Society of Automotive Engineers (SAE) Standard J1100a) shall have not less than three designated seating positions, unless the seat design or vehicle design is such that the center position cannot be used for seating.

designated service
Railroad. Exclusive operation of a locomotive under the following conditions: a) the locomotive is not used as an independent unit or the controlling unit is a consist of locomotives except when moving for the purpose of servicing or repair within a single yard area; b) the locomotive is not occupied by operating or deadhead crews outside a single yard area; and c) the locomotive is stenciled "Designated Service-DO NOT OCCUPY."

designated uses
Those water uses identified in state water quality standards which must be achieved and maintained as required under the Clean Water Act. Uses can include cold water fisheries, public water supply, agriculture, etc.

designer bugs
Popular term for microbes developed through biotechnology that can degrade specific toxic chemicals at their source in toxic waste dumps or in groundwater.

desmalgia
Pain in a ligament.

desorption
The process of removing an adsorbed material from the solid in which it is adsorbed and retained.

desorption efficiency
The fraction of a known quantity of analyte that is recovered from a spiked solid sorbent media blank.

desquamation
The sloughing off of the epidermal layer of the skin.

destabilizing pressure
A pressure that produces comprehensive stresses in a pressurized structure or pressure component.

destination
(1) For travel period trips, the destination is the farthest point of travel from the point of origin of a trip of 75 miles or more one way. For travel day trips, the destination is the point at which there is a break in travel. (2) The place/location in which the cargo was unloaded and/or the transit terminated.

destruction and removal efficiency (DRE)
An expression of hazardous waste incinerator efficiency stated as the percentage of incoming principal organic hazardous components destroyed during incineration.

destructive test
A procedure in testing product quality in which the material or product being tested is either partially or totally destroyed.

desulfurization
Removal of sulfur from fossil fuels to reduce pollution.

desynchronize
(1) Change the electroencephalogram from a low frequency, high amplitude rhythm to a higher frequency, lower amplitude rhythm. (2) Change a biological rhythm from a normal or typical phase relationship to another.

desynchronosis
See ***jet lag***.

detachment of the retina
Separation of the inner layers of the retina from the pigment layer, which remains attached to the choroid. The onset of symptoms may be gradual or sudden, depending on the cause, size, and location of the area involved. The victim may see flashes of light and then days or weeks later notice cloudy vision or the loss of central vision, Another common symptom is the sensation of spots or moving particles in the field of vision. In severe retinal detachment, there may be complete loss of vision.

detectability
One or more qualities of a signal, display, or other stimulus which affect its probability of being perceived, either in isolation or against a background.

detection limit
The lowest amount that can be distinguished from the normal electronic noise of an analytical instrument. For Hazardous Ranking System purposes, the detection limit used is the method detection limit (MDL) or, for real-time field instruments, the detection limit of the instrument as used in the field.

detection threshold
See ***threshold***.

detector
(1) *General.* The portion of an instrument that is responsive to the material being measured. (2) *Radiation.* A device which converts ionizing radiation energy to a form more amenable to measurement. For example, an ionization detector, scintillation detector, etc.

detector tube
An air sampling device used to measure the concentration of various air contaminants. Consists of a glass tube filled with a solid chemical that changes color when it reacts with the air contaminant being sampled in combination with a hand-held pump device to

draw air through the tube at a measured rate. Also known as *colorimetric tube.*

detent
A releasable element used to restrain a part before or after its motion; detents are common arming mechanisms. For example, safe and arm (S&A) device safing pins use a spring-loaded detent to secure the pin in the device.

detention time
The period of time that a volume of liquid remains in a tank.

detergent
Synthetic washing agent that helps to remove dirt and oil. Some contain compounds which kill useful bacteria and encourage algae growth when they are in wastewater that reaches receiving waters.

deterioration
(1) The process or state of growing worse. (2) Disintegration or wearing away. (3) With regard to a commodity, consists of a constitutional hurt or impairment, involving some degeneration in the substance of the thing, such as that arising from decay, corrosion, or disintegration. With respect to values or prices, a decline.

determinate errors
Errors which occur that are correctable if their cause can be determined.

determination
Transit. A document signed by the Administrator of the General Services Administration, setting forth the decision to establish an Interagency Fleet Management Center at a specific location.

determinism
The doctrine that the will is not free but is absolutely determined by psychic and physical conditions.

deterministic
Pertaining to those data which can be explained or predicted with reasonable accuracy via some explicit solvable mathematical relationship.

detonating cord
A flexible fabric tube containing a filler of high explosive material intended to be initiated by an electroexplosive device; often used in destruction and separation functions.

detonation
An exothermic chemical reaction that propagates with such rapidity that the rate of advance of the reaction zone into the unreacted material exceeds the velocity of sound. The rate of advance of the reaction zone is termed *detonation velocity.* When this rate of advance attains such a value that it will continue without diminution through the unreacted material, it is termed the *stable detonation velocity.* When the detonation velocity is equal to or greater than the stable detonation velocity of the explosive, the reaction is termed a *high-order detonation;* when it is lower, the reaction is termed a *low-order detonation.*

detonation velocity
See ***detonation***.

detonator
An explosive device (usually an electroexplosive device) that is the first device in the explosive train and is designed to transform an input (usually electrical) into an explosive reaction.

detoxification
The destruction of toxic properties of a substance, a major function of the liver.

detresfa
Distress Phase. The code word used to designate an emergency phase wherein there is reasonable certainty that an aircraft and its occupants are threatened by grave and imminent danger or require immediate assistance.

detrimental deformation
Term used to indicate a type of deformation including all structural deformations, deflections, or displacements that prevent any portion of the structure from performing its intended function or that reduce the probability of successful completion of the mission.

detritus
(1) Decaying organic matter such as root hairs, stems, and leaves usually found on the bottom of a water body. (2) Grit or fragments of rock or minerals.

detritus tank
Square tank grit chamber incorporating a revolving rake to scrape settled grit to a sump for removal.

deutan
A person with anomalous color vision, marked by derangement or loss of the red-green sensory mechanism.

deuteranomaly
A color vision deficiency involving a reduced ability to discriminate green in colors.

deuteranopia
A form of color blindness involving an inability to discriminate the green content of colors. Also referred to as *green-blindness*.

deuterium
The mass two isotope of hydrogen, symbol ^2H or D; it is available as a gas or heavy water (deuterium oxide) and is used as a tracer or indicator in studying fat and amino acid metabolism.

deuteron
The nucleus of a deuterium atom.

devanning
The unloading of a container or cargo van.

developer
A person, government unit, or company that proposes to build a hazardous waste treatment, storage, or disposal facility.

development
(1) *General.* Gradual growth or expansion, especially from a lower to a higher stage of complexity. (2) Under the Federal Antarctic Protection Act of 1990: Any activity, including logistic support, which takes place following exploration, the purpose of which is the exploitation of specific mineral resource deposits, including processing, storage, and transport activities.

development test
A test to provide design information that may be used to check the validity of analytic technique and assumed design parameters, to uncover unexpected system response characteristics, to evaluate design changes, to determine interface compatibility, to prove qualification and acceptance procedures and techniques, or to establish acceptance and rejection criteria.

development time
The temporal period required to design, engineer, and prepare the manufacturing documentation for some device.

developmental age
An index of growth using an age equivalent determined by standardized observations. May include body measures, mental, emotional, social, and mental observations. *See also* ***chronological age*** *and* ***mental age***.

developmental anthropometry
The study of growth in size and/or proportions of the human body.

developmental quotient
The value of the ratio of the developmental age and the chronological age.

developmental RfD
An estimate (with uncertainty spanning perhaps an order of magnitude or greater) of an exposure level for the human population, including sensitive subpopulations, that is likely to be without an appreciable risk of developmental effects. Developmental RfDs are used to evaluate the effects of a single event (generally one day) exposure.

developmental toxic effect
Harmful effect to the embryo, or fetus, such as embryotoxicity, fetotoxicity, or teratogenicity.

deviant
(1) Varying from a determinable standard. (2) A person with characteristics varying from what is considered standard or normal.

deviation
(1) An alternate method of compliance with the intent of specific requirements. A departure from established or usual conduct or ideology. (2) The amount by which a score or other measure differs from the mean, or other descriptive statistic. (3) *Aviation.* a) A departure from a current clearance, such as an off-course maneuver to avoid weather or turbulence. b) Where specifically authorized in the FARs and requested by the pilot, Air Traffic Control (ATC) may permit pilots to deviate from certain regulations.

device
(1) According to the Federal Food, Drug, and Cosmetic Act, an instrument, apparatus, implement, machine, contrivance, implant, *in vitro* reagent, or other similar or related article, including any component, part or accessory, which is a) recognized in the official National Formulary, or the United States

Pharmacopoeia, or any supplement to them; b) intended for use in the diagnosis of disease or other conditions, or in the cure, mitigation, treatment or prevention of disease, in man or other animals; or c) intended to affect the structure of any function of the body of man or other animals and which does not achieve its primary intended purposes through chemical action within or on the body of man or other animals and which is not dependent upon being metabolized for the achievement of its primary intended purposes. (2) A unit of an electrical system which is intended to carry but not utilize electric energy.

device driver
*See **driver**.*

device-independent
An operation or procedure which has similar functions may be executed or performed on a variety of pieces of equipment which may differ in structure, method of operation, and/or appearance.

dew
Water droplets that form on cool surfaces following condensation of atmospheric water vapor.

dew point
The temperature at which air at a constant pressure and constant water-vapor content will be saturated. Cooling below the dew point usually results in frost or dew. When this temperature is below 0°C, it is sometimes called the *frost point*.

dexterity
The degree of manipulative ability via perceptual-motor coordination.

dextrality
Preferring the right-hand over the left.

dextrose
A sugar, also called glucose or grape sugar, containing six carbon atoms. Dextrose is considered one of the most important carbohydrates because it makes up 80% of all simple sugar absorbed into the blood. It is present in the juice of many sweet fruits and in the blood of all animals. Through the process of metabolism, dextrose is used by the body to provide energy or, in excess, it is converted to fat. The liver cells convert glucose to glycogen, so that it can be stored until needed. When the blood sugar drops below normal, there is increased production of epinephrine, which causes glycogen to be changed back into glucose and used for the production of energy.

DF
*See **direction finder**.*

DFT
*See **dry film thickness**.*

DH
*See **decision height**.*

DHEW
Department of Health, Education, and Welfare (United States).

diabetes
Inordinate and persistent increase in the urinary secretions, especially diabetes mellitus.

diabetes insipidus
A metabolic disorder resulting from decreased activity of the posterior lobe of the pituitary gland. Re-absorption of water from the renal tubules is promoted by vasopressin, or antidiuretic hormone, a hormone from the posterior pituitary lobe. A deficiency of this hormone leads to the symptoms of diabetes insipidus which include excessive thirst and the passage of large amounts of urine with no excess of sugar.

diabetes mellitus
A disorder of carbohydrate metabolism in which the ability to oxidize and utilize carbohydrates is lost as a result of disturbances in the normal insulin mechanism. A serious disruption of carbohydrate metabolism leads to abnormalities of protein and fat metabolism. The oxidation of fat is accelerated in diabetes, and thus there is an accumulation of the end products of fat metabolism in the blood and the development of the symptoms of ketosis, acidosis, and coma. Factors leading to disturbances in the normal insulin mechanism and the onset of diabetes mellitus include insufficient production of insulin from the beta cells of the islands of Langerhans in the pancreas, an increase in the insulin requirement by the tissue cells, or a decrease in the effectiveness of insulin due to one or more insulin antagonists which can deactivate insulin. Any of these factors can produce the symptoms of diabetes mellitus. Because diabetics are un-

able to utilize the carbohydrates in their blood, they are improperly nourished no matter how much food they consume. The accumulation of unused glucose leads to weakness, fatigue, and a spilling over of sugar into the urine. The high levels of sugar in the blood make the diabetic particularly susceptible to infection. In a prolonged severe diabetic condition, the raised fat and glucose level of the blood may cause damage to blood vessels and to tissues and organs containing blood vessels. The resulting poor circulation may be a factor leading to other complications such as gangrene of the hands or feet. Also, the heart and kidneys may suffer damage, there can be difficulty with vision (up to and including total blindness), or the nervous system may be affected. *See also* ***insulin***.

diagnose
To isolate or recognize a disease or the cause of an illness.

diagnosis
The art or method of identifying or recognizing a disease. A medical term, meaning the discovery of the source of a patient's illness or the determination of the nature of his/her disease from a study of its symptoms.

Diagnostic Rhyme Test
A forced-choice test in which an individual is required to select the word he/she believes was spoken from two rhyming options.

diagnostic study
A preliminary investigation or study of some operation, process, individual, or group in an attempt to learn the causes of problems. May also be referred to as a *diagnostic survey*.

diagnostic survey
See ***diagnostic study***.

diagram
A geometric drawing used to explain a fact, a process, the sequence of an activity, or the composition of an element as associated with an accident or incident.

dial caliper
A caliper which uses a rotary dial to indicate the distance between two points.

dial-a-ride
See demand response.

dial up access terminal (DUAT)
The capability for direct user access terminals to file flight plans into the National Airspace System (NAS) and access weather information from the National Graphic Weather Display System.

dialogue
The content of a structured sequence of steps in an interaction between a user and a computer.

dialogue box
A pop-up display window which requests user input regarding some computer system function.

dialysis
The passage of solute molecules through a semipermeable membrane from higher concentration to a solution of lower concentration.

diameter
The length of a straight line passing through the center of a circle and connecting the opposite points on its circumference; hence the distance between two specified opposite points on the periphery of a structure.

diamide
A double amide

diamine
A double amine.

diaphragm
(1) In anatomy, the musculomembranous partition separating the abdominal and thoracic cavities that plays an important role in the respiration process. (2) A thin septum dividing a cavity. (3) A disk with a fixed or flexible opening, mounted in relation to a lens, by which part of the light may be excluded from the area.

diaphysis
The shaft or central portion of a long bone.

diarrhea
Fecal discharge of an abnormal frequency and liquidity. A condition often associated with foodborne illnesses and one of the most common symptoms of the gastroenteritis syndrome.

diastolic blood pressure
The minimum arterial blood pressure occurring during that portion of the cardiac cycle

when the heart relaxes and the ventricle fills with blood.

diatom
A unicellular algae with a yellowish brown color and siliceous shell.

diatomaceous earth
Also known as *diatomite*, a chalk-like material (fossilized diatoms) used to filter out solid waste in wastewater treatment plants, also used as an active ingredient in some powdered pesticides and in swimming pool filtration systems.

diatomaceous earth filter
Water treatment filter that uses a layer of diatomaceous earth as the filter medium.

diatomite
See ***diatomaceous earth***.

diazinon
An insecticide. In 1986, EPA banned its use on open areas such as sod farms and golf courses because it posed a danger to migratory birds who gathered on them in large numbers. The ban did not apply to its use in agriculture, or on lawns of homes and commercial establishments.

dichoptic
Pertaining to viewing conditions in which the visual display to each of the two eyes differs with respect to some property of the stimulus.

dichotic
Pertaining to listening conditions in which differential stimulation of the two ears occurs according to some definable physical property of the stimulus such as duration, frequency, phase, intensity, or bandwidth.

dichromat
One who has dichromatopsia.

dichromatopsia
A form of color blindness involving an inability to discriminate one of the three primary colors: red, green, or blue (in other words, the person is capable of seeing two of the three colors). *See also* ***deuteranopia*** *and* ***protanopia***.

dicofol
A pesticide used on citrus fruits.

dielectric heating
The heating of a nominally insulating material due to its own dielectric losses when the material is placed in a varying electric field.

diesel-electric plant
A generating station that uses diesel engines to drive its electric generators.

diesel fuel
A fuel composed of distillates obtained in a petroleum refining operation or blends of such distillates with residual oil used in motor vehicles. The boiling point and specific gravity are higher for diesel fuels than for gasoline. *See also* ***diesel fuel system***.

diesel fuel, No. 1
A volatile distillate fuel oil with a boiling range between 300 and 575 degrees Fahrenheit and used in high-speed diesel engines generally operated under wide variations in speed and load. Includes type C-B diesel fuel used for city buses and similar operations. Properties are defined in American Society for Testing and Materials (ASTM) Specification D 975.

diesel fuel, No. 2
A gas-oil type distillate of lower volatility with distillation temperatures at the 90 percent point between 540 and 640 degrees Fahrenheit for use in high-speed diesel engines generally operated under uniform speed and load conditions. Includes Type R-R diesel fuel used for railroad locomotive engines, and Type T-T for diesel engine trucks. Properties are defined in American Society for Testing and Materials (ASTM) Specification D 975.

diesel fuel system
Diesel engines are internal combustion engines that burn diesel oil rather than gasoline. Injectors are used to spray droplets of diesel oil into the combustion chambers, at or near the top of the compression stroke. Ignition follows due to the very high temperature of the compressed intake air, or to the use of "glow plugs," which retain heat from previous ignitions (spark plugs are not used). Diesel engines are generally more fuel efficient than gasoline engines, but must be stronger and heavier due to high compression ratios. *See also* ***carburetor, diesel fuel,*** *and* ***fuel injection***.

diet
(1) The total food consumed by an individual. (2) A prescription of food required or permitted to be eaten by a patient. Also called therapeutic diet.

dietary supplement
Under the Federal Food, Drug, and Cosmetic Act: A product (other than tobacco) intended to supplement the diet that bears or contains one or more of the following dietary ingredients: 1) a vitamin; 2) a mineral; 3) an herb or other botanical; 4) an amino acid; 5) a dietary substance for use by man to supplement the diet by increasing the total dietary intake; 6) a concentrate, metabolite, constituent, extract, or combination of any ingredient described in 1), 2), 3), 4), or 5). A dietary supplement is also a product that is 1) intended for ingestion in a form described in the Federal Food, Drug, and Cosmetic Act; 2) is not represented for use as a conventional food or as a sole item of a meal or the diet; and 3) is labeled as a dietary supplement. It does include an article that is approved as a new drug under the Act, certified as an antibiotic under the Act, or licensed as a biologic under the CAA, and was, prior to such approval, certification, or license, marketed as a dietary supplement or as a food unless the Secretary of Health and Human Services has issued a regulation, after notice and comment, finding that the article, when used as or in a dietary supplement under the conditions of use and dosages set forth in the labeling for such dietary supplement, is unlawful under the Act.

difference limen
*See **difference threshold**.*

difference spectrum
That spectrum which is obtained by the subtraction of one spectrum from another.

difference threshold
The degree or intensity by which two suprathreshold stimuli must differ if a difference is to be noted on a specified percentage of the trials. Also referred to as *differential threshold* and *difference limen*. *See also **just noticeable difference**.*

differential
(1) An amount added or deducted from base rate to make a rate to or from some other point or via another route. (2) *Standard type.* The gear assembly on the drive axle that permits the wheels to turn at different speeds; no-slip or limited-slip type: a gear assembly on the drive axle that will not permit one wheel to spin while the other is motionless.

differential absorption ratio
The ratio of concentration of an isotope in a given organ or tissue to the concentration that would be obtained if the same administered quantity of this isotope were uniformly distributed throughout the body.

Differential Aptitude Test (DAT)
A commonly used test for determining verbal, abstract, and mechanical reasoning ability, spatial relations, clerical speed and accuracy, and grammatical/spelling skills.

differential equation
Any equation containing one or more derivatives of a mathematical function.

differential piecework
That form of compensation in which the piece rate is variable, and based on the total number of pieces produced during a specified period.

differential pressure
The difference in the static pressure between two locations.

differential threshold
*See **difference threshold**.*

differential timing
The use of subtraction or simultaneous equations for obtaining the time value of an extremely short duration work element in a time study by combining the time values of elements preceding and following the element in successive cycles.

differentiate
(1) Distinguish between one or more conditions. (2) Mathematically determine the ratio of a small change in a dependent variable as a function of change in the independent variable.

differentiation
The process by which single cells grow into particular forms of specialized tissue (e.g., root, stem, leaf).

diffraction
The bending or breaking of a ray of light into its individual parts.

diffuse
Re-direct or scatter energy transmission in multiple directions over a region.

diffuse lighting
That light which is not incident from any particular direction and is of approximately the

same intensity within the volume of consideration.

diffuse reflectance
The value of the ratio of diffused flux leaving a surface to the incident flux.

diffuse reflection
The distribution of an incident energy flux in many directions from a surface on which it is incident.

diffuse sound field
A sound field in which the sound energy will flow in all directions with equal probability. This type of noise environment exists in a reverberation room and can be used to test sound absorption materials.

diffuse transmission
The passage of an incident energy flux through a material with a wide distribution either internally or on emergence from that material.

diffuse transmittance
The value of the ratio of the flux passed as diffused to the incident flux.

diffused air
A type of aeration that forces oxygen into sewage by pumping air through perforated pipes inside a holding tank and bubbling it through the sewage.

diffusers and grilles
Components of the ventilation system that distribute and diffuse air to promote air circulation in the occupied space. Diffusers supply air and grilles return air.

diffusing medium
*See **diffusers and grilles**.*

diffusion
The continual movement and intermingling of molcculcs in liquids and gases. These movements are random and are caused by thermal agitation. In the body fluids, the molecules of water, gases, and the ions of substances in solution are in constant motion. As each molecule moves about, it bounces off other molecules and loses some of its energy to each molecule it hits, but at the same time, it gains energy from the molecules that collide with it. The rate of diffusion is influenced by the size of the molecules; larger molecules move less rapidly because they require more energy to move about. Molecules of a solution of higher concentration move less rapidly toward those of lesser concentration. In other words, the rate of movement from higher to lower concentration is greater than the movement in the opposite direction.

diffusion detector
A passive type detection device which utilizes the principle of diffusion as the means to transport airborne contaminants to the detector. No mechanical means is employed to transport the sampled air from the surroundings to the detector. This is often referred to as a *passive sampler* or *passive sampling*.

diffusion rate
The rate at which a gas or vapor disperses into or mixes with another vapor. A measure of the tendency of a gas or vapor to disperse into or mix with another gas or vapor.

diffusive sampling
*See **passive sampling**.*

dig
A bubble defect which lies on the surface of a transparent material or window.

digester
In wastewater treatment, a closed tank; in solid waste conversion, a unit in which bacterial action is induced and accelerated to break down organic matter and establish the proper carbon to nitrogen ratio.

digestion
(1) The biochemical decomposition of organic matter, resulting in partial gasification, liquefacation, and mineralization of pollutants. (2) The conversion of materials into simpler compounds, physically or chemically, especially the breaking down of food into substances that can be absorbed into the blood and utilized by the body tissues. Digestion is accomplished by physically breaking down, churning, diluting, and dissolving the food substances, and also by splitting them chemically into simpler compounds. Carbohydrates are broken down to monosaccharides (simple sugars); proteins are broken down into amino acids; and fats are absorbed as fatty acids and glycerol (glycerin). The digestive process takes place in the *alimentary canal*. The salivary glands, liver, gallbladder, and pancreas are located outside the alimentary canal, but they are considered accessory organs of di-

gestion because their secretions provide essential enzymes.

digestive tract
: *See **alimentary canal**.*

digit
: (1) Any of the fingers or toes. A convention is to number the digits with Roman numerals, beginning with the thumb and big toe, for example, the thumb is digit I, index finger is digit II, etc. (2) A numerical symbol, having a potential range of the integers from 0 through 9. The actual range may vary with the number base being used.

digital brite radar indicator tower equipment (DBRITE)
: Alphanumeric display systems for control towers using digital scan converter systems in a radar scope-type presentation.

digital dermatoglyph
: *See **fingerprint**.*

digital display
: *See **numerical display**.*

digital-to-analog conversion
: The process of changing numerical data from a sequence of bits to a continuous graphical curve. Typically done with time series data.

digital-to-analog converter
: That electrical or electromechanical equipment used for digital-to-analog conversion.

Digital Versatile Disc (DVD)
: Resembles a standard compact disc (CD), but has much higher capacity. Also called *Digital Video Disc*.

dihedral angle
: That angle between two planes.

dike
: (1) A low wall that can act as a barrier to prevent a spill from spreading. (2) A construction, usually of piling or stone and usually at right angles to the current, for the purpose of diverting the river current away from the banks and toward the channel. A dike serves the same purpose as a wingdam. Dike pilings are usually visible at normal water stages but are often submerged in high water and constitute a navigational hazard. *See also **hurdle**.*

dike light
: A light installed on the end of a dike, normally a portable, 90-mm, battery-operated, light.

dilatant
: Property of a liquid whose viscosity increases as agitation is increased.

dilation
: The process of expanding or enlarging.

diligence
: (1) Vigilant activity; attentiveness; or care, of which there are infinite shades, from the slightest momentary thought to the most vigilant anxiety. Attentive and persistent in doing a thing; steadily applied; active. (2) The attention and care required of a person in a given situation and is the opposite of negligence. *See also **due diligence**.*

diluent
: Material used to reduce the concentration of an active material to achieve a desirable effect.

diluent gas narcosis
: *See **inert gas narcosis** and **nitrogen narcosis**.*

dilute
: To reduce the concentration of a material in the air or in a liquid.

dilution
: The process of increasing the proportion of the solvent or diluent in a mixture.

dilution factor
: The volumetric ratio of solvent to solute.

dilution ratio
: The relationship between the volume of water in a stream and the volume of incoming water. It affects the ability of the stream to assimilate waste.

dilution ventilation system
: A system of airspace ventilation that relies on the mixing of contaminated air with uncontaminated air for the purpose of controlling potential air borne health hazards. Also referred to as *general exhaust ventilation*.

dilution weight
: Parameter in the Hazardous Ranking System surface water migration pathway that reduces the point value assigned to targets as the flow or depth of the relevant surface water body increases.

dim
To reduce the light intensity from a source.

dimension
(1) Any orthogonal spatial axis, typically representing length, width, or depth. (2) An aspect of a picture, concept, or other entity for consideration.

dimmability
Having the capability of reducing light intensity without turning off one or more sources.

dimmer
Any device whose purpose is to cause dimming.

DIN color system
A color ordering system based on the relative importance of hue (T), saturation (S), and darkness (D) using a standard daylight (D_{65}) and CIE tristimulus values.

dinner bucket boat
A boat operating without benefit of a cook house.

dinocap
A fungicide used primarily by apple growers to control summer diseases. In 1986, EPA proposed restrictions on its use when laboratory tests found it caused birth defects in rabbits.

dinoseb
A herbicide that is also used as a fungicide and insecticide. It was banned by EPA in 1986 because it posed the risk of birth defects and sterility.

dioctyl phthalate
A colorless liquid that can be used to generate particles of uniform size (0.3 micrometers diameter) for use in testing the efficiency of filter media.

diopters
A measure of the power of a lens, equal to the reciprocal of the focal length in meters.

diotic
Pertaining to listening conditions in which both ears are stimulated by identical stimuli.

dioxide
An oxide with two oxygen atoms.

dioxin
Any of a family of compounds known chemically as dibenzo-*p*-dioxins. Their potential toxicity and their ability to contaminant commercial products have been the cause of some concern in recent years. Laboratory tests on animals indicate that it is one of the more toxic manmade compounds.

dip
Seismology. The angle between a geologic surface (for example, a fault plane) and the horizontal. The direction of dip can be thought of as the direction a ball, if placed upon the tilted surface, would roll. Thus, a ball placed on a north-dipping fault plane would roll northward. The dip of a surface is always perpendicular to the strike of that surface. See also *strike*.

dip slip
Seismology. Fault movement (slip) that is parallel to the dip of the fault. This can describe both normal slip and reverse slip.

diphtheria
An acute, contagious disease in children that normally affects the membranes of the throat. The causative agent *Corynebacterium diphtheriae* is spread from the nose, mouth, and throat of infected persons.

diplacusis binauralis
A condition in which a single tone, presented to both ears, is perceived as having a different pitch in each ear.

diplegia
Paralysis of like parts on either side of the body.

diploe
The spongy bone between the inner and outer layers of the flat skull bones.

diplopia
An eyesight defect in which an object appears double. Commonly referred to as *double vision*.

dipsomania
A mental disease characterized by an uncontrollable desire for intoxicating drinks. An irresistible impulse to indulge in intoxication, either by alcohol or other drugs.

dipthong
A vowel sound which involves some articulator movement.

direct
Aviation. Straight-line flight between two

navigational aids, fixes, points, or any combination thereof. When used by pilots in describing off-airway routes, points defining direct route segments become compulsory reporting points unless the aircraft is under radar contact.

direct altitude and identity readout (DAIR) system
The DAIR system is a modification to the AN/TPX42 Interrogator System. The Navy has two adaptations of the DAIR System-Carrier Air Traffic Control Direct Altitude and Identification Readout System for Aircraft Carriers and Radar Air Traffic Control Facility Direct Altitude and Identity Readout System for land-based terminal operations. The DAIR detects, tracks, and predicts secondary radar aircraft targets. Targets are displayed by means of computer-generated symbols and alphanumeric characters depicting flight identification, altitude, ground speed, and Right plan data. The DAIR System is capable of interfacing with ARTCCs.

direct anthropometric measurement
The measurement of some anthropometric dimension using one or more tools in physical or near physical contact with the body.

direct assistance
Transportation and other relief services provided by a motor carrier or its driver(s) incident to the immediate restoration of essential services (such as electricity, medical care, sewer, water, telecommunications, and telecommunication transmissions) or essential supplies (such as food and fuel). It does not include transportation related to long-term rehabilitation of damaged physical infrastructure or routine commercial deliveries after the initial threat to life and property has passed.

direct component
That portion of an energy flux which arrives at a given location on a path from the source, without reflection.

direct cost
A cost due to or for supporting direct labor.

direct current (DC)
A non-oscillating current flow, traveling only in one direction.

direct damages
Those that follow immediately upon the act done. Damages which arise naturally or ordinarily from breach of contract. They are damages which, in the ordinary course of human experience, can be expected to result from a breach.

direct discharger
A municipal or industrial facility which introduces pollution through a defined conveyance or system; a point source.

direct filtration
Filtration process that does not include flocculation or sedimentation pretreatment.

direct glare
A type of glare experienced when a bright light source is within an individual's field of view.

direct injury
A wrong which directly results in the violation of a legal right and which must exist to permit a court to determine the constitutionality of an act of Congress.

direct labor
(1) That effort expended on a product or service which advances that product or service toward its specifications of completion. Also referred to as *productive labor*. (2) Any effort which is readily identified with and chargeable to a specific product or project.

direct lighting
That illumination environment in which approximately 90% or more of the luminous flux is directed onto a work or other surface.

direct loss
One which results immediately and proximately from an occurrence and not remotely from some of the consequences of effects thereof.

direct manipulation
A user-computer interface in which the entity being worked is continuously displayed, the communication involves button clicks and movements instead of text-like commands, and changes are quickly represented and reversible.

direct manipulation control
Having command of an object or cursor via the use of a direct manipulation device.

direct manipulation device
Any device intended for use in controlling a cursor or other responding object on a display.

Also referred to as *cursor control device* or simply *control device*.

direct manipulation dialogue
The manipulation of symbols in display via a cursor.

direct radiation effect
Any of those cellular effects in which radiation damage is caused by ionization of the DNA molecules without an intermediate step.

direct ratio
The value of the ratio of the luminous flux actually reaching a given surface to the luminous flux emitted from a luminaire.

direct-reading instrument
An apparatus providing a direct readout of the contaminant level without further off-site laboratory analysis.

direct transit passengers
Passengers stopping temporarily at a designated airport and departing on an aircraft with the same flight number. They are counted only once.

direct user access terminal system (DUATS)
An automated pilot self-briefing and flight plan filing system. For pilots with access to a computer, modem, and touch telephone, the system provides direct access to a national weather database and the ability to file flight plans without contact with a flight service station.

direct viewing
Having an object, especially one being manipulated, within sight of the unaided eye. Synonymous with *direct vision*.

direct worker
An employee involved in direct labor.

directed verdict
In a case in which the party with the burden of proof has filed to present a *prima facie* case for jury consideration, the trial judge may order the entry of a verdict without allowing the jury to consider it because, as a matter of law, there can be only one such verdict.

direction finder (DF)
A radio receiver equipped with a directional sensing antenna used to take bearings on a radio transmitter. Specialized radio direction finders are used in aircraft as air navigation aids. Others are ground-based, primarily to obtain a "fix" on a pilot requesting orientation assistance or to locate downed aircraft. A location "fix" is established by the intersection of two or more bearing lines plotted on a navigational chart using either two separately located Direction Finders to obtain a fix on an aircraft or by a pilot plotting the bearing indications of his Direction Finder (DF) on two separately located ground-based transmitters both of which can be identified on his chart.

direction finder (DF) approach procedure
Used under emergency conditions where another instrument approach procedure cannot be executed. Direction finder (DF) guidance for an instrument approach is given by Air Traffic Control (ATC) facilities with DF capability.

direction finder (DF) fix
The geographical location of an aircraft obtained by one or more direction finders.

direction finder (DF) guidance
Headings provided to aircraft by facilities equipped with direction finding equipment. These headings, if followed, will lead the aircraft to a predetermined point such as the direction finder (DF) station or an airport. DF guidance is given to aircraft in distress or to other aircraft which request the service. Practice DF guidance is provided when workload permits.

directional lighting
That lighting exposing an object or surface primarily from a given direction.

directional microphone
A microphone whose response/sensitivity varies significantly by design with the direction of incident sound.

directional response
A usually graphical description of a transducer response as a function of the direction of the emitted/incident energy in a specified plane and/or at a specified frequency.

directional route miles
The mileage in each direction over which public transportation vehicles travel while in revenue service. It is computed with regard to direction but without regard to the number of traffic lanes or rail tracks existing in the ROW.

director's and officer's (D&O) liability insurance
Insures corporate directors and officers against claims based on negligence, failure to disclose, and to a limited extent, other defalcations. Such insurance provides coverage against expenses and to a limited extent fines, judgments, and amounts paid in settlement. *See also* ***insurance***.

dirt
Any material or substance which causes an unclean condition. *See also* ***adhesive dirt, attractive dirt,*** *and* ***inert dirt***.

dirt depreciation
A reduction in light transmission or reflection due to dirt accumulation. *See also* ***luminaire dirt depreciation*** *and* ***room surface dirt depreciation***.

disability
An impairment or defect of one or more organs or body members. In more general terms, the want of legal capability to perform an act. The term is more typically used to indicate an incapacity for the full enjoyment of ordinary legal rights. For *workers' compensation* purposes, the following categories are generally used to determine the level of benefit to be awarded: (1) *Permanent partial.* A permanent physical impairment (loss of an eye, hand, etc.) that restricts the ability of the worker to perform certain jobs. Benefits are normally based upon the percentage of disability incurred. (2) *Permanent total.* A disability that is so extensive it prevents the worker from obtaining or competing for a job. (3) *Temporary partial.* A condition that leaves the employee capable of performing some work and will probably improve to pre-injury or illness status over time and with treatment. (4) *Temporary total.* A disability that renders the worker incapable of working, but from which he or she is expected to recover fully.

disability clause
A provision in an insurance policy calling for the waiver of premiums during a period of disability.

disability compensation
Payments from public or private funds to one during a period of disability and incapacity from work, e.g., social security or workers' compensation disability benefits.

disability glare
A viewing condition in which glare interferes with visual clarity, thus reducing visual performance.

disability insurance
Insurance coverage purchased to protect the insured financially during periods of incapacity from working. Often purchased by professionals.

disable
Ordinarily, to take away the ability of, to render incapable of proper and effective action.

disabled person
A person who lacks the legal capacity to act or one who is mentally or physically disabled from acting in his/her own behalf or from pursuing any occupation.

disabling damage
Damage which precludes departure of a motor vehicle from the scene of the accident in its usual manner in daylight after simple repairs.

disabling injury
Bodily harm resulting in death, permanent disability, or any degree of temporary total disability. It is an injury which prevents a person from performing a regular established job for at least one full day beyond the day the injury occurred.

disabling injury frequency rate
The total number of days lost per million employee-hours of exposure. More commonly referred to as *lost-time injury rate.*

disaster
A subjective term used to describe a loss, or the degree of loss, resulting from a given event or occurrence. It may include the loss of life, or serious (life-threatening) injuries, or property loss, or any combination of these or other resultant losses.

disbarment
Act of a court in suspending an attorney's license to practice law. A disbarment proceeding is neither a civil nor criminal action. It is a special proceeding peculiar to itself, disciplinary in nature, and of summary character resulting from inherent power of courts over their officers.

disc brake
(1) A car friction braking system that forces a pad against a metal disc attached to the wheel or axle to produce a retarding force. (2) A brake used primarily on rail passenger cars that uses brake shoes clamped by calipers against flat steel discs.

disc screen
A screening device consisting of a circular disc fitted with wire mesh that rotates on a horizontal axis.

discharge
(1) In general, a setting free, or liberation. A material or force set free, as electric energy, or an excretion or substance evacuated. (2) According to the Federal Water Pollution Control Act: Includes, but is not limited to, any spilling, leaking, pumping, pouring, emitting, emptying, or dumping. (3) Under the Federal Oil Pollution Act of 1990: Any emission (other than natural seepage), intentional or unintentional, and includes, but is not limited to, any spilling, leaking, pumping, pouring, emitting, emptying or dumping.

discharge incidental to the normal operation of a vessel
Under the Federal Water Pollution Control Act: 1) A discharge, including graywater, bilge water, cooling water, weather deck run-off, ballast water, oil water separator effluent, and any other pollutant discharge from the operation of a marine propulsion system, shipboard maneuvering system, crew habitability system, or installed major equipment, such as an aircraft carrier elevator or a catapult, or from a protective, preservative, or absorptive application to the hull of the vessel. 2) A discharge in connection with the testing, maintenance, and repair of a system described above whenever the vessel is waterborne.

discipline
A branch of knowledge or learning, such as physics, chemistry, industrial hygiene, safety engineering, etc.

disclosure guidelines
As pertains to corporate environmental violations: Guidelines issued in 1991 by the United States Department of Justice which are designed to encourage companies to engage in self-auditing, self-policing, and voluntary disclosure of regulatory violations by employees. Complete title is *Factors in Decisions on Criminal Prosecutions for Environmental Violations in the Context of Significant Voluntary Compliance or Disclosure Efforts by the Violator*. See also ***federal sentencing guidelines for organizations***.

discometry
The study or process of measuring the pressure in the nucleus pulposus of an intervertebral disk.

discomfort
A state other than well-being due to the presence of one or more undesirable environmental stressors.

discomfort glare
A viewing condition in which glare from one or more high-intensity sources within the field of view causes an observer to experience visual pain or annoyance.

discomfort index
A method for estimating effective temperature as a heat stress measure.

discomfort threshold
That stimulus intensity at which, in a specified proportion of the trials and/or in a specified proportions of individuals, will sufficiently activate a sensory system to cause a reported change from a typical sensation for a given modality to a sensation of being uncomfortable. Sometimes referred to as *threshold of discomfort*.

disconnecting means
A device, or group of devices, or other means by which the conductors of a circuit can be disconnected from their source of supply.

disconnecting switch
As pertains to systems over 600 volts (nominal) a mechanical switching device used for isolating a circuit or equipment from a source of power.

discontinued operations income (loss)
Transit. Gain or loss from disposal of investor controlled companies or nontransport ventures. Does not include earnings or losses from discontinued transport or transport-related operations.

discontinuous timing
See ***repetitive timing***.

discover
To uncover that which was hidden, concealed, or unknown from every one. To get first sight or knowledge of; to get knowledge of what has existed but has not previously been known to the discoverer.

discovery
(1) *General.* Ascertaining that which was previously unknown; the disclosure or coming to light that which was previously hidden. (2) *Trial Practice.* The pre-trial devices that can be used by one party to obtain facts and information about the case from the other party in order to assist the party's preparation for trial. Under the Federal Rules of Civil Procedure (and in states which have adopted rules patterned on such), tools of discovery include depositions upon oral or written questions, written interrogatories, production of documents or things, permission to enter upon land or other property, physical and mental examinations, and requests for admission.

discrete
Having separate, clearly distinguishable components.

discrete code
Aviation. As used in the Air Traffic Control Radar Beacon System (ATCRBS), any one of the 4096 selectable Mode 3/A aircraft transponder codes except those ending in zero. (e.g., discrete codes: 0010, 1201, 2317, 7777; nondiscrete codes: 0100, 1200, 7700). Nondiscrete codes are normally reserved for radar facilities that are not equipped with discrete decoding capability and for other purposes such as emergencies (7700), visual flight rules (VFR) aircraft (1200), etc. *See also* ***code*** *and* ***decoder***.

discrete frequency
Aviation. A separate radio frequency for use in direct pilot-controller communications in air traffic control which reduces frequency congestion by controlling the number of aircraft operating on a particular frequency at one time. Discrete frequencies are normally designated for each control sector in en route/terminal Air Traffic Control (ATC) facilities. Discrete frequencies are listed in the Airport/Facility Directory and the Department of Defense (DOD) FLIP Instrument Flight Rules (IFR) En Route Supplement.

discrete particle settling
Phenomenon referring to sedimentation of particles in a suspension of low solids concentration.

discrete spectrum
A presentation of the amount of energy in a complex waveform at each frequency present in the waveform.

discrete variable
A variable which can assume only a specified, finite number of values.

discrete word recognition
See ***word recognition***.

discretion
Having the freedom to make decisions.

discriminate
(1) Distinguish reliability between conditions, stimuli, or divisions on a measurement scale. (2) Treat differently based on some attribute of a person.

discrimination reaction time
The temporal interval required to discriminate between two or more stimuli and decide if a response is appropriate.

disease
A deviation from normal health status associated with a characteristic sequence of signs and symptoms and caused by a specific etiologic agent.

disequilibrium
A loss of balance accompanied by swaying of the body and tremors sometimes experienced by workers exposed to whole body vibrations above 2 Hz.

disfigurement
That which impairs or injures the beauty, symmetry, or appearance of a person or thing; that which renders unsightly, misshapen, or imperfect, or deforms in some manner.

disinfect
Destroy most or all disease-causing microorganisms, except viruses.

disinfectant
(1) Chemicals used to reduce or kill microorganisms present on inanimate objects or surfaces. (2) One of three groups of antimicrobials registered by EPA for public health uses. The EPA considers an antimicrobial to be a disinfectant when it destroys or irreversibly

inactivates infectious or other undesirable organisms, but not necessarily their spores. The EPA registers three types of disinfectant products based upon submitted efficacy data: limited, general or broad spectrum, and hospital disinfectant. (3) A chemical or physical process that kills pathogenic organisms in water. Chlorine is often used to disinfect sewage treatment plant effluent, water supplies, wells, and swimming pools.

disinfection
The act or process of destroying organisms that may cause disease.

disinfection byproduct (DBP)
Byproducts that occur or are anticipated to occur, from the addition of commonly used water treatment disinfectants, including chlorine, chloramine, chlorine dioxide, and ozone.

disintegrated
Excessive degree of separation or decomposition into fragments with complete loss of the original form of the material.

disintegration
In radiation, the process of spontaneous breakdown of a nucleus of an atom resulting in the emission of a particle and/or photon.

disintegration constant
*See **decay constant**.*

disinterested
Not concerned, with respect to possible gain or loss, in the result of the pending proceedings or transactions. Not having any interest in the matter referred to or in controversy; free from prejudice or partiality; impartial or fair method; without pecuniary interest.

disinterested witness
One who has no interest in the cause or matter in issue, and who is lawfully competent to testify.

disk
(1) A round, flat magnetic or optical medium for storage of digital data. (2) *See **intervertebral disk**.*

disk memory
*See **virtual memory**.*

dislocated shoulder
An injury in which the head or the humerus has been forced out of the glenoid cavity of the scapula.

dislocation
Displacement of a bone from a joint. The most common dislocations are those involving a finger, thumb, or shoulder. Less common are those of the mandible, elbow, knee, or hip. Symptoms include loss of motion, temporary paralysis of the involved joint, pain and swelling, and sometimes shock. A dislocation is usually caused by a blow or fall, although unusual physical effort may lead to this condition. Some dislocations, especially of the hip, are congenital, usually resulting from a faulty construction of the joint.

dismantling allowance
*See **tear-down allowance**.*

dismemberment
Amputation of an extremity, usually designating separation other than through a joint.

disorientation
The loss of the normal recognition of time, place, or persons.

dispatch point
A location where arrangements may be made for the short-term or trip rental use of an Interagency Fleet Management System (IFMS) vehicle.

dispensary
A place for the dispensation of free or low-cost medical treatment. It is also the name typically given to an on-site medical and/or first aid facility located at remote worksites and usually offering no-cost treatment to employees who have suffered occupational injuries or illnesses.

dispersant
A chemical agent that is used to break up or disperse concentrations of a material, such as an oil spill in water.

dispersion
(1) The mixing and movement of contaminants in their surroundings (e.g., air) with the resultant effect of diluting the contaminant. (2) The spread of scores or other quantitative results in a given sample or frequency distribution. *See also **measure of dispersion** and **variability**.* (3) An indication of the rate of change of the refraction index on the various wavelengths of energy passing through a transparent medium; the spread of white light into its different component wavelengths.

dispersion rate
A diffusion parameter of gas plumes or stack effluents.

dispersion staining
A particle identification technique in which the material of interest is immersed in a liquid media, such as an oil of specific index of refraction, and examined microscopically (e.g., by polarized light microscopy) for identification.

displaced threshold
Aviation. A threshold located at a point on the runway other than at the beginning of the usable runway pavement. The displaced area is available for takeoff.

displacement
(1) *General.* Removal to an abnormal location or position. (2) *Psychology.* Unconscious transference of an emotion from its original object onto a more acceptable substitute. (3) *Vibration.* The change in distance or position of an object relative to a reference point. (4) *Maritime.* The weight, in tons of 2,240 pounds, of the vessel and its contents. Calculated by dividing the volume of water displaced in cubic feet by 35, the average density of sea water.

displacement joystick
See ***isotonic joystick***.

display
The presentation of data and/or graphics from a system or device in a format designed for human perception through one or more of the senses.

display-control layout
An aspect of workstation design involving both the location and grouping of an integrated layout involving both controls and displays for the human operator. Also referred to as *control-display layout.*

display density
The proportion of the total screen area which is used to present information or data.

display format
That arrangement of the data, command areas, messages, and other features on a display.

display layout
The grouping of displays at a workplace. *See also* ***display-control layout***.

disposable
An item which is intended for use only once.

disposable income
Personal income less personal tax and non-tax payments.

disposal
(1) Final placement or destruction of toxic, radioactive, or other wastes; surplus or banned pesticides or other chemicals; polluted soils; and drums containing hazardous materials from removal actions or accidental releases. Disposal may be accomplished through use of approved secure landfills, surface impoundments, land farming, deep well injection, ocean dumping, or incineration. (2) Under the Federal Solid Waste Disposal Act: The discharge, deposit, injection, dumping, spilling, leaking, or placing of any solid waste or hazardous waste into or on any land or water so that such solid waste or hazardous waste or any constituent thereof may enter the environment or be emitted into the air or discharged into any waters, including groundwaters.

disposal date
Transit. The date a vehicle is disposed of and no longer included in the inventory.

disposal package
According to the Federal Nuclear Waste Policy Act of 1982: The emplacement in a repository of high-level radioactive waste, spent nuclear fuel, or other highly radioactive material with no foreseeable intent of recovery, whether or not such emplacement permits the recovery of such waste.

disposal site
With regard to ocean dumping of wastes, an interim or finally approved and precise geographical area within which ocean dumping of wastes is permitted under conditions specified in permits issued under the Marine Protection, Research, and Sanctuaries Act of 1972. Such sites are identified by boundaries established by either coordinates of latitude and longitude for each corner, or by coordinates of latitude and longitude for the center point and a radius in nautical miles from that point. Boundary and coordinates shall be identified as precisely as is warranted by the accuracy with which the site can be located with existing navigational aids or by the implementation of

transponders, buoys, or other means of marking the site.

disposal storage
Transit. An inventory accountability category of vehicles not in use and scheduled for disposal.

disposal storage date
Transit. The date a vehicle is taken out of service and placed in disposal storage.

disposal system
With regard to radiation protection, any combination of engineered and natural barriers that isolate spent nuclear fuel or radioactive waste after disposal.

disqualification
(1) The suspension, revocation, cancellation, or any other withdrawal by a state of a person's privileges to drive a commercial motor vehicle. (2) A determination by the Federal Highway Administration (FHWA), under the rules of practice for motor carrier safety contained in 49 CFR 386, that a person is no longer qualified to operate a commercial motor vehicle under 49 CFR 391. (3) The loss of qualification which automatically follows conviction of an offense listed in 49 CFR 383.51.

dissipating stage
The final stage in the development of an air mass thunderstorm when downdrafts exist throughout the cumulonimbus cloud.

dissipative muffler
A type of acoustic muffler that is typically used for reducing noise emissions from a source, such as large engines. The muffler housing is lined with a sound-absorbing material.

dissociation
(1) *General.* Separation into parts or elements. (2) *Psychology.* A mental disorder in which ideas are split off from the personality and are buried in the unconscious. (3) *Chemistry.* The separation of a molecule into two or more constituents as a result of added energy (e.g., heat) or the effect of a solvent on a dissolved polar compound.

dissolve
To liquefy by means of a solvent.

dissolved air flotation (DAF)
The clarification of flocculated material by contact with minute bubbles of air causing the air/floc mass to be buoyed to the surface, leaving behind a clarified water.

dissolved nitrogen flotation (DNF)
The clarification of flocculated material by contact with minute bubbles of nitrogen causing the air/floc mass to be buoyed to the surface, leaving behind a clarified water.

dissolved organic carbon (DOC)
The fraction of total organic carbon that is dissolved in a water sample.

dissolved oxygen
The oxygen freely available in water. Dissolved oxygen is vital to fish and other aquatic life and for the prevention of odors. Traditionally, the level of dissolved oxygen has been accepted as the single most important indicator of a water body's ability to support desirable aquatic life. Secondary and advanced waste treatment are generally designed to protect dissolved oxygen in waste-receiving waters.

dissolved solids
Disintegrated organic and inorganic material contained in water. Excessive amounts render water unsafe to drink or use in industrial processes.

distal
Remote; a point or region which is farther from the trunk or point of attachment than some reference point.

distance
Expression of the linear measurement of space separating two specified points.

distance measuring equipment (DME)
Airborne and ground equipment used to measure, in nautical miles, the slant range distance of an aircraft from the distance measuring equipment (DME) navigational aid.

distance measuring equipment (DME) fix
A geographical position determined by reference to a navigational aid which provides distance and azimuth information. It is defined by a specific distance in nautical miles and a radial, azimuth, or course (i.e., localizer) in degrees magnetic from that aid.

distance measuring equipment (DME) separation
Spacing of aircraft in terms of distances (nautical miles) determined by reference to distance measuring equipment.

distance weight
Parameter in the Hazardous Ranking System air migration, groundwater migration, and soil exposure pathways that reduces the point value assigned to targets as their distance increases from the site.

distillate
A liquid product condensed from vapor during distillation.

distillate fuel oil
A general classification for one of the petroleum fractions produced in conventional distillation operations. It is used primarily for space heating, on and off highway diesel engine fuel (including railroad engine fuel and fuel for agricultural machinery), and electric power generation. Included are products known as No. 1, No. 2, and No. 4 fuel oils and No. 1, No. 2, and No. 4 diesel fuels. No. 1 distillate is a petroleum distillate which meets the specifications for No. 1 heating or fuel oil as defined in American Society for Testing and Materials (ASTM) D 396 and/or the specifications for No. 1 diesel fuel as defined in ASTM Specification D 975. No. 2 distillate is a petroleum distillate which meets the specifications for No. 2 heating oil or fuel oil as defined in ASTM D 396 and/or the specifications for No. 2 diesel fuel as defined in ASTM Specification D 975.

distillation
The act of purifying liquids through boiling so that the steam condenses to a pure liquid and the pollutants remain in a concentrated residue.

distilled water
Water which has been heated to its boiling point or above to form steam, then condensed with cooling, the process intending to remove minerals and other materials.

distort
Cause a (usually undesirable) change in the natural shape or form of a physical entity, image, information, or energy waveform.

distract
Divert attention from, prevent concentration on, or inhibit a timely or correct response on some task.

distractor
Any environmental feature which distracts.

distress
(1) A condition of being threatened by serious and/or imminent danger and of requiring immediate assistance. (2) The state of being in peril, to any degree, for a person and/or property.

distressed unit
A person and/or property in peril to any degree.

distributary
A branch of a river that flows away from the main stream and does not return to it.

distributed control
Having controlling mechanisms or subsystems at other than a central location.

distributed control system (DCS)
A collection of modules, each having a specific function, interconnected to carry out an integrated data acquisition and control operation.

distributed practice
A training or experimental procedure in which practice periods are separated by rest periods or periods of different activity. Synonymous with *spaced practice.*

distribution
The movement of a chemical substance or foreign material from entry site and throughout the body.

distribution coefficient (K_d)
Measure of the extent of partitioning of a substance between geologic materials (e.g., soil, sediment, rock) and water. The distribution coefficient is used in the Hazardous Ranking System to evaluate the mobility of a substance for the groundwater migration pathway. It is measure as ml/g. Also referred to as *partition coefficient.*

distribution into commerce
According to the Federal Toxic Substances Control Act (TSCA), either the introduction, holding, or selling of a chemical substance, mixture, or article into commerce.

distribution main
Generally, mains, services, and equipment that carry or control the supply of gas from the point of local supply to and including the sales meters.

distribution temperature (T_d)
That temperature of a blackbody radiator whose relative spectral power distribution is essentially the same as that of the radiation source being considered.

distributor
A company primarily engaged in the sale and delivery of natural and/or supplemental gas directly to consumers through a system of mains.

district attorney
The law enforcement lawyer, usually a prosecutor, for a region of a state such as a county. Also known as *county attorney* or a *state's attorney*.

district commander
The district commander of the Coast Guard or his authorized representative, who has jurisdiction in the particular geographical area.

disturbance input
An undesired input affecting the value of an output signal for which control is being attempted.

disuse osteoporosis
An osteoporotic condition induced by lack of use rather than a metabolic dysfunction.

disversant
A chemical agent used to break up concentrations of organic material, such as spilled oil.

ditch light
Spotlight aimed at the right side of a road.

diuresis
Secretion of urine; often used to indicate increased function of the kidney.

diuretic
(1) Causing diuresis. (2) A substance that stimulates the flow of urine. Certain common substances such as tea, coffee, and water act as diuretics.

diurnal
Occurring during a 24-hour period.

divergence
(1) *Anatomy*. An outward rotation of both eyes to focus on a point further away from the observer. (2) *Meteorology*. An atmospheric condition that exists when the winds cause a horizontal net outflow of air from a specific region.

divergent phoria
A tendency for an observer to fixate behind a stationary target.

diverse vector area
Aviation Safety. In a radar environment, that area in which a prescribed departure route is not required as the only suitable route to avoid obstacles. The area in which random radar vectors below the Minimum Vectoring Altitude/Minimum Instrument Flight Rules Altitude (MVA/MIA), established in accordance with the Terminal Instrument Procedures criteria for diverse departures obstacles and terrain avoidance, may be issued to departing aircraft.

diversion chamber
A chamber used to divert all or part of a flow to various outlets.

diversity index
A mathematical expression that depicts the diversity of a species in quantitative terms.

diverticulitis
Inflammation of the diverticula, small blind pouches that form in the lining of the colon. Weakness of the muscles of the colon, sometimes produced by chronic constipation, leads to the formation of diverticula. Inflammation may occur as a result of collections of bacteria or other irritating agents trapped in the pouches. Symptoms include muscle spasms and cramp-like pains in the abdomen, especially in the lower left quadrant. Diagnosis is confirmed by barium enema in which the diverticula are clearly shown.

divided attention
That form of attention in which an individual must perform two or more separate tasks concurrently, all of which require attention. Also referred to as *division of work*. *See also* ***attention***.

divided highway
A multi-lane facility with a curbed or positive barrier median, or a median that is 4 feet (1.2 meters) or wider.

division
Separation into parts.

division of labor
The separation of a job into smaller tasks. Also referred to as *division of work* or *divided attention*.

division of work
See ***divided attention***.

divulsor
An instrument for forcible dilation or separation of body parts.

DM respirator
Dust and mist respirator.

DME
See ***distance measuring equipment***.

DMF respirator
Dust, mist, and fume respirator.

DNA
See ***deoxyribonucleic acid***.

DNA hybridization
Use of a segment of DNA, called a DNA probe, to identify its complementary DNA; used to detect specific genes. This process takes advantage of the ability of a single strand of DNA to combine with a complementary strand.

DNA identification
DNA profiling or fingerprinting is an analysis of deoxyribonucleic acid resulting in the identification of an individual's patterned chemical structure of genetic information. A method of determining distinctive patterns in genetic material in order to identify the source of a biological specimen, such as blood, tissue, or hair.

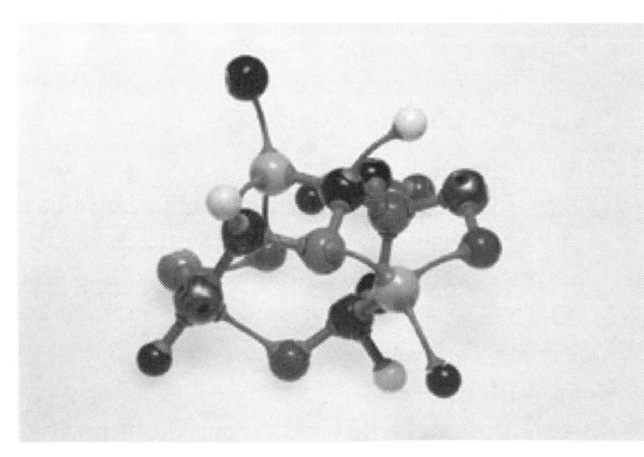

A model of a DNA segment

DNF
See ***dissolved nitrogen flotation***.

do (DO)
A physical basic work element in which a worker performs some operation which results in a change in the form, physical condition, or chemical composition of a product.

DOC
See ***dissolved organic carbon***.

dock
(1) *Transportation*. Move a vehicle adjacent to another compatible vehicle or a compatible facility and join the two. (2) *Maritime*. For ships, a cargo handling area parallel to the shoreline; for land transportation, a loading or unloading platform at an industrial location or carrier terminal.

dock receipt
(1) Written acknowledgment showing that goods have been delivered and received at a dock or warehouse or ocean liner. (2) A receipt used to transfer accountability when the export item is moved by the domestic carrier to the port of embarkation and left with the international carrier for export.

dock walloper
One who loads and unloads vehicles and handles freight on the dock.

dockage
Charge assessed against a vessel for berthing at a wharf, pier, bulkhead structure, or bank or for mooring to a vessel so berthed.

docket
A minute, abstract, or brief entry, or the book containing such entries. A formal record, entered in brief, of the proceedings in a court of justice.

doctor blade
A scraping device used to remove or regulate the amount of material on a belt, roller, or other moving or rotating surface.

document retention
See ***records retention policy***.

documented yacht
A vessel of five or more net tons owned by a citizen of the United States and used exclusively for pleasure with a valid marine document issued by the Coast Guard. Documented vessels are not numbered.

A "documented yacht" according tithe U.S. Coast Guard

documents against acceptance
Instructions given by a shipper to a bank indicating that documents transferring title to goods should be delivered to the buyer (or drawee) only upon the buyer's payment of the attached draft.

DOE
Department of Energy (United States).

dog
Transit (slang). A truck with little power.

dog chart
Railroad. With respect to rail operations, a diagrammatic representation of the mechanical locking of an interlocking machine, used as a working plan in making up, assembling and fitting the locking. *See also* ***locking dog***.

dog tracks
Transit (slang). Unit or straight truck that runs out of line.

DOL
Department of Labor (United States).

doldrums
The region near the equator that is characterized by low pressure and very light, shifting winds.

dolly
Transit (trucking). An auxiliary axle assembly having a fifth wheel used for purpose of converting a semitrailer to a full trailer.

dolomite
A natural mineral consisting of calcium carbonate and magnesium carbonate. The chemical formula is $CaMg(CO^3)^2$.

dolomite lime
Lime containing 35-40% magnesium oxide.

dolphin
An isolated cluster of piles used as a support for mooring devices or marker lights. *See also* ***mooring cell***.

domestic
(1) *General.* Produced in the United States, including the Outer Continental Shelf (OCS).
(2) *Transportation.* Traffic (passengers and freight) performed between airports located within the same country or territory.

Domestic Air Defense Identification Zone
An Air Defense Identification Zone (ADIZ) within the United States along an international boundary of the United States.

domestic air operator
Aviation. Commercial air transportation within and between the 50 United States and the District of Columbia. Includes operations of certificated route air carriers, Pan American, local service, helicopter, intra-Alaska, intra-Hawaii, all-cargo carriers, and other carriers. Also included are trans-border operations conducted on the domestic route segments of U.S. air carriers. Domestic operators are classified based on their operating revenue as follows: Major (over $1 billion); National ($100-1,000 million); Large Regional ($10-99.9 million); Medium Regional ($0-9.99 million).

domestic airspace
Airspace which overlies the continental land mass of the United States plus Hawaii and U.S. possessions. Domestic airspace extends to 12 miles offshore.

domestic fleet
All reportable agency-owned motor vehicles, operated in any State, Commonwealth, Territory or possession of the United States.

domestic freight
All waterborne commercial movements between points in the United States, Puerto Rico, and the Virgin Islands, excluding traffic with the Panama Canal Zone. Cargo moved for the military in commercial vessels is reported as ordinary commercial cargo; military cargo moved in military vessels is omitted.

domestic intercity trucking
Trucking operations within the territory of the United States, including intra-Hawaiian and

intra-Alaskan, which carry freight beyond the local areas and commercial zones.

domestic operation
In general, operations within and between the 50 States of the United States, the District of Columbia, American Samoa, Caroline Islands, Guam Island, Johnston Island, Marianna Islands, Midway Island, Puerto Rico, U.S. Virgin Islands, and Wake Island.

domestic operations
All air carrier operations having destinations within the 50 United States, the District of Columbia, Puerto Rico, and the U.S. Virgin Islands.

domestic passenger
Any person traveling on a public conveyance by water between points in the United States, Puerto Rico, and the Virgin Islands.

domestic transportation
Transportation between places within the United States other than through a foreign country.

domestic wastewater
Wastewater originating from sanitary conveniences in residential dwellings, office buildings, and institutions. Also called *sanitary wastewater*.

dominant eye
The preference for the use of one eye over the other when given a choice (may be subconscious).

dominant wavelength ($_d$)
The visual wavelength represented on a chromaticity diagram by the point of intersection with the spectrum locus of an extended straight line from a sample chromaticity through the achromatic point.

domino effect
The descriptive term used to illustrate the cause and effect relationship one event may have to another. As the term implies, one failure event may result in a sequence of additional failure events unless other forces (such as barriers) are in place to prevent or interfere with this process.

Donaldson scale
A scoring system based on a large number of variables for judging how well an individual is capable of performing the activities of daily living, of caring for himself, and of mobility.

donut area
The area outside of the Federal Highway Administration (FHWA) approved adjusted boundary of one or more urbanized areas but within the boundary of a National Ambient Air Quality Standards (NAAQS) nonattainment area.

doodle bug
A small tractor used to pull two axle dollies in a warehouse.

door
A structure commonly having a thickness much less than its length and width, and which is attached on one side by hinges for use in closing off one volume from another.

door sill step
On trucks, any step normally protected from the elements by the cab door when closed.

doorway
A short passageway surrounded by a frame, in which a door may be mounted.

Doppler effect
An observed change in pitch or frequency due to a difference in relative velocity between an energy source and a receiver. Also referred to as the *Doppler shift*.

Doppler radar
A radar that determines the velocity of falling precipitation either toward or away from the radar unit using the *Doppler shift* or *Doppler effect*.

Doppler shift
See ***Doppler effect***.

dorsal
Directed toward or situated on the back surface; opposite of ventral.

dorsal flexor
See ***dorsiflexor***.

dorsal hand skinfold
The thickness of a skinfold at the middle of the back of the hand and parallel to the long axis of the hand.

dorsiflexion
A motion involving raising the toes and upper part of the foot.

dorsiflexor
Any muscle which raises the toes and upper foot about the ankle joint.

DOS
Disk operating system. The basic programming required to operate a personal computer system.

dosage
A specific quantity of a substance applied to a unit quantity of liquid to obtain a desired effect.

dose
(1) *General.* The amount of a substance to which an organism is exposed. (2) *Radiation.* A quantity (total or accumulated) of ionizing (or nuclear) radiation. Exposure dose, expressed in roentgens, is a measure of the total amount of ionization that a quantity of radiation could produce in air. Absorbed dose, expressed in reps or rads, represents the energy absorbed from the radiation per gram of body tissue. Biological dose, expressed in Rems, is a measure of the biological effectiveness of the radiation exposure. (2) *Toxicology.* The total amount of a toxicant, drug, or other chemical administered to the organism.

dose-effect relationship
The relationship between the dose given and the occurrence and severity of the effect produced.

dose equivalent
In radiation protection, the product of absorbed dose and appropriate factors to account for differences in biological effectiveness due to the quality of radiation and its spatial distribution in the body.

dose limit
See ***maximum permissible dose***.

dose rate
(1) *Industrial Hygiene.* The dose of a hazardous agent (chemical, physical, biological) delivered or taken into the body per unit time. (2) *Radiation.* The amount of ionizing radiation to which an individual would be exposed to that he or she would receive per unit of time.

dose ratemeter
An instrument which measures ionizing radiation dose rate.

dose-response assessment
The determination of the relation between the magnitude of the exposure and the probability of the occurrence of an adverse health effect.

dose-response curve
A graphical representation of the response of an animal or individual to increasing doses of a substance.

dose-response relationship
The relationship between dose administered and resulting response. Simplistically, an increase in the dose results in an increase in the response. It is actually a complex relationship dependent upon many factors.

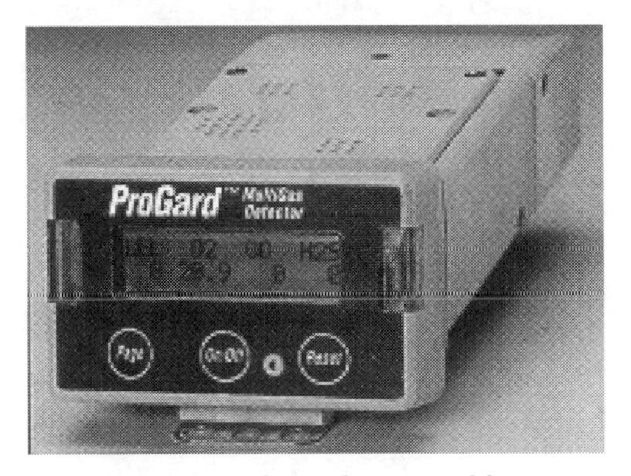

A type of dosimeter, a multi-gas detector used for monitoring combustible gases, carbon monoxide, hydrogen sulfide, and decreases in oxygen (by MSA, Pittsburgh, PA)

dosimeter
An instrument or device used to determine the amount of exposure to an agent or toxic chemical, usually over a period of time. Typically used to measure exposures to noise, radiation, and chemicals.

dosing siphon
A siphon that automatically discharges liquid onto a trickling filter bed or other wastewater treatment device.

dosing tank
A tank into which raw or partly treated wastewater is accumulated and held for subsequent discharge and treatment at a constant rate.

DOT
See ***Department of Transportation***.

dote
See ***decay*** *(4)*.

double
(1) *Transit.* A combination of two trailers pulled by a power unit. Usually refers to a

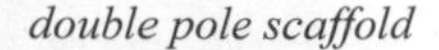

power unit pulling two 28-foot trailers. *See also* ***Rocky Mountain double*** *and* ***turnpike double***. (2) *Maritime*. The maneuver whereby a towboat with barges in tow must break the tow and push half of it into the lock chamber, lock that part through, and then enter the remaining barges with the towboat. In other words, two distinct lockages must be made to pass the entire tow of barges and towboat. Also called *double lockage*.

double-blind
An experimental condition in which neither the administrator nor the subject knows the true experimental treatment on a given trial.

double block and bleed
A method to isolate a piece of equipment, vessel, confined space, etc. from a line, duct, or pipe by locking or tagging closed two valves in series with each other in the line, duct, or pipe, and locking or tagging open to the outside atmosphere a bleed in the line between the two closed valves.

double bottom
Transit. A truck unit consisting of a tractor, semitrailer, and full trailer. Also called *twin trailers* and *doubles*.

double click
Press a button on a computer input device, such as a mouse, two times within a specified brief time period to command two operations at once, such as specify and open a file. *See also* ***click***.

double clutching
Shifting the gears of a truck transmission without clashing them.

double deck bus
A bus with two separate passenger compartments, one above the other.

double-decked buses
High-capacity buses having two levels of seating, one over the other, connected by one or more stairways. Total bus height is usually 13 to 14.5 feet, and typical passenger seating capacity ranges from 40 to 80 people.

double heading cock
Railroad. A manually operated valve by means of which the control of brake operation is transferred to the leading locomotive.

Typical double decked bus

double indemnity
Payment of twice the basic benefit in the event of a loss resulting from specified causes or under specified circumstances. A provision in life insurance contracts requiring payment of twice the face amount of the policy by the insurer in the event of death by accidental means.

double jeopardy
A Fifth Amendment guarantee, enforceable against the states through the Fourth Amendment, which protects against a second prosecution for the same offense after acquittal or conviction, and against multiple punishments for the same offense. The "evil" sought to be avoided is double trial and double conviction, not necessarily double punishment.

double lockage
See ***double***.

double pole scaffold
A scaffold supported from the base by a double row of uprights, independent of support from the walls and constructed of uprights, ledgers, horizontal platform bearers, and diagonal bracing. Sometimes referred to as *independent pole scaffold*.

double shift
The working of two shifts during a 24-hour period.

double suction pump
A centrifugal pump with suction pipes connected to the casing from both sides.

double trip
The maneuver necessary when a towboat has more barges in tow than the power of the boat can handle in certain areas of swift current or conditions at the lock will permit. A tow will tie off below the swift water a portion of his tow, push the others above the questionable area, tie them off to the bank, and go back for the remainder of the tow. This is also at times a necessary maneuver in ice.

double underline
A highlighting technique in which two horizontal lines are drawn below a line of text or used as a graphic.

double vision
See ***diplopia***.

down shape of (revetment, shore, etc.)
Running the shape of the shore, staying approximately the same distance off the shore at all times.

Down's syndrome
A congenital condition characterized by physical malformations and some degree of mental retardation. The disorder is also known as *mongolism* because the person's facial characteristics resemble those persons of the Mongolian race, and *trisomy 21* because the disorder is concerned with a defect in the twenty-first chromosome.

downburst
A weather phenomenon described as a severe localized downdraft that can be experienced beneath a severe thunderstorm.

downcomer
A pipe directed downward.

downdraft
The natural tendency for the river current to pull objects downstream.

downgrade
(1) A dilution or reduction of the skill level required for a task or job. (2) The lowering of a particular job in such aspects as responsibility, scope, degree of difficulty, or wage category.

download
To receive a file from another computer via modem or network interface card.

downtime
(1) The time during which an operation cannot proceed or a piece of equipment or instrumentation cannot be used productively due to maintenance, breakdown, lack of materials, or other causes. (2) The amount of time a vehicle or equipment is out of service for repair.

downtown people mover
A type of automated guideway transit vehicle operating on a loop or shuttle route within the central business district of a city.

DOX
Dissolved organic halogen.

dpm
Disintegrations per minute.

Draeger tube
See ***detector tube***.

draft
(1) *General*. A first version of a document of product not intended for final release or sale. (2) *Ventilation*. The movement of air in a manner which results in discomfort to persons exposed to it due to its velocity, temperature, or other cause. It also refers to the difference in pressure between the inside and outside of a structure due to a combustion process (e.g., furnace, boiler, etc.). The draft causes the products of combustion to flow from the combustion process to the outside atmosphere. A *back draft* can result if there is insufficient air to sustain the combustion process. (3) *Marine Navigation*. Cross-current tows will drift to the right or left depending on the draft (usually qualified as out draft, or left- or right-handed draft). Also, the depth of water a vessel draws, loaded or unloaded.

draft tube
A centrally located vertical tube used to promote mixing in a sludge digester or aeration basin.

drag
(1) *General*. To draw across a surface. (2) *Computers*. To move a computer input device

such as a mouse such that a screen element or cursor moves across a display; a direct manipulation operation.

drag down
In a manual transmission vehicle, the act of shifting too slowly to lower gears.

drag tank
A rectangular sedimentation basin that uses a chain and flight collector mechanism to remove dense solids.

drain
(1) *Noun.* A trench or ditch to convey water from wet land; a channel through which water may flow off. The word has no technical legal meaning. Any hollow space in the ground, natural or artificial, where water is collected and passed off is considered a ditch or drain. (2) *Verb.* To conduct water from one place to another, for the purpose of drying the former. To make dry; to draw off water; to rid the land of its superfluous moisture by adapting or improving natural water courses and supplementing them when necessary by artificial ditches. *See also* ***public drainage system***.

drain tile
Short lengths of pipes laid in underground trenches to collect and carry away excess groundwater, or to discharge wastewater into the ground.

drainage basin
The land drained by a river system.

drainage district
A political subdivision of the state, created for the purpose of draining and reclaiming wet and overflowed land, as well as to preserve the public health and convenience.

drainage rights
A landowner may not obstruct or divert the natural flow of a watercourse or natural drainage course to the injury of another. In urban areas, "natural drainage course" is narrowly interpreted to include only streams with well-defined channels and banks. In rural areas, the term is more broadly construed, apparently including the flow and direction of diffused surface waters.

drainage water
Ground-, surface-, or stormwater collected by a drainage system and discharged into a natural waterway.

Draize Test
An animal test procedure for assessing the potential irritation or corrosive effect of a material on the skin or eyes.

draw back
Repayment in whole or in part of duties or taxes paid on imported merchandise that is re-exported.

draw down
The procedure of spilling water through one dam prior to the arrival of excessive water from the upper reaches of the river. This maneuver is used when flash floods are expected or have occurred or where tributary streams are emptying excessive amounts of water into the main streams.

draw span
The movable portion of a bridge deck.

drawbridge
A bridge that pivots or lifts so as to let a boat through.

drawer
A structure which is usually open on one side and closed on all other sides and the bottom and which is designed to slide into and out of a cabinet, rack, or other housing.

drayage
Charge made for local hauling by dray or truck.

DRE
See ***destruction and removal efficiency***.

dredge
To remove sediment or sludge from rivers or estuaries to maintain navigation channels.

dredge material
Under the Federal Marine Protection, Research, and Sanctuaries Act of 1972: Any material excavated or dredged from the navigable waters of the United States.

dredged cut
(1) One pass made by a dredge in a channel within the confines of the riverbed for the purpose of maintaining the proper depth of water. (2) A dredged channel.

dredged shipping lane
Lane that has been dug out to provide an adequate depth of water for navigation.

dredging
Removal of mud from the bottom of water bodies using a scooping machine. This process disturbs the ecosystem and causes silting that can kill aquatic life. Dredging of contaminated mud can expose aquatic life to heavy metals and other toxics. Dredging activities may be subject to regulation under Section 404 of the Clean Water Act.

dredging spoil
The discharge from a dredge.

drift
(1) *General.* Includes any buoyant material that, when floating in navigable waters, may cause damage to a commercial or recreational vessel. (2) *Instrumentation.* The gradual and unintentional deviation of a given variable. It is the gradual change in readout due to component aging, variation in power supply, characteristics of the detector, temperature effect on the detection system, etc. (3) *Maritime.* The motion of a boat floating with no mechanical aid. Also, colloquially used as a synonym for currents. Also referred to as *current, set,* or *draft.*

drift barrier
An artificial barrier designed to catch driftwood or other floating material.

drift test
A part of the emissions certification process in which the continuous emissions monitoring system must operate unattended for some period of time without the analyzers drifting out of calibration.

drifting
Marine Navigation. Underway, but proceeding over the bottom without use of engines, oars or sails; being carried along only by the tide, current, or wind.

drilling mud
A fluid, often containing bentonite, used to cool and lubricate a drilling bit and to remove cuttings from the bit and carry them to the well's surface.

drinking water
Water safe for human consumption, or for the use in the preparation of food or beverages, or for cleaning articles used in the preparation of food or beverages.

drinking water cooler
Under the Federal Safe Drinking Water Act: Any mechanical device affixed to drinking water supply plumbing which actively cools water for human consumption.

drinking water equivalent level (DWEL)
The lifetime exposure level at which adverse health effects are not anticipated to occur, assuming 100% exposure from drinking water.

Drinking Water Priority List (DWPL)
A 1988 list of drinking water contaminants that may pose a health risk and warrant regulation.

drinking water supply
Under the Federal Safe Drinking Water Act: Any raw or finished water source that is or may be used as public water or as drinking water by one or more individuals.

drip irrigation
A micro-irrigation water management technique used primarily for landscaping in which drips of water are emitted near the base of the plant either continuously (in especially dry climates) or on a programmed schedule.

drip proof
Designation for a motor enclosure with ventilating openings constructed so that drops of liquids or solids falling on the motor will not enter the unit directly or by running along an inwardly inclined surface.

drive
(1) To maneuver or control a vehicle designed for essentially 2-dimensional travel, as on the ground or a relatively hard, fixed surface. (2) The motivation required to complete a process, task, or ambition. (3) *Computing.* The mechanism (peripheral device) that allows computer users to access stored or important information.

driveaway-towaway
Refers to a carrier operation, such as a fleet of tow trucks, used to transport other vehicles, when some or all wheels of the vehicles being transported touch the road surface.

driveaway-towaway operation
Any operation in which a motor vehicle constitutes the commodity being transported and one or more set of wheels of the vehicle being transported are on the surface of the roadway during transportation.

driver
(1) *Computing.* In a Windows environment, it is software that Windows loads at startup. Drivers give Windows specific instructions about your video card and printer that Windows and Windows-based applications use to display information on the screen and to print information on your printer. (2) *Transit.* A person who operates a motorized vehicle. If more than one person drives on a single trip, the person who drives the most miles is classified as the principal driver. Also, an occupant of a vehicle who is in physical control of a motor vehicle in transport or, for an out-of-control vehicle, an occupant who was in control until control was lost.

driver applicant
An individual who applies to a state to obtain, transfer, upgrade, or renew a commercial driver's license (CDL).

driver's license
A license issued by a state or other jurisdiction to an individual which authorizes the individual to operate a motor vehicle on the highways.

driving a commercial motor vehicle while under the influence of alcohol
Committing any one or more of the following acts in a commercial motor vehicle (CMV): driving a CMV while the person's alcohol concentration is 0.04 percent or more; driving under the influence of alcohol, as prescribed by state law; or refusal to undergo such testing as is required by any state or jurisdiction in the enforcement of 49 CFR 383.51b)2)(i)a) or b), or 49 CFR 392.5a)2).

driving piece
A crank secured to a locking shaft by means of which horizontal movement is imparted to a longitudinal locking bar.

driving under the influence (DUI)
The driving or operating of any vehicle or common carrier while drunk or under the influence of liquor or narcotics.

driving while intoxicated (DWI)
An offense committed by one who operates a motor vehicle while under the influence of intoxicating liquor or drugs. A showing of complete intoxication is not required. State statutes specify levels of blood alcohol content at which a person is presumed to be under the influence of intoxicating liquor. *See also* ***blood alcohol count, blood test evidence,*** *and* ***breathalyzer test.***

drizzle
Small drops between 0.2 and 0.5 mm in diameter that fall slowly and reduce visibility more than light rain.

drop delivery
The simple release of an object after being transported to some location where it is to be transported further, stored, disposed of, or processed.

drop it on the nose
(slang). Uncoupling a tractor from a semitrailer without lowering the landing gear to support the trailer's front end.

drop the body
Unhook and drive a tractor away from a parked semi.

dropping out line
A line used in dropping a barge out of a tow.

droplet
Liquid particle suspended in air, and which settles out quite rapidly.

drought
An extended period of dry weather which, as a minimum, can result in a partial crop failure or an inability to meet normal water demands.

drug
According to the Federal Food, Drug, and Cosmetic Act: 1) articles recognized in the official United States Pharmacopoeia, official Homeopathic Pharmacopoeia of the United States, or official National Formulary, or any supplement to any of them; and 2) articles intended for use in the diagnosis, cure, mitigation, treatment, or prevention of disease in man or other animals; and 3) articles (other than food) intended to affect the structure or any function of the body of man or other animals; and 4) articles intended for use as a component of any article specified in 1), 2), or 3) above.

drug abuse
Legally defined as a state of chronic or periodic intoxication detrimental to the individual and to society, produced by the repeated consumption of a drug, natural or synthetic.

drug dependence
Habituation to, abuse of, and/or addiction to a chemical substance.

drug tolerance
The progressive decrease in susceptibility of the body to a drug's effects resulting from repeated administrations or addiction.

drum
A flat-ended or convex-ended cylindrical packaging made of metal, fiberboard, plastic, plywood, or other suitable materials. This definition also includes packaging of other shapes made of metal or plastic (e.g., round taper-necked packaging or pail-shaped packaging) but does not include cylinders, jerricans, wooden barrels, or bulk packaging.

drum pulverizer
A rotating cylinder used to shred solid waste by the intermingling action of internal baffles acting on the wetted solid waste.

drum screen
A cylindrical screening device used to remove floating and suspended solids from water or wastewater.

dry adiabatic rate
The rate of change of temperature in a rising or descending unsaturated air parcel. The rate of adiabatic cooling or warming is 10°C per 1000 meters (5.5°F per 1000 feet).

dry-bulb temperature (DBT)
The temperature derived from a thermal sensor or a thermometer that is shielded from direct radiant energy. It is used for estimating comfort conditions and is also one of three ambient indices used for heat-stress analysis.

dry bulk cargo
Cargo which may be loose, granular, free-flowing or solid, such as grain, coal, and ore, and is shipped in bulk rather than in package form. Dry bulk cargo is usually handled by specialized mechanical handling equipment at specially designed dry-bulk terminals.

dry-bulk container
A container constructed to carry grain, powder and other free-flowing solids in bulk. Used in conjunction with a tilt chassis or platform.

dry cargo
Cargo that does not require temperature control.

dry chemical
An extinguishing agent composed of very small particles of chemicals such as, but not limited to, sodium bicarbonate, potassium bicarbonate, urea-based potassium bicarbonate, potassium chloride, or monoammonium phosphate supplemented by special treatment to provide resistance to packing and moisture absorption (caking) as well as to provide proper flow capabilities. Dry chemical does not include dry powders.

dry cleaning wastes
Wastewater from laundry cleaning operations that use non-aqueous chemical solvents to clean fabrics.

dry film thickness (DFT)
Thickness of a dried paint or coating, usually expressed in mils.

dry gas
Natural gas from which the entrained liquids and nonhydrocarbon gases have been removed by lease facilities and/or plant processing. This is the gas that the interstate pipeline purchased, or expects to purchase, to serve its annual requirements.

dry-gas meter
A secondary air flow calibration device, similar to a domestic gas meter, that can be used for determining the flow rate of air sampling pumps.

dry hole
An exploratory or development well found to be incapable of producing either oil or gas in sufficient quantities to justify completion as an oil or gas well.

Dry Ice™
A trademark for solid carbon dioxide. It is frequently used as a refrigerant that vaporizes without passing through a liquid state.

dry line
A boundary that separates warm, dry air from warm, moist air. It usually represents a zone of instability along which thunderstorms form.

dry location
See ***location***.

dry powder
A compound used to extinguish class D fires. *See also* ***class D fire***.

dry run
A trial run without the use of hazardous materials, or the operation of a process at less than design, in order to identify problems, verify operating characteristics/parameters, test procedures, etc.

dry weather flow (DWF)
The flow of wastewater in a sanitary sewer during dry weather. The sum of wastewater and dry weather infiltration.

dry well
(1) A dry compartment in a pumping station where pumps are located. (2) A well that produces no water.

drydock
An artificial basin fitted with gate or caisson into which a vessel may be floated and from which the water may be pumped out to expose the bottom of the vessel.

drydock gate
Gate at the entrance to a drydock.

dual capacity doctrine
If an employer acts in a capacity other than that strictly of employer in a "dual capacity State" and an employee is injured, then the employer may be sued for negligence arising out of its dual capacity role.

dual drive
Also referred to as *tandem* drive. Box axles have drive mechanisms and are connected to engine power output. There are two common types: a) pusher tandem: only the rearmost axle is driving type and the forward unit is free rolling, also called "dead axle;" b) tag axle: forward unit of tandem is driving type while rear unit is free rolling.

dual flow screen
A traveling water screen arranged in a channel so that water enters through both the ascending and descending wire mesh panels and exits through the center of the screen.

dual media filter
Granular media filter utilizing two types of filter media, usually silica sand and anthracite.

dual shift
An operating mode in which workers are working two shifts, usually with the employees divided into two teams.

duals
(slang). A pair of tires mounted together.

DUAT
See ***dial up access terminal.***

DUATS
See ***direct user access terminal system.***

duck water
Slack water. Smooth water generally found on the inside shore of a river bend, under a point, under a bar, etc.

duckweed
See ***Lemnaceae.***

duct
A conduit used for conveying air at low pressure.

duct velocity
(1) The air velocity within a duct carrying that air to some location. (2) The air velocity through a duct cross-section.

ductile fracture
A type of failure mode in structural materials generally preceded by large amounts of plastic deformation and in which the fracture surface is inclined to the direction of the applied stress.

ductless fume hood
A hood which returns filtered air to the area where it is located. This type hood is to be used only with nontoxic chemicals. Also referred to as a *ductless lab hood.*

ductless lab hood
See ***ductless fume hood.***

dudding
The process of permanently degrading an electroexplosive initiator to a state where it cannot perform its designed function.

due care
Just, proper, and sufficient care, so far as circumstances demand; the absence of negligence. That degree of care that a reasonable person can be expected to exercise to avoid harm reasonably foreseeable if such care is not taken. *See also* ***ordinary care*** *and* ***reasonable care.***

due compensation
Term used in eminent domain and refers to the value of land taken and the damages, if any, which result to the owner as a conse-

quence of the taking without considering either general benefits or injuries. *See also **just compensation**.*

due diligence
Such a measure or prudence, activity, or assiduity, as is properly to be expected from, and ordinarily exercised by, a reasonable and prudent person under the particular circumstances; not measured by any absolute standard, but depending on the relative facts of the special case.

due diligence steps
Under the Federal Sentencing Guidelines (FSGs): Good faith efforts to prevent statutory violations. The seven steps or elements of an effective compliance program under the FSGs are considered due diligence steps.

due process of law
Law in it regular course of administration through the courts of justice.

due process rights
All rights which are of such fundamental importance as to require compliance with due process standards of fairness and justice.

DUI
The crime of driving under the influence of alcohol or drugs. *See **driving while intoxicated**.*

dumb barge
Slang term for a barge that does not have its own power.

dumb terminal
A CRT and keyboard having no local processing capability other than simple input/output.

dummy
*See **mannequin**.*

dummy variable
A discrete variable in regression analysis which is not continuously distributed and has at least two distinct values.

dump
A site used to dispose of solid wastes without environmental controls.

dump body
Truck body of any type which can be tilted to discharge its load.

dumping
A disposition of material.

Dunn cell
A glass device formerly used to contain an aliquot of the dust collecting media in which the airborne dust sample was collected and which enabled counting of the dust so that a determination of its concentration could be made.

dunnage
Pieces of wood placed against the sides and bottom of the hold of a vessel, to preserve the cargo from the effect of leakage, according to its nature and quality.

duplex pump
A reciprocating pump having two side-by-side cylinders and connected to the same suction and discharge lines.

duplex stainless steel
A high-strength stainless steel containing two forms of iron, typically austenite and ferrite.

duplicate samples
Provide information about the precision of a laboratory's results by providing a check to determine if the correct sampling technique or method was used; may be a mandatory requirement of some regulatory agencies. Duplicate samples should be collected at locations where suspected contaminant levels are believed to be at their highest concentrations.

dura matter
The tough, outermost membrane which covers the surface of the brain and spinal cord.

duration of exposure
The period of time during which exposure to a hazardous substance or physical agent occurs, or how long a time one works with a substance or in the environment where the agent is used.

dust
Small airborne or settled solid particles usually formed by abrasion or arising from soil, bedding, or from surfaces such as floors and walls.

dust collector
An air-cleaning device for removing particulates from air being discharged to the environment.

dust devil
Common term used to describe a *whirlwind* which is a small but rapidly rotating wind made visible by dust, sand, and debris it picks up from the surface. It develops on clear, dry, hot afternoons.

dust explosion
A dust combustion process so confined as to result in an appreciable rise in pressure.

dust mask
A semipermeable facial mask used to protect the wearer from respirable dust particles that are airborne in the work environment. There are many types; most are disposable and should be replaced once breathing becomes difficult due to the saturation of the mask material by dust particulates.

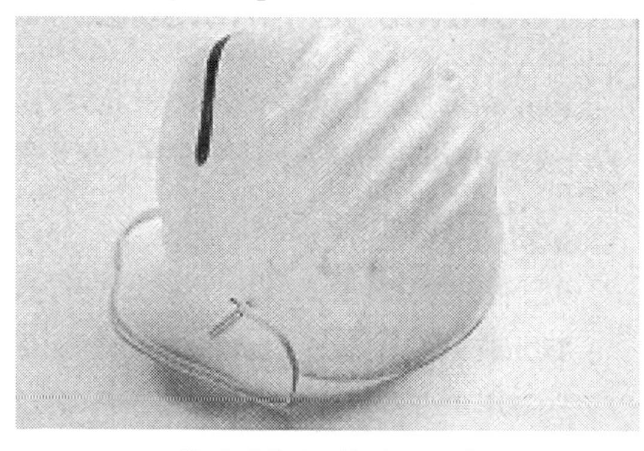

Typical disposable dust mask

dustfall jar
An open container used to collect large particles from the air for measurement and analysis.

dusting
Transit (slang). Driving with wheels on road shoulder, thereby causing a cloud of dust.

dustproof
Constructed such that dust will not interfere with its operation.

duty
(1) *Law.* A human action which is exactly conformable to the laws which require us to obey them. Legal or moral obligation. An obligation that one has by law or contract. (2) *Immigration.* A tax imposed by a government on imports

DVD
See ***Digital Versatile Disc.***

DVFR
See ***defense visual flight rules.***

Dvorak keyboard
See ***QWERTY keyboard.***

DWEL
See ***drinking water equivalent level.***

dwell time
(1) That length of time for which the eye is fixated on a given point or within a specified region. (2) That period of time which an aircraft of other vehicle is capable of staying at or over its destination/target before having to return.

DWF
See ***dry weather flow.***

DWI
See ***driving while intoxicated.***

DWPL
See ***Drinking Water Priority List.***

dwt
Deadweight tons.

dynamic
Involving motion or progress; not static.

dynamic action
Any muscle contraction or elongation. *See also* ***isotonic action, isoinertial action, isokinetic action, eccentric action,*** *and* ***concentric action.***

dynamic anthropometry
Study and/or measurement of the changes in body dimensions during motion. Also called *functional anthropometry.*

dynamic display
Any display containing one or more screen structures which are updated at or near real time.

dynamic equilibrium
The ability to maintain and control body position while in motion through the integrated involvement of the cristae in the semicircular ducts, vision, and the cerebellum and muscle activity. *See also* ***static equilibrium.***

dynamic flexibility
The ability to perform extension flexibility rapidly and repetitively.

dynamic loss
See ***turbulence loss.***

dynamic measurement
An aspect of anthropometry involving the correct location of controls, tools, and other items requiring worker manipulation.

dynamic modulus
The ratio of stress to strain under vibro-acoustic conditions.

dynamic moment
See ***angular acceleration***.

dynamic muscle work
See ***dynamic work***.

dynamic response index model (DRI)
A model representing the human torso as a single-degree-of-freedom system for predicting probability of spinal injury for a given $+g_z$ acceleration time history, assuming a restrained seated crew member in an ejection seat.

dynamic routing
In demand-response transportation systems, the process of constantly modifying vehicle routes to accommodate service requests received after the vehicle began operations, as distinguished from predetermined routes assigned to a vehicle.

dynamic strength
A measure of the ability to apply force through a range of motion.

dynamic vision
The ability to interpret moving visual stimuli.

dynamic visual acuity
A measure of the ability to resolve detail in a changing or moving stimulus. *See also* ***visual acuity***.

dynamic work
The work performed when one or more muscle lengths change, producing external motion. Also referred to as ***dynamic muscle work***.

dynamics
The study of the body in motion, whether due to internal generation or external forces.

dynamite the brakes
Transit (slang). An emergency stop using every brake on the unit.

dynamograph
See ***oscillograph, kymograph,*** *and* ***polygraph***.

dynamometer
A device for measuring external force or torque, especially that generated by human muscular contraction. *See also* ***Asmussen dynamometer*** *and* ***Cybex dynamometer***.

dyne
The unit of force that, when acting upon a mass of 1 gram, will produce an acceleration of 1 centimeter per second per second.

dysarthria
Imperfect articulation of speech due to disturbances of muscular control which result from damage to the central nervous system or peripheral nervous system.

dysbaric osteonecrosis
A form of decompression sickness resulting in bone lesions, especially near joints. Probably due to air embolism. Also known as *aseptic bone necrosis*.

dysbarism
Chemical effects resulting from exposure to an atmospheric pressure different from that of the total gas pressure within the body.

dyscrasia
A morbid condition caused by poisons in the blood.

dysentery
A lower intestinal infection caused by bacteria, protozoa, or virus, and associated with diarrhea and cramps. Dysentery is less prevalent today than in years past because of improved sanitary facilities throughout the world; it was formerly a common occurrence is crowded parts of the world and it particularly plagued army camps. In dysentery, there is an unusually fluid discharge of stool from the bowels, as well as fever, stomach cramps, and spasms of involuntary straining to evacuate, with passage of little feces. The stool is often mixed with pus and mucus and may be streaked with blood.

dysfunction
Any impaired function of some body part or of the body as a whole.

dyskinesia
Any of a variety of abnormal involuntary movements, generally due to some pathology in the extrapyramidal system. *See also* ***tremor, athetosis, chorea, ballism,*** *and* ***movement disorder***.

dyslalia
Any speech impairment due to some defect in the speech-generating structures, especially the tongue.

dyslexia
A difficulty in reading, or an inability to learn how to read.

dysmenorrhea
Painful menstruation, characterized by cramp-like pains in the lower abdomen, and sometimes accompanied by headache, irritability, mental depression, malaise, and fatigue.

dysmetria
An inability to perform accurate control of range of voluntary movement, especially of the hand.

dyspepsia
Impairment of the function of digestion.

dysphagia
Difficulty in swallowing.

dysphasia
Impairment of ability to understand and use the symbols of language, both spoken and written.

dysplastic
A body type which cannot be readily classified as any of Kretschmers' standard athletic, asthenic, or pyknic somatotypes; misshapen. *See also **Kretschmer somatotype**.*

dyspnea
Shortness of breath, or the sensation of it, due to labored breathing.

dystonia
A movement disorder involving lack of normal muscle tone.

dystrophic lakes
Shallow bodies of water with a high humus and/or organic matter content, that contain many plants but few fish and are highly acidic.

dystrophy
Any disorder caused by defective nutrition.

dysuria
Painful or difficult urination.

E

ear

A structure within and external to the side of the head consisting of three major aspects (external ear, middle ear, and inner ear), which is used for hearing and equilibrium. The outer ear consists of the auricle, or pinna, and the external acoustic meatus. The auricle collects sound waves and directs them to the external acoustic meatus which conducts them to the tympanum (the cavity of the middle ear). The tympanic membrane (eardrum) separates the outer ear from the middle ear. In the middle ear are the three ossicles: the malleus (referred to as the "hammer" because of its shape), the incus (or "anvil") and the stapes (or "stirrup"). These three small bones form a chain across the middle ear from the tympanum to the oval window in the membrane separating the middle ear from the inner ear. The middle ear is connected to the nasopharynx by the Eustachian tube, through which the air pressure on the inner side of the eardrum is equalized with the air pressure on its outside surface. The middle ear is also connected with the cells in the mastoid bone just behind the outer ear. Two muscles attached to the ossicles contract when loud noises strike the tympanic membrane, limiting its vibration and thus protecting it and the inner ear from damage. In the inner ear (or labyrinth) is the cochlea, containing the nerves that transmit sound to the brain. The inner ear also contains the semicircular canals, which are essential to the sense of balance. When sound strikes the ear, it causes the tympanic membrane to vibrate. The ossicles function as levers, amplifying the motion of the tympanic membrane, and passing the vibrations on to the cochlea. From there, the vestibulocochlear (eighth cranial) nerve transmits the vibrations, translated into nerve impulses, to the auditory center in the brain. *See also* ***external ear, middle ear,*** *and* ***inner ear***.

ear breadth

The horizontal linear distance from the most anterior point to the most posterior point of the external ear. Measured with the head level and the scalp and facial muscles relaxed.

ear clearing

The process of equalizing pressure between the middle ear and the external environment. Commonly accomplished by holding one's mouth closed, pinching the nostrils closed, and gently blowing through the nose until the pressure is equalized.

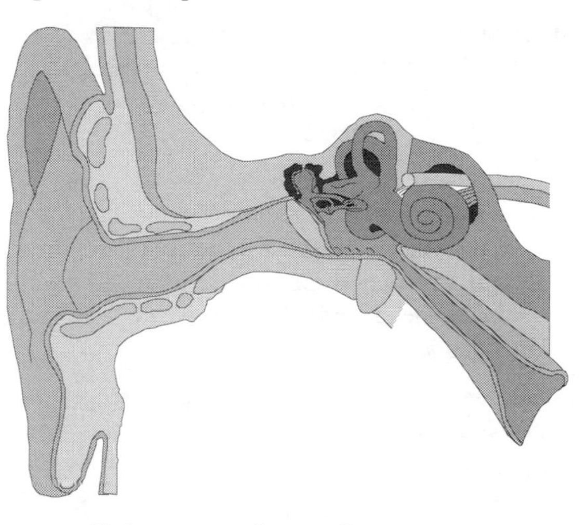

The human ear and surrounding components

ear defender

Outdated term for devices, such as earplugs, earmuffs, canal caps, etc., that are used by individuals to provide personal hearing protection from noise.

ear insert

A hearing protective device that is designed to be inserted into the ear canal in order to reduce the level of noise reaching the hearing sensitive part of the ear.

ear length

The vertical distance between the highest point of the upper rim and the most inferior point of the ear lobe of the external ear. Measured with the head level and the scalp and facial muscles relaxed.

ear length above tragion

The vertical distance along the axis of the auricle from tragion to the level of the upper rim.

ear mark

A mark put upon a thing to distinguish it from another. Originally, and literally, a mark upon the ear; a mode of marking sheep and other animals.

ear mark rule
Through the process of commingling money or deposits with funds already in a bank, the money or deposits lose their "identity," with the resultant effect of defeating the right of preference over general creditors.

ear protector
Any device designed to reduce the level of noise passing through a person's auditory system (ear muffs, ear plugs, etc.).

ear protrusion
The horizontal distance from the bony eminence directly behind the auricle to the most lateral protrusion of the auricle. Measured with the head level and the scalp and facial muscles relaxed.

ear squeeze
See ***barotalgia***.

ear witness
In the law of evidence, one who attests or can attest anything as heard by himself/herself.

earache
Pain in the ear. Medically referred to as *otalgia*.

earblock
The failure of the middle ear to equalize pressure with the external environment due to blockage of the Eustachian tube.

earcon
The auditory counterpart of the visual icon.

earcup
The cavity on the lateral interior structure of a helmet, headphone, or other headgear, into which the pinna is expected to fit when the headgear is worn.

eardrum
The tympanic membrane that separates the outer ear from the middle ear.

earflap
Any piece of cloth, fur, or other soft material designed into headwear for protecting the auricle from cold, sun, or other environmental stressors.

earlier maturity rule
The rule under which bonds maturing first are entitled to priority when the sale of a security is not sufficient to satisfy all obligations.

earlobe
The fleshy tissue at the base of the auricle.

earmuffs
Devices worn to protect against hearing loss in high-noise environments or to protect against exposure to cold. *See also* ***circumaural protectors***.

earn
To acquire by labor, service, or performance. To merit or deserve, as for labor or service.

earned income
Income from services (e.g., salaries, wages, or fees); distinguished from passive, portfolio, and other unearned income.

earned income credit
A refundable tax credit on earned income up to a certain amount for low income workers who maintain a household for dependent children. The amount of the credit is reduced dollar for dollar if earned income (or adjusted gross income) is greater than a specified amount.

earned premium
In insurance, that portion of the premium properly allocable to a policy which has expired. An "earned premium" is the difference between the premium paid by the insured and the portion returnable to him/her by the insurance company on cancellation of the policy during its term.

earned surplus
Retained earnings. That species of surplus which has been generated from profits as contrasted with paid-in surplus. The term relates to the net accumulation of profits. It is a part of the surplus that represents net earnings, gains or profits, after deduction of all losses, but has not been distributed as dividends, or transferred to stated capital or capital surplus, or applied to other purposes permitted by law.

earned time
The standard time, in a specified time unit (usually hours), which is credited to one or a group of personnel on completion of one or more jobs.

earnest money
A sum of money paid by a buyer at the time of entering a contract to indicate the intention and ability of the buyer to carry out the con-

tract. Normally, such earnest money is applied against the purchase price. Often, the contract provides for forfeiture of this sum if the buyer defaults.

earning capacity
Refers to the capability of a worker to sell his/her labor or services in any market reasonably accessible to him/her, taking into consideration his/her general physical functional impairment resulting from his/her accident, any previous disability, his/her occupation, age at the time of injury, nature of the injury, and his/her wages prior to and after the injury. The term does not necessarily mean the actual earnings that one who suffers an injury was making at the time the injuries were sustained, but refers to that which, by virtue of the training, the experience, and the business acumen possessed, an individual is capable of earning.

earnings
Income. That which is earned (i.e., money earned from the performance of labor, services, sale of goods, etc.). *Gross earnings*. Total income from all sources without considering deductions, personal exemptions, or other reductions of income in order to arrive at taxable income. *Net earnings*. The excess of gross income over expenses incurred in connection with the production of such income.

earnings and profits
A tax concept peculiar to corporate taxpayers which measures the economic capacity to make a distribution to shareholders that is not a return of capital. Such a distribution will result in dividend income to the shareholders to the extent of the corporation's current and accumulated earnings and profits.

earnings per share
One common measure of the value of common stock. The figure is computed by dividing the net earnings for the year (after interest and prior dividends) by the number of shares of common stock outstanding.

earnings profile
An individual's anticipated future annual income from employment. Use often during litigation proceedings of wrongful death claims to determine the lifetime earnings the deceased would have been expected to accumulate had the fatality never occurred.

earphone
An electro-acoustic transducer intended to be closely coupled acoustically to the ear.

earplugs
Any device which fits into the external auditory canal for the purpose of reducing the acoustic intensity reaching the eardrum. Usually constructed of a soft, sponge-like material allowing for "one-size-fits-all." *See also **aural insert protectors** and **ear protector**.*

earring
A piece of jewelry worn on or about the earlobe. It can be fastened by a mechanical clip or pierced directly through the lobe or other portion of the external ear.

earth surface
The outermost surface of the land and waters of the planet.

earthquake
*See **plate tectonics**.*

earwax
Cerumen.

EAS
*See **Employee Aptitude Survey**.*

ease
To provide or obtain comfort, consolation, contentment, enjoyment, happiness, pleasure, satisfaction.

easement
A right of use over the property of another. Traditionally the permitted kinds of uses were limited, the most important being rights of way and rights concerning flowing waters. The easement was normally for the benefit of adjoining lands, no matter who the owner was (an *easement appurtenant*), rather than for the benefit of a specific individual (*easement in gross*). The land having the dominant tenement and the land which is subject to the easement is known as the *servient tenement*.

easterly wave
A migratory wavelike disturbance in the tropical easterlies. Easterly waves occasionally intensify into tropical cyclones.

Eastern California Shear Zone (ECSZ)
A region of increased seismic activity which stretches from the San Andreas fault near In-

dio, north-northwest across the Mojave and northward into Owens Valley. It may accommodate as much as 10 to 20 percent of the relative motion between the North American and Pacific Plates.

EAT
Earnings after taxes.

eat inde sine die
Law (Latin). Words used on the acquittal of a defendant, or when a prisoner is to be discharged, *that he may go thence without a day,* (i.e., be dismissed without any further continuance or adjournment).

eaves-drip
The drip or dropping of water from the eaves of a house on the land of an adjacent owner; the easement of having the water so drip, or the servitude of submitting to such drip.

eavesdropping
Knowingly and without lawful authority: a) entering into a private place with intent to listen surreptitiously to private conversations or to observe the personal conduct of any other person or persons therein; or b) installing or using outside a private place any device for hearing, recording, amplifying, or broadcasting sounds originating in such place, which sounds would not ordinarily be audible or comprehensible outside, without the consent of the person or persons entitled to privacy therein; or c) installing or using any device or equipment for the interception of any telephone, telegraph, or other wire communication without the consent of the person in possession or control of the facilities for such wire communication. Such activities are regulated by state and federal statutes, and commonly require a court order.

ebonation
The removal of loose pieces of bone from a wound.

ebriety
Drunkenness; alcoholic intoxication.

Ebstein's anomaly
A malformation of the tricuspid valve, usually associated with an atrial septal defect.

EC_{50}
Concentration in which a given effect (e.g., death, incoordination) is observed in 50 percent of exposed organisms. The effective concentration for 50 percent of exposed organisms is usually reported along with the duration of exposure (e.g., 96 hours EC_{50}).

eccentric action
A dynamic muscle action which involves muscle lengthening with an increase in muscle tension. Also referred to as *eccentric contraction* and *eccentric muscle contraction.*

eccentric contraction
See ***eccentric action***.

eccentric muscle contraction
See ***eccentric action***.

eccentricity
Personal or individual peculiarities of mind and disposition which markedly distinguish the subject from the ordinary, normal, or average types of men/women, but do not amount to mental unsoundness or insanity.

ecchymosis
A small hemorrhagic spot in the skin or mucous membrane forming a non-elevated blue or purplish spot.

eccrine gland
A sweat gland whose ducts terminate on the free skin surface. *See also* ***apocrine gland***.

ECD
Electron capture detector. Used in gas chromatography primarily to analyze halogenated organics.

ECG
See ***electrocardiogram***.

echo
(1) To display on a computer screen the character or other symbol typed on a keyboard. (2) An acoustic or electromagnetic reflected energy signal which has sufficient magnitude and delay to be distinguishable from the original emitted signal.

echoacousia
The subjective hearing of repetition of sound after the stimuli producing it have ceased.

echocardiogram
The record produced by echocardiography.

echocardiography
The recording of the position and motion of the heart borders and valves by reflected echoes of ultrasonic waves transmitted through the chest wall.

echoencephalogram
The record produced by echoencephalography.

echoencephalography
The mapping of intracranial structures by means of reflected echoes of ultrasound transmitted through the skull.

echography
See ***sonography***.

echoic memory
A sensory memory associated with the auditory system.

echovirus
A group of viruses, the name of which was derived from the first letters of the description "enteric cytopathogenic human orphan." At the time of the isolation of the viruses, the diseases they caused were not known, hence the term "orphan." But it is now known that these viruses produce many different types of diseases, including forms of meningitis, diarrhea, and various respiratory diseases.

ecological impact
The effect that a manmade or natural activity has on living organisms and their non-living (abiotic) environment.

ecological stress vector
See ***environmental stressor***.

ecology
The relationship of living things to one another and their environment, or the study of such relationships.

economic discrimination
Any form of discrimination within the field of commerce such as a boycott of a particular product or price fixing.

economic duress
A legal defense of "economic duress," or business compulsion, arises where one individual, acting upon another's fear of impending financial injury, unlawfully coerces the latter to perform an act in circumstances which prevent his/her exercise of free will.

economic impact analysis
A corporate analysis which assesses direct and indirect costs of a rule or policy and examines how it will affect the local, regional, and national economies and what economic sectors will bear the burden of costs. It estimates the magnitude and distribution of the financial burden but does not assess whether or not the rule or policy is nonetheless worthwhile.

economic life
That period of time which either minimizes an asset's total equivalent annual cost or maximizes an asset's equivalent annual net income. Also referred to as *minimum cost life* and *optimum replacement interval*.

economic loss
In a products' liability action, recovery of damages for "economic loss" includes recovery for costs of repair and replacement of defective property which is the subject of the transaction, as well as commercial loss for inadequate value and consequent loss of profits or use.

economic obsolescence
Loss of desirability and useful life of property due to economic developments (e.g., deterioration of neighborhood or zoning change) rather than deterioration (functional obsolescence).

economic poisons
Chemicals used to control pests and to defoliate cash crops such as cotton.

economic strike
Refusal to work because of a dispute over wages, hours or working conditions, or other conditions of employment. An economic strike is one neither prohibited by law nor by collective bargaining agreement nor caused by employer unfair labor practices, but is typically for the purpose of enforcing employer compliance with union collective bargaining demands, and economic strikers possess more limited reinstatement rights than unfair labor practice strikers.

economic waste
An overproduction or excessive drilling of oil or gas.

economies of scale
Cost reductions or productivity efficiencies achieved through size-optimization in relation to operational circumstances. For example, commodity freight rates usually decline as the volume of cargo tonnage shipped increases.

economy
(1). *General*. Frugal management of money, materials, resources, and the like. Also, the

practical administration of the material resources of a country, community, or establishment. (2) *Transportation.* Transport service established for the carriage of passengers at fares and quality of service below that of coach service.

economy of scale factor
The ratio of the change in investment cost to the change in capacity.

ecosphere
The "bio-bubble" that contains life on earth, in surface waters, and in the air.

ecosystem
The interacting system of a biological community and its non-living environmental surroundings.

ECSZ
See ***Eastern California Shear Zone***.

ectocanthic breadth
The horizontal linear distance from the ectocanthus of the right eye to the ectocanthus of the left eye. Measured with the individual sitting or standing erect, and the facial musculature relaxed. May also be referred to as *biocular breadth* or *bicanthic diameter.*

ectocanthus
The junction of the most lateral parts of the upper and lower eyelids, with the eyelids open normally. May also be referred to as the *external canthus* or *lateral canthus*.

ectocanthus to back of head
The horizontal linear distance from ectocanthus to the back of the head. Measured with the individual standing or sitting erect and looking straight ahead, and the facial musculature relaxed. Equivalent to *ectocanthus to wall.*

ectocanthus to otobasion
The horizontal linear distance from ectocanthus to otobasion superior. Measured with the individual sitting or standing erect, with the facial musculature relaxed.

ectocanthus to top of head
The vertical linear distance from ectocanthus to the vertex level of the head. Measured with the individual standing or sitting erect, with the facial musculature relaxed.

ectocanthus to wall
The horizontal distance from ectocanthus to a reference wall. Measured with the individual standing erect with his/her back and head against the wall, looking straight ahead, and the facial musculature relaxed. Equivalent to *ectocanthus to back of head.*

ectoderm
The outermost of the three primitive germ layers of the embryo; from it are derived the epidermis and epidermic tissues, such as the nails, hair, and glands of the skin, the nervous system, external sense organs (eye, ear, etc.), and mucous membrane of the mouth and anus.

ectomorph
A Sheldon somatotype having characteristics of a thin, frail-appearing body build with little fat or muscle, small bones, and thin chest.

ecuresis
Production of absolute dehydration of the body by excessive urinary excretion in relation to the intake of water.

ECW
See ***extracellular water***.

eczema
Generalized term for an inflammatory process involving the epidermis and marked by itching, weeping, and crusting.

ED
See ***effective dose***.

ED_{10}
Ten percent effective dose. Estimated dose associated with a 10% increase in response over control groups. For Hazard Ranking System purposes, the response considered is cancer. It is measured as milligrams of toxicant per kilogram body weight per day (mg/kg-day).

ED_{50}
Dose in which a given effect (e.g., death, incoordination) is observed in 50 percent of exposed organisms. The effective dose for 50 percent of the exposed organisms is usually reported along with the duration of exposure (e.g., 80 hours ED_{50}).

EDB
See ***ethylene dibromide***.

EDCT
See ***expected departure clearance time***.

EDD
See ***enforcement decision document***.

eddy
(1) A small volume of air (or any fluid) that behaves differently from the larger flow in which it exists. (2) A current running contrary to the main current, causing water turbulence, e.g., below the bridge pier where a swift current is passing through, or below a bar or point.

eddy above and below
Maritime. Channel report term meaning that eddies should be expected both above and below the object mentioned in the marks, such as dikes, top and bottom of crossings, sunken obstructions, etc.; tricky water.

eddy extends way out
Maritime. Term meaning that an eddy extends from the shore or the dike into or across the range formed by this set of marks, or extends one-third or more across the river.

eddy makes out from right (left) shore
Maritime. Term indicating that one should watch for eddy along the shore designated in this set of marks.

eddy viscosity
The internal friction produced by turbulent flow. *See also* ***molecular viscosity***.

edema
A condition in which body tissues contain an excessive amount of fluid. Edema can be caused by a variety of factors, including hypoproteinemia in which a lowered concentration of plasma proteins decreases the osmotic pressure, thereby permitting passage of abnormal amounts of fluid out of the blood vessels and into the tissue spaces. Some other causes are poor lymphatic drainage, increased capillary permeability (as in inflammation), and congestive heart failure. Local edema due to inflammation or poor drainage through the lymph vessels may be relieved by elevation of the part and application of cold to the area. Generalized edema is treated by the administration of diuretics, which increase the loss of certain salts and thereby increase removal of tissue fluids, which are eliminated as urine.

EDF
Environmental Defense Fund.

edge lease
One located on the edge of an oil-bearing structure.

edit
To manually change the data or information in a file, document, or other form of textual or graphic material.

eductor
See ***ejector***.

EEC
European Economic Community.

EEG
See ***electroencephalograph*** and ***electroencephalogram***.

EEL
See ***emergency exposure limit***.

EEO
See ***Equal Employment Opportunity***.

effect
That which is produced by an agent or cause; result; outcome; consequence.

effective assistance of counsel
The conscientious, meaningful representation wherein the accused is advised of his/her rights and honest, learned, and able counsel is given a reasonable opportunity to perform the task assigned to him/her.

effective compliance program
Under the Federal Sentencing Guidelines (FSGs), an organization may take advantage of mitigating factors if it has an effective program to prevent and detect violations of law. An effective program includes a minimum of seven due diligence steps which the organization must have in place to receive reduced fines at the time of an offense. The organization must generally:

1. have established compliance standards and procedures,
2. have a specific individual within high level personnel of the organization to oversee compliance with such standards and procedures,
3. use due care not to delegate substantial discretionary authority to individuals who had a propensity to engage in illegal activities,

4. take steps to communicate the above to all employees and agents,
5. take reasonable steps to achieve compliance with its standards,
6. consistently enforce standards through disciplinary mechanisms, and
7. respond appropriately to the offense and prevent further similar offenses.

The size of the organization, the likelihood that certain offenses may occur because of the nature of its business, and the prior history of the organization are relevant factors to be considered. *See also* ***Federal Sentencing Guidelines***.

effective dose (ED)
The amount of a toxicant (or drug) required to bring about a given functional change in an intact organism, at a biochemical site, or in an isolated tissue. Expressed in a proportion to the population affected (ED_{50}, for example).

effective intensity
That intensity of a light in candela as defined by the Illuminating Engineering Society's Guide for Calculating the Effective Intensity of Flashing Signal Lights, November, 1964.

effective locking device
Railroad. A manually operated switch or derail which is a) vandal resistant; b) tamper resistant; and c) capable of being locked and unlocked only by the class, craft, or group of employees for whom the protection is being provided.

effective stack height
The sum of the actual stack height and the rise of the plume after emission from the stack.

effective sound pressure
The root mean square value of the pressure exerted at a given location by an acoustical waveform over a complete cycle. Also referred to as *root mean square sound pressure,* or, simply, *sound pressure*.

effective temperature
The combination of the dry-bulb and wet-bulb temperature of slowly moving air which produces immediate sensations of warmth and coolness. The combinations of dry-bulb and wet-bulb temperature and air movement are located on an effective temperature chart from which the effective temperature can be read.

effective temperature index
An arbitrary index which combines into a single value the effect of temperature, humidity, and air movement on the sensation of warmth or cold felt by the human body. A sensory index, developed by ASHRAE, of the degree of warmth that a person, stripped to the waist and engaged in light activity, would experience upon exposure to different combinations of air temperature, humidity, and air movement. This index is applicable to work situations where light activity is performed over a several-hour period. A revised effective temperature chart has been developed for sedentary type work situations, as well as one where radiant heat is a concern.

effective thermal insulation value of clothing
See ***total thermal insulation value of clothing***.

effectively grounded
As pertains to systems over 600 volts (nominal), permanently connected to earth through a ground connection of sufficiently low impedance and having sufficient ampacity that ground fault current which may occur cannot build up to voltages dangerous to personnel.

effectiveness
The ability to produce a specific result or to exert a specific measurable influence.

effectus sequitur causam
Law (Latin). The effect follows the cause.

efferent
Conveying information away from a central point, pertaining especially to neural signals.

efferent nerve
A collection of one or more axons which conducts signals primarily from the central nervous system to the periphery.

efficacy
The capacity or ability to produce the desired effect.

efficiency
The effectiveness of some process, usually measured with respect to the amount of output compared to energy, cost, or other measure input.

efficient cause
The working cause; that cause which produces effects or results. An intervening

cause, which produces results which would not have come to pass except for its interposition, and for which, therefore, the person who set in motion the original chain of causes is not responsible. That cause of an injury to which legal liability is attached.

efficient intervening cause
An intervening efficient cause is a new and independent force, which breaks the causal connection between the original wrong and the injury, and is the proximate and immediate cause of the injury. This means that the original negligent actor is not liable for an injury that could not have been foreseen or reasonably anticipated as the probable consequence of his/her negligent act, and would not have resulted from it had not the intervening efficient cause interrupted the natural sequence of events, turned aside their own course, and produced the injury.

efflorescence
A rash or eruption. Any skin lesion.

effluent
Wastewater, treated or untreated, that flows out of a treatment plant, sewer, or industrial outfall. Generally refers to wastes discharged into surface waters.

effluent limitation
Restrictions established by a state or the EPA on quantities, rates, and concentrations in wastewater discharges.

effort
(1) That point of force application on a lever. (2) The expenditure of physical and/or mental energy in the performance of some task.

effort arm
That portion of a lever arm from the fulcrum to the point at which an effort is applied. Also referred to as *force arm*.

effort-controlled cycle
See ***self-paced work***.

effort rating
See ***performance rating***.

effort time
That part of the cycle time during which an employee is required to use his/her skill and effort.

egestion
The elimination from the body of waste products and residue of ingested nutrients.

EGG
See ***electrogoniogram*** *and* ***electrogoniography***.

ego
In psychoanalytic theory, one of the three major parts of the personality, the others being the id and the superego.

egregious policy
OSHA's fining strategy implemented in 1990 which allowed the agency to fine employers for multiple violations of the same standard as if each were a separate and distinct violation. This allowed the assessment of huge fines against employers found to be in violation of the same requirement in several different instances (or at several different company locations) during an OSHA inspection.

egress
To exit from a region or space. The path or opening by which a person goes out; exit. The means or act of going out.

EHF
See ***extremely high frequency***.

EH&S
Environmental Health and Safety.

EHS
Extremely hazardous substance.

EIA
See ***environmental impact assessment***. *See also* ***Energy Information Administration***.

Eiband tolerance curve
A graph developed from both human and animal data illustrating the likelihood and severity of injuries based on uniform acceleration of short duration (an older concept).

eidoptometry
A measurement of the acuteness of visual perception.

eight hour laws
Statutes which establish eight hours as the length of a day's work, prohibited work beyond this period, and required payment of overtime for work in excess of this period.

Eighteenth Amendment
The amendment to the U.S. Constitution added in 1919 which prohibited the manufacture, sale, transportation, and exportation of intoxicating liquors in all the States and Territories of the United States and which was repealed in 1933 by the Twenty-first Amendment.

Eighth Amendment
The amendment to the U.S. Constitution added in 1791 which prohibits excessive bail, excessive fines, and cruel and unusual punishment.

einsteinium
A chemical element, atomic number 99, atomic weight 254, symbol Es.

EIS
See ***environmental impact statement****.*

ejection
Refers to occupants being totally or partially thrown from a vehicle as a result of an impact or rollover.

ejection seat
A seat structure which uses rockets or explosive devices to propel a crew member from a high performance aircraft in a life-threatening, emergency situation.

ejector
An air-moving device employing compressed air to create a vacuum as it is passed through a venturi or straight pipe, which then induces air to flow. Often used when contaminant air could corrode a fan if it were passed through it. Ejectors are not very efficient air-moving devices but do have application in special situations. Sometimes referred to as *eductors*.

Ekman spiral
An idealized description of the way the wind-driven ocean currents vary with depth. In the atmosphere, it represents the way the winds vary from the surface up through the friction layer.

El Niño
A condition that generally develops about every eight years or so just before Christmas off the coast of Peru when the ocean water turns warm as upwelling diminishes. El Niño means "little boy" in Spanish; when capitalized, it refers to the Christ child. This innocent-sounding name originated in the 19th century when Peruvian sailors noticed that every few years around Christmas, coastal waters warmed up and the current shifted southward. El Niño occurs when weather patterns in the tropical Pacific shift violently. Normally, strong, westward-blowing trade winds of South America push surface water toward Asia. These trade winds almost literally "pile" warm water against the coastlines of Australia, Indonesia, and the Philippines. Above the warm water, moist air rises, lowering atmospheric pressure and triggering the tropical downpours that nourish the rain forests of Asia. Meanwhile, high-altitude winds travel back toward South America. There, the now cooled air sinks, raising atmospheric pressure and suppressing rain along most of the Pacific coast, making it one of the driest regions in the world. When El Niño strikes, this pattern reverses. Atmospheric pressure in the western Pacific rises, setting the stage for drought from Australia to India. The trade winds decrease, or in extreme years, reverse to blow eastward. With no wind to push it toward Asia, some of the huge mass of warm water flows back toward South America, spawning storms from Chile to California. Meanwhile over the Pacific, towering ten-mile high thunderheads further heat the atmosphere, fueling a stronger-than-normal jet stream, which often splits into two. One branch veers north, warming the Pacific Northwest, central Canada, and Alaska. Another branch surges south, producing heavy rains in the U.S. gulf states and southwest.

elapsed time
The temporal interval from the beginning point of some activity to a specified or current point of that activity.

elastic
Susceptible of being stretched, compressed, or distorted, and then tending to assume its original shape.

elastic limit
The level of physical deformation beyond which damage to a structure occurs and/or the structure will not return to its original condition.

elasticity
The property of a material to return to its original shape after being distorted by the application of an external force.

elastomer
A rubber or rubber-like materials, for example, a synthetic polymer with rubber-like characteristics.

elation
Emotional excitement marked by the acceleration of mental and bodily activity.

elbow
(1) The joint between the upper arm and the forearm. It joins the large bone of the upper arm, or humerus, with the two smaller bones of the lower arm, the radius and ulna. The elbow is one of the body's most versatile joints, with a combined hinge and rotating action allowing the arm to bend and the hand to make a half turn. The flexibility of the elbow and shoulder joints together permits a nearly infinite variety of hand movements. The action of the elbow is controlled primarily by the biceps and the triceps muscles. When the biceps contracts, the arm bends at the elbow. When the triceps contracts, the arm straightens. In each action, the opposite muscle exerts a degree of opposing tension, moderating the movement so that it is smooth and even instead of sudden and jerky. The *funny bone* is not a bone but the ulnar nerve, a vulnerable and sensitive nerve that lies close to the surface near the point of the elbow. Hitting causes a tingling pain or sensation that may be felt all the way to the fingers. (2) That joint in a robotic arm capable of planar motion and corresponding by analogy to the human elbow in function.

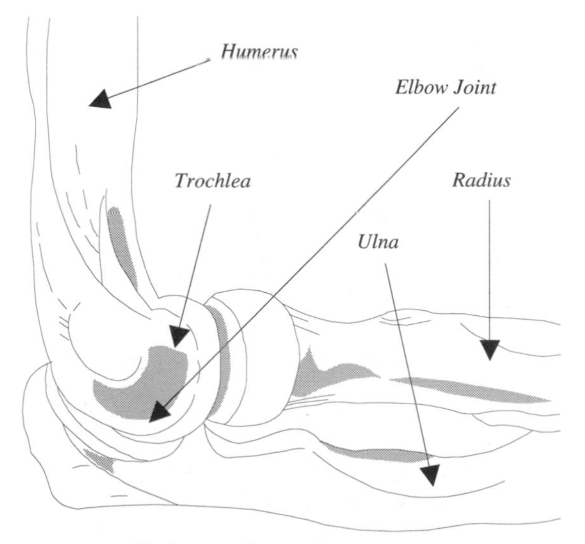

The human elbow and its components

elbow breadth
The horizontal linear distance between the medial and lateral epicondyles of the humerus. Also referred to as *humeral breadth*. Measured with the flesh compressed, the individual standing erect, and the arms hanging naturally at the sides in the anatomical position.

elbow circumference, flexed
The surface distance around the flexed elbow over the olecranon prominence and through the elbow crease. Measured with the elbow flexed 90°, the shoulder flexed 90° laterally such that the upper arm is horizontal, and the hand clenched into a fist.

elbow circumference, fully bent
The surface distance around the olecranon prominence and the crease of the elbow. Measured with the elbow maximally flexed and the fingers extended touching the shoulder.

elbow - elbow breadth
The horizontal distance across the body from the lateral surface of the left elbow to the lateral surface of the right elbow. Also called *elbow-to-elbow breadth*. Measured with the individual sitting erect, the elbows flexed 90°, and resting lightly against the body.

elbow - fingertip breadth
See ***forearm - hand length***.

elbow - grip length
The horizontal distance from the posterior tip of the elbow to the center of the clenched fist. Measured with the elbow flexed 90°.

elbow height
The vertical distance from the floor or other reference surface to the height of radiale. Also called *radiale height*. Measured with the individual standing erect and the arms hanging naturally at the sides.

elbow rest height, sitting
The vertical distance from the sitting surface to the bottom tip of the elbow. Also called elbow rest height. Measured with the individual sitting erect, the upper arm resting vertically at his/her side, and the elbow flexed 90°.

elbow - wrist length
The horizontal linear distance from the posterior tip of the elbow flexed 90° to the tip of

the styloid process of the radius. Measured with the individual sitting or standing erect, the upper arm vertical, and the palm facing medially.

electoral college
The college or body of electors of a State chosen to elect the president and vice-president; also, the whole body of such electors, composed of the electoral colleges of the several states.

electric arc
The visible effect of an undesired electrical discharge between two electrical connections; produces burned spots or fused metal.

electric discharge lamp
A source of radiant electromagnetic energy within or near the visible spectrum resulting from the passage of electrical current through one or more materials in the gaseous state.

electric lock
Rail Operations. A device to prevent or restrict the movement of a lever, a switch or a movable bridge, unless the locking member is withdrawn by an electrical device such as an electromagnet, solenoid, or motor.

clcctric locking
Rail Operations. The combination of one or more electric locks and controlling circuits by means of which levers of an interlocking machine, or switches or other units operated in connection with signaling and interlocking, are secured against operation under certain conditions.

electric shock
Effect caused by electric current passing through the body. The longer the contact with electricity, the smaller the chance of survival. The victim's breathing may stop, and his/her body may appear stiff.

electric sign
A fixed, stationary, or portable self-contained, electrically illuminated utilization equipment with words or symbols designed to convey information or attract attention.

electric system
Physically connected generation, transmission, and distribution facilities operated as an integrated unit under one central management or operating supervision.

electric utility steam generating unit
Under the Clean Air Act: (1) Any fossil fuel-fired combustion unit of more than 25 megawatts that serves a generator that produces electricity for sale. (2) A unit that cogenerates steam and electricity and supplies more than one-third of its potential electric output capacity and more than 25 megawatts electrical output to any utility power distribution system for sale shall be considered an electric utility steam-generating unit.

electrical component
A component such as a switch, fuse, resistor, wire, capacitor, or diode in an electrical system.

electrical current
In all Systems of Units, the basic unit of electrical current is the *ampere,* which has been defined to be that constant flow of electricity which, if maintained in two straight parallel conductors of infinite length, each having negligible circular cross-section, and placed 1.0 meter apart in a vacuum, would produce-between these conductors and normal to the direction in which these conductors are positioned, a repulsive force equal to 2×10^{-7} newtons per meter of conductor length.

electrical ground
An electrical reference point or return path for current flow. Also referred to simply as *ground.*

electrical hygrometer
See ***hygrometer***.

electrical impedance (Z)
The total opposition to an alternating current in an electrical circuit. Also called *impedance.*

electrical muscle stimulation (EMS)
The stimulation of muscles or muscle tissue with electrical current/voltage.

electrical resistance (R)
A measure of the opposition to electric current flow. Also called *resistance. See also* ***electrical impedance***.

electrical resistance thermometer
Thermometer that uses electrical conducting wires (or *thermistors*) whose electrical resistance changes with the temperature. It is used in *radiosondes.*

electrical shock
The passage of electrical current/voltage through the body, resulting in the abnormal stimulation of muscles and nerves.

electrical skin resistance (ESR)
See ***skin resistance response****.*

electrical stimulation
Any form of artificial activation of nerves, muscles, or other materials by the application of electrical current/voltage.

electricity
A form of energy generated by friction, induction, or chemical change that is caused by the presence and motion of elementary charged particles of which matter consists. *See also* ***gigawatt, kilowatt,*** *and* ***megawatt****.*

electro-pneumatic switch
A switch operated by an electro-pneumatic switch-and-lock movement.

electro-pneumatic valve
A valve electrically operated which, when operated, will permit or prevent passage of air.

electro-silence
The absence of measurable electrical potentials in biological tissues.

electroaffinity
The tenacity with which the ions of an element hold their charges.

electroanalysis
Chemical analysis by means of electric current.

electrobiology
The science of the relationship between electricity and living organisms.

electrocardiogram (ECG)
A graphical record or other visual display of the electrical activity of the heart as recorded from various points on the body surface, usually consisting of a P wave, a QRS wave complex, and a T wave, depending on the recording locations. Often referred to as *EKG*.

electrocardiograph
The instrumentation used to obtain a graphical recording of heart electrical activity.

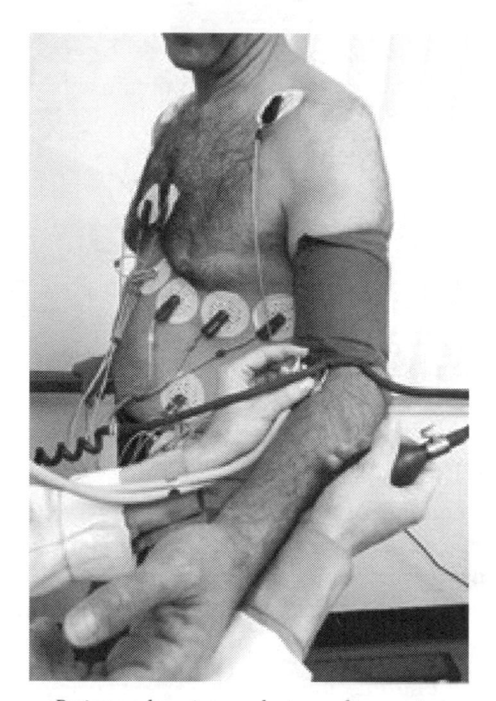

Patient undergoing an electrocardiogram test

electrocardiography
The study, measurement, recording, analysis, and/or interpretation of the electrical activity of the heart.

electrochemical detector
A detector that operates on the principle of electrochemical oxidation or reduction of a specific chemical in an electrolyte or galvanic cell. The electrons produced in the chemical reaction are proportional to the contaminant concentration.

electrochemistry
The science that deals with the use of electrical energy to bring about a chemical reaction or with the generation of electrical energy by means of chemical action.

electrode
Any electrically conductive device used for sensing or applying electrical current/voltage.

electrodialysis
A process that uses electrical current applied to permeable membranes to remove minerals from water. Often used to desalinize salty or brackish water.

electroencephalogram (EEG)
A graphical recording or other visual display of the electrical potentials generated by the brain and measured by electrodes attached to the scalp or implanted within the brain itself.

electroencephalograph (EEG)
The instrumentation used to obtain a graphical recording or the graphical recording itself of brain electrical activity.

electroencephalography
The study, measurement, recording, analysis, and/or interpretation of electrical activity from the brain.

electrogoniogram (EGG)
The electronic display or hardcopy record of changes in a joint angle using a potentiometer-equipped or other type of electrical goniometer.

electrogoniography (EGG)
The measurement, study, or analysis of changes in joint angles using potentiometer-equipped or other type of electrical goniometers.

electrogoniometer
An electromechanical goniometer, normally using changes in electrical resistance across a potentiometer to indicate the joint angle.

electroluminescence
The emission of light due to the application of an electromagnetic field to certain materials, and which is not due to heating effects alone.

electrolyte
A chemical substance that breaks down into electrically charged particles (ions) when dissolved or melted.

electromagnetic field (EMF)
Any combination of an electric field and a magnetic field which occur as a result of natural or artificially generated electromagnetic radiation.

electromagnetic interference (EMI)
A disturbance of some system due to the presence of electromagnetic fields.

electromagnetic radiation
A traveling wave motion resulting from changing electric or magnetic fields. The length of these waves can be relatively short (x-rays and gamma rays) or relatively long (ultra-violet, visible, and infrared through to radar and radiowaves). All electromagnetic radiation travels with the speed of light in a vacuum. Generally speaking, the shorter the wavelength, the more penetrating the radiation.

electromagnetic spectrum
The range of frequencies and wavelengths emitted by atomic systems. The spectrum includes radiowaves as well as the short cosmic rays.

electromagnetic susceptibility
Degraded performance of an instrument caused by an electromagnetic field.

electromagnetic waves
*See **radiant energy**.*

electromyogram (EMG)
A graphical recording or other visual display of the electrical potentials generated by a muscle, muscle group, or a large segment of muscle tissue and measured by electrodes placed in or over the tissues involved.

electromyographic kinesiology
The use of electromyography in the analysis of human motion. Also referred to as *correlative kinesiology*.

electromyography (EMG)
The study, measurement, recording, analysis, and/or interpretation of the electrical activity of muscles. Also referred to as *myography*.

electron
A negatively charged particle that is a fundamental constituent of all atoms. A unit of negatively charged electricity found in orbit around the nucleus of the atom. It has a negative electric charge of 1.60210 E-19 coulombs, and can exist as a constituent of an atom or in the free state (e.g., a beta particle).

electron capture
As pertains to ionizing radiation, a mode of radioactive decay in which an orbital electron merges with a proton in the nucleus. The process is followed by emission of an electron or photon.

electron capture detector
A type of detector employed in gas chromatography.

electron microscopy
An analytical method which utilizes a beam of electrons for the analysis of materials. This methodology is used for the identification of asbestos and other materials.

electron volt
A unit of energy equivalent to that gained by an electron in passing through a potential dif-

ference of 1 volt. Often expressed in large units such as keV (thousand electron volts), MeV (million electron volts), BeV (billion electron volts.

electronystagmogram (ENG)
A graphical recording or other visual display of the electrooculogram during nystagmus. *See also **electrooculogram**.*

electrooculogram (EOG)
A graphical display or recording of eye movements as detected by surface electrodes positioned on the skin around the eye socket, which is due to the relative orientations between the eyeball (corneo-retinal potential) and the electrodes.

electrooculography
The study, measurement, recording, analysis, and/or interpretation of the electrical activity associated with eye movements.

electrophoresis
The movement of charged particles suspended in a liquid on various media (e.g., paper, starch, agar), under the influence of an applied electric field. The method is used to analyze the plasma protein content in order to diagnose certain diseases.

electrophysiological kinesiology
The use of electrophysiological techniques in biomechanical and kinesiological research and training.

electrophysiology
The study of any form of electrical activity of the body, either associated with natural processes or due to external stimulation.

electroretinography (ERG)
The study, measurement, recording, analysis, and/or interpretation of the electrical potentials from the retina.

electrostatic discharge
A spontaneous or enticed release of static electricity.

electrostatic precipitator (ESP)
An air pollution control device that removes particles from a gas stream (smoke) after combustion occurs. The ESP imparts an electrical charge to the particles causing them to adhere to metal plates inside the precipitator. Rapping on the plates causes these particles to fall into a hopper for disposal.

electrotherapy
The use of various aspects of non-ionizing electromagnetic radiation or conduction in an attempt to heal, reduce pain, or create other beneficial effects.

element
(1) A pure substance that cannot be broken down into a simpler substance by chemical change but whose atoms will disintegrate in simpler particles through physical decomposition when exposed to drastic bombardment with high-energy particles. (2) A basic division of work, whether for man or machine, consisting of one or more basic, describable, and quantifiable motions or processes.

element breakdown
A descriptive listing of work elements, with or without certain parameters for each.

element time
That period of time required or allowed to perform a specified work element or other portion of a process or task.

elemental motion
*See **therblig**.*

elements
The forces of nature. Violent or severe weather. The ultimate undecomposable parts which unite to form anything. Popularly: fire, air, earth, and water.

elements of crime
Those constituent parts of a crime which must be proved by the prosecution to sustain a conviction.

elephantiasis
Massive subcutaneous edema, with accompanying thickening of the skin, the result of lymphatic obstruction. The disease derives its name from the symptoms, particularly swelling of the legs which makes them look like those of an elephant. The condition is usually caused by a slender, threadlike parasite, the filarial worm which enters the lymphatic system, causing an obstruction to drainage. The disease is transmitted by mosquitoes or flies which carry blood infected with filaria larva. The first visible signs are inflammation of the lymph nodes, with temporary swelling in the affected area, red streaks along the leg or arm, pain, and tenderness.

elevated on fill
Rail Operations. Rail transit way above the surface level fill. Transition segments above surface level on fill are included.

elevated on structure
Rail Operations. Rail transit way above surface level on structure. Transition segments above surface level on structures are included.

elevated rail subway
Includes elevated and subway trains in a city.

elevated temperature material
Transit. A material which, when offered for transportation or transported in a bulk packaging is a) in a liquid phase and at a temperature at or above 100°C (212°F); b) in a liquid phase with a flash point at or above 37.8°C (100°F) that is intentionally heated and offered for transportation or transported at or above its flash point; or c) in a solid phase and at a temperature at or above 240°C (464°F).

eleven contiguous western states
According to the Federal Land Policy and Management Act of 1976: Arizona, California, Colorado, Idaho, Montana, Nevada, New Mexico, Oregon, Utah, Washington, and Wyoming.

Eleventh Amendment
The Amendment to the U.S. Constitution, added in 1798, which provides that the judicial power of the U.S. shall not extend to any suit in law or equity, commenced or prosecuted against one of the United States by citizens of another State, or by citizens of any foreign state.

ELF
Extremely low frequency range of rf radiation (3 to 3,000 Hz). *See* ***extremely low frequency***.

ELF EM field
Extremely low frequency electromagnetic field.

eligible costs
The construction costs for wastewater treatment works upon which EPA grants are based.

elimination
(1) The removal of a chemical substance from the body by metabolism or excretion. Also, the removal of health or physical hazard risk through control, substitution, or some other means. (2) Defecation or urination. (3) The reduction in the use or importance of an impaired process as proficiency in an alternate process is developed.

Elkins Act
Federal Act (1903) which strengthened the Interstate Commerce Act by prohibiting rebates and other forms of preferential treatment to large shippers.

ELP
See ***Environmental Leadership Program***.

elutriation
Purification of a substance by dissolving it in a solvent and pouring off the solution, thus separating it from the undissolved foreign material.

elutriator
An air-sampling device that uses gravitational force to remove non-respirable dust from the air sample. It separates particles according to mass and aerodynamic size by maintaining laminar flow through it, thereby permitting particles of greater mass to settle out rapidly with the smaller particles depositing at greater distances from the entry point of the elutriator.

eluviation
The movement of soil caused by excessive water in the soil.

ELT
See ***emergency locator transmitter***.

emaciation
A wasted, lean appearance due to extreme weight loss.

embankment
A raised structure of earth, ground, etc.

embedded measure
A hidden process, operation, or test which an individual completes as a subset of a regular job or task, and which is intended to provide another individual or group with information about that person's performance.

embezzlement
(1) The fraudulent appropriation of property by one lawfully entrusted with its possession. (2) To "embezzle" means willfully to take, or convert to one's own use, another's money or

property, of which the wrongdoer acquired possession lawfully, by reason of some office or employment or position or trust.

embolism
A blockage of a blood vessel by some substance.

embolus
A mass of undissolved material, usually part or all of a thrombus, carried in the blood stream and frequently causing obstruction of a vessel (i.e., an embolism).

embracery
The crime of attempting to influence a jury corruptly to one side or the other, by promises, persuasions, entreaties, entertainment, and the like. The person guilty of this offense if called an "embraceor." This is both a state and federal crime, and is commonly included under the offense of "obstructing justice."

embryo
(1) *Anatomy.* An organism in an early stage of development. (2) *Meteorology.* In cloud physics, a tiny ice crystal that grows in size and becomes an ice nucleus.

embryotoxicity
The toxic effect of a substance on the embryo.

embryotoxin
A material that is harmful to the developing embryo. Substances that act during pregnancy to cause adverse effects on the fetus.

emergency
(1) *General.* A deviation from normal operation, a structural failure, or severe environmental conditions that probably would cause harm to people or property. (2) *Department of Transportation.* Any hurricane, tornado, storm (e.g., thunderstorm, snowstorm, ice storm, blizzard, sandstorm, etc.), high water, wind-driven water, tidal wave, tsunami, earthquake, volcanic eruption, mud slide, drought, forest fire, explosion, blackout or other occurrence, natural or manmade which interrupts the delivery of essential services (such as, electricity, medical care, sewer, water, telecommunications, and telecommunication transmissions) or essential supplies (such as, food and fuel) or otherwise immediately threatens human life or public welfare, provided such hurricane, tornado, or other event results in: a) a declaration of an emergency by the President of the United States, the Governor of a State, or their authorized representatives having authority to declare emergencies; by the Regional Director of Motor Carriers for the region in which the occurrence happens; or by other Federal, State or local government officials having authority to declare emergencies, or b) a request by a police officer for tow trucks to move wrecked or disabled vehicles. (3) *Chemical.* A situation created by an accidental release or spill of hazardous chemicals which poses a threat to the safety of workers, residents, the environment, or property. (4) *Confined Spaces.* Any occurrence (including any failure of hazard control or monitoring equipment) or event internal or external to the permit space that could endanger entrants. (5) *Law.* A sudden unexpected happening; an unforeseen occurrence or condition; perplexing contingency or complication of circumstances; a sudden unexpected occasion for action; exigency; pressing necessity.

emergency action plan
A plan for a workplace, or parts thereof, describing what procedures the employer and employees must take to ensure employee safety from fire or other emergencies.

emergency brake
A mechanism designed to stop a motor vehicle after a failure of the service brake system.

emergency brake system
A mechanism designed to stop a vehicle after a single failure occurs in the service brake system of a part designed to contain compressed air or brake fluid or vacuum (except failure of a common valve, manifold brake fluid housing, or brake chamber housing).

emergency button
A type of emergency stop consisting of a pushbutton installed on or near a piece of equipment which is capable of quickly shutting off electricity to that equipment.

emergency contingency vehicles
Revenue vehicles placed in an inactive contingency fleet for energy or other local emergencies after the revenue vehicles have reached the end of their normal minimum useful life. The vehicles must be properly stored and maintained, and the Emergency Contingency Plan must be approved by FTA. Sub-

stantial changes to the plan (10% change in fleet) require re-approval by FTA.

Emergency Court of Appeals
Court created during World War II to review orders of the Price Control Administrator. It was abolished in 1953 but reestablished in 1970 under Section 211 of the Economic Stabilization Act to handle primarily wage and price control matters.

emergency doctrine
Under the doctrine variously referred to as the "emergency," "imminent peril," or "sudden peril" doctrine, when one is confronted with a sudden peril requiring instinctive action, he/she is not, in determining a course of action, held to the exercise of the same degree of care as when he/she had time for reflection, and in the event that a driver of a motor vehicle suddenly meets with an emergency which naturally would overpower the judgement of a reasonably prudent and careful driver, so that momentarily he/she is thereby rendered incapable of deliberate and intelligent action, and as a result injures a third person, he/she is not negligent, provided he/she has used due care to avoid meeting such an emergency and, after it arises, exercises such care as a reasonably prudent and capable driver would use under the unusual circumstances. In an emergency situation when medical service is required for an adult who by virtue of his/her physical condition is incapable of giving consent, or with respect to a child, whose parent or other guardian is absent, and thus incapable of giving consent, the law implies the consent required to administer emergency medical services. This is a good defense to an action of tort for an alleged battery.

emergency episode
*See **air pollution episode**.*

emergency escape route
The route that employees are directed to follow in the event they are required to evacuate the workplace or seek a designated refuge area.

emergency exposure limit (EEL)
The concentration of an air contaminant to which, it is believed, an individual can be exposed in an emergency without experiencing permanent adverse health effects but not necessarily without experiencing temporary discomfort or other evidence of irritation or intoxication.

emergency lighting
A system for providing adequate illumination automatically in the event of interruption of the normal lighting system. The emergency lighting should provide, throughout a means of egress, not less than one foot-candle of illumination for a period of one and one-half hours.

emergency locator transmitter (ELT)
A radio transmitter attached to the aircraft structure which operates from its own power source on 121.5 mHz and 243.0 mHz. It aids in locating downed aircraft by radiating a downward sweeping audio tone, 2-4 times per second. It is designed to function without human action after an accident.

emergency mover
A skeletal muscle which may be used to assist a prime mover when a very high force level is required.

emergency opening window
Rail. That segment of a side-facing glazing location which has been designed to permit rapid and easy removal during a crisis situation.

emergency procedure
An action plan to be implemented in the event of an emergency. It typically describes, as a minimum, roles and responsibilities, types of emergency situations to be expected, emergency notification and/or communication procedures, public relations procedures during an emergency, and any other contingency plans applicable to the facility and its processes.

emergency relief
Transit. An operation in which a motor carrier or driver of a commercial motor vehicle is providing direct assistance to supplement state and local efforts and capabilities to save lives or property or to protect public health and safety as a result of an emergency.

emergency respirator use
The use of a respirator when a hazardous atmosphere develops suddenly and requires its immediate use for escape or for responding to the emergency in locations, areas, or operations where the hazardous situation may exist or arise.

Emergency Response Planning Guides (ERPG)
Concentration ranges, developed by the American Industrial Hygiene Association (AIHA) committee, above which adverse health effects could reasonably be expected to occur if exposures exceed the time limit established for the guides. Different effects are identified for exposure periods of one hour in ERPG-1, ERPG-2, and ERPG-3.

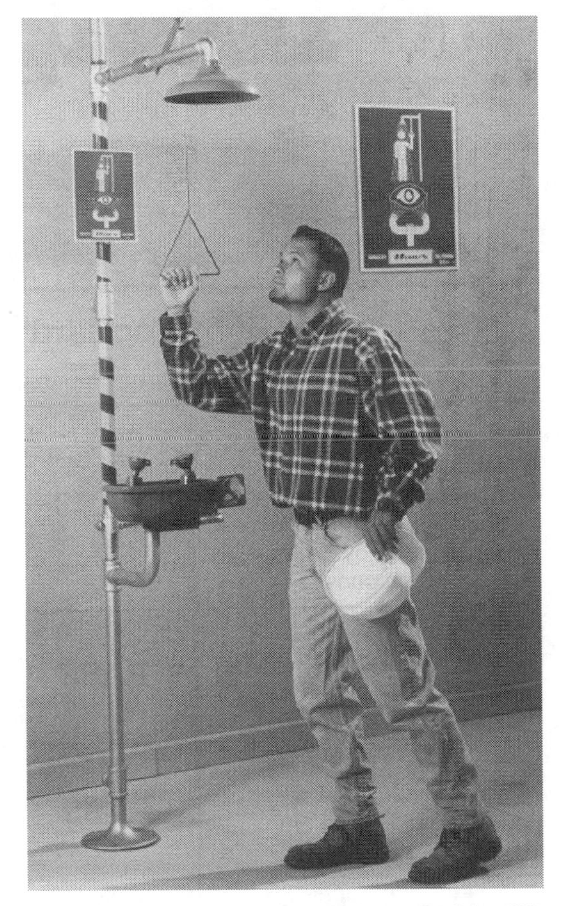

Emergency Shower (Drench Shower, Haws Corp., Berkeley, CA)

emergency shower
A water shower designed and located for use if an employee or other individual contacts a material that must be removed promptly in order to prevent an adverse health effect. Typically, it is recommended that such showers be capable of providing a continuous flow of deluge water for a period of not less than 15 minutes.

emergency stop
(1) A pushbutton, switch, or other control device installed in or on a piece of equipment which is capable of quickly cutting power to that equipment in an emergency. (2) A rapid cessation of the forward motion of a vehicle to avoid undesirable consequences.

emergency switch
A type of emergency stop consisting of a switch located in some readily accessible position for quickly shutting down a system in an emergency.

emergency temporary standard (ETS)
*See **Section 6 (c) standard**.*

emetic
An agent that induces or causes vomiting.

EMF
(1) Electromotive force. (2) Electromagnetic force. (3) *See **electromagnetic field**.* (4) electric and magnetic field.

EMG
*See **electromyogram** and **electromyography**.*

EMI
*See **electromagnetic interference**.*

eminent domain
Government taking or forced acquisition of private land for public use, with compensation paid to the landowner. The power to take private property for public use by the state, municipalities, and private persons or corporations authorized to exercise functions of public character. In the United States, the power of eminent domain is founded in both federal (Fifth Amendment) and state constitutions. The Constitution limits the power to taking for a public purpose and prohibits the exercise of the power of eminent domain without just compensation to the owners of the property which is taken. The process of exercising the power of eminent domain is commonly referred to as "condemnation," or "expropriation."

emission
Pollution discharged into the atmosphere from smokestacks, other vents, and surface areas of commercial or industrial facilities; from residential chimneys, and from motor vehicle, locomotive, or aircraft exhausts.

emission factor
The relationship between the amount of pollution produced and the amount of raw material processed. For example, an emission factor for a blast furnace making iron would

be the number of pounds of particulates per ton of raw materials.

emission inventory

A listing, by source, of the amount of air pollutants discharged into the atmosphere of a community. It is used to establish emission standards.

emission standard

(1) Standards for the levels of pollutants emitted from automobiles and trucks. Congress established the first standards in the Clean Air Act of 1963. Currently, standards are set for four vehicle classes: automobiles, light trucks, heavy duty gasoline trucks, and heavy-duty diesel trucks. (2) The maximum amount of air polluting discharge legally allowed from a single source, mobile or stationary.

emissions trading

EPA policy that allows a plant complex with several facilities to decrease pollution from some facilities while increasing it from others, so long as total results are equal to or better than previous limits. Facilities where this is done are treated as if they exist in a bubble in which total emissions are averaged out. Complexes that reduce emissions substantially may "bank" their "credits" or sell them to other industries. Also referred to as *bubble policy*.

emissivity

The ratio of the radiation intensity from a surface to the radiation intensity of the same wavelength from a black body at the same temperature. The emissivity of a perfect black body is 1.

emmetrope

One who has normal refractive vision.

emmetropia

A condition of normal optical vision in which parallel light rays are brought to an accurate focus on the retina without the need for accommodation.

emotion

A feeling or state of mental excitement that is usually accompanied by physical changes in the body.

emotional insanity

The species of mental aberration produced by a violent excitement of the emotions or passions, though the reasoning faculties may remain unimpaired. A passion, effecting for a space of time, complete derangement of a person's intellect, or an impulse, which his/her mind is not able to resist, to do an act.

empathy

Intellectual understanding of something in another person which is foreign to oneself.

emphasizing facts

A jury instruction is said to emphasize facts which may contain sufficient facts to authorize a verdict, but nevertheless some fact or facts are selected from the evidence and mentioned in such a way as to indicate to the jury that they have especial importance when that is not justified.

emphysema

Overdistention of the alveolar sacs of the lungs. A condition of the lungs in which there is dilation of the air sacs, resulting in labored breathing and increased susceptibility to infection.

empiric

A practitioner in medicine or surgery, who proceeds on experience only, without science or legal qualification; a quack.

empirical

Derived from practical experience or relying on observations or experimental results as opposed to theory.

empirical distribution

A distribution of sampled events or data.

empirical probability

When many possible outcomes can result, including a desired outcome, the probability of occurrence of such outcomes is referred to as empirical and requires statistical evaluation to determine the likelihood of expected results based upon past performance.

empirical workplace design

The evolutionary design of the working environment based on a combination of human factors engineering and experience.

emplead

To indict; to prefer a charge against; to accuse.

employ

To engage in one's service; to hire; to use as an agent or substitute in transacting business;

to commission and entrust with the performance of certain acts or functions or with the management of one's affairs; and, when used with respect to a servant or hired laborer, the term is equivalent to hiring, which implies a request and a contract for compensation.

employed

(1) Performing work under an employer-employee relationship. The term signifies both the act of doing a thing and the being under contract or orders to do it. (2) To give employment to or to have employment.

employee

(1) *General.* The person taking the direction from the employer. An individual who has an agreement to work for an employer and is compensated by that employer for his/her time and/or effort. (2) *Transit.* a) A driver of a commercial motor vehicle (including an independent contractor while in the course of operating a commercial motor vehicle); b) a mechanic; c) a freight handler; d) any individual who is employed by an employer and who in the course of his or her employment directly affects commercial motor vehicle safety, but such term does not include an employee of the United States, any State, any political subdivision of a State, or any agency established under a compact between States and approved by the Congress of the United States who is acting within the course of such employment; e) an individual who is compensated by the transit agency and whose expense is reported in object class 501 labor. (3) *Law.* A person in the service of another under any contract of hire, express or implied, oral or written, where the employer has the power or right to control and direct the employee in the material details of how the work is to be performed.

Employee Aptitude Survey (EAS)

A commonly used test for determining symbolic, verbal, and numeric reasoning abilities, word fluency and comprehension, spatial visualization, visual pursuit, speed and accuracy abilities, and manual speed and accuracy.

employee hours

(1) *General.* The total number of hours worked by all employees in a facility or company. May also be referred to as *exposure hours*. (2) *Transit.* The number of hours worked by all employees of the railroad during the previous calendar year.

employee human factor

Railroad. Includes any of the accident causes signified by the rail equipment accident/incident cause codes listed under "Train Operation-Human Factors" in the current "Federal Railroad Administration (FRA) Guide for Preparing Accident/Incident Reports," except for Cause Code 506. *See also* ***human factor***.

employee not on duty

Railroad. A railroad employee who is on railroad property for a purpose connected with his or her employment or with other railroad permission, but who is not en d in rail operations for financial or other compensation. Two classifications or categories are used: a) *Employee on duty (Class A):* Those persons who are en d in the operation of a railroad. Ordinarily the fact that the employee is or is not under pay will determine whether he or she is "on duty." However, employees on railroad property while on rest or meal periods, "training time," or doing work which they are expected to do, but actually perform before pay starts, must be considered as "employees on duty." b) *Employee on duty (Class B):* Those employees who are on railroad property for purposes connected with their employment or with other railroad permission, but who are not "on-duty" as defined above.

employee participation team

See ***quality circles***.

Employee Retirement Income Security Act (ERISA)

A government regulation with the intent of guaranteeing employees' pensions if they leave a company before retirement age and that sufficient funds will exist to pay pensions when due.

Employee Stock Ownership Plan (ESOP)

A type of qualified profit sharing plan that invests in securities of the employer. Such plans acquire shares of the employer corporation for the benefit of employees, usually through contributions of the employer to the plan. In a contributory ESOP, the employer usually contributes its shares to a trust and receives a deduction for the fair market value of such stock. Generally, the employee recognizes no income until the stock is sold after its distribution to him/her upon retirement or other separation from service. Special tax benefits are provided to companies with such benefits.

employer
(1) *General.* The person who has the authority to direct and control the activities of another. Also, the person who supervises the employee on a day to day basis is usually considered the employer. This means that temporary and part-time workers may be considered "employees." (2) *Transit.* Any person engaged in a business affecting interstate commerce who owns or leases a commercial motor vehicle in connection with that business, or assigns employees to operate it, but such terms do not include the United States, any State, any political subdivision of a State, or an agency established under a compact between States approved by the Congress of the United States. (3) *Law.* One who employs the services of others; one for whom employees work and who pays their wages or salaries. The correlative of "employee."

employers' liability acts
Statutes, such as the Federal Employer's Liability Act and Workers' Compensation Acts, defining or limiting the occasions and the extent to which public and private employers shall be liable in damages (compensation) for injuries to their employees occurring in the course of their employment, and particularly abolishing the common-law rule that the employer is not liable if the injury is caused by fault or negligence of a fellow servant, and also the defenses of contributory negligence and assumption of risk.

employers' liability insurance
In this form of insurance, the risk insured against is the liability of the insured to make compensation of pay damages for an accident, injury, or death occurring to a servant or other employee in the course of his/her employment, either at common law or under statutes imposing such liability on employers. It is coverage which protects the employer as to claims not covered under workers' compensation insurance. *See also* ***insurance***.

employment
The act of employing or the state of being employed. That which engages or occupies; that which consumes time and attention; also an occupation, profession, trade, post, or business.

employment agency
A business operated by a person, firm, or corporation engaged in procuring, for a fee, employment for others and employees for employers. The fee may be paid by either the employer or the employee, depending upon the terms of the agreement.

employment at will
This doctrine provides that, absent the express agreement to the contrary, either the employer of the employee may terminate their relationship at any time, for any reason. Such employment relationship is one which has no specific duration, and such a relationship may be terminated at will by either the employer or the employee, for or without cause. *See also* ***at-will employment***.

employment contract
An agreement or contract between employer and employee in which the terms and conditions of one's employment are provided.

emporiatrics
That branch of medicine particularly concerned with the health problems of travelers about the world.

empower
(1) To give an individual the challenge or opportunity to show creativity, demonstrate personal responsibility, and provide quality work. (2) A grant of authority rather than a command of its exercise.

emptor
Law (Latin). A buyer or purchaser. Used in the maxim "caveat emptor," meaning let the buyer beware (i.e., the buyer of an article must be on guard and take the risks of his/her purchase). *See also* ***caveat emptor***.

empty car mile
Rail Operations. A mile run by a freight car without a load. In the case of intermodal movements, the car miles generated will be loaded or empty depending on whether the trailers/containers are moved with or without a waybill, respectively.

empty chair doctrine
Under this doctrine, a trial justice may charge a jury that it may infer from the litigant's unexplained failure to produce an available witness who would be expected to give material testimony in the litigant's behalf that the witness, had he/she occupied the empty chair, would have testified adversely to the litigant.

empty field myopia
The condition of eye accommodation for near, as opposed to far, vision when viewing a homogeneous field.

empyema
The presence of pus in a body cavity, particularly the presence of a purulent exudate within the pleural cavity (pyothorax). It occurs as an occasional complication of pleurisy or some other respiratory disease. Symptoms include dyspnea, coughing, chest pain on one side, malaise, and fever.

EMS
See ***electrical muscle stimulation***.

EMU
See ***extravehicular mobility unit***.

emulsifier
A surface-active agent that promotes the dispersion of one liquid in another, such as small fat globules in water.

en route
Aviation. One of three phases of flight services (terminal, en route, oceanic). En route service is provided outside of terminal airspace and is exclusive of oceanic control.

en route air traffic control service
Air traffic control service provided for aircraft on Instrument Flight Rules (IFR) flight plans, generally by Air Route Traffic Control Center (ARTCC), when these aircraft are operating between departure and destination terminal areas. When equipment capabilities and controller workload permit, certain advisory or assistance services may be provided to Visual Flight Rules (VFR) aircraft.

en route center
An Air Route Traffic Control Center.

en route descent
Descent from the en route cruising altitude which takes place along the route of flight.

en route facility activity
Total Instrument Flight Rules aircraft handled; (2 x departures) + Domestic and Oceanic overflights.

en route high altitude charts
Provide aeronautical information for en route instrument navigation (IFR) in the high altitude stratum. Information includes the portrayal of jet routes, identification and frequencies of radio aids, selected airports, distances, time zones, special uses airspaces, and related information.

en route low altitude charts
Provide aeronautical information for en route instrument navigation (IFR) in the low altitude stratum. Information includes the portrayal of airways, limits of controlled airspace, position identification and frequencies of radio aids, selected airports, minimum en route and minimum obstruction clearance altitudes, airway distances, reporting points, restricted areas, and related data. Area charts, which are a part of this series, furnish terminal data at a larger scale in congested areas.

en route minimum safe altitude warning
A function of the National Airspace System (NAS) Stage A en route computer that aids the controller by alerting him when a tracked aircraft is below or predicted by the computer to go below a predetermined minimum Instrument Flight Rules (IFR) altitude (MIA).

en route spacing program
A program designed to assist the exit sector in achieving the required in trail spacing.

enable
To give power to do something; to make able. In the case of a person under a disability as to dealing with another, "enable" has the primary meaning of removing that disability; not of giving a compulsory power that can be used against another person.

Enabling Act
A term referring to the foundation statute creating an agency and giving it jurisdiction and authority, usually also establishing some standards and procedures for it to follow. *See also* ***enabling statute***.

enabling clause
That portion of a statute or constitution which gives to governmental offices the power and

authority to put it into effect and to enforce such.

enabling statute

Term applied to any statute enabling persons or corporations, or agencies to do what before they could not. It is applied to statutes which confer new powers. *See also **Enabling Act** and **enabling clause**.*

enact

To establish by law; to perform or effect; to decree. The common introductory formula in making statutory laws is "Be it enacted."

enacting clause

A clause at the beginning of a statute which states the authority by which it is made. That part of a statute which declares its enactment and serves to identify it as an act of legislation proceeding from the proper legislative authority.

enactment

The method or process by which a bill in the legislature becomes a law.

enamel

The calcified tissue of ectodermal origin covering the crown of a tooth.

encapsulant

A material that can be applied to a solid or semisolid material to prevent the release of a component(s), such as fibers from an ACM.

encapsulation

The process of coating an asbestos-containing material, manmade mineral fiber, lead-containing or other material from which release of a contaminant is to be controlled by the encapsulating material. An example is the coating of asbestos-containing material with a bonding or sealing agent to prevent the release of fibers.

encephalitis

Inflammation of the brain and the coverings (the meninges) producing persistent drowsiness, delirium, and rarely, coma. There are several different forms, a few of which are occasionally epidemic in limited areas of the United States. The epidemic forms are caused by a virus transmitted to man by the bite of mosquitoes and ticks. The condition can also occur as a rare complication of some other virus disease, and it is occasionally produced by contact with a toxic substance, such as lead.

encephalopathy

Any degenerative disease of the brain.

enclosed

Surrounded by a case, housing, fence, or walls which will prevent persons from accidentally contacting energized parts.

enclosed structure

A structure with a roof or ceiling and at least two walls which may present fire hazards to employees, such as accumulations of smoke, toxic gases and heat, similar to those found in buildings.

enclosing hood

A hood that encloses the source of contamination.

enclosure

(1) *General.* The case or housing of an apparatus, or the fence or walls surrounding an installation, to prevent personnel from accidentally contacting energized parts, or to protect the equipment from physical damage. (2) *Asbestos.* A tight structure around an area of asbestos-containing material to prevent the release of fibers into the surrounding area.

encoder

Any device for coding one or more values for use by another device or computer.

encourage

Law. In criminal law, to instigate; to incite to action; to give courage to; to inspirit; to embolden; to raise confidence; to make confident; to help; to forward; to advise.

encroach

To enter by gradual steps or stealth into the possessions or rights of another; to trespass or intrude. To gain or intrude unlawfully upon the lands, property, or authority of another.

encroachment

An illegal intrusion in a highway or navigable river, with or without obstruction. An encroachment upon a street or highway is a fixture, such as a wall or fence, which illegally intrudes into or invades the highway or encloses a portion of it, diminishing its width or area, but without closing it to public travel.

encumbrance

Any right to, or interest in, land which may subsist in another to diminution of its value, but consistent with the passing of the fee by conveyance. A claim, lien, charge, or liability attached to and binding real property (e.g., a mortgage, judgement lien; mechanics' lien; lease; security interest; easement or right of way; accrued and unpaid taxes). If the liability relates to a particular asset, the asset is encumbered. While encumbrances usually relate to real property, a purchaser of personal property is provided with a warranty of title against unknown encumbrances.

end effector
A remote mechanical latching device for gripping, holding, and/or performing work.

end facing glazing location
Railroads. With regard to safety glazing on rail car windows: Any location where a line perpendicular to the plane of the glazing material makes a horizontal angle of 50 degrees or less with the centerline of the locomotive, caboose, or passenger car. Any location which, due to curvature of the glazing material, can meet the criteria for either a front facing location or a side facing location shall be considered a front facing location.

end item
The final manufactured product, typically built to certain requirements or specifications.

end plate
(1) A specialized region of muscle cell membrane in which an axon terminates with extensive branching. Also referred to as *motor end plate*. (2) A layer of cartilage at the top and bottom of each intervertebral disk.

end-plate potential (EPP)
A prolonged potential change from the resting potential across the membrane of a muscle cell which may or may not result in a muscle action potential.

end-use energy consumption
DOE. (1) Primary end-use energy consumption is the sum of fossil fuel consumption by the four end-use sectors (residential, commercial, industrial, and transportation) and generation of hydroelectric power by non-electric utilities. Net end-use energy consumption includes electric utility sales to those sectors but excludes electrical system energy losses. Total end-use energy consumption includes both electric utility sales to the four end-use sectors and electrical system energy losses. (2) The sum of fossil fuel consumption by the four end-use sectors (residential, commercial, industrial, and transportation) plus electric utility sales to those sectors and generation of hydroelectric power by non-electric utilities. Net end-use energy consumption excludes electrical system energy losses. Total end-use energy consumption includes electrical system energy losses.

end-use sectors
The residential, commercial, industrial, and transportation sectors of the economy.

endangered assessment
A study conducted to determine the nature and extent of contamination at a site on the National Priorities List and the risk posed to public health or the environment. EPA or the state conduct the study when a legal action is to be taken to direct potentially responsible parties to clean up a site or pay for the cleanup. An endangered assessment supplements a remedial investigation.

endangered species
Under the Federal Endangered Species Act of 1973: Any species which is in danger of extinction throughout all or a significant portion of its range other than a species of the Class Insecta determined by the Secretary of the Interior or the Secretary of Commerce to constitute a pest whose protection under the provisions of the Federal Endangered Species Act of 1973 would present an overwhelming and overriding risk to man.

endarteritis
Inflammation of the innermost coat of an artery.

endeavor
To exert physical and intellectual strength toward the attainment of an object. A systematic or continuous effort.

endemic
Refers to diseases or infectious agents in the human population within a given geographic area that are constantly present or usually prevalent.

ending milepost
Transit. The continuous milepost notation, to the nearest 0.01 mile that marks the end of any road or trail segment.

endocanthic breadth
The horizontal linear distance between the right and left endocanthi. Also referred to as *interocular breadth*.

endocanthus
The junction of the most medial parts of the upper and lower eyelids, with the eyelids open normally. May be referred to as *internal canthus* or *medial canthus*.

endocarditis
An inflammation of the inner lining membrane of the heart, usually involving the heart valves. Bacterial endocarditis is an acute or subacute, febrile, systemic disease characterized by bacterial infection of the heart valves or irregular areas on the endocardium, with the formation of bacteria-laden vegetation on these areas.

endocardium
The membrane lining the chambers of the heart and covering the cusps of the various valves.

endocrine
(1) Secreting internally. (2) Pertaining to internal secretion.

endocrine gland
Gland that regulates body activity by special secretions, the hormones, which are delivered directly into the blood. Each of the glands within the endocrine system has one or more specific functions, but they are all dependent upon other glands in the system for maintenance of a normal hormonal balance in the body.

endocrinology
The study of the glands of internal secretions.

endogenous
Originating within an organ or part.

endolymph
The fluid within the semicircular ducts, the utricle, saccule, and cochlear duct of the inner ear.

endometriosis
A condition in which tissue, more or less perfectly resembling the uterine mucous membrane, occurs aberrantly in various locations in the pelvic cavity. The condition may be characterized by pelvic pain, abnormal uterine or rectal bleeding, dysmenorrhea, and symptoms of pressure within the pelvic cavity. Sterility and dyspareunia also may be present.

endometrium
The mucous membrane lining the uterus.

endomorph
A Sheldon somatotype characterized generally by a soft, rounded body, with greater amounts of fatty tissue, little muscle, and an abdominal protrusion.

endorsement
(1) *Insurance*. An amendment to an insurance policy. (2) *Transit*. An authorization to an individual's commercial driver's license (CDL) required to permit the individual to operate certain types of commercial motor vehicles.

endoscope
An instrument used for direct visual inspection of hollow organs or body cavities.

endoskeleton
The framework of hard structures, embedded in and supporting the soft tissues of the body of higher animals, derived principally from the mesoderm.

endospore
A thick-walled structure formed within the cells of certain bacteria that allows the organism to withstand adverse environmental conditions, such as drying.

endothermic
Refers to a reaction in which the products contain more energy than the reacting materials, causing the absorption of energy as heat.

endothoracic
Within the thorax; situated internal to the ribs.

endotoxin
A heat-stable toxin that is present in the bacterial cell but not in cell-free filtrates of cultures of intact bacteria.

endpoint
See ***breakpoint***.

endurance
A measure of the ability to maintain some specific level of effort, usually represented in units of time. May be referred to as *capacity*.

enema
(1) The introduction of fluid into the rectum. (2) A solution introduced into the rectum to promote evacuation of feces or as a means of administering nutrient or medicinal substances, anesthetics, or opaque material in roentgen examination of the lower intestinal tract.

energy
(1) *General*. The capacity for doing work or the amount of work done. The product of power (watts) and time duration (seconds) where one watt-second equals one joule. Forms of energy include chemical, nuclear, kinetic, and others. (2) *DOE*. The capacity for doing work as measured by the capability of doing work (potential energy) or the conversion of this capability to motion (kinetic energy). Energy has several forms, some of which are easily convertible and can be changed to another form useful for work. Most of the world's convertible energy comes from fossil fuels that are burned to produce heat that is then used as a transfer medium to mechanical or other means to accomplish tasks. Electrical energy is usually measured in kilowatt hours, while heat energy is usually measured in British Thermal Units (BTUs).

energy average level
A quantity calculated by taking ten times the common logarithm of the arithmetic average of the antilogs of one-tenth of each of the levels being averaged. The levels may be of any consistent type, such as maximum sound levels, sound exposure levels, and day-night sound levels.

energy capacity
Measured in kilowatt hours. The energy delivered by the battery, when tested at C/3 discharge rate, up to termination of discharge specified by the battery manufacturer. The required acceleration power must be delivered by the battery at any point up to 80% of the battery's energy capacity rating.

energy consumption
The use of energy as a source of heat or power or as an input in the manufacturing process.

energy efficiency
In reference to transportation, the inverse of energy intensiveness. The ratio of outputs from a process to the energy inputs, for example, miles traveled per gallon of fuel (mpg).

energy efficient motors
Are also known as "high-efficiency motors" and "premium motors." They are virtually interchangeable with standard motors, but differences in construction make them more energy efficient.

energy expenditure
*See **metabolic rate**.*

energy facilities
Under the Federal Coastal Zone Management Act of 1972: Any equipment or facility which is or will be used primarily in the exploration for, or the development, production, conversion, storage, transfer, processing, or transportation of any energy resource; or for the manufacture, production, or assembly of equipment, machinery, products, or devices which are involved in any such activity. The term includes, but is not limited to electric generating plants; petroleum refineries and associated facilities; gasification plants; facilities used for the transportation, conversion, treatment, transfer, or storage of liquefied natural gas; uranium enrichment or nuclear fuel processing facilities; oil and gas facilities, including platforms, assembly plants, storage depots, tank farms, crew and supply bases, and refining complexes; facilities including deepwater ports, for the transfer of petroleum; pipelines and transmission facilities; and terminals which are associated with any of the foregoing.

energy flow
Under ISO 14000, input flow to or output flow from a unit process or product system measured in units of energy.

Energy Information Administration (EIA)
An independent agency within the U.S. Department of Energy that develops surveys, collects energy data, and analyzes and models energy issues. The Agency must meet the requests of Congress, other elements within the Department of Energy, Federal Energy Regulatory Commission, the Executive Branch, its own independent needs, and assist the general public, or other interest groups, without taking a policy position.

energy intensity
In reference to transportation, the ratio of energy inputs to a process to the useful outputs form that process; for example, gallons of fuel per passenger-mile or BTU per ton mile.

energy management
The allocation or use of energy.

Energy Research and Development Administration (ERDA)
The part of the now defunct Atomic Energy Commission (AEC) that became the reactor development section and was subsequently incorporated into the Department of Energy.

energy source
A substance, such as petroleum, natural gas, or coal, that supplies heat or power. In Energy Information Administration reports, electricity and renewable forms of energy, such as biomass, geothermal, wind, and solar, are considered to be energy sources.

energy summation of levels
A quantity calculated by taking ten times the common logarithm of the sum of the antilogs of one-tenth of each of the levels being summed. The levels may be of any consistent type, such as day-night sound level or equivalent sound level.

energy trace and barrier analysis (ETBA)
A system safety analytical technique used to evaluate the flow of energy through a system and analyze the effectiveness of existing barriers within the system which are intended to prevent unwanted transfers of that energy flow.

enfleshment
The use of volumes surrounding body segments or links in human computer modeling to stimulate the presence of body tissues.

enforcement
(1) *Law.* The act of putting something such as a law into effect; the execution of a law; the carrying out of a mandate or command. (2) *Environmental.* EPA, state, or local actions to obtain compliance with environmental laws, rules, regulations, or agreements and/or obtain penalties or criminal sanctions for violations. Enforcement procedures may vary, depending on the specific requirements of different environmental laws and related implementing regulatory requirements.

enforcement decision document (EDD)
A document that provides an explanation to the public of EPA's selection of the cleanup alternatives at enforcement sites on the National Priorities List. Similar to a record of decision.

enforcement powers
The 13th, 14th, 15th, 19th, 23rd, 24th, and 26th Amendments to the U.S. Constitution; each contains clauses granting to Congress the power to enforce by appropriate legislation the provisions of such Amendments.

ENG
See ***electronystagmogram****.*

engage
To employ or involve oneself; to take part in; to embark on.

engaged in commerce
To be "engaged in commerce" for purposes of Fair Labor Standards Act and Federal Employers' Liability Act, an employee must be actually engaged in the movement of commerce or the services he/she performs must be so closely related thereto as to be for all practical purposes an essential part thereof, rather than an isolated local activity.

engaged in employment
To be rendering service for an employer under the terms of employment, and is more than being merely hired to commence work.

engine
A locomotive propelled by any form of energy and used by a railroad.

engine classification
A 2-digit numeric code identifying vehicle engines by the number of cylinders.

engine displacement
The volume in inches, through which the head of the piston moves, multiplied by the number of cylinders in the engine. Also known as cubic inch displacement (CID), may also be measured in liters. *See also* ***cylinder*** *and* ***engine size****.*

engine retarder
Electronic equipment which governs engine speed control.

engine size
The total volume within all cylinders of an engine, when pistons are at their lowest positions. The engine is usually measured in "liters" or "cubic inches of displacement (CID)." Generally, larger engines result in greater engine power, but less fuel efficiency. There are 61.024 cubic inches in a liter. *See also* ***cylinder*** *and* ***engine displacement****.*

engineer
(1) An individual qualified by education, training, and/or experience to practice in one or more fields of engineering. (2) A person responsible for operating and maintaining the power system on a vessel.

engineered barriers
Under the Federal Nuclear Waste Policy Act of 1982: Manmade components of the disposal system designed to prevent the release of radionuclides into the geologic medium involved. The term includes the high-level radioactive waste form, high-level radioactive waste canisters, and other materials placed over and around such canisters.

engineered performance standard
See ***standard time***.

engineering
A discipline in which knowledge of the mathematical and natural sciences, gained by some combination of education, training, and practical experience, is integrated with various natural materials and forces to shape the environment.

engineering anthropometry
The application of anthropometric data for designing products to be used by humans. *See also* ***human factors engineering***.

engineering controls
Measures taken to prevent or minimize hazard exposure through the application of controls such as improved ventilation, noise reduction techniques, chemical substitution, equipment and facility modifications, etc.

engineering model
A full-size structural model which is functionally identical to and dimensionally corresponds with the intended or actual final production item.

engineering psychology
See ***human factors engineering***.

engineering tolerance
The maximum degree of variation permitted or allowed on a given specification, drawing, or part. Also referred to as *tolerance, tolerance specification,* and *tolerance limits*.

English System
A nearly obsolete system of measurement, used only in the United States, whose primary units are essentially "non-metric" in nature (e.g., feet, inches, yards, miles, gallons, etc.). *See also* ***basic units*** *and* ***English Units***.

English Units
The term "English" refers to the United States legislative interpretation of the units as defined in a document prepared by the National Institute of Standards and Technology (NIST), U.S. Department of Commerce, Special Publication 330. Commonly used English units in Highway Performance Monitoring System (HPMS) are miles, feet, and inches. *See also* ***English System*** *and* ***base units***.

engram
A postulated neural pathway representing the trace of a memory in the brain.

engulfment
As pertains to confined spaces, the surrounding and effective capture of a person by a liquid or finely divided solid substance that can be aspirated to cause death by filling or plugging the respiratory system or that can exert enough force on the body to cause death by strangulation, constriction, or crushing.

enhancement coding
Any technique for increasing the chances that a particular item will stand out against a background. Examples include color coding, blinking, and bolding.

enjoin
To require; command; positively direct. To require a person, by writ of injunction, to perform, or to abstain or desist from, some act.

enplaned passenger
The total number of revenue passengers boarding aircraft.

enplaned revenue tons of freight and mail
The number of revenue tons of freight and mail loaded on an aircraft including originating and transfer tons.

enplanement
Domestic, territorial, and international revenue passengers who board an aircraft in the states in scheduled and non-scheduled service of aircraft in intrastate, interstate, and foreign commerce and includes intransit passengers (passengers on board international flights that transit an airport in the US for non-traffic purposes).

enrichment

The addition of nutrients (e.g., nitrogen, phosphorus, carbon compounds) from sewage effluent or agricultural runoff to surface water. This process greatly increases the growth potential for algae and aquatic plants.

enrolled bill

The final copy of a bill or joint resolution which has passed both houses of a legislature and is ready for signature. In legislative practice, a bill which has been duly introduced, finally passed by both houses, signed by the proper officers of each, approved by the governor (or president) and filed by the secretary of state.

enter

A user operation which signifies the end of a sequence of keystrokes or other operations and directs the computer to take action based on the content of that sequence.

enteric

Pertaining to the intestines.

entering judgements

The formal entry of the judgement on the rolls or records (e.g., civil docket) of the court, which is necessary before bringing an appeal or an action on the judgement. The entering of judgement is a ministerial act performed by the clerk of court by means of which permanent evidence of a judicial act in rendering judgement is made a record of the court.

enteritis

An inflammation of some portion of the intestines. A general condition that can be produced by a variety of causes. Bacteria and certain viruses may irritate the intestinal tract and produce symptoms of abdominal pain, nausea, vomiting, and diarrhea. Similar effects may result from poisonous foods such as mushrooms and berries, or from a harmful chemical present in food or drink. Enteritis may also be the consequence of overeating, alcoholic excesses, or emotional tension.

enterocolitis

Inflammation of the small intestine and colon.

enteromegaly

Enlargement of the intestines.

enterotoxin

(1) A toxin specific for the cells of the intestinal mucosa. (2) A toxin arising in the intestine. (3) An exotoxin that is protein in nature and relatively heat-stable, produced by staphylococci and causing food poisoning.

enterprise

A business venture or undertaking.

enterprise liability

Imposition of liability upon each member in industry who manufactures or produces a product which causes injury or harm to a consumer and apportions liability of each member of industry by reference to that member's share of the market for the product.

enthalpy

Heat function at constant pressure. Enthalpy is sometimes also called the heat content of the system.

entire loss of sight

In legal terms, with respect to one eye or both, means substantial blindness, not necessarily absolute.

entitlement

*See **apportionment**.*

entity

(1) One of the more basic graphical elements, such as a line, arc, or circle. (2) An individual, organism, or other object having existence.

entraining agent

Any event, signal, or cue which is a driver for maintaining periodicity in biological rhythms. Also referred to as *Zeitgeber* and *synchronizer*.

entrainment

The mixing of environmental air into a preexisting air current or cloud so that the environmental air becomes part of the current or cloud.

entrant

A person who has been authorized by their employer to enter a permit-required confined space.

entrapment

(1) *Law.* The act of officers or agents of the government in inducing a person to commit a crime not contemplated by him/her, for the purpose of instituting a criminal prosecution against him/her. (2) *Vehicle Safety.* Refers to persons being partially or completely in the vehicle and mechanically restrained by a

damaged vehicle component. Jammed doors and immobilizing injuries, by themselves, do not constitute entrapment. Occupants pinned by cargo shift are not considered to be entrapped. Occupants who are completely or partially ejected and subsequently become pinned by their own vehicle and any surface other than their own vehicle are not considered entrapped. An occupant whose seat belt buckle release mechanism is jammed as a result of a crash is not considered entrapped.

entrepreneur
One who, on his/her own, initiates and assumes the financial risks of a new enterprise and who undertakes its management.

entropy
A measure of the degree of disorder in a system, wherein every change that occurs and results in an increase of disorder is said to be a positive change in entropy. All spontaneous processes are accompanied by an increase in entropy. The internal energy of a substance that is attributed to the internal motion of the molecules.

entrust
To give something over to another after a relation of confidence has been established.

entry
(1) *Law.* The act of making or entering a record; a setting down in writing of particulars; or that which is entered; an item. (2) *Confined Spaces.* The act of passing through an opening into a confined space and the ensuing work in the space. An entry occurs when any part of the body breaks the plane of an opening of what is classified as a confined space. An alternate definition is any action resulting in any part of the face of the employee breaking the plane of any opening of a confined space as well as any ensuring work inside the space.

entry loss
Loss in pressure caused by air flowing into a duct or hood opening.

entry permit
The written authorization of the employer for entry into a confined space under defined conditions for a stated purpose during a specified time.

entry point
The point at which an aircraft transitions from an offshore control area to oceanic airspace.

entry supervisor
As pertains to confined spaces, the person (such as the employer, foreman, or crew chief) responsible for determining if acceptable entry conditions are present at a permit space where entry is planned, for authorizing entry and overseeing entry operations, and for terminating entry. An entry supervisor may also serve as an *attendant* or as an *authorized entrant,* as long as that person is trained and equipped as required by OSHA for each role he or she fills. Also, the duties of entry supervisor may be passed from one individual to another during the course of any operation.

enumerated
This term is often used in law as equivalent to "mentioned specifically," "designated," or "expressly named or granted," as in speaking of "enumerated" governmental powers, items of property, or articles in a tariff schedule.

enumerated powers
The powers specifically delegated by the Constitution to some branch or authority of the national government, and which are not denied to that government or reserved to the States or to the people. The powers specifically given to Congress are enumerated in Article I of the U.S. Constitution.

envelope
A specified volume as determined by some methodology or required function.

environment
The sum of all external conditions affecting the life, development, and survival of an organism. Includes water, air, land, and all plants and man and other animals living therein, and the interrelationships which exist among these.

environmental anthropometry
The measurement or study of changes in an individual's anthropometry due to his/her physical environment.

environmental aspect
Under ISO 14000, the element of an organization's activities, products, or services that can interact with the environment.

environmental assessment
A written environmental analysis which is prepared pursuant to the National Environmental Policy Act to determine whether a federal action would significantly affect the environment and thus require preparation of a more detailed environmental impact statement. Also referred to as *environmental impact assessment (EIA)*.

environmental audit
(1) An independent assessment of the current status of a party's compliance with applicable environmental requirements. (2) An independent evaluation of a party's environmental compliance policies, practices, and controls. (3) Auditing an organization's policies and procedures to bring industrial operations and practices into compliance with environmental laws and regulations, its permits, and any agreements with government agencies before they trigger enforcement action.

environmental audit privilege statutes
State legislation enacted by many states which insulates companies from abuse of their self-policing efforts. Although the statutes of the individual states do vary, generally the elements include a) documentation using Environmental Audit Report, b) immunity or reduction in penalties for voluntary disclosure, c) waiver of privilege, d) loss of privilege in certain cases, and e) a burden of proof in proving the privilege and due diligence toward compliance. *See also* ***State Audit Immunity Statutes***.

environmental control
The regulation or alteration of the environment to maintain certain conditions.

environmental due diligence
The process used to investigate a commercial or industrial property (usually prior to completion of a real estate transaction) for contamination by hazardous wastes or hazardous substances.

environmental factors
Conditions other than indoor air contaminants that cause stress, comfort and/or health problems (e.g., humidity extremes, drafts, lack of air circulation, noise, and overcrowding).

environmental fate
Term used to describe the transport and transformation processes which occur to a chemical in the environment.

Environmental Guidelines for Sentencing Organizations (Draft)
A proposed Chapter 9 of the Federal Sentencing Guidelines (FSGs) for the sentencing of organizations for environmental crimes developed by the Advisory Working Group on Environmental Sanctions (March 3, 1993). Not yet adopted by the Federal Sentencing Commission.

environmental health
(1) The body of knowledge concerned with the prevention of disease through the control of biological, chemical, or physical agents in air, water, and food. Also concerned with the control of *environmental factors* that may have an impact on the well-being of people. (2) The activities necessary to ensure that the health of employees, customers, and the public is adequately protected from any health hazards associated with a company's operations.

environmental impact
Under ISO 14000, any change to the environment, whether adverse or beneficial, wholly or partially resulting from an organization's activities, products, or services.

environmental impact assessment (EIA)
A report prepared by an applicant for a discharge permit which identifies and analyzes the impact of a new source of emission to the environment and discusses possible alternatives.

environmental impact statement (EIS)
A document required of federal agencies by the National Environmental Policy Act for major projects or legislative proposals significantly affecting the environment. A tool for decision making, it describes the positive and negative effects of the undertaking and lists alternative actions.

environmental impairment liability
A type of insurance coverage carried by hazardous waste generators and others involved in hazardous waste handling and disposal. The coverage typically provides funds for remediating environmental impairment or paying for damages resulting from the impairment. Not all such insurance polices include the same types of coverage; however, some have specific exclusion for certain types of occurrences or releases.

environmental inputs
The economic, social, psychological, managerial, mechanical, and climatic variables which cause an individual to respond, either physiologically or behaviorally.

environmental labeling or declaration
A tool of environmental management which is a claim indicating the environmental aspects of a product or service that may take the form of statements, symbols, or graphics on product or package labels, product literature, technical bulletins, advertising, publicity, etc. An element of ISO 14000.

environmental lapse rate
The distribution of the temperature vertically. It is most often measured with a radiosonde. Also called the *lapse rate*.

environmental leadership program (ELP)
An EPA positive incentives program which earns a company a degree of trust by that agency and public recognition once the company has met the very highest standards of compliance. Applicant companies must be held to a high standard of performance at the time of entry into the program and must have addressed any outstanding problems with either state or federal officials. The company's own internal self-evaluation system would serve as evidence of its continuous compliance. *See also* ***carrot and stick approach*** *and* ***positive incentives***.

environmental monitoring
The systematic collection, analysis, and evaluation of environmental samples, such as from air, to determine the contaminant levels to which workers are exposed.

environmental noise
Under the Federal Noise Control Act of 1972, the intensity, duration, and the character of sounds from all sources.

environmental objective
Under ISO 14000, the overall environmental goal, arising from the environmental policy, that an organization sets itself to achieve, and which is quantified where practicable.

environmental performance
Under ISO 14000, the measurable results of the environmental management system, related to an organization's control of its environmental aspects, based on its environmental policy, objectives, and targets.

environmental policy
Under ISO 14000, a statement by an organization of its intentions and principles in relation to its overall environmental performance which provides a framework for action and for setting of its environmental objectives and targets.

Environmental Protection Agency (EPA)
Established in 1970 by Presidential Executive Order (President Nixon), the EPA is the primary federal agency charged with ensuring the protection and preservation of environmental resources in the United States. It is responsible for pollution control and abatement, including programs for air, water, pollution, solid and toxic waste, pesticide, control, noise abatement, and other pollution sources and concerns.

Environmental Protection Agency Certification Files
Computer files produced by Environmental Protection Agency (EPA) for analysis purposes. For each vehicle make, model and year, the files contain the EPA test Miles Per Gallon (MPG) (city, highway, and 55/45 composite). These MPGs are associated with various combinations of engine and drivetrain technologies (e.g., number of cylinders, engine size, gasoline or diesel fuel, and automatic or manual transmission). These files also contain information similar to that in the Department of Energy (DOE)/EPA Gas Mileage Guide, although the MPGs in that publication are adjusted for shortfall.

Environmental Protection Agency Composite Mile Per Gallon (MPG)
The harmonic mean of the Environmental Protection Agency (EPA) city and highway MPG, weighted under the assumption of 55 percent city driving and 45 percent highway driving.

environmental response team
EPA experts located in Edison, New Jersey and Cincinnati, Ohio who can provide around-the-clock technical assistance to EPA regional offices and states during all types of emergencies involving hazardous waste sites and spills of hazardous substances.

environmental restoration
Restitution for the loss, damage, or destruction of natural resources arising out of the accidental discharge, dispersal, release, or escape into or upon the land, atmosphere, watercourse, or body of water of any commodity transported by a motor carrier. This shall include the cost of removal and the cost of necessary measures taken to minimize or mitigate damage to human health, the natural environment, fish, shellfish, and wildlife.

environmental risk
The probability of a human health effect resulting from some environmental state or circumstance.

environmental sampling
The taking of samples from the environment for analysis. Also called *sampling*. *See also* ***environmental monitoring***.

environmental stressor
Any condition in the environment which produces stress in an organism, whether climatological, biological, chemical, mechanical, or particulate. Also referred to as *ecological stress vector*.

environmental target
Under ISO 14000, the detailed performance requirement, quantified where practicable, applicable to the organization or parts thereof, that arises from the environmental objectives and that needs to be set and met to achieve those objectives.

environmentally sensitive area
An area of environmental importance which is in or adjacent to navigable waters.

enzyme
An organic compound, frequently a protein, that accelerates (catalyzes) specific transformations of material, as in the digestion of foods.

EOG
See ***electrooculogram***.

E. P. Tox
EP Toxicity or Extraction Procedure Toxicity; an analytical laboratory characterization using extraction procedures for determining primarily toxic metal concentrations and/or leaching potential. Recently updated as a series of combined tests now called *TCLP* or *toxicity characterization leaching procedure*.

EPA
See ***Environmental Protection Agency***.

EPCRA
Emergency Planning and Community Right-to-Know Act of 1986 (Federal). It is Title III of the Superfund Amendments and Reauthorization Act (SARA) of 1986.

ephedrine
An alkaloid obtained from the shrub *Ephedra equisetina* or produced synthetically; used, in the form of ephedrine hydrochloride or ephedrine sulfate, as a sympathomimetic, as a pressor substance, to relieve bronchial spasm and as a central nervous system stimulant. It may be administered orally, topically, intramuscularly, or intravenously.

epicardia
The lower portion of the esophagus, extending from the esophageal hiatus to the cardia, the upper orifice of the stomach.

epicardium
The layer of the pericardium that is in contact with the heart.

epicenter
The point on the earth's surface directly above the (subterranean) point of origin (the *hypocenter*) of an earthquake. Only two measurements, latitude and longitude, are need to locate it.

epicondyle
A bony protrusion at the distal end of bones such as the humerus, radius, and femur.

epicondylitis
A cumulative trauma disorder (CTD) characterized by inflammation or infection in the general area of the elbow, such as tennis elbow.

epidemic
The occurrence of cases that are of similar nature in human populations in a particular geographic area and that are clearly in excess of the usual incidence.

epidemiologist
A person who applies epidemiological principles and methods to the prevention and control of diseases.

epidemiology
The study of the distribution and determinants of disease causation in human populations.

Examines the frequency of occurrence and distribution of a disease throughout a population, often with the purpose of determining the cause. To the industrial hygienist, it is the determination of statistically significant relationships of specific diseases of specific organs of the human body in selected organs of the human body in selected occupational groups (cohorts) in comparison with selected controls.

epidermis
The outer, non-vascular, non-sensitive layer of the skin that covers the *true skin.*

epiglottis
A large piece of cartilage at the top of the larynx which closes the tracheal entrance when swallowing to prevent food from entering.

epilation
The removal of hair by the roots. Loss of body hair.

epilepsy
A disruption of the normal rhythm of the brain. An occasional, periodic, excessive and disorderly discharge of nerve cells in the brain. The discharge is chemical-electrical in nature. While the discharge itself is hidden, it manifests itself in various forms of visible activity called seizures. The type of seizures will vary according to the location of the discharge in the brain, and the spread of the charges from cell to cell. In many cases, seizures are so mild (a brief twitch, a momentary attention loss) that they are not recognized. Even when they are, they have a minimal effect. A major convulsion which the public tends to associate immediately with epilepsy is only one of a number of seizure types.

epinephrine
A catecholamine which may act as a neurotransmitter or hormone, depending on the location and source. More commonly referred to as *adrenaline*.

epiphyseal separation
Not a bone fracture in true sense, but a separation of the fibers and cartilaginous tissues which attach the epiphysis to the femur.

epiphyseitis
Inflammation of an epiphysis (a process of bone attached for a time to another bone by cartilage).

epiphysis
The region at the end of a long bone having an expanded cross-section.

episode
(1) *Epilepsy*. With regard to grand mal epilepsy, a seizure event. (2) *Air Pollution*. An incident within a given region as a result of a significant concentration of an air pollutant with meteorological conditions such that the concentration may persist and possibly increase with the likelihood that there will be a significant increase in illnesses and possibly deaths, particularly among those who have a preexisting condition that may be aggravated by the pollutant.

epistaxis
Hemorrhage from the nose; a nosebleed.

epithelial
Pertaining to or comprised of epithelium.

epithelioma
Tumor derived from epithelium.

epithelium
Refers to cells that line all canals and surfaces that have contact with external air, and also cells that are specialized for secretion in certain organs such as the liver and kidneys.

eponychium
The thin layer of tissue which overlaps the lunula at the base of a fingernail or toenail.

EPP
*See **end-plate potential**.*

EPRI
Electric Power Research Institute.

EP toxic waste
A waste with certain toxic substances present at levels greater than limits specified by regulation.

Equal Access to Justice Act
This 1980 Act entitles certain prevailing parties to recover attorney and expert witness fees, and other expenses, in actions involving the United States, unless the government action was substantially justified.

Equal Employment Opportunity (EEO)
A series of government regulations intended to prevent discrimination in hiring, firing, and promotion of minorities and women.

Equal Employment Opportunity Commission (EEOC)
The EEOC was created by Title VII of the Civil Rights Act of 1964 (78 Stat. 241; 42 U.S.C.A. § 2000a), and became operational July 2, 1965. The purposes of the Commission are to end discrimination based on race, color, religion, age, sex, or national origin in hiring, promotion, firing, wages, testing, training, apprenticeship, and all other conditions of employment; and to promote voluntary action programs by employers, unions, and community organizations to put equal employment opportunity into actual operation.

equal-energy white point
See ***achromatic point***.

equal-interval scale
A measurement scale which meets the criteria for an ordinal scale and which items can be classified by value on a linear magnitude measure, with equal distances between measures, but providing no information as to the absoluteness of the magnitudes. May be referred to as *interval scale*.

Equal Pay Act
Federal law which mandates the same pay for all persons who do the same work without regard to sex, age, etc. For work to be "equal" within the meaning of the Act, it is not necessary that the jobs be identical but only that they be substantially equal.

equal protection clause
That provision in the 14th Amendment to the U.S. Constitution which prohibits a state from denying to any person within its jurisdiction the equal protection of the laws. This clause requires that persons under like circumstances be given equal protection in the enjoyment of personal rights and the prevention and redress of wrongs.

equal protection of the law
The constitutional guarantee of "equal protection of the laws" means that no person or class of persons shall be denied the same protection of the laws which is enjoyed by other persons or other classes in like circumstances in their lives, liberty, property, and in their pursuit of happiness.

equalizing reservoir
Rail. An air reservoir connected with and adding volume to the top portion of the equalizing piston chamber of the automatic brake valve, to provide uniform service reductions in brake pipe pressure regardless of the length of the train.

equilibrium
(1) *Physiology.* A state in which the body maintains desired posture or retains control in body movement through continuous sensory monitoring and the balancing of muscle tensions. *See also* ***static equilibrium*** *and* ***dynamic equilibrium***. (2) *Radiation.* The state at which the radioactivity of consecutive elements within a radioactive series is neither increasing nor decreasing.

equilibrium vapor pressure
The necessary vapor pressure around liquid water that allows the water to remain in equilibrium with its environment. Also called *saturation vapor pressure*.

equinoxes
The two periods of the year (vernal equinox about March 21st, and the autumnal equinox about September 22nd) when the time from the rising of the sun to its setting is equal to the time from its setting to its rising.

equinus
A deformity where the foot is continuously plantar-flexed.

equipment
A general term including material, fittings, devices, appliances, fixtures, apparatus, and the like, used as a part of, or in connection with, an electrical installation.

equipment code
Transit. A six-digit numeric code used to classify equipment by its usage characteristics (passenger carrying, cargo hauling, etc.), gross weight rating, and equipment configuration (panel truck, pick-up, stake body, dump etc.).

equipment consist
Rail. An equipment consist is a train, locomotive(s), cut of cars, or a single car not coupled to another car or locomotive.

equipment damage
Rail. All costs, including labor and material, associated with the repair or replacement-in-

kind of on-track rail equipment. Trailers and/or container on flat cars are considered to be lading and damage to these is not to be included in on-track equipment damage. Damage to a flat car carrying a trailer/container is to be included in reportable damage.

equipment-type flow process chart
A flow process chart which provides a plan or usage record for equipment.

equity
A legal doctrine which emphasizes fairness as opposed to law in resolving disputes. Sometimes referred to as *balancing of equities*; for instance, when a court decides whether or not to issue an injunction.

equivalent airspeed
The calibrated airspeed of an aircraft corrected for adiabatic compressible flow for the particular altitude. Equivalent airspeed is equal to calibrated airspeed in standard atmosphere at sea level.

equivalent diameter
*See **aerodynamic diameter**.*

equivalent form
Any of two or more forms of some test which are very similar in content and difficulty and which are expected to yield similar means and variability for a given group.

equivalent groups method
*See **matched groups design**.*

equivalent mean luminance
The transformed luminance output by a flickering light compared to an equivalent steady light.

equivalent method
Any method of sampling or analyzing for air pollution which has been demonstrated to the EPA Administrator's satisfaction to be, under specific conditons, an acceptable alternative to the normally used reference methods.

equivalent sound level
The level, in decibels, of the mean-square A-weighted sound pressure during a stated time period, with reference to the square of the standard reference sound pressure of 20 micropascals. It is the level of the sound exposure divided by the time period and is abbreviated as L_{eq}.

equivalent weight
The weight of an element that combines chemically with 8 grams of oxygen or its equivalent.

erbium
A chemical element, atomic number 68, atomic weight 167.26, symbol Er.

ERDA
*See **Energy Research and Development Administration**.*

erect
Pertaining to a standing posture in which the individual's shoulders are back and the neck is fully extended.

erg
A unit of work equal to the force of one dyne acting through a distance of one centimeter.

ERG
*See **electroretinography**.*

ergograph (Kelso-Hellebrandt)
A device used for measuring muscle work output in a series of repetitive movements.

ergometer
Any device which permits some determination of the work performed by an individual over a period of time.

ergonometrics
*See **physiological work measurement**.*

ergonomic analysis
*See **human factors analysis**.*

ergonomic design of jobs
*See **job design**.*

ergonomic job analysis
*See **human factors analysis**.*

ergonomic lifting calculator
A sliding rule device distributed by the National Safety Council for determining whether or not a lifting task is acceptable.

ergonomics
A multi-disciplinary activity that concentrates on the interactions between the human and their total working environment with consideration for the stressors that may be present in that environment such as atmospheric heat, illumination, and sound as well as all the tools and equipment used in the work place. Also referred to as *human factors* and *human factors engineering*.

ergonomist
An individual trained in health, behavioral, and technological sciences and who is competent to apply those fields to the industrial environment to reduce stress on personnel and thereby prevent work strain from developing to pathological levels or producing fatigue, careless workmanship, or high employee turnover.

ERISA
See ***Employee Retirement Income Security Act.***

Erlanger-Gasser classification
A method for classifying motor neurons, based on conduction velocity, into three primary groups: A, B, and C, with the A group being further divided into four subgroups: α, β, γ, and δ.

ERMAC
Electromagnetic Radiation Management Advisory Council.

erosion
The wearing away of land surface by wind or water. Erosion occurs naturally from weather or runoff but can be intensified by land-clearing practices related to farming, residential or industrial development, road building, or timber-cutting.

ERPG
See ***Emergency Response Planning Guides.***

erroneous
Involving error; deviating from the law. This term is not generally used as designating a corrupt or evil act.

erroneous judgement
One rendered according to course and practice of court, but contrary to law, upon mistaken view of law, or upon erroneous application of legal principles.

error
(1) The difference between the true or actual value to be measured and the value to be measured and the value indicated by the measuring system. Any deviation of an observed value from the true value. (2) An inappropriate response by a system, whether of commission, omission, inadequacy, or timing. (3) A mistaken judgment or incorrect belief as to the existence or effect of matters of fact, or a false or mistaken conception or application of the law.

error in exercise of jurisdiction
Error in determination of questions of law or fact on which the court's jurisdiction in a particular case depends.

error in fact
Error in fact occurs when, by reason of some fact which is unknown to the court and not apparent on the record (e.g., infancy, or death of one of the parties), it renders a judgement void. Such occurs when some fact which really exists is unknown, or some fact is supposed to exist which really does not.

error in law
An error of the court in applying the law to the case on trial (e.g., in ruling on the admission of evidence, or in charging the jury.

error rate
The number of errors per division, in which the division may be time, number of products output, motions, or other quantifiable variable.

errors and omissions (O&E) insurance
A type of insurance that indemnifies the insured for any loss sustained because of an error or oversight on his/her part. *See also* ***insurance.***

ERV
See ***expiratory reserve volume.***

erysipelas
An inflammation of the skin marked by red patches with sharp border lines, usually due to Group A hemolytic streptococci. The visible symptoms or erysipelas, a form of cellulitis, are round or oval patches on the skin that promptly enlarge and spread, becoming swollen, tender, and red. The affected skin is hot to the touch, and, occasionally, the adjacent skin blisters. Headache, vomiting, fever, and sometimes complete prostration can occur.

erythema
A abnormal redness of the skin, due to distention of the capillaries with the blood. It can be caused by a various agents such as heat, certain drugs, ultraviolet rays, and ionizing radiation.

erythemal region
The electromagnetic spectrum in the ultraviolet region from 2800 angstroms to 3200 angstroms.

erythemal threshold
That level at which erythema becomes apparent. Also referred to as *minimal perceptible erythema*.

erythrasma
A chronic infection of the skin, marked by the development of red or brownish patches on the inner side of the thigh, on the scrotum, and in the axilla.

erythroblastemia
The presence in the peripheral blood of abnormally large numbers of nucleated red cells.

erythrocyte
A red blood cell which contains hemoglobin and transports oxygen to body tissues.

erythromycin
An antibiotic obtained from *Streptomyces erythreus*. It is effective against a wide variety of organisms, including gram-negative and gram-positive bacteria and many rickettsial and viral infectious agents. It may be administered orally or parenterally.

escalator clause
(1) In *union contracts,* a provision that wages will rise or fall depending on some standard like the cost of living index. (2) In a *lease,* a provision that rent may be increased to reflect an increase in real estate taxes, operating costs, and even increases in Consumer Price Index. (3) In *construction contracts,* a clause authorizing a contractor to increase his/her contract price should costs of labor or materials increase.

escape clause
A provision in a contract, insurance policy, or other legal document permitting a party or parties to avoid liability or performance under certain conditions.

eschar
Damage created to the skin and underlying tissue from a burn or as a result of contact with a corrosive material.

esophagus
That portion of the digestive system composed of the passageway extending from the lower part of the pharynx to the stomach. The hollow muscular tube extending from the pharynx to the stomach, consisting of an outer fibrous coat, a muscular layer, a submucous layer and an inner mucous membrane. The junction between the stomach and esophagus is closed by a muscular ring known as the cardiac sphincter, which opens to allow the passage of food into the stomach. In an adult the esophagus is usually 10 to 12 inches long.

esophoria
A condition in which the eyes tend to turn inward, preventing binocular vision.

ESP
See ***electrostatic precipitator***.

ESR
Electrical skin resistance. *See* ***skin resistance response***.

EST
See ***ex-ship's tackle***.

establishment
(1) *According to OSHA:* a) A single physical location where business is conducted or where services or industrial operations are performed. Examples include a factory, mill, store, hotel, restaurant, movie theater, farm, ranch, bank, sales office, warehouse, or central administrative office. When distinctly separate activities are performed at a single physical location (such as contract construction activities operated from the same physical location as a lumbar yard, each activity shall be treated as a separate establishment. b) For firms engaged in activities such as agriculture, construction, transportation, communications, and electric, gas, and sanitary services, which may be physically dispersed, records may be maintained at a place to which employees report each day. c) Records of personnel who do not primarily report or work at a single establishment, and who are generally not supervised in their daily work, such as traveling sales personnel, technicians, and engineers, shall be maintained at the location from which they are paid or the base from which personnel operate to carry our their activities. (2) *According to FRA:* A single physical location where business is conducted or where services or industrial operations are performed. Examples of railroad establishments include, but are not limited to

an operating division, general office, and a major installation such as a locomotive or car repair or construction facility. For employees who are engaged in dispersed operations, such as track maintenance workers, the "establishment" is the location where these employees report for work assignments. (3) *Law.* An institution or place of business, with its fixtures and organized staff.

establishment list
A list that contains the names of particular plants located within the territorial jurisdiction of the local OSHA Area Office that are of the types of industries that have been noted on the industry rank report. *See **also industry rank report**.*

esthesiometer
An instrument for measuring touch sensitivity.

esthetic
Pertaining to the senses, especially when pleasuring to the senses.

estimate ratio
The ratio of two population aggregates (totals). For example, "average miles traveled per vehicle" is the ratio of total miles driven by all vehicles, over the total number of vehicles, within any subgroup. There are two types of ratio estimates; those computed using aggregates for vehicles and those computed using aggregates for households. Also referred to as *aggregate ratio. See also **mean** and **ratio estimate**.*

estimated arrival time
The time the flight is estimated to arrive at the gate (scheduled operators) or the actual runway on times for nonscheduled operators.

estimated elapsed time
The estimated time required to proceed from one significant point to another.

estimated en route time
Aviation. The estimated flying time from departure point to destination liftoff to touchdown).

estrogen
(1) An estrus-producing substance. (2) A general name for the principal female sex hormones. These hormones are manufactured in the ovaries and, though each has a slightly different function, they are closely related and are usually referred to collectively as estrogen.

estuarine sanctuary
A research area which may include any part or all of an cstuary and any island, transitional area, and upland in, adjoining, or adjacent to such estuary, and which constitutes to the extent feasible a natural unit, set aside to provide scientists and students the opportunity to examine over a period of time the ecological relationships within the area.

estuary
(1) That part of a river or stream or other body of water having unimpaired connection with the open sea, where the sea water is measurably diluted with fresh water derived from land drainage. The term includes estuary-type areas of the Great Lakes and the Chesapeake Bay. (2) Associated aquatic ecosystems and those portions of tributaries draining into the estuary up to the historic height of migration of anadromous fish or the historic head of tidal influence, whichever is higher.

ET
Effective temperature.

ETA
*See **explosive transfer assembly**.* Also, an acronym for estimated time of arrival.

ETBA
*See **energy trace and barrier analysis**.*

ethanol
Otherwise known as ethyl alcohol, alcohol, or grain-spirit. A clear, colorless, flammable oxygenated hydrocarbon with a boiling point of 78.5°C in the anhydrous state. In transportation, ethanol is used as a vehicle fuel by itself (E100), blended with gasoline (E85), or as a gasoline octane enhancer and oxygenate (10% concentration).

Ethernet
Computing. A software protocol for building networks.

ethics
That moral code practiced by an individual or groups, typically referring to a moral code involving honesty, integrity, and other qualities generally judged to be good.

ethmocarditis
Inflammation of the connective tissue of the heart.

ethmoid bone
A relatively complex, irregularly shaped bone within the anterior medial region of the skull behind the nose.

ethnic group
A group of people who either maintains affiliation due to strong racial and/or cultural ties or is descended from a certain race or culture.

ethylene dibromide (EDB)
A chemical used as an agricultural fumigant and in certain industrial processes. Extremely toxic and found to be a carcinogen in laboratory animals, EDB has been banned for most agricultural uses in the United States.

ethylenediamine
A volatile, colorless liquid with an ammonia odor that is used as a solvent and in organic synthesis.

etiologic agents
Infectious microorganisms, viruses, or parasitic agents capable of producing infection and/or disease in a susceptible host.

etiology
The study or theory of the causation of disease; the sum of knowledge regarding disease causes.

ETS
Emergency temporary standard. *See* ***Section 6(c) standard***.

euphoria
The absence of pain or distress. An exaggerated sense of well-being.

eustachian tube
A hollow, tubular structure connecting the middle ear with the nasal/oral cavity.

eutrophic lakes
Shallow, murky bodies of water that have excessive concentrations of plant nutrients causing excessive algae production. An increase in mineral and organic nutrients reduces the dissolved oxygen, producing an environment that favors plant over animal life.

eutrophication
The slow aging process during which a lake, estuary, or bay evolves into a bog or marsh and eventually disappears. During the later stages of eutrophication, the water body is choked by abundant plant life as the result of increased amounts of nutritive compounds such as nitrogen and phosphorus. Human activities can accelerate the process.

EVA
See ***extravehicular activity***.

evaporation
The change of a substance from the solid or liquid phase to the gaseous or vapor phase.

evaporation fog
Fog produced when sufficient water vapor is added to the air by evaporation. The two common types are *steam fog,* which forms when cold air moves over warm water, and *frontal fog,* which forms as warm raindrops evaporate in a cool air mass.

evaporation ponds
Areas where sewer sludge is dumped and allowed to dry out.

evaporation rate
The rate at which a material will vaporize (evaporate) as compared to the known rate of a standard material (such as normal-butyl acetate). It is the ratio of the time required to evaporate a measured amount of a liquid to the time required to evaporate the same amount of a reference liquid under ideal test condition. Normal-butyl acetate has typically been used as the reference standard.

evaporative heat loss
The dissipation of body heat through perspiration, indicated by an equation of the form:

$$H = kA(P_s - P_a)$$

where:
H = evaporative heat loss
k = evaporative coefficient
A = body surface area
P_s = saturated vapor pressure of water at skin temperature
P_a = ambient water vapor pressure

evaporative heat transfer coefficient
The value of the ratio of the permeability index to the total thermal insulation value of clothing. Also referred to as *coefficient of evaporative heat transfer* and *evaporative transmissibility*.

evapotranspiration
The loss of water from the soil both by evaporation and by transpiration from the plants growing in the soil.

evasé
A gradual enlargement at the outlet of an exhaust system to reduce the air discharge velocity efficiently so that velocity pressure can be regained instead of being wasted as occurs when air is discharged directly from a fan housing.

evasive answer
One which consists of refusing either to admit or to deny a matter in a direct, straightforward manner as to which the person is necessarily presumed to have knowledge. An evasive answer is considered and treated as a failure to answer, for which a party may on motion seek a court compelling answers to discovery questions.

evening person
Slang term for an individual who generally likes to go to sleep late at night, likes to sleep late, and has trouble waking early in the morning.

evening shift
See ***second shift***.

event
(1) A collection of one or more sample points. (2) The consequence of anything; the issue or outcome of an action as finally determined.

event recorder
Rail Operations. A device, designed to resist tampering, that monitors and records data on train speed, direction of motion, time, distance, throttle position, brake applications and operations (including train brake, independent brake, and, if so equipped, dynamic brake applications and operations) and, where the locomotive is so equipped, cab signal aspect(s), over the most recent 48 hours of operation of the electrical system of the locomotive on which it is installed.

event tree
A graphic depiction of system or operational events as they are related to the top event or failure condition.

event tree analysis
A system safety analysis method, similar to fault tree analysis, used to examine different system or operational responses to various positive or negative conditions which occur during system operation.

eversion
A turning of the bottom of the foot outward such that the more sagittal portions are also elevated slightly.

evertor
Any muscle which is involved in eversion of the foot.

evidence
Any species of proof, or probative matter, legally presented at the trial of an issue, by the act of the parties and through the media of witnesses, records, documents, exhibits, concrete objects, etc. for the purpose of inducing belief in the minds of the court or jury as to their contention.

evidence by inspection
Such evidence as is addressed directly to the senses without intervention of testimony. Tangible, physical evidence.

evidence codes
Statutory provisions governing admissibility of evidence and burden of proof at hearings and trials.

evidence rules
Rules which govern the admissibility of evidence at hearing and trials (e.g., Federal Rules of Evidence, Uniform Rules of Evidence).

evoked potential (EP)
An electrophysiological response recorded from the brain or scalp which is time-linked to peripheral sensory stimulation. Synonymous with *evoked response*.

evoked response
See ***evoked potential***.

ex quay
Maritime. The seller makes the goods available to the buyer on the quay (wharf) at the destination named in the sales contract. The seller has to bear the full cost and risk involved in bringing the goods there.

ex ship
Maritime. The seller will make the goods available to the buyer on board the ship at the destination named in the sales contract. The seller bears all costs and risks involved in bringing the goods to the destination.

ex ship's tackle (EST)
Maritime. Similar to Cost, Insurance and Freight, but seller is responsible for loss and

damage until goods are delivered on dock at port of destination. Seller has to insure goods up to this point. Also called *ex ship*.

ex warehouse
Buyer is responsible for all charges to destination and has to arrange insurance to cover the goods from the time they leave the warehouse at the place of shipment until their arrival at final destination. Also called *ex works*.

ex works
See ***ex warehouse***.

exa
Prefix indicating 1 E+18.

examine (E)
A mental basic work element involving examination of a part or product.

exceedance
Violation of environmental protection standards by exceeding allowable limits or concentration levels.

excess
Any property under the control of a federal Agency which that agency determines is not required for its needs or for the discharge of its responsibilities.

excess air
A quantity of air in excess of the theoretical amount required to completely combust a material, such as a fuel, waste, etc. Also referred to as excess combustion air and is expressed as a percentage (e.g., 20% excess air).

excess baggage revenue
Revenues from the transportation by air of passenger baggage in excess of the free allowance.

excess liability damages
A cause of action in tort by an insured against his/her liability carrier for the negligent handling of settlement negotiations which result in a judgment against the insured in excess of his/her policy limits.

excess work allowance
A special time allowance given a worker for additional work required beyond that specified in his normal task or job or due to some alteration from usual working conditions. Also referred to as ***additional work allowance***.

excessive damages
Those damages awarded by a jury which are grossly in excess of the amount warranted by law on the facts and circumstances of the case; unreasonable or outrageous damages.

exchange rate
A tradeoff for an increased sound pressure level above recommended limits for a proportionately reduced period of time.

excitation
The addition of energy to a system, thereby transferring it from its ground state to an excited state.

excitation purity (p_e)
The distance between a color sample and neutral white in the 1931 CIE chromaticity diagram relative to the distance between neutral white and the spectrum locus or the purple boundary in the same direction.

excited state
An atom with an electron at a higher energy level than it normally occupies. This principal is employed in the use of thermoluminescent dosimeters (TLDs) for determining exposure to ionizing radiation with this type device.

exclusion zone
An area surrounding a Liquefied Natural Gas (LNG) facility in which an operator or government agency legally controls all activities in accordance with 49 CFR 193.2057 and 49 CFR 193.2059 for as long as the facility is in operation.

exclusionary
Any form of zoning ordinance that tends to exclude specific classes of persons or businesses from a particular district or area.

exclusive event
As pertains to fault tree analysis (FTA) and/or the Management Oversight and Risk Tree (MORT), a conditional event which places specific restrictions upon the occurrence of other events. Represented graphically as an oval. *See also* ***conditional event***.

exclusive right-of-way
A highway or other facility that can only be used by buses or other transit vehicles. *See also* ***controlled access rights-of-way***.

exclusive rights-of-way
Roadways or other right-of-way reserved at all times for transit use and/or other high occupancy vehicles. The restriction must be sufficiently enforced so that 95 percent of vehicles using the right-of-way are authorized to use it.

Excobedo Rule
Under this rule, when police investigation begins to focus on a particular suspect, the suspect is in custody, the suspect requests and is denied counsel, and the police have not warned him/her of his/her right to remain silent, the accused will be considered to have been denied assistance of counsel and no statement elicited during such interrogation may be used in a criminal trial.

excretion
The removal of a substance or its metabolites from the body in urine, feces, or expired air.

excursion
A movement or deviation from the norm. In industrial hygiene, it is the deviation above the norm that is of concern.

excursion limit
The amount by which an exposure limit can be exceeded, and the number of times in an exposure period it can be exceeded without causing an adverse health effect, narcosis, discomfort, impairment of self rescue, or reducing work efficiency.

Executive Order
A document promulgated by the President of the United States or the Governor of a state binding federal or state agencies, usually instructing them how to carry out or coordinate policies or programs. There are federal Executive Orders on flood plains and wetlands, for example.

executive transportation
Any use of an aircraft by a corporation, company, or other organization for the purposes of transporting its employees and/or property not for compensation or hire, and employing professional pilots for the operation of the aircraft.

exempt carrier
A for-hire interstate operator which transports commodities or provides types of services that are exempt from federal regulation, could also operate within exempt commercial zones.

exempt intracity zone
The geographic area of a municipality or the commercial zone of that municipality described by the Interstate Commerce Commission (ICC) in 49 CFR 1048, revised as of October 1, 1975. The descriptions are printed in Appendix F to Subchapter B of 49 CFR, Chapter III. The term "exempt intracity zone" does not include any municipality or commercial zone in the State of Hawaii. For the purposes of 49 CFR 390.3(g), a driver may be considered to operate a vehicle wholly within an exempt intracity zone notwithstanding any common control, management, or arrangement for a continuous carriage or shipment to or from a point without such zone.

exempt motor carrier
A person engaged in transportation exempt from economic regulation by the Interstate Commerce Commission (ICC) under 49 U.S.C. 10526. Exempt motor carriers are subject to the safety regulations set forth in 49 CFR, Chapter III, Subchapter B.

exempt solvent
Specific organic compounds that are not subject to requirements of regulation because they have been deemed by the EPA to be of negligible photochemical reactivity.

exempted aquifer
Underground bodies of water defined in the Underground Injection Control program as aquifers that are sources of drinking water (although they are not being used as such) and that are exempted from regulations barring underground injection activities.

exemption
A temporary or permanent grant, license, or form of legal permission given by an agency to deviate from a regulation or provision of law administered by that agency. Issued in response to a petition for relief submitted by an individual or company.

executive branch of government
That branch of government consisting of the chief executive (i.e., the President), and those offices and positions held under its control.

executive privilege
This privilege, based on constitutional doctrine of separation of powers, exempts the executive from disclosure requirements applicable to the ordinary citizen or organization where such exemption is necessary to the discharge of highly important executive responsibilities involved in maintaining governmental operations, and extends not only to military and diplomatic secrets but also to documents integral to an appropriate exercise of the executive's domestic decision and policy making functions, that is, those documents reflecting the frank expression necessary in intragovernmental advisory and deliberative communications.

exemplary damages
Damages on an increased scale, awarded to the plaintiff over and above what will barely compensate for his/her property loss, where the wrong done to him/her was aggravated by circumstances of violence, oppression, malice, fraud, or wanton and wicked conduct on the part of the defendant, and are intended to solace the plaintiff for mental anguish, laceration of his/her feelings, shame, degradation, or other aggravations of the original wrong, or else to punish the defendant for his/her evil behavior or to make an example of him/her, for which reason they are also called *punitive* or *punitory damages* or *vindictive damages.* Unlike compensatory or actual damages, exemplary or punitive damages are based upon an entirely different public policy consideration: that of punishing the defendant or of setting an example for similar wrongdoers. In cases in which it is proved that a defendant has acted willfully, maliciously, or fraudulently, a plaintiff may be awarded exemplary damages in addition to compensatory or actual damages.

exemption
Freedom from a general duty or service; immunity from a general burden, tax, or charge.

exercise
The use of muscular exertion to maintain conditioning, train for an athletic event, or in an attempt to maintain health.

exercise physiology
The study of the metabolic activities and changes ongoing during exercise, including the aerobic and anaerobic mechanisms, and respiratory, neuromuscular, and cardiovascular mechanisms.

exfiltration
The flow of air from inside a building to the outside due to the existence of negative pressure outside the building surface.

exfoliation
The peeling or flaking off of the skin.

exhalation
(1) The expulsion of air or other vapor from the lungs. (2) Escape in the form of vapor. (3) Vapor escaping from a body or substance.

exhaust air
That air rejected to the outside from a ventilation system.

exhaust grill
Fixture in the wall, floor, or ceiling through which air is exhausted from a space.

exhaust hood
A structure to enclose or partially enclose a contaminant-producing operation or process, or to guide air flow in an advantageous manner to capture a contaminant and is connected to a duct/pipe or channel for removing the contaminant from the hood.

exhaust rate
The volumetric flow rate at which air is removed by a ventilation system.

exhaust system
(1) The combination of components which provides for the enclosed flow of exhaust gas from the engine exhaust port to the atmosphere. (2) Any constituent components of the combination that conducts exhaust gases and which are sold as separate products. (3) A system for removing contaminated air from a space, comprising one or more of the elements including an exhaust hood, duct work, air-cleaning equipment, exhauster, and stack.

exhaust ventilation
Mechanical removal of air from a portion of a building (e.g., piece of equipment, room or general area).

exhausting work
That level of work activity which has a gross metabolic cost of over 380 calories per square meter of skin surface per hour in young men.

exhaustion of administrative remedies
A legal doctrine stipulating the need for a party to pursue all available, possibly fruitful

appeals within an agency before challenging that agency in court.

existing source
Under the Clean Air Act, any stationary source other than a new source.

exit
That portion of a means of egress which is separated from all other spaces of the building or structure by construction or equipment to provide a protected way of travel to the exit discharge. *See also* ***means of egress*** *and* ***exit discharge***.

exit access
That portion of a means of egress which leads to an entrance to an exit. *See also* ***means of egress***.

exit discharge
That portion of a means of egress between the termination of an exit and a public way. *See also* ***means of egress***.

exogenous
Derived or developed from external causes.

exophoria
A condition in which the eyes tend to turn outward, preventing binocular vision.

exoskeleton
An external hard framework that supports and protects the soft tissues of lower animals, derived from the ectoderm. In vertebrates, the term is sometimes applied to structures produced by the epidermis, as hair, nails, hoofs, teeth, etc.

exosphere
The outermost portion of the atmosphere.

exothermic
When applied to reactions, describes those that produce substances that have less energy than the reaction materials resulting in a release of energy as heat.

exotoxin
A microbial toxin (i.e., a toxin excreted by a microorganism into a surrounding medium).

expandable
Flatbed trailer which can be expanded beyond its regular length to carry larger shipments.

expect altitude at time or fix
Air traffic control terminology. Used under certain conditions to provide a pilot with an altitude to be used in the event of two-way communications failure. It also provides altitude information to assist the pilot in planning.

expect further clearance
Air traffic control terminology. Used to inform a pilot of the routing he can expect if any part of the route beyond a short range clearance limit differs from that filed.

expectancy damages
As awarded in actions for nonperformance of contract, such damages are calculable by subtracting the injured party's actual dollar position as a result of the breach from that party's projected dollar position had performance occurred. The goal is to ascertain the dollar amount necessary to ensure that the aggrieved party's position after the award will be the same (to the extent money can achieve this identity as if the other party had performed as expected).

expectation
A mental set in which an individual anticipates a certain outcome in a given situation.

expected attainment
See ***fair day's work***.

expected departure clearance time (EDCT)
Air traffic control terminology. The runway release time assigned to an aircraft in a controlled departure time program and shown on the flight progress strip as an EDCT.

expected work pace
The rate of work output required to achieve a certain level of earnings or production standards.

expectorate
To cough up and eject from the mouth by spitting.

expedite
Used by Air Traffic Control (ATC) when prompt compliance is required to avoid the development of an imminent situation.

expenditure
All amounts of money paid out by a government, net of recoveries and other correcting transactions, other than retirement of debt, investment in securities, extension of credit, or agency transactions. Federal expenditures are also referred to as outlays.

expenditures
Funds spent for energy purchased and paid for or delivered to a manufacturer during a calendar year. The expenditure dollar includes state and local taxes and delivery charges.

experience
The verifiable, objective history of one's work performance.

experience curve
A graphical plot of a worker's performance over time, especially in the learning phase of a job.

experience rating
A method for adjusting workers' compensation rates using a three-year history of the employer's claim experience. *See also* ***merit rating (2)***.

experimental aircraft
An aircraft which does not have a type design or does not meet other certification standards. The "experimental" designation is one of several "Special Airworthiness Certificates" which allows the aircraft to operate in U.S. airspace. None may be used for commercial purposes. Experimental aircraft are divided into three groups: a) Amateur Built: an aircraft, built by one or more persons who undertake the effort for the purpose of recreation and education; d) Exhibition: a unique (one-of-a-kind) aircraft, a replica, a foreign or U.S. military surplus aircraft which may be used for exhibition purposes, movie and television productions, or sanctioned, organized events where the unique or unusual characteristics of the aircraft can be displayed; c) Other: includes experimental aircraft that are not amateur or exhibition. This includes aircraft involved in research and development, crew training, market surveys, air racing, those used to show compliance with regulations, and the like.

experimental variable
See ***independent variable*** *and* ***dependent variable***.

experimenter
One who designs, supervises, and/or conducts research.

experimenter error
Any error resulting from an experimenter's inappropriate action or inaction, regardless of its nature.

expert
An individual who (a) possesses certain knowledge, wisdom, and/or skills in a particular subject not likely to be possessed by ordinary persons, (b) acquired such knowledge, wisdom, and/or skills by study, investigation, and/or experience, (c) is capable of reasoning, inference, and drawing conclusions based on hypothetical facts relating to that subject, and (d) can offer reasonable opinions regarding one or more situations dealing with that particular subject.

expert evidence
Any testimony given by an expert witness based on objective data or information, or information derived directly from such objective data or information. *See also* ***expert testimony***.

expert opinion
A statement of belief by an expert witness, based on a given situation.

expert system
A decision-making job aid, generally developed in consultation with experts in a given field and which typically contains a computer-based model and database generated from that human expertise.

expert testimony
The opinion of a person skilled in a particular art, science, or profession, having demonstrated special knowledge through experience and education, beyond that which is normally considered common for that art, science, or profession. *See also* ***expert, expert evidence,*** *and* ***expert witness***.

expert witness
(1) *General.* A witness qualified as a subject expert based upon their knowledge, skill, experience, training, or education. Unlike other witnesses, an expert's testimony may be in the form of an opinion. *See also* ***expert testimony***. (2) *Law.* One who by reason of education or specialized experience possesses superior knowledge with respect to a subject about which persons having no particular training are incapable of forming an accurate opinion or deducing correct conclusions.

expiration
(1) Exhaling of the lungs caused by the relaxation of the diaphragm and rib muscles which causes decreased chest cavity space,

thus forcing air out through the trachea. (2) Cessation; termination from mere lapse of time, as the expiration date of a lease, insurance policy, statute, and the like. Coming to a close; termination or end.

expiratory flow rate
The maximum rate at which air can be expelled from the lungs.

expiratory reserve volume (ERV)
The maximum amount of air that can be forcibly expired after a normal expiration.

exploration
Under the Federal Antarctic Protection Act of 1990: Any activity, including logistic support, the purpose of which is the identification or evaluation of specific mineral resource deposits. The term includes exploratory drilling, dredging, and other surface or sub-surface excavations required to determine the nature and size of mineral resource deposits and the feasibility of their development.

explosimeter
A device for detecting the presence of, and measuring the concentration of, gases or vapors that can reach explosive concentrations.

explosion
A rapid build-up and release of pressure caused by chemical reaction or by an overpressurization within a confined space leading to a massive rupture of the pressurized container.

explosion/detonation
Railroad Operations. An accident/incident caused by the detonation of material carried by or transported by rail. A detonation occurs when a shock wave exceeds the speed of sound. Explosions/detonations resulting from mishaps during loading or unloading operations, and those caused by fire aboard on-track equipment are included in this definition.

explosion-proof
The design of a device or equipment to eliminate the possibility of its igniting volatile materials. A type of construction that is designed to contain an explosion and prevent its propagation to the atmosphere outside the device/equipment.

explosion-proof apparatus
An apparatus enclosed in a case that is capable of withstanding an explosion of a specified gas or vapor which may occur within it and of preventing the ignition of a specified gas or vapor surrounding the enclosure by sparks, flashes, or explosion of the gas or vapor within, and which operates at such external temperature that it will not ignite a surrounding flammable atmosphere.

explosive
Any chemical compound, mixture, or device, the primary or common purpose of which is to function by explosion (i.e., with substantially instantaneous release of gas and heat), unless such compound, mixture, or device is otherwise specifically classified by a governing authority, such as the U.S. Department of Transportation (DOT). The term "explosive" shall include all material which is classified as Class A, Class B, and Class C by the DOT and includes, but is not limited to, dynamite, black powder, pellet powders, initiating explosives, blasting caps, electric blasting caps, safety fuse, fuse lighters, fuse ignitors, squibs, cordeau detonate fuse, instantaneous fuse, ignitor cord, ignitors, small arms ammunition, small arms ammunition primers, smokeless propellant, cartridges for propellant-actuated power devices, and cartridges for industrial guns. Commercial explosives are those explosives which are intended to be used in commercial industrial operations.

explosive-actuated power devices
Any tool or special mechanized but not including propellant-actuated power devices. Examples of explosive-actuated power devices are jet tappers and jet perforators.

explosive atmosphere
An atmosphere containing a mixture of vapors or gases which is within the explosive or flammable range. Also referred to as an *explosive mixture*.

explosive decompression
A rapid and significant decrease in barometric pressure.

explosive limit
See ***lower flammable limit*** *and* ***upper flammable limit***. Also referred to as *explosive limit*.

explosive mixture
See ***explosive atmosphere***.

explosive quantity distance site plan
A formal plan for explosives facilities and areas detailing the quantity of explosives, operating and storage limits and restrictions, and resultant distance clearance requirements.

explosive range
See ***flammability range***.

explosive strength
That force expended in a very short burst of intense muscular activity.

explosive train
See ***explosive transfer assembly***.

explosive transfer assembly (ETA)
An arrangement of explosive or combustible elements used to perform or transfer energy to an end function. Also referred to as *explosive train*.

exponent
A number conventionally placed to the right and above a base number, representing the power to which the base number is raised for evaluation.

exponential decay
As pertains to ionizing radiation, a mathematical expression describing the rate at which radioactive materials decay.

exponential distribution
A distribution having the probability distribution function of

$$f(x) = ae^{-ax}$$

where:
a = 1/mean, and $a > 0$ for $x > 0$
$f(x)$ = 0 for $x \leq 0$

export license
A government document permitting designated goods to be shipped out of the country as specified.

exports
(1) Outbound international freight, including re-export of foreign merchandise. (2) Shipments of goods from the 50 States and the District of Columbia to foreign countries and to Puerto Rico, the Virgin Islands, and other U.S. possessions and territories.

exposed
(1) *General.* Having come into close contact with something that may cause adverse physical or mental effects. (2) *Wiring Methods.* Where the circuit is in such a position that in case of failure of supports or insulation, contact with another circuit may result. Also, wires installed on or attached to the surface or behind panels designed to allow access.

exposed pipeline
A pipeline where the top of the pipe is protruding above the seabed in water less than 15 feet deep, as measured from the mean low water.

exposure
(1) *General.* a) A measure representing some combination of the amount of time an individual or object has been located in some environment and the severity of that environment. b) The amount of radiation or pollutant present in an environment which represents a potential health threat to the living organisms in that environment; the amount of biological, physical, or chemical agent that reaches a target population. c) The proximity to a condition that may produce injury or damage. (2) *Physiology.* Contact of an organism with a chemical, biological, or physical agent. Exposure is quantified as the amount of the agent available at the exchange boundaries of the organism (skin, lungs, etc.) and available for absorption; also, the route by which an organism comes in contact with a toxicant (inhalation, ingestion, dermal absorption, injection).

exposure assessment
(1) The defining of exposure pathways and the calculation of the potential magnitude of exposure. (2) The determination or estimation (qualitative or quantitative) of the magnitude, frequency, duration, and route of exposure. (3) Estimation of the amount of chemicals that may be ingested, inhaled, or absorbed through the skin by people living, working, or recreating in areas where air, water, soil, etc. may be contaminated.

exposure dose
A measure of the x-radiation or gamma radiation at a certain place, based upon the ability of the radiation to produce ionization. The unit of measure is the roentgen (R).

exposure dose rate
The radiation exposure dose per unit time expressed as R/unit time.

exposure event
An incident of contact with a chemical or physical agent. An exposure event can be defined by time (e.g., day, hour) or by the incident (e.g., eating a single meal of contaminated fish).

exposure hours
*See **employee hours**.*

exposure limit
A somewhat outdated term for the maximum vibration acceleration as a function of frequency and duration.

exposure pathway
The course a chemical or physical agent takes from the source to the exposed organism. An exposure pathway describes a unique mechanism by which an individual or population is exposed to chemicals or physical agents at or originating from the site. Each exposure pathway includes a source or release from a source, an exposure point, and an exposure route. If the exposure point differs from the source, a transport/exposure medium (e.g., air) or media (in cases of intermedia transfer) also is included.

exposure point
A point of potential contact between an organism and a chemical or physical agent.

exposure point concentration
The concentration of a chemical at the exposure point.

exposure route
The way a chemical or physical agent comes in contact with an organism (i.e., by ingestion, inhalation, injection, or dermal contact).

express body
Open box truck body.

express bus
A bus that operates a portion of the route without stops or with a limited number of stops.

expressway
A divided highway for through traffic with full or partial access control and including grade separations at all or most major intersections.

extend
Move adjacent body segments connected by a common joint such that the angle between the segments increases in the direction opposite to that of maximum flexion.

extended duty hours
*See **extended work hours**.*

extended functional reach
*See **thumb-tip reach, extended**.*

extended hours
*See **extended work hours**.*

extended over-water operations
(1) With respect to aircraft other than helicopters, an operation over water at a horizontal distance of more than 50 nautical miles from the nearest shoreline. (2) With respect to helicopters, an operation over water at a horizontal distance of more than 50 nautical miles from the nearest shoreline and more than 50 nautical miles from an off-shore heliport structure.

extended source
Any energy source whose dimensions are significant relative to the distance between the source and the point of observation. (Note: "significant" usually refers to greater than about 10' of arc for visual work).

extended work hours
That working time beyond the normal workday hours. Also referred to as *extended duty hours* or *extended hours*.

extension ladder
A non-self-supporting portable ladder adjustable in length. It consists of two or more sections traveling in guides or brackets so arranged as to permit length adjustment. Its size is designated by the sum of the lengths of the section measured along the side rails.

extension trestle ladder
A self-supporting portable ladder, adjustable in length, consisting of a trestle ladder base and a vertically adjustable single ladder, with suitable means for locking the ladders together. The size is designated by the length of the trestle ladder base.

extensor
Any muscle whose contraction normally causes joint extension.

extensor retinaculum
A membranous band of fibers in the posterior hand/wrist which forms the carpal tunnel through which the finger extensor tendons pass. Also referred to as *transverse dorsal ligament*.

extent flexibility
The ability to twist, stretch, bend, or reach out with one or more parts of the body on a one-time basis.

external
Beyond the outer or surface portion of the body or a body segment.

external auditory canal
The tubular structure leading from the external environment to the tympanic membrane. Also referred to as the *external auditory meatus*.

external auditory meatus
See ***external auditory canal***.

external canthus
See ***ectocanthus***.

external combustion engine
An engine in which fuel combustion takes place outside the cylinder, turbine, or the like and in which energy is turned into mechanical force; for example, a steam engine.

external ear
The visible, most lateral aspects of the ear, including the auricle, external auditory canal, and the tympanic membrane. Sometimes referred to as the *outer ear*.

external element
Any work element in a progress or operation which is performed by the operator outside the machine- or process-controlled time. *See also* ***external work***.

external load
Aviation. A load that is carried, or extends, outside of the aircraft fuselage.

external load attaching
The structural components used to attach an external load to an aircraft, including externally loaded containers, the backup structure at the attachment points, and any quick-release device used to jettison the external load.

external mechanical environment
The manmade physical environment, consisting of tools, equipment, etc.

external naris
The entrance from the exterior to the air passageway of the nose. Also referred to more commonly as *nostril*.

external occipital protuberance
See ***inion***.

external pacing
Pertaining to externally paced work.

external radiation
Ionizing radiation in which the source is located outside the body and the radiation penetrates into deeper tissues.

external time
That amount of time required to perform manual work elements when a machine is not in operation.

external viewing
Having the capability for seeing outside a vehicle, either to view the vehicle itself or the surrounding environment.

external work
Any work element or combination of work elements in a process or operation which is performed by the operator outside the machine- or process-controlled time. Also referred to as *outside work*. *See also* ***external element***.

externally operable
Capable of being operated without exposing the operator to contact with live parts.

externally paced element
A work element whose completion is beyond a worker's control. Also referred to as *restricted element*.

externally paced work
Any manual or human/machine work in which the work pace and/or output is at least in part beyond a worker's control. Also called *restricted work*.

exteroceptor
Any sensory receptor at the body surface which receives information about the external environment.

extinguisher classification
The letter classification given an extinguisher to designate the class or classes of fire on which an extinguisher will be effective.

extinguisher rating
The numerical rating given to an extinguisher which indicates the extinguishing potential of the unit based on standardized tests developed by Underwriters' Laboratories, Inc.

extinguishing agent
Any substance capable of performing a fire extinguishing function.

extorsion
A rotation of one or both eyes about their vertical axes away from the midline (opposite of *intorsion*).

extortion
The obtaining of property from another induced by wrongful use of actual or threatened forces, violence, or fear, or under control of an official right. A person is guilty of theft by extortion if he/she purposely obtains property of another by threatening to: a) inflict bodily injury on anyone or commit any other criminal offense; or, b) accuse anyone of a criminal offense; or c) expose any secret tending to subject any person to hatred, contempt, ridicule, or to impair his/her credit or business repute; or d) take or withhold action as an official, or cause an official to take or withhold action; or e) bring about or continue to strike, boycott, or other collective unofficial action, if the property is not demanded or received for the benefit of the group in whose interest the actor purports to act; or f) testify or provide information or withhold testimony or information with respect to another person's legal claim or defense; or g) inflict any other harm which would not benefit the actor. *See also* ***blackmail***.

extra allowance
That additional time allowed for the completion of work which is not specified in the standard allowance.

extracanthic diameter
The horizontal linear distance between endocanthus and ectocanthus of one eye.

extracellular water (ECW)
That bodily water external to the cells. *See also* ***total body water***.

extrafusal fiber
The contractile fiber of muscle tissue which is capable of generating motion or tension. *See also* ***intrafusal fiber***.

extraocular muscle
Any of the six voluntary muscles which are capable of positioning the eyeball within the orbit.

extraordinary flood
A flood whose unexplained occurrence is not foreshadowed by the usual course of nature, and whose magnitude and destructiveness could not have been anticipated or provided against by the exercise of ordinary foresight. One such unusual occurrence is that it could not have been foreseen by men of ordinary experience and prudence.

extraordinary grand jury
Such a jury is limited in the scope of its investigation and may not go beyond terms of executive proclamation, and examination of witness must be confined within those terms, and must not be used as a means of disclosing or intermeddling with extraneous matters.

extraordinary hazards
One not commonly associated with a job or undertaking. If hazards are increased by what other employees do, and an injured employee has no part in increasing them, they are considered to be "extraordinary."

extraordinary items income (loss)
Income or loss which can be characterized as material, unusual, and of infrequent occurrence.

extraordinary nuclear occurrence
Under the Federal Atomic Energy Act of 1954: Any event causing a discharge or dispersal of source, special nuclear, or byproduct material from its intended place of confinement in amounts off site, or causing radiation levels off site, which the Nuclear Regulatory Commission or the Secretary of Energy, as appropriate, determines to be substantial, and which the Nuclear Regulatory Commission or the Secretary of Energy, as appropriate, determines has resulted or will probably result in substantial damages to persons off site or property off site. Any determination by the Nuclear Regulatory Commission or the Secretary of Energy, as appropriate, that such an event has, or has not, occurred shall be final

and conclusive, and no other official or any court shall have power or jurisdiction to review any such determination. The Nuclear Regulatory Commission or the Secretary of Energy, as appropriate, shall establish criteria in writing setting forth the basis upon which such determination shall be made.

extraordinary risk
A risk lying outside of the sphere of the normal, arising out of conditions not usual in the business. It is one which is not normally and necessarily incident to the employment, and is one which may be obviated by the exercise of reasonable care by the employer.

extrapolate
To estimate a value beyond current knowledge by using known current values and a predictor.

extrapolation
Using known data to predict or estimate unknown outcomes. A calculation, based on limited data from natural or experimental observation of humans or other organisms exposed to a substance, that aims to estimate the dose-effect relationship outside the range of the available data.

extrapyramidal system
A collection of subcortical neural structures involved in skeletal muscle activities which generally have more central integration, are slower than and supportive of pyramidal system motor function, and have involvement with postural motions.

extrasystole
A premature heartbeat.

extratropical cyclone
A cyclonic storm that most often forms along a front in middle and high latitudes. It is not a tropical storm or hurricane.

extravehicular activity (EVA)
That activity outside a support or transport vehicle, especially referring to space flight which requires a space suit.

extravehicular mobility unit (EMU)
An enclosed and self-contained clothing set for protecting the occupant outside a protective vehicle in a hazardous environment.

Astronaut performing an EVA wearing an EMU

extreme value projection
In system safety, a risk projection technique used to provide information about potential losses (i.e., in the future) that are more severe than those occurring in the past.

extremely hazardous substances
Any of 406 chemicals identified by the EPA on the basis of toxicity, and listed under SARA Title III. The list is subject to revision.

extremely high frequency (EHF)
That portion of the electromagnetic spectrum consisting of radiation frequencies between 30 GHz and 300 GHz.

extremely low frequency (ELF)
That portion of the electromagnetic spectrum consisting of radiation frequencies below 300 Hz.

extremely low frequency magnetic field
A magnetic field with a frequency in the range of 0 to 3000 hertz that results from current flowing in electrical conductors.

extremity
Term referring to either an upper limb (arms) or a lower limb (legs).

extrinsic
Pertaining to a structure or mechanism which originates outside the structure on which it acts.

eye
(1) *Anatomy.* The organ of vision. The total of all structures and tissues enclosing and enclosed within the eyeball. (2) *Meteorology.* A region in the center of a hurricane (tropical storm) where the winds are light and skies are clear to partly cloudy.

eye blink
A brief closure and re-opening of both eyelids. Also referred to simply as *blink.*

eye blink rate
The number of occasions within a specified temporal interval that an individual executes an eye blink.

eye dominance
See ***ocular dominance****.*

eye height, sitting
The vertical distance from the upper seat surface to endocanthus. Measured with the individual seated erect and looking straight ahead.

eye height, standing
The vertical distance from the floor or other reference surface to endocanthus. Measured with the individual standing erect, looking straight ahead, and his/her weight balanced evenly on both feet.

eye movement
Any active or passive, conscious or unconscious movement of the eyeball relative to the orbit.

eye protector
A device worn by a person or affixed to equipment to deter harmful substances from contact with the human eye. The use of safety glasses, splash goggles, or other protective eye wear that will reduce the potential for eye contact with a hazardous material being used, handled, or processed is required under OSHA 29 CFR 1910.132. Such eyewear must meet or exceed the specification of the American National Standards Institute (ANSI Z-87). Normally, eye protection is provided at no cost to the worker when it is determined that such protection is required in a particular work environment.

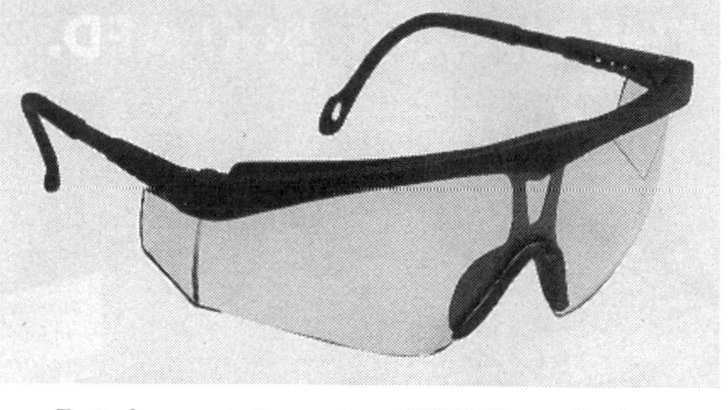

Typical eye protector meeting ANSI Z-87 specifications

eye scan
To scan the visual field by eye movement alone, not allowing or using any head movements.

eye sensitivity curve
See ***spectral luminous efficiency function****.*

eye wall
A wall of dense thunderstorms that surrounds the eye of a hurricane.

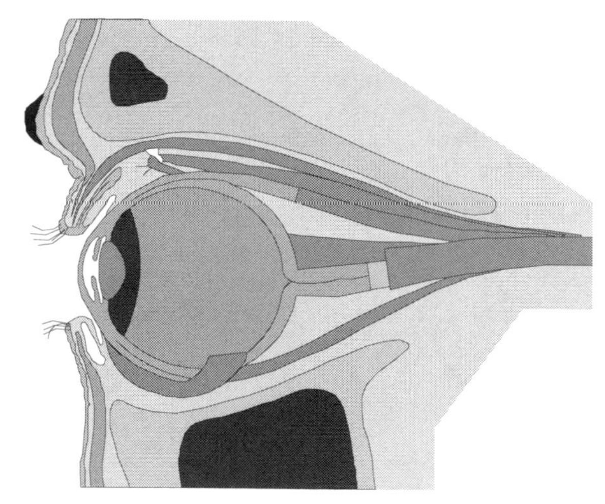

The human eyeball and surrounding components

eyeball
The approximately spherical portion of the eye, including the sclera, cornea, pupil/iris, retina, intraocular fluids, lens, and blood vessels. The *cornea* is the clear transparent layer on the front of the eyeball. It is a continuation of the sclera (the white of the eye), the tough outer coat that helps protect the delicate mechanism of the eye. The *choroid* is the middle layer and contains blood vessels. The third layer, the *retina,* contains rods and cones, which are specialized cells that are sensitive to light. Behind the cornea and in front of the lens is the iris, the circular pigmented band around the pupil. The iris works

much like the diaphragm in a camera, widening or narrowing the pupil to adjust to different light conditions.

eyebrow
The supraorbital ridge with its associated overlying tissues and hairs.

eyecup
A small vessel for the application of cleansing or medical solution to the exposed area of the eyeball.

eyeflush
The process of rinsing fluid over the conjunctiva and anterior eyeball with water or eyewash.

eyeground
The fundus of the eye.

eyelash
A short, curved hair embedded in the free edges of the eyelids, usually in two or three separate rows.

eyelid
A thin, soft, movable structure which overlies the anterior portion of the eyeball, is capable of closure to protect the eyeball from certain stimuli, is lined on its posterior surface by the conjunctiva, and contains various glands, a muscle, and the eyelashes.

eyepiece
The lens or system of lenses of a microscope nearest the eye of the observer when the instrument is in use.

eyestrain
A visuo-motor fatigue resulting from a prolonged period of muscle tension to focus to overcome glare or any other vision-interfering conditions. Also referred to as *visual strain*.

eyewash
A solution for flushing the eyes.

eyewash fountain
A device used to irrigate and flush the eyes in the event of eye contact with a hazardous substance. Generally speaking, water deluge from the fountain must be capable of providing a continuous flow for a period of not less than 15 minutes.

eyewear
Any type of eye covering, whether for eye protection or for improving vision. *See also* ***eye protector***.

eyewitness
A person who can testify as to what he/she has seen from personal observation. One who saw the act, fact, or transaction to which he/she testifies. Distinguished from *earwitness*.

F

F

A variable obtained from computing the *F* ratio and used in tests of statistical significance.

°F

Degrees Fahrenheit. *See* ***Fahrenheit temperature scale***.

***F* distribution**

That frequency distribution obtained by taking repeated random pairs of independent samples and calculating the *F* ratio.

***F* ratio**

The ratio of two chi squares divided by their respective degrees of freedom.

***F* test**

The use of an obtained *F* value with the degrees of freedom for each of the mean squares in an *F* distribution to indicate the probability that the samples are from the same population. Also referred to as *variance ratio test*.

FAA

See ***Federal Aviation Administration***. Also, in maritime insurance means "Free of all average," denoting that the insurance is against total loss only.

fabric filter

A cloth device that catches dust particles from industrial emissions.

fabric softener

Any of a class of cationic amine compounds of substituted fatty acids which act to reduce wrinkling and increase fluffiness while retaining moisture to reduce static electricity/cling. Also called *textile softener*.

fabricated evidence

(1) Evidence manufactured or arranged after the fact, and either wholly false or else warped and discolored by artifice and contrivance with a deceitful intent. To fabricate evidence is to arrange or manufacture circumstances or *indicia* (after the fact committed) with the purpose of using them as evidence, and of deceitfully making them appear as if accidental or undesigned. (2) To devise falsely or contrive by artifice with the intention to deceive. Such evidence may be wholly forged and artificial, or it may consist of so warping and distorting real facts as to create an erroneous impression in the minds of those who observe them and then presenting such impression as true and genuine.

fabricated fact

In the law of evidence, a fact existing only in statement, without any foundation in truth. An actual or genuine fact to which a false appearance has been given by design; a physical object placed in a false connection with another, or with a person on whom it is designed to cast suspicion.

Fabry's syndrome

A genetically transmitted disorder characterized by remittent attacks of fever, lightning pains and burning dysesthesia of the extremities, proteinuria and hematuria, and cutaneous lesions.

face

(1) The anterior portion of the head, from crinion to menton, and from right otobasion to left otobasion. (2) The surface of anything, especially the front, upper, or outer part or surface. That which particularly offers itself to the view of a spectator. The words of a written paper in their apparent or obvious meaning, as, the face of a note, bill, bond, check, draft, judgment record, or contract. The face of a judgment for which it was rendered exclusive of interest.

face amount

The amount of an instrument is that shown by the mere language employed, and excludes any accrued interest.

face breadth

See ***bizygomatic breadth***.

face line

A line used from head of boat to the tow.

face shield

A protective device designed to prevent hazardous materials, dusts, sharp objects, and other materials from contacting the face. A device worn in front of the eyes and a portion of, or all of, the face. It supplements the eye protection afforded by a primary protective device, such as safety glasses.

face up

To make-up the towboat to the tow (i.e., maneuver barges into position and secure for towing).

face validity
Having apparently relevant or appropriate measure, statement, or data.

face velocity
The average air velocity in the plane of an opening into an enclosure, such as a hood, through which air moves, usually expressed in feet per minute or meters per second.

face wires
Heavy cables securing boat to tow (i.e., pusher to barge).

facepiece
That part of a respirator which covers the wearer's nose, mouth, and in a full facepiece, the eyes.

facet
A smooth, generally flat surface on a bone.

facial angle
That angle formed by the intersection of a line connecting nasion and gnathion with the Frankfort plane of the head.

facial breadth
See ***bizygomatic breadth***.

facial disfigurement
That which impairs or injures the beauty, symmetry, or appearance of a person. That which renders unsightly, misshapen, or imperfect, or deforms in some manner.

facial hair policy
Respirators are not to be worn when conditions prevent a good facial seal. Such conditions may include the presence of a beard, long sideburns, mustache, or other facial hair growth. A facial hair policy is one which does not permit the presence of facial hair that could prevent a good respirator-to-face seal on personnel who may be required to wear such devices. Some facilities do not permit such facial hair on anyone who comes on the site.

facial height
The vertical linear distance between crinion and menton in the midsagittal plane. Also referred to as *facial length*. *See also* ***facial height, total***.

facial height, total
The sellion-menton length. *See also* ***facial height***.

facial index
The ratio of the facial length to the face breadth.

facial length
See ***facial height*** *and* ***facial height, total***.

facial nerve
A cranial nerve having both motor and sensory aspects, and which is involved in facial expressions, cutaneous sensations, and taste.

facies
The front aspect of the head.

facility
(1) *General.* Any building, plant, factory, office complex, or other structure where work or other designated activity or activities occur on a regularly scheduled or unscheduled basis. Also, all or any portion of buildings, structures, sites, complexes, equipment, roads, walks, passageways, parking lots, or other real or personal property, including the site where the building, property, structure, or equipment is located. (2) *CERCLA.* Broadly defined under Superfund to include any structure, installation, equipment, landfill impoundment, storage vessel, vehicle, or any site or area where hazardous substances have been deposited or otherwise have come to be located. (3) *OSHA.* The buildings, containers, or equipment which contain a process.

Facility Index System (FINDS)
An informational database that provides the EPA with an inventory of almost 500,000 facilities. FINDS contains both facility information and pointers to other sources of information that contain more detailed information about the facility.

facing movement
Rail. The movement of a train over the points of a switch which face in a direction opposite to that in which the train is moving.

facing point lock plunger
Rail. That part of a facing point lock which secures the lock rod to the plunger stand when the switch is locked.

facing point switch
Rail. A switch, the points of which face traffic approaching in the direction for which the track is signaled.

fact
A thing done; an action performed or an incident transpiring; an event or circumstance; an actual occurrence; an actual happening in time or space or an event mental or physical; that which has taken place.

fact question
Those issues in a trial or hearing which concern facts or events and whether such occurred and how they occurred as contrasted with issues and questions of law. Fact questions are for the jury, unless the issues are presented at a bench trial, while law questions are decided by the judge. Fact questions and their findings are generally not appealable though rulings of law are subject to appeal.

factor
(1) A set of related variables as determined by factor analysis. (2) An agent or element that contributes to the production of a result. (3) *See* ***variable***.

factor analysis
A statistical data treatment in which variable scores are analyzed and rotated to obtain orthogonality and achieve a summary in terms of a minimum number of factors.

factor loading
A calculated measure of the degree of generalization between variables and factors in a factor analysis.

factor of safety
(1) *Ultimate.* The ratio of the ultimate stress to the maximum calculated stress based on limit loads, as follows:

$$\textit{Ultimate Factor of Safety} = \frac{\textit{Ultimate Strength}}{\textit{Limit Load Stress}}$$

(2) *Yield.* The ratio of the yield stress to the maximum calculated stress based on limit loads, as follows:

$$\textit{Yield Factor of Safety} = \frac{\textit{Yield Strength}}{\textit{Limit Load Stress}}$$

factorial design
A type of experimental design in which two or more independent variables are examined as part of the same process to permit the study of both their independent and interaction effects on a dependent variable.

factory acts
Laws enacted for the purpose of regulating the hours of work, and the health and safety conditions.

factory investigative audit
Maritime. The presence of the Officer in Charge of Marine Inspection (OCMI) and other Coast Guard personnel at a manufacturing facility to gather information and evidence to prove or disprove violations of the statutes, or to investigate potential defects which may present substantial risks of personal injury.

Factory Mutual Association
An industrial fire protection, engineering, and inspection bureau established and maintained by mutual insurance companies. The Factory Mutual laboratories test and list fire protection equipment for approval, assist in the development of standards, and conduct research in fire protection. Approvals by the Factory Mutual Association carry the designation FM.

facts in issue
Those matters of fact on which the plaintiff proceeds by his/her action, and which the defendant controverts in his/her defense. Under civil rule practice in the federal courts, and in most state courts, the facts alleged in the initial complaint are usually quite brief, with the development of additional facts being left to discovery and pretrial conference.

factum probandum
(*Latin*) In the law of evidence, the fact to be proved; a fact which is at issue, and to which evidence is to be directed.

facultative
Capable of adaptation to different conditions.

facultative anaerobe
Microorganisms that can multiply either in the presence or in the absence of oxygen. They can obtain energy either by respiration or by fermentation and do not require oxygen for biosynthesis.

facultative saprophytes
Organisms which can only survive on dead organic matter.

faculty
(1) A normal power or function, especially of the mind. (2) The teaching staff of an educational institution or organization.

FAF
See ***final approach fix.***

Fahrenheit temperature scale
The scale of temperature in which 212 degrees is the boiling point of water at 760 mm mercury pressure and 32 degrees is the freezing point. Abbreviated °F.

fail
Fault, negligence, or refusal. To fall short; be unsuccessful or deficient. Also, fading health.

fail operational
A design characteristic which allows continued operation of a system or subsystem despite a discrete failure.

fail operational, fail safe
A fail operational design which also remains acceptably safe. *See also* ***fail operational*** *and* ***fail safe.***

fail passive
A system or component design feature that, under failure conditions, will have no effect on the operation of the overall system.

fail safe
A system or component design feature that, under failure conditions, will permit the failed component or system to revert to a safe mode and not present an unacceptable hazard risk or flow of energy due to the failure condition.

failure
The inability of a component or system to perform its designed function within specified limits.

failure analysis
See ***failure mode and effect analysis.***

failure assessment
The process in which the cause, effect, responsibility, and cost of a failure are determined and reported.

failure condition
As pertains to fault tree analysis (FTA) and/or the Management Oversight and Risk Tree (MORT), the top event, or that primary event subject to a failure analysis through an event tree.

failure management
The planning, decision-making, and policy implementation which attempt to identify and eliminate potential failures or apply corrective policies/procedures after a failure occurrence.

failure mechanism
See ***fault.***

failure mode
The status in which or process during which a piece of equipment failed.

failure mode and effect analysis (FMEA)
An in-depth analysis of possible failures and their resulting effects related to system function and performance *(functional FMEA)* or system hardware and components *(hardware FMEA).*

failure of consideration
As applied to notes, contracts, conveyances, etc., this term does not necessarily mean a want of consideration, but implies that a consideration, originally existing and good, has since become worthless or has ceased to exist or been extinguished, partially or entirely.

failure of proof
Inability or failure to prove the cause of action or defense in its entire scope and meaning.

failure to state cause of action
Failure of the plaintiff to allege sufficient facts in the complaint to maintain action. In other words, even if the plaintiff proved all the facts alleged in the complaint, the facts would not establish a cause of action entitling the plaintiff to recover against the defendant. The motion to dismiss for failure to state a cause of action is sometimes referred to as a) a demurrer or b) a failure to state a claim upon which relief can be granted.

failure to testify
In a criminal trial, defendant is not required to testify and such failure may not be commented on by judge or prosecution because of protection under the Fifth Amendment of the U.S. Constitution.

failure tolerance
The ability of a system to experience one or more failures and still maintain some functional capability.

faint
The temporary loss of consciousness as a result of a reduced supply of blood to the brain. Also referred to as *syncope. See also* ***unconsciousness.***

fair
(1) Having the qualities of impartiality and honesty; free from prejudice, favoritism, and

self-interest. (2) A gathering of buyers and sellers for the purpose of exhibiting and sale of goods; usually accompanied by amusements, contests, entertainment, and the like.

fair and impartial jury
Jury chosen to hear evidence and render verdict without any prior fixed opinion concerning the guilt, innocence, or liability of defendant. Means that every member of the jury must be a fair and impartial juror.

fair and impartial trial
A hearing by an impartial and disinterested tribunal; a proceeding which hears before it condemns, which proceeds upon inquiry, and renders judgment only after consideration of evidence and facts as a whole. A basic constitutional guarantee contained implicitly in the Due Process Clause of the Fourteenth Amendment of the U.S. Constitution.

fair comment
A form of qualified privilege applied to news media publications relating to discussion of matters which are of legitimate concern to the community as a whole, because they materially affect the interests of all the community.

fair day's work
A concept of the amount of daily work output expected by management from qualified employee(s), assuming no processing limitations. Also referred to as *expected attainment*.

Fair Labor Standards Act (FLSA)
A comprehensive federal employment regulation providing employer requirements such as equal pay, overtime, minimum wage, employment of minors, and recordkeeping. Sometimes referred to as *Wage and Hour Law*.

fair market value
Automotive Industry. The value of a vehicle as stated by the National Automotive Dealers Association (NADA) or other sale publication. For vehicles under the 3-year replacement cycle, fair market value is the average loan indicated in the appropriate NADA publication.

fair preponderance of evidence
Evidence sufficient to create in the minds of the triers of fact the conviction that the party upon whom is the burden of proof has established its case. The greater and weightier the evidence, the more convincing the evidence. This term is not a technical term, but simply means that the evidence outweighs that which is offered to oppose it, and does not necessarily mean the greater number of witnesses.

fair representation
Refers to the duty of a union to represent fairly all its members, both in the conduct of collective bargaining and in the enforcement of the resulting agreement, and to serve the interests of all members without hostility or discrimination toward any and to exercise its discretion with complete good faith and honesty and to avoid arbitrary conduct.

fairly close
Maritime. As close to the shore, dike, or light as practicable (approximately 150 feet off).

FAK
*See **freight all kinds***.

fall streaks
Falling ice crystals that evaporate before reaching the ground. They appear as streaks of grayish blue in the sky. *See also **virga***.

fall time
The time interval between an initial response in an instrument and a specified percent decrease (e.g., 90%) after a decrease in the inlet concentration.

fall wind
A strong, cold wind that blows downslope off snow-covered plateaus.

fallen skier
Coast Guard. A person who has fallen off their water skies.

falling river
The river condition when gauge readings are decreasing day by day.

Fallot's tetralogy
A combination of congenital cardiac defects, namely, pulmonary stenosis, ventricular septal defects, dextroposition of the aorta, so that such combination overrides the interventricular septum and receives venous as well as arterial blood, and right ventricular hypertrophy.

fallout
Radioactive debris from a nuclear detonation which becomes airborne, or has deposited on the earth. It is the dust and other particulate material which contain radioactive fission products from a nuclear explosion.

false alarm
An indication of a problem when no operational problem exists other than in the sensing mechanism. *See also* ***type I error***.

false chokes
A choking sensation or cough due to breathing 100% oxygen for an extended period of time, which results in dry lung tissues.

false statement
Statement knowingly false, or made recklessly without honest belief in its truth, and with purpose to mislead or deceive. The federal criminal stature governing false statements applies to three distinct offenses: falsifying, concealing, or covering up a material fact by any trick, scheme, or device; making false, fictitious, or fraudulent false documents or writing.

falsifying a record
It is a crime, under state and federal statutes, for a person, knowing that he/she has no privilege to do so, to falsify or otherwise tamper with public records with the purpose of deceiving or injuring anyone or concealing any wrongdoing.

falx
A sickle-shaped structure.

fan
A mechanical device which physically moves air and creates static pressure.

fan, airfoil
A type of backward inclined blade fan with blades that have an airfoil cross-section.

fan, axial
A fan in which airflow is parallel to the fan shaft and air movement is induced by a screw-like action of the fan blade.

fan, backward inclined blade
A centrifugal fan with blades inclined opposite to fan rotation.

fan, centrifugal
A fan in which the air leaves the fan in a direction perpendicular to the direction of entry.

fan curve
A curve relating the pressure versus volume flow rate of a given fan at a fixed fan speed (rpm).

fan, forward curved blade
A centrifugal fan with blades inclined in the direction of fan rotation.

fan laws
Statements and equations that describe the relationship between fan volume, pressure, brake horsepower, size, and any other changes made in fan operation. For example, volume varies directly as fan speed and horsepower varies as the cube of the fan speed.

fan, paddle wheel
A centrifugal fan with radial blades.

fan, propeller
An axial fan employing a propeller to move air.

fan, radial blade
A centrifugal fan with radial blades extending out radially from the fan wheel shaft.

fan rating table
Tables published by fan manufacturers presenting the range of capacities of a particular fan model along with the static pressure developed and the fan speed within the limits of the fan's construction.

fan, squirrel cage
A centrifugal blower with forward curved blades.

fan static pressure
The static pressure added to that of the ventilation system due to the presence of the fan. It equals the sum of pressure losses in the system minus the velocity pressure in the air at the fan inlet.

fan, tube axial
An axial fan mounted in a duct section.

fan, vane axial
An axial flow fan mounted in a duct section with vanes to straighten the airflow and increase static pressure.

FAP
See ***final approach point***.

FAR
See ***Federal Aviation Regulation***.

far field
In acoustics, the uniform sound field which is free and undisturbed by bounding surfaces and other sources of sound and in which the sound pressure level obeys the inverse-square

law relationship and decreases 6 dB for each doubling of distance from the source. Also referred to as a *free sound field*.

far infrared
That portion of the infrared radiation spectrum with wavelengths ranging from about 5000 nm to 1 mm. Also referred to as *long wavelength infrared*.

far ultraviolet
That portion of the ultraviolet radiation spectrum consisting of wavelengths from about 100 to 200 nm.

far vision
The ability to see the distant physical environment.

farad (F)
A unit of capacitance; that amount of capacitance between two conductors separated by a dielectric with a potential difference of one volt and charged by one coulomb.

fare
The required payment for a ride on a public transportation vehicle. It may be paid by any acceptable means, for example, cash, token, ticket, transfer, fare card, voucher, or pass or user fee.

fare evasion
The unlawful use of transit facilities by riding without paying the applicable fare.

fare recovery ratio
The ratio of fare revenue to operating expenses.

farm-to-market agricultural transportation
The operation of a motor vehicle controlled and operated by a farmer who: a) is a private motor carrier of property; b) is using the vehicle to transport agricultural products from a farm owned by the farmer, or to transport farm machinery or farm supplies to or from a farm owned by the farmer; and c) is not using the vehicle to transport hazardous materials of a type or quantity that requires the vehicle to be placarded in accordance with 49 CFR 177.823.

farm vehicle driver
A person who drives only a motor vehicle that is a) controlled and operated by a farmer as a private motor carrier of property; b) being used to transport either agricultural products, or farm machinery, farm supplies, or both, to or from a farm; c) not being used in the operation of a for-hire motor carrier; c) not carrying hazardous materials of a type or quantity that requires the vehicle to be placarded in accordance with 49 CFR 177.823, and d) being used within 150 air-miles of the farmer's farm.

farmer's lung disease
A syndrome that consists initially of chills and fever, followed by impairment of lung function. It is normally caused by chronic exposure to moldy hay or other moldy organic material. It is also known as *thresher's lung*.

farmer's skin
See ***sailor's skin***.

farsightedness
See ***hyperopia***.

FAS
See ***free alongside ship***.

fast file
Aviation. A system whereby a pilot files a flight plan via telephone that is tape recorded and then transcribed for transmission to the appropriate air traffic facility. Locations having a fast file capability are contained in the Airport/Facility Directory.

fast meter response
The "fast" response of the sound level meter shall be used. The fast dynamic response shall comply with the meter dynamic characteristics in the American National Standard Institute (ANSI) specification for Sound Level Meters.

fast twitch muscle
See ***white muscle***.

fastenings
With regard to ladders, a device to attach a ladder to a structure, building, or equipment.

fat
(1) The adipose or fatty tissue of the body. (2) An oily substance consisting of glycerin (a form of alcohol called glycerol) and a group of fatty acids, chiefly palmitic, stearic and oleic acids, combined as glycerin esters. Fats consist of carbon, hydrogen, and oxygen in most foods, especially in meats and dairy products. Fats may be solid, such as butter, or liquid such as olive oil.

fat body mass
That portion of the body mass which is due to fat.

fat-free body
A physical/metabolic state in which an individual has only the minimal amount of fat stored in his/her body.

fat-free mass
*See **lean body mass**.*

fat-free weight
*See **lean body weight**.*

fat patterning
The distribution of subcutaneous fat throughout the body.

fatal accident
(1) *General.* An accident causing the death of one or more persons in or as a direct result of that accident. (2) *National Safety Council.* An accident that results in one or more deaths within one year. (3) *Transportation.* a) A motor vehicle traffic accident resulting in one or more fatal injuries. b) An accident for which at least one fatality was reported. c) Statistics reported to the Federal Highway Administration (FHWA) shall conform to the 30-day rule, i.e., a fatality resulting from a highway vehicular accident is to be counted only if death occurs within 30 days of the accident.

fatal accident rate
Transportation. The fatal accident rate is the number of fatal accidents per 100 million vehicle miles of travel.

fatal alcohol involvement crash
A fatal crash is alcohol related or alcohol involved if either a driver or a non-motorist (usually a pedestrian) had a measurable or estimated blood alcohol concentration (BAC) of 0.01 grams per deciliter (g/dl) or above.

fatal crash
Transit. A police-reported crash involving a motor vehicle in transport on a traffic-way in which at least one person dies within 30 days of the crash.

fatal injury
(1) *Law-Insurance.* A term embracing injuries resulting in death, which, as used in accident and disability insurance policies is distinguished from "disability," which embraces injuries preventing the insured from performing the work in which he/she is usually employed, but not resulting in death. (2) Any injury which results in death within 7 days of the accident. (3) *National Transportation Safety Board.* Any injury which results in death within 30 days of the accident.

fatal plus nonfatal injury accidents
The sum of all fatal accidents and nonfatal-injury accidents.

fatality
(1) *General.* A death due to any cause. (2) *American Gas Association.* Death resulting from the failure or escape of gas. (3) *Highway Transit.* Those deaths a) which result from motor vehicle accidents that occurred during the relevant calendar year, and b) those in which the injured person(s) died within 30 days of the accident. Also, a transit-caused death confirmed within 30 days of a transit incident. Also, a death as the result of a crash that involves a motor vehicle in transport on a traffic-way and in which at least one person dies within 30 days of the crash. For purposes of statistical reporting on transportation safety, fatality shall be considered a death due to injuries in a transportation accident or incident that occurs within 30 days of that accident or incident. (4) *Rail Operations.* The death of a person resulting from an injury incurred during railroad operations or resulting from an occupational illness, if death occurs within 365 days of initial diagnosis. Also, a death confirmed within 30 days after an incident which occurs under the collision, derailment, personal casualty, or fire categories.

fatality/injury
Refers to the average number of fatalities and injuries which occurred per one hundred accidents. Frequently used as an index of accident severity.

fatality rate
Transit. (1) The average number of fatalities which occurred per accident or per one hundred accidents. (2) The fatality rate is the number of fatalities per 100 million vehicle miles of travel.

fatfold
*See **skinfold**.*

fatigue
(1) *Physiological.* The condition of being extremely tired as a result of some physical and/or mental exertion. A state characterized by lack of motivation, interest, and/or an inability to maintain normal, consistent productivity and quality due to recent physical or mental exertion. (2) *Structural.* The progressive localized permanent structural change that occurs in a material subjected to constant or variable amplitude loads at stresses having a maximum value less than the ultimate strength of the material.

fatigue allowance
That additional time which is added to the normal time to permit a worker to rest.

fatigue-decreased proficiency
A decrease in performance due to prolonged whole-body vibration exposure (an older term).

fatigue-decreased proficiency boundary
Those limits of human whole-body vibration exposure for certain time durations at specified frequencies which are intended to maintain a basic performance level (an older term).

fatigue life
The number of cycles of stress or strain of a specified character that a given material sustains before failure of a specified nature occurs. Also referred to as *cycle life*.

fauces
The opening between the posterior mouth and the oropharynx.

fault
(1) *General.* A manifestation of an error. Any condition which may or will cause a system to fail. (2) *Seismology.* A fracture or zone of fractures along which there has been displacement of the sides relative to one another, parallel to the fracture. (3) *Law.* Negligence; an error or defect of judgement or of conduct; any deviation from prudence, duty, or rectitude; any shortcoming, or neglect of care or performance resulting from inattention, incapacity, or perversity; a wrong tendency, course, or act; bad faith or mismanagement; neglect of duty.

fault hazard analysis (FHA)
A system safety analysis method, usually an extension of the failure mode and effect analysis, that evaluates the overall effect of functional failures on other subsystems or the overall system itself. Synonymous with *functional hazard analysis*.

fault of operator
Coast Guard. Speeding; overloading; improper loading, not properly seating occupants of a boat; no longer lookout; carelessness; failure to heed weather warnings; operating in a congested area; not observing the Rules of the Road; unsafe fueling practices; lack of experience; ignorance of aids to navigation; lack of caution in an unfamiliar area of operation; improper installation or maintenance of hull, machinery, or equipment; poor judgment; recklessness; overpowering the boat; panic; proceeding in an unseaworthy craft; operating a motorboat near persons in the water; starting engine with clutch engaged or throttle advanced; irresponsible boat handling such as quick, sharp turns.

fault stand
In seismology, an individual fault of a set of closely spaced parallel or sub-parallel faults of a fault system.

fault tolerance
The built-in ability of a system to provide continued correct operation in the presence of a specified number of faults or failures.

fault tree analysis (FTA)
A system safety analysis technique used as an inductive method (top down, from the known to the unknown) to evaluate fault or failure events in a system or process.

FBI
See ***Federal Bureau of Investigation***.

FBSA
See ***Federal Boating Safety Act***.

FBT
See ***full berth terms***.

fc
See ***foot-candle***.

FCC
See ***Federal Communications Commission***.

FCDC
Flexible, confined detonating cord.

FCLT
See ***freeze calculated landing time***.

FDA
See ***Food and Drug Administration***.

FDAAL
See ***Food and Drug Administration Action Level***.

feasibility study
(1) *OSHA*. A study performed by OSHA to determine if a proposed standard is practical for the exposure under consideration as well as from an implementation perspective. (2) *EPA*. An analysis of the practicability of a proposal; e.g., a description and analysis of the potential cleanup alternatives for a site or alternatives for a site on the National Priorities List. The feasibility study usually recommends selection of a cost-effective alternative. It usually starts in tandem with the Remedial Investigation (RI). Performed together, the process is commonly referred to as the RI/FS. The term can apply to a variety of proposed corrective or regulatory actions.

feasible
A measure that is practical and capable of being accomplished or brought about.

feasor
The doer or maker. Also used in the compound term "tort-feasor," meaning one who commits or is guilty of a tort.

featherbedding
The name given to employee practices which create or spread employment by unnecessarily maintaining or increasing the number of employees used, or the amount of time consumed, to work on a particular job. Most of these practices stem from a desire on the part of employees for job security in the face of technological improvements.

feathered propeller
Aviation. A propeller whose blades have been rotated so that the leading and trailing edges are nearly parallel with the aircraft flight path to stop or minimize drag and engine rotation. Normally used to indicate shutdown of a reciprocating or turboprop engine due to malfunction.

febella
A sesamoid fibrocartilage in the gastrocnemius muscle.

fecal coliform bacteria
Bacteria found in the intestinal tracts of mammals. Their presence in water or sludge is an indicator of pollution and possible contamination by pathogens.

feces
The collective excretions normally passing through the anus, including undigested and unabsorbed food and intestinal secretions.

Fechner's Law
A proposed logarithmic relationship between stimulus intensity and sensory strength, having the form

$$S = k \log I_s$$

where:
S = sensory strength
k = constant depending on the units of measurement and modality
I_s = stimulus intensity

fecundity
The physiological ability to reproduce.

federal
Belonging to the general government or union of the states. Founded on or organized under the Constitution of the United States. Of or constituting a government in which power is distributed between a central authority (i.e., federal government) and a number of constituent territorial units (i.e., states).

federal agency
Any executive department, military department, government corporation, government-controlled corporation or other establishment in the executive branch of government including the Executive Office of the President or any independent regulatory agency.

federal-aid highways
Those highways eligible for assistance under Title 23 U.S.C. except those functionally classified as local or rural minor collectors.

Federal-Aid Primary Highway System
The Federal-Aid Highway System of rural arterials and their extensions into or through urban areas in existence on June 1, 1991, as described in 23 U.S.C. 103(b) in effect at that time.

Federal Aid Secondary Highway System
This existed prior to the ISTEA [Intermodal Surface Transportation Efficiency Act] of 1991 and included rural collector routes.

Federal Aid Urban Highway System
This existed prior to the ISTEA [Intermodal Surface Transportation Efficiency Act] of 1991 and included urban arterial and collector routes, exclusive of urban extensions of the Federal-Aid Primary system.

Federal Aviation Administration (FAA)
Formerly the Federal Aviation Agency, the Federal Aviation Administration was established by the Federal Aviation Act of 1958 (49 U.S.C. 106) and became a component of the Department of Transportation in 1967 pursuant to the Department of Transportation Act (49 U.S.C. app. 1651 note). The Administration is charged with a) regulating air commerce in ways that best promote its development and safety and fulfill the requirements of national defense; b) controlling the use of navigable airspace of the United States and regulating both civil and military operations in such airspace in the interest of safety and efficiency; c) promoting, encouraging, and developing civil aeronautics; d) consolidating research and development with respect to air navigation facilities; e) installing and operating air navigation facilities; f) developing and operating a common system of air traffic control and navigation for both civil and military aircraft; and g) developing and implementing programs and regulations to control aircraft noise, sonic boom, and other environmental effects of civil aviation.

Federal Aviation Regulations (FAR)
The set of regulatory obligations contained in Title 14 of the Code of Federal Regulations which FAA is charged to enforce in order to promote the safety of civil aviation both domestically and internationally.

Federal Boating Safety Act (FBSA)
Enacted by Congress on 10 August 1971, it gave the Coast Guard the authority to establish comprehensive boating safety programs, authorized the establishment of national construction and performance standards for boats and associated equipment and created a more flexible regulatory authority concerning the use of boats and associated equipment. Amended by the Recreational Boating Safety and Facilities Improvement Act of 1980, also known as The Recreational Boating Fund Act of 1980 (The Biaggi Act) which provided financial assistance, in part through motorboat fuel taxes, for state recreational boating safety programs. Now re-codified as Chapter 43 of Title 46, United States Code.

Federal Bureau of Investigation (FBI)
The FBI (established in 1908) is charged with investigating all violations of federal laws with the exception of those which have been assigned by legislative enactment or otherwise to some other federal agency. The FBI's jurisdiction includes a wide range of responsibilities in the criminal, civil, and security fields. Among these are espionage, sabotage, and other subversive activities; kidnapping; extortion; bank robbery; interstate transportation of stolen property; civil rights matters; interstate gambling violations; fraud against the Government; and assault or killing the President or a Federal officer. Cooperative services of the FBI for other duly authorized law enforcement agencies include fingerprint identification, laboratory services, police training, and the National Crime Information Center.

Federal Communications Commission (FCC)
The FCC was created by the Communications Act of 1934 to regulate interstate and foreign communications by wire and radio in the public interest. It was assigned additional regulatory jurisdiction under the provisions of the Communications Satellite Act of 1962. The scope of its regulatory powers includes radio and television broadcasting, telephone, telegraph, and cable television operation; two-way radio and radio operators; and satellite communication.

federal crimes
Those acts which have been made criminal by federal law. There are no federal common-law crimes though many federal statutes have incorporated the elements of common-law crimes. Most federal crimes are codified in Title 18 of the United States Code; though other Code Titles also include specific crimes.

Federal Emergency Management Agency (FEMA)
An independent agency that advises the President on meeting civil emergencies and provides assistance to individuals and public entities that suffered property damage in emergencies and disasters when recommended by the President.

Federal Employees' Compensation Act
Type of workers' compensation plan for federal employees by which payments are made for death or disability sustained in performance of duties of employment.

Federal Employer's Liability Act
Federal workers' compensation law which protects employees of railroads engaged in interstate and foreign commerce. Payments are made for death or disability sustained in performance of duties of employment.

Federal Energy Regulatory Commission (FERC)
The federal agency with jurisdiction over interstate electricity sales, wholesale electric rates, hydroelectric licensing, natural gas pricing, oil pipeline rates, and gas pipeline certification. Federal Energy Regulatory Commission (FERC) is an independent regulatory agency within the Department of Energy and is the successor to the Federal Power Commission.

Federal Facility Compliance Act (FFCA) of 1992
Amendment to the Resource Conservation and Recovery Act (RCRA). Ensures that there is a complete and unambiguous waiver of sovereign immunity with regard to fines and penalties against Federal Facilities. Act allows State environmental agencies and the EPA to impose civil penalties and administrative fines on Federal Facilities under RCRA for violations of federal, state, and local solid and hazardous waste laws.

Federal Highway Administration (FHWA)
Became a component of the Department of Transportation in 1967 pursuant to the Department of Transportation Act (49 U.S.C. app. 1651 note). It administers the highway transportation programs of the Department of Transportation under pertinent legislation and the provisions of law cited in section 6a) of the act (49 U.S.C. 104). The Administration encompasses highway transportation in its broadest scope seeking to coordinate highways with other modes of transportation to achieve the most effective balance of transportation systems and facilities under cohesive federal transportation policies pursuant to the act. The Administration administers the Federal-Aid Highway Program; is responsible for several highway-related safety programs; is authorized to establish and maintain a national network for trucks; administers a coordinated federal lands program; coordinates varied research, development and technology transfer activities; supports and participates in efforts to fund research and technology abroad; plus a few additional programs.

Federal Implementation Plan (FIP)
Implemented by the EPA when a state fails to implement their own plan for the establishment, regulation, and enforcement of air pollution standards.

Federal Maritime Commission
Regulates the waterborne foreign and domestic offshore commerce of the United States, assures that United States international trade is open to all nations on fair and equitable terms, and guards against unauthorized monopoly in the waterborne commerce of the United States. This is accomplished through maintaining surveillance over steamship conferences and common carriers by water; assuring that only the rates on file with the Commission are charged; approving agreements between persons subject to the Shipping Act; guaranteeing equal treatment to shippers and carriers by terminal operators, freight forwarders, and other persons subject to the shipping statutes; and ensuring that adequate levels of financial responsibility are maintained for indemnification of passengers or oil spill cleanup.

Federal Mediation and Conciliation Service
The Federal Mediation and Conciliation Service helps prevent disruptions in the flow of interstate commerce caused by disputes between labor and management by providing mediators to assist disputing parties in the resolution of their differences. The Service can intervene on its own motion or by invitation of either side in a dispute. Mediators have no law enforcement authority and rely wholly on persuasive techniques. The Service also helps provide qualified third-party neutrals as fact finders or arbitrators.

Federal Motor Carrier Safety Regulations (FMCSR)
The regulations are contained in the Code of Federal Regulations, Title 49, Chapter III, Subchapter B.

Federal Power Act

Enacted in 1920, amended in 1935, the Act consists of three parts. The first part incorporated the Federal Water Power Act administered by the former Federal Power Commission, whose activities were confined almost entirely to licensing non-Federal hydroelectric projects. Parts II and III were added with the passage of the Public Utility Act. These parts extended the Act's jurisdiction to include regulating the interstate transmission of electrical energy and rates for its sale, at wholesale rates, in interstate commerce. The Federal Energy Regulatory Commission is now charged with the administration of this law.

federal preemption

The U.S. Constitution and acts of Congress have given to the federal government exclusive power over certain matters such as interstate commerce. Sedition, to the exclusion of state jurisdiction, is also given to federal courts. When such preemptions are invoked, they are commonly described as involving a "federal question."

Federal Railroad Administration (FRA)

The FRA was created pursuant to section 3(e)(1) of the Department of Transportation Act of 1966 (49 U.S.C. app. 1652). The purpose of the Federal Railroad Administration is to promulgate and enforce rail safety regulations, administer railroad financial assistance programs, conduct research and development in support of improved railroad safety and national rail transportation policy, provide for the rehabilitation of Northeast Corridor (NEC) rail passenger service, and consolidate government support of rail transportation activities.

Federal Register (FR)

The official daily publication of the United States government that provides a uniform system for publishing Presidential and federal agency documents.

Federal Rules of Civil Procedure

Body of procedural rules which governs all civil actions in the U.S. District Courts and after which most states have modeled their own rules of civil procedure. These rules were promulgated by the U.S. Supreme Court in 1938 under power granted by Congress, and have since been frequently amended. Such rules also govern adversary proceedings in the bankruptcy courts; and Supplemental Rules, in addition to main body of rules, govern admiralty and maritime actions.

Federal Rules of Evidence

Rules which govern the admissibility of evidence at trials in the Federal District Courts and before U.S. Magistrates. Many states have adopted Evidence Rules patterned on these federal rules.

Federal Sentencing Guidelines for Organizations (Section 8)

Guidelines that went into effect on November 1, 1991 designed to promote uniformity of sentencing for corporations convicted in federal cases, such as violations of securities or antitrust laws, kickbacks, or bribery. Mitigating factors lessen penalties if an organization has an Effective Compliance Program. Guidelines do not apply to environmental offenses. *See also* ***effective compliance program*** *and* ***Environmental Guidelines for Sentencing Organizations***.

Federal Standard 595a

A color ordering system developed by the U.S. Government for standardizing colors used by federal agencies according to a 5-digit code and a gloss/luster criterion. *See also* ***color ordering system***.

Federal Tort Claims Act

The government of the United States may not be sued in tort without its consent. That consent was given in the Federal Tort Claims Act (1946), which largely abrogated the federal government's immunity from tort liability and established the conditions for suits and claims against the federal government. The Act (28 U.S.C.A §§ 1346(b), 2674) preserves governmental immunity with respect to the traditional categories of intentional torts, and with respect to acts or omissions which fall within the "discretionary function or duty" of any federal agency or employee.

Federal Trade Commission

Agency of the federal government created in 1914. The Commission's principal functions are to promote free and fair competition in interstate commerce through the prevention of general trade restraints such as price-fixing agreements, false advertising, boycotts, illegal combinations of competitors and other unfair methods of competition.

Federal Transit Administration (FTA)
Formerly the Urban Mass Transportation Administration, it operates under the authority of the Federal Transit Act, as amended (49 U.S.C. app. 1601 et seq.). The Federal Transit Act was repealed on July 5, 1994, and the Federal transit laws were codified and re-enacted as Chapter 53 of Title 49, United States Code. The Federal Transit Administration was established as a component of the Department of Transportation by section 3 of Reorganization Plan No. 2 of 1968 (5 U.S.C. app.), effective July 1, 1968. The missions of the Administration are a) to assist in the development of improved mass transportation facilities, equipment, techniques, and methods, with the cooperation of mass transportation companies both public and private; b) to encourage the planning and establishment of area-wide urban mass transportation systems needed for economical and desirable urban development, with the cooperation of mass transportation companies both public and private; c) to provide assistance to state and local governments and their instrumentalities in financing such systems, to be operated by public or private mass transportation companies as determined by local needs; and d) to provide financial assistance to state and local governments to help implement national goals relating to mobility for elderly persons, persons with disabilities, and economically disadvantaged persons.

Federal Water Pollution Control Act (FWPCA)
Law passed in 1970 and amended in 1972 giving the Coast Guard a mandate to develop, among other things, marine sanitation device regulations.

fee damages
Damages sustained by and awarded to an abutting owner of real property occasioned by the construction and operation of an elevated railroad in a city street, are so called because compensation is made to the owner for the injury to, or deprivation of, his/her easements of light, air, and access, and these are parts of the fee.

feed
A mechanism which introduces material to a machine for processing.

feedback
The return of meaningful information within a closed-loop system so that system performance can be appropriately modified. Also referred to as *knowledge of results*.

feedback control system
See ***closed-loop system***.

feedback mechanism
A process whereby an initial change in an atmospheric process will tend to either reinforce the process (*positive feedback*) or weaken the process (*negative feedback*).

feeder
All circuit conductors between the service equipment, or the generator switchboard of an isolated plant, and the final mechanical rather than electrical function.

feeder bus
A bus service that picks up and delivers passengers to a rail rapid transit station or express bus stop or terminal.

feeder fix
Aviation. The fix depicted on Instrument Approach Procedure Charts which establishes the starting point of the feeder route.

feeder route
Aviation. A route depicted on instrument approach procedure charts to designate routes for aircraft to proceed from the en route structure to the initial approach fix (IAF).

feeder vessel
A vessel which transfers containers to a "mother ship" for an ocean voyage.

feedlot
A relatively small, confined area for the controlled feeding of animals that tends to concentrate large amounts of animal wastes that cannot be absorbed by the soil and, hence, may be carried to nearby streams or lakes by rainfall runoff.

feedstock energy
Under ISO 14000, the gross combustion heat of raw material inputs, which are not used as energy sources, to a byproduct system.

fellow servant
One who works for and is controlled by the same employer; a co-worker. Employees who derive authority and compensation from the

same employer, and are engaged in the same general business.

fellow servant rule
A common-law doctrine, now generally abrogated by workers' compensation acts and Federal Employers' Liability Act, that in an action for damages brought against an employer by an injured employee the employer may allege that the negligence of another fellow employee was partly or wholly responsible for the accident resulting in the injury, and thus reducing or extinguishing his/her own liability.

felon
A person who commits or has committed a felony.

felony
A crime of a graver or more serious nature than those designated as misdemeanors. For example, an aggravated assault (a felony) is contrasted with simple assault (a misdemeanor). Under many state statutes, any offense punishable by death or imprisonment for a term exceeding one year. The federal and many state criminal codes define felony status crimes, and in turn also have various classes of felonies (e.g., Class A, B, C, etc.) or degrees (e.g., first, second, third) with varying sentences for each class.

Fels index
An estimate for the percentage of body fat and nutritional status of the body.

FEMA
See ***Federal Emergency Management Association***.

femoral breadth
See ***knee breadth***.

femur
The long bone in the thigh extending from the pelvis to the knee. It is the longest and straightest bone in the body. Its proximal end articulates with the acetabulum, a cup-like cavity in the pelvic girdle. The greater and lesser trochanters are the two processes (prominences) at the proximal end of the femur.

fen
A type of wetland that accumulates peat deposits. Fens are less acidic than bogs, deriving most of their water from groundwater rich in calcium and magnesium. *See also* ***wetlands***.

FERC
See ***Federal Energy Regulatory Commission***.

Feret's diameter
The distance between the extreme boundaries of a particle.

fermentation
The breakdown of organic substance by microorganisms with a resulting release of energy.

ferrel cell
The name given to the middle latitude cell in the 3-cell model for general circulation.

ferries
Establishments primarily engaged in operating ferries for the transportation of passengers or vehicles.

ferruginous bodies
Bodies formed by fibers that have entered the lungs. These bodies can be formed by any kind of durable fiber including asbestos, fiberglass, and vegetable fibers of siliceous origin.

ferry boat
A boat providing fixed-route service across a body of water.

ferry crossing
Route used to transport traffic between two points separated by water.

ferry flight
Aviation. A flight for the purpose of a) returning an aircraft to base, b) delivering an aircraft from one location to another, or c) moving an aircraft to and from a maintenance base. Ferry flights, under certain conditions, may be conducted under terms of a special flight permit.

ferry vessel
A vessel which is limited in its use to the carriage of deck passengers or vehicles or both, operates on a short run on a frequent schedule between two points over the most direct water route, other than in ocean or coastwise service, and is offered as a public service of a type normally attributed to a bridge or tunnel.

ferryboats
Vessels for carrying passengers and/or vehicles over a body of water. The vessels are

generally steam- or diesel-powered conventional ferry vessels. They may also be hovercraft, hydrofoil and other high speed vessels.

fertility toxin
A substance which reduces male or female fertility.

fertilizer
Materials such as nitrogen and phosphorous that provide nutrients for plants. Commercially sold as fertilizers and may contain other chemicals or may be in the form of processed sewage sludge.

fetotoxicity
Harmful effects exhibited by a fetus, due to exposure to a toxic substance, that may result in death, reduced birth weight, or impairment of growth and physiological dysfunction.

fetotoxin
A substance which is toxic to the fetus.

fetus
Later stage of development in unborn organisms, following the embryonic stage.

FEV
Forced expiratory volume.

FEV-1
*See **forced expiratory volume-one second**.*

fever
A condition in which the body temperature is above normal.

FFCA
*See **Federal Facility Compliance Act of 1992**.*

FFDCA
Federal Food, Drug, and Cosmetic Act.

FHA
*See **fault hazard analysis**.*

FHWA
*See **Federal Highway Administration**.*

fiber
(1) *General.* A particle having a length to diameter/width ratio of greater than 3 to 1. (2) *PCM Method.* Particulate at least 5 micrometers in length with an aspect ratio (length to width ratio) of at least 3 to 1. A rod-like structure having a length at least three times its diameter. (3) *EPA-TEM Method.* Structure greater than or equal to 0.5 micrometers in length with an aspect ratio of 5 to 1, or greater, and having substantially parallel sides.

fiber optics
A system of flexible quartz or glass fibers with internal reflective surfaces that can transmit light.

A fiber optic lead

fiberglass
A commercial, nonflammable fiber that is made from spun glass primarily used for insulation. Fibers of this material can penetrate the skin causing dermatitis in some people and, when airborne, can affect the lungs of some people. Fiberglass is resistant to most chemicals and solvents.

fiberglass (plastic) hull
Hulls of fiber reinforced plastic. The laminate consists of two basic components, the reinforcing material (glass filaments) and the plastic or resin in which it is embedded.

fiberscope
A flexible instrument for direct visual examination of the interior of hollow organs or body cavities, constructed of fibers having special optical properties.

fibrillation
(1) A transitory muscular contraction resulting from spontaneous activation of single muscle cells or fibers. (2) Rapid and uncoordinated contractions of the heart. (3) The quality of being made up of fibers.

fibrinogen
A blood protein which precipitates out to form fibers during the clotting process.

fibroblast
Connective tissue cell.

fibroma
A tumor composed mainly of fibrous or fully developed connective tissue.

fibrosis
The formation and accumulation of fibrous tissue, especially in the lungs. Also the chronic collagenous degeneration of the pulmonary parenchyma.

fibrosis producing dust
A dust which, when inhaled, deposited, and retained in the lungs, can produce fibrotic growth that may result in pulmonary disease.

fibrous
A material which contains fibers.

fibula
The smaller, more lateral bone of the lower leg.

fibular height
The vertical distance from the floor or other reference surface to the superior tip of the fibula. Measured with the individual standing erect and the weight distributed evenly on both feet.

FID
*See **flame ionization detector**.*

fidelity
(1) The degree to which a system's input is reflected in its output. (2) The degree of realism in a simulation.

field
(1) That portion of the interlaced display which is represented by every other horizontal scan line. Two fields make a frame on an interlaced video display. (2) A limited area, such as the area of a slide visible through the lens system of a microscope.

field and gathering pipelines
A network of pipelines (mains) transporting natural gas from the individual wells to a compressor station, processing point, or main trunk pipeline.

field area
Energy. A geographic area encompassing two or more pools that have a common gathering and metering system, the reserves of which are reported as a single unit. This concept applies primarily to the Appalachian region. *See also **pool**.*

field blank sample
Sampling media, such as a charcoal tube, filter cassette, or other device, which is handled in the field in the same manner as are other sampling media of the same type but through which no air is sampled. These are used in sampling and analysis procedures to determine the contribution to the analytical result from the media plus any contamination which may have occurred during handling in the field, shipping, and storage before analysis. Often referred to as a *blank sample*.

field-constructed tanks
Vertical cylinders with a capacity of greater than 50,000 gallons.

field duplicate
A sample that is collected concurrently with another sample of the same type, and in the same location for the same duration. It is an extra field sample that helps to ensure quality control.

field of view
The solid angle within the visual field for which the eye or other optical sensor provides useful data.

field sampling plan
Provides guidance for all fieldwork by defining in detail the sampling and data-gathering methods to be used on a project.

field separation facility
A surface installation designed to recover lease condensate from a produced natural gas stream usually originating from more than one lease and managed by the operator of one or more of these leases.

field study
An investigation in which subjects are observed or measured in their natural environments.

FIFRA
Federal Insecticide, Fungicide, and Rodenticide Act.

Fifteenth Amendment
Amendment to the U.S. Constitution, ratified by the States in 1870, guaranteeing all citizens the right to vote regardless of race, color, or previous condition of servitude. Congress was given the power to enforce such rights by appropriate legislation.

Fifth Amendment
Amendment to the U.S. Constitution providing that no person shall be required to answer for a capital or otherwise infamous offense unless on indictment or presentment of a grand jury in military cases; that no person will suffer double jeopardy; that no person will be compelled to be a witness against himself/herself; that no person shall be deprived of life, liberty, or property without due process of law and that private property will not be taken for public use without just compensation.

fifth wheel
(1) A device mounted on a truck tractor or similar towing vehicle (e.g., converter dolly) which interfaces with and couples to the upper coupler assembly of a semitrailer. (2) Load-supporting plate mounted to frame of vehicle. Pivot mounted, it contains provisions for accepting and holding the kingpin of a semitrailer, providing a flexible connection between the tractor and the trailer.

fighting words doctrine
The First Amendment doctrine that holds that certain utterances are not constitutionally protected as "free speech" if they are inherently likely to provoke a violent response from the audience. Words which by their very utterance inflict injury or tend to incite an immediate breach of the peace, having direct tendency to cause acts of violence by the persons to whom, individually, the remark is addressed. The test is what persons of common intelligence would understand to be words likely to cause an average addressee to fight. Certain racial slurs may fall into this category. The "freedom of speech" protected by the Constitution is not absolute at all times and under all circumstances and there are well-defined and narrowly limited classes of speech, the prevention and punishment of which do not raise any constitutional problems, including the lewd and obscene, the profane, the libelous, and the insulting or "fighting words" which by their very utterance inflict injury or tend to incite an immediate breach of the peace.

figure
Any drawing, graphical display, photograph, or similar entity composed of more than just text in a document.

Filar micrometer
A microscopic attachment used for determining the size of particles.

file
A collection of information or data which is stored as a single unit or within a specified restricted location (can be electronic, paper, or other type of media).

File Transfer Protocol (FTP)
A type of Internet site for file downloads.

filed
Aviation. Normally used in conjunction with flight plans, meaning a flight plan has been submitted to Air Traffic Control.

filed en route delay
Aviation. Any of the following preplanned delays at points/areas along the route of flight which require special flight plan filing and handling techniques: a) *Terminal Area Delay.* A delay within a terminal area for touch and go, low approach, or other terminal area activity. b) *Special Use Airspace Delay.* A delay within a Military Operating Area, Restricted Area, Warning Area, or Air Traffic Control (ATC) Assigned Airspace. c) *Aerial Refueling Delay.* A delay within an Aerial Refueling Track or Anchor.

fill material
Any material used for the primary purpose of replacing an aquatic area with dry land or of changing the bottom elevation of a body of water.

fill up work
*See **internal work**.*

filling
Depositing dirt and mud or other materials into aquatic areas to create more dry land, usually for agriculture or commercial devel-

opment. Such activities often damage the ecology of the area.

film
(1) A thin layer or coating. (2) A thin sheet of material (e.g., gelatin, cellulose acetate) specially treated for use in photography or radiography; also used to designate the sheet after exposure to the energy to which it is sensitive.

film analysis
A systematic frame-by-frame study of an activity from motion picture film.

film analysis chart
See ***film analysis record***.

film analysis record
A record generated from a film analysis, containing sequential elemental motions or operations, the beginning and ending clock times, and some type of descriptive symbol. Also referred to as *film analysis chart*.

film badge
A pack of photographic film used for approximate measurement of radiation exposure for personnel monitoring purposes. Also called a *film dosimeter*.

film dosimeter
See ***film badge***.

film loop analysis
A film analysis with a cut and spliced segment of film to form a contiguous loop for repeated viewing. *See also* ***cassette loop analysis***.

film ring
A film ring badge in the form of a finger ring that is typically worn by personnel whose hands may be exposed to ionizing radiation during use of a radiation source, (e.g., operation of an x-ray diffraction unit).

filter
(1) *General.* Any device which removes undesired materials, noise, signal, or information. (2) *Respirator.* The media component of a respirator which removes particulate materials, such as dusts, fumes, fibers, and/or mists from inspired air. (3) *Sample.* Sampling media for collection of airborne particulate contaminants in order to determine the concentration of the material in the air. Filter media may be made of cellulose fibers, glass fibers, mixed cellulose esters (membrane filter), polyvinyl chloride, Teflon, polystyrene, or other material.

filter efficiency
The efficiency of a filter media expressed as collection efficiency (percentage of total particles collected), or as penetration (percent of particles that pass through the filter).

filtration
(1) *Wastewater Treatment.* A treatment process, under the control of qualified operators, for removing solid (particulate) matter from water by passing the water through porous media such as sand or manmade filters. The process is often used to remove particles that contain pathogenic organisms. (2) *Sampling.* The process of collecting a contaminant on an appropriate filter media for determining its composition and concentration in the sampled air, as well as determining if the exposure level is acceptable or whether exposure controls must be developed and implemented. (3) *Respiratory Protection.* The process of removing a contaminant from air being inhaled.

final
Aviation. Commonly used to mean that an aircraft is on the final approach course or is aligned with a landing area.

final approach
Aviation. That part of an instrument approach procedure which commences at the specified final approach fix or point, or where such a fix or point is not specified: a) at the end of the last procedure turn, base turn or inbound turn of a racetrack procedure, if specified; or b) at the point of interception of the last track specified in the approach procedure, and ends at a point in the vicinity of an aerodrome from which a landing can be made, or a missed approach procedure is initiated.

final approach course
Aviation. A published Microwave Landing System (MLS) course, a straight line extension of a localizer, a final approach radial/bearing, or a runway centerline all without regard to distance.

final approach fix (FAF)
Aviation. The fix from which the final approach Instrument Flight Rule (IFR) to an airport is executed and which identifies the beginning of the final approach segment. It is designated on government charts by the Mal-

tese Cross symbol for non-precision approaches and the lightning bolt symbol for precision approaches; or when Air Traffic Control directs a lower-than-published glideslope/path Intercept Altitude, it is the resultant actual point of the glideslope/path intercept.

final approach point (FAP)
Aviation. The point, applicable only to a non-precision approach with no depicted final approach fix (FAF) (such as an on-airport VOR), where the aircraft is established inbound on the final approach course from the procedure turn and where the final approach descent may be commenced. The final approach point (FAP) serves as the FAF and identifies the beginning of the final approach segment.

final approach segment
Aviation. That segment of an instrument approach procedure in which alignment and descent for landing are accomplished.

final controller
See ***air traffic controller***.

final monitor aid
Aviation. A high resolution color display that is equipped with the controller alert system hardware/software which is used in the precision runway monitor (PRM) system. The display includes alert algorithms providing the target predictors, a color change alert when a target penetrates or is predicted to penetrate the no transgression zone (NTZ), a color change alert if the aircraft transponder becomes inoperative, synthesized voice alerts, digital mapping, and like features contained in the PRM system.

final product
Under ISO 14000, a product which requires no additional transformation prior to its use.

finding of no significant impact (FNSI)
A document prepared by a federal agency that presents the reasons why a proposed action would not have a significant impact on the environment and thus would not require preparation of an Environmental Impact Statement. An FNSI is based on the results of an environmental assessment. Also referred to as *FONSI*.

FINDS
See ***Facility Index System***.

fine
A money sanction ordered by a government agency or court, sometimes loosely used to include civil penalties but more properly applied only to criminal fines.

finger
Any of the structures on the hand composed of three phalanges and the surrounding tissues of a digit.

finger dexterity
The ability to make rapid, coordinated finger movements using one or both hands to manipulate small objects.

finger diameter
The maximum medial-lateral cross-sectional diameter of a finger. Measured by a determination of the smallest diameter hole into which the finger can be inserted (specifying the digit involved).

finger-shaping
Providing the alternating troughs and ridges on a handle or gripping structure to accommodate the fingers and the gaps between them.

fingernail
The harder elastic tissue covering the dorsal portion of the terminal phalanges of the hand.

fingerprint
The pattern of unique whorls and ridges on the pad of the distal phalanx of each finger. Also known as ***digital dermatoglyph***.

fingertip height
See ***dactylion height***.

finished aviation gasoline
All special grades of gasoline for use in aviation reciprocating engines, as given in American Society for Testing and Materials (ASTM) Specification D910 and Military Specification MIL-G-5572. Excludes blending components that will be used for blending or compounding into finished aviation gasoline. *See also* ***gasoline***.

finished gasohol motor gasoline
A blend of finished motor gasoline (leaded or unleaded) and alcohol (generally ethanol, but sometimes methanol) in which 10 percent or more of the product is alcohol. *See also* ***gasohol*** *and* ***gasoline***.

finished leaded gasoline
Contains more than 0.05 gram of lead per gallon or more than 0.005 gram of phosphorus per gallon. Premium and regular grades are included, depending on the octane rating. Includes leaded gasohol. Blend-stock is excluded until blending has been completed. Alcohol that is to be used in the blending of gasohol is also excluded.

finished leaded premium motor gasoline
Motor gasoline having an antiknock index, calculated as (R+M)/2, greater than 90 and containing more than 0.05 gram of lead per gallon or more than 0.005 gram of phosphorus per gallon.

finished leaded regular motor gasoline
Motor gasoline having an antiknock index, calculated as (R+M)/2, greater than or equal to 87 and less than or equal to 90 and containing more than 0.05 gram of lead or 0.005 gram of phosphorus per gallon.

finished motor gasoline
(1) A complex mixture of relatively volatile hydrocarbons, with or without small quantities of additives, blended to form a fuel suitable for use in spark-ignition engines. Specification for motor gasoline, as givcn in American Society for Testing and Materials (ASTM) Specification D439-88 or Federal Specification VV-G-1690B, include a boiling range of 122 degrees to 158 degrees Fahrenheit at the 10 percent point to 365 degrees to 374 degrees Fahrenheit at the 90 percent point and a Reid vapor pressure range from 9 to 15 psi. "Motor gasoline" includes finished leaded gasoline, finished unleaded gasoline, and gasohol. Blendstock is excluded until blending has been completed. Alcohol that is to be used in the blending of gasohol is also excluded. (2) Motor gasoline that is not included in the reformulated or oxygenated categories.

finished unleaded gasoline
Contains not more than 0.05 gram of lead per gallon and not more than 0.005 gram of phosphorus per gallon. Premium and regular grades are included, depending on the octane rating. Includes unleaded gasohol. Blendstock is excluded until blending has been completed. Alcohol that is to be used in the blending of gasohol is also excluded.

finished unleaded midgrade motor gasoline
Motor gasoline having an antiknock index, calculated as (R+M)/2, greater than or equal to 88 and less than or equal to 90 and containing not more than 0.05 gram of phosphorus per gallon.

finished unleaded premium motor gasoline
Motor gasoline having an antiknock index, calculated as (R+M)/2, greater than 90 and containing not more than 0.05 gram of lead or 0.005 gram of phosphorus per gallon.

finished unleaded regular motor gasoline
Motor gasoline having an antiknock index, calculated as (R+M)/2, of 87 containing not more than 0.05 gram of lead per gallon and not more than 0.005 gram of phosphorus per gallon.

finite element
A small segment of a large object obtained by some standard division process.

finite element analysis
The use of finite elements to model force components on a large object or complex structure and draw conclusions about that object or structure as a whole.

FIP
*See **Federal Implementation Plan**.*

FIR
*See **flight information region**.*

fire
(1) Uncontrolled combustion by flame or smoke resulting in evidence of charring, melting, or other evidence of ignition. (2) The process of rapid oxidation that generally produces both heat and light. May also be referred to generally as *combustion*.

fire alarm
Any fire protection device or system which indicates the presence of a fire.

fire brigade
An organized group of employees who are knowledgeable, trained, and skilled in at least basic fire fighting operations. May also be referred to as a *private fire department* or an *industrial fire department*.

fire classification
A division of fires by the types of materials being burned. Briefly: *Class A* – Ordinary combustible materials (e.g., wood, paper);

Class B – Flammable liquid or gas (e.g., oil, paint, grease); *Class C* – Energized electrical circuits (e.g., electrical wiring, equipment); *Class D* – Combustible metals (e.g., magnesium, sodium, lithium). Portable fire extinguishers are also based on these fire classifications (i.e., they are classed for the type of fire they are capable of extinguishing). *See also* ***Class A Fire, Class B Fire, Class C Fire,*** *and* ***Class D Fire***.

fire detection
The use of any fire protection device or system intended to determine that a fire is present. Usually sensitive to heat, smoke, or flame.

fire door
Any door which has been designed, tested, and rated for preventing the spread of fire.

fire/explosion, fuel
Accidental combustion of vessel fuel, liquids, including their vapors, or other substances, such as wood or coal.

fire/explosion, other
Accidental burning or explosion of any material on board except vessel fuels or their vapors.

fire insurance
A contract of insurance by which the underwriter, in consideration of the premium, undertakes to indemnify the insured against all losses in his/her houses, buildings, furniture, ships in port, or merchandise by means of accidental fire happening within a prescribed period. *See also* ***insurance***.

fire or violent rupture
Rail. An accident or incident caused by combustion or violent release of material carried by or transported by rail. Examples of this type include fuel and electrical equipment fires, crankcase explosions, and violent release of liquefied petroleum or anhydrous ammonia.

fire point
The minimum temperature to which a material must be heated to sustain combustion after ignition by an external source.

fire prevention
The study and/or implementation of measures specifically designed to control ignition and fuel sources.

fire protection
The implementation of measures for preventing, detecting, controlling, and extinguishing fire to protect life and property.

fire resistant
(1) Pertaining to a normally non-combustible material which will withstand the effects of a fire. (2) With respect to sheet or structural members, it means the capacity to withstand the heat associated with fire at least as well as aluminum alloy in dimensions appropriate for the purpose for which they are used. (3) With respect to fluid-carrying lines, fluid system parts, wiring, air ducts, fittings, and powerplant controls, it means the capacity to perform the intended functions under the heat and other conditions likely to occur when there is a fire at the place concerned.

fire resistive
The ability of a structure or material to provide a predetermined degree of fire resistance, usually rated in hours.

fire retardant
Any material or substance which slows the progress of a fire through reduced combustibility.

fire triangle
The recognition that three elements must be present in the right proportion for a fire to exist. These are oxygen (or an oxidizing agent), fuel (or a reducing agent), and heat. Keeping the three elements of the fire triangle apart is the key to preventing fires, and removing one or more of these elements is the key to extinguishing fires that do start.

Fire triangle concept
(remove any one leg and a fire cannot start or an existing fire will go out)

fire wall

Any self-supporting vertical structure designed to resist the horizontal spread of a fire from one enclosed region to another.

firefighting vehicle

A vehicle designed exclusively for the purpose of fighting fires.

Fireman's Rule

Doctrine which holds that professionals, whose occupations by nature expose them to particular risks, may not hold another negligent for creating the situation to which they respond in their professional capacity.

fireproof

According to commonly accepted terminology, the word "fireproof" is technically not an accurate term since few materials are actually incapable of "total" resistance to flame or fire. However, as defined in 14 CFR 1, the term "fireproof" means: (1) with respect to materials and parts used to confine fire in a designated fire zone, the capacity to withstand at least as well as steel in dimensions appropriate for the purpose for which they are used, the heat produced when there is a severe fire of extended duration in that zone; and (2) with respect to other materials and parts, the capacity to withstand the heat associated with fire at least as well as steel in dimensions appropriate for the purpose for which they are used.

fireworks

See ***pyrotechnics***.

firing circuit

The current path between the power source and the initiating device.

firmware

Computer programs and data loaded in a class of memory that cannot be dynamically modified by the computer during processing. For System Safety purposes, firmware is to be treated as software.

first aid

(1) Any emergency care provided to an ill or injured person in order to relieve pain, counteract shock, or prevent death or further injury until better medical care becomes available. (2) Under OSHA 29 CFR 1904.12(e), any one-time treatment, and any follow-up visit for the purpose of observation, of minor scratches, cuts, burns, splinters, and so forth, which do not ordinarily require medical care. Such one-time treatment, and follow-up visit for the purpose of observation, are considered first aid even though they may be provided by a physician or registered professional personnel.

first aid injury

Any injury requiring first aid treatment only; considered by OSHA to be non-recordable for recordkeeping purposes. *See also* ***first aid***.

First Amendment

Amendment to the U.S. Constitution guaranteeing basic freedoms of speech, religion, press, and assembly and the right to petition the government for redress of grievances. The various freedoms and rights protected by the First Amendment have been held applicable to the states through the due process clause of the Fourteenth Amendment. *See also* ***fighting words doctrine***.

first-class lever

A lever in which the fulcrum is located between the effort and resistance.

first-class passenger revenue

Aviation. Revenues from the air transportation of passengers moving at either standard fares, premium fares, or at reduced fares not predicated upon the use of aircraft space specifically separated from first class, and for whom standard or premium quality services are provided.

first-class service

Aviation. Transport service established for the carriage of passengers moving at either standard fares or premium fares, or at reduced fares not predicated upon the operation of specifically allocated aircraft space, and for whom standard or premium quality services are provided.

first degree murder

A murder committed with deliberately premeditated malice aforethought, or with extreme atrocity or cruelty, or in the commission or attempted commission of a crime punishable with death or imprisonment for life, is murder in the first degree.

first draw

The water that immediately comes out when a tap is first opened. This water is likely to have the highest level of lead contamination from plumbing materials.

first harmful event
Highway Transit. (1) A first harmful event is the first event during a traffic accident that causes an injury (fatal or nonfatal) or property damage. (2) The first event during a crash that caused injury or property damage.

first phalanx length
The linear distance of the most proximal segment of a finger. Measured across the surfaces from the distal tip of the third metacarpal to the proximal tip of the second phalanx while the hand is held in a fist (while specifying the digit involved).

first piece time
The time permitted or required for the production of the first complete item in the starting sequence of several complete items.

first-order content
*See **rate control**.*

first shift
A day work shift of about 8 hours' duration, approximately between 7 A.M. and 5 P.M. Also called day shift or A-shift.

first tier center
Aviation. The air route traffic control center immediately adjacent to the impacted center.

fish
According to the Federal Outer Continental Shelf Lands Act Amendments of 1978: Fin fish, mollusks, crustaceans, and all other forms of marine animal and plant life other than marine mammals, birds, and highly migratory species.

Fishberg concentration test
A laboratory test used to determine the ability of the kidneys to concentrate urine. Samples of urine are collected and tested for specific gravity.

fishy back
Transportation (slang). The movement of loaded truck trailers by barge or ferry.

fission
A type of nuclear reaction occurring in very heavy atoms in which the nucleus, following bombardment by neutrons or other atomic particles, splits into two nuclei of nearly comparable mass, accompanied by the release of energy. Also referred to as *atomic fission* and *nuclear fission.*

fission products
The products produced as a result of the splitting (fission) of a substance.

fissionable material
A material that can be split (fission) into other nuclei by any process. *See also **fission**.*

fissure
A narrow slit or cleft.

fist
A hand posture consisting of a maximal flexion of the hand in which the phalanges of digits II – V (the fingers) are tightly collapsed into the palm with the metacarpals and phalanges of digit I (the thumb) flexed to overlie the fingers.

fist circumference
The surface distance around the fist over the thumb and the knuckles. Measured with the thumb lying across the end of the fist.

The human hand, in fist orientation

fistula
Any abnormal, tubelike passage within body tissue.

fit

(1) The adequacy, suitability, and/or appropriateness of some individual, equipment, object, or structure with consideration of size, shape, conditioning, or other aspects to perform some function or fulfil a need or use. (2) A sudden, brief exhibition of emotion or motor activity.

fit check

*See **fit test**.*

fit factor

The value of the ratio of the outside concentration of a substance to the concentration of that substance inside a respirator/face mask during a fit test.

fit test

The testing of a prototype item on either a sample or potentially the population as a whole to verify that a design is acceptable, appropriate, or the best option for the environment. The term usually refers to clothing or personal protective equipment. Also referred to as *fit check*.

fitting

An accessory such as a lock nut, bushing, or other part of a wiring system that is intended primarily to perform a mechanical rather than an electrical function.

Fitts' Law

A rule for movement time prediction, in which the average movement time in a response is a function of the target separation distance and the width of the target. *See also **index of difficulty***. Expressed as

$$MT = a + b \log_2 \left(\frac{2A}{W}\right)$$

where:

MT = movement time
A = distance to target
W = width of target

fix

Aviation. A geographical position determined by visual reference to the surface, by reference to one or more radio navigational aids (NAVAIDs), by celestial plotting, or by another navigational device.

fix balancing

Aviation. A process whereby aircraft are evenly distributed over several available arrival fixes reducing delays and controller workload.

fixation

(1) The focusing and convergence of the eyes on some point or object at a distance. (2) Having a particular attachment for one technique for performing some task.

fixation disparity

A condition in which the visual axes intersect at some point other than in the desired fixation plane.

fixation distance

That distance at which the visual axes intersect.

fixation muscle

*See **fixator**.*

fixation plane

That fixation surface which is at such a distance from the observer that the arc may be assumed for practical purposes to be planar. *See also **fixation surface**.*

fixation point

That location in a normal individual's line of sight at which the eyes' visual axes intersect. Also referred to as *point of fixation*.

fixation reflex

An ocular reflex mechanism which tends to orient the eyes toward a stationary light or object or to keep the eyes oriented toward a light or object which is in motion relative to the observer.

fixation surface

That curved surface which is perpendicular to the observer's line of sight and which contains the fixation point of the eyes. *See also **fixation plane**.*

fixative

A chemical, such as alcohol or formaldehyde, used for the preservation of biological materials.

fixator

A muscle which undergoes an isometric contraction to steady a body part or segment against some other muscle contraction or against an external force. Also referred to as *fixation muscle* and *stabilizer*.

fixed collision barrier

A flat, vertical, unyielding surface with the following characteristics: 1) The surface is sufficiently large that when struck by a tested vehicle, no portion of the vehicle projects or passes beyond the surface; 2) The approach is a horizontal surface that is large enough for the vehicle to attain a stable attitude during its approach to the barrier, and that does not restrict vehicle motion during impact; 3) When struck by a vehicle, the surface and its supporting structure absorb no significant portion of the vehicle's kinetic energy, so that a performance requirement described in terms of impact with a fixed collision barrier must be met no matter how small an amount of energy is absorbed by the barrier.

fixed crane
A crane whose principal structure is mounted on a permanent or semipermanent foundation.

fixed dam
A dam which does not permit the passage of marine traffic and requires the use of a lock in contrast to movable dams which, during periods of high water, are lowered to allow traffic to pass directly over the dam. Also, any dam that has a fixed height without adjustment such as a concrete spillway throughout the length of the dam exclusive of the lock chamber.

fixed extinguishing system
A permanently installed system that either extinguishes or controls a fire at the location of the system.

fixed function key
A keyboard key which directs a computer to perform some unchangeable, specific function when pressed.

fixed guideway system
A system of vehicles that can operate only on its own guideway constructed for that purpose (e.g., rapid rail, light rail). Federal usage in funding legislation also includes exclusive right-of-way bus operations, trolley coaches and ferryboats as "fixed guideway" transit.

fixed linkage mechanism
See ***link***.

fixed object
Stationary structures or substantial vegetation attached to the terrain.

fixed operating cost
Transit. In reference to passenger car operating cost, refers to those expenditures that are independent of the amount of use of the car, such as insurance costs, fees for license and registration, depreciation and finance charges.

fixed route
Service provided on a repetitive, fixed-schedule basis along a specific route with vehicles stopping to pick up and deliver passengers to specific locations; each fixed-route trip serves the same origins and destinations, unlike demand response and taxicabs.

fixed route system
A system of transporting individuals (other than by aircraft), including the provision of designated public transportation service by public entities and the provision of transportation service by private entities, including, but not limited to, specified public transportation service, on which a vehicle is operated along a prescribed route according to a fixed schedule.

fixed shift
A work shift in which the working hours remain the same over time.

fixture
(1) Any device at a workplace used for positioning or holding materials being assembled, worked on, or used. (2) *See* ***lighting fixture***.

fixture hand
That hand being used to hold an object while the other hand performs some work on the object.

fl
See ***footlambert***.

flag
Aviation. A warning device incorporated in certain airborne navigation and flight instruments indicating that a) instruments are inoperative or otherwise not operating satisfactorily, or b) signal strength or quality of the received signal falls below acceptable values.

flag drop charge
The charge for an initial distance (usually specified by regulation) for taxi service. It is actually the minimum fare.

flame
The electromagnetic radiation from a fire, typically referring to the visible range.

flame arrester
Device used in gas vent lines, and other similar locations, to arrest or prevent the passage

of flame into an enclosed space, such as a container or flammable liquid storage cabinet.

flame ionization detector (FID)
A carbon detector which relies on the detection of ions formed when carbon-containing material, such as a volatile or gaseous hydrocarbon, is burned in a hydrogen-rich flame. This detector is commonly used in a gas chromatograph to detect and quantify organic compounds. It is also employed in some portable instruments.

flame photometric detector
A detection system based on the luminescent emissions between 300 and 425 nanometers when sulfur compounds are introduced into a hydrogen-rich flame. An optical filter system is used to differentiate the sulfur compounds present from other materials. This detector finds application in gas chromatography.

flame propagation
The spread of a flame throughout an entire volume of a vapor-air mixture from a single source of ignition.

flame resistant
(1) Not susceptible to combustion to the point of propagating a flame, beyond safe limits, after the ignition source is removed. (2) The property of materials, or combinations of component materials, to retard ignition and restrict the spread of flame.

flameout
Unintended loss of combustion in turbine engines resulting in the loss of engine power.

flammability range
The difference between the lower and upper flammable limits, expressed in terms of percentage of a vapor or gas in air or oxygen by volume. *See also* ***flammable range*** *and* ***upper flammability limit (UFL)***.

flammable
(1) Any substance that is easily ignited and burns, or has a rapid rate of flame spread. (2) Capable of being ignited and burning. (3) With respect to a fluid or gas, means susceptible to igniting readily or to exploding.

flammable atmosphere
A surrounding gaseous environment which contains a mixture of gases or vapors within their flammable range(s).

flammable limits
The percent by volume limits (i.e., upper and lower flammable limits) of the concentration of a flammable gas at normal temperature and pressure in air above and below at which flame propagation does not occur on contact with a source of ignition. *See also* ***flammability range, lower flammable limit,*** *and* ***upper flammable limit***.

flammable liquid
(1) *Class I flammable liquid.* Any liquid having a flash point below 100°F (37.8°C). (2) *Class II flammable liquids.* Any liquid having a flash point above 100°F (37.8°C) but below 140°F (60°C).

flammable mixture
Any combination of flammable vapor or gas and an appropriate oxidizing agent within the flammable range.

flammable solid
A solid material that is easily ignited and that burns rapidly.

flanged hood
A barrier placed around the periphery of a chemical hood to reduce air turbulence and hood entry pressure loss by keeping the hood from drawing air from behind the hood face.

flanking buoy
Buoy tied to the corner of a tow so pilot can tell when tow has been checked.

flanking maneuver
Maritime. Maneuvering action of a tow (when down-bound) approaching at an angle (usually 30 to 45 degrees) at bridges or locks or in sharp bends. Only the current is utilized for headway, and the engines and rudders are used to maintain the angle until just before the lead barges reach the bridge span, at which time the engines are backed and the head of the tow is swung gently in line with the opening. Then full power is applied to drive through the opening. This is the safest way that a heavy tow can make tight passages.

flanking rudder
Maritime. A rudder installed forward of the screw, used for maneuvering when the propellers are turning a stem regardless of the direction of actual movement of the towboat. *Also called* ***backing rudders***.

flap extended speed
Aviation. The highest speed permissible with wing flaps in a prescribed extended position.

flash
(1) A sudden, great increase in brightness for a short period of time. (2) A highlighting technique in which a selected portion of a display momentarily increases in brightness.

flash blindness
A temporary inability to see detail of objects having poor illumination following a brief exposure to very intense light.

flash burn
(1) An inflammation of the lens of the eye due to excessive exposure to ultraviolet radiation, usually from a welding arc. (2) Any injury to tissue from sudden intense heat radiation.

Direct (unprotected) exposure of the eye to a welding arc or other ultraviolet light source is the most common cause of flash burn injuries

flash point
(1) The lowest temperature of a liquid at which there are sufficient vapors given off to form a combustible mixture in the air near the surface of the liquid. (2) That temperature at which a material, liquid or solid, will provide sufficient quantity of vapors to ignite in the presence of an ignition source.

flash rate
The number of times a highlighted portion of a display increases in brightness within a specified temporal interval.

flash resistant
Not susceptible to burning violently when ignited.

flashback arrester
A mechanical device utilized on a vent of a flammable liquid or gas storage container to prevent flashback into the container, when a flammable or explosive mixture ignites outside the container.

flasher
In rail systems, the flashing light at railroad grade crossings that warns motorists, bicyclists, and pedestrians of approaching trains.

flat
(1) *General.* A smooth, level surface. (2) Having little or no gloss. (3) *Maritime-General.* A small barge with flat top used for transporting fuel or other miscellaneous cargo. (4) *Maritime-Navigation.* A place covered with water too shallow for navigation with vessels ordinarily used for commercial purposes. The space between high and low water marks along the edge of an arm of the sea, bay, tidal river, etc.

flat car
A rail car without a roof and walls.

flat face
Cab over engine configuration.

flat pool
The normal stage of water in the area between two dams that is to be maintained by design when little or no water is flowing; hence the pool flattens out.

flat rate manual
A manual published by an equipment manufacturer or an independent publisher that indicates the length of time required for performing specific mechanical tasks such as installing a clutch. Normally, the costs of parts required for a specific job are also listed.

flatbed
A truck or trailer without sides and top.

flatboat
A rectangular, flat-bottomed boat used on the western rivers during the 18th and 19th centuries.

flatulence
Having gas in the gastrointestinal tract.

flatus
Gas or air expelled from the gastrointestinal tract.

flaw
An imperfection or unintentional discontinuity that is detectable by nondestructive examination.

fleet
The vehicles in a transit system. Usually, "fleet" refers to highway vehicles and "rolling stock" to rail vehicles.

fleet management center (FMC)
A formally approved element of the Interagency Fleet Management System (IFMS) responsible for the administrative control of Interagency Fleet Management System (IFMS) vehicles in a specified geographic area as defined in the determination that is approved by the Administrator of General Services.

fleet management subcenter
A formally approved element of the Interagency Fleet Management System (IFMS) the Fleet Management Center is physically detached from the central or main Fleet Management Center.

fleet management system (FMS)
The automated inventory and control system used by the Interagency Fleet Management System (IFMS) to track vehicle assignments, vehicle utilization, and provide direct input to the Finance Division to bill customer agencies for the use of IFMS vehicles.

fleet policy insurance
Type of blanket insurance covering a number of vehicles of the same insured (e.g., covers pool or fleet vehicles owned by a business). *See also* ***insurance***.

fleet vehicles
(1) Private fleet vehicles: ideally, a vehicle could be classified as a member of a fleet if it is operated in mass by a corporation or institution, operated under unified control, or used for non-personal activities; however, the definition of a fleet is not consistent throughout the fleet industry. Some companies make a distinction between cars that were bought in bulk rather than singularly, or whether they are operated in bulk, as well as the minimum number of vessels that constitute a fleet (i.e., 4 or 10). (2) Government fleet vehicles: includes vehicles owned by all federal (GSA) state, county, city, and metro units of government, including toll road operations.

fleeting
Storing of barges (loaded or unloaded) until they can be moved to the unloading area or until the owner can pick them up.

flex
To move adjacent body segments connected by a common joint so that the angle formed by the joint and the two segments is decreased.

flexibility
(1) The capability for adjusting to varying conditions. (2) A measure of the mobility of a joint or a series of joints. Quantified as the range of motion, reach.

flexibility of closure
The ability to discover a known pattern masked by the background material.

flexible work schedule
See ***flextime***.

flexion
Movement in which the angle between two bones connecting to a common joint is reduced.

flexor
Any muscle which causes joint flexion.

flexor muscles
Muscles which when contracted, decrease the angle between limb segments.

flexor retinaculum
The ligament which forms the carpal tunnel in the wrist through which the finger flexor tendons and the median nerve pass.

flextime
A work schedule in which an employee has the freedom within certain limits to choose his work starting and stopping times, but which usually includes a period of time within a given shift during which the employee must be present. Also referred to as *flextime* and *flexible work schedule*.

flicker
A perceptible temporal variation of brightness or movements occurring several times per

second in a display or other source within the visual field.

flicker-free display
A visual display unit with a refresh rate greater than 60 Hz.

flicker fusion
The perception of a regular, intermittent visual stimulus as continuous or steady by the eye or video sensor. May be referred to more commonly as *fusion*.

flicker fusion frequency (fff)
The frequency at which flicker fusion occurs. Also referred to as *critical flicker frequency, critical fusion frequency,* and *critical fusion frequency*.

flicker photometry
The use of a single field of view and rapidly alternating light sources of different colors to determine equal-appearing intensity.

flight advisory service
A service specifically designed to provide, upon pilot request, timely weather information pertinent to his type of flight, intended route of flight, and altitude. The flight service stations providing this service are listed in the Airport/Facility Directory. *See also* ***flight watch***.

flight check
A call-sign prefix used by Federal Aviation Administration (FAA) aircraft engaged in flight inspection/certification of navigational aids and flight procedures. The word "recorded" may be added as a suffix; (e.g., "flight check 320 recorded" to indicate that an automated flight inspection is in progress in terminal areas). *See also* ***flight inspection***.

flight crew member
A pilot, flight engineer, or flight navigator assigned to duty in an aircraft during flight time.

flight deck
(1) That region of an aircraft or spacecraft in which the flight controls and instrumentation, the pilot, and others involved in operating the vehicle are based. (2) That region of an aircraft-carrying ship on which air-support and/or ground-support operations, including launching and landing, take place.

flight equipment
The total cost of property and equipment of all types used in the in-flight operations of aircraft and construction work in progress.

flight information region (FIR)
An airspace of defined dimensions within which Flight Information Service and Alerting Service are provided.

flight information service
A service provided for the purpose of giving advice and information useful for the safe and efficient conduct of flights.

flight inspection
In-flight investigation and evaluation of a navigational aid to determine whether it meets established tolerances. *See also* ***flight check***.

flight level
A level of constant atmospheric pressure related to a reference datum of 29.92 inches of mercury. Each is stated in three digits that represent hundreds of feet. For example, flight level 250 represents a barometric altimeter indication of 25,000 feet; flight level 255, an indication of 25,500 feet. *See also* ***cardinal altitude***.

flight line
A term used to describe the precise movement of a civil photogrammetric aircraft along a predetermined course(s) at a predetermined altitude during the actual photographic run.

Flight Management System
A computer system that uses a large database to allow routes to be preprogrammed and fed into the system by means of a data loader. The system is constantly updated with respect to position accuracy by reference to conventional navigation aids. The sophisticated program and its associated database ensures that the most appropriate aids are automatically selected during the information update cycle.

Flight Management System Procedure
An arrival, departure, or approach procedure developed for use by aircraft with a slant (/G) equipment suffix.

flight path
A line, course, or track along which an aircraft is flying or intending to be flown. *See also* ***bearing, course***.

flight plan

Specified information, relating to the intended flight of an aircraft, that is filed orally or in writing with air traffic control.

flight plan area

The geographical area assigned by regional air traffic divisions to a flight service station for the purpose of search and rescue for Visual Flight Rule (VFR) aircraft, issuance of NOTAMs, pilot briefing, in-flight services, broadcast, emergency services, flight data processing, international operations, and aviation weather services. Three letter identifiers are assigned to every flight service station and are annotated in AFDs and Order 7350.6 as tie-in-facilities.

flight recorder

A general term applied to any instrument or device that records information about the performance of an aircraft in flight or about conditions encountered in flight. Flight recorders may make records of airspeed, outside air temperature, vertical acceleration, engine RPM, manifold pressure, and other pertinent variables for a given flight.

Flight Service Station (FSS)

Air traffic facilities offering pilot briefings, en route communications, and VFR search and rescue services. Additionally, the FSS assists lost aircraft and aircraft in emergency situations; relay ATC clearances; originate Notices to Airmen; broadcast aviation weather and NAS information; receive and process IFR flight plans; monitor radio Navigation Aids (NAVAIDS). Also, at selected locations, FSSs take weather observations, issue airport advisories, and advise Customs and Immigration of transborder flight.

flight services

The sum of flight plans originated and pilot briefs, multiplied by two, plus the number of aircraft contacted.

flight simulator

A flight trainer with computer-driven functional displays and controls, possibly including motion.

flight stage

The operation of an aircraft from takeoff to landing.

Flight Standards District Office

A Federal Aviation Administration (FAA) field office serving an assigned geographical area and staffed with Flight Standards personnel who serve the aviation industry and the general public on matters relating to the certification and operation of air carrier and general aviation aircraft. Activities include general surveillance of operational safety, certification of airmen and aircraft, accident prevention, investigation, enforcement, etc.

flight test

A flight for the purpose of a) investigating the operation/flight characteristics of an aircraft or aircraft component, and/or b) evaluating an applicant for a pilot certificate or rating.

flight time

The time from the moment the aircraft first moves under its own power for the purpose of flight until the moment it comes to rest at the next point of landing. Also known as *block to block time.*

flight trainer

A ground-based pilot training device containing a representation of an aircraft cockpit for familiarization and basic training purposes.

flight visibility

The average forward horizontal distance, from the cockpit of an aircraft in flight, at which prominent unlighted objects may be seen and identified by day and prominent lighted objects may be seen and identified by night.

flight watch

A shortened term for use in air-ground contacts to identify the flight service station providing En Route Flight Advisory Service (e.g., "Oakland Flight Watch"). *See also **flight advisory service**.*

float

(1) *General.* To refrain from or prevent sinking. (2) *Transit.* A flatbed semitrailer. (3) *Computing.* The amount of slack in a network. (4) *See **bank**.*

float light

A 10-foot wooden platform mounted on pontoons supporting a battery-operated light. Used exclusively on the Upper Mississippi River in a certain area.

float scaffold
A scaffold hung from overhead supports by means of ropes and consisting of a substantial platform having diagonal bracing underneath, resting upon and securely fastened to two parallel plank bearers at right angles to the span. Also called *ship scaffold*.

floater
Transit (slang). A driver without a steady job.

floater insurance
A form of insurance that applies to movable property whatever its location, if within the territorial limits imposed by the contract. *See also* ***insurance***.

floating crane
A crane mounted on a barge or pontoon which can be towed or self-propelled from place to place.

floating kidney
A condition in which the kidney does not remain fixed in its normal position. Nephroptosis refers to a dropping of the kidney from its normal position. Surgical correction, by nephropexy, is necessary when the condition interferes with normal kidney function.

floating pin
A mooring pin or timber head attached to a floating tank in a lock chamber set in a guided recess in the lock walls, for mooring tows within the lock chamber whereby a short mooring line suffices without an attendant.

floating the gears
Transportation (slang). Shifting gears without using the clutch.

floats
Large single tires, instead of dual tires.

floc
A dump of solids formed in sewage by biological or chemical action.

flocculate
See ***agglomeration***.

flocculation
The process by which clumps of solids in water or sewage are made to increase in size by biological or chemical action so that they can be separated from the water.

flood
(1) *General.* An uncontrolled overrun of a liquid (usually water) into an area where it is not normally expected to be in such great quantities. (2) *Computing.* To send multiple messages to the viewing screen, with clearing or scrolling of the screen before all can be read.

flood insurance
Insurance indemnifying against loss by flood damage. Required by lenders in areas designated as potential flood areas. The insurance is privately issued but federally subsidized. *See also* ***insurance***.

flood plain
An area which is subject to periodic flooding. *See also* ***floodplain***.

flood stage
Condition of the river when it rises above a stage predetermined by the Corps of Engineers to be designated as flood stage. Also, the stage at which some part of the main bank may be overflowed, but not necessarily all of it.

floodgate
Gate placed across/along a channel to control floodwater or a gate across a roadway in levee. *See also* ***gate and tidegate***.

flooding
Filling with water, regardless of method of ingress, but retaining sufficient buoyancy to remain on the surface.

floodplain
Land adjacent to rivers which, because of its level topography, floods when a river overflows. *See also* ***flood plain***.

floodwater
Waters which escape from a stream or other body of water and overflow adjacent territory, under conditions which do not usually occur.

floor hole
An opening measuring less than 12 inches but more than 1 inch in its least dimension, in any floor, platform, pavement, or yard, through which materials but not persons may fall; such as a belt hole, pipe opening, or slot opening.

floor opening

An opening measuring 12 inches or more in its least dimension, in any floor, platform, pavement, or yard, through which persons may fall; such as a hatchway, stair or ladder opening, pit, or large manhole. Floor openings occupied by elevators, dumb waiters, conveyors, machinery, or containers are excluded from this definition (OSHA 29 CFR 1910.21(a)(2).

floor reference plane

(1) That plane through the floor reference point and perpendicular to the local vertical axis. (2) That point on a floor or other base surface which provides an origin for representing all other coordinates within the volume of interest.

floor sweep

A vapor collection system designed to capture vapors which are heavier than air and which collect along the floor.

floppy disk

A flexible plastic disk used as a common form of external storage for information in a microcomputer system.

flow analysis

An examination of the progressive sequence of activities and locations of personnel, equipment, and materials involved in the performance of a particular task or operation.

flow control

Aviation. Measures designed to adjust the flow of traffic into a given airspace, along a given route, or bound for a given aerodrome (airport) so as to ensure the most effective utilization of the airspace.

flow diagram

A scaled graphic/pictorial representation of the layout and locations of activities or operations and the flow paths of materials between activities in a process.

flow line

See ***flow path***.

flow meter

A gauge that shows the speed of wastewater moving through a treatment plant. Also used to measure the speed of liquids moving through various industrial processes.

flow path

The route(s) taken by personnel, equipment, and materials involved in production as the manufacturing process continues. Also called *flow line* or *line of flow*.

flow process chart

A graphic/symbolic representation using standardized symbols for the manipulations involved for an item through each of the various steps required. *See also* ***process chart, process chart symbol, worker-type flow process chart, material-type flow process chart,*** *and* ***equipment-type flow process chart***.

flow rate

The volume per time unit (e.g., liters per minute, etc.) given to the flow of air or other fluid by the action of a pump, fan, etc.

flowchart

A diagram consisting of standardized symbols which enclose text and/or other symbols and are governed by specific layout rules for describing the steps involved in a given operation.

FLSA

See Fair Labor Standards Act.

FLSC

Flexible linear-shaped charge.

flue

A pipe or other channel through which combustion air, smoke, steam, or other material is vented to the atmosphere.

flue gas

The air coming out of a chimney after combustion in the burner it is venting. It can include nitrogen oxides, carbon oxides, water vapor, sulfur oxides, particles, and many chemical pollutants.

flue gas desulfurization

A technology which uses a sorbent, usually lime or limestone, to remove sulfur dioxide from the gases produced by burning fossil fuels. Flue gas desulfurization is currently the state-of-the-art technology in use by major sulfur dioxide emitters, such as power plants.

fluence

The number of particles or photons passing per unit area, usually square centimeter. Also referred to as *radiation fluence*.

fluid
(1) A material that flows readily in the natural state; a liquid or gas. (2) Composed of elements that yield to pressure without disruption of the mass. (3) One of the ultimate states of matter, being composed of molecules that can move about within limits, permitting change in the shape of the mass without disruption of the substance.

fluid balance
A physiological state in which water intake equals water loss. Also referred to as *water balance*.

flume
A natural or manmade channel that diverts water.

fluorence
A hue similar to fluorescent materials.

fluorescence
Phenomenon involving the absorption of radiant energy by a substance (usually a crystal) and its re-emission as visible or near-visible light.

fluorescent lamp
A light source which operated by passing an electrical current through a closed tube containing mercury vapor and one or more suitable fluorescing powders coating the interior surface of the tube.

fluorescent screen
A sheet of material coated with a substance (usually calcium tungstate or zinc sulfide) that will emit light when irradiated with ionizing radiation.

fluoridation
The addition of a fluoride, a chemical salt containing fluorine, to drinking water. This has been found to reduce the occurrence of dental caries in children by one-half. Minute traces of fluoride are found in almost all food, but the quantity apparently is too small to meet the requirements of the body in building tooth enamel that resists cavities.

fluorides
Gaseous, solid, or dissolved compounds containing fluorine that result from industrial processes. Excessive amounts in food can lead to fluorosis.

fluorine
A chemical element, atomic number 9, atomic weight 18.998, symbol F.

fluorocarbon
Any of a number of organic compounds analogous to hydrocarbons in which one or more hydrogen atoms are replaced by fluorine.

fluoroscope
A screen mounted in front of an x-ray tube used for indirect visualization of internal body organs or internal structures of inanimate objects.

fluorosis
An abnormal condition caused by excessive intake of fluorine, characterized chiefly by mottling of the teeth.

flush
(1) To open a cold water tap to clear out all the water which may have been sitting for a long time in the pipes. In new homes, to flush a system means to send large volumes of water gushing through the unused pipes to remove loose particles of solder and flux. (2) To force large amounts of water through liquid to clean out piping or tubing, storage or process tanks.

flush-mounted
Pertaining to any piece of equipment which is embedded within a structure such that the exposed surface of the equipment is level with the structure surface.

flutter
(1) Any deviation in frequency of the reproduced sound from the original sound. (2) Any low-frequency vibration of an object capable of such vibration. (3) A tremulous, generally ineffective movement.

flux
(1) *Electromagnetic Radiation.* The number of visible-light photons, gamma-ray photons, neutrons, particles, or energy crossing a unit surface area per unit time. The units of flux are the number of particles (or energy, etc.) per square centimeter per second. (2) *Soldering.* A substance used to clean the surface and promote fusion in a soldering procedure.

fly
(1) To control an aircraft or spacecraft in flight, generally including takeoff and land-

ing. (2) A two-winged insect that is often the vector of organisms causing disease.

fly ash
Noncombustible residual particles from the combustion process, carried by flue gas.

fly-by-wire
A technique for controlling aircraft in which a digital signal carried by wire to hydraulic actuators in the wings and tail which move the flight control surfaces.

fly-fix-fly
A description of the early approach to system safety, with reference to the aviation industry, that focused upon an after-the-fact method of designing safe systems.

fly heading (degrees)
Aviation. Informs the pilot of the heading he should fly. The pilot may have to turn to, or continue on, a specific compass direction in order to comply with the instructions. The pilot is expected to turn in the shorter direction to the heading unless otherwise instructed by ATC.

flyaway value
Aviation. Includes the cost of the airframe, engines, electronics, communications, armament, and other installed equipment.

flyback method
See ***repetitive timing***.

flybar
A system which provides airspeed, turn, and bank indications via auditory signals, instead of the conventional visual flight instruments.

flyer
Transit (slang). A run in which the driver takes a trailer to a distant terminal, leaves it there and immediately pulls another trailer back to his home terminal.

flying operations expenses
Aviation. Expenses incurred directly in the in-flight operation of aircraft and expenses related to the holding of aircraft and aircraft operational personnel in readiness for assignment for an in-flight status.

flying orders
Transit. Trip instructions issued to a driver by his/her dispatcher.

FM
See ***Factory Mutual Association***.

FMC
See ***fleet management center***.

FMCSR
See ***Federal Motor Carrier Safety Regulations***.

FMEA
See ***failure mode and effect analysis***.

FMS
See ***fleet management system***.

FNSI
See ***finding of no significant impact***. Also referred to as *FONSI*.

FO
See ***free out***.

foam
(1) A fluid mixture of bubbles which floats on or flows over a surface. (2) A stable aggregation of small bubbles which flow freely over a burning liquid surface and form a coherent blanket which seals combustible vapors and thereby extinguishes the fire.

foamed buoy
A buoy whose interior is filled with styrofoam for the purpose of improving flotation when in a damaged condition.

FOB
See ***free on board***.

focal mechanism
In seismology, the direction and sense of slip on a fault plane at the point of origin (see *hypocenter*) of an earthquake, as inferred from the first seismic waves which arrive at various locations. Often, they are drawn on maps with a "baseball-like" symbol. The dark areas denote compression, the white areas denote dilation. The fault plane which moved is parallel to one of the two planes dividing the sphere in half.

focus
(1) The point of convergence of light rays or sound waves. (2) *See* ***hypocenter***.

focused attention
See ***selective attention***.

Foehn
See ***Chinook***.

fog
A term loosely applied to visible aerosols, less than 40 microns in diameter, that are liquid;

formation by condensation is sometimes implied. Basically, fog is a cloud with its base at the earth's surface. It reduces visibility to below 1 kilometer.

fog lamps
Car lamps, installed just above the front bumper, designed to give better lighting during foggy weather.

fogging
Applying a pesticide by rapidly heating the liquid chemical so that it forms very fine droplets that resemble smoke or fog. It may be used to destroy mosquitoes, black flies, and similar pests.

FOIA
*See **Freedom of Information Act**.*

folliculitis
The inflammation of follicles, particularly hair follicles.

follower
Any selected object on a display which is moved by manipulation of a control.

fomites
Intimate personal articles, such as clothing, a drinking glass, a handkerchief, etc.

font
The size and style of written type.

font change
A highlighting technique in which a different font, different pitch, point size, or representation of the same font, or some other alteration is used.

food
According to the Federal Food, Drug, and Cosmetic Act: (1) articles used for food or drink for man or other animals, (2) chewing gum, and (3) articles used for components of any such article.

food additive
According to the Federal Food, Drug, and Cosmetic Act: Any substance the intended use of which results or may reasonably be expected to result, directly or indirectly, in its becoming a component or otherwise affecting the characteristics of any food (including any substance intended for use in producing, manufacturing, packing, processing, preparing, treating, packaging, transporting, or holding food; and including any source of radiation intended for any such use), if such substance is not generally recognized among experts qualified by scientific training and experience to evaluate its safety as having been adequately shown through scientific procedures (or, in the case of a substance used in food prior to January 1, 1958, through either scientific procedures or experience based on common use in food) to be safe under the conditions of its intended use.

Food and Drug Administration (FDA)
Agency within the Department of Heath and Human Services established to set safety and quality standards for foods, drugs, cosmetics, and other household substances sold as consumer products. Among the basic tasks of the FDA are research, inspection, and licensing of drugs for manufacturing and distribution. This agency is in charge of administering the Food, Drug, and Cosmetic Act.

Food and Drug Administration Action Level (FDAAL)
Under the Federal Food, Drug, and Cosmetic Act, as amended, concentration of a poisonous or deleterious substance in human food or animal feed at or above which FDA will take legal action to remove adulterated products from the market. Only FDAALs established for fish and shellfish apply in the Hazard Ranking System.

food chain
A sequence of organisms, each of which uses the next, lower member of the sequence as a food source.

Food, Drug Cosmetic Act
Federal Act of 1938 prohibiting the transportation in interstate commerce of adulterated or misbranded food, drugs, and cosmetics. The Act is administered by the Food and Drug Administration.

food engineering
The implementation of food science and technology in the manufacturing, processing, and packaging of food items.

food poisoning
A broad term including foodborne illnesses caused by the ingestion of foods containing microbial toxins or chemical poisons. *See also **foodborne disease**.*

food waste
The organic residues generated by the handling, storage, sale, preparation, cooking, and serving of foods. Commonly called *garbage.*

foodborne disease
Any disease that is transmitted through food contaminated with bacteria, viruses, fungi, parasites, or even some toxic chemicals.

foot
(1) *Anatomy.* That bodily structure composed of the phalanges, metatarsal, cuneiform, navicular, cuboid, talus, and calcaneus bones with their associated, surrounding tissues. (2) *Measurement.* A unit of length in the English system; equal to 12 inches.

foot acre
One acre of coal one foot thick.

foot breadth
The maximum width of the foot measured perpendicular to its longitudinal axis. Measured with the individual standing and his/her weight distributed evenly on both feet.

foot-candle (fc)
The illumination resulting from the uniform distribution of a flux of one lumen (lm) on a surface area of one square foot. Hence, one foot candle equals one lumen per square foot.

foot control
Any control device intended for normal operation using a foot.

foot lambert
The unit of photometric brightness equal to the uniform brightness of a perfectly diffusing surface emitting or reflecting one lumen per square foot.

foot-leg
Involving both the foot and the leg, generally pertaining to sensory or other external influences on both the foot and the leg.

foot length
The maximum length of the foot measured parallel to its long axis, from the back of the heel to pternion. Measured without tissue compression, with the individual standing erect and his/her weight evenly distributed on both feet.

foot-pound
An English system measure of torque; equal to one pound of force acting at a distance of one foot from the fulcrum.

foot restraint
A platform structure which serves to immobilize one or both feet to hold an individual in position for performing a task.

footbar
A rod or molded tube which serves as a footrest for a chair when the seat pan of the chair is too high for the feet to reach the floor or another surface.

footfall
The striking of the bottom of the foot or footwear on a surface in human gait.

footlambert (fl)
A unit of luminous intensity; the luminance of a surface which receives 1.0 lumen per square foot (an outdated measure).

footprints
In the law of evidence, impressions made upon the earth, snow, or other surface by the feet of persons, or by their shoes, boots, or other foot coverings.

footrest
Any structure on which the foot may rest, usually when seated.

footrest angle
The angle between a footrest having a flat surface and the lower leg link.

footring
A tube or rod attached in a circular pattern about the legs of a stool or chair as a footrest when the seat pan is too high for the feet to reach the floor or other base surface.

footstool
A short structure which is easily portable and may be stood upon to improve one's vertical reach.

footswitch
Any type of switch which closes when the foot or some portion of the foot makes contact with the floor or ground.

footwall
Seismology. Of the two sides of a non-vertical fault, the side below the fault plane. It is called the footwall because where faults have been "filled in" with mineral deposits and then mined, this is the side on which miners walk. *See also* ***hanging wall.***

footwear
Any type of material or covering worn over the foot.

for hire
Refers to a vehicle operated on behalf of or by a company that provides transport services to its customers.

for-hire carriage
Transportation of property by motor vehicle except when a) the property is transported by a person engaged in a business other than transportation; and b) the transportation is within the scope of, and furthers a primary business (other than transportation) of, the person.

for-hire motor carrier
A person engaged in the transportation of goods or passengers for compensation.

foramen
A natural opening or passage; used as a general term in anatomic nomenclature to designate such a passage, especially on or into a bone.

forbidden or not acceptable explosives
Any explosives which are forbidden or not acceptable for transportation by common carriers by rail freight, rail express, highway, or water in accordance with the regulations of the U.S. Department of Transportation.

force
(1) The push or pull that tends to impart motion to a body at rest, or to increase or diminish speed, or to change the direction of a body already in motion. (2) A physical influence exerted on an object which tends to cause a change in velocity.

force arm
See ***effort arm***.

force feedback
Any means of providing information to an operator about the forces involved on a remote or teleoperated end effector.

force joystick
See ***isometric joystick***.

force majeure
The title of a standard clause found in marine contracts exempting the parties for nonfulfillment of their obligations by reasons of occurrences beyond their control, such as earthquakes, floods, or war.

force plate
A system consisting of a cover plate and one or more transducers for measuring the forces or accelerations of an object either positioned on the cover or as the object strikes the cover plate. Also referred to as *force platform* and *reactance platform*.

force reflection
Providing an operator or system with tactile information about the forces/torques experienced by a remote device.

force-velocity curve
A graphical plot showing a characteristic of concentric muscular contractions in which the velocity of a muscular contraction is inversely related to the force of the contraction. May also be referred to as *force-velocity relationship*.

forced choice
An experimental methodology in which a subject must make a selection from one of the available choices.

forced convection
On a small scale, a form of mechanical stirring taking place when twisting eddies of air are able to mix hot surface air with the cooler air above. On a larger scale, it can be induced by the lifting of warm air along a front (frontal uplift) or along a topographic barrier (orographic uplift). *See also* ***convection***.

forced draft
The positive pressure created by air being blown into a furnace or other combustion equipment by a fan or blower.

forced expiratory volume-one second (FEV-1)
The maximum volume of air that can be forced from an individual's fully inflated lungs in one second.

forced grasping
A movement disorder in the adult in which the victim grasps any object which touches his/her hand, frequently with great strength. (Note: Different from the normal grasp reflex.)

forced vital capacity (FVC)
The volume of air that can be forcibly expelled from the lungs after a full inspiration of air.

ford
The shallow part of a river which can be easily crossed.

Fordyce's disease
A congenital condition characterized by minute yellowish white papules on the oral mucosa.

fore and aft
The direction on a vessel parallel to the centerline

fore and aft line
A line used to secure two barges end to end.

fore bay
An enclosure of the river, usually above a dam.

forearm
The radius, ulna, and all other organized tissues comprising that part of the arm from the elbow to the wrist.

forearm circumference
The surface distance around the forearm at the level at which the maximum value is obtained. Measured with the individual standing erect, the shoulder slightly abducted, and the hand relaxed with the fingers extended. May be referred to as *arm circumference.*

forearm circumference, elbow flexed
The maximum surface distance around the forearm with the elbow flexed 90 degrees. Measured with the shoulder flexed 90 degrees laterally (so that the upper arm is horizontal), and the fist clenched.

forearm circumference, relaxed
The maximum surface distance around the forearm. Measured with the elbow flexed 90 degrees and the hand relaxed.

forearm – forearm breadth, sitting
The horizontal linear distance from the most lateral point on the right forearm, across the body to the most lateral point on the left forearm. Measured without tissue compression, with the individual seated erect, the upper arms hanging naturally at the sides, and the elbows flexed 90 degrees while resting lightly against the torso

forearm – hand length
The distance from the posterior elbow to the tip of the longest finger. Also referred to as *elbow – fingertip length.* Measured with the individual seated erect, the upper arm vertical at the side, the forearm and hand horizontal, and the fingers maximally extended.

forearm length
See ***radiale – stylion length***.

forearm skinfold
The thickness of a vertical skinfold on the posterior midline of the forearm at the level of the forearm circumference. Measured with the individual standing erect and the arms relaxed naturally at the sides.

forefinger
See ***index finger***.

forefinger length
See ***index finger length***.

forefoot
The anterior portion of the foot, including the phalanges, metatarsals, cuneiform, and cuboidal bones and the soft tissues surrounding them.

forehead
That superior portion of the face from the supraorbital ridges upward and between the maximum lateral bulges of the brow ridges near the ends of the eyebrow. May be referred to as *brow* or *frons*. *See also* ***eyebrow***.

forehead breadth
See ***frontal breadth (maximum),*** *and* ***frontal breadth (minimum)***.

foreign
With regard to commerce, refers to outside the fifty United States and the District of Columbia.

foreign air carrier
Any person other than a citizen of the United States, who undertakes directly, by lease or other arrangement, to engage in air transportation. *See also* ***foreign flag air carrier***.

foreign air commerce
The carriage by aircraft of persons or property for compensation or hire, or the carriage of mail by aircraft, or the operation or navigation of aircraft in the conduct or furtherance of a business or vocation, in commerce between a place in the United States and any place outside thereof; whether such commerce moves wholly by aircraft or partially by aircraft and partially by other forms of transportation.

foreign air transportation
The carriage by aircraft of persons or property as a common carrier for compensation or hire, or the carriage of mail by aircraft, in commerce between a place in the United States and any place outside of the United States, whether that commerce moves wholly by aircraft or partially by aircraft and partially by other forms of transportation.

foreign body
Usually refers to any material that has entered and/or become imbedded in a body part (such as the eye). Examples of foreign bodies can include splinters, slivers, dirt, etc.

foreign current
A term applied to stray electric currents which may affect a signaling system, but which are not a part of the system.

foreign element
A work element which is not normally part of the work cycle and provides an interruption to it, usually with a random/unpredictable frequency of occurrence.

foreign exchange gains and losses
Gains or losses resulting from nonroutine abnormal changes in the rates of foreign exchange.

foreign flag air carrier
(1) An air carrier other than a United States flag air carrier in international air transportation. "Foreign air carrier" is a more inclusive term than "foreign flag air carrier," including those non-U.S. air carriers operating solely within their own domestic boundaries. In practice, the two terms are used interchangeably. (2) An air carrier other than a United States flag air carrier providing international air transportation. Certificated in accordance with Federal Aviation Regulations Part 129. *See also* ***foreign air carrier***.

foreign fleet
All reportable agency-owned motor vehicles, operated outside any State, Commonwealth, Territory or possession of the United States.

foreign freight
Movements between the United States and foreign countries and between Puerto Rico, the Virgin Islands, and foreign countries. Trade between U.S. territories and possessions (e.g., Guam, Wake, American Samoa) and foreign countries is excluded. Traffic to or from the Panama Canal Zone is included.

foreign freight forwarder
An independent business which makes shipments for exporters for a fee.

foreign mail
Mail transported outside the United States by U.S. flag carriers on behalf of any foreign government.

foreign trade
The exchange of waterborne commodity movements (imports and exports) between the United States and its territories, and foreign countries.

foreign trade zone
An isolated area, attached to a port, where facilities for dockage and unloading are provided, and where foreign merchandise may be stored or manipulated pending sale or reshipment without limitation as to time and without compliance with the customs laws and regulations relating to the entry of merchandise. Most such privileges are equally available at other regular ports of entry by arrangement with U.S. Customs Bureau.

forensic
Belonging to the courts of justice.

forensic engineering
The application of the principles and practices of engineering to the elucidation of questions before courts of law. Practiced by legally qualified professional engineers who are experts in their field, by both education and experience, and who have experience in the courts and an understanding of jurisprudence. A forensic engineering engagement may require investigation, studies, evaluations, advice to counsels, reports, advisory opinions, depositions and/or testimony to assist in the resolution of disputes relating to life or property in cases before courts, or other lawful tribunals.

forensic medicine
That science which teaches the application of every branch of medical knowledge to the purposes of the law. Hence its limits are, on the one hand, the requirements of the law, and, on the other hand, the whole range of medicine. Anatomy, physiology, medicine, surgery, chemistry, physics, and botany lend

their aid as necessity arises; and in some cases all these branches of science are required to enable a court of law to arrive at a proper conclusion on a contested question affecting life or property.

forensic pathology
That branch of medicine dealing with diseases and disorders of the body in relation to legal principles and cases.

forensic psychiatry
That branch of medicine dealing with disorders of the mind in relation to legal principles and cases.

foreseeability
A concept in which an individual may be held liable for actions resulting in injury or damage only if he/she could be reasonably expected to foresee the risk or danger.

foreseeable damages
Loss that the party in breach had reason to know of when the contract was made.

foreshock
Any earthquake which is followed, within a short time span, by a larger earthquake in the exact same location can be labeled a "foreshock." (In the case of an earthquake *swarm,* this terminology is not generally applied.)

foreshore
The part of a seashore between high-water and low-water marks.

forestall
As applied to an automatic train stop or train control device, to prevent an automatic brake application by operation of an acknowledging device or by manual control of the speed of the train.

forklift truck
A high-powered vehicle equipped with hydraulic-driven protruding metal blades, that is used to raise and lower unitized freight and/or handle other material(s).

form
A display or hardcopy with organized categories for the user or operator to fill in.

form analysis chart
See ***form process chart.***

form process chart
A flow process chart for one or more paperwork forms. Also referred to as *information process analysis, functional form analysis, form analysis chart, paperwork flow chart,* and *procedure flow chart.*

formal rulemaking
The process of promulgating rules based upon the formal procedures established in the Administrative Procedures Act (APA) of 1946 requiring (most notably) hearings, substantiation of evidence, and the cross-examination of witnesses.

formaldehyde
A colorless, pungent, irritating gas (CH_2O) used chiefly as a disinfectant and preservative and in synthesizing other compounds and resins.

formant
A resonance which is associated with vocal tract reflections in the production of sound.

formation flight
More than one aircraft which, by prior arrangement between the pilots, operate as a single aircraft with regard to navigation and position reporting.

Separation between aircraft within the formation is the responsibility of the flight leader and the pilots of the other aircraft in the flight. This includes transition periods when aircraft within the formation are maneuvering to attain separation from each other to effect individual control and during join-up and breakaway.

A *standard formation* is one in which a proximity of no more than 1 mile laterally or longitudinally and within 100 feet vertically from the flight leader is maintained by each wingman.

Nonstandard formations are those operating under any of the following conditions: a) when the flight leader has requested and Air Traffic Control (ATC) has approved other than standard formation dimensions; b) when operating within an authorized altitude reservation (ALTRV) or under the provisions of a letter of agreement; c) when the operations are conducted in airspace specifically designed for a special activity. *See also* ***altitude reservation.***

F-18 Hornets flying in formation flight

formed elements
The enclosed structures within the blood, consisting of erythrocytes, leukocytes, and platelets.

formite
Any substance that may harbor or transmit pathogenic organisms.

formulation
The substance or mixture of substances which is comprised of all active and inert ingredients in a pesticide.

formulation time
The temporal period required for the end-user and manufacturer to determine what characteristics a desired system should have.

Fortran
A high-level computer language designed for scientific and mathematical use with the name of Formula Translator and the acronym, Fortran.

forward chaining
A reasoning or control strategy in which the starting point is selected and all possible resulting states are derived from that point.

forward control
Transportation. (1) A configuration in which more than half of the engine length is rearward of the foremost point of the windshield base and the steering wheel hub is in thc forward quarter of the vehicle length. (2) Vehicle with driver controls (pedals, steering wheel, instruments) located as far forward as possible. Supplied with or without body, the controls are stationary mounted as opposed to the special mountings of tilt cabs.

forward masking
A form of temporal masking in which the masking stimulus just precedes the test stimulus.

forward wing
Aviation. A forward-lifting surface of a canard configuration or tandem-wing configuration airplane. The surface may be a fixed, movable, or variable geometric surface, with or without control surfaces.

fossa
A depression in the surface of a bone.

fossil fuel
(1) Fuel, such as natural gas, petroleum, coal, etc., that originated from the remains of plant, animal, and sea life of previous geological eras. (2) Any naturally occurring organic fuel, such as petroleum, coal, and natural gas.

fossil water
See ***connate water***.

foul bill of lading
A receipt for goods issued by a carrier with an indication that the goods were damaged when received.

foul ground
An area identified as a danger to maritime navigation where the holding qualities for an anchor are poor, or where danger exists of striking or fouling the ground or other obstructions.

fouling section
Rail Operations. The section of track between the switch points and the clearance point in a turnout.

foundation
A structural, knowledge, or economic base which enables further growth or development.

foundation garment
Underwear (an older term).

four banger
Transit (slang). Term used to describe a four-cylinder engine.

four by four
Four-speed transmission and 4-speed auxiliary transmission.

four ps
In evidence collection following an accident, the phrase given to the four common categories of people, parts, papers, and positions.

Fourier analysis
The mathematical decomposition of a complex periodic waveform into its sinusoidal

components. Often used with non-periodic waveforms to get frequency components.

Fourteenth Amendment
Amendment to the U.S. Constitution, ratified in 1868, creates or at least recognizes for the first time a citizenship of the United States, as distinct from that of the states; forbids the making or enforcement by any state of any laws abridging the privileges and immunities of citizens of the United States; and secures all "persons" against any state action which results in either deprivation of life, liberty, or property without due process of law, or, in denial of equal protection of the laws. This Amendment also contains provisions concerning the apportionment of representatives in Congress.

Fourth Amendment
Amendment of the U.S. Constitution guaranteeing people the right to be secure in their homes and protect their property against unreasonable searches and seizures and providing that no warrants shall issue except upon probable cause and then only as to specific places to be searched and persons and things to be seized.

fovea
A depressed region within the macula lutea of the posterior retina at which cone density is highest and the greatest visual acuity occurs. Also referred to as *fovea centralis*.

foveal blindness
The lack of visual capability in the center of the visual field, due to damage or other problem with the fovea or macula lutea. Also referred to as *central visual field blindness*.

foveal vision
That photopic sensory stimulation mediated by the fovea.

FR
See ***Federal Register***.

FRA
See ***Federal Railroad Administration***.

fractionation
Any of several processes, apart from radioactive decay, that result in change in the composition of radioactive debris.

fracture
A sudden break or crack in a bone or other solid material. It may be caused by trauma, by twisting due to muscle spasm or indirect loss of leverage, or by disease that results in decalcification of the bone.

fracture control
The application of design philosophy, analysis method, manufacturing technology, quality assurance, and operating procedures to prevent premature structural failure due to the propagation of cracks or crack-like flaws during fabrication, testing, transportation and handling, and service.

fracture mechanics
An engineering concept used to predict flaw growth of materials and structures containing cracks or crack-like flaws; an essential part of a fracture control plan to prevent structure failure due to flaw propagation.

fracture toughness
A generic term for measures of resistance to extension of a crack.

frame
One complete scan or image on a CRT, videotape, motion picture film, or other type of display.

frame counter
Any electrical, mechanical, or electromechanical device which is used to determine and/or display a count of the number of frames displayed on a film or video medium.

frame rate
The number of frames recorded or displayed per unit time.

frangible navigational aid
Aviation. A navigational aid whose properties allow it to fail at a specified impact load.

Frankfort plane
An imaginary plane through the head, used for head orientation purposes, established by the lateral extensions of a line between tragion and the lowest point of the orbit. Also referred to as *Frankfort horizontal plane*.

fraud
(1) An intentional perversion of truth for the purpose of inducing another in reliance upon it to part with some valuable thing belonging to him/her or to surrender a legal right. (2) A false representation of a matter of fact, whether by words or by conduct, by false or misleading allegations, or by concealment of

that which should have been disclosed, which deceives and is intended to deceive another so that he/she shall act upon it to his/her legal injury. (3) Anything calculated to deceive, whether by a single act or a combination, or by suppression of the truth, or suggestion of what is false, whether it be by direct falsehood or innuendo, by speech or silence, word of mouth, or look or gesture.

free alongside ship (FAS)
Maritime. A price quotation under which the exporter quotes a price that includes delivery of the goods to the vessel's side and within reach of its loading tackle. Subsequent risks and expenses are for the account of the buyer.

free alongside ship (FAS) value
Maritime. The value of a commodity at the port of exportation, generally including the purchase price plus all charges incurred in placing the commodity alongside the carrier at the port of exportation in the country of exportation.

free chlorine residual
Portion of the total residual chlorine remaining at the end of a specific contact time which will react as hypochlorous acid or hypochlorite ion.

free convection
See ***convection***.

free field
See ***free sound field***.

free-field room
An enclosed volume which provides essentially a free sound field. *See also* ***anechoic room***.

free float
That calculated additional time available for an activity from the earliest possible completion time of that activity and the earliest possible beginning of the next activity linked to it in a network.

free in
Maritime. A pricing term indicating that the party who charters a vessel is responsible for the cost of loading goods onto the vessel.

free in and out
Maritime. Terms under which cost of loading and discharging cargo is borne by parties other than the vessel owner or operator.

free of particular average
A marine insurance term meaning that insurer will not allow payment for partial loss or damage to a foreign shipment.

free on board (FOB)
(1) A price quotation under which the exporter quotes a price that includes delivery of the goods on board the vessel. Subsequent risks and expenses are for the account of the buyer. (2) A transaction whereby the seller makes the product available within an agreed-upon period at a given port at a given price. It is the responsibility of the buyer to arrange for the transportation and insurance.

free on board (FOB) airport
FOB airport is based on the same principle as the ordinary FOB term. The seller's obligation includes delivering the goods to the air carrier at the airport of departure. The risk of loss or damage to the goods is transferred from the seller to the buyer when the goods have been so delivered. *See also* ***free on board***.

free out (FO)
Terms under which the owner of goods is responsible for discharging costs.

free port
A restricted area at a seaport for the handling of duty-exempted import goods.

free radical
An atom or a chemically combined group of atoms which have a free electron and are very chemically reactive.

free-running rhythm
A biological rhythm without the use of entrainment cues, often resulting in a slight change of period.

free silica
Silica in the form of cristobalite, tridymite, or alpha quartz.

free sound field
A sound field in which the boundary effects are negligible over the frequencies of interest. *See also* ***far field***.

free time
The amount of time that a carrier's equipment may be used without incurring additional charges.

free trade zone
A port designated by the government of a country for duty-free entry of any non-prohibited goods. Merchandise may be stored, displayed, used for manufacturing, etc. within the zone and re-exported without duties.

Freedom of Information Act (FOIA)
Allows all U.S. citizens and residents to request any records in possession of the executive branch of the federal government. The term "records" includes documents, papers, reports, letters, films, photographs, sound recordings, computer tapes and disks. An object that cannot be reproduced is not considered a record in this case. The federal Freedom of Information Act (FOIA) covers the President's cabinet agencies, independent agencies, regulatory commissions, and government-owned corporations. Congress is exempt, as are federal courts and state and local governments. Some states and municipalities have laws modeled after the federal FOIA. The federal act includes nine exemptions that agencies may claim as a basis for withholding information. An administrative appeal can be filed that argues for disclosure based on benefits to the public vs. privacy. If a good argument is made, appellate reviewers may waive an exemption.

freeway
An expressway with full control of access.

freeze
A condition occurring over a widespread area when the surface air temperature remains below freezing for a sufficient time to damage certain agricultural and ornamental crops. A freeze most often occurs as cold air is advected into a region, causing freezing conditions to exist in a deep layer of surface air. Also called *advection frost*.

freeze calculated landing time (FCLT)
Aviation. A dynamic parameter number of minutes prior to the meter fix calculated time of arrival for each aircraft when the tentative calculated landing time (TCLT) is frozen and becomes an actual calculated landing time (ACLT) i.e., the vertex time of arrival (VTA) is updated and consequently the TCLT is modified as appropriate until freeze calculated landing time (FCLT) minutes prior to meter fix calculated time of arrival, at which time updating is suspended and an ACLT and a frozen meter fix crossing time (MFT) is assigned.

freeze protected deluge shower
A deluge shower that is designed to operate at temperatures which would normally freeze water in the system.

freeze trap
A method to collect gases/vapors by cooling the sampled air to a temperature at which the substance(s) of interest condense, and thus collect.

freezing drizzle
*See **freezing rain**.*

freezing nuclei
Any particle that has a shape similar to that of an ice crystal and allows rapid freezing of supercooled water. Such particles include certain clay minerals, meteoric dust, and ice crystals themselves.

freezing rain
Rain or drizzle that falls in liquid form and then freezes upon striking a cold object or the ground. Both can produce a coating of ice on objects, which is called *glaze*.

freight
(1) Property (other than express and passenger baggage) transported by air, rail, truck, seafaring vessel, or other commercial transport means. (2) Any commodity being transported.

freight agent
An establishment that arranges the transportation of freight and cargo for a fee. Revenue for freight agents (also known as shipping agents or brokers) represents commissions of fees and not the gross charges for transporting goods.

freight all kinds (FAK)
Goods classified FAK are usually charged higher rates than those marked with a specific classification and are frequently in a container which includes various classes of cargo.

freight and other transportation services
Forwarding: Includes establishments that provide forwarding, packing, and other services incidental to transportation. Also included are horse-drawn cabs and carriages for hire.

freight container
A reusable container having a volume of 64 cubic feet or more, designed and constructed to permit being lifted with its contents intact and intended primarily for containment of packages (in unit form) during transportation.

freight forwarder
(1) An individual or company that accepts less-than-truckload (TLT) or less-than-carload (LCL) shipments from shippers and combines them into carload or truckload lots. Designated as a common carrier under the Interstate Commerce Act. (2) A broker that functions as an intermediary between shippers (consignors/consignees) and carriers. Functions performed by a freight forwarder may include receiving small shipments (e.g., less than container load) from consignors, consolidating them into larger lots, contracting with carriers for transport between ports of embarkation and debarkation, conducting documentation transactions, and arranging delivery of shipments to the consignees.

freight forwarding
Establishments primarily engaged in undertaking the transportation of goods from shippers to receivers for a charge covering the entire transportation, and in turn making use of the services of various freight carriers in effecting delivery. Establishment pays transportation charges as part of its costs of doing business and assumes responsibility for delivery of the goods. There are no direct relations between shippers and the various freight carriers performing the movement.

freight revenue
Revenues from the transportation by air of property other than passenger baggage.

freight service operating expenses
The sum of operating expenses directly assignable to freight service and an apportionment of expenses common to both freight and passenger service.

freight service revenue
Revenue from the transportation of freight, switching of freight train cars, water transfers of freight, vehicles and livestock, movement of freight trains at a rate per train mile or for a lump sum, storage of freight, demurrage, grain elevators, stockyards, and miscellaneous services and facilities in connection with the transportation of freight.

frequency
The number of cycles, revolutions, or vibrations completed per unit of time. In sound, for example, the frequency describes the rate at which complete cycles of high- and low-pressure peaks are produced. The unit of measurement is cycles per second or hertz (Hz). The normal human ear has a frequency range of 20 to 20,000 Hz at moderate sound pressure levels.

frequency distribution
The tabulation of data from the lowest to the highest, or highest to the lowest, along with the number of times each of the values was observed or occurred in the distribution.

frequency domain
The expression of a function in terms of frequency.

frequency function
See ***frequency distribution***.

frequency masking
See ***simultaneous masking***.

frequency of exposure
The number of times per shift, day, year, etc. that an individual is exposed to a harmful substance or physical agent.

frequency of lift
The number of times a specified mass is raised and/or lowered within a unit time. (Note: The most common time interval is one minute.)

frequency of use principle
A rule that states the most frequently used controls and displays should be placed in optimal locations.

frequency polygon
A graphical representation in which the ordinal values corresponding to abscissa values are plotted in a coordinate system and connected by straight lines.

frequency rate
Relates the injuries that occur to the hours worked during the period and expresses them in terms of a million man-hour unit.

frequency response
That range of frequencies which a system is capable of producing or a sensor is capable of detecting.

frequency response curve
A graph of the input frequency spectrum vs. output frequency spectrum for a system.

frequency spectrum
A description of the frequency components and associated amplitudes of a time series waveform.

frequency-time spectrum
*See **compressed spectral array**.*

frequent
In terms of probability of hazard or mishap occurrence, a hazard or event likely to occur numerous times during the life of an item.

fresh water
Water that generally contains less than 1,000 milligrams per liter of dissolved solids.

friable
Refers to materials that have a tendency to crumble easily. Most often used to describe the condition that exists when asbestos fibers can potentially be released and become airborne presenting a respiratory hazard.

friable asbestos-containing material
Any asbestos-containing material applied on ceilings, walls, structural members, piping, duct work, or any other part of a building which when dry may be crumbled, pulverized, or reduced to powder by hand pressure. The term includes non-friable asbestos-containing material after such previously non-friable material becomes damaged to the extent that when dry it may be crumbled, pulverized, or reduced by hand pressure.

fricative
A consonant produced by the steady frictional or turbulent passage of air through a narrowing of a segment within the vocal tract. Also called *spirant*.

friction
A force which opposes the motion of a body or tends to hold a stationary body in place. *See also **static friction** and **kinetic friction**.*

friction layer
The atmospheric layer near the surface usually extending up to about 1 km (3300 feet) where the wind is influenced by friction of the earth's surface and objects on it.

friction loss
The pressure loss in a ventilation system due to friction of the moving air on the ductwork.

Friedman two-way analysis of variance
A non-parametric statistical test using matched sample rank data to test the null hypothesis.

fringe benefit
That compensation to an employee which is not in the form of wages, salary, or bonuses.

fringe parking
An area for parking usually located outside the Central Business District (CBD) and most often used by suburban residents who work or shop downtown.

frit
The porous section at the end of a glass tube which is employed in a glass flask to breakup an air stream into small bubbles, thereby improving the absorption of air contaminants by the sorbent as air is sampled through it. Often referred to as a glass frit.

fritted bubbler
A glass frit. *See **frit**.*

frivolous
Pertaining to a lawsuit with no basis in fact, and which is based on nonsensical legal theory or intended to harass the defendant or grandstand in court.

frog
Rail Operations. A track component used at the intersection of two running rails to provide support and guidance for the wheels. It allows wheels on each rail to cross the other rail.

from
A shipping term under which price quoted applies only at the point of origin, such as ex-mill, ex-rail car, ex-barge, and the seller agrees to place the goods at the disposal of the buyer at the agreed place within a fixed period of time.

from a little open
Transit (slang). An expression meaning to depart on a new course from a point 50 yards or less from a defined object.

from foot of dike
From the end of the dike where it is attached to the shore.

from (lower) end of dike
From the outward or channel end of a dike.

frons
See ***forehead***.

front
The transition zone between two distinct air masses.

front-end analysis
The process of determining whether or not a problem exists. Also referred to as *needs assessment* and *discrepancy analysis*.

frontal
Pertaining to the anterior portion of the body or of a body part, or the frontal plane.

frontal arc, minimum
The minimum surface distance across the forehead to the temporal crests at their points of maximum indentation. Measured with the individual sitting or standing erect and the facial muscles relaxed.

frontal bone
The flat bone making up the forehead and superior frontal portion of the skull.

frontal breadth, maximum
The horizontal linear distance between the maximum lateral bugles of the brow ridges near the ends of the eyebrow. Also referred to as *forehead breadth*.

frontal breadth, minimum
The horizontal linear distance across the forehead from the points of greatest indentation of the temporal crests.

frontal fog
See ***evaporation fog***.

frontal lobe
The most anterior portion of the cerebral hemisphere, extending from the frontal pole to the central sulcus.

frontal plane
Any vertical plane at right angle to the midsagittal and horizontal planes which divides the body into anterior and posterior portions. Often referred to as *coronal plane*.

frontal thunderstorms
Thunderstorms that form in response to forced convection (forced lifting) along a front. Most go through a cycle similar to those of air mass thunderstorms.

frontal wave
A wavelike deformation along a front in the lower levels of the atmosphere. Those that develop into storms are termed *unstable waves,* while those that do not are called *stable waves*.

frontogenesis
A meteorological term for the formation, strengthening, or regeneration of a front.

frontolysis
A meteorological term for the weakening or dissipation of a front.

frost
A covering of ice produced by deposition (sublimation) on exposed surfaces when the air temperature falls below the frost point (the dew point is below freezing). Also called *hoarfrost*.

frost point
See ***dew point***.

frostbite
The destruction of tissue resulting from exposure to extreme cold or contact with extremely cold objects.

frozen dew
The transformation of liquid dew into tiny beads of ice when the air temperature drops below freezing.

frozen section
A specimen of tissue that has been quick-frozen, cut by microtome, and stained immediately for rapid diagnosis of possible malignant lesions. A specimen processed in this manner is not satisfactory for detailed study of the cells, but it is valuable because it is quick and gives the surgeon immediate information regarding the malignancy of a piece of tissue.

fructose
A colorless or white crystalline sugar; also called *levulose* and *fruit sugar*. It is used in solution as a fluid and nutrient replenisher.

FSG
See ***Federal Sentencing Guidelines***.

FSS
*See **Flight Service Station**.*

FTA
*See **fault tree analysis**. See also **Federal Transit Administration**.*

FTP
*See **File Transfer Protocol**.*

fuel
The primary fuel or energy source delivered to a residential site. It may be converted to some other form of energy at the site. Electricity is included as a fuel. Other primary fuels are coal, fuel oil, kerosene, liquefied petroleum gas (LPG), natural gas, wood, and solar.

fuel cell
(1) A device for converting chemical energy into electrical energy. (2) A device that produces electrical energy directly from the controlled electrochemical oxidation of the fuel. It does not contain an intermediate heat cycle, as do most other electrical generation techniques.

fuel code
A 2-digit numeric code that identifies the type of fuel used. The code identifies regular (gasoline and diesel) fuels, alternative fuels such as natural gas and methanol, and vehicles able to operate on a combination of these fuels (regular and alternative).

fuel dumping
Aviation. Airborne release of usable fuel. This does not include the dropping of fuel tanks. *See also **jettisoning of external stores**.*

fuel economy standard
The Corporate Average Fuel Economy Standard (CAFE) which went into effect in 1978. It was meant to enhance the national fuel conservation effort by slowing fuel consumption through a miles-per-gallon requirement for motor vehicles.

fuel fire/explosion
Accidental combustion of vessel fuel, liquids, including their vapors, or other substance such as wood or coal.

fuel injection
A fuel delivery system whereby gasoline is pumped to one or more fuel injectors under high pressure. The fuel injectors are valves that, at the appropriate times, open to allow fuel to be sprayed or atomized into a throttle bore or into the intake manifold ports. The fuel injectors are usually solenoid-operated valves under the control of the vehicle's on-board computer (thus the term "electronic fuel injection"). The fuel efficiency of fuel injection systems is less temperature-dependent than carburetor systems. Diesel engines always use injectors. *See also **carburetor** and **diesel fuel system**.*

fuel oil
A liquid petroleum product less volatile than gasoline, used as an energy source. Fuel oil includes distillate fuel oil (No. 1, No. 2, and No. 4), residual fuel oil (No. 5 and No. 6), and kerosene.

fuel oil, No. 1
A light distillate fuel oil intended for use in vaporizing pot-type burners. ASTM Specification D 396 specifies for this grade maximum distillation temperature of 400 degrees Fahrenheit at the 10 percent point and 550 degrees Fahrenheit at the 90 percent point, and kinematic viscosity between 1.4 and 2.2 centistoke at 100 degrees Fahrenheit.

fuel oil, No. 2
A distillate fuel oil for use in atomizing type burners for domestic heating or for moderate capacity commercial/industrial burner units. ASTM Specification D 396 specifies for this grade distillation temperature at the 90 percent point between 540 and 640 degrees Fahrenheit, and kinematic viscosity between 2.0 and 3.6 centistoke at 100 degrees Fahrenheit.

fuel oil, No. 4
A fuel oil for commercial burner installations not equipped for preheating facilities. It is used extensively in industrial plants. This grade is a blend of distillate fuel oil and residual fuel oil stocks that conform to ASTM Specification D 396 or Federal Specification VV-F-815C; its kinematic viscosity is between 5.8 and 26.4 centistoke at 100 degrees Fahrenheit. Also included is No. 4-D, a fuel oil for lower and medium speed diesel engines that conforms to ASTM Specification D 975.

fuel remaining
Aviation. A phrase used by either pilots or controllers when relating to the fuel remaining on board until actual fuel exhaustion. When transmitting such information in response to either a controller question or pilot-initiated cautionary advisory to air traffic control, pilots will state the appropriate number of minutes the flight can continue with the fuel remaining. All reserve fuel should be included in the time stated, as should an allowance for established fuel gauge system error. *See also **minimum fuel**.*

fuel siphoning
The unintentional release of fuel caused by overflow, puncture, loose cap, etc.

fuel tank
A tank other than a cargo tank, used to transport flammable or combustible liquid, or compressed gas for the purpose of supplying fuel for propulsion of the transport vehicle to which it is attached, or for the operation of other equipment on the transport vehicle.

fuel tank fitting
Any removable device affixed to an opening in the fuel tank with the exception of the filler cap.

fueling
Any stage of the fueling operation; primarily concerned with introduction of explosive or combustible vapors or liquids on board.

fugitive emissions
Emissions that are not caught by a capture system. The release of airborne contaminants into the surrounding air other than through a stack, such as the sealing mechanisms of sources including pumps, compressors, flanges, valves, and other type seals. Thus, fugitive emissions result from an equipment leak and are characterized by a diffuse release of materials such as VOCs, hydrocarbons, etc. into the atmosphere. The EPA defines fugitive emissions as those emissions that do not occur as part of the normal operation of the plant.

fugitive releases
Under ISO 14000 criteria, emissions to air, water, or land that are not controlled.

fulcrum
A fixed point representing the axis about which a lever may operate.

full berth terms (FBT)
Maritime. Terms under which cost of loading and discharge is included in the steamship rate quoted. Ship owner pays loading and discharge costs.

full double
Maritime. The maximum tow that can be locked.

full facepiece respirator
A respirator which covers the wearer's entire face from the hairline to the chin.

full hearing
Embraces not only the right to present evidence, but also a reasonable opportunity to know the claims of the opposing party, and to meet them.

full radiator
*See **blackbody**.*

full scale
The maximum measurement value or maximum limit for a given range on an instrument.

full shift
The regularly scheduled work period, typically of 8 hours duration.

full-time employment
Having a job consisting of about 35 or more hours per week on a regular basis.

full trailer
(1) Any motor vehicle other than a pole trailer which is designed to be drawn by another motor vehicle and so constructed that no part of its weight, except for the towing device, rests upon the self-propelled towing unit. A semitrailer equipped with an auxiliary front axle (converter dolly) shall be considered a full trailer. (2) A truck-trailer with front and rear axles. The load weight is distributed over both the front axle(s) and rear axle(s).

fullmount
A smaller vehicle mounted completely on the frame of either the first or last vehicle in a saddlemount combination.

fumble
An unintentional sensory-motor error.

fume
Small solid particles generated following the volatilization of a metal or plastic when their gaseous state condenses quickly upon contact with cooler air. Welding, for example, causes the volatilization of metals into a gas followed by condensation upon contact with cooler air. This creates welding fumes typically on the order of 0.1-1 micrometer in diameter. In popular usage, the word *fume* is often incorrectly used to describe virtually any type of air contaminant.

fume cupboard
British term for laboratory fume hood.

fume fever
See ***metal fume fever***.

fumigant
A pesticide that is vaporized to kill pests. Used in buildings and greenhouses.

function
(1) *General*. That activity which a product or system is to carry out. (2) *Computing*. A software-supported capability to aid the user in performing a task or operation.

function area
Computing. A portion of a screen display reserved by a given application for a specific purpose.

function key
A key which directs the computer to perform some specific function when pressed. *See also* ***fixed function key***.

Functional Analysis System Technique (FAST)
A diagramming process which permits a hierarchy of two-word function definitions derived from a product's consequences and cause.

functional anatomy
The study of the body and its component parts, relating them to biomechanical and/or physiological function.

functional anthropometry
See ***dynamic anthropometry***.

functional capacity level rating scale
A seven-point classification for grouping individuals, especially the elderly, according to their ability to perform the activities of daily living.

functional deafness
See ***psychogenic deafness***.

functional electrical stimulation
See ***electrical stimulation***.

functional equivalent
Term used to describe EPA's decision-making process and its relationship to the environmental review conducted under the National Environmental Policy Act (NEPA). A review is considered functionally equivalent when it addresses the substantive components of a NEPA review.

functional flow logic diagram
A technique for determining what operations or processes are necessary to achieve certain objectives from a system.

functional form analysis
See ***form process chart***.

functional hazard analysis
See ***fault hazard analysis***.

functional impact
A purposeful impact in fulfilling a useful task. *See also* ***beneficial impact***.

functional impairment
A reduced ability to perform certain functions. May also be called *functional limitation*.

functional injury
A form of trauma not readily detectable by visual examination, but which is indicated by one or more variables measuring a functional limitation.

functional leg length
The linear distance from the back at waist level to the heel, measured along the longitudinal axis of the leg. Measured with the individual sitting erect on the edge of a chair and the knee fully extended.

functional limitation
See ***functional impairment***.

functional principle
See ***functionality principle***.

functional reach
See ***thumb-tip reach***.

functional residual capacity (FRC)
That volume of air which remains in the lungs after a normal exhalation.

functional vibration
An intentional vibration generated to accomplish some end.

functionality principle
A rule stating that displays and controls which have related functions should be grouped together.

fundamental frequency
In the study of acoustics, the lowest periodic frequency component present in a complex spectrum.

fundamental motion
*See **therblig**.*

fundus
The bottom or base of anything; used in anatomic nomenclature as a general term to designate the bottom or base of an organ, or the part of a hollow organ farthest from its mouth.

fungi
*See **fungus**.*

fungicide
Pesticides which are used to control, prevent, or destroy fungi.

fungus
A general term used to describe the diverse morphological forms of yeast, rust, mildew, and mold. Any non-chlorophyll-bearing thallophyte (that is, any non-chlorophyll-bearing plant of a lower order than mosses and liverworts). *Fungi* (plural) are *heterotrophs* and obtain nourishment by absorption usually from dead or decaying organic matter. Some fungi are beneficial in foods and pharmaceutical development while other can cause pulmonary diseases. Fungi are found in soil, water, and air. Also referred to as *mold*.

funnel chest
A deformity of the front of the chest wall, characterized by a funnel-shaped depression with its apex over the lower end of the sternum. Also called *pectus excavatum*.

funnel cloud
A rotating cone-like cloud that extends downward from the base of a thunderstorm. When it reaches the surface it is called a *tornado*. If it touches a body of water (lake, pond, ocean, etc.) it is referred to as a *waterspout*.

furniture van body
Truck body designed for the transportation of household goods; usually a van of drop-frame construction.

further clearance time
Aviation. The time a pilot can expect to receive clearance beyond a clearance limit.

furuncle
A focal, suppurative inflammation of the skin and subcutaneous tissue, enclosing a central slough or "core." *See also **boil**.*

fuscin
A brown pigment of the retinal epithelium.

fuse
(1) *Electrical.* Pertaining to systems over 600 volts (nominal), an overcurrent protective device with a circuit-opening fusible part that is heated and severed by the passage of overcurrent through it. A fuse comprises all the parts that form a unit capable of performing the prescribed functions. It may or may not be the complete device necessary to connect it into an electrical circuit. (2) *Explosives.* A system used to initiate an explosive train.

fusiform neuron
*See **gamma motor neuron**.*

fusion
(1) *Nuclear Energy.* A nuclear reaction characterized by the joining together of light nuclei to form heavier nuclei. *See also **nuclear fusion, binocular fusion,** and **flicker fusion**.* (2) *Anatomy.* The combining or blending of distinct bodies into one, such as the fusion into a single image of the separate impressions received by the two eyes, or the surgical process of making a formerly movable structure (joint) immovable.

fusus
(1) A spindle-shaped structure. (2) A minute air vesicle in a hair shaft.

future damages
Those sums awarded to an injured party for, among other things, residuals or future effects of an injury which have reduced the capability of an individual to function as a whole person, future pain and suffering, loss or impairment of earning capacity, and future medical expenses.

fuzzy logic
The use of approximations in reasoning rather than exact, discrete points or information.

FVC
See ***forced vital capacity***.

FWPCA
See ***Federal Water Pollution Control Act***.

G

g
Abbreviation for gram(s).

g force
That force experienced on the body due to acceleration(s) from vehicular or other motion. Usually expressed as some multiple or fraction of g.

g force syndrome
See ***acceleration syndrome*** *(an older term).*

g-induced loss of consciousness
See ***gravity-induced loss of consciousness.***

g-load
That loading imposed on the body due to gravity of other accelerations.

g-tolerance
A measure of the ability to withstand positive acceleration(s) without a system failure or blackout.

GAC
See ***granular activated carbon treatment.***

GADO
See ***General Aviation District Office.***

gadolinium
A chemical element, atomic number 64, atomic weight 157.25, symbol Gd.

gag
(1) A surgical device for holding the mouth open. (2) To retch, or strive to vomit. (3) Something placed in or around the mouth with the specific intent of preventing speech (usually placed against a person's will).

gag order
An unruly defendant at trial may be constitutionally bound and gagged to prevent further interruptions in the trial. The term may also refer to an order by the court, in a trial with a great deal of notoriety, directed to attorneys and witnesses, to not discuss the case with reporters (such an order being felt necessary to assure the defendant of a fair trial). Term also refers to order of the court directed to reporters to not report court proceedings, or certain aspects thereof.

gain
(1) In instrumentation, the ratio of the signal output to input. Gain is frequently referred to as *span*. (2) The constant multiplier in the numerator of a transfer function.

gain sharing
Any means through which an employee receives benefit in wages from his greater than standard production rates.

gainful employment
In general, any calling, occupation, profession or work which one may or is able to profitably pursue. Within a disability clause of an insurance policy, the term means ordinary employment of the particular insured, or such other employment, if any, as the insured may fairly be expected to follow.

gait
The mobility style using an individual's or robotic legs. Many clinical types of gaits have been identified. *See also* ***walk, run, jog,*** *and* ***limp.***

gait analysis
The study of gait. Usually with the intent to determine mechanisms or quantify disorders.

gal
Common abbreviation for gallon(s).

galactic cosmic radiation
That cosmic background radiation, consisting of extremely high energy particles, which comes from outside the solar system.

galactose
A monosaccharide derived from lactose.

gallbladder
A small sac-like organ located below the liver. It serves as a storage place for bile. The gallbladder may be subject to such disorders as inflammation and the formation of gallstones. Acute inflammation of the gallbladder causes severe pain and tenderness in the right upper abdomen, accompanied by fever, nausea, prostration, and sometimes jaundice. If the inflammation does not subside quickly, the gallbladder must be removed before it becomes gangrenous and ruptures. Chronic inflammation of the gallbladder may cause habitual indigestion, accompanied by flatulence, and nausea. The indigestion is most evident after heavy meals or meals of fatty foods. There may also be repeated at-

tacks of pain in the right upper abdomen. These may be very brief or may last as long as several hours. Gallstones are often present. The condition may respond to conservative treatment with diet and medications or it may require surgical removal of the gallbladder, especially if there are gallstones.

galley
That location on certain ships in which food is prepared for consumption. *See also* ***kitchen***.

gallon
A volumetric measure equal to 4 quarts (231 cubic inches) used to measure fuel oil. One barrel equals 42 gallons. *See also* ***barrel***.

gallstone
A stone-like mass, called a *calculus,* that forms in the gallbladder. The presence of gallstones is known medically as *cholelithiasis*. Their cause is unknown, although there is evidence of a connection between gallstones and obesity. They are most common in women after pregnancy and in men and women after the age of 35. Gallstones may be present for years without causing trouble. The usual symptoms, however, are vague discomfort and pain in the upper abdomen. There may be indigestion and nausea, especially after eating fatty foods. Either directly or by use of a dye introduced into the gallbladder, x-rays will generally reveal the presence of gallstones. The most common complication of gallstones occurs when one of the stones escapes from the gallbladder and travels along the common bile duct, where it may lodge, blocking the flow of bile to the intestine and causing obstructive jaundice. This condition must be corrected by surgery before liver damage occurs. When a gallstone travels through or obstructs a bile duct it can cause severe biliary colic, probably the most severe pain that can be experienced. The pain is located in the upper right quadrant of the abdomen and radiates through to the scapula.

galoshes
A type of waterproof footwear worn external to the shoes.

galvanic cell
An electrolytic cell brought about by the difference in electrical potential between two dissimilar metals.

galvanic current
Direct current from an electricity source, usually a battery (an outdated term).

galvanic skin reflex
See ***skin resistance response***

galvanic skin response
See ***skin resistance response***.

galvanometer
An electrical instrument for measuring small electric currents.

galvo
See ***metal fume fever***.

game fish
Species like trout, salmon, or bass, caught for sport. Many of them show more sensitivity to environmental change than "rough" fish.

gamma
A unit of magnetic field strength.

gamma angle
The angle formed by the intersection of the optical axis and the visual axis (line of sight), usually about 4°.

gamma efferent
See ***gamma motor neuron***.

gamma globulin
A plasma protein developed in the lymphoid tissues and reticuloendothelial system in response to invasion by harmful agents such as bacteria, viruses, and toxins.

gamma motor neuron
An A-class motor neuron in the Erlanger-Gasse classification system having a medium conduction velocity which innervates muscle spindle intrafusal fibers and is involved in regulating muscle activity. Also referred to as *gamma efferent* and *fusiform neuron*.

gamma ray
Electromagnetic radiation of high energy originating in atomic nuclei and accompanying many nuclear reactions (fission, radioactivity, and neutron capture). Physically, gamma rays are identical to x-rays of high energy; however, x-rays do not originate from atomic nuclei. Gamma rays are true rays of energy in contrast to alpha and beta radiation. The properties are similar to x-rays and other electromagnetic waves. They are the most penetrating waves of radiant nuclear energy

but can be blocked by dense materials such as lead. *See also* ***x-ray***.

gamma ray irradiation
A process to reduce pathogens in solid waste by irradiating sludge with gamma rays from certain isotopes.

gang chart
A multiple activity process chart used for coordinating work crews.

ganged controls
A set of controls which are grouped or stacked on a single axis, usually having a different outside diameter.

ganglion
(1) A mass of human or animal tissue-containing nerve cells (neurons). (2) A knot or knot-like mass; used in anatomic nomenclature as a general term to designate a group of nerve cell bodies located outside the central nervous system. (3) A form of cystic tumor occurring on an aponeurosis or tendon, as in the wrist.

gangrene
(1) An infection caused by an anaerobic bacteria resulting in the destruction of body tissue. (2) The death and putrefaction of body tissue, caused by the stoppage of circulation to an area, often caused by infection or injury. There are three types of gangrene: moist, dry, and gas gangrene. Moist and dry gangrene result from loss of blood circulation due to various causes; gas gangrene occurs in wounds infected by species of *Clostridium* that break down tissue by gas production and by toxins.

gangway
Any ramp-like or stair-like means of access provided to enable personnel to board or leave a vessel, including accommodation ladders, gangplanks, and brows.

gantry
A frame structure raised on side supports so as to span over or around something.

gantry crane
A crane-hoisting machine mounted on a frame or structure spanning an intervening space. Used primarily in modern container-handling ports.

Gantt chart
A two-dimensional graphical representation of the planned activities and the dates/times at which each of those activities should be completed over the duration of a project or other activity.

Gantt task and bonus plan
A wage incentive plan in which employees are rewarded with a percentage bonus for higher than normal performance.

ganzfeld
A homogeneous, uniformly illuminated, formless visual field.

GAO
See ***General Accounting Office***.

gap
Low point or opening between hills or mountains or in a ridge or mountain range.

gaps analysis
An assessment of an organization's current management methods, techniques, and systems to determine its conformance to standards and other requirements, such as ISO 14000.

garage
A space large enough to accommodate a car, with a door opening at least six feet wide and seven feet high. "Attached" means it is under part or all of the house or it shares part of a wall in common with the house. Not included are carports, barns, or buildings (not connected to the house) or storage space for golf carts or motorcycles.

garbage
See ***food waste***.

garbage and trash collection
Establishments that are primarily engaged in collecting and transporting garbage, trash, and refuse, within a city, town, or other local area, including adjoining towns and suburban areas.

garbage in/garbage out (GIGO)
Computing. A phrase indicating that if errors are made in computer input, errors will be present in the output, even if the programming and logic are correct.

garment
Any piece of clothing intended for wear over one or more body parts.

garment design
The development of a garment, ideally with consideration given to size, style, color, patterns, fabric types, layering, and insulation value.

gas
(1) A thin fluid, like air, capable of indefinite expansion but convertible by compression and cooling into a liquid and eventually a solid. Gases may be either elements (such as argon) or compounds (such as carbon dioxide). (2) A state of matter in which the material has very low density and viscosity, and can expand and contract greatly in response to changes in temperature and pressure. A gas easily diffuses into other gases, readily and uniformly distributing itself throughout any container. (3) Except when designated as inert, natural gas, other flammable gas, or gas which is toxic or corrosive. (4) A non-solid, non-liquid combustible energy source that includes natural gas, coke-oven gas, blast-furnace gas, and refinery gas.

gas amplification
As applied to gas-ionization radiation-detection instruments, the ratio of the charge collected to the charge produced by the initial ionizing event.

gas chromatograph-mass spectrometer
Refers to both an analytical method, as well as the apparatus used in the analysis. The gas chromatograph serves to separate the components of the sample and the mass spectrometer serves to identify then by exposing the eluted components to a beam of electrons which causes ionization to occur. The ions produced are accelerated by an electric impulse, passed through a magnetic field, separated, and identified on their mass. Often referred to simply as *GC-MS* or *GC-Mass Spec*.

gas chromatography (GC)
An analytical chemical procedure involving passing a sample through a column of specific make-up to separate the components of the sample, enabling them to elute, or pass out of the column separately and be detected and quantified by one or more detectors such as a flame ionization detector, thermal conductivity detector, electron capture detector, etc.

gas discharge lamp
A lamp which produces light at specific wavelengths of the spectrum by electrical excitation of the gas within the lamp. Also called *gaseous discharge lamp*.

gas distribution company
Company which obtains the major portion of its gas operating revenues from the operation of a retail gas distribution system, and which operates no transmission system other than incidental connections within its own system or to the system of another company.

gas exchange
The diffusion of gases through a membrane or other porous material.

gas-forming bacteria
Organisms that ferment lactose in foods or other carbohydrates producing both acid and gas, which may render a food product as unacceptable.

gas free
A tank, compartment, or other type containment or area is considered to be gas free when it has been tested, using appropriate instruments, and found to be sufficiently free, at the time of the test, of toxic or explosive gases or vapors for a specified purpose.

gas frit
A sintered or fritted glass surface which is designed to break up an air stream into small bubbles in order to increase the contact of the air with a liquid sorbent, thereby improving the absorption of specific gaseous contaminants present in the air. *See also **frit**.*

gas guzzler tax
Originates from the 1978 Energy Tax Act (Public Law 95418). A new car purchaser is required to pay the tax if the car purchased has a combined city/highway fuel economy rating that is below the standard for that year. For model years 1986 and later, the standard is 22.5 mpg.

gas laser
A type of laser in which the laser action takes place in a gas medium, such as carbon dioxide.

gas law
The thermodynamic law applied to a perfect gas that relates the pressure of the gas to its density and absolute temperature.

gas mask
A full-face respirator equipped with an air-purifying cartridge or canister that removes contaminants and renders air breathable to the user. (Not for use in oxygen deficient atmospheres.)

gas pressure
The force, generally designated in pounds per square inch (psi), that is exerted by a gas on its surroundings.

gas sorption
Devices used to reduce levels of airborne gaseous compounds by passing air through materials that extract gases. The performance of a solid sorbent is dependent on the airflow rate, concentration of the pollutants, presence of other gases or vapors, and other factors.

gas tension
The partial pressure of a gas.

gas test
An analysis of the air to detect unsafe concentrations of toxic or explosive gases and/or vapors.

gas/vapor detection instrument
An assembly of electrical, mechanical, and often chemical components that senses and responds to the presence of a gas/vapor in air mixtures.

gaseous agent
A fire extinguishing agent which is in the gaseous state at normal room temperature and pressure. It has low viscosity, can expand or contract with changes in pressure and temperature, and has the ability to diffuse readily and to distribute itself uniformly throughout an enclosure.

gasification
Conversion of solid material such as coal into a gas for use as a fuel.

gasohol
A blend of finished motor gasoline (leaded or unleaded) and alcohol (generally ethanol but sometimes methanol) limited to 10 percent by volume of alcohol. Gasohol is included in finished leaded and unleaded motor gasoline. *See also* ***fuel*** *and* ***gasoline***.

gasoline
(1) A blend of light hydrocarbon fractions of relatively high antiknock value, with proper volatility, clean burning characteristics, additives to prevent rust and oxidation, and sufficiently high octane rating to prevent knocking. Gasoline typically contains some benzene. (2) A complex mixture of relatively volatile hydrocarbons, with or without small quantities of additives, obtained by blending appropriate refinery streams to form a fuel suitable for use in spark ignition engines. Motor gasoline includes both leaded or unleaded grades of finished motor gasoline, blending components, and gasohol. *See also* ***fuel, gasohol,*** *and* ***kerosene***.

gasoline aviation/gasoline blending components
Naphthas that will be used for blending or compounding into finished aviation gasoline (e.g., straight-run gasoline, alkylate, reformate, benzene, toluene, and xylene). Excludes oxygenates (alcohols, ethers), butane, and pentanes plus. Oxygenates are reported as other hydrocarbons, hydrogen, and oxygenates.

gastric
Pertaining to the stomach.

gastritis
Chronic or acute inflammation of the stomach.

gastrocolitis
Inflammation of the stomach and colon.

gastrocnemius muscle
The large voluntary skeletal muscle in the posterior lower leg which forms a majority of the calf.

gastroenteritis
Inflammation of the mucous membrane of the stomach and intestines that may be caused by various bacteria or viruses. Symptoms may include diarrhea, abdominal cramps, nausea, vomiting, fever, malaise, muscle ache, and fatigue.

gastrointestinal tract
The system consisting of the stomach, intestines, and related organs. Commonly referred to as the *GI tract*.

gastroscope
An endoscope especially designed for passage into the stomach to permit examination of its interior.

GATB
See ***general aptitude test battery***.

gate
A structure that may be swung, drawn, or lowered to block an entrance or passageway.

gate dam
A type of opening in a dam whereby the water passes over the top.

gate hold procedure
Procedures at selected airports to hold aircraft at the gate or other ground location whenever departure delays exceed or are anticipated to exceed 16 minutes. The sequence for departure will be maintained in accordance with initial call-up unless modified by flow control restrictions. Pilots should monitor the ground control and clearance delivery frequency for engine start/taxi advisories or new proposed start/taxi time if the delay changes.

gateway
In the context of travel activities, gateway refers to a major airport or seaport. Internationally, gateway can also mean the port where customs clearance takes place.

gathering line
A pipeline 219.1 mm (8 5/8 inches) or less nominal diameter that transports petroleum from a production facility.

gauge
Marine Navigation. A scale graduated in tenths of a foot which indicates the water level or river stage.

gauge pressure
(1) The pressure with respect to atmospheric pressure, or above atmospheric pressure as indicated on the appropriate pressure gauge. (2) The difference between two absolute pressures, one of which is usually atmospheric pressure

gauging station
A structure used to measure the characteristics of a hydrographic feature.

gauss
The centimeter-gram-second electromagnetic unit of magnetic flux density, equal to one Maxwell per square centimeter (an older term). Also referred to as *abtesla*. *See also* ***Maxwell***.

Gaussian distribution
Pertaining to or having the appearance of a normal distribution. *See* ***normal distribution***.

Gaussian noise
See ***white noise***.

gavage
Dosing an animal by introducing a test material through a tube into the stomach.

GAWR
See ***gross axle weight rating***.

gaze
To look in one direction for an extended period of time.

GB
Gigabyte – approximately one billion bytes.

GC
See ***gas chromatography***. Also abbreviation for gas chromatograph.

GCA
See ***ground controlled approach***.

GC-ECD
Gas chromatography-electron capture detector.

GC-FID
Gas chromatography-flame ionization detector.

GC-FPD
Gas chromatography-flame photometric detector.

GC-MS
See ***gas chromatograph-mass spectrometer***.

GC-PID
Gas chromatography-photoionization detector.

GC-TCD
Gas chromatography-thermal conductivity detector.

GCW
See ***gross combination weight***.

GCWR
See ***gross combination weight rating***.

GDP
See ***gross domestic product***.

gear banger
Transit (slang). Driver who grinds gears when shifting.

gear jammer
Transit (slang). One who constantly clashes the gears.

gear ratio
The number of revolutions a driving gear requires to turn a driven gear one revolution. For a pair of gears, the ratio is found by dividing the number of teeth on the driven gear by the number of teeth on the driving gear.

Geiger counter
An electrical device that detects the presence of certain types of radioactivity. It consists of a needle-like electrode inside a hollow metallic cylinder filled with gas which, when ionized, sets up a current in an electrical field.

Geiger-Mueller counter
A refined version of the Geiger counter that has an amplifying system and is used for detecting and measuring radioactivity.

GEMI
*See **Global Environmental Management Institute**.*

gender
(1) Referring to feminine, masculine, or neuter terms in a language. (2) A classification for the male or female of the species. Preferred by some to the term *sex* when referring to the male and female. *See also* ***sex***.

gene
(1) A functional unit of heredity that occupies a specific location on a chromosome that is capable of producing itself exactly at each cell division, and can direct the formation of an enzyme or other protein. (2) A length of DNA that directs the synthesis of a protein.

gene library
A collection of DNA fragments from cells or organisms. So far, no simple way of sorting the contents of gene libraries has been devised. However, DNA pieces can be moved into bacterial cells where sorting according to gene function becomes feasible.

General Accounting Office (GAO)
The GAO of the federal government has the following basic purposes: Assist the Congress, its committees, and its members to carry out their legislative and oversight responsibilities, consistent with its role as an independent nonpolitical agency in the legislative branch; carry out legal, accounting, auditing, and claims settlement functions with respect to federal government programs and operations as assigned by the Congress; and make recommendations designed to make government operations more efficient and effective. The GAO is under the control and direction of the Comptroller General of the United States and the Deputy Comptroller General of the United States, appointed by the President with the advice and consent of the Senate for a term of 15 years.

general administration
Transit. All activities associated with the general administration of the transit system, including transit system development, injuries and damages, safety, personnel administration, legal services, insurance, data processing, finance and accounting, purchasing and stores, engineering, real estate management, office management and services, customer services, promotion, market research, and planning.

General and Administrative (G&A) expenses
Those expenses of a general corporate nature and expenses incurred in performing activities which contribute to more than a single operating function such as general financial accounting activities, purchasing activities, representation at law, and other general operational administration not directly applicable to a particular function. In the Transportation Industry, for example, passenger service, aircraft and traffic servicing, and promotion and sales expenses are included for certain small air carriers.

general aptitude test battery (GATB)
A commonly used test for determining general intelligence, numerical, verbal, and spatial skills, motor coordination, finger and manual dexterity, and clerical perception.

general average
A general loss voluntarily incurred to save all interest involved in a common maritime adventure from an impending peril including hull, cargo, and freight at risk.

general aviation

(1) Movements of aircraft and helicopters belonging to companies with an air taxi or air work license; an individual, a flying club or a company whose main objective is not to provide revenue passenger transport. (2) All civil flying except that of air carriers. (3) That portion of civil aviation which encompasses all facets of aviation except air carriers. (4) All civil aviation activity except that of air carriers certificated in accordance with Federal Aviation Regulations (FAR) Parts 121, 123, 127, and 135. The types of aircraft used in general aviation activities cover a wide spectrum from corporate multi-engine jet aircraft piloted by professional crews to amateur-built single engine piston acrobatic planes, balloons, and dirigibles.

general aviation airport

Any airport which is used or to be used for public purposes, under the control of a public agency, the landing area of which is publicly owned.

General Aviation Crashworthiness Project

An effort sponsored by the National Transportation Safety Board which was intended to improve the crashworthiness of small airplanes.

General Aviation District Office (GADO)

A Federal Aviation Administration (FAA) field office serving a designated geographical area and staffed with Flight Standards personnel who have the responsibility for serving the aviation industry and the general public on all matters relating to the certification and operation of general aviation aircraft.

general aviation operations

Takeoffs and landings of all civil aircraft, except those classified as air carriers or air taxis. *See also* ***general aviation***.

general cargo

(1) General cargo consists of those products or commodities such as timber, structural steel, rolled newsprint, concrete forms, agricultural equipment that are not conducive to packaging or unitization. Break-bulk cargo (e.g., packaged products such as lubricants and cereal) are often regarded as a subdivision of general cargo. (2) The tonnes of cargo assessed at the general rate of tolls as defined in the St. Lawrence Seaway Tariff of Tolls.

general cargo ship

A ship configured to accommodate general, break-bulk, and containerized cargoes. Cargo handling operations are labor intensive and conducted with either ship's cranes or jib cranes onshore. These ships traditionally have numerous holds located on several decks, have smaller hatches than bulk carriers or containerships, and are usually equipped with a boom or crane positioned at each hatch cover.

general circulation of the atmosphere

Large-scale atmospheric motions over the entire earth.

general contractor

One who contracts for the construction of an entire building or project, rather than for a portion of the work. The general contractor hires subcontractors (e.g., plumbing, electrical, carpentry, etc.), coordinates all the work, and is responsible for ensuring payment to the subcontractors. Also called *prime contractor*. *See also* ***contractor***.

general damages

Damages that are the immediate, direct, and proximate result of the wrongful act that is subject to a complaint.

General Duty Clause

Refers to Section 5(a)(1) of the Occupational Safety and Health Act of 1970, which states: *"Each employer shall furnish to each of his employees employment and a place of employment which are free from recognized hazards that are causing or are likely to cause death or serious physical harm to his employees, and shall comply with the occupational safety and health standards promulgated under this Act."* It is often cited by OSHA to cover hazards for which a specific Standard or Regulation does not yet exist.

general duty clause violation

Under the Occupational Safety and Health Act, a violation of the general duty clause exists when OSHA can show that the hazard is a recognized hazard, the employer failed to render its workplace free from the recognized hazard, the occurrence of an accident or adverse health effect was reasonably foreseeable, the likely consequence of the incident (accident or adverse effect) was death or a form of serious physical harm, and there exists feasible means to correct the hazard.

general environment
With regard to ionizing radiation, the total terrestrial, atmospheric, and aquatic environment outside sites within which any activity, operation, or process authorized by a general or special license is performed.

general exhaust ventilation
A mechanical system for exhausting air from a work area thereby reducing the contaminant concentration by dilution. *See also* ***dilution ventilation system***.

general export license
Authorization to export without specific documentary approval.

general freight carrier
(1) A carrier which handles a wide variety of commodities. (2) Trucking company engaged in shipping packaged, boxed, and palletized goods that can be transported in standard, enclosed tractor-trailers, generally 40 to 48 feet in length.

general hearing
The ability to detect sound and/or discriminate between sounds over a wide range of pitch and loudness (an older term).

General Industry Standard (GIS)
See ***OSHA General Industry Standard***.

general law
A law that affects the community at large. A general law, as distinguished from one that is special or local, is a law that embraces a class of subjects or places, and does not omit any subject or place naturally belonging to such class.

general license
As pertains to ionizing radiation, a license issued by the Nuclear Regulatory Commission (NRC), or an Agreement State, for the possession and use of certain radioactive materials, often for small quantities, for which a specific license is not required. Individuals are automatically licensed when they buy or obtain a radioactive material from a vendor who has a license from the NRC to sell products containing small amounts of some radioactive materials.

general lighting
The approximately uniform background illumination within a specific area or volume.

general permit
A permit applicable to a class or category of dischargers.

general-use snap switch
See ***switch (2)***.

general-use switch
See ***switch (1)***.

general utility, stage I airport
This type of airport serves all small airplanes. Precision approach operations are not usually anticipated. This airport is designed for airplanes in Airport Reference Code B-II.

general utility, stage II airport
This type of airport serves large airplanes in Aircraft Approach Categories A and B and usually has the capability for precision approach operations. This airport is normally designed for Airport Reference Code B-III.

general ventilation
This term is used synonymously with dilution ventilation. General ventilation is used typically for the control of temperature, humidity, or odors.

general warehousing and storage
Those establishments primarily engaged in the warehousing and storage of a general line of goods.

generality
See ***coefficient of determination***.

generation
(1) The process of begetting offspring. (2) A successive step or degree in natural descent, or, the average period between any two such successive steps (about thirty years for human beings). (3) Any group of individuals born at or about the same time. (4) The act or process of producing solid waste.

generator
A individual, facility, or mobile source that emits or causes or contributes to the emission of pollutants into the air or releases hazardous wastes into the water or soil.

generic name
A nonproprietary name, such as the chemical identity of a material or product rather than identification by a registered trade name.

genetic
See ***gene***.

genetic defect
A defect in a living organism as a result of a deficiency in the genes of the original reproductive cells from which the organism was conceived.

genetic effects
Inheritable changes, chiefly mutations, produced by the absorption of ionizing radiation, exposure to certain chemicals, ingestion of some medications, and from other causes.

genetic engineering
A process of inserting new genetic information into existing cells to modify any organism for the purpose of changing one of its characteristics.

genetic mutation
A change in a gene which is reflected in body structure and/or function.

genetics
Inheritable changes, chiefly mutations, produced by the absorption of ionizing radiation, exposure to certain chemicals, ingestion of some medications, and from other causes.

genome
A total set of chromosomes derived from one parent.

genotoxic
(1) Refers to the ability of a chemical to adversely affect the genome of living cells, such that upon duplication, a mutagenic or carcinogenic event is expressed due to the alteration of the genome molecular structure. (2) Chemical or radioactive substance known to cause or suspected of causing damage to the DNA in individual cells, thus causing mutations or cancer.

genotoxic teratogen
A substance which includes malformations in a developing embryo via genetic damage (i.e., mutations).

genotoxin
A substance that is toxic to genetic material.

geo map
Aviation. The digitized map markings associated with the Airport Surveillance Radar (ASR-9) System.

geodesic line
The shortest line which connects two points on a curved surface.

Geographical Information System (GIS)
A system of hardware, software, and data for collecting, storing, analyzing, and disseminating information about areas of the Earth. For Highway Performance Monitoring System (HPMS) purposes, Geographical Information System (GIS) is defined as a highway network (spatial data which graphically represent the geometry of the highways, an electronic map) and its geographically referenced component attributes (HPMS section data, bridge data, and other data including socioeconomic data) that are integrated through GIS technology to perform analyses. From this, GIS can display attributes and analyze results electronically in map form.

geomedicine
The branch of medicine dealing with the influence of climatic and environmental conditions on health.

geometric mean
The median in a lognormal distribution. Expressed as the nth root of a product of n numbers:

$$GM = \sqrt[n]{x_1 \cdot x_2 \cdot \ldots \cdot x_n}$$

geometric progression
A sequence of values corresponding to the form, a, ar^1, ar^2, ar^3, ...

geometric series
An infinite series having the form $a + ar^1 + ar^2 + ar^3 + \ldots$.

geometrical access
See ***optical axis***.

geometry
The study of size and shape.

geophysical techniques tests
Used to locate buried metallic objects, such as Underground Storage Tanks (USTs), and to map groundwater pathways. Testing methodologies include magnetometer, surveys, ground penetrating radar, electrical resistance, and seismic refraction.

geostationary satellite
A satellite that orbits the earth at the same rate that the earth rotates and thus remains over a fixed place above the equator.

geostrophic wind
The horizontal wind blowing in a straight path, parallel to the isobars or contours, at a constant speed. The geostrophic wind results when the Coriolis force exactly balances the horizontal pressure gradient force.

geothermal resources
Under the Federal Geothermal Energy Research, Development, and Demonstration Act of 1974: (1) all products of geothermal processes, embracing indigenous steam, hot water, and brines; (2) steam and other gases, hot water and hot brines, resulting from water, gas, or other fluids artificially introduced into geothermal formations; and (3) any byproduct derived from them.

GERDDA
Geothermal Energy Research, Development, and Demonstration Act of 1974 (federal).

geriatrics
The study of aging and any diseases associated with aging.

germ
A general term for a microorganism.

germ cell
The cells of an organism whose function it is to reproduce the kind (i.e., an ovum or spermatozoon). The cells of an organism whose function is reproduction.

German measles
A contagious virus disease, most common in children between the ages of 3 and 12 years. Also called *rubella,* or *3-day measles*.

germanium
A chemical element, atomic number 32, atomic weight 72.59, symbol Ge.

germicidal effectiveness
See ***bactericidal effectiveness***.

germicidal lamp
See ***bactericidal lamp***.

germicide
(1) Any compound that kills disease-causing microorganisms. (2) Any substance that kills microbes, or an agent that destroys pathogenic microorganisms.

gerontology
The study of aging processes and their associated problems.

get
To pick up and acquire control of an object (may include several therbligs).

GFCI
See ***ground-fault circuit-interrupter***.

GFF
Glass fiber filter.

GFI
Ground fault interrupter. *See* ***ground-fault circuit interrupter***.

GHG
Greenhouse gas.

GHz
Gigahertz, 1 E+9 Hz.

GI tract
See ***gastrointestinal tract***.

giant nuclei
See ***condensation nuclei***.

giga–
(prefix) 10^9 or 1 billion times the base unit.

gigabyte (GB)
Approximately one billion bytes.

gigawatt (GW)
One billion watts or one thousand megawatts. *See also* ***electricity, kilowatt,*** *and* ***megawatt***.

gigawatt electric (GWE)
One billion watts of electric capacity.

gigawatt hour (GWH)
One billion watt-hours.

GIGO
See ***garbage in/garbage out***.

Gilbreth basic element
See ***therblig***.

gimbal
A device with two mutually perpendicular and intersecting axes of rotation which permits orientation or motion in two directions.

gingiva
The mucous membrane and other fibrous tissue covering the upper and lower jaws and bases of the teeth within the mouth.

gingival septum
That portion of the gingiva which lies between two teeth.

gingival sulcus
The groove between the gingiva and the tooth surface. Also called *gingival crevice.*

gingivitis
Inflammation of the gums of the mouth. Bleeding is the primary symptom. Other symptoms include swelling, redness, pain, and difficulty in chewing.

girth
The distance around an approximately circular object or cross-section of a structure.

GIS
See ***Geographical Information System****. See also* ***OSHA General Industry Standard****.*

glabella
The most anterior point of the forehead between the brow ridges in the midsagittal plane.

glabella – inion length
The horizontal linear distance from glabella to inion in the midsagittal plane. Measured with the individual standing erect and looking straight ahead.

glabella to back of head
See ***glabella – inion length****.*

glabella to top of head
The vertical distance from the most anterior point of the forehead between the brow ridges to the level of the top of the head. Measured with the individual standing erect.

glabella to wall
The horizontal linear distance from a wall to the most anterior point of the forehead between the brow ridges. Measured with the individual standing erect with his/her back and head against the wall and looking straight ahead.

glaciated cloud
A cloud or portion of a cloud where only ice crystals exist.

glad hands
Transit (slang). Air hose brake system connections between tractor and trailer.

gland
A structure, ranging from a cell to an organ in size, which manufactures, stores, and/or secretes one or more substances for bodily use.

glare
The sensation produced by luminance within the visual field that are sufficiently greater than the luminance to which the eyes are adapted. This causes annoyance, discomfort, or loss of visual performance and acuity, a concern especially for individuals using video display terminals for extended periods.

glare sensitivity
The ability to see objects despite the presence of glare or strong ambient lighting.

glare shield
Any transparent structure which can be used to reduce glare.

glass blower's cataract
An opacity of the rear surface of the lens in the eye caused by excessive exposure of the eyes to luminous radiation, primarily visible and infrared. Found in those occupationally exposed to furnaces or other hot devices for extended periods of time.

glass cockpit
Aviation (slang). An aircraft cockpit in which the use of multifunctional and computerized displays replaces many of the dedicated gages and instruments.

glass frit
See ***frit****.*

glaucoma
An abnormally high pressure in the eyeball. It is caused by an increase in fluid pressure inside the eye, somewhat like an overfilled water balloon. Most cases result from the inability of the eye fluid to drain.

glaze
The coating of ice that forms on cold objects that have been exposed to rain or drizzle. *See also* ***freezing rain****.*

GLC
Ground level concentration.

glenoid cavity
The depression in the scapula inferior to acromion which articulates with the head of the humerus to comprise the shoulder joint.

glide
A speech sound generally considered as being between a vowel and a consonant, and which is produced by movement or gliding from an articulatory position to an adjacent sound.

glidepath
A descent profile determined for vertical guidance during a final approach.

glider
A heavier-than-air aircraft, that is supported in flight by the dynamic reaction of the air against its lifting surfaces and whose free flight does not depend principally on an engine.

glideslope
Provides vertical guidance for aircraft during approach and landing. The glideslope or "glidepath" is based on the following: a) electronic components emitting signals which provide vertical guidance by reference to airborne instruments during instrument approaches such as Instrument Landing System (ILS)/Microwave Landing System (MLS); or b) visual ground aids, such as Visual Approach Slope Indicator (VASI), which provide vertical guidance for a Visual Flight Rules (VFR) approach or for the visual portion of an instrument approach and landing; or c) used by Air Traffic Control (ATC) to inform an aircraft making a Precision Approach Radar (PAR) approach of its vertical position (elevation) relative to the descent profile. *See also **instrument landing system, intercept glideslope altitude, localizer, middle marker,** and **outer marker**.*

Global Environmental Management Institute (GEMI)
Established by some major U.S. companies to provide leadership to other companies in developing tools and strategies to help business achieve environmental, health, and safety excellence and economic success.

Global Positioning System (GPS)
A space-based radio positioning, navigation, and time transfer system being developed by the Department of Defense. The first satellite in the system was launched from Cape Canaveral Air Station (CCAS) in Florida on February 14, 1989 aboard a Delta-II expendable launch vehicle (ELV). When fully deployed, the system is intended to provide highly accurate position and velocity information, and precise time, on a continuous global basis, to an unlimited number of properly equipped users. The system will be unaffected by weather, and will provide a worldwide common grid reference system. The Global Positioning System (GPS) concept is predicated upon accurate and continuous knowledge of the spatial position of each satellite in the system with respect to time and distance from a transmitting satellite to the user. The GPS receiver automatically selects appropriate signals from the satellites in view and translates these into a three-dimensional position, velocity, and time. Predictable system accuracy for civil users is projected to be 100 meters horizontally. Performance standards and certification criteria have not yet been established.

global scale
*See **planetary scale**.*

global warming
*See **greenhouse effect**.*

globe temperature
A thermal value representing the composite of the dry-bulb temperature, radiation heating, and convection/wind effects. Measured with the thermometer in the center of a 6" sphere which is assumed to be a blackbody radiator or represents the material being tested. *See **globe thermometer**.*

globe thermometer
A dry-bulb thermometer suspended in the center of a sphere that has been painted flat black and is used to measure radiant heat.

globulin
Any of a group of proteins found in animal and vegetable tissues that can be precipitated from serum or plasma.

gloss
An attribute of a surface which results in a shiny appearance.

gloss trap
A cavity or other structure designed to absorb specular reflections from incident light.

glossal
Pertaining to the tongue.

glossitis
Inflammation of the tongue.

glossmeter
A photometer for measuring the gloss of a material in the general direction of specular reflection.

glossopharyngeal nerve
A nerve having both motor and sensory components, and generally involved in salivation, muscular control of the pharynx, and taste. May also be called the *ninth cranial nerve*.

glossy
Word (adjective) used to describe a polished surface with a mirror-like finish.

glottis
The opening between the vocal cords.

glove
An article of clothing which has separate appendages for covering the digits and the remainder of the hand, as well as possibly covering the wrist and some portion of the distal forearm. Although there are some social uses that are nothing more than cosmetic, the general intent is to protect tissue from some undesirable or hazardous environment.

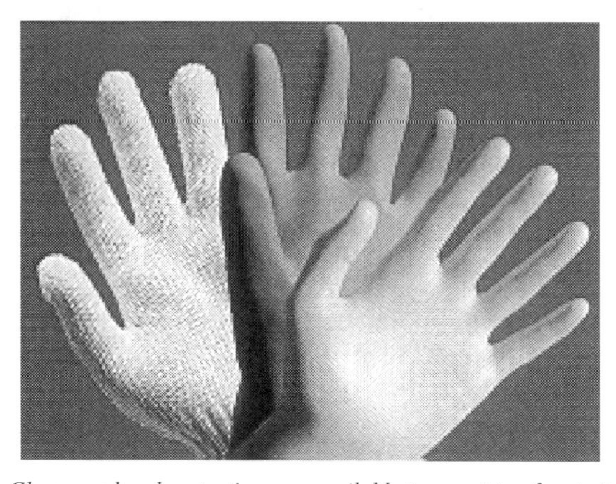

Gloves, as hand protection, are available in a variety of materials to ensure protection against skin exposure to virtually any type of chemical or physical substance

glove box laboratory hood
An enclosed, six-sided hood with arm-length gloves provided at the front or side of the hood for access. An air-lock pass-through port is often provided to insert and remove materials from the hood.

glove controller
A lightweight glove-like device which is equipped with transducers and can transmit information about arm, hand, and finger position to a computer for controlling another device.

glovebag
A plastic bag which is placed around a pipe or other structure from which the removal of a material, such as asbestos, is to be carried out without its release to the atmosphere.

GLP
Good laboratory practice.

glucose
A 6-carbon monosaccharide (blood sugar). The most common type of sugar and the primary metabolic energy sources. It forms the basis for the *glycemic index*. All carbohydrates are eventually converted to glucose by the body.

glutamine
A nitrogen compound occurring in body tissues and having a part in the production of ammonia by the kidney.

glutaraldehyde
A compound used as a disinfectant and as a tissue fixative for light and electron microscopy because of its preservation of fine structural detail and localization of enzyme activity.

gluteal arc
That portion of the posterior body surface represented primarily by the curvature of the buttock.

gluteal arc length
The surface distance over the buttock from the gluteal furrow to the posterior waist level. Measured with the individual standing erect and the back/hip/leg muscles relaxed except as necessary to maintain posture.

gluteal furrow
The crease at the inferior junction of the buttock and superior portion of the posterior thigh.

gluteal furrow height
The vertical distance from the floor or other reference surface to the gluteal furrow. Measured with the individual standing erect and the back/hip/leg muscles relaxed except as necessary to maintain posture.

glycemic index
A measure of the speed at which a carbohydrate is assimilated. The rapidness of assimilation is determined by the structure of the carbohydrate as well as its fiber content.

glycerin
A clear, colorless, syrupy liquid, used as an emollient and as a solvent for drugs; a product, along with fatty acids, of the hydrolysis of ingested fats.

glycogen
A polysaccharide (i.e., carbohydrate) molecule, containing glucose and water, which is stored in various body tissues as a quick reserve source of sugar/energy. It is converted to glucose when additional energy is required. Glycogen is the primary way that carbohydrates are stored in skeletal muscle and the liver. Also referred to as *animal starch*.

glycolysis
The breakdown of carbohydrates in bodily metabolism.

GMT
See ***Greenwich Mean Time***.

GNP
See ***Gross National Product***.

go ahead
Aviation. Aircraft-to-tower communication protocol meaning *proceed with your message*. The term cannot be used for any other purpose.

go around
Aviation. Instructions for a pilot to abandon his approach to landing. Additional instructions may follow. Unless otherwise advised by Air Traffic Control (ATC), a Visual Flight Rules (VFR) aircraft or an aircraft conducting visual approach should overfly the runway while climbing to traffic pattern altitude and enter the traffic pattern via the crosswind leg. A pilot on an Instrument Flight Rules (IFR) flight plan making an instrument approach should execute the published missed approach procedure or proceed as instructed by ATC. *See also* ***missed approach***.

go/no-go display
(slang). A display which provides information from which the user can make only one of two opposing responses.

go/no-go reaction
(slang). One of a set of responses open to an individual in which he/she either responds (go) or withholds (no-go) depending on a stimulus, display, or other input.

go well over
Maritime Navigation. A term applied in making a crossing meaning to go well over near the shore on the opposite side before turning out to either shape the shore or pass an easy distance off before coming up on the next set of marks.

goal
An objective for which some activity is initiated and sustained.

goal gradient
The influence of the nearness to reaching a goal on the energy expended toward achieving that goal.

goal-oriented problem solving
See ***backward chaining***.

goals, operators, methods, and selection rules (GOMS)
A method for analyzing and/or modeling the knowledge required for interface use.

goat 'n' shoat man
Transit (slang). Driver of a livestock carrier.

goggle
A tight-fitting device worn over the eyes to provide splash and/or impact protection.

going rate curve
A relationship between the evaluation of jobs and their rates of pay in the labor market.

goiter
Enlargement of the thyroid gland, causing a swelling in the front part of the neck.

gold
A chemical element, atomic number 79, atomic weight 196.967, symbol Au.

Gold Book
Common name for an EPA publication known as the Quality Criteria for Water which was developed as a means of ensuring some level of minimum consistency between the states. The EPA has established minimum criteria for 137 specific pollutants based upon identifiable effects of each pollutant on the public health and welfare, aquatic life, and recreation.

Golgi tendon organ
A stretch receptor located primarily near the tendon-muscle junction which measures muscle tension and provides feedback to the nerv-

ous system. Also called *neurotendinous spindle*.

GOMS
See ***goals, operators, methods, and selection rules***.

gonad
A primary sex gland, consisting of an ovary in the female or testis in the male.

gonial angle
The point on the lower jaw at which the posterior lower portion of the ramus and lower body of the mandible meet.

goniometer
An apparatus for measuring the limits of flexion (bending) and extension of the joints of the fingers. The goniometer arms are normally aligned with the bones of adjacent body segments, and the angle read from the pivot point.

goniophotometer
An instrument for measuring the quantity of light emitted/reflected in various directions to determine the spatial distribution of light.

goniophotometric curve
A graph or function showing the light emitted/reflected from an object at varying angles of view with a fixed angle of incidence.

good condition classification
No corrective maintenance is needed at time of an inspection. The facility is serving the purpose for which it was constructed.

good faith
An intangible and abstract quality with no technical meaning or statutory definition, and it encompasses, among other things, an honest belief, the absence of malice, and the absence of design to defraud or to seek an unconscionable advantage, and an individual's personal good faith is a concept of his/her own mind and inner spirit and, therefore, may not conclusively be determined by his/her protestations alone.

Good Samaritan doctrine
One who sees a person in imminent and serious peril through negligence of another cannot be charged with contributory negligence, as a matter of law, in risking his/her own life or serious injury in attempting to effect a rescue, provided the attempt is not recklessly or rashly made. Under this doctrine, negligence of a volunteer rescuer must worsen the position of the person in distress before liability will be imposed. The protection from liability is provided by statute in most states.

goodness of fit
A measure of how well a sample or model approximates a prescribed curve.

goose pimples
Also known as "goose flesh," a skin conditions marked by numerous small elevations around the hair follicles caused by the action of the *arrectores pilorium* ("raisers of hair") muscles.

gout
A disease in which uric acid appears in excessive quantities in the blood and may be deposited in the joints and other tissues. During an acute gout attack, there is swelling, inflammation and extreme pain in a joint, frequently the big toe. After several years of attacks, the chronic form of the disease may set in, permanently damaging and deforming joints and destroying cells of the kidney. About 95% of all cases occur in men and the first attack rarely occurs before the age of 30. The causes of gout are not fully understood. It is a disorder of the metabolism of purines. These nitrogenous substances are found in high-protein foods and the net product of their metabolism is uric acid. For unknown reasons, the uric acid, normally expelled in the urine, is retained in the blood in excess amounts. Uric acid crystals are deposited in the joints and in cartilage, where they form lumps called tophi. The uric acid crystals also predispose to the formation of calculi in the kidney (kidney stones) and lead to permanent damage of the kidney cells.

governing element
A work element which requires a longer time than any other element being performed concurrently in a work cycle.

government aid cargo
The tonnes of cargo assessed at the government aid rate of tolls as defined in the St. Lawrence Seaway Tariff of Tolls.

government fleet vehicle
Includes vehicles owned by all federal General Services Administration (GSA), state,

county, city, and metro units of government, including toll road operations.

government leased vehicle
A vehicle obtained by an executive agency by contract or other source for a period of 60 continuous days or more.

government light
A colloquial term applied to an aid to navigation maintained by the Coast Guard.

government-owned contractor-operated vehicle
A vehicle that is owned or leased by the federal government but used by a contractor under a cost reimbursement contract with a federal agency.

government owned vehicle
A vehicle that is owned by the federal government.

government tort
A wrong perpetrated by the government through an employee or agent or instrumentality under its control which may or may not be actionable depending upon whether there is governmental tort immunity. Tort actions against the federal government are governed by the Federal Tort Claims Act.

government transportation expenditures
Those expenditures that are the final actual costs for capital goods and operating services covered by the government transportation program.

government transportation revenue
The transportation revenue estimates contained in this report consist of those funds identified as government transportation-related user charges, taxes, or fees in the various data sources. Therefore, general revenue is not included.

governor
(1) *General.* One who governs. The elected chief executive of any State in the United States. An official appointed to administer a province, territory, etc. (2) *Automotive Mechanics.* A device which limits the speed of an engine. A governor is also a part on an automatic transmission which signals internal transmission components to shift to a higher gear.

gpm
Gallons per minute.

GPS
See ***Global Positioning System***.

grab bar
With regard to ladders, individual handholds placed adjacent to or as an extension above ladders for the purpose of providing access beyond the limits of the ladder.

grab one
Transit (slang). To shift into a lower gear as a means of gaining power when driving uphill.

grab sample
(1) To collect an air sample for a short period of time to test for the presence of contaminants in a work or other environment. (2) In industrial hygiene application, a type of air sample in which the air is admitted into a bag, vessel, or instrument instantaneously for subsequent analysis.

graben
An elongated part of the Earth's crust bounded by faults on its long sides and relatively down-dropped compared to or relative to its surroundings.

gradability
The ability of a vehicle to negotiate a given grade at a specified Gross Combination Weight Rating (GCWR) or Gross Vehicle Weight Rating (GVWR). It is the measure of the starting and grade climbing ability of a vehicle, and is expressed in percent grade (1 percent is a rise of 1 foot in a horizontal distance of 100 feet).

grade
(1) One level in a series of defined sequential levels according to a set of criteria. (2) The angle of an incline, either up or down from horizontal. (3) To segregate a quantity of some product by quality.

grade crossings
An intersection of highway roads, railroad tracks, or dedicated transit rail tracks that run either parallel or across mixed traffic situations with motor vehicles, light rail, commuter rail, heavy rail, trolley bus, or pedestrian traffic. Collisions at grade crossings involving transit vehicles apply only to light rail, commuter rail, heavy rail, or trolley bus. *See also*

at grade, highway-rail crossing, *and* ***rail-highway grade crossing***.

Grade D breathing air
Breathing air which meets the specifications of the Compressed Gas Association (CGA) Commodity Specification for Grade D air. It must have between 19.5 and 23% oxygen content and must contain maximums of 5 mg/m^3 condensed hydrocarbons, 20 ppm carbon monoxide, and 1000 ppm carbon dioxide; and it must have no pronounced odor.

gradient
The rate of increase or decrease in magnitude of a variable or response.

gradient wind
A wind that blows parallel to curved isobars or contours.

gradually pull down
Marine Navigation. To swing slowly to a new course on a mark further downstream.

gradually pull down shape of bend
Marine Navigation. Term used in crossings meaning to keep well out until tow is well down, then alter course to follow the shore shape of the bend.

gradually pull down shore
Marine Navigation. Term used in crossings meaning, when well over, to gradually swing the vessel's head downstream along the shore.

Graham's law
The rate of diffusion of a gas through porous membranes varies inversely with the square root of its density.

grain
(1) A unit of weight equal to 64.8 milligrams. (2) The seed of cereal plants.

grain body
Low side, open top truck body designed to transport dry fluid commodities.

grain cargo
The tonnes of cargo assessed at the Food or Feed Grains rate of tolls as defined in the St. Lawrence Seaway Tariff of Tolls.

grain loading
The rate at which particles are emitted from a pollution source. Measurement is made by the number of grains per cubic foot of gas emitted.

gram
The basic unit of mass in the metric system. One gram is equal to 15.432 grains.

gram-atomic weight
A mass in grams numerically equal to the atomic weight of the element.

gram mole
See ***gram molecular weight***.

gram-molecular weight
Mass in grams numerically equal to the molecular weight of a substance.

Gram's stain
A stain for bacteria, used as one means of identifying unknown bacterial organisms.

Grand Jury
A tribunal which is part of a criminal procedure to which the prosecutor submits evidence from investigations and which determines whether or not there is probable cause to believe a crime was committed and by whom. If so, next follows an *indictment*.

grand mal
A major epileptic seizure attended by the loss of consciousness and convulsive movements, as distinguished from petit mal, a minor seizure.

grandfather clause
Provision in a new law or regulation exempting those already in or a part of the existing system which is being regulated.

Grandfathering Provision
Under the Clean Water Act, any new source which has been constructed to meet current BADT standards will not be subjected to any additional more stringent standards of performance for as much as 10 years into the future.

grants
A federal financial assistance award making payment in cash or in kind for a specified purpose. The federal government is not expected to have substantial involvement with the state or local government or other recipient while the contemplated activity is being performed. The term "grants-in-aid" is commonly restricted to grants to states and local governments.

granular activated carbon treatment (GAC)
A filtering system often used in small water systems and individual homes to remove or-

ganics. GAC can be highly effective in removing elevated levels of radon from water.

granulocytes
Any cell containing granules, especially a leukocyte containing neutrophil, basophil, or eosinophil granules in its cytoplasm.

granulocytosis
An abnormally large number of granulocytes in the blood.

granuloma
A tumorlike mass or nodule of vascular tissue due to a chronic inflammation process associated with an infectious disease.

Granz rays
X-rays produced at voltages of 5 to 20 kilovoltage peak (KVP).

graph
A plot of some function or distribution using a coordinate system.

graphic
A pictorial hardcopy or display representing an object or a dataset which involves more than simple straight or curved lines.

graphic display
A graphic presented on a CRT, flat panel, or other graphics-capable monitor.

Graphical User Interface (GUI)
The use of direct manipulation and icons or other graphical symbols on a display to interact with a computer.

grapple
To close a device on the end effector of a robotic or teleoperated arm to gain control of an object.

grasp
(1) To position the required number of digits and/or the palm to enable an individual to move, pick up, or hold an object. (2) A therblig; to flex the hand and fingers around an object to gain control of that object.

grasp reflex
A grasping motion which occurs on stimulation of the palm or sole of the foot.

graticule
See ***reticle***.

graupel
See ***snow pellets***.

graveyard
See ***burial ground***.

graveyard shift
See ***third shift***.

gravimetric method
An analytical method for determining the concentration of a substance based on determination of the weight of the material collected on a filter, absorbed in a sorbent, or formed in a subsequent analytical procedure.

gravitational field
That vector field due to gravity extending through space which would cause the source and any object entering that field to be mutually attracted to each other. One of the basic fields in nature.

gravitational force
See ***gravity***.

gravitational physiology
The study of the effects of different gravity levels on the body's structure and function.

gravity
A force which causes objects to attract each other as a function of their masses and the distance between them.

gravity feed
The process of using gravitational force to pass materials from one location to another, lower location.

gravity-induced loss of consciousness (g-LOC)
That loss of consciousness due to high positive g-force maneuvers with the resulting reduction in cranial blood supply in high performance aircraft. *See also* ***grayout*** *and* ***blackout***.

gray
(1) *General.* An achromatic color between total white and total black. (2) The unit of absorbed radiation dose. One *gray* is equal to one joule per kilogram.

gray scale
A series of achromatic shades with varying proportions of white and black, to give the full range between total whiteness and total blackness.

gray water
The term given to domestic wastewater composed of washwater from sinks, kitchen sinks, bathroom sinks and tubs, and laundry tubs.

graying of vision
See ***grayout***.

grayout
A condition in which the visual field begins to narrow and decrease in brightness. Also referred to as *graying of vision*. *See also* ***gravity-induced loss of consciousness*** *and* ***blackout***.

grazing permit and lease
Under the Federal Land Policy and Management Act of 1976: Any document authorizing use of public lands or lands in National Forests in the eleven contiguous western states for the purpose of grazing domestic livestock.

great bodily injury
Bodily injury which involves a substantial risk of death, serious permanent disfigurement, or protracted loss or impairment of function of any part of an organ of the body.

great care
Law. Great care is such as persons of ordinary prudence usually exercise about affairs of their own which are of great importance; or it is that degree of care usually bestowed upon the matter in hand by the most competent, prudent, and careful persons having to do with the particular subject. Highest degree of care and utmost care have substantially the same meaning. "Highest degree of care" only requires the care and skill exacted of persons engaged in the same or similar business. It means the highest degree required by law where human safety is at stake, and the highest degree known to the usage and practice of very careful, skillful, and diligent persons engaged in the same business by similar means or agencies.

Great Lakes-St. Lawrence Seaway Freight Transportation
Establishments primarily engaged in the transportation of freight on the Great Lakes and St. Lawrence Seaway, either between U.S. ports or between U.S. and Canadian ports.

Great River Environmental Action Team
A multi-agency planning group organized to develop a resource management plan for the Upper Mississippi River. Operates under the auspices of the Upper Mississippi River Basin Commission.

greater hazard defense
A well-established Occupational Safety and Health Review Commission (OSHRC) doctrine that, on some occasions, allows employers to escape sanctions for violations of otherwise applicable safety regulations because the act of abating the violation would itself pose an even greater threat to the safety and health of their employees.

greater multiangular bone
See ***trapezium***.

greater trochanter
A large lateral projection of the proximal femur.

green
A primary color, corresponding to that hue apparent to the normal eye when stimulated only with electromagnetic radiation approximately between 495 to 575 nm wavelength.

green blindness
See ***deuteranopia***.

green development
An emerging practice in real estate development which emphasizes research and incorporation of environmentally and economically sound measures into real estate projects.

green flash
A small, green color that occasionally appears on the upper part of the sun as it rises or sets.

Greenburg-Smith impinger
A relatively large impinger that has been employed for the collection of airborne dust samples. Requires a sample rate of 1 cubic foot per minute.

greenhouse effect
The warming of the earth's atmosphere caused by a build-up of carbon dioxide or other trace gases; it is believed by many scientists that this build-up allows light from the sun's rays to heat the earth but prevents a counterbalancing loss of heat. Also known as *global warming* or, simply *atmospheric effect*.

greening
Incorporating environmental performance and sustainability into overall corporate strategies and environments.

Greenwich Mean Time (GMT)
A world time standard; the mean solar time at the Greenwich (England) Meridian.

grid
A flat section of a region which is subdivided into smaller, usually square, sections.

grievance
Any dissatisfaction with working conditions or pay which is expressed by one or more employees to management. Such employees are typically, but not necessarily, represented by a collective bargaining agreement.

grievance committee
A group of workers, usually in a union shop, who have been chosen by their fellow workers (of the same union) to represent employees to management during grievance proceedings.

grievance procedures
Any sequence of steps which should be followed in pursuing an employee's grievance through an organization in an attempt to obtain resolution.

grille
Component of a ventilation system through which air is returned to the system from the space to which it was supplied.

grind
A process using an abrasive disk rotating at high speed.

grinder pump
A mechanical device which shreds solids and raises the fluid to a higher elevation through pressure sewers.

grinder's asthma
Asthmatic symptoms related to the inhalation of fine particles generated in the grinding of metals. Also called *grinder's rot*.

grip
(1) To hold firmly; *see also* ***grasp***. (2) That portion of a tool or other device which is normally held by the operator for carrying or operating the tool. The grip design typically attempts to conform to the shape of the hand and fingers.

grip diameter, inside
The diameter of the widest level of a cone which an individual can grasp with his/her thumb and middle finger (digit III) touching. Measured at the level of the thumb crotch.

grip diameter, outside
The linear distance between the joint of the 1st and 2nd phalanges of the thumb and the metacarpal-phalangeal joint of the middle finger (digit III). Measured with the hand held around a cone at the widest level at which the thumb and middle finger (digit (III) can still touch.

grip strength
The amount of force which may be applied when grasping or squeezing an object under specified conditions.

gristle
See ***cartilage***.

grit
Coarse nuisance dust particles that are larger than 75 microns in diameter.

groin
That region between the thighs at the apex of the pubic crotch.

grooving
The practice of designing a tool with grooves to accommodate the user's fingers.

gross adjustment
See ***primary positioning movement***.

gross alpha particle activity
The total activity, commonly measured in picocuries, due to emission of alpha particles. Generally used as a screening measurement for naturally occurring radionuclides.

gross anatomy
That portion of anatomy which involves the bodily features apparent to the naked eye.

gross axle weight rating (GAWR)
Value specified by the vehicle manufacturer as the load carrying capacity of a single axle system, as measured at the tire-to-ground interfaces.

gross beta particle activity
The total activity, commonly measured in picocuries, due to emission of beta particles. Used as a screening measurement for human-made radionuclides.

gross body coordination
The ability to integrate motion of the body segments while the entire body is in motion.

gross body equilibrium
A measure of the ability to retain or acquire one's balance, regardless of bodily position or motion.

gross combination weight (GCW)
The maximum allowable fully laden weight of a tractor and its trailer(s).

gross combination weight rating (GCWR)
The value specified by the manufacturer as the loaded weight of a combination (articulated) vehicle. In the absence of a value specified by the manufacturer, GCWR will be determined by adding the Gross Vehicle Weight Rating (GVWR) of the power unit and the total weight of the towed unit and any load thereon.

Gross Domestic Product (GDP)
The total value of goods and services produced by labor and property located in the United States. As long as the labor and property are located in the United States, the supplier (that is, the workers and, for property, the owners) may be either U.S. residents or residents of foreign countries.

gross head
A dam's maximum allowed vertical distance between the upstream's surfacc water (headwater) forebay elevation and the downstream's surface water (tailwater) elevation at the tail-race for reaction wheel dams or the elevation of the jet at impulse wheel dams during specified operation and water conditions.

gross horsepower
The power of a basis engine at a specified revolution per mile without alternator, water pumps, fan, etc. Gross horsepower is the figure commonly given as the horsepower rating of an engine.

gross metabolic cost
The total amount of energy expended to perform some specific activity. *See also **net metabolic cost**.*

Gross National Product (GNP)
A measure of monetary value of the goods and services becoming available to the nation from economic activity. Total value at market prices of all goods and services produced by the nation's economy. Calculated quarterly by the Department of Commerce, the Gross National Product is the broadest available measure of the level of economic activity.

gross registered tonnage (GRT)
(1) The capacity of a vessel in cubic feet of the spaces within the hull and of the enclosed spaces above the main deck available for cargo, stores, and crew, divided by 100. (2) The gross registered tonnage of a vessel according to the country of registry.

gross ton mile
The number of tons behind the locomotive (cars and contents, company service equipment, and cabooses) times the distance moved in road freight trains.

gross tonnage
The gross tonnage of a vessel is the internal cubic capacity of all spaces in and on the vessel which is permanently enclosed, with the exception of certain permissible exemptions. It is expressed in tons of 100 cubic feet.

gross vehicle weight (GVW)
(1) The maximum allowable weight in pounds or tons that a truck is designed to carry. (2) The weight of the empty vehicle plus the maximum anticipated load weight.

gross vehicle weight rating (GVWR)
(1) The maximum loaded weight in pounds of a single vehicle. Vehicle manufacturers specify the maximum gross vehicle weight rating (GVWR) on the vehicle certification label. (2) The maximum rated capacity of a vehicle, including the weight of the base vehicle, all added equipment, driver and passengers, and all cargo loaded into or on the vehicle. Actual weight may be less than or greater than GVWR.

gross weight
(1) Entire weight of goods, packing, and container ready for shipment. (2) The weight of a packaging plus the weight of its contents. *See also **net weight**.*

ground
(1) *General.* The surroundings of a figure or object which are perceived as behind or not belonging directly to the figure or object of interest. Also, the surface of the earth. (2) *Electricity.* A conducting connection, whether intentional or accidental, between an electrical circuit or equipment and the earth, or to some conducting body that serves in place of the

earth. *See also* ***electrical ground***. (3) To restrict from certain activities, especially flying.

ground blizzard
See ***blizzard***.

ground controlled approach (GCA)
Aviation. A radar approach system operated from the ground by air traffic control personnel transmitting instructions to the pilot by radio. The approach may be conducted with airport surveillance radar (ASR) only or with both surveillance and precision approach radar (PAR). Usage of the term "GCA" by pilots is discouraged except when referring to a Ground Controlled Approach (GCA) facility. Pilots should specifically request a "PAR" approach when a precision radar approach is desired or request an "ASR" or "surveillance" approach when a non-precision radar approach is desired. *See also* ***airport surveillance radar*** *and* ***precision approach radar***.

ground cover
Plants grown to keep soil from eroding.

ground current
Any current passing to or through the earth from electrical equipment.

ground delay
The amount of delay attributed to Air Traffic Control (ATC), encountered prior to departure, usually associated with a Controlled Departure Time (CDT) program.

ground-fault circuit-interrupter (GFCI)
A device whose function is to interrupt the electric circuit to the load when a fault current to ground exceeds some predetermined value that is less than that required to operate the overcurrent protective device of the supply circuit.

ground-fault interrupter (GFI)
See ***ground-fault circuit interrupter***.

ground fog
See ***radiation fog***.

ground potential
See ***electrical ground***.

ground property, equipment and other
The total cost of ground property and equipment and land.

ground speed
The speed of an aircraft relative to the surface of the earth.

ground stop
Aviation. Normally, the last initiative to be utilized; this method mandates that the terminal facility will not allow any departures to enter the Air Route Traffic Control Center (ARTCC) airspace until further notified.

ground state
The lowest energy level of an atom.

ground surface
The land surface of the earth, both exposed and underwater.

ground visibility
Prevailing horizontal visibility near the earth's surface as reported by the United States National Weather Service or an accredited observer.

grounded conductor
A system or circuit conductor that is intentionally grounded.

grounded, effectively
See ***effectively grounded***.

grounding
(1) *Electronics.* The practice of eliminating the difference in voltage potential between an object and ground. Procedure involves connecting the object to an effective ground (metal to metal) by an appropriate wire. (2) *Maritime Navigation.* Running aground of a vessel, striking or pounding on rocks, reefs, or shoals; stranding.

grounding conductor
A conductor used to connect equipment or the grounded circuit of a wiring system to a grounding electrode or electrodes.

grounding conductor, equipment
The conductor used to connect the non-current-carrying parts of equipment, raceways, and other enclosures to the system grounded conductor and/or the grounding electrode conductor at the service equipment or at the source of a separately derived system. *See also* ***grounding electrode conductor***.

grounding electrode conductor
The conductor used to connect the grounding electrode to the equipment grounding conductor and/or to the grounded conductor of the circuit at the service equipment or at the source of a separately derived system.

groundwater
The supply of fresh water found beneath the Earth's surface, usually in aquifers, which is often used for supplying wells and springs. Because groundwater is a major source of drinking water, there is a growing concern over areas where leaching agricultural or industrial pollutants or substances from leaking underground storage tanks are contaminating groundwater.

group
Two or more persons having some common relationship or interest.

group II railroad
Railroads, excluding Class I, with an annual accumulation of over 400,000 employee hours worked.

group dynamics
The interactions between the members of a group or their functioning as a unit.

group incentive plan
An incentive plan under which a number of workers are collectively rewarded based on the results of the entire group's behavior.

group technology
A concept which holds that the similarities of part geometric shapes or processes can be grouped to reduce manufacturing costs.

growing degree-day
A form of the degree-day used as a guide for crop planting and for estimating crop maturity dates.

growth
(1) An increase in the number of cells and/or cell size. (2) An expansion in consciousness or value.

growth curve
A graphic representation of the pattern of increase in some measure.

growth rate
A measure of the rapidity in some aspect of individual or entity growth.

GRT
See ***gross registered tonnage****.*

guard
(1) A person whose primary function is to restrict entry to a certain facility and observe that facility for hazards or violations. (2) A physical device to prevent undesired contact with a source of energy between people, equipment, materials, and the environment. (3) Any structure designed to restrict or limit entry into some hazardous region of a piece of equipment for preventing injuries.

guard rail
Transit. A strong fence or barrier to prevent vehicles from leaving the roadway, or for people's safety. *See also* ***guardrail****.*

guard wall
The river wall of a lock which prevents boats from being drawn into the dam.

guarded
Covered, shielded, fenced, enclosed, or otherwise protected by means of suitable covers, casings, barriers, rails, screens, mats, or platforms to remove the likelihood of approach to a point of danger or contact by persons or objects.

guardrail
A rail secured to uprights and erected along the exposed sides and ends of platforms to protect persons from otherwise exposed openings through which they may fall.

GUI
See ***Graphical User Interface****.*

guide wall
The extension of the inner lock wall on the upper and lower side of the lock chamber to assist navigators in guiding vessels or tows into the lock chamber. It is usually 600 feet in length, although some are now 1,200 feet long.

guideline
A recommended practice or other non-binding suggestion issued by an agency, without the force of law. Contrasted to a *regulation*.

guideway
In transit systems, a track or other riding surface (including supporting structure) that supports and physically guides transit vehicles specifically designed to travel exclusively on it.

Gulf Intracoastal Waterway
The system of that name extending from St. Marks, Florida to Brownsville and Harlingen, Texas and including the Pearl River, Tombigbee River, Apalachicola River, Flint River,

and such other navigable tributaries to which barge operations extend.

Gulf of Mexico and its inlets
The waters from the mean high water mark of the coast of the Gulf of Mexico and its inlets open to the sea (excluding rivers, tidal marshes, lakes, and canals) seaward to include the territorial sea and Outer Continental Shelf (OCS) to a depth of 15 feet, as measured from the mean low water.

Gulf Stream
A warm, swift, narrow ocean current flowing along the east coast of the United States.

gum
See ***gingiva***.

gum ball machine
Transit (slang). The rotating warning light on top of an emergency vehicle.

gust front
A boundary that separates a cold downdraft of a thunderstorm from warm, humid surface air. On the surface its passage resembles that of a cold front.

gustation
The sense of taste.

gut
The intestines.

GVW
See ***gross vehicle weight***.

GVWR
See ***gross vehicle weight rating***.

GW
See ***gigawatt***.

GWE
See ***gigawatt electric***.

GWH
See ***gigawatt hour***.

gypsy
Transit (slang). (1) An independent truck operator who drives his own truck and secures freight wherever he can. (2) One who trip-leases to authorized carriers.

gyre
A large, circular, surface ocean current pattern.

gyrodyne
A rotorcraft whose rotors are normally engine driven for takeoff, hovering, and landing, and for forward flight through part of its speed range, and whose means of propulsion, consisting usually of conventional propellers, is independent of the rotor system.

gyroplane
A rotorcraft whose rotors are not engine driven, except for initial starting, but are made to rotate by action of the air when the rotorcraft is moving; and whose means of propulsion, consisting usually of conventional propellers, is independent of the rotor system. *See also* ***helicopter***.

H

H point
The mechanically hinged hip point of a manikin which simulates the actual pivot center of the human torso and thigh, described in Society of Automotive Engineers (SAE) Recommended Practice J826, "Manikins for Use in Defining Vehicle Seating Accommodations," November 1962.

habeas corpus
(*Latin*). Meaning "you have the body."

Haber's Rule
States that a toxic effect is dependent upon the product of exposure time and the contaminant concentration. Thus, exposure at a higher concentration for a short period would be equivalent to exposure at a lower concentration for a longer period in direct proportion to the product of exposure concentration and time. This reportedly, however, holds only for short exposure periods. Also referred to as *Haber's Law.*

habilitate
Bring to an initial state of fitness or capability, as in overcoming a congenital handicap. *See also* ***rehabilitate***.

habit
An acquired, well-practiced behavior pattern which is carried out with minimal or no conscious direction.

habitability
A measure of the interaction quality of an individual or group with their physical, social, and psychological environment to produce certain working and living conditions.

habitable volume
That volume which is suitable for living, containing breathable air and necessary or reasonable accommodations.

habitat
The place where population (e.g., human, animal, plant, microorganism) lives and its surroundings, both living and non-living.

habitual criminal
A legal category created by statute in many states by which more severe penalties can be imposed on offenders who have multiple felony convictions. The criminal history of a defendant is an important factor in imposing sentence under federal sentencing guidelines.

habituation
A decline in response or conscious sensitivity to repeated or maintained exposure to one or more environmental stimuli.

habituation error
The tendency to keep making the same response, even if the stimulus or conditions change.

haboob
A dust or sandstorm that forms as cold downdrafts from a thunderstorm turbulently lift dust and sand into the air.

habutai
A soft, lightweight, plain weave silk.

hacking
A massaging technique in which the medial edge of the open hand is brought repeatedly against the body surface.

Hadley cell
A thermal circulation pattern first proposed by George Hadley to explain the movement of the trade winds. It consists of rising air near the equator and sinking air near 30° latitude.

hafnium
A chemical element, atomic number 72, atomic weight 178.49, symbol Hf.

hahnium
A chemical element, atomic number 105, atomic weight 260, symbol Ha.

hailstones
Transparent or partially opaque particles of ice that range in size from that of a pea to that of golf balls. Some may even reach larger proportions, such as that of a softball, but such development is very uncommon.

hair
(1) The collective hair shafts growing in various portions of the body, such as the scalp, face, or pubic region. (2) A single keratinized shaft growing from a hair root within the skin. (3) A threadlike structure, especially the specialized epidermal structure developing from a papilla sunk in the corium, produced only by mammals and characteristic of that group of animals.

hair esthesiometer
A device developed by von Frey to determine skin touch sensitivity, consisting of a filament attached to some type of holder. *See also* ***von Frey filament***.

hair follicle
That structure surrounding the root of a hair in the skin.

hair hygrometer
See ***hygrometer***.

hairball
A concentration of hair sometimes found in the stomach or intestines of man or other animals.

HAL
See ***height above landing***.

half-life
(1) *Biological.* The time required for the body to eliminate, by natural biological means, half of the material taken into it. (2) *Effective.* The time in which the quantity of a radioactive isotope in the body will decrease to half as a result of *both* radioactive decay and biological elimination. (3) *Radioactive.* The time for the activity of a given radioactive isotope to decrease to half of its initial value, due to radioactive decay. The half-life is a characteristic property of each radioactive isotope and is independent of its amount or condition.

half-mask respirator
Respirator which covers half the face, from the bridge of the nose to below the chin.

half-thickness
See ***half-value layer***.

half-value layer (HVL)
The thickness of a specified material which, when introduced into the path of a given beam of ionizing radiation, reduces the exposure rate by one-half. Also referred to as the *half-thickness*.

halfway-to-hip circumference
The surface distance around the torso at a level midway between the waist height and the trochanteric height levels. Measured with minimal tissue compression.

halide
A compound of a halogen with an element or radical.

halide meter
An instrument used for the direct measurement of halogenated hydrocarbons.

halitosis
A condition in which one's breath is offensive to others.

hallucination
An apparently real sensory perception (auditory or visual) without any real external stimuli to cause it; commonly experienced by psychotics.

hallucinogen
A psychedelic agent; a compound that produces changes in perception, thought, or mood without causing major disturbances in the nervous system (autonomic). An example is LSD.

halo
A ring or arc that appears to encircle the sun or moon when seen through an ice crystal cloud or a sky filled with falling ice crystals. Halos are produced by refraction of light.

halo effect
A tendency for an evaluator to be overly influenced by an individual's ratings on one trait or due to some past outstanding achievement.

halogen
Any of a group of five chemically related nonmetallic elements that includes bromine, fluorine, chlorine, iodine, and astatine.

halogenation
The process whereby halogens are used for disinfecting purposes.

Halon 1211
A colorless, faintly sweet smelling, electrically nonconductive liquefied gas, chemical formula CBrC1F(2), which is a medium for extinguishing fires by inhibiting the chemical chain reaction of fuel and oxygen. It is also known as bromochlorodifluoromethane.

Halon 1301
A colorless, odorless, electrically nonconductive gas, chemical formula CBrF(3), which is a medium for extinguishing fires by inhibiting the chemical chain reaction of fuel and oxygen. It is also known as bromotrifluoromethane.

halons
Bromine-containing compounds, normally used in firefighting methodologies, with long atmospheric lifetimes whose breakdown in the stratosphere is thought to cause ozone depletion.

halothane
A colorless, mobile, non-flammable, heavy liquid used by inhalation to produce anesthesia.

hamarthritis
Arthritis of all the joints.

hamate bone
One of the distal group of bones on the wrist.

Hamman's disease
Spontaneous interstitial emphysema of the lungs.

Hamman-Rich syndrome
Diffuse interstitial pulmonary fibrosis.

hammer provision
Common term for the automatic promulgation of required standards by Congress, usually invoked when a regulatory agency such as the EPA fails to promulgate the required standard by an established deadline.

hammermill
A high-speed machine that hammers and cutters use to crush, grind, chip, or shred solid wastes.

hamstring
The tendon for the hamstring muscles.

hamstring muscles
A group of muscles in the posterior thigh, consisting of the biceps femoris, semitendinosus, and semimembranosus muscles. Sometimes simply referred to as *hamstrings*.

hand
(1) *Measurement.* A measure of length equal to four inches, usually used in measuring the height of horses. (2) *Anatomy.* The metacarpal and phalangeal bones and other associated tissues normally existing distal to the wrist. *See also* ***metacarpal bone***.

hand-arm
Involving both the hand and the arm, generally pertaining to sensory or other external influences on both the hand and the arm. *See also* ***arm-hand***.

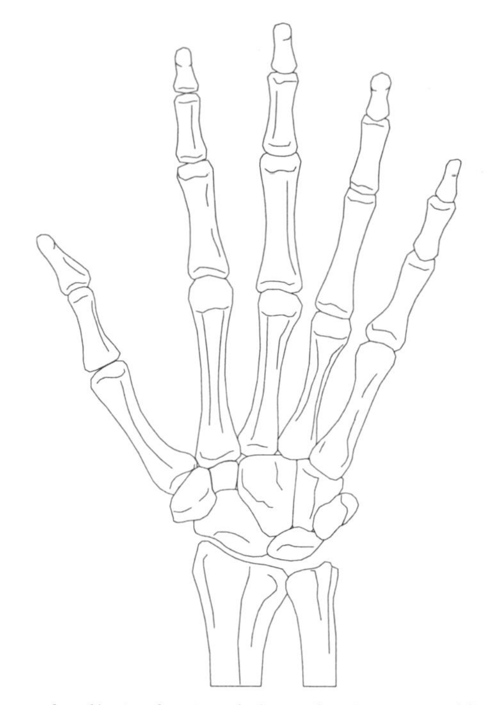

Human hand/wrist showing phalangeal and metacarpal bones

hand breadth, metacarpal
The maximum linear width of the hand across the distal ends of the metacarpal bones. Measured with the fingers extended and adducted.

hand breadth, thumb
The maximum width of the hand at the level of the distal end of the first metacarpale of the thumb. Measured with the fingers extended and adducted, and the thumb adducted to the side of the plan.

hand circumference
The surface distance around digits II – V at the metacarpal-phalangeal level. Measured with the hand flat and the fingers extended.

hand circumference, overthumb
The surface distance around the hand, in a plane at right angles to the long axis of the hand, passing over the metacarpals and the metacarpal-phalangeal joint of the thumb; also hand circumference including thumb. Measured with the hand flat, the fingers extended, and the thumb aligned with the index finger.

hand control
Any control on a panel or other structure which is used for controlling some process and is normally designed for positioning by the hand.

hand feed
That portion of a machine at which the materials or operating portion are fed for processing at a pace determined by the worker.

hand-held drench shower
A flexible hose connected to a water supply and used to irrigate and flush eyes, face, and body areas in the event of contact with a hazardous material that is corrosive, irritating, absorbed through the skin, etc.

hand hole
A slot in the side or end of a container used for carrying items.

hand length
The linear distance from the plane where the base of the hand/thumb joins the wrist in the fleshy tip of the middle finger (digit III) parallel to the longitudinal axis of the hand. Measured with the fingers extended and adducted, the wrist rotated/supinated into the anatomical position.

hand-operated switch
A non-interlocked switch which can only be operated manually.

hand protection
Gloves, or other type hand protection which will prevent the harmful exposure of the wearer to hazardous materials.

hand rail
A single bar or pipe supported on brackets from a wall or partition, as on a stairway or ramp, to furnish persons with a handhold in case of tripping.

hand steadiness
A measure of the ability to sustain a fixed position of the hand and/or finger with minimal tremor. Also referred to as *manual steadiness*. *See also* ***arm-hand steadiness***.

hand thickness, metacarpale III
The thickness of the metacarpo-phalangeal joint of the middle finger (digit III). Measured with the hand flat, fingers extended and adducted.

hand tool
Any small tool capable of being held and used easily by one or both hands for manufacturing, servicing, or other activities.

handcontroller
A small device, usually grasped by or fitting the hand, which responds to axial and/or rotational movements for allowing an operator to control a larger/stronger/ remote system.

handedness
A preference for using one arm-hand or the other, or a combination of the two.

handhold
A structure consisting of a segment which normally is an elliptical- or rod-shaped cross-section and of suitable outside perimeter and length to permit a hand to grasp it for carrying, for assistance in remaining in a desired position, or for mobility.

handicap
(1) A compensating factor which attempts to equalize performance levels on one or more aspects in some activity. (2) A physical or mental condition which prevents an individual from functioning at a normal performance level, especially referring to those functions such as activities of daily living.

handle
(1) A structure designed for gripping an object. (2) To move an object or material from one location to another, via a suitable combination of motions.

handling
Physically moving cargo between point-of-rest and any place on the terminal facility, other than the end of ship's tackle.

handling aid
See ***job aid***.

handling structures
Structures such as beams, plates, channels, angles, and rods assembled with bolts, pins, and/or welds. Includes lifting, supporting, and manipulating equipment such as lifting beams, support stands, spin tables, rotating devices, and fixed and portable launch support frames.

handling time
The period of time required to move parts or materials to or from a work area or operation.

handoff
Aviation. A Controller action taken to transfer the radar identification of an aircraft from one Controller to another if the aircraft will

enter the receiving Controller's airspace and radio communications with the aircraft will be transferred.

handwear
Any form of clothing worn over the hand.

handwheel
A large control device intended for rotation when a mechanism requires a greater amount of torque than can be applied by a knob.

handy line
A small line used to throw between separated barges or boat and shore, (i.e., heaving line).

hanging wall
Of the two sides of a fault, the side above the fault plane. It is called the hanging wall because where faults have been "filled in" with mineral deposits and then mined, this is the side on which miners can hang their lanterns. *See also* ***footwall***.

hangnail
A shred of epidermis at one side of a nail. Hangnail is prevented by gently pushing the cuticle instead of cutting it, and it is treated by clipping off the shred of skin and applying antiseptic to the area to prevent infection.

HANS™
See ***Head And Neck Support***.

happiness sheet
(slang) A written survey obtained from students at the end of a course or training session to provide feedback to the instructor regarding various aspects of the training.

HAPS
Hazardous air pollutants.

haptic
Pertaining to the sensation of pressure.

haptophore
Anatomy. The specific group of atoms in a toxin molecule by which it attaches itself to another molecule. It is capable of neutralizing antitoxin and of acting as an antigen to stimulate specific antitoxin production by body cells.

harassment of wildlife
Any act of pursuit, torment, or annoyance which has the potential to injure wildlife or has the potential to disturb wildlife by causing disruption of behavioral patterns, including, but not limited to, migration, breathing, nursing, breeding, feeding, or sheltering.

harbor
An area of water where ships, planes, or other watercraft can anchor or dock. *See also* ***port***.

harbor line
See ***permit line***.

harbor master
Maritime Safety. An officer who attends to the berthing, etc. of ships in a harbor.

hard hat
A safety helmet maintained in position on the head by straps, for protecting the wearer from being injured by falling objects.

Typical hard hat (or "safety helmet")

hard light
A light source which causes objects to cast well-defined shadows.

hard palate
The anterior portion of the roof of the mouth, backed by the maxilla and palatine bones and covered by mucous membranes. *See also* ***soft palate***.

hard soap
Any soap made with sodium hydroxide and packaged in bar form.

hard water
Alkaline water containing dissolved salts that interfere with some industrial processes and prevent soap from lathering.

hardcopy
A paper or other sheeted material display.

hardware
Computers. The physical equipment used in processing. The equipment or components made of physical materials, often referring to

electronics and structural portion of a computer.

hardwired
A system in which there is a direct connection of components by electrical wires or cables.

harelip
Congenitally cleft lip.

harmful
Term indicating the potential for an agent or condition to produce injury or an adverse health effect.

harmful quantities of oil discharge
Any discharge that violates a water quality standard or causes a film or sheen upon the surface of the water.

harmful quantity
With regard to oil and hazardous substances, those which may be harmful to the public health or welfare and includes harm to fish, shellfish, wildlife, and public and private property, shorelines, and beaches. EPA further defines a hazardous quantity of oil as an amount which either violates applicable water quality standards, or causes a surface film or sheen, or a discoloration of the water or adjoining shoreline.

harmless error doctrine
The doctrine that minor or harmless errors during a trial do not require reversal of the judgement by an appellate court. An error which is trivial or formal or merely academic and was not prejudicial to the substantial rights of the party assigning it, and in no way affected the final outcome of the case.

harmonic
In acoustics, a tone in the harmonic series of overtones that are produced by the fundamental tone. A frequency component at a frequency that is an integer multiple of the fundamental frequency. *See also **harmonic series**.*

harmonic motion
*See **simple harmonic motion**.*

harmonic series
A set of overtones whose frequencies are separated by integral multiples of the fundamental frequency.

harmonic vibration
*See **simple harmonic motion**.*

Harmonized System of Codes
An international goods classification system for describing cargo in international trade under a single commodity coding scheme. This code is a hierarchically structured product nomenclature containing approximately 5,000 headings and subheadings.

harness
Any combination of straps intended to hold an occupant of a vehicle in his/her seat, especially those straps holding the torso against the seatback.

Harrison Antinarcotic Act
A federal law, enacted March 1, 1915, that regulates the possession, sale, purchase, and prescription of opium and cocoa and all their preparations, natural and synthetic derivatives, and salts. These include the drugs cocaine, morphine, codeine, and papaverine. Laws patterned after the Harrison Antinarcotic Act in some states prohibit the possession or sale of derivatives of barbituric acid except under proper licenses, so that they may not be dispensed without a prescription.

Hashimoto's disease
A progressive disease of the thyroid gland with degeneration of its epithelial elements and replacement by lymphoid and fibrous tissue. Also called *struma lymphomatosis*.

hashish
The stalks and leaves of the hemp plant, cannabis, with narcotic properties similar to those of marijuana.

hat
Any head covering made largely of soft materials, but having a rigid shape.

HAT
*See **height above touchdown**.*

hatch
A full-body or materials passageway through some solid structure which may be sealed to separate different fluids or pressures.

haulage cost
Cost of loading ore at a mine site and transporting it to a processing plant.

hauling post holes
Transit (slang). Driving an empty truck or trailer.

have numbers
Aviation. Communication protocol. Term used by pilots to inform Air Traffic Control (ATC) that they have received runway, wind, and altimeter information only.

HAVS
Hand arm vibration syndrome.

hawser
A tow line.

Hawthorne effect
A phenomenon in which employee-perceived interest by the employer proved to be a factor in productivity and employee moral. Specifically, intentional variations in physical work environment variables (such as lighting, ventilation, noise, etc.) were examined in an experiment and resulted in conclusions that were opposite to those expected (i.e., employees worked harder when environmental conditions were made worse) thereby confounding the experiment. Apparently, it did not matter what the employer did to affect the environmental conditions. The employees viewed these variations in environmental conditions as interest by the employer in their work. They therefore worked harder to impress their employers. This phenomenon is based on a study conducted at the Western Electric Company Hawthorne Works plant in Chicago, Illinois. It is often generalized to apply to confounded results from unconsidered variables in experiments.

hay fever
An allergy characterized by sneezing, itchy and watery eyes, running nose, and burning palate and throat. Like all allergies, hay fever is caused by sensitivity to certain substances—most commonly pollens and the spores of molds. Pollen is the fertilizing element of flowering plants. It is a fine dust, easily airborne, that enters the body by inhalation. Hay fever deserves to be recognized as more than a mere nuisance. By causing lack of sleep and loss of appetite, it can lower the body's resistance to disease. It can cause inflammation of the ears, sinuses, throat, and bronchi. A number of hay fever sufferers develop asthma.

hazard
(1) *General.* A risky, perilous, or dangerous condition or situation that could result in the exposure of individuals to unnecessary physical or health risks. Hazards can be biological, chemical, physical, mechanical, human-made, or naturally occurring. (2) *Safety.* A dangerous condition, potential or inherent, that can interrupt or interfere with the expected orderly progress of an activity. It is any real or potential condition which either has previously caused or could reasonably be expected to cause personal injury or property damage. (3) *System Safety.* A condition or situation that exists within the working environment capable of causing an unwanted release of energy resulting in physical harm, property damage, or both. (4) *Toxicology.* Potential for harm to humans or what they value. For toxic substances, hazard refers to the probability that injury will occur under stated sets of exposure conditions. (5) *Industrial Hygiene.* A material poses a hazard if it is likely that an individual will encounter a harmful exposure to it. Hazard is the estimated potential of a chemical, physical agent, ergonomic stress, or biologic organism to cause harm based on the likelihood of exposure, the magnitude of exposure, and the toxicity or effect. (6) *Law.* A risk or peril, assumed or involved, whether in connection with contract relation, employment, personal relation, sport, or gambling. A danger or risk lurking in a situation which by change or fortuity develops into an active agency of harm. Exposure to the chance of loss or injury. (7) *Insurance.* The risk, danger, or probability that the event insured against may happen, varying with the circumstances of the particular case.

hazard analysis
The analysis of systems, processes, and/or procedures to determine potential hazards and recommended actions to eliminate or control those hazards.

hazard and operability study (HAZOP)
A formal, structured investigative system for examining potential deviations of operations from design conditions that could create process-operating problems and hazards.

hazard classification
Designation of relative accident potential based on the likelihood that an accident will occur.

Hazard Communication Standard
A regulatory requirement, as promulgated by OSHA (29 CFR 1910.1200) that establishes requirements for the evaluation of the hazards of chemicals used in industry, labeling of chemical containers, preparation of material safety data sheets (MSDS), training of employees and the provisions of employee access to information about the potential hazards of chemicals they handle. Also known as "Right to Know" or "Worker Right to Know."

hazard correction
The elimination or control of a workplace hazard in accord with the requirements of applicable federal or state statutes, regulations, or standards.

hazard elimination
The removal of a known, already existing hazard.

hazard identification
*See **hazardous identification**.*

hazard pay
*See **hazardous duty pay**.*

hazard probability
A measure of the likelihood that a condition or set of conditions will exist or occur in a given situation or operating environment.

hazard proof
A method of making electrical equipment safe for use in hazardous locations; these methods include explosion proofing, intrinsically safe, purged, pressurized, and nonincendiary, and must be rated for the degree of hazard present.

hazard quotient
The ratio of a single substance exposure level over a specified period of time (e.g., chronic) to a reference dose for that substance derived from a similar exposure period.

hazard ranking system
*See **hazardous ranking system**.*

hazard recognition
In terms of OSHA compliance, a concept based upon the premise that hazardous conditions cannot be eliminated or controlled until they are first recognized as such by the employer. An important concept since employers cannot be held in violation of a requirement if they did not recognize that the hazardous condition existed.

hazard risk index
*See **risk assessment code**.*

hazard severity
A categorical description of hazard level or degree, based upon real or perceived potential for resulting in harm, injury, and/or damage caused by a given hazard condition.

hazard to navigation
Marine Safety. For the purpose of 49 CFR 195, a pipeline where the top of the pipe is less than 12 inches below the seabed in water less than 15 feet deep, as measured from the mean low water.

hazard warning signal
Transit. Lamps that flash simultaneously to the front and rear, on both the right and left sides of a commercial motor vehicle, to indicate to an approaching driver the presence of a vehicular hazard.

hazard zone
(1) *DOT.* One of four levels of hazard (Hazard Zones A through D) assigned to gases, as specified in 49 CFR 173.116(a), and one of two levels of hazards (Hazard Zones A and B) assigned to liquids that are poisonous by inhalation, as specified in 49 CFR 173.133(a). A hazard zone is based on the LC_{50} value for acute inhalation toxicity of gases and vapors, as specified in 49 CFR 173.133(a). (2) *Maritime Navigation.* An area identified as a danger to maritime navigation.

hazardous
Exposed to or involving danger; perilous; risky; involving risk of loss.

hazardous air pollutants
Air pollutants which are not covered by ambient air quality standards but which, as defined in the Clean Air Act, may reasonably be expected to cause or contribute to irreversible illness or death. Such pollutants include asbestos, beryllium, mercury, benzene, coke oven emissions, radionuclides, and vinyl chloride.

Hazardous and Solid Waste Amendments of 1984 (HSWA)
A set of statutory amendments that expanded and strengthened the Resource Conservation and Recovery Act (RCRA) of 1976.

hazardous area reporting service

Flight monitoring for Visual Flight Rules (VFR) aircraft crossing large bodies of water, swamps, and mountains. This service is provided for the purpose of expeditiously alerting Search and Rescue (SAR) facilities when required. Radio contacts are desired at least every 10 minutes. If contact is lost for more than 15 minutes, SAR will be alerted.

hazardous assessment

Help to define the potential adverse health or environmental effects associated with chemicals on site, the potential magnitude to exposure, and the frequency of exposure.

hazardous atmosphere

(1) *General.* Any atmosphere which is oxygen deficient or contains toxic or other types of health hazards at concentrations exceeding established exposure limits. It is also considered to be an atmosphere that may expose personnel to the risk of death, incapacitation, or impairment of one's ability for self-rescue, injury, or illness. (2) As pertains to *confined spaces,* an atmosphere that may expose employees to the risk of death, incapacitation, impairment of ability to self-rescue (that is, escape unaided from a permit space), injury, or acute illness from one or more of the following causes:

(1) Flammable gas, vapor, or mist in excess of 10 percent of its lower flammable limit (LFL).

(2) Airborne combustible dust at a concentration that meets or exceeds it LFL. This concentration may be approximated as a condition in which the dust obscures vision at a distance of 5 feet (1.52 m) or less.

(3) Atmospheric oxygen concentration below 19.5% or above 23.5%.

(4) Atmospheric concentration of any substance for which a dose or a permissible exposure limit is published in Subpart G, *Occupational Health and Environmental Control,* or in Subpart Z, *Toxic and Hazardous Substances,* of OSHA 29 CFR 1910 and which could result in employee exposure in excess of its dose or permissible exposure limit. An atmospheric concentration of any substance that is not capable of causing death, incapacitation, impairment of ability to self-rescue, injury, or acute illness due to its health effects is not covered by this provision.

(5) Any other atmospheric condition that is immediately dangerous to life or health. For air contaminants for which OSHA has determined a dose or permissible exposure limit, other sources of information, such as material safety data sheets that comply with the hazard communication standard (OSHA 29 CFR 1910.1200), published information, and internal documents can provide guidance in establishing acceptable atmospheric conditions.

hazardous chemical

According to OSHA (29 CFR 1910.1200), any chemical that is a health or physical hazard and for which there is statistically significant evidence that acute or chronic health effects may occur in exposed individuals.

hazardous condition

Circumstances which are causally related to an exposure to a hazardous material.

hazardous employment

High risk and extra perilous work. When used in the context of workers' compensation, it refers to employment which requires the employer to carry workers' compensation coverage or its equivalent regardless of the number of employees.

hazardous duty pay

The additional monetary compensation given to workers performing dangerous tasks.

hazardous goods

The categories of hazardous goods carried by inland waterways are those defined by the European Provisions concerning the International Carriage of Dangerous Goods by Inland Waterways.

hazardous identification

(1) The identification of those chemicals that may pose a threat to human health or the environment. (2) An initial evaluation of media (air, water, soil, etc.) that may be contaminated and the chemicals that are most likely to present a public health threat.

Hazardous In-flight Weather Advisory Service (HIWAS)

Continuous recorded hazardous in-flight weather forecasts broadcasted to airborne pilots over selected very high frequency omnidirectional (VOR) outlets defined as an HIWAS BROADCAST AREA.

hazardous insurance
Insurance effected on property which is in unusual or peculiar danger of destruction by fire, or on the life of a person whose occupation exposes him/her to special or unusual perils.

hazardous LBB
A pressure vessel that exhibits a *leak before burst,* or *LBB,* failure mode and contains a hazardous material.

hazardous liquid
DOT. (1) Liquefied Natural Gas (LNG) or a liquid that is flammable or toxic. (2) Petroleum, petroleum products, or anhydrous ammonia.

hazardous material (HAZMAT)
(1) *General.* Any substance or compound that has the ability to produce an adverse health effect in a worker. (2) *DOT.* A substance or material which has been determined by the Secretary of Transportation to be capable of posing an unreasonable risk to health, safety, and property when transported in commerce, and which has been so designated. The term includes hazardous substances, hazardous wastes, marine pollutants, and elevated temperature materials as defined in this section, materials designated as hazardous under the provisions of 49 CFR 172.101 and 172.102, and materials that meet the defining criteria for hazard classes and divisions in 49 CFR 173. *See also **highly volatile liquid** and **marine pollutant**.*

hazardous material employee
A person who is employed by a HAZMAT employer and who in the course of employment directly affects hazardous materials transportation safety. This term includes an owner-operator of a motor vehicle which transports hazardous materials in commerce. This term includes an individual, including a self-employed individual, employed by a HAZMAT employer who, during the course of employment: a) loads, unloads, or handles hazardous materials; b) tests, reconditions, repairs, modifies, marks, or otherwise represents containers, drums, or packaging as qualified for use in the transportation of hazardous materials; c) prepares hazardous materials for transportation; d) is responsible for safety of transporting hazardous materials; or e) operates a vehicle used to transport hazardous materials.

hazardous material employer
A person who uses one or more of its employees in connection with transporting hazardous materials in commerce; causing hazardous materials to be transported or shipped in commerce; or representing, marking, certifying, selling, offering, reconditioning, testing, repairing, or modifying containers, drums, or packaging as qualified for use in the transportation of hazardous materials. This term includes an owner-operator of a motor vehicle which transports hazardous materials in commerce. This term also includes any department, agency, or instrumentality of the United States, a State, a political subdivision of a State, or an Indian tribe engaged in an activity described in the first sentence of this definition.

hazardous material residue
The hazardous material remaining in a packaging, including a tank car, after its contents have been unloaded to the maximum extent practicable and before the packaging is either refilled or cleaned of hazardous material and purged to remove any hazardous vapors.

Hazardous Material Transportation Act of 1974 (HMTA)
This statute provides the United States Department of Transportation with the authority to issue and enforce requirements for the packaging, labeling, and transporting of all hazardous materials including wastes. These requirements cover transportation by air, water, rail, or highway.

hazardous materials
(1) *General.* Liquids, gases, or solids that may be toxic, reactive, or flammable or that may cause oxygen deficiency either by themselves or in combination with other materials. (2) *DOT.* Any toxic substance, explosive, corrosive material, combustible material, poison, or radioactive material that poses a risk to the public's health, safety, or property when transported in commerce.

Hazardous Materials Incident Report System HMIRS
HMIRS contains hazardous material spill incidents reported to the Department of Transportation.

hazardous pressure systems
Systems used to store and transfer hazardous fluids such as cryogens, flammables, combustibles, hypergols, etc.

hazardous ranking system (HRS)
The principle screening tool used by the EPA to evaluate risks to public health and the environment associated with abandoned or uncontrolled hazardous wastes sites. The HRS calculates a score based on the potential of hazardous substances spreading from the site through the air, surface water, or groundwater and on other factors such as nearby populations. This score is the primary factor in deciding if the site should be on the National Priorities List and, if so, what ranking it should have compared to other sites.

hazardous ranking system factor
Primary rating elements internal to the hazardous ranking system (HRS). *See also **hazardous ranking system**.*

hazardous ranking system factor category
Set of HRS factors, such as likelihood of release (or exposure), waste characteristics, targets. *See also **hazardous ranking system**.*

hazardous ranking system migration pathways
HRS groundwater, surface water, and air migration pathways. *See also **hazardous ranking system**.*

hazardous ranking system pathway
Set of HRS factor categories combined to produce a score to measure relative risks posed by a site in one of four environmental pathways (i.e., groundwater, surface water, soil, and air). *See also **hazardous ranking system**.*

hazardous ranking system site score
A composite of the four HRS pathway scores. *See also **hazardous ranking system** and **hazardous ranking system pathway**.*

hazardous secondary materials
As defined by the Resource Conservation and Recovery Act (RCRA), any spent materials, sludges, byproductss, commercial chemical products, and scrap metals.

hazardous substance
(1) *General.* Any material that poses a threat to human health and/or the environment. Typical hazardous substances are toxic, corrosive, ignitable, explosive, or chemically reactive. (2) *EPA.* Any substance designated by the EPA to be reported if a designated quantity of the substance is spilled in the waters of the United States or if otherwise emitted to the environment. (3) *DOT.* A material, and its mixtures or solutions, that a) is identified in the appendix to 49 CFR 172.101; b) is in a quantity, in one package, which equals or exceeds the reportable quantity (RQ) listed in Appendix A to 49 CFR 172.101; and c) when in a mixture or solution which, for radionuclides, conforms to paragraph 6 of Appendix A, or, for other than radionuclides, is in a concentration by weight which equals or exceeds the concentration corresponding to the Reportable Quantity (RQ) of the material, as shown in the table appearing in 49 CFR 171.8. This definition does not apply to petroleum products that are lubricants or fuels.

hazardous waste
(1) Byproducts of society that can pose a substantial or potential hazard to human health or the environment when improperly managed. Possesses at least one of four characteristics (ignitability, corrosivity, reactivity, or toxicity), or appears on any special EPA list. (2) A hazardous material generated as the result of an industrial, research, commercial, domestic, or institutional process for which no intended further use or reuse is anticipated. (3) A solid waste or a combination of solid wastes which, because of its quantity, concentration, or physical, chemical, or infectious characteristics may cause, or significantly contribute to an increase in mortality or an increase in or pose a substantial present or potential hazard to human health or the environment when properly treated, stored, transported, or disposed of, or otherwise managed. (4) Any material that is subject to the hazardous waste manifest requirements of the Environmental Protection Agency (EPA) specified in 40 CFR 262 or would be subject to these requirements absent an interim authorization to a State under 40 CFR 123, subpart F.

hazardous waste management
According to the Federal Solid Waste Disposal Act: The systematic control of the collection, source separation, storage, transportation, processing, treatment, recovery, and disposal of hazardous wastes.

hazardous waste stream
Material containing hazardous substances, as defined by CERCLA, that are deposited, stored, disposed, or placed in, or that otherwise migrated to, a source.

hazards analysis
The procedure involved in identifying potential sources of release of hazardous materials from fixed facilities or transportation accidents; determining the vulnerability of a geological area to a release of hazardous materials; and comparing hazards to determine which present greater or lesser risks to a community.

hazards identification
(1) Providing information on which facilities have extremely hazardous substances, what those chemicals are, and how much there is at each facility. The process also provides information on how the chemicals are stored and whether they are used at high temperatures. (2) The process of determining whether or not exposure to an agent can cause an increase in the incidence of a particular adverse health effect (e.g., cancer, birth defect) and whether the adverse health effect is likely to occur in humans. *See also* ***hazardous identification***.

HazCAT
A procedure or set of procedures developed for systematically performing simple, on-site analytical tests for determining the chemical nature (generically) of unknown, potentially hazardous materials.

haze
(1) Fine dry or wet dust or salt particles dispersed through a portion of the atmosphere. Individually these are not visible but cumulatively they can diminish visibility. (2) A cloudiness in a surface or coating.

HAZMAT
See ***hazardous material***.

HAZOP
See ***hazard and operability study***.

Hb
See ***hemoglobin***.

HBAO
See ***high boiling aromatic oils***.

HbCO
See ***carboxyhemoglobin***.

HbO_2
See ***oxyhemoglobin***.

HBV
See ***Hepatitis B Virus***.

H_c
See ***hue composition***.

HCI
See ***human-computer interface***.

HCP
Hearing conservation program. *See* ***hearing conservation***.

HDL
See ***high density lipoprotein***.

HDTV
See ***high-definition television***.

head
(1) *Pressure Systems.* Term used for indicating pressure such as a head of one inch water gauge. (2) *Hydrology.* The product of the water's weight and a usable difference in elevation gives a measurement of the potential energy possessed by water. (3) *Military.* Marine restroom facility. (4) *Anatomy.* That part of the human body superior to the neck when standing erect, including the skull and facial bones, skin, brain, and other associated tissues. Also, a point of origin, as in a muscle.

Head And Neck Support (HANS™)
A head, neck, and upper torso restraint modeling system, consisting of a helmet and tethers, for minimizing neck injuries in a vehicular crash.

head breadth
The maximum linear side-to-side width of the head superior to the auricles. Measured at whatever level provides the maximum, with minimal tissue compression.

head circumference
The maximum surface distance around the head, including the hair, at a level just above, but not including, the brow ridges. Measured with hair compression. Also referred to as *occipitofrontal circumference*.

head diagonal, inion to pronasale
The linear distance from inion to provasale. Measured with the face and scalp muscles relaxed, without tissue compression.

head diagonal, maximum, menton to occiput
The maximum linear distance from menton to occiput. Measured with the face and scalp muscles relaxed, without tissue compression.

head diagonal maximum, nuchale to pronasale
The maximum linear distance from nuchale to pronasale. Measured with the face and scalp muscles relaxed, without tissue compression.

head-down display
A display, generally located on a control panel, which requires the operator to lower his/her normal line of sight to obtain the desired information.

head height
The vertical distance between tragion or the lowest point on the inferior orbit and the horizontal plane which intersects the vertex in the midsagittal plane. This uses a restricted definition of "head."

head impact area
Automotive Safety Design. All nonglazed surfaces of the interior of a vehicle that are statically contactable by a 6.5-inch diameter spherical head form of a measuring device having a pivot point to "top-of-head" dimension infinitely adjustable from 29 to 33 inches in accordance with the procedure explained in 49 CFR 390.5.

Head Injury Criterion (HIC)
A measure for determining the tolerance to concussion in a head impact, based on the duration and acceleration involved. An HIC value of 1000 with a duration of less than 15 msec is an acceptable tolerance.

$$HIC = [t_2 - t_1]\left[\frac{1}{t_2 - t_1}\int_{t_1}^{t_2} a(t)dt\right]^{2.5}$$

where:
t_1 = start time of impact
t_2 = end time of impact
$a(t)$ = acceleration function (in g units)

head lamps
Lamps used to provide general illumination ahead of a motor vehicle.

head length
See ***glabella – inion length***.

head length, maximum
The horizontal linear distance between pronasale and inion in the midsagittal plane.

head log
Maritime. The heavily reinforced section at each end of the barges and at the bow of the towboat to take the pressure of pushing the entire tow.

head-mounted display (HMD)
Any system which can be attached to the head, neck, and/or shoulders for enabling presentation of a head-up display.

head movement
Any motion of the head as a unit, relative to the torso.

head of bend
Maritime. The top or upstream beginning of a bend.

head of navigation
The furthest (upriver) location on a river deep enough for navigation.

head of passes
A point near the mouth of the Mississippi River where the three principal distributary passes diverge. It is the point from which river distances are measured.

head-on collision
(1) *General Transit.* Refers to a collision where the front end of one vehicle collides with the front-end of another vehicle while the two vehicles are traveling in opposite directions. (2) *Rail Operations.* A collision in which the trains or locomotives involved are traveling in opposite directions on the same track.

head on landing
Maritime. Landing in which the bow of the boat only is made fast.

head scan
Scan through the visual environment using head movements, allowing for accompanying eye movements.

head-up display (HUD)
A display in which information is presented on a nearby transparent surface such that the operator is capable of viewing both the infor-

mation and the external world with his/her normal line of sight.

headache
A pain or ache in the head. One of the most common ailments of man, it is a symptom rather than a disorder in itself. It accompanies many diseases and conditions, including emotional distress. Although recurring headache may be an early sign of serious organic disease, relatively few headaches are caused by disease-induced structural changes. Most result from vasodilation of blood vessels in tissues surrounding the brain, or from tension in the neck and scalp muscles.

headache rack
Transit (slang). Heavy bulkhead that extends over cab from trailers, usually made of pipe and used in steel hauling.

header bar
The rear cross piece on open top trailer.

header board
A protective shield at the front end of a flat-bottom trailer to prevent freight from shifting forward.

headform
An object whose shape resembles that of the human head for sizing, modeling, or simulation purpose.

headgear
Any protective structure worn on the head to protect the individual from possible injury due to hazards, usually from impacts.

headgear retention
A measure of the ability of a piece of headgear to remain in place during an impact and any post-impact events.

headgear retention assembly
Any combination of chinstraps, internal form fitting, or other techniques to aid in headgear retention.

headline
A mooring line used in combination to hold a fleet or barge "in."

headrest
Any padded structure which provides support to the head when sitting or reclining.

headroom
That distance available to accommodate an individual's head, generally referring to that distance between the vertex of an individual's head and a roof, passageway, or other limiting environmental feature when standing, sitting, walking, or other motion/posture as the situation requires.

headset
A device having one or a pair of transducers for converting electrical energy to sound and having a spring mechanism or other device over the head, under the jaw, or around the neck to hold it/them in place.

headwaters
The upper part of a river system, denoting the upper basin and source streams of a river.

headwear
Any form of clothing worn only on or around the head, such as a hat, cap, or helmet.

headway
The time interval between transit revenue vehicles passing a specified location.

healing
The restoration of structure and function of injured or diseased tissues. The healing processes include blood clotting, tissue mending, scarring, and bone healing.

health
A state in which an individual's and/or population's mental, physical, physiological, and social conditions are within normal limits. *See also **mental health** and **physical health**.*

health and safety study
As defined by TSCA, any study of any effect of a chemical substance or mixture on health and/or the environment, including underlying epidemiological studies, studies of occupational exposure, toxicological, clinical, and ecological studies of a chemical substance or mixture.

health assessments
Under the Federal Solid Waste Disposal Act: Includes preliminary assessments of the potential risk to human health posed by individual sites and facilities subject to the Federal Solid Waste Disposal Act, based on such factors as the nature and extent of contamination, the existence of a potential for pathways of human exposure (e.g., ground or surface water contamination, air emissions, and food chain contamination), the size and potential susceptibility of the community

within the likely pathways of exposure, the comparison of expected human exposure levels to the short-term and long-term health effects associated with identified contaminants and any available recommended exposure or tolerance limits for such contaminants, and the comparison of existing morbidity and mortality data on diseases that may be associated with the observed levels of exposure. The assessment includes evaluation of the risks to the potentially affected population from all sources of such contaminants, including known point or non-point sources other than the site or facility in question. A purpose of such preliminary assessments shall be to help determine whether full-scale health or epidemiological studies and medical evaluations of exposed populations shall be taken.

health care facilities
Buildings or protons of buildings and mobile homes that contain, but are not limited to, hospitals, nursing homes, extended care facilities, clinics, and medical and dental offices, whether fixed or mobile.

health hazard
A property of a chemical, mixture of chemicals, physical stress, pathogen, or ergonomic factor for which there is statistically significant evidence, based on at least one test or study conducted in accordance with established scientific principles, that acute or chronic adverse health effects may occur among workers exposed to the agent.

health index
Any qualitative or quantitative measure for describing the relative or absolute health of an individual or a population.

health insurance
A program which includes come percentage of payment or reimbursement for medical, dental, vision, counseling, and/or other care beyond a specified deductible limit. Often a fringe benefit paid at least in part by employers and generally used to provide financial protection in the event of a major family health problem.

health physicist
An individual trained in radiation (ionizing) physics, its associated health hazards, the means to control exposures to this physical hazard, and in establishing procedures for work in radiation areas.

health physics (HP)
The branch of radiological science dealing with the protection of personnel from harmful effects of ionizing radiation.

Health Physics Society (HPS)
Professional society of persons active in the field of health physics, the profession devoted to the protection of people and their environment from radiation hazards.

health standard
Those standards that generally prescribe requirements for worker exposure to hazards presented by toxic substances. Such hazards usually involve the potential for long-term adverse health effects (such as those posed by exposure to lead, noise, asbestos, silica, radiation, vibration, etc.).

healthy worker effect
A phenomenon observed in studies of occupational diseases in which workers exhibit lower death rates than the general population because hospitalized, severely ill, and many disabled persons have been excluded from employment and those that are employed are generally healthy.

hearing
(1) *Anatomy.* The physiological process of sound perception. That specialized sense through which sound is perceived, by conversion of sound waves into nerves impulses, which are then interpreted by the brain. (2) *Legal.* A legal proceeding convened at an announced time and place for a governmental purpose; for instance, to entertain new legislation, consider promulgating new regulations, hear an applicant for a permit or license, consider revoking or amending such an approval, present evidence, hear motions by parties, or announce decisions. Some hearings are *public hearings* required by law to be conducted with an opportunity for the public to attend. Others are *adjudicatory hearings* where only the interested parties and their representatives and witnesses attend. A *public meeting* at which a board or other tribunal deliberates is not necessarily a *public hearing.*

hearing aid
A device which amplifies sound intensity or filters noise, typically for use by persons with hearing impairments.

hearing conservation
The prevention or minimizing of occupational noise-induced hearing defects through the combined use of hearing protectors, training, the use of engineering and administrative control measures, annual audiometric testing, and the establishment of a written program. The written program is referred to as a *hearing conservation program (HCP)*.

hearing impaired
A person with a hearing loss sufficient to affect their efficiency in the course of everyday living.

hearing impairment
(1) Loss of the ability to hear, either partially or completely. (2) The deviation of an individual's absolute auditory threshold in decibels using a calibrated audiometer or by comparison to the absolute auditory threshold of a person with normal hearing. Also called *hearing loss*.

hearing level
The deviation, in decibels, of an individual's hearing threshold at various test frequencies as determined by an audiometric test based on an accepted standard reference level. *See also* ***hearing threshold***.

hearing loss
See ***hearing impairment***.

hearing protection
See ***hearing protective device***.

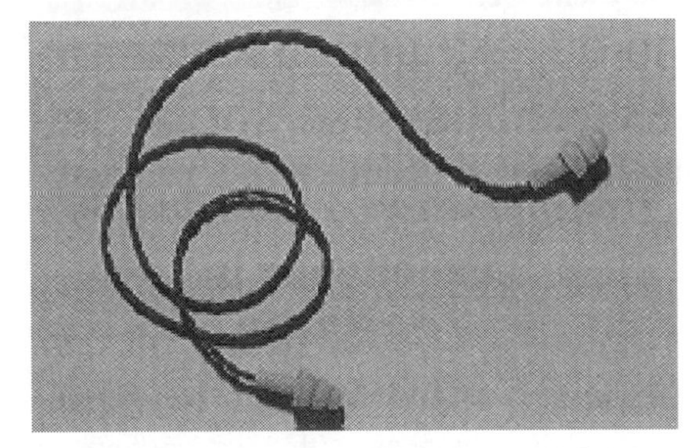

Common earplugs, a type of hearing protective device

hearing protective device (HPD)
Any device or material, capable of being worn on the head or in the ear canal, that is sold wholly or in part on the basis of its ability to reduce the level of sound entering the ear. This includes devices of which hearing protection may not be the primary function, but which are nonetheless advertised as providing hearing protection to the user.

hearing scotoma
See ***tonal gap***.

hearing test
Any method of evaluating hearing capabilities. *See also* ***audiometry*** *and* ***tuning fork test***.

hearing threshold
The weakest or minimally perceived sound, in decibels, that an individual can detect during an audiometric test at a particular time.

hearsay
A term applied to that species of testimony given by a witness who relates, not what he/she knows personally, but what others have told him/her, or what he/she has heard said by others.

heart
The multi-chambered muscular organ within the thorax which pumps blood through the circulatory system.

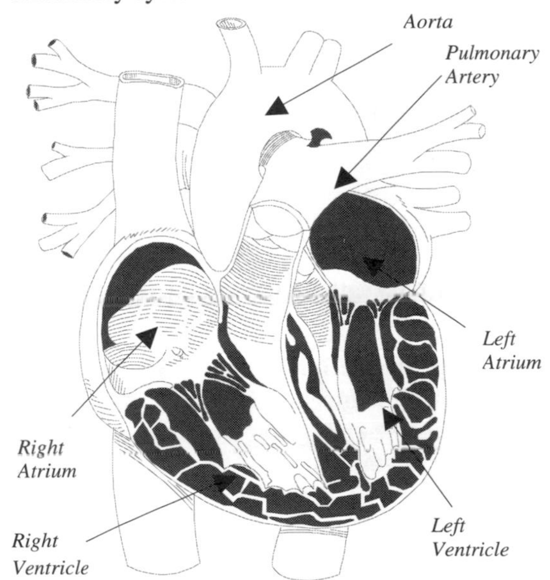

Cross-section of the human heart, showing its four chambers

heart block
A condition in which the atria and ventricles of the heart contract independently, causing interference in the rate or regularity of the heartbeat.

heart failure
Inability of the heart to perform its proper function of expelling blood from the ventricles.

heart-lung machine
A mechanical device that temporarily takes over the functions of the heart and lungs. It is used as an aid during surgery.

heart murmur
Any sound in the heart region other than normal heart sounds. A murmur may be caused by several different factors, including changes in the valves of the heart or blood leaking through a disease-scarred valve that does not close properly.

heart rate
The number of complete heart contraction cycles per minute. Synonymous with *pulse rate.*

heartbeat
The cycle of contraction of the heart muscle, during which the chambers of the heart contract. The beat begins with a rhythmic impulse in the sinoatrial node, which serves as a pacemaker for the heart.

heartburn
(slang) A burning sensation in the esophagus, or below the sternum in the region of the heart. It is one of the common symptoms of indigestion. *See also* ***reflux esophagitis***.

heat
The energy associated with a mass because of random motions of its molecules. It is a form of energy that is transferred between systems by virtue of their individual temperature differences.

heat acclimatization
A physiological adjustment to living or working at higher external temperatures and/or humidity.

heat balance
The difference between the heat produced by the body and that which is given off to the environment.

heat capacity
That heat energy absorbed by an object under given conditions for each degree rise in temperature. *See also* ***specific heat***.

heat collapse
See ***heat exhaustion***.

heat conduction
Heat transfer from one entity to another via direct contact.

heat conservation
Any mechanism such as peripheral vasoconstriction, piloerection, or reduction in sweating which may be used to retain heat within the body. Also referred to simply as *heat retention.*

heat convection
Heat transfer from one entity to another or within an entity via a fluid capable of storing heat, such as air.

heat cramps
A condition related to work and/or exercise in hot environments that causes painful muscle spasms due to heavy swelling and the consumption of large amounts of water without adequate salt intake and adequate exercise-rest balance.

heat disorder
Any condition resulting from exposure to heat or hot work environments that results in an adverse effect on the health of the exposed individual. Such disorders include heat cramps, heat exhaustion, heat stress, and heat stroke. Also may be referred to as *heat stress.*

heat drying
A process to reduce pathogens in solid waste by drying dehydrated sludge cake by direct or indirect contact with hot gases, and reducing moisture content to ten percent or lower.

heat exchanger
Device for transferring heat from one fluid or body to another for the purpose of heating or cooling.

heat exhaustion
A potentially dangerous condition caused by work and exertion in high-temperature environments marked by mild elevation in body temperature, weak pulse, pale complexion, dizziness, fainting, profuse sweating, headache, low blood pressure, and cool, moist skin. Synonymous with *heat collapse* and *heat prostration.*

heat index
See ***heat stress index***.

heat island effect
A "dome" of elevated temperatures over an urban area caused by structural and pavement heat fluxes, and pollutant emissions from the area below the dome.

heat lightning
Distant lightning that illuminates the sky but is too far away for its thunder to be heard.

heat loss
The release of heat from the body to the environment via conduction, convection, radiation, or evaporation.

heat of fusion
The heat released by a liquid freezing to a solid, or that gained by a solid melting to a liquid without a change in temperature.

heat prostration
See ***heat exhaustion***.

heat pyrexia
See ***heat stroke***.

heat radiation
The transfer of heat via electromagnetic radiation. Also called *thermal radiation*.

heat rash
See ***miliaria***.

heat ray cataract
An opacity in the lens of the eye which occurs in occupations requiring long exposures to high temperatures and glare. *See also* ***glass blower's cataract***.

heat regulation
See ***thermoregulation***.

heat retention
See ***heat conservation***.

heat strain predictive system
A method for predicting heat stress based on variable clothing effect.

heat stress
Thermal stress upon the body from the surrounding environment, including heat stroke, heat cramps, and heat exhaustion, caused by the body's inability to rid itself of excessive heat. *See also* ***heat disorder***.

heat stress index (HSI)
Any number of estimators for body heat stress which may be based on temperature, humidity, air velocity, workload, clothing, and their interactions. *See also* ***Belding-Hatch heat stress index***.

heat stroke
A serious, potentially life-threatening condition marked by a rapid rise in body temperature, hot dry skin, mental confusion, loss of consciousness, convulsions, coma, and the absence of sweating. The condition is caused by excessive physical exertion in hot environments by unacclimatized individuals and dehydration. Recent intake of alcohol may expedite the onset of the condition.

heat syncope
A condition marked by fainting while standing erect and immobile in hot environments caused by the pooling of blood in dilated vessels of the skin and lower part of the body.

heated wire anemometer
See ***thermoanemometer***.

heater
Any device or assembly of devices or appliances used to heat the interior of any motor vehicle. This includes a catalytic heater which must meet the requirements of 49 CFR 177.834(1) when flammable liquid or gas is transported.

Heath-Carter somatotype
A body type classification system which uses a combination of anthropometric measures (such as stature, weight, skinfolds, girths, and breadths) for determining or modifying the basic classifications.

heating degree-day
A form of the degree-day used as an index for fuel consumption.

heating equipment
According to OSHA, 29 CFR 1910.306(g), the term includes equipment used for heating purposes if heat is generated by induction or dielectric methods.

heating, ventilation, and air conditioning (HVAC) system
The system that is in place to provide ventilation, heating, cooling, dehumidification, humidification, control of odors, and cleaning of the air for maintaining comfort, safety, and health of the occupants of a building, workplace, etc.

heavy duty scaffold
A scaffold designed and constructed to carry a working load not to exceed 75 pounds per square foot.

heavy duty truck
Truck with a gross vehicle weight generally in excess of 19,500 pounds (class 6-8). Other minimum weights are used by various law or government agencies.

heavy hauler trailer
A trailer with one or more of the following characteristics: a) its brake lines are designed to adapt to separation or extension of the vehicle frame; or b) its body consists only of a platform whose primary cargo carrying surface is not more than 40 inches above the ground in an unloaded condition, except that it may include sides that are designed to be easily removable and a permanent "front-end structure" as that term is used in 49 CFR 393.106.

heavy ion
An ion having a normal atomic mass equal to or greater than that of carbon.

heavy lifts
Maritime. Freight too heavy to be handled by regular ship's tackle.

heavy metal
Metals such as arsenic, barium, cadmium, chromium, lead, mercury, and silver that do not rapidly break down in the body or the environment and thus can exert toxic effects because of their cumulative or residual properties.

heavy rail
(1) An electric railway with the capacity for a "heavy volume" of traffic and characterized by exclusive rights-of-way, multi-car trains, high speed and rapid acceleration, sophisticated signaling, and high platform loading. (2) High-speed, passenger rail cars operating singly or in trains of two or more cars on fixed rails in separate rights-of-way from which all other vehicular and foot traffic is excluded. *See also* ***heavy rail (rapid rail)***.

heavy rail passenger cars
Rail cars with motive capability, driven by electric power taken from overhead lines or third rails, configured for passenger traffic and usually operated on exclusive rights-of-way.

heavy rail (rapid rail)
Transit service using rail cars powered by electricity which is usually drawn from a third rail and usually operated on exclusive rights-of-way. It generally uses longer trains and has longer spacing between stations than light rail.

heavy work
That level of work activity which involves the entire body and has a gross metabolic cost of 280 – 380 calories per square meter of skin surface per hour.

hedonic damages
Damages awarded in some jurisdictions for the loss of enjoyment of life, or for the value of life itself, as measured separately from the economic productive value that an injured or deceased person would have had. It should be noted that many courts hold that such loss is included in damages for disability and pain and suffering.

heel
The calcaneus and surrounding soft tissue of the inferior and posterior portion of the foot.

heel – ankle circumference
The surface distance around the foot under the tip of the heel and over the instep at the junction of the foot and anterior lower leg. Measured with minimal tissue compression, minimal weight on the foot being measured, and the foot muscles relaxed.

heel breadth
The maximum medial to lateral linear width of the heel behind the vertical projection downward from the ankle bones. Measured with the individual's weight equally distributed on both feet and with minimal tissue compression.

HEG
See ***homogeneous exposure group***.

height
The straight-line vertical distance from the floor or other reference surface to the level of the referenced body part or the top of an object.

height above airport
The height of the Minimum Descent Altitude above the published airport elevation. This is published in conjunction with circling minimums. *See also* ***minimum descent altitude***.

height above landing (HAL)
The height above a designated helicopter landing area used for helicopter instrument approach procedures.

height above touchdown (HAT)
The height of the Decision Height or Minimum Descent Altitude above the highest runway elevation in the touchdown zone (first 3,000 feet of the runway). Height above touchdown (HAT) is published on instrument approach charts in conjunction with all straight in minimums.

height/decision altitude
A specified altitude or height (A/H) in the precision approach at which a missed approach must be initiated if the required visual reference to continue the approach has not been established. *Note 1:* Decision altitude is referenced to mean sea level and decision height is referenced to the threshold elevation. *Note 2:* The required visual reference means that section of the visual aids or of the approach area which should have been in view for sufficient time for the pilot to have made an assessment of the aircraft position and rate of change of position, in relation to the desired flight path.

height velocity
The rate at which stature increases during physical maturation.

helicopter
(1) A rotorcraft that, for its horizontal motion, depends principally on its engine-driven rotors. (2) A rotary-wing aircraft which depends principally for its support and motion in the air upon the lift generated by one or more power-driven rotors, rotating on substantially vertical axes. A helicopter is a V/STOL (vertical/short takeoff and landing) aircraft. (3) A heavier-than-air aircraft supported in flight chiefly by the reactions of the air on one or more power-driven rotors on substantially vertical axes. *See also* ***gyroplane***.

helipad
A small, designated area, usually with a prepared surface, on a heliport, airport, landing/takeoff area, apron/ramp, or movement area used for takeoff, landing, or parking of helicopters.

heliport
(1) An area of land, water, or structure used or intended to be used for the landing and takeoff of helicopters and includes its buildings and facilities if any. (2) An area, either at ground level or elevated on a structure, that is used for the landing and takeoff of helicopters and includes some or all of the various facilities useful to helicopter operations such as helicopter parking, hangar, waiting room, fueling, and maintenance equipment. *See also* ***aerodrome, aircraft facility,*** *and* ***airport***.

helistop
A minimum facility heliport, either at ground level or elevated on a structure for the landing and takeoff of helicopters, but without such auxiliary facilities as waiting room, hangar parking, etc.

helium
A chemical element, atomic number 2, atomic weight 4.003, symbol He.

helix
(1) *Anatomy.* The rolled outer portion of the auricle. (2) *Geometry.* A spiraling geometrical pattern.

helmet
(1) A piece of headgear with a hard exterior covering and internal cushioning designed to fit over the top of or enclose the entire head to protect the head from impacts or other hazards. (2) A head protective device consisting of a rigid shell, energy absorption system, and chin strap intended to be worn to provide protection for the head or portions thereof, against impact, flying or falling objects, electric shock, penetration, heat and flame.

helmet-mounted display (HMD)
A display projected within or on the visor of a user's helmet such that both the information presented and the external environment are simultaneously within the line of sight.

Helmholtz resonator
A passive acoustical filter consisting of a cavity with a narrow neck and an enlarged interior.

Helmholtz-Kohlrausch effect
A tendency for apparent brightness to increase as color saturation increases.

help
An online software user assistance feature.

hemangioma
A benign tumor made up of newly formed blood vessels, clustered together. Hemangioma may be present at birth in various parts of the body, including the liver and bones. In the majority of cases, however, it appears as a network of small blood-filled capillaries near the surface of the skin, forming a reddish or purplish birthmark. These marks are not malignant.

hematemesis
The vomiting of blood. The appearance of the vomitus depends on the amount and character of the gastric contents at the time blood is vomited and on the length of time blood has been in the stomach. Gastric acids change the bright red blood to a brownish color and the vomitus is often described as "coffee-ground" color. Bright red blood in the vomitus indicates a fresh hemorrhage and little contact of the blood with gastric juices. The most common causes are peptic ulcer, gastritis, esophageal lesions or varices, and cancer of the stomach. Benign tumors, traumatic postoperative bleeding and swallowed blood from points in the nose, mouth, and throat can also produce hematemesis.

hematocrit
The percent by volume of erythrocytes in whole blood.

hematologist
An individual trained in the science encompassing the generation, anatomy, physiology, pathology, and therapeutics of blood.

hematology
The branch of medical science concerned with the generation, anatomy, physiology, and therapeutics of blood. The study of the form and structure of blood-forming organs.

hematoma
(1) An enclosed volume of blood in tissue external to the circulatory system, from whatever cause. (2) A tumor-like mass produced by coagulation of extravasated blood in a tissue or cavity. Contusions (bruises) and black eyes are familiar forms of hematoma that are seldom serious. Hematomas can occur almost anywhere on the body; they are almost always present with a fracture and are especially serious when they occur inside the skull, where they may produce local pressure on the brain. In minor injuries the blood is absorbed unless infection develops. *See also* ***bruise***.

hematopoietic
Pertaining to or affecting the formation of blood cells.

hematopoietic changes
Changes in the formation of blood cells.

hematotoxicity
The toxic effects of various substances and physical agents in blood and blood-forming organs.

hematuria
Appearance of blood in the urine.

heme
The non-protein, iron-containing part of the hemoglobin molecule that carries oxygen and accounts for the color of blood.

hemi-
(prefix) Meaning half; pertaining to one side of the body.

hemianopsia
A unilateral or bilateral blindness in one half of the visual field.

hemiballismus
A unilateral form of ballismus.

hemiplegia
A condition in which one side of the body (especially both limbs) is affected by paralysis. Usually caused by a brain lesion, such as a tumor, or by a cerebral vascular accident. The paralysis occurs on the side opposite the brain disorder.

hemisphere
Half of a spherical or roughly spherical structure or organ.

hemochromatosis
A disorder of iron metabolism with excess deposition of iron in the tissues, skin pigmentation, cirrhosis of the liver, and decreased carbohydrate tolerance.

hemodynamics
The study of the physical principles of blood and its circulation.

hemoglobin (Hb)
The red pigment protein matter in the red blood corpuscles that carries oxygen from the lungs to the tissues and carbon dioxide from the tissue to the lungs. Hemoglobin is a chromoprotein, that is, a protein combined with a colored pigment. The protein is globin; the pigment is heme, which is red. When erythrocytes are broken down, degradation of hemoglobin releases the pigment bilirubin which is converted into pigments responsible for the characteristic color of bile. Heme is a complex molecule containing iron. Hemoglobin has the property of combining chemically with certain gases to form various substances. One of the most important is oxyhemoglobin, formed by the combination of oxygen and hemoglobin. This function of hemoglobin is important in respiration because it provides a means of transporting oxygen from the lungs to the tissues. The oxygen combined with hemoglobin in arterial blood is responsible for its bright red color; venous blood is a darker color because of its lower oxygen content. It is important to note that hemoglobin has an affinity towards carbon monoxide approximately 200 times greater than that for oxygen. This explains why carbon monoxide poisoning can occur so quickly.

hemoglobinuria
Excretion of hemoglobin in the urine.

hemolysis
Breakdown of red blood cells or erythrocytes with the release of hemoglobin into the blood plasma. This results in hemoglobinuria.

hemophilia
A condition characterized by impaired coagulation capability of the blood, and a strong tendency to bleed. The classic disease is hereditary, and limited to males, being transmitted always through the female to the second generation, but many similar conditions attributable to the absence of different factors from the blood are now recognized.

hemoptysis
Expectoration of blood or of blood-stained sputum.

hemorrhage
The loss of blood from blood vessels and/or capillaries.

hemorrhoid
An enlarged blood vessel in the anal or rectal wall that causes pain, itching, discomfort, and bleeding.

hemostasis
The stoppage of blood flow or loss.

hemothorax
A collection of blood in the pleural cavity.

hemotoxin
Any substance that causes destruction of red blood cells.

henry (H)
The inductance of a closed circuit in which a potential of one volt is produced when the electric current in the circuit is uniform at one ampere per second.

Henry's Law Constant
Measure of the volatility of a substance in a dilute solution of water at equilibrium. It is the ratio of the vapor pressure exerted by a substance in the gas phase over a dilute aqueous solution of that substance to its concentration in the solution at a given temperature. For hazardous ranking system (HRS) purposes, use the value reported at or near 25°C [atmospheric-cubic meters per mol (atm-m^3/mol)].

HEP
*See **human error probability**.*

HEPA
*See **high-efficiency particulate air filter**.*

Hepatitis B Virus (HBV)
A virus that causes inflammation of the liver. Can also occasionally be caused by toxic agents other than viral.

hepatotoxic
Refers to an agent that produces damage to the liver.

Heptachlor
An insecticide that was banned in some food products in 1975 and all of them in 1978. It was allowed for use in seed treatment until 1983. More recently it was found in milk and other dairy products in Arkansas and Missouri, as a result of illegally feeding treated seed to dairy cattle.

herb
A plant which may be used as food flavoring or for medicinal purposes.

herbicide
A chemical pesticide designed to control or destroy plants, weeds, or grasses.

herbivore
An animal that feeds on plants.

here she comes
Term used when another boat appears around a bend.

hereditary mutagenicity
The ability of a chemical to cause an inheritable change in the genetic material (i.e., DNA) or organisms.

heredity
The transmission of characteristics from parent to offspring.

Hering opponent process theory
See ***opponent process theory***.

hermetically sealed
Closed by fusion, gasketing, crimping, or equivalent means so that no gas or vapor can enter or escape.

hernia
A failure/rupture or weakness in the wall of a bodily structure, usually a rupture of the abdominal wall or an intervertebral disk which results in the protrusion of part of an organ or tissue through the failure.

herniate
The creation of a hernia.

herniated disk
A protrusion of the nucleus pulposus of an intervertebral disk into or through the annulus. Commonly referred to as a *slipped disk.*

heroin
Narcotic drug which is a derivative of opium and whose technical name is diacetylmorphine. It is classified as a Class A substance for criminal purposes and the penalty for its unlawful manufacture, distribution, sale, or possession is severe.

Hertz (Hz)
(1) A measure of frequency in cycles per second (cps). (2) The standard radio equivalent of frequency in cycles per second of an electromagnetic wave. Kilohertz (kHz) is a frequency of one thousand cycles per second. Megahertz (mHz) is a frequency of one million cycles per second.

hetero-
(prefix) Relating to combinations of different entities.

heterochromatic
Combining or pertaining to two or more different colors.

heterochromatic flicker photometry
A technique used for measuring an observer's relative sensitivity to light of different wavelengths by comparing a fixed-luminance reference light alternating with a light of a different wavelength to minimize the sensation of flicker.

heteromodal
See ***multisensory***.

heterophoria
A tendency to position the eyes such that binocular vision cannot be used.

heterosphere
The region of the atmosphere above about 85 km where the composition of the air varies with height.

heterotroph
An organism that must obtain food energy by ingesting other organic material.

heuristic
Pertaining to a learning or problem-solving technique which uses certain empirical rules or guidelines to ultimately reach a solution.

heuristic program
A set of instructions which directs a computer to use a heuristic approach to problem solving.

hexadecimal
Pertaining to a numbering system based on 16, using the alphanumerics zero through nine and A through F.

HHC
See ***highly hazardous chemical***.

HIC
See ***Head Injury Criterion***.

hiccough
Spasmodic involuntary contraction of the diaphragm that results in uncontrolled breathing in of air. The peculiar noise of hiccoughs is

produced by the attempt to inhale while the air passages are partially closed. Also called *singultus* and *hiccup*.

Hick-Hyman Law
A rule that the choice reaction time is linearly related to the logarithmic transformation of the amount of stimulus information presented. Represented as follows:

$$CRT = d + t_b H$$

where:

CRT	=	average choice reaction time
d	=	summed time required for all non-decision-making activities, e.g., stimulus transmission time plus motor response time, assumed to be a constant
t_b	=	time required to process one bit of information, assumed to be a constant
H	=	amount of information in bits (=$\log_2 N$), often taken as the number of available choices.

hidden digit test
See ***Stilling test***.

hidden line
A graphic line not displayed on a model, especially a wire-frame model, which would not be visible from a particular view if the model were solid.

hidden window
A display window partially or completely covered by another.

hierarchical decomposition
The breakdown of a high-level task into smaller, lower level steps.

hierarchical menu
A menu structure or format in which each item on a given menu has another menu consisting of a subset of additional selections until the lowest level menu is reached.

high
(1) A general feeling of euphoria produced most commonly by artificial means, such as with drugs or other influencing substances. However, certain naturally occurring influencing factors, such as exercise and laughter, have also been reported to produce a similar feeling presumably due to a release of endorphins. (2) A meteorological phenomenon. See ***anticyclone***.

high blood pressure
A disorder of the circulatory system marked by excessive pressure of the blood against the walls of the arteries. *See also* ***hypertension***.

high boiling aromatic oils (HBAO)
These are high boiling components produced during catalytic cracking and thermal cracking of petroleum streams, and also during the extraction of lube base stocks. They contain complex mixtures of hydrocarbons in the boiling range of 500-1000°F and have demonstrated carcinogenic potential in animal testing. These are also referred to as *aromatic process oils*.

high-definition television (HDTV)
A video medium with a resolution of approximately 1200 lines.

high degree of care
See ***reasonable care***.

high density lipoprotein (HDL)
A substance present in blood which functions to return cholesterol to the liver for reprocessing and elimination. Often referred to as the "good cholesterol" because of its ability to reduce overall cholesterol levels of the blood. It is assumed that the higher the HDL level, the healthier the outlook.

high-density polyethylene
A material that produces toxic fumes when burned. Used to make plastic bottles and other products.

high-efficiency particulate air (HEPA) filter
A filter capable of removing 99.97% of all particles with a mean aerodynamic diameter of 0.3 micron. Often used to filter air in air-purifying respirators, vacuum systems, exhaust systems, and fans.

high-energy heavy ion (HZE)
A high-velocity particle consisting of an ionized heavy atom.

high fidelity
Pertaining to an audio or graphic (including photographic) reproduction which is comparable with the original.

high frequency
The frequency band between 3 and 30 mHz.

high frequency communications
High radio frequencies (HF) between 3 and 30 mHz used for air-to-ground voice communication in overseas operations.

high frequency loss
In acoustics, refers to the hearing loss in frequency bands of 2000 Hz and above. Also referred to as *high frequency hearing loss.*

high-hazard contents
Those contents which are liable to burn with extreme rapidity or from which poisonous fumes or explosions are to be feared in the event of fire.

high inversion fog
A fog that lifts above the surface but does not completely dissipate because of a strong inversion (usually subsidence) that exists above the fog layer.

high-level oversight structure
With regard to regulatory compliance, full support by a company's board of directors and senior management of the company's compliance program.

high-level radioactive waste (HLW)
Under the Federal Nuclear Waste Policy of 1982: The aqueous waste resulting from the operation of the first cycle solvent extraction system, or equivalent, and the concentrated waste from subsequent extraction cycles, or equivalent, in a facility for reprocessing irradiated reactor fuels, or irradiated fuel from nuclear power reactors. Includes liquid waste produced directly in reprocessing and any solid material derived from such liquid waste that contains fission products in sufficient concentrations.

high-mileage households
Households with estimated aggregate annual vehicle mileage that exceeds 12,500 miles.

high occupancy vehicle (HOV)
Vehicles that can carry two or more persons. Examples of high occupancy vehicles are buses, vanpools, and carpools.

high occupancy vehicle (HOV) lane
An exclusive road or traffic lane limited to buses, vanpools, carpools, and emergency vehicles.

high-order detonation
*See **detonation**.*

high-speed cinematography
The sampling of activities using motion picture film with a frame rate much higher than the normal projection rate.

high-speed rail
(1) A rail service having the characteristics of intercity rail service which operates primarily on a dedicated guideway or track not used, for the most part, by freight, including, but not limited to, trains on welded rail, magnetically levitated (MAGLEV) vehicles on a special guideway, or other advanced technology vehicles designed to travel at speeds in excess of those possible on other types of railroads. (2) A rail transportation system with exclusive right-of-way which serves densely traveled corridors at speeds of 124 miles per hour and greater.

high-speed taxiway
A long radius taxiway designed and provided with lighting or marking to define the path of aircraft, traveling at high speed (up to 60 knots), from the runway center to a point on the center of a taxiway. Also referred to as long radius exit or turn-off taxiway. The high speed taxiway is designed to expedite aircraft turning off the runway after landing, thus reducing runway occupancy time.

high task
*See **incentive pace**.*

high type road surface
Highly flexible, composite, rigid, etc. (Surface/Pavement Type Codes 61, 62, 71-76 and 80).

high volume air sampler
Sampling device used for the collection of particulates in the ambient air. One type is employed for collecting PM 10 (particulate materials equal to or less than 10 micrometers in diameter), and another for collecting all suspended particulates to determine the total suspended particulate concentration.

high volume area
Maritime Emergency Response. Area where an oil pipeline having a nominal outside diameter of 20 inches or more crosses a major river or other navigable waters which, because of the velocity of the river flow and vessel traffic on the river, would require a more rapid response in case of a worst case dis-

charge or substantial threat of such a discharge from the oil pipelione.

high water buoy
Small unlighted buoys permanently secured to the end of dikes, lock walls, bear traps, and other river front structures such as mooring cells for the purpose of marking them during high water when they are submerged.

high water station
The location to which lights or buoys are moved when a river is at or near its flood stage, the purpose being a) to guide navigation in the high water and b) to locate the light in a position of security against loss.

highest degree of care
That degree of care that a very careful and prudent person would use under the same or similar circumstances. *See also **great care** and **care**.*

highlight
Use some feature different from the background to attract a user's attention to some portion of a display.

highly hazardous chemical
Chemicals listed in Appendix A of the OSHA standard related to process safety management of highly hazardous chemicals (OSHA 29 CFR 1910.119). They are substances possessing toxic, reactive, flammable, or explosive properties.

highly volatile liquid
A hazardous liquid which will form a vapor cloud when released to the atmosphere and which has a vapor pressure exceeding 276 kPa (40 psia) at 37.8°C (100°F). *See also **hazardous material**.*

highpass filter
A device which allows frequencies higher than the cutoff frequency to exit from the device unattenuated, while the intensity of frequencies lower than the cutoff frequency is attenuated.

highway
(1) Any public street, public alley, or public road. (2) Any road, street, parkway, or freeway/expressway that includes rights-of-way, bridges, railroad-highway crossings, tunnels, drainage structures, signs, guardrail, and protective structures in connection with highways. The highway further includes that portion of any interstate or international bridge or tunnel and the approaches thereto (23 U.S.C. 101a).

highway capacity manual
A publication of the Institute of Transportation Engineers defining level of service criteria to determine peak hour traffic congestion.

highway construction project
A project financed in whole or in part with federal-aid or federal funds for the construction, reconstruction or improvement of a highway or portions thereof, including bridges and tunnels.

highway mode
Consists of public roads and streets, automobiles, vans, trucks, motorcycles, and buses (except local transit buses) operated by transportation companies, other businesses, governments, and households, garages, truck terminals, and other facilities for motor vehicles.

Highway Performance Monitoring System (HPMS)
The state/federal system used by the FHWA to provide information on the extent and physical condition of the nation's highway system, its use, performance, and needs. The system includes an inventory of the nation's highways including traffic volumes.

highway-rail crossing
A location where one or more railroad tracks intersect a public or private thoroughfare, a sidewalk, or a pathway. *See also **at grade** and **grade crossings**.*

highway-rail crossing accident/incident
An impact between on track railroad equipment and a highway user (e.g., an automobile, bus, truck, motorcycle, bicycle, farm vehicle, pedestrian, or other highway user) at a designated crossing site. Sidewalks, pathways, shoulders, and ditches associated with the crossing are considered to be part of the crossing site. The term "highway user" includes pedestrians, cyclists, and all other modes of surface transportation.

Highway Research Information Service (HRIS)
A computer-based information storage and retrieval system developed by the Transportation Research Board with financial support from the state highway and transportation de-

partments and the Federal Highway Administration. It consists of summaries of research projects in progress and abstracts of published works.

highway trust fund
The federal account established by law to hold receipts collected by the government and earmarked for highway programs and a portion of the federal mass transit program. It is supported by the federal gasoline tax and other user taxes.

highway user fee or tax
A charge levied on persons or organizations based on the use of public roads. Funds collected are usually applied toward highway construction, reconstruction, and maintenance. Examples include vehicle registration fees, fuel taxes, and weight-distance taxes.

hinge joint
The type of joint which permits only a single degree of freedom, as rotation about a pivot point within a plane. In anatomy, the elbow and knee are examples of a hinge joint.

hip
The coxal bone, its joints with the sacrum and femurs, and all the associated surrounding tissues.

hip bone
See ***coxal bone***.

hip breadth, sitting
The maximum horizontal linear distance across the widest portion of the hips. Measured with the individual sitting erect, knees flexed at 90°, knees and thighs together, and feet flat on the floor.

hip breadth, standing
The maximum horizontal linear distance across the lower torso in the hip region. Measured with the individual standing erect, feet together, and his/her weight distributed evenly on both feet.

hip circumference at trochanterion
The surface distance around the hip at the trochanteric height. Measured with the individual standing erect and his/her weight equally distributed on both feet.

hip circumference, sitting
See ***buttock circumference, sitting***.

hip circumference, standing
See ***buttock circumference***.

hip joint
The joint composed of the junction of the femur head and the coxal bone.

Hippocrates
The late 5th century B.C. "Father of Medicine." The son of a priest-physician, he was born on the island of Cos. By stressing that there is a natural cause for disease, he did much to dissociate the care of the sick from the influence of magic and superstition. His carefully kept records of treatment and solicitous observation of the ill provided a foundation for clinical medicine in the case report; and by also reporting unsuccessful methods of treatment, he anticipated the modern scientific attitude. A moral code for medicine has been established by his ideals of ethical conduct and practice as embodied in the Hippocratic Oath.

hiring at will
A general or indefinite hiring with the right to terminate such at the will of the employer.

histamine
A substance produced by the breakdown of histidine, a common amino acid derived from protein that occurs naturally in the body. Histamine is found in all tissues of the body. Although histamine was discovered in 1909, its role is still not fully understood. Histamine normally functions as a stimulant to the production of gastric juice. It also dilates the small blood vessels, as part of the regular adaptation of the body to changing inner and exterior conditions. An excess of histamine can dilate blood vessels to the extent that extravasation occurs. This appears as the reddening and swelling known as inflammation. Continued extravasation causes edema.

histidine
A naturally occurring amino acid, essential for optimal growth of infants.

histogram
A graphical representation of two or more amplitude measures using rectangular shapes along either a discrete or continuous dimension. More commonly referred to as a bar graph or bar chart.

histology
The study of the structure of tissues.

histolysis
Process whereby tissue is broken down.

histopathology
Pathologic histology or the change in the function of tissues as a result of a disease.

histoplasmosis
Bacterial infection resulting from the inhalation of the spores of *Histoplasma Capsulatum*. Occupations at risk are those associated with the raising and processing of fowl.

historic site
Any building, structure, area, or property that is significant in the history, architecture, archeology or culture of a state, its communities, or the nation and has been so designated pursuant to a statute. Such structures or properties are commonly statutorily protected and cannot be altered without permission of the appropriate authorities.

historical data
That data which have been previously collected in a given work situation and serve as a standard reference for performance. Typically refers to historical time, but not necessarily restricted to that.

histotoxic hypoxia
An inability of the tissues to use oxygen, even though it is present in amounts equal to or greater than normal.

hit and run
A hit-and-run occurs when a motor vehicle in transport, or its driver, departs from the scene after being involved in a crash prior to police arriving on the scene. Fleeing pedestrians and motor vehicles not in transport are excluded from the definition. It does not matter whether the hit-and-run vehicle was striking or struck. *See also* ***crash***.

HIV
See ***human immune deficiency virus***. Also called human immunodeficiency virus.

HIWAS
See ***Hazardous In-flight Weather Advisory Service***.

HLSC
See ***human life cycle safe concentration***.

HLW
See ***high-level radioactive waste***.

HMD
See ***head-mounted display*** *and/or* ***helmet-mounted display***.

HMI
See ***human-machine interface***.

HMTA
See ***Hazardous Materials Transportation Act of 1974***.

hoarfrost
See ***frost***.

hobo
Transit (slang). A tractor that is shifted from terminal to terminal.

Hodgkin's disease
A painless, progressive, and fatal condition characterized by pruritus and enlargement of the lymph nodes, spleens, and lymphoid tissue generally, which often begins in the neck and spreads through the body. Although Hodgkin's disease can occur at any age, it affects primarily those between the ages of 20 and 40 and is almost twice as frequent among men as among women. The first sign of the disease is often swelling of the lymph nodes, usually those of the neck, armpit, or groin, but sometimes those lying deep within the chest or abdomen. Severe itching is often an early sign of the disorder. As the disease progresses, it is usually marked by sweating, weakness, fever, and loss of weight and appetite. It spreads through the lymphatic system, involving other lymph nodes elsewhere in the body as well as the spleen, liver, and bone marrow. The lymph nodes and the spleen and liver may swell, and by obstructing other organs may cause coughing, breathlessness, or enlargement of the abdomen. The patient often becomes anemic, and because of blood changes the body becomes less able to combat infections. Also called *malignant granuloma* and *lymphogranuloma*.

hoist angle
An angle at which the load line is pulled during a hoisting operation.

hoistway
Any shaftway, hatchway, well hole, or other vertical opening or space in which an elevator or dumbwaiter is designed to operate.

hold
A therblig; a work element in which an object is held in a fixed orientation and location by the hand or other body member.

hold for release
Aviation. Communication Protocol. Used by Air Traffic Control (ATC) to delay an aircraft for traffic management reasons, i.e., weather, traffic volume, etc. Hold for release instructions (including departure delay information) are used to inform a pilot or a controller (either directly or through an authorized relay) that an Instrument Flight Rules (IFR) departure clearance is not valid until a release time or additional instructions have been received.

hold harmless agreement
A contractual arrangement whereby one party assumes the liability inherent in a situation, thereby relieving the other party of responsibility. *See also* ***indemnification agreement***.

hold open
Maritime Navigation. To hold below or above an object (i.e., wide of the mark) being steered on, depending on direction. Upstream tows normally hold above, downstream tows below, the object.

holdfire
An interruption of the ignition circuit of a launch vehicle.

holding agency
A federal agency having accountability for motor vehicles owned by the government. This term applies when a federal agency has authority to take possession of, assign, or reassign motor vehicles regardless of which agency is using the motor vehicles.

holding mark
Maritime Navigation. An object, usually an aid to navigation, on which the pilot of a tow will steer.

holding on
Steering steadily on a mark or object.

holding pond
A pond or reservoir, usually made of earth, built to store polluted runoff.

Holocene
The most recent geologic era; from approximately 10,000 years ago to the present. The Holocene is the latest epoch of the *Quaternary period*.

homatropine
A chemical which dilates the pupil and paralyzes accommodation when applied to the eye surface. Usually applied during an eye examination to permit viewing of the eyeball interior.

home base
The location where a vehicle is usually parked when not in use or on the road.

home row
The row of letters in a typewriter or computer keyboard on which the fingertips normally rest when typing in a standard mode.

home rule
A legal doctrine (usually found in a state constitution) whereby municipalities such as cities and towns are authorized to enact legislation in the form of *bylaws* or *ordinances* on certain subjects which the state legislation could authorize them to do, without the need to wait for an *Enabling Act* or other state approval. In some states, municipalities have been granted home rule authority to varying degrees over financial affairs, taxation, and exercise of the police power.

home signal
A roadway signal at the entrance to a route or block to govern trains in entering and using that route or block.

homeostasis
A state of physiologic equilibrium within the body.

homeothermy
The ability of some species to regulate body temperature within narrow limits, despite large temperature fluctuations in the environment. Commonly referred to as *warm-blooded*.

homesickness
A strong desire to be home such that one becomes sluggish and performance is affected, with possible psychosomatic or other symptoms if prolonged.

homicide
The killing of one human being by the act, procurement, or omission of another. A person is guilty of criminal homicide if he/she

purposely, knowingly, recklessly, or negligently causes the death of another human being.

homing
The procedure of using the direction finding equipment of one radio station with the emission of another radio station, where at least one of the stations is mobile, and whereby the mobile station proceeds continuously toward the other station.

homogeneous exposure group (HEG)
A group of employees who experience exposures similar enough so that monitoring the exposure of any of the group will provide exposure data that are useful for predicting the exposures of the remainder of the group.

homogeneous menu hierarchy
A menu hierarchy having the same number of options in each menu.

homogeneous radiation
A beam or flux consisting of radiation of the same kind and energy.

homogeneous reactor
A nuclear reactor in which the fissionable material and the moderator (if used) are combined in a mixture such that an effectively homogeneous medium is presented to the neutron.

homograph
A word which is spelled the same as another, but which has a different origin, pronunciation, and/or meaning.

homologous
Having the same structural relationship.

homologous motion
A movement which can be achieved in more than one way.

homoscedastcity
A condition in which each distribution has the same variance.

homosphere
The region of the atmosphere below about 85 km where the composition of the air remains fairly constant.

homunculus
A representation of the human body mapped onto the surface of the brain cortex. *See also* ***sensory homunculus*** *and* ***motor homunculus***.

HON
Hazardous Organic NESHAP. *See also* ***NESHAPS***.

hood
A shaped inlet designed to capture contaminated air and conduct it into an exhaust duct and/or exhaust fan.

hood capture efficiency
The emissions from a process which are captured by a hood and directed into the control device, expressed as a percent of all emissions.

hood entry loss
The pressure loss from turbulence and friction as air enters a ventilation system hood.

hood lifter
Transit (slang). A garage mechanic.

hood static pressure
The static pressure near a hood in the duct serving the hood; measured static pressure about 2-5 duct diameters downstream in the duct near a hood. It represents the suction that is available to draw air into the hood.

hook
A point within software at which additional steps or code can be easily added at a later time.

hook grip
A type of grip where only the fingers flex around an object, with the thumb not being used.

hookworm
A parasitic worm that infests people and causes debilitation. Major infestations can cause anemia and retardation of mental and physical development. Adult hookworms feed on blood and tissue from the wall of the intestine. Eggs pass out in the feces, undergo a period of development in soil, and the larvae enter a new host by burrowing through the skin, usually through the sole of the foot. The first sign of the disease may appear on the skin as small eruptions that develop into pus-filled blisters; this condition is sometimes called *ground itch*. Meanwhile, the hookworms enter blood vessels and are carried by the blood into the lungs. They leave the lungs, propel themselves up the trachea, are swallowed and washed through the stomach and end up in the intestines. Here, if left

alone, they will make a permanent home, using their host's body as a source of nourishment. By the time they reach the intestines, about 6 weeks after they enter the body as larvae, the worms are full-grown adults. Each worm now attaches itself by its host's blood by contraction and expansion of its gullet. If large number of worms are present, they can cause considerable loss of blood and severe anemia. The symptoms include pallor and loss of energy while the appetite may increase. The thousands of eggs laid every day by each female worm pass out of the body in the stool, in which they can easily be seen. If the stool is not properly disposed of, the larvae that hatch from the eggs may infect other persons.

hopper
A top-loading, funnel-shaped structure for temporary storage of loose materials, which will be dispensed from the bottom.

hopper barge
An open-compartment barge used for dry bulk cargo that does not require protection from the weather.

hopper body
Truck body capable of discharging its load through a bottom opening without tilting.

horizon
The apparent boundary line between the earth's surface and the sky.

horizontal axis of Helmholtz
The horizontal axis connecting the centers of rotation of the two eyes.

horizontal disparity
See ***binocular disparity***.

horizontal fault
Seismology. A fault with no dip. Still theoretical, this sort of fault should only exist within a region of strong compression or extension where the tectonic forces required for such movement could be present.

horizontal job enlargement
See ***job enlargement***.

horizontal leg room
See ***knee well width*** *and* ***knee well depth***.

horizontal plane
Any plane parallel to the floor, ground, or other reference surface.

horizontal scroll
Move the cursor sufficiently to the left or right under operator control such that the display changes to present information not visible before.

horizontal standard
An OSHA standard that essentially has application across a number of different industries, such as the Hazard Communication Standard and other General Industry Standards.

hormone
A chemical substance found in one organ or part of the body and carried in the blood to another part. Hormones can alter the function and sometimes the structure of one or more organs. Hormones act as chemical messengers to body organs, stimulating certain life processes and retarding others. Growth, reproduction, sexual attributes, and even mental conditions and personality traits are dependent on hormones. Hormones are produced by various organs and body tissues, but mainly by the endocrine glands (such as the pituitary and gonads).

Horner's syndrome
Sinking in of the eyeball, ptosis of the upper eyelid, slight elevation of the lower lid, constriction of the pupil, narrowing of the palpebral fissure and anhidrosis caused by paralysis of the cervical sympathetic nerve supply.

horology
The study of time measurement, including the principles and technologies involved in the time-measuring devices.

horopter
The locus of points in space which produce images falling on the corresponding points of both eyes with a constant amount of convergence such that a single image is seen.

horse
Transit (slang). A tractor or power unit.

horse latitudes
The belt of latitude at about 30° to 35° where winds are predominantly light and weather is hot and dry.

horse light
Spotlight mounted on cab to reveal open-range livestock.

horse scaffold
A scaffold for light or medium duty, composed of wooden or metal horses supporting a work platform.

horse van body
Truck designed for the transportation of valuable horses (livestock).

horsepower (hp)
(1) A unit of measure of work done by a machine equal to 745.7 watts or 33,000 foot-pounds per minute. (2) The amount of work that an engine can perform within a given time.

hose mask
Respiratory protective device that supplies air to the wearer from an uncontaminated source through a hose that is connected to the facepiece.

host
(1) In genetics, an organism, simple or complex and including humans, that is capable of being infected by a specific agent. (2) In medicine, an animal infected by another organism.

host factors
The personal characteristics of individuals who harbor or nourish a parasite.

hostility
An outwardly directed expression of anger, animosity, or antagonism toward another entity.

hot-deck imputation
A statistical procedure for deriving a probable response to a questionnaire item concerning a household or vehicle, where no response was given during the survey. To perform the procedure, the households or vehicles are sorted by variables related to the missing item. Thus, a series of "sort categories" are formed, which are internally homogeneous with respect to the sort variables. Within each category, households or vehicles for which the questionnaire item is not missing are randomly selected to serve as "donors" to supply values for the missing item of "recipient" households or vehicles.

hot flow
A flow of a hazardous commodity in a newly assembled system to normally passivate system walls and components and to remove residual, nonactive contaminants or flushing fluid. The hot flow is not intended for leak checks because of the potential hazards due to leaks.

hot lines
A confidential telephone service used by employees or agents of an organization for internal or external reporting of law or corporate policy. It may be in-house or an outside service. *See **self-reporting system**.*

hot load
Transit (slang). A rush shipment of cargo.

hot-wire anemometer
A device, also known as a *thermal anemometer,* used to measure air velocity by the cooling effect of moving air over a heated element.

Welding is just one example of "hot work"

hot work
Mechanical or other work that involves a source of heat, sparks, or other source of ignition that is sufficient to cause ignition of a flammable material. Work involving sources of ignition or temperatures high enough to cause the ignition of a flammable mixture. Examples include welding, burning, soldering, use of power tools, operating engines, sandblasting, electric hot plates, explosives, open fires, portable electrical equipment which has not been tested and classified as

intrinsically safe, and other sources of ignition.

hot work permit
The employer's written authorization to perform operations (for example, riveting, welding, cutting, burning, and heating) capable of providing a source of ignition.

hour
A unit of time, corresponding to 1/24 of the time required for the earth to rotate about its axis.

household
A group of persons whose usual place of residence is a specific housing unit; these persons may or may not be related to each other. The total of all U.S. households represents the total civilian noninstitutionalized population. Does not include group quarters (i.e., 10 or more persons living together, none of whom are related).

household trip
One or more household members traveling together.

household vehicle
A motorized vehicle that is owned, leased, rented or company owned and available to be used regularly by household members during the travel period. Includes vehicles used solely for business purposes or business-owned vehicles if kept at home and used for the home to work trip (e.g., taxicabs, police cars, etc.) which may be owned by, or assigned to, household members for their regular use. Includes all vehicles that were owned or available for use by members of the household during the travel period even though a vehicle may have been sold before the interview. Excludes vehicles that were not working and not expected to be working within 60 days, and vehicles that were purchased or received after the designated travel day.

housekeeping
The maintenance of the orderliness and cleanliness of an area or facility.

housing unit
A house, apartment, a group of rooms, or a single room occupied or intended for occupancy as separate living quarters. Separate living quarters are those in which the occupants do not live and eat with any other persons in the structure and which have either a) direct access from the outside of the building or through a common hallway intended to be used by the occupants of another unit or by the general public, or b) complete kitchen facilities for the exclusive use of the occupants. The occupants may be a single family, one person living alone, two or more families living together, or any other group of related or unrelated persons who share living arrangements.

HOV
*See **high occupancy vehicle**.*

hover check
Used to describe when a helicopter/vertical takeoff and landing (VTOL) aircraft requires a stabilized hover to conduct a performance/power check prior to hover taxi, air taxi, or takeoff. Altitude of the hover will vary based on the purpose of the check.

hover taxi
Used to describe a helicopter/vertical takeoff and landing (VTOL) aircraft movement conducted above the surface and in ground effect at airspeeds less than approximately 20 knots. The actual height may vary, and some helicopters may require hover taxi more than 26 feet above ground level (AGL) to reduce ground effect turbulence or provide clearance for cargo slingloads.

how do you hear me?
Aviation. Communication Protocol. A question relating to the quality of the transmission or to determine how well the transmission is being received.

hp
*See **horsepower**.*

HP
*See **health physics**.*

HPD
*See **hearing protective device**.*

HPLC
High performance liquid chromatography.

HPMS
*See **Highway Performance Monitoring System**.*

HPS
*See **Health Physics Society**.*

HRI
Hazard risk index. *See* ***risk assessment code***.

HRIS
See ***Highway Research Information Service***.

HRS
See ***hazardous ranking system***.

HS
See ***hydrogen sulfide***.

HSI
Heat stress index. *See* ***Belding-Hatch heat stress index.***

HSWA
See ***Hazardous and Solid Waste Amendments of 1984***.

html
Computing. See ***hypertext markup language***.

http
Computing. Acronym for *hypertext transfer protocol.*

HUD
(1) Department of Housing and Urban Development (United States). (2) *See* ***head-up display***.

hue
A perceptual attribute of color determined primarily by the wavelength of the light entering the eye.

hue composition (H_c)
An expression of hue as percentages of the components.

hue contrast
See ***chromatic contrast***.

hull inspector
Maritime Navigation (slang). Colloquial river term for any large piece of drift or submerged piling, log, rock, etc.

human-computer dialogue
The interchange of data, commands, or information in those activities between a human and computer.

human-computer interaction
The total of the relationship and activities occurring between a human operator and a computer or terminal.

human-computer interface (HCI)
The total of the relationship and activities occurring between a human operator and a computer or terminal. Also referred to as *man-computer interface, user-computer interface, computer-human interface (CHI).*

human describing function
See ***human transfer function***.

human ecology
The study of the relationships of individuals with each other and with their community's environment.

human engineering
See ***human factors engineering***.

human-environment interface
Any region of contact between man and his surroundings.

human error
The end result of multiple factors which influence human performance in a given situation. An often overused causal factor finding which, by itself, is not entirely descriptive of a true accident cause. Human error is considered more a symptom than a cause. *See also* ***human factor***.

human error probability (HEP)
A measure of the likelihood of occurrence of a human error under special conditions:

$$HEP = error\ count/number\ of\ possibilities$$

human factor
Any one of a number of underlying circumstances or conditions which directly or indirectly affect human performance. These include physical as well as psychological factors that can potentially lead a person to make an error in judgment or action (human error) resulting in an accident. *See also* ***ergonomics***.

human factors analysis
A systematic study of those elements involving a human-machine interface or other situation with the intent of improving working conditions, operations, or an individual's well-being. Also referred to as *ergonomic analysis*.

human factors engineer
One who has the appropriate education, training, and experience to be capable of properly performing human factors engineering activities.

human factors engineering
The use of information derived from human factors research, theory, and modeling for the specification, design, development, testing, analysis, and evaluation of products or systems for human use. *See also* ***ergonomics***.

human factors specialist
An individual who has the necessary educational, training, and experimental background to have a working understanding of human factors principles and is capable of research or other work toward achieving human factors goals.

human immune deficiency virus (HIV)
Also called the human "immunodeficiency" virus, HIV is the virus that causes acquired immune deficiency syndrome (AIDS).

human life cycle safe concentration (HLSC)
The highest concentration of a substance which will not cause an adverse effect when humans are exposed continuously over their life times.

human-machine interface (HMI)
Any region or point at which a person interacts with a machine.

human-machine system
A system in which the functions of both man and machine are interrelated, both being necessary for proper system operation.

human modeling
The use of any system which is capable of modeling one or more human structures or other characteristics for education, research, or engineering purposes.

human operator
An individual who is involved in the routine control, function, or support of a system or subsystem, but is specifically not involved in any maintenance on that system.

human performance
(1) The degree to which an individual's skill or ability is implemented in a specific task. (2) Any result from the measurement of human activity under specified conditions.

human performance technology
The use of people, systems, and/or programs to influence behavior and accomplishment.

human reliability
The probability that an individual or group will adequately perform a given task at the appropriate time.

human resources engineering
The process of using human skill resources as factors in design tradeoffs.

human tolerance
The ability of the human body and/or psyche to withstand physical and/or mental stresses without permanent injury or damage.

human transfer function
A mathematical description of what output(s) the human operator would produce as a function of specific input(s).

humane
Under the Federal Mammal Protection Act of 1972: In the context of taking a marine mammal, means that method of taking which involves the least possible degree of pain and suffering practicable to the mammal involved.

Humanscale
A manual modeling system for estimating body link, strength, postures, and other aspects for use in human factors engineering.

humectant
Any chemical which absorbs and helps retain moisture.

humeral breadth
See ***elbow breadth***.

humerus
The bone in the upper arm, extending from shoulder to elbow. It consists of a shaft and two enlarged extremities. The proximal end has a smooth round head that articulates with the scapula to form the shoulder joint. Just below the head are two rounded processes called the greater and lesser tubercles. Just below the tubercles is the surgical neck, so named because of its liability to fracture. The distal end of the humerus has two articulating surfaces: the trochlea, which articulates with the ulna, and the capitulum, which articulates with the radius.

humidify
Increase the water vapor content of the atmosphere.

humidistat
A device for measuring and/or controlling humidity levels.

humidity
The amount of moistness or dampness in the air. *See also* ***relative humidity***.

humiture
An index that relates air temperature and relative humidity to how hot it feels.

humor
Any fluid or semifluid in the body.

humping
Rail Operations. The process of connecting a moving rail car with a motionless rail car within a rail classification yard in order to make up a train. The cars move by gravity from an incline or "hump" onto the appropriate track.

humus
Decomposed organic material.

hunchback
A rounded deformity, or hump, of the back, or a person with such a deformity. The condition is also called kyphosis and is the result of an abnormal backward curvature of the spine.

hunger
The feeling of a need for food to satisfy an empty feeling in the stomach.

Hunter Lab color system
A color ordering system which is defined from a simple relationship to the CIE X, Y, and Z tristimulus values and is specified by lightness (L), redness or greenness (a), and yellowness or blueness (b).

Huntington's chorea
A hereditary type of chorea which develops in adults and is accompanied by mental deterioration. Also referred to as *adult chorea*. *See also* ***Sydenham's chorea***.

hurdle
A colloquial term for a dike. *See also* ***dike***.

hurricane
A severe tropical cyclone having winds in excess of 64 knots (74 mph). Such storms originate in warm tropical waters of the Atlantic Ocean, Caribbean Sea, or Gulf of Mexico and have circulation about their centers. According to the Saffir/Simpson Scale, there are five Categories of hurricanes which are dictated by sustained wind speeds, as follows:

CATEGORY	WIND/MPH
I	74 – 95 storm surge 4-5 feet above normal
II	96 – 110 storm surge 6-8 feet above normal
III	110 – 130 storm surge 9-12 feet above normal
IV	131 – 155 storm surge 13-18 feet above normal
V	156+ storm surge greater than 18 feet above normal

The tremendous power of a hurricane as seen from space

HVAC
An air handling system designed primarily for temperature, humidity, odor control, and air quality. *See* ***heating, ventilation and air conditioning system***.

HVL
See ***half-value layer***.

hybrid
A cell or organism resulting from a cross between two unlike plant or animal cells or organisms.

HYBRID
One of a series of anthropomorphic dummies developed for use in automotive and aircraft crash testing.

HYBRID II
An instrumented anthropomorphic dummy used in automobile head-on collision research.

HYBRID III
An instrumented anthropomorphic dummy used in automobile head-on collision research. Also referred to as *Part 572 dummy*.

hybrid rulemaking
A process of rulemaking that has elements of both formal and informal rulemaking procedures.

hybridoma
A hybrid cell that produces monoclonal antibodies in large quantities.

hydrargyria
Chronic mercury poisoning.

Hydraset
The trade name for a closed circuit hydraulically operated instrument installed between a crane hook and load that allows precise control of lifting operations and provides an indication of applied load; a precision load positioning device.

hydration
The process of absorbing or combining with water; the chemical addition of water to a compound.

hydraulic
Operated by water or any other liquid under pressure, including all hazardous fluids as well as typical hydraulic fluids that are normally petroleum based.

hydraulic head
The distance between the respective elevations of the upstream water surface (headwater) above and the downstream surface water (tailwater) below a hydroelectric power plant.

hydrocarbon
A compound composed solely of the two elements hydrogen and carbon. The simplest and lightest forms of hydrocarbon are gaseous. With greater molecular weights, they are liquid, while the heaviest are solids.

hydrochloric acid (HCL)
A normal constituent of gastric juice in man and other animals. The absence of free hydrochloric acid in the stomach, called achlorhydria, may be found with chronic gastritis, gastric carcinoma, pernicious anemia, pellagra, and alcoholism. This condition is also referred to as gastric anacidity.

hydrodynamic element
A modeling fluid which is governed by pressure and volume laws.

hydrogen
A chemical element, atomic number 1, atomic weight 1.00797, symbol H.

hydrogen embrittlement
A mechanical environmental failure process that results from the initial presence or absorption of excessive amounts of hydrogen in metals, usually in combination with residual or applied tensile stresses.

hydrogen sulfide (HS)
Gas emitted during organic decomposition. Also a byproduct of oil refining and burning. In heavy concentrations, HS can cause illness.

hydrogenation
The addition of hydrogen to a gaseous substance by the use of gaseous hydrogen combined with a catalyst.

hydrogeology
The geology of groundwater, with particular emphasis on the chemistry and movement of water.

hydrologic cycle
A model that illustrates the movement and exchange of water among the earth, atmosphere, and oceans.

hydrology
The study of the distribution and movement of water.

hydrolysis
The formation of an acid and a base from a salt by the ionic dissociation of water.

hydrometer
An instrument used for determining the specific gravity of liquids.

hydrophobic
The ability to resist the condensation of water vapor. Usually used to describe "water-repelling" condensation nuclei.

hydrophylic
Materials that absorb water which results in their swelling and forming reversible gels.

hydrostatic equilibrium
The state of the atmosphere when there is a balance between the vertical pressure gradient force and the downward pull of gravity.

hydrostatic pressure
Pressure created by water at rest equally at any point within a confined area.

hydrostatic weighing
A part of one technique for estimating body composition by weighing an individual completely submerged under water to determine body volume.

hydrothermal
The generic term that refers to any geologic process involving heated or superheated water.

hygiene
Refers to the science of health and the preservation of well-being (named for the Greek God Hygeia).

hygrometer
An instrument used for the detection of atmospheric moisture. The sensing part of the instrument can be hair (*hair hygrometer*), a plate coated with carbon (*electrical hygrometer*), or an infrared sensor (*infrared hygrometer*).

hygrometry
The determination of the water vapor content of the air.

hygroscopic
Refers to substances that absorb water from the atmosphere. Usually used to describe "water-seeking" condensation nuclei.

hygrothermograph
A recording instrument which provides a simultaneous reading of ambient temperature and humidity.

hyoid bone
A U-shaped bone in the neck which is connected by ligaments to the temporal bone and which supports the tongue. Unique in that it does not articulate directly with any other bone.

hyper-
(prefix) Greater than normal, excessive.

hyperabduct
To abduct a joint beyond the normal joint range of motion limits, with or without injury.

hyperactivity
A disorder characterized by prolonged generally excessive movement, but which may be voluntarily controlled. Also referred to as *hyperkinesis, hyperkinesia,* and *hyperkinetic syndrome*.

hyperbaric
Air pressure in excess of that at sea level.

hyperbaric oxygen therapy
A treatment using pure oxygen at greater than atmospheric pressures in a pressure chamber to treat decompression sickness, lesions or sores that resist healing, and other pathologies.

hyperbarism
A condition resulting from exposure to atmospheric pressure that exceeds the pressure within the body.

hypercapnia
An excessive amount of CO_2 in the blood.

hyperemia
An excess of blood in tissue, organ, or other part of the body.

hyperextend
To extend a joint beyond its normal range of motion or comfortable working limits, with or without injury.

hyperflex
To flex a joint beyond its normal range of motion or comfortable working limits, with or without injury.

hypergolic
Ignites spontaneously upon contact, such as certain rocket fuels and oxidizers.

hyperkeratosis
Hypertrophy of the horny layer of the skin.

hyperkinesis
See ***hyperactivity***.

hyperkinetic syndrome
See ***hyperactivity***.

hypermetropia
See ***hyperopia***.

hyperopia
A refraction disorder in the eye in which the focal point of the parallel light rays from the distant object come to a focus posterior to the retina under relaxed accommodation. Commonly referred to as *farsightedness*.

hyperoxia
A condition in which the partial pressure of oxygen is greater than that found in a standard atmosphere.

hyperparathyroidism
Abnormally increased activity of the parathyroid glands, causing loss of calcium from the bones and excessive secretion of calcium and phosphorus by the kidney. Among the symptoms are kidney stones, back pain, joint pains, thirst, nausea, and vomiting. The conditions also makes bones more susceptible to fracture.

hyperplasia
The abnormal multiplication or increase in the number of normal cells in normal arrangement in tissue.

hyperpnea
An increased depth and rate of respiration.

hypersensitivity
A state of altered reactivity in which the body reacts to a foreign agent more strongly than normal; anaphylaxis and allergy are forms of hypersensitivity.

hypersensitivity diseases
Diseases characterized by allergic responses to animal antigens. The hypersensitivity diseases most clearly associated with indoor air quality are asthma, rhinitis, and hypersensitivity pneumonitis. Hypersensitivity pneumonitis is a rare but serious disease that involves progressive lung damage as long as there is exposure to the causative agent.

hypersensitivity pneumonitis
See ***building-related illness***.

hypersonic
Traveling at or pertaining to a velocity equal to or greater than five times the velocity of sound.

hypertension
(1) A state in which an individual chronically maintains an arterial blood pressure higher than optimal levels, generally ≥90 mm Hg diastolic and/or ≥140 mm Hg systolic. (2) A state in which a muscle is overly tensed.

hypertensive
(1) Characterized by or causing increased tension or pressure, as abnormally high blood pressure. (2) A person with abnormally high blood pressure.

hypertext markup language (html)
A standard language for creating documents on the World Wide Web.

hyperthermia
A marked sustained increase in body temperature due to the inability of the body to dissipate excessive heat generated through metabolic activity that can result in severe cellular damage and death if not treated promptly.

hyperthyroidism
Excessive functional activity of the thyroid gland. The condition is called also *thyrotoxicosis,* and is often accompanied by *goiter*. Symptoms include profuse sweating, dislike of heat, palpitation, insomnia, nervousness, and excitability. The basal metabolic rate is increased. Sometimes there is diarrhea. There may also be bulging of the eyes, in which case the condition may be referred to as *exophthalmic goiter,* or *Grave's disease.*

hypertonia
Having an above normal muscle tension.

hyperventilation
Abnormally prolonged, rapid, and deep breathing. This results in reduced carbon dioxide in the blood (acapnia) and consequent apnea (intermittent cessation of breathing). Symptoms include faintness (or impaired consciousness without actual loss of consciousness).

hyphenated point
Transportation. Basically, two or more neighboring communities which, in terms of authorization shown in a carrier's Certificate of Public Convenience and Necessity, are treated as a single community.

hypnotic
Substance that induces sleep or sleepiness.

hypo-
(prefix) Less than normal.

hypoallergenic
Having a low probability of stimulating allergic reactions.

hypobaric
Air pressure below that which exists at sea level.

hypocenter
The point of origin of an earthquake. It can be expressed with no fewer than three measurements: latitude, longitude, and depth. Also known as the *focus*.

hypochondriac
(1) Pertaining to the hypochondrium. (2) A person affected with hypochondriasis.

hypochondriasis
An abnormal concern about one's health. The hypochondriac exaggerates trivial symptoms and often believes that he/she is suffering from some serious ailment. True hypochondriasis is a type of neurosis caused by an unresolved conflict in the person's unconscious mind.

hypochondrium
The abdominal region on either side, just below the thorax.

hypodynamia
The lack of gravitational loading on the skeleton.

hypoglossal nerve
A cranial nerve which regulates part of the motor activity of the tongue.

hypoglycemia
An abnormally low level of sugar (glucose) in the blood. The condition may result from an excessive rate of removal of glucose from the blood or from decreased secretion of glucose into the blood. Overproduction of insulin from the islands of Langerhans in the pancreas or an overdose of exogenous insulin can lead to increased utilization of glucose, so that glucose is removed from the blood at an accelerated rate.

hypogravity
A state in which a significantly reduced gravitational force is experienced, generally with reference to the accepted standard gravitational force of the earth at its surface.

hypokinesia
An abnormally reduced capacity for voluntary muscular movement while having full, normal consciousness.

hypokinetic hypoxia
A hypoxic condition due to the reduced flow of blood.

hypoparathyroidism
A disorder caused by underproduction of the parathyroid hormone. It most often occurs as a result of accidental removal of, or damage to, one or all of the parathyroids during thyroid surgery. Insufficiency of parathyroid hormone causes lowering of the calcium content of the blood and may result in the calcium content of the blood and may result in tetany, of which most obvious sign is spasm of the muscles, especially those of the fingers and toes.

hypoplasia
The incomplete development of an organ so that it fails to reach adult size.

hyposensitivity
(1) Abnormally decreased sensitivity. (2) The specific or general ability to react to a specific allergen reduced by repeated and gradually increasing doses of the offending substance.

hypotension
Diminished tension; lowered blood pressure. A consistently low blood pressure with a systolic pressure less than 100 mm of mercury is no cause for concern. In fact, low blood pressure is associated with long life and an old age free of illness. An extremely low blood pressure is occasionally a symptom of a serious condition. Hypotension may be associated with Addison's disease and inadequate thyroid function, but in both cases the primary disease produces so many other symptoms that the hypotension is considered comparatively unimportant.

hypotensive
Condition in which there is a lack of oxygen supply to the tissues.

hypothalamus
The portion of the brain that controls body temperature and produces hormones that affect the pituitary gland.

hypothenar
Pertaining to the fleshy mass on the medial/ulnar side of the palm.

hypothenar eminence
The fleshy protrusion on the medial ulnar side of the palm.

hypothermia
Loss of body heat and decreased temperature due to extensive exposure to cold.

hypothesis
An assumption which may be accepted or rejected, based on experimental findings, such as by statistical tests of significance.

hypothesis testing
The conducting of a properly controlled experiment, including any supporting statistical analyses, to determine the likelihood of a hypothesis being true.

hypothetical question
A form of question framed in such a manner as to call for an opinion from an expert based on a series of assumptions claimed to have been established as fact by the evidence, and a ground for inferring guilt of innocence, as the case may be, or as indicating a probable or possible motive for the crime.

hypothyroidism
Deficiency of thyroid gland activity, with underproduction of thyroxine, or the condition resulting from it.

hypotonia
A condition involving decreased muscle tone.

hypotonic
(1) Having an abnormally reduced tonicity or tension. (2) Having an osmotic pressure lower than that of the solution with which it is compared.

hypoventilation
Decrcasc of air in the lungs below the normal amount.

hypoxemia
Deficient oxygenation of the blood.

hypoxia
A condition experienced by humans when the brain does not receive sufficient oxygen. Anemic hypoxia is the reduction of the oxygen-carrying capacity of the blood as a result of a decrease in the total hemoglobin or as the result of an alteration of the hemoglobin constituents.

hypsokinesis
A backward swaying or falling in erect posture, seen in paralysis agitans and other neurologic disorders.

hysteresis
The maximum difference in output for any given input when the value is approached first with increasing input signal then with decreasing input signal. The nonuniqueness in the relationship between two variables as a parameter increase or decrease.

hysteresis error
The difference in response output when increasing a variable as opposed to decreasing that variable.

hysteretic damping
Damping due to the internal mechanical properties of materials.

hysteria
A highly emotional state. A form of psychoneurosis in which the individual converts anxiety created by emotional conflict into physical symptoms that have no organic basis.

hyzone
An unstable, triatomic form of hydrogen, H_3.

Hz
See ***Hertz***.

HZE
See ***high-energy heavy ion***.

I

I say again
Aviation. Communication Protocol. The message will be repeated.

IADL
See ***instrumental activities of daily living***.

IAEA
International Atomic Energy Agency.

IAP
See ***intra-abdominal pressure***.

IAQ
See ***indoor air quality***.

IARC
International Agency for Research on Cancer.

IATA
See ***International Air Transportation Association***.

IC
Integrated circuit. Also *ion chromatography*. See also ***inspiratory capacity***.

ICAO
See ***International Civil Aviation Organization***.

ICAO word list
A standard word list in which the first letter of each word represents the corresponding sequence of letters in the alphabet (e.g., alpha, bravo, charlie, delta, echo, foxtrot, gulf, hotel, india, juliet, kilo, lima, etc.)

ICC
Interstate Commerce Commission (United States).

ice action on bridge piers
The force required to break ice, transmitted to bridge piers and other structures in the river. Such a force could damage the structures.

Ice Age
See ***Pleistocene epoch***.

ice clause
Maritime Law. A standard clause in the chartering of ocean vessels. It dictates the course a vessel master may take if the ship is prevented from entering the loading or discharging port because of ice, or if the vessel is threatened by ice while in the port. The clause establishes rights and obligations of both vessel owner and charterer if these events occur.

ice crystal process
A process that produces precipitation. The process involves tiny ice crystals in a supercooled cloud growing larger at the expense of the surrounding liquid droplets. Also called the *Wegener-Bergeron-Findeisen process*.

ice fog
A type of fog composed of tiny suspended ice particles that forms at very low temperatures.

ice gorge
A conglomeration of ice solidly packed from bank to bank which is obstructing the flow of the river and marine traffic.

ice nuclei
Particles that act as nuclei for the formation of ice crystals in the atmosphere.

ice pellets
See ***sleet***.

ice pier
A heavily constructed cluster of piling or concrete behind which towboats moor or shelter from running ice.

ice shelf
Seaward extension of an ice sheet, floating but attached to the land on at least one side and bounded on the seaward side by a steep cliff rising 2 to 50 m or more above sea level.

iceberg
A large mass of detached land ice in the sea or stranded in shallow water.

Icelandic low
The subpolar low-pressure area that is centered near Iceland on charts that show mean sea level pressure.

ICHCA
See ***International Cargo Handling Coordination Association***.

ichthyismus
Disease caused by eating rancid fish or poisonous fish.

ichthyosis
Dryness, roughness, and scaliness of the skin, resulting from the failure of shedding of the keratin produced by the skin cells.

icon
(1) A graphical, nonlinguistic representation of an object or action. (2) A small picture that represents a function, file, or program. In Windows, for example, users can run programs by choosing icons rather than having to remember the program name and type a command.

iconic memory
A sensory memory associated with the visual system.

ICPES
See ***inductively coupled plasma emission spectroscopy***.

ICRP
International Commission on Radiological Protection.

icterus
Jaundice due to the deposition of bile pigment in the skin and mucous membranes with a resulting yellow appearance of the individual.

ICW
See ***intracellular water***.

i.d.
Inside diameter.

id
(1) A Freudian term used to describe that part of the personality which harbors the unconscious, instinctive impulses that lead to immediate gratification of primitive needs such as hunger, the need for air, the need to move about and relieve body tension, and the need to eliminate. Id impulses are physiological and body processes, as opposed to the ego and superego, which are psychological and social processes. The id is dominated by the pleasure principle and some gratification of the id impulses is necessary for survival of a person's personality. (2) A skin eruption occurring as an allergic reaction to an agent causing primary lesions elsewhere.

ideal blackbody
See ***blackbody***.

ideal radiator
See ***blackbody***.

ideal spectrum
A frequency distribution in which a pure tone appears as a vertical line due to perfectly sharp filtering.

ideation
The mental process(es) through which ideas are formed.

ideational fluency
The ability to generate a number of ideas on a given topic.

ident
Aviation. Communication Protocol. A request for a pilot to activate the aircraft transponder identification feature. This will help the controller to confirm an aircraft identity or to identify an aircraft.

ident feature
A special feature in the Air Traffic Control Radar Beacon System (ATCRBS) equipment. It is used to immediately distinguish one displayed beacon target from other beacon targets.

identification
(1) A mental mechanism by which an individual unconsciously takes as his or her own characteristics, postures, achievements, or other identifying traits of other persons or groups. (2) The official legends "For Official Use Only" and "U.S. Government," and other legends showing either the full name of the department, establishment, corporation, or agency by which it is used, if such title readily identifies the department, establishment, corporation, or agency concerned.

identification lamps
Lamps used to identify certain types of commercial motor vehicles.

identified
As used in reference to a conductor or its terminal, means that such conductor or terminal can be readily recognized as grounded.

idiopathic
A disease of unknown origin or cause.

idiosyncratic error
A type of human error due to peculiarities of an individual's characteristics, such as attitudes, social problems, or emotional state.

idle thrust
The jet thrust obtained with the engine power control level set at the stop for the least thrust position at which it can be placed.

idle time
A temporal interval, excluding standby time, during which a worker, a piece of equipment, or a system is at the workplace, but not producing output, regardless of the cause. See also ***delay time***.

IDLH
*See **immediately dangerous to life and health**.*

IEEE
Institute of Electrical and Electronics Engineers.

IEMG
*See **integrated electromyogram**.*

IES
Illuminating Engineering Society.

IF
*See **intermediate fix**.*

if no transmission received for (time)
Aviation. Communication Protocol. Used by Air Traffic Control (ATC) in radar approaches to prefix procedures which should be followed by the pilot in the event of lost communications.

IFR
*See **instrument flight rules**.*

IFS
*See **in-flight survey**.*

ignitable
Capable of burning or causing fire.

ignitable waste
A waste that poses a fire hazard during routine storage, handling, or disposal.

ignition
The introduction of some external spark, flame, or glowing object that initiates self-sustained combustion.

ignition temperature
(1) The lowest temperature that will cause a gas/vapor to ignite and burn independent of the heating source. (2) The lowest temperature at which sustained combustion for a volatile substance will occur when heated in air or another specified oxidizing environment.

ignitor
A device containing a specifically arranged charge of ready burning composition, usually black powder, used to amplify the initiation of a primer.

IH
Industrial hygienist or industrial hygiene.

ileitis
Inflammation of the ileum, or lower portion of the small intestine. It may result from infection, obstruction, severe irritation, or faulty absorption of material through the intestinal walls. A specific type of inflammation of unknown cause involving the small and large intestines is known as regional ileitis, regional enteritis, or Crohn's disease. The advanced stage is marked by hardening, thickening, and ulceration of parts of the bowel lining. An obstruction may cause the development of a fistula. A common symptom of ileitis is pain in the lower right quadrant of the abdomen or around the umbilicus. Other symptoms include loss of appetite, loss of weight, anemia, and diarrhea, which may alternate with periods of constipation.

ileus
Intestinal obstruction, especially failure of peristalsis. The condition frequently accompanies peritonitis and usually results from disturbances in neural stimulation of the bowel.

iliac crest
The lateral, superior rim of the coxal bone.

iliac spine
A projection from the coxal bone at the anterior portion of the iliac crest.

iliocristale height
The vertical distance from the floor or other reference surface to the highest point of the iliac crest in the midaxillary plane. Measured with the individual standing erect and his/her weight equally balanced on both feet.

iliospinale
The most anterior point on the iliac spine.

iliospinale height
The vertical distance from the floor or other reference surface to iliospinale. Measured with the individual standing erect and his/her weight evenly distributed between both feet.

illegally obtained evidence
Evidence which is obtained in violation of a defendant's rights because officers had no

warrant and no probable cause to arrest or because the warrant was defective and no valid grounds existed for seizure without a warrant.

illiteracy
Having no ability to read and write.

illness
(1) A condition or pronounced deviation from the normal health state; sickness. Illness can be the result of disease or injury. (2) Sickness, disease, or disorder of body or mind.

illness incident rate
The number of annual occupational illnesses experienced by a company in one year, based on 100 full-time employees. Expressed as:

$$IIR = \frac{No.\ of\ illnesses\ x\ 200{,}000}{No.\ of\ man\text{-}hours\ worked}$$

illuminance
The amount of light falling on a surface. Illuminance is expressed in units of foot-candles or lux.

illuminance category
An alphabetic character, ranging from A through H, representing illumination ranges for various types of work such that the further the letter is from A, the brighter the light.

illuminance meter
A device, composed of a photodetector, filter, and electronic circuitry, for measuring the luminous flux incident on a plane.

illuminance threshold
That lowest luminance level which the eye or other image sensor is capable of detecting, given a specified luminance contrast, position within the field of view, dark adaptation, flicker rate, source dimensions, and color.

illuminant
Any light source or combination of flight sources.

illuminant A
A standard CIE illuminant corresponding to a typical tungsten filament incandescent lamp.

illuminant B
A standard CIE illuminant corresponding to direct sunlight.

illuminant C
A standard CIE illuminant corresponding to average daylight.

illuminant D
A series of standard CIE illuminants corresponding to a daylight which measures beyond the normal visible spectrum.

illuminate
To distribute or provide light to an area or region.

illumination
The density of light flux incident upon a surface.

illusion
A perceptual misinterpretation of a stimulus.

ILO
International Labor Organization (of the United Nations).

IM
See ***inner marker***.

image
(1) The sum of the perceptions by an individual, group, or population about itself or another entity. (2) An electronic or photographic representation of onc or more entities. (3) A subjective sensory experience, especially in the visual modality.

image analysis
Any computer or other electronic processing to quantify an image, usually with the intent of deriving some statistically based conclusions.

image enhancement
That electronic or other processing to improve the resolution, features, or other quality of an electronic or photographic image.

image processing
Any type of computer-based alteration of the data representing an image, including enhancement, analysis, and reconstruction.

image reconstruction
The process of re-working data for image enhancement.

IMC
See ***instrument meteorological conditions***.

immaterial evidence
Evidence which lacks probative weight and is unlikely to influence the tribunal in resolving

the issue before it. Such evidence is commonly objected to by opposing counsel, and disallowed by the court.

immaterial facts
Those which are not essential to the right of action or defense.

immaterial issue
In pleading, an issue taken on an immaterial point; that is, a point not proper to decide the action.

immediate cause
The last of a series or chain of causes tending to a given result, and which, of itself, and without the intervention of any further cause, directly produces the result or event. A cause may be immediate in this sense, and yet not "proximate," and, conversely, the proximate cause (that which directly and efficiently brings about the result) may not be immediate. The familiar illustration is that of a drunken man falling into the water and drowning. His intoxication is considered the proximate cause of his death, if it can be said that he would not have fallen into the water when sober; but the immediate cause of death is suffocation by drowning. *See also* ***proximate cause***.

immediate danger
The definition of "immediate danger" as part of the humanitarian doctrine contemplates that there be some inexorable circumstance, situation, or agency bearing down on the plaintiff with reasonable probability of danger prior to the negligent act of the defendant.

immediately
Aviation. Communication Protocol. Used by Air Traffic Control (ATC) when such action compliance is required to avoid an imminent situation.

immediately dangerous to life and health (IDLH)
The maximum level to which a healthy individual can be exposed to a chemical for thirty minutes and escape without suffering irreversible health effects or impairing symptoms. For example, some materials such as hydrogen fluoride gas and cadmium vapor may produce immediate transient effects that, even if severe, may pass without medical attention, but are followed by sudden, possibly fatal collapse 12-72 hours after exposure. The victim "feels normal" after recovery from transient effects until collapse. Such materials in hazardous quantities are considered to be "immediately" dangerous to life or health.

immersion foot
That damage to the skin, blood, vessels, and nerves of the feet resulting from prolonged exposure to water at temperatures between freezing and approximately 60°F.

imminent danger
Any conditions or practices in a place of employment which are such that danger exists which could reasonably be expected to cause death or serious physical harm immediately or before the imminence of such danger can be eliminated.

imminent hazard
(1) *General.* A hazardous situation, condition, or circumstance the nature of which poses a serious and imminent threat to human health or the environment. If actions are not taken to immediately correct or stop the hazard cause, the results could be catastrophic. (2) *Federal Insecticide, Fungicide, and Rodenticide Act.* A situation which exists when the continued use of a pesticide during the time required for cancellation proceedings would be likely to result in unreasonable adverse effects on the environment or will involve unreasonable hazard to the survival of a species declared endangered or threatened by the Secretary of Agriculture pursuant to the Endangered Species Act of 1973.

imminently hazardous chemical substance or mixture
A chemical substance or mixture which presents an imminent and unreasonable risk of serious or widespread injury to health or to the environment. Such a risk to health or the environment shall be considered imminent if it is shown that the manufacture, processing, distribution in commerce, use, or disposal of the chemical substance or mixture, or that any combination of such activities, is likely to result in such injury to health or the environment.

immiscible
Not capable of being uniformly mixed or blended.

immune
Not affected or responsive. Not susceptible to a particular disease.

immunity
Not susceptible. Biologically, immunity is usually to a specific infectious agent and is one result of infection. The quality or condition of being immune. An inherited, acquired, or induced condition to a specific pathogen. The power of the body to successfully resist infection and the effects of toxins.

immunoassay
The measurement of an antigen-antibody interaction.

immunodeficient
Lacking in the ability to produce antibodies in response to an antigen.

immunoglobulin
Serum globulin having antibody activity. Most of the antibody activity apparently resides in the gamma fraction of globulin.

immunotoxin
An antibody to the toxin of a microorganism, zootoxin (spider or bee toxin), or phytotoxin (toxin from a plant) which combines specifically with the toxin, resulting in the neutralization of its toxicity.

IMO
See ***International Maritime Organization***.

impact
A rapid transmission of physical momentum from one object to another in a mechanical system.

impact acceleration
An acceleration lasting less than one second.

impact acceleration profile
A graphical display or plot of the deceleration sequence experienced by a vehicle in a crash.

impact analysis
A subjective technique for attempting to quantify the positive and negative aspects of a system or plan.

impact attenuation
The reduction in impulsive forces due to cushioning or other means of spreading out the forces in space or time.

impact biodynamics
See ***biodynamics***.

impact load
A force implemented by a rapid blow.

impact noise
Variations in the noise level such that the maximum noise level occurs at intervals of greater than one second.

impact strength
The impulse energy required to fracture a material.

impact velocity
The velocity at which one object strikes another.

impaction
The forcible contact of particles with a surface. The cascade impactor is a device that operates on this principle.

impactor
An object which makes contact with another body or structure.

impairment
Any dysfunction in which one or more body systems or subsystems are not capable of functioning to the degree considered normal.

impartial expert
A witness appointed by tribunal for an unbiased opinion on a matter addressed to the court.

impedance
Obstruction or opposition to passage or flow, as of an electric current or other form of energy. *See also* ***electrical impedance***.

imperial gallon
A British gallon, slightly larger than the U.S. gallon (an older term).

impermeable
Not capable of being permeated or not allowing substances to pass through the openings or interstices of the material.

impetigo
A skin disease characterized by pustules and caused by streptococci, often in association with staphylococci. The disease occurs most frequently in children, especially in very young infants because of their low resistance. It is spread by direct contact with the moist discharges of the lesions.

impingement
The process by which particulate material in air is collected by passing the air through a

nozzle or jet and impinging the air-particle mixture onto a surface that is immersed in a liquid, such as water. The particles are retained in the liquid. The midget and Greenburg-Smith impingers are examples of instruments using this principle of dust collection.

impinger
A sampling device used to collect airborne particulates. The midget impinger and the Greenburg-Smith impinger were widely used types.

implementation allowance
That time allowance provided for workers in beginning new techniques or changing to a different method to prevent them from losing income during the change.

implosion
A violent inward collapse of an item, such as an evacuated glass vessel.

importance
A subjective rating of greater worth, necessity, or regard relative to other items or functions.

importance principle
A rule stating that displays and controls with the greatest operational importance should be placed in optimum locations with regard to convenient access and visibility.

imports
Receipts of goods into the 50 states and the District of Columbia from foreign countries and from Puerto Rico, the Virgin Islands, and other U.S. possessions and territories.

impounding space
A volume of space formed by dikes and floors which is designed to confine a spill of hazardous liquid.

impounding system
Includes an impounding space, including dikes and floors for conducting the flow of spilled hazardous liquids to an impounding space.

impoundment
A body of water or sludge confined by a dam, dike, floodgate, or other barrier.

imprecision
That variance due to measurement error from repeated measurements within a short period of time, and which are attributed to measurement process only.

impression
(1) An indentation or dent. (2) A negative copy or counterpart of some object made by bringing into contact with the object, with varying degrees of pressure, some plastic material that later becomes solidified. (3) An effect on the mind or senses produced by external objects.

improbable
In terms of probability of hazard or mishap occurrence, a hazard or event whose occurrence is so unlikely during the life of an item or system, it can be assumed that the hazard will not occur.

improper loading
Maritime Safety. Loading, including weight shifting, of a vessel causing instability, limited maneuverability, or dangerously reduced freeboard.

improper lookout
Maritime Safety. No proper watch; the failure of the operator to perceive danger because no one was serving as lookout, or the person so serving failed in that regard.

impulse
(1) A human urge based more on emotional than cognitive factors and without significant consideration of possible consequences. (2) The area under the curve of a force for the brief time duration of the force application.

impulsive force
See ***impact***.

impulsive noise
An acoustic event characterized by very short rise time and duration.

impurity
Chemicals. (1) A chemical that remains in a product that is distributed in commerce. (2) A chemical substance which is unintentionally present in another chemical substance.

imputed negligence
The negligence of one person may be chargeable to another depending upon the relationship of the parties. For example, the negligence of an agent acting within the scope of his/her employment is chargeable to the principal.

in
Inch.

in^2
Square inch or square inches.

in^3
Cubic inch or cubic inches.

in bulk
The transportation, as cargo, of property, except Class A and B explosives and poison gases, in containment systems with capacities in excess of 3,500 water gallons.

in draft
Maritime Navigation. Current moving across the lock entrance toward the shore.

in-out racks
Maritime. Dry land boat storage on a vertical rack system.

in phase
Pertaining to waveforms having the same frequency and which are at the same point in their respective cycles at the same time.

in shape
Maritime Navigation. Term used when a tow is properly aligned for entering a lock or passing through a narrow channel or opening between bridge piers.

in situ
In its original place.

in-stream use
Water use taking place within a stream channel, e.g., hydroelectric power generation, navigation, water quality.

in the marks
Maritime Navigation. Proceeding along the channel line as described in the channel report. Well on the line (imaginary) running from one mark to the other or from one light to the other.

in-use mile per gallon (mpg)
A miles per gallon (mpg) that was adjusted for seasonal fluctuations and annual miles traveled.

in vitro
(1) "In glass" (a test-tube culture). (2) Any laboratory test using living cells taken from an organism. Refers to an experiment or procedure that is observable with a test tube, other laboratory equipment, or an artificial environment.

in vivo
In the living body of a plant or animal, in vivo tests are those laboratory experiments carried out on whole animals or human volunteers.

inaccessible
Incapable of being reached or entered by a human, a human body part, a remotely operated system, or a tool for retrieval or repair of a system or subsystem.

inactive aircraft
All legally registered civil aircraft which flew zero hours.

inactive window
In computing, an open, perceptually and functionally available window which must be activated before the user may work within it.

inadequate damages
Damages are referred to as "inadequate" (within the rule that an injunction will not be granted where adequate damages at law could be recovered for the injury sought) when such a recovery at law would not compensate the parties and place them in the position in which they formerly stood.

inapparent infection
Infection without recognizable clinical signs or symptoms.

inboard-outboard
U.S. Coast Guard. Regarded as inboard because the power unit is located inside the boat. Also referred to as *inboard/outdrive.*

incandescence
The emission of light and other forms of electromagnetic energy due solely to heating a source material.

incandescent lamp
A light source derived from incandescence, usually from electrical heating of a filament within a sealed bulb.

incapacitated person
Any person who is impaired by reason of mental illness, mental deficiency, physical illness or disability, advanced age, chronic use of drugs, chronic intoxication, or other cause (except minority) to the extent that he/she lacks sufficient understanding or capacity to

make or communicate responsible decisions concerning his/her person.

incendiary
A material that is primarily used to start fires.

incendive spark
A spark of sufficient temperature and energy to ignite a flammable vapor/gas.

incentive
Any condition which motivates behavior to obtain a reward or avoid punishment.

incentive operators
Those employees whose wages are determined either entirely or in part by the quality and/or quantity of their output.

incentive pace
The performance level of a worker under incentive conditions and without excess fatigue.

incentive plan
Any procedure by which an organization attempts to promote increased productivity.

incerfa (uncertainty phase)
Aviation. A situation wherein uncertainty exists as to the safety of an aircraft and its occupants.

inch
A unit of length in the English system, equal to 2.54 cm in the metric system.

inch of mercury
A unit used in measuring or expressing pressure. One inch of mercury pressure is equivalent to 0.491 pounds per square inch.

inches of water
A pressure term. One inch of water is equal to 0.0735 inches of mercury, or 0.036 pounds per square inch (psi). Atmospheric pressure at standard conditions is 407 inches water gauge (w.g.).

incidence
Number of new cases of diseases within a specified period of time.

incidence (or incident) rate
For OSHA recordkeeping purposes, the number of injuries, illnesses, or lost workdays related to a common exposure base of 100 full-time workers (working 40 hours per week, 50 weeks per year).

incident
(1) *General.* An occurrence, happening, or energy transfer that results from either positive or negative influencing events and may be classified as an accident, mishap, near-miss, or none of them, depending on the level and degree of the negative or positive outcome. (2) *Transportation.* Collisions, derailments, personal casualties, fires, and property damage in excess of $1000, associated with transit agency revenue vehicles; all other facilities on the transit property; and service vehicles, maintenance areas, and rights-of-way.

incident reporting thresholds
Transit. For an incident to be reportable, it must involve a transit vehicle or occur on transit property, and result in death, injury, or property damage in excess of $1,000.

incidental damages
Any commercially reasonable charges, expenses, or commissions incurred as a result of the stopping of a delivery; in the transportation, care and custody of goods after the buyer's breach; in connection with the return or resale of the goods; or otherwise resulting from the breach. Also, such damages, resulting from a seller's breach of contract, include expenses reasonably incurred in inspection, receipt, transportation, and care and custody of goods rightfully rejected, any commercially reasonable charges, expenses, or commissions in connection with effecting cover and any other reasonable expense incident to the delay or other breach.

incidental element
See ***irregular element.***

incidental learning
The acquisition of information or skills as a byproduct of one's simple presence or through other, unrelated activities.

incidental vibration
Any unintended vibration (an older term).

incineration
(1) Burning of certain types of solid, liquid, or gaseous materials. (2) A treatment technology destruction of waste by controlled burning at high temperatures, e.g., burning sludge to remove the water and reduce the remaining residues to a safe, nonflammable ash which can be disposed of safely on land, in some waters, or in underground locations.

incineration at sea
Disposal of waste by burning at sea on especially designed incinerator ships.

incineration vessel
According to CERCLA: Any vessel which carries hazardous substances for the purpose of incineration of such substances, so long as such substances or the residues of such substances are on board.

incinerator
A furnace for burning wastes under controlled conditions.

incipient fire stage
A fire which is in the initial or beginning stage and which can be controlled or extinguished by portable fire extinguishers, Class II type standpipe, or small hose systems without the need for protective clothing or breathing apparatus.

incisor
Any one of the four front teeth of either jaw.

inclination
A sloping or leaning; the angle of deviation from a particular line or plane of reference.

incline railway
Rail Operations. A railway used to traverse steep slopes.

inclined manometer
A manometer, used in pressure measurement, that amplifies the vertical movement of the water column through the use of an inclined leg.

inclined plane
Rail Operations. Railway operating over an exclusive right-of-way on steep grades with unpowered vehicles propelled by moving cables attached to the vehicles and powered by engines or motors at a central location not on board the vehicle.

inclined plane vehicles
Rail Operations. Special type of passenger vehicles operating up and down slopes on rails via a cable mechanism.

inclusion
(1) Any unintended or undesirable foreign particle in a finished object. (2) Enclosure within something else. (3) Anything that is enclosed.

incombustible
Incapable of burning.

incompatible
(1) Describes materials that may cause dangerous, violent, or lethal reactions when coming into direct contact with each other. (2) Not suited for harmonious coexistence or simultaneous administration; not to be combined in the same preparation or taken concomitantly.

incompetence
An inadequacy for performing a certain function, regardless of cause.

inconsistent
Mutually repugnant or contradictory. Contrary, the one to the other, so that both cannot stand, but the acceptance or establishment of the one implies the abrogation or abandonment of the other.

incontinence
An inability to control the elimination of feces and/or urine.

incorporation by reference
The inclusion of specifications, requirements, regulations, or other information into a given document simply by referring to a second document which already contains the desired information.

incremental threshold
*See **difference threshold**.*

incubate
(1) To provide proper conditions for growth and development, as to maintain optimal temperature for the growth of bacteria. (2) Material that has been incubated.

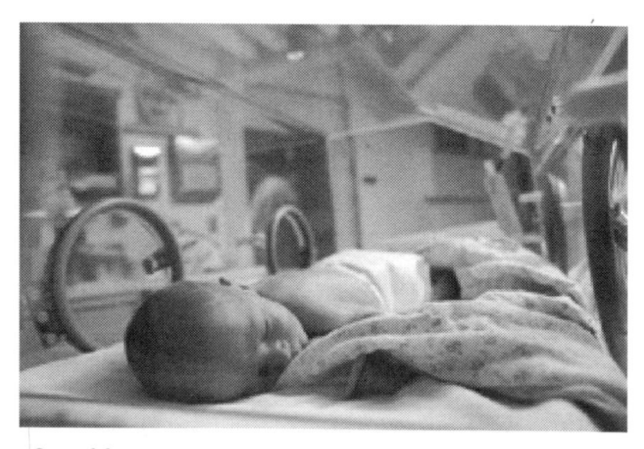

One of the more common uses of incubation, a premature infant in an incubator is provided the care it will need to survive

incubation
The growth and development of microorganisms.

incubation period
The time interval between effective exposure of a susceptible host to an agent (infection) and onset of clinical signs and symptoms of disease in that host. Incubation periods of some common communicable diseases are noted in the table below.

Disease Name	Incubation Period*	
	Average	Range
Amoebic dysentery	21 – 21	8 – 90
Anthrax	1 – 4	1 – 7
Bacillary dysentery	2 – 4	1 – 7
Brucellosis	14	6 – 30+
Chancroid	3 -5	1 – 12
Chickenpox	14	12 – 21
Cholera	3	1 – 5
Dengue	5 – 6	3 – 15
Diphtheria	2 – 5	2 – 5
Erysipelas	0 – 2	0 – 2
Food Poisoning:		
Staphylococcus	2 – 4 hr.	1 – 6 hr.
Salmonella	12 hr.	6 – 48 hr.
Botulinus	18 – 24 hr	2 – 48 hr
German measles	16 – 18	10 – 21
Gonorrhea	3 – 5	1 – 14
Hepatitis, infectious	25	15 – 35
Hepatitis, serum	80 – 100	60 – 180
Impetigo contagiosa	5	5
Infectious keratoconjunctivitis	5 – 7	5 – 7
Influenza	1 – 3	1 – 3
Malaria	10 – 17	up to 35+
Measles	9 – 14	9 – 14
Meningitis, meningococci	7	2 – 10
Mumps	18	12 – 26
Paratyphoid	1 – 10	1 – 10
Pertussis	5 – 9	2 – 21
Plague	3 – 6	3 – 6
Pneumonia, bacterial	1 – 3	1 – 3
Puerperal infection	1 – 3	1 – 3
Rabies	14 – 42	10 - 180
Relapsing fever (tick)	3 – 6	2 – 12
Relapsing fever (louse)	7	5 – 12
Rocky Mountain spotted fever	3 – 10	3 – 10
Scabies	1 – 2	1 – 2
Scarlet fever	2 – 5	2 – 5
Smallpox	12	7 – 21
Syphilis	21	10 – 90
Tetanus	4 – 21	4 – 21
Tuberculosis	variable	variable
Tularemia	3	1 – 10
Typhoid fever	7 – 14	3 – 38
Yellow Fever	3 – 6	3 – 6

* *"Average" and "Range" are in DAYS unless otherwise noted*

incubus
(1) A nightmare. (2) A heavy mental burden.

incurable disease
(1) Any disease which has reached an incurable stage in the patient afflicted therewith, according to general state of knowledge of the medical profession. (2) A disease for which there is no known cure.

incurred risk
A defense to a claim of negligence, separate and distinct from a defense of contributory negligence. It contemplates acceptance of a specific risk of which the plaintiff has actual knowledge.

incus
The middle bone of the auditory ossicles in the middle ear.

indemnification agreement
A written promise by one party that it will not hold another party liable. Also called a ***hold harmless clause***.

indemnify
To restore the victim of a loss, in whole or in part, by payment, repair, or replacement.

indemnity insurance
Insurance which provides indemnity against loss, in contrast to contracts which provide for indemnity against liability. The latter are known as liability contracts or policies, and the former as indemnity contracts or policies. *See also* ***insurance***.

independent
Not capable of being influenced by other systems.

independent audit
One conducted by an outside person or firm not connected in any way with the company or person being audited. *See also* ***audit***.

independent contractor
Generally, one who, in the exercise of an independent employment, contracts to do a piece of work according to his/her own methods and is subject to his/her employer's control only as to the end product or final result of his/her work.

independent pole scaffold
See ***double pole scaffold***.

Independent Private Sector Inspector General (IPSIG)
Legal auditor or investigator who reviews the risk management factors in a company. This

may include attesting to the adequacy of a compliance program, independent audits, and hot line operation.

independent psychomotor abilities
A set of movement capabilities reportedly determined by factor analysis to be independent of one another and which may be used for task and job analyses, performance measurement, etc.

independent surveillance
Aviation. A system which requires no airborne compatible equipment.

independent variable
A variable which can be either set to a desired value or controlled by the experimenter, or matched or observed as it occurs naturally.

indeterminate errors
Errors that occur randomly and whose cause is not determinable and thereby cannot be corrected.

index finger
The three phalanges and surrounding tissues of digit II of the hand.

index finger length
The linear distance from the thumb crotch to the tip of the index finger. Measured with the index finger fully extended. This definition is not consistent with other finger or finger segment lengths, since it includes a portion of the metacarpal length.

index of difficulty
An indication of the amount of information required to generate a movement. *See also* ***Fitts' law***.

$$ID = \log_2 \left(\frac{2A}{W}\right)$$

where:
A = distance to the target
W = width of the target

index of forecasting efficiency
That reduction in prediction error obtained by using the correlation between two variable.

$$E = 1 - \sqrt{1 - r^2}$$

where:
r = the correlation between the variables

index of physiological effects
A measure of heat stress.

index of refraction
The value of the ratio of the velocity of electromagnetic radiation in one medium relative to another medium. A constant for a given pair of media and a given wavelength.

index of relative strain
A measure of heat stress based on clothing insulation and clothing effects on evaporation.

index of thermal stress
An indicator of the degree of heat stress which predicts the sweating rate required to cool the body based on the heat load combined with the effects of clothing and humidity levels.

Indian reservation
A part of public domain set aside by proper authority for use and occupation of a tribe or tribes of Indians, and under superintendence of the government which retains title to the land.

Indian summer
An unreasonably warm spell of weather with clear skies near the middle of autumn. Usually follows a substantial period of cool weather.

indicated airspeed
The speed of an aircraft as shown on its pitot static airspeed indicator calibrated to reflect standard atmosphere adiabatic compressible flow at sea level uncorrected for airspeed system errors. *See also* ***airspeed***.

indicating thermometer
A non-recording thermometer that allows the user to measure the temperature, generally on the Fahrenheit scale.

indication
(1) The response or evidence from the application of a nondestructive examination, including visual inspection. (2) The information conveyed by the aspect of a signal.

indication locking
Rail Operations. Electric locking which prevents manipulation of levers that would result in an unsafe condition for a train movement if a signal, switch, or other operative unit fails to make a movement corresponding to that of its controlling lever, or which directly prevents

the operation of a signal, switch, or other operative unit, in case another unit which should operate first fails to make the required movement.

indicator
(1) *Instrumentation.* Any device for displaying information. (2) *Biology.* An organism, species, or community whose characteristics show the presence of specific environmental conditions.

indicator compounds
Chemical compounds, such as carbon dioxide, whose presence at certain concentrations may be used to estimate certain building conditions (e.g., airflow, presence of sources).

indictment
A formal written accusation originating with a prosecutor and issued by a grand jury against a party charged with a crime.

indigestion
Failure of the digestive function; dyspepsia. Among the symptoms of indigestion are heartburn, nausea, flatulence, cramps, a disagreeable taste in the mouth, belching, and sometimes vomiting or diarrhea. Ordinary indigestion can result from eating too much or too fast; from eating when tense, tired, or emotionally upset; from food that is too fatty or spicy; and from heavy fried food or food that has been badly cooked or processed. Indigestion and its symptoms may also accompany other disorders such as allergy, migraine, influenza, typhoid fever, food poisoning, peptic ulcer, inflammation of the gallbladder (chronic cholecystitis), appendicitis, and coronary occlusion (heart attack).

indirect anthropometric measurement
A bodily measurement obtained by remote or noncontact techniques, such as stereometric anthropometry.

indirect cause
A contributing causal factor other than direct cause associated with an incident.

indirect discharge
Introduction of pollutants from a nondomestic source into a publicly owned waste treatment system. Indirect dischargers can be commercial or industrial facilities whose wastes go into the local sewers.

indirect employment
In respect to waterways industry, not necessarily engaged directly in river activities, but dependent upon the river.

indirect labor
That work which is a part of indirect operations.

indirect lighting
That illuminated environment in which approximately 90% or more of the luminous flux is directed toward a continuous solid structure away from a task.

indirect material
Any of the materials not used in direct operations.

indirect operations
Those administrative, management, or other functions within an organization necessary to support the manufacture or output of a product but which are not directly involved in producing a product or service for sale in the marketplace and which do not add value to that product.

indirect point source discharges
Discharge by industries of pollutants indirectly into U.S. waters through publicly owned treatment works (POTW).

indirect radiation effect
Any of those cellular effects causing damage to DNA by first creating radicals in other bodies or cellular materials, which in turn affect the DNA.

indirect source
Under the Clean Air Act, any facility, building, structure, installation, real property, road, or highway which attracts, or may attract, mobile sources of pollution.

indirect viewing
The use of video or other aids to view a scene or object being manipulated when direct viewing is not practical or possible.

indirect vision
Peripheral vision.

indirect worker
An employee involved in indirect operations.

indium
A chemical element, atomic number 49, atomic weight 114.82, symbol In.

individual incentive plan
An incentive plan in which each worker is rewarded based on his/her own efforts.

individual-rung ladder
A fixed ladder, each rung of which is individually attached to a structure, building, or equipment.

indolent
A person who is not inclined to work. An habitually lazy person.

indoor air
The breathing air inside a habitable structure or conveyance.

indoor air pollution
The presence of chemical, physical, or biological contaminants in indoor air in concentrations that could have an adverse effect on human health.

indoor air quality (IAQ)
General term that applies to the assurance or the evaluation and assessment of indoor air pollution to determine if contaminant levels exceed established standards for a particular pollutant or set of pollutants.

indoor climate
Temperature, humidity, lighting, and noise levels in a habitable structure or conveyance. Indoor climate can affect indoor air pollution.

indoors work
*See **inside work (1)**.*

induced draft
Negative pressure created by the action of a fan or ejector located between a combustion chamber and a stack/exhaust vent.

induced environment
That environment imposed upon an object or system from manmade conditions.

induced radioactivity
Radioactivity produced in certain materials as a result of nuclear reactions that involve the formation of unstable nuclei.

induction
(1) The generation of an electrical current by a change in magnetic flux in a conductor. (2) The alteration of a perception by indirect stimulation. (3) The process or act of inducing, or causing to occur. (4) *See **inductive reasoning**.*

inductive reasoning
The ability to integrate specific, diverse bits of information to arrive at a general conclusion.

inductively coupled plasma emission spectroscopy (ICPES)
A method typically used for the simultaneous analysis of many heavy metals.

inductor
Rail Operations. A track element consisting of a mass of iron, with or without a winding, that stimulates the train control, train stop, or cab signal mechanisms on the rail vehicle.

industrial anthropometry
The use of anthropometry for designing and constructing equipment for human use in the industrial environment. *See also **human factors**.*

industrial dermatitis
An inflammation of the skin surface caused by contact with industrial compounds and a subsequent allergic reaction. *See also **dermatitis**.*

industrial disease
*See **occupational illness**.*

industrial engineer
One who is qualified by education, training, and experience to practice the discipline of industrial engineering.

industrial engineering
That engineering discipline concerned with the design, development, installation, and improvement of integrated systems of people, materials, equipment, and energy in the industrial environment.

industrial ergonomics
Human factors applied to an industrial setting.

industrial hygiene (IH)
The art and science of anticipating, recognizing, evaluating, and controlling occupational and environmental health hazards in the work place and the surrounding community.

industrial hygienist
An individual who possess a degree from an accredited university in industrial hygiene, chemistry, physics, medicine, or other physical or biological science, and who, by virtue of specialized studies and training, has acquired competence in industrial hygiene.

industrial medicine
See ***occupational medicine***.

industrial psychology
That field of study and practice involving the testing, development of criteria and predictors for personnel selection and human performance in the workplace.

industrial radiography
The examination of the macroscopic structure of materials by nondestructive methods using sources of ionizing radiation.

industrial robot
A programmable manipulator for moving or operating on materials, components, products, or other objects in the industrial environment.

industrial safety
See ***occupational safety***.

industrial solid waste
The solid waste generated by industrial processes and manufacturing.

industrial special
Aviation. Any use of an aircraft for specialized work allied with industrial activity, excluding transportation and aerial application, e.g., pipeline patrol, survey, advertising, photography, helicopter hoist, etc.

industrial ventilation (IV)
The equipment or operation associated with the supply or exhaust of air, by natural or mechanical means, to control airborne hazards in the industrial setting.

industrial sector
Construction, manufacturing, agricultural, and mining establishments.

industrial terminal
A specialized terminal whose primary purpose is manufacturing, not transportation services.

industrial track
Rail Operations. A switching track serving industries, such as mines, mills, smelters, and factories.

industry briefing
Maritime. The appearance of Coast Guard Standards-trained personnel before a gathering of boat and/or associated equipment manufacturers and/or dealers.

industry rank report
A report from OSHA's National Office in Washington, DC supplied to each local Area Office that ranks industries (such as automotive, petroleum refining, transportation, etc.) according to their lost workday injury (LWDI) rate. *See also* ***establishment list***.

industry standards
With regard to issues of compliance: Organizations must know what the relevant industry and regulatory standards are and make sure their own programs' features meet those standards. *See also* ***effective compliance program***.

industry track
Rail Operations. A switching track, or series of tracks, serving the needs of a commercial industry other than a railroad.

ineffective time
That part of the elapsed time spent on any activity which is not a specified part of the task or job, excluding check time.

inert
Not chemically reactive at normal temperature and pressures.

inert atmosphere
The atmosphere of a confined space that has been made non-flammable, non-explosive, or otherwise chemically non-reactive and, therefore, also generally incapable of supporting or sustaining human life.

inert condition
A tank or other enclosure is in an inert condition when the oxygen content of the atmosphere throughout the enclosed space has been reduced to 8% or less by volume through the addition of an inert gas.

inert dirt
Any form of dirt which has no inherent attraction to any surface except through gravitation.

inert dust
Dusts which have a long history of little or no adverse effect on lungs and do not produce significant organic disease or toxic effect when enclosures are kept under reasonable control. Such dusts are often called biologically inert dusts. *See also* ***nuisance dust***.

inert gas
A non-reactive gas such as argon, helium, neon, or krypton. These are gases that will not burn or support combustion, and are not toxic. Nitrogen is often used as an inert gas in process operations for reducing the risk of fire and/or explosion.

inert gas narcosis
A toxic effect of the diluting or carrier gas in a breathing mixture at increased pressures, characterized by euphoria, diminished cognitive function, and impaired coordination. Also referred to as *diluent gas narcosis*. *See also* ***nitrogen narcosis***.

inert ingredient
An ingredient which is not active.

inertia
The tendency of a body at rest to remain at rest or a body in motion to stay in motion in a straight line unless distributed by an external force.

inertial frame
A reference frame to which the law of inertia applies.

inertial navigation system
Aviation. An Area Navigation (RNAV) system which is a form of self-contained navigation.

inertial separator
A device that uses centrifugal force to separate waste particles.

inerting
The displacement of the atmosphere in a permit space by a noncombustible gas (such as nitrogen) to such an extent that the resulting atmosphere is noncombustible. This produces an oxygen-deficient atmosphere that is immediately dangerous to life and health.

inevitable accident
Law. An unavoidable accident. One produced by an irresistible physical cause. An accident which cannot be prevented by human skill or foresight, but results from natural causes, such as lightning or storms, perils of the sea, inundation or earthquake, or sudden death or illness. In legal terms, an accident is termed "inevitable" so as to preclude recovery on the grounds of negligence, if the person by whom it occurs neither has nor is legally bound to have sufficient power to avoid it or prevent its injuring another person.

inexcusable neglect
Such neglect which will preclude the setting aside of default judgement and implies something more than the unintentional inadvertence or neglect common to all who share the ordinary frailties of mankind.

infant
A child less than two years of chronological age.

infant formula
Under the Federal Food, Drug and Cosmetic Act: A food which purports to be or is represented for special dietary use solely as a food for infants by reason of its simulation of human milk or its suitability as a complete or partial substitute for human milk.

infant mortality
(1) *General.* The death of an infant. (2) *Production.* The failure of a system in the early portion of its projected useful life.

infant mortality rate
The reported death rate for infants under one year of age per 1000 reported live births in a calendar year for a specified region.

infarct
A localized area of ischemic necrosis produced by occlusion of the arterial supply or the venous drainage of the part.

infarction
The development or presence of an infarct.

infected person
A person who harbors an infectious agent, whether or not the infection is accompanied by disease.

infection
The entry and multiplication of an infectious agent that occurs in the body tissues of a human or animal and that results in cellular injury. Several factors are necessary for the development of an infection. The microorganisms must enter the body in sufficient number and they must be virulent, or capable of destroying healthy tissues. The host must be susceptible to the disease. If the host has developed immunity to the disease, either by having had the disease or by having undergone immunization, he/she will not be af-

fected by the microorganisms. Some persons have greater natural resistance to infections than others. Finally, the disease must be transmitted through the proper route. Infection may be transmitted by *direct contact,* by *indirect contact,* or by *vectors*. Direct contact may be with body excreta such as urine, feces, or mucous, or with drainage from an open sore, ulcer, or wound. Indirect contact refers to transmission via inanimate objects such as bed linens, doorknobs, drinking glasses, or eating utensils. Vectors are flies, mosquitoes, or other insects capable of harboring and spreading the infectious agent. Synonymous with the term *infectious disease.*

infectious
Capable of invading a susceptible host, replicating, and causing an altered host reaction, such as disease.

infectious agent
An organism, usually a microorganism, that is capable of producing infection or infectious disease.

infectious disease
A disease of humans or animals resulting from the invasion of the body by pathogenic agents and the reaction of the tissue to these agents and/or the toxins they may produce. *See* ***infection***.

infectious waste
(1) Equipment, instruments, utensils, and formites of a disposable nature from the rooms of patients who are suspected to have or have been diagnosed as having a communicable disease and must, therefore, be isolated as required by public health agencies. (2) Laboratory wastes, such as pathological specimens (e.g., all tissues, specimens of blood elements, excreta, and secretions obtained from patients or laboratory animals) and disposable formite (any substance that may harbor or transmit pathogenic organisms) attendant thereto, and similar disposable materials from outpatient areas and emergency rooms.

inference
The conclusion resulting from the inductive reasoning process.

inference space
Those limits within which the results of an experiment may be applied.

inferential statistics
A technique for inferring something and drawing conclusions from data or information obtained from a representative sample taken from a population. It provides a means of drawing conclusions about a larger body or population based on sample data from that population.

inferior
(1) Of less than acceptable quality or performance. (2) Lower than or beneath some reference structure in position.

inferior angle of scapula
The thick lowermost portion of the scapula.

inferior mirage
See ***mirage***.

inferior nasal concha
A bone forming part of the lateral wall of the nasal cavity.

inferior oblique muscle
A voluntary extraocular muscle extending beneath the eyeball. Principally for rotation of the upper part of the eye laterally about the optical axis.

inferior rectus muscle
A voluntary extraocular muscle parallel to the optical axis beneath the eyeball. Involved in the anterior downward pitch/rotation of the eye.

infestation
The lodgment, development, and reproduction of anthropods such as mites, ticks, or fleas on the surface of the body, in clothing, or in dwellings.

infiltration
(1) The penetration of water through the ground surface into subsurface soil or the penetration of water from the soil into sewer or other pipes through defective joints, connections, or manhole walls. (2) A land application technique where large volumes of wastewater are applied to land, allowed to penetrate the surface and percolate through the underlying soil. *See also* ***percolation***. (3) Air leakage into a space through cracks and interstices, and through ceilings, floors, and walls.

inflammable
See ***flammable***.

inflammation

Normal tissue response to cellular injury or foreign material invasion, characterized by dilation of small blood vessels (capillaries) and mobilization of defense cells. The injury may be caused by a physical blow, or by exposure to an excessive amount of radiation from sunlight, x-rays, or an ultraviolet lamp; or it may be caused by corrosive chemicals, burns, extreme heat or cold, or foreign objects. Inflammation is also the usual response to a bacterial infection. The physiological changes that take place during the inflammatory process include vascular dilation, leukocytosis, and fluid exudation. The vascular changes occur at the site of the injury to the tissues. There is automatic dilation of the capillaries and arterioles so that a greater supply of blood is brought to the area. The speed of circulation is decreased with the result that leukocytes leave the blood vessels and enter the tissues spaces. The vascular changes are responsible for the redness that accompanies inflammation. The injured tissues release chemicals that attract the leukocytes to the site of the injury. There, they ingest or surround and destroy the cause of the inflammation. Body fluids also collect at the site. This increase of fluids is called exudation. The exudate brings immune bodies (antibodies) and special enzymes, and also helps in the removal of dead bacteria, destroyed tissue cells, and blood cells. The four classic symptoms of inflammation are redness (rubor), swelling (tumor), heat (calor), and pain (dolor), Loss of function of the affected part may also occur.

inflection point

A point on a curve such that the following are true: (a) the curve changes from concave to convex, (b) the mathematical derivative of the curve is increasing on one side of the point and decreasing on the other side, and (c) the second derivative changes sign.

in-flight survey (IFS)

The in-flight survey is administered to United States (U.S.) and foreign travelers departing the U.S. as a means of providing data on visitor characteristics, travel patterns and spending habits, and for supplying data on the U.S. international travel dollar accounts as well as to meet balance of payments estimation needs. The IFS covers about 70% of U.S. carriers and 35% of foreign carriers who voluntarily choose to participate.

inflow

Entry of extraneous rain water into a sewer system from sources other than infiltration, such as basement drains, manholes, storm drains, and street washing.

influent

Water, wastewater, or other liquid flowing into a reservoir, basin, or treatment plant.

influenza

An acute infectious epidemic disease caused by a filterable virus. Four main types of the virus have been recognized, arbitrarily labeled by researchers as types A, B, C, and D, and sometimes subdivided into A_1 and A_2. The A_2 virus is a comparatively new strain that first emerged in 1957. The disease it produces is often called the Asian flu. Influenza has a brief incubation period. The symptoms appear suddenly and though the virus enters the respiratory tract it soon affects the entire body. The symptoms include fever, chills, headache, sore throat, cough, gastrointestinal disturbances, muscular pain, and neuralgia.

informal contract

A contract that does not require a specified form or formality for its validity. Generally refers to an oral contract as contrasted with a written contract or specialty instrument.

informal factory visit

A visit by Coast Guard personnel to a manufacturing facility to acquaint the manufacturer with the existence of the law, regulations, general administrative requirements affecting him/her, and possible penalties for violations.

informal rulemaking

Also known as *notice and comment rulemaking,* requires OSHA provide "interested parties an opportunity to participate in the rulemaking through submission of written data, views, or arguments with or without opportunity for oral presentation." It does not require a hearing, although OSHA may hold one if it so desires. It allows the agency to look beyond any hearing records in making rules. Also, when courts review OSHA's actions under informal rulemaking, OSHA is not held to the "substantial evidence" test required under formal proceedings. Rather, the agency must only prove that their decisions and determinations are not "arbitrary" or "capricious."

information
(1) A meaningful collection of facts, figures, and/or data. (2) That which reduces uncertainty; typical unit is the *bit*.

information aid
Any work aid which provides the worker with text, numbers, figures, or other details appropriate for performing in the working environment.

information area
Any region of a display containing useful general-purpose information.

information file
In the Superfund program, a file that contains accurate, up-to-date documents on a Superfund site. The file is usually located in a public building such as a school, library, or city hall that is convenient for local residents.

information ordering
The ability to correctly follow a set of rules in arranging items.

information process analysis
See ***form process chart***.

information request
Aviation. A request originated by an Flight Service Station (FSS) for information concerning an overdue Visual Flight Rules (VFR) aircraft.

information theory
That aspect of communications dealing with the coding of messages and with the content and amount of information conveyed. *See also* ***information*** *and* ***bit***.

informed consent
A person's agreement to allow something to happen (such as surgery) that is based on a full disclosure of the facts needed to make the decision intelligently. *See also* ***voluntary informed consent***.

informer's privilege
The government's privilege to withhold from disclosure the identity of persons who furnish information on violations of the law to officers charged with the enforcement of that law.

infra-
(prefix) Under, below, or less than.

infradian rhythm
A biological rhythm having less than one cycle per day, or a period longer than one day.

infrared
Electromagnetic radiation of wavelength between the longest visible red (7000 Angstroms or 7 x 10^{-4} millimeter) and about 1 millimeter.

infrared detector
A measurement technique in which infrared radiation is passed through a cell containing the sampled material. The absorption of the IR energy at a wavelength which coincides with the absorption band of the analyte (contaminant) and it's proportional to the amount of contaminant present. This principle can also be applied to the determination of materials present in air drawn through a cell through which a beam of IR radiation is passed.

infrared gas analyzer
A real-time air sampling device that measures the absorbency of inorganic and organic gases and vapors.

infrared hygrometer
See ***hygrometer***.

infrared lamp
A lamp which emits its primary radiation in the infrared portion of the electromagnetic spectrum, and any radiation in the visible portion of the spectrum is not normally of interest.

infrared radiation
Electromagnetic radiation with wavelengths between about 0.7 and 1000 μm. This radiation is longer than visible radiation but shorter than microwave radiation.

infrared radiometer
An instrument designed to measure the intensity of infrared radiation emitted by an object. Also called *infrared sensor*.

infrared sensor
See ***infrared radiometer***.

infrared touchscreen
A display having a frame with embedded infrared transmitters and receivers which uses blockage of the infrared beam to indicate a touch location.

infrasonic
At a frequency below the audio frequency range. Also called *subsonic*.

infrasound
A mechanical vibration at frequencies below those normally heard by the human ear, generally below about 16 to 20 Hz. *See also **infrasonic**.*

infrastructure
(1) In transit systems, all the fixed components of the transit system, such as rights-of-way, tracks, signal equipment, stations, park-and-ride lots, bus stops, maintenance facilities. (2) In transportation planning, all the relevant elements of the environment in which a transportation system operates.

infusion
(1) Steeping of a substance in water to obtain its soluble principles. (2) A solution obtained by steeping a substance in water. (3) The introduction of a solution into a vein by gravity. (Note: an *infusion* flows by gravity, an *injection* is forced in by a syringe, an *instillation* is dropped in, an *insufflation* is blown in, and an *infection* slips in unnoticed.)

Ingersoll glarimeter
An early instrument for measuring gloss using polarized light.

ingestant
A substance capable of entering the body through the mouth or digestive system.

ingestion
(1) The process of taking substances into the body by mouth. (2) The taking in of substances, especially via the mouth.

ingredient statement
Under the Federal Insecticide, Fungicide, and Rodenticide Act: A statement which contains a) the name and percentage of each active ingredient, and the total percentage of all inert ingredients, in the pesticide; and b) if the pesticide contains arsenic in any form, a statement of the percentages of total and water-soluble arsenic, calculated as elementary arsenic.

ingress
To enter a region or space. The right or act of entering.

ingress point
The location for entering a region or space.

ingrown nail
An overlapping of the anterior corners of a nail by the flesh of the digit, causing pain, inflammation, and possible infection. The condition occurs most frequently in the great toe, and is often caused by pressure from tight-fitting shoes. Another common cause is improper cutting of the toenails, which should be cut straight across or with a curved toenail scissors so that the sides are a little longer than the middle.

inguinal
Pertaining to the groin.

inguinal crease
The groove at the junction of the anterior-medial thigh and the torso.

inguinal hernia
Hernia occurring in the groin; protrusion of intestine or omentum, or both, either directly through a weak point in the abdominal wall (direct inguinal hernia) or downward into the inguinal banal (indirect inguinal hernia).

inhalable dust
*See **respirable dust**.*

inhalable fraction
The mass fraction of total airborne particulates that is inhaled through the nose and mouth.

inhalant
(1) A substance which is inhaled. (2) A gaseous substance that is or may be taken into the body by way of the nose and trachea (through the respiratory system).

inhalation
The breathing in of a substance, such as air or a contaminant in the atmosphere.

inherent defect
Fault or deficiency in a thing, no matter the use made of such, which is not easily discoverable and which is fixed in the object itself and not from without.

inherent delay
*See **delay time**.*

inherent right
One which abides in a person and is not given from something or someone outside itself. A right which a person has because he/she is a person.

inhibit
An independent and verifiable mechanical and/or electrical device that prevents a hazardous event from occurring. The device has direct control and is not the monitor of such a device.

inhibitor
An agent that arrests or slows chemical action or a material used to prevent or retard rust or corrosion.

inion
The most posterior protuberance of the occipital bone. Also referred to as *external occipital protuberance*.

initial approach fix
Aviation. The fixes depicted on instrument approach procedure charts that identify the beginning of the initial approach segment(s).

initial approach segment
Aviation. That segment of an instrument approach procedure between the initial approach fix and the intermediate approach fix or, where applicable, the final approach fix or point.

initial crack size
A crack dimension determined by nondestructive examination methods or proof test logic.

initial flaw
A flaw in a structural material before the application of load and/or environmental stressors.

initial impact point
Transit. The first impact point that produced property damage or personal injury, regardless of "first" or "most harmful event."

initial luminance
That luminance reaching the work surface from a given luminaire when new.

initial terminal
Rail Operations. The starting point of a locomotive for a trip.

initiate
To begin something. Also, a mental activity preceding a psychomotor task.

initiation
An irreversible genetic change in the cell, which is believed to be the first stage in the development of cancer.

initiator
Includes low voltage electroexplosive devices and high voltage exploding bridge wire devices.

injection
(1) Introduction of a fluid substance into the body, usually by means of a syringe or other device connected to a hollow needle. (2) The solution so administered.

injection well
A well into which fluids are injected for purposes such as waste disposal, improving the recovery of crude oil, or solution mining.

injection zone
A geological formation, group of formations, or part of a formation receiving fluids through a well.

injunction
A type of court order compelling a party in civil litigation to do something or not to do something. Thus, injunctions are usually *mandatory* or *prohibitory,* or a combination of both. There are three types of injunctions: the *temporary restraining order, the preliminary injunction,* and the *permanent injunction.*

injure
(1) To violate the legal right of another or inflict an actionable wrong. (2) To do harm, damage, or impair. (3) To hurt or wound, as the person; to impair the soundness of, as health.

injurious exposure
Such an exposure as will render the employer liable for occupational disease of the employee. For example, a concentration of a toxic material which would be sufficient to cause disease in the event of prolonged exposure to such concentration regardless of the length of exposure required to actually cause the disease.

injury
(1) Physical harm or damage to a person. (2) An incident involving lost time or other than on-site medical treatment. (3) Bodily injury resulting from a motor vehicle accident. To qualify as an "injury," the injured person must require and receive medical treatment away from the accident scene. (4) Harm to a person resulting from a single event, activity, occur-

rence, or exposure of short duration. (5) Any physical damage or harm to a person requiring medical treatment, or any physical damage or harm to a person reported at the time and place of occurrence. For employees, an injury includes incidents resulting in time lost from duty or any definition consistent with a transit agency's current employee injury reporting practice. (6) Physical harm or damage to the body resulting from an exchange, usually acute, of mechanical, chemical, thermal, or other environmental energy that exceeds the body's tolerance. (7) Any physical disturbance to, damage to, or destruction of one or more body structures which prevents/impairs normal functioning or appearance. (8) Any wrong or damage done to another, either in his/her person, rights, reputation, or property. The invasion of any legally protected interest of another. (9) A specific impairment of body structure or function caused by an outside agent or force, which may be physical, chemical, or psychic.

injury accident
An accident for which at least one injury, but no fatalities, were reported.

injury crash
A police-reported crash that involves a motor vehicle in transport on a traffic-way in which no one died but at least one person was reported to have a) an incapacitating injury; b) a visible but not incapacitating injury; c) a possible, not visible injury; or d) an injury of unknown severity.

injury incident rate
The number of injuries experienced by a company based on a year's work for 100 full-time employees

injury index
NTSB. Refers to the highest degree of personal injury sustained as a result of the accident.

injury potential
A potential difference across a membrane, generally of about 30-40 mv, between regions of normal and injured tissue.

injury rate
The average number of nonfatal injuries per accident or per one hundred accidents.

injury severity
Transportation. The police-reported injury severity of the occupant, pedestrian, or pedalcyclist (e.g., severe or fatal; killed or incapacitating; minor or moderate; evident, but not incapacitating; complaint of injury; injured, severity unknown; no injury).

inland
Means transit to and from inland ports connected by water routes made navigable by one or more lock structures.

inland and coastal channels and waterways
These terms include the Atlantic Coast Waterways, the Atlantic Intracoastal Waterway, the New York State Barge Canal System, the Gulf Coast Waterways, the Gulf Intracoastal Waterway, the, Mississippi River System (including the Illinois Waterway), Pacific Coast Waterways, the Great Lakes, and all other channels (waterways) of the United States, exclusive of Alaska, that are usable for commercial navigation.

inland area
The area shoreward of the boundary lines defined in 46 CFR 7, except that in the Gulf of Mexico, it means the area shoreward of the lines of demarcation (COLREG lines) defined in 33 CFR 80.740-80.850. The inland area does not include the Great Lakes.

inland bill of lading
A bill of lading used in transporting goods overland to the exporter's international carrier. Although a through bill of lading can sometimes be used, it is usually necessary to prepare both an inland bill of lading and an ocean bill of lading for export shipments.

inland carrier
A transportation line that hauls export or import traffic between ports and inland points.

inland marine insurance
Originally, a form of insurance protection for goods transported other than on the ocean. Now, the term applies to a variety of coverage on floating personal property and to general liability as a bailee. *See also* ***insurance***.

inland navigation facility
Aviation. A navigation aid on a North American Route at which the common route and/or the non-common route begins or ends.

inland waters of the United States
Those waters of the United States lying inside the baseline from which the territorial sea is measured and those waters outside such baseline which are a part of the Gulf Intracoastal Waterway. *See also* ***inland waterway of the United States***.

inland waterway convoy
One or more non-powered inland waterways transport (IWT) vessels which are towed or pushed by one or more powered IWT vessels.

inland waterway journey
Any movement of an inland waterways transport (IWT) vessel from a specified point of origin to a specified point of destination.

inland waterway of the United States
Any improved waterway, the improvements to which are primarily for the use of vessels other than ocean going vessels. *See also* ***inland waters of the United States***.

inland waterway transport (IWT)
Any movement of goods and/or passengers using an IWT vessel on a given inland waterways network.

inland waterways cabotage transport
National inland waterways transport (IWT) performed by an IWT vessel registered in another country.

inland waterways fleet
Number of inland waterways transport (IWT) vessels registered at a given date in a country and authorized to use inland waterways open for public navigation.

inland waterways on national territory traffic
Any movement of an inland waterways transport (IWT) vessel within a national territory irrespective of the country in which the vessel is registered.

inland waterways passenger
Any person who makes a journey on board of an inland waterways transport (IWT) vessel. Service staff assigned to IWT vessels are not regarded as passengers.

inland waterways passenger-kilometer
Unit of measure representing the transport of one passenger by inland waterway over one kilometer.

inland waterways passenger transport link
The combination of the place of embarkment and the place of disembarking of the passenger conveyed by inland waterways whichever itinerary is followed.

inland waterways traffic
Any movement on an inland waterways transport (IWT) vessel on a given network.

inland waterways transit
Inland waterways transport (IWT) through a country between two places (a place of loading and a place of unloading) both located in another country or in other countries provided the total journey within the country is by an IWT vessel and that there is no loading and unloading in that country.

inland waterways transport (IWT) enterprise
An enterprise carrying out in one or more places activities for the production of IWT services using IWT vessels and whose main activities according to the value added are inland waterway transport and services allied to inland waterway transport.

inland waterways transport (IWT) freight vessel
A vessel with a carrying capacity of not less than 20 tons designed for the carriage of freight by navigable inland waterways.

inland waterways transport (IWT) passenger vessel
Vessel designed exclusively or primarily for the public carriage or passengers by navigable inland waterways.

inland waterways transport (IWT) vessel
A floating craft designed for the carriage of goods or public transport of passengers by navigable inland waterways.

inland zone
The environment inland of the coastal zone excluding the Great Lakes, Lake Champlain, and specified ports and harbors on inland rivers. The term inland zone delineates an area of federal responsibilities for response actions. Precise boundaries are determined by agreements between the Environmental Protection Agency (EPA) and the United States Coast Guard (USCG) and are identified in Federal Regional Contingency Plans.

inlet
An opening of the sea into the land or of a lake into its shore.

INM
See ***Integrated Noise Model.***

INMARSAT
See ***International Maritime Satellite Organization.***

innage
The height of a liquid in a tank from the bottom datum plate of the tank to the liquid surface.

innate
Due to one's genetic make-up.

innavigable
(1) As applied to streams, not capable of or suitable for navigation; impassable by ships or vessels. (2) As applied to vessels in the law of maritime insurance, it means unfit for navigation; so damaged by misadventures at sea as to be no longer capable of making a voyage.

inner-approach obstacle free zone
Aviation. The inner-approach obstacle free zone (OFZ) is a defined volume of airspace centered on the approach area. The inner-approach OFZ applies only to runways with an approach lighting system. The inner-approach OFZ begins 200 feet from the runway threshold at the same elevation as the runway threshold and extends 200 feet beyond the last light unit in the approach lighting system. The width of the inner approach OFZ is the same as the runway OFZ and rises at a slope of 50 (horizontal) to 1 (vertical) from the beginning. *See also* ***obstacle free zone.***

inner ear
That portion of the ear embedded in the temporal bone and consisting of the vestibule, cochlea, and semicircular canals. Also referred to as the *internal ear* or the *labyrinth.*

inner marker (IM)
Aviation. A marker beacon used with an Instrument Landing System (ILS) (CAT II) precision approach located between the middle marker and the end of the ILS runway, transmitting a radiation pattern keyed at six dots per second and indicating to the pilot, both aurally and visually, that he is at the designated decision height (DH), normally 100 feet above the touchdown zone elevation, on the ILS CAT II approach. It also marks progress during a CAT III approach. *See also* ***outer marker.***

inner packaging
DOT. A packaging for which an outer packaging is required for transport. It does not include the inner receptacle of a composite packaging.

inner receptacle
DOT. A receptacle which requires an outer packaging in order to perform its containment function. The inner receptacle may be an inner packaging of a combination packaging or the inner receptacle of a composite packaging.

inner-transitional obstacle free zone
Aviation. The inner transitional surface obstacle free zone (OFZ) is a defined volume of airspace along the sides of the runway and inner-approach OFZ and applies only to precision instrument runways. The inner-transitional surface OFZ slopes 3 (horizontal) to 1 (vertical) out from the edges of the runway OFZ and inner-approach OFZ to a height of 150 feet above the established airport elevation. *See also* ***obstacle free zone.***

innervation
The distribution of nerves or neurons to all or some portion of the body.

innocent
The term applied to a defendant in a criminal prosecution who is determined to be not guilty. More loosely applied to persons who did not know or have reason to know about a violation or problem, such as an innocent landowner ignorant of real estate contamination.

innocent purchaser
One who, by an honest contract or agreement, purchases property or acquires an interest therein, without knowledge, or means of knowledge sufficient to charge him/her in law with knowledge, of any infirmity in the title of the seller.

innocuous
Harmless, or having no adverse effects.

innominate bone
See ***coxal bone.***

inoculation
(1) Introduction of pathogenic microorganisms into the body to stimulate the production of antibodies and immunity. (2) Introduction of infectious material into culture medium in an effort to produce growth of the causative organism.

inoculum
(1) Bacterium placed in compost to start biological action. (2) A medium containing organisms which is introduced into cultures of living organisms.

inorganic
(1) Having no organs. (2) Not of organic origin.

inorganic chemicals
Chemical substances of mineral origin, not basically of carbon structure.

inorganic compound
Chemical compounds of mineral origin, not of basically carbon structure.

input
That information, signal, or form of energy which enters a system.

input device
Any piece of equipment or instrumentation used to provide the human with an interface for providing input to a system.

input/output (I/O)
Any activity which inputs to or receives output from a computer.

input point
The physical location at which some information or signal can enter some system.

input storage
The temporary placement of data in a computer file until time for processing.

inquest
The inquiry by a medical examiner or coroner, sometimes with the aid of a jury, into the manner of the death of any one who has been killed, or has died suddenly under unusual or suspicious circumstances, or by violence, or while in prison.

inrolling nip point
A system in which two or more rollers rotate parallel to each other, but in opposite directions, and which can grab and pull on such items as loose clothing, and ties.

insanity
Severe mental disorder that may make a person irresponsible, unreasonable, and unable to function normally in society. His/her thoughts and actions are distinctly different from accepted patterns of behavior. The term is a legal rather than a medical one, and includes different kinds of mental illness. A person who is judged to be insane by a court is not held legally responsible for his/her actions and may have to be institutionalized.

insect
Any of the numerous small invertebrate animals generally having the body more or less obviously segmented, for the most part belonging to the class insecta, comprising six-legged, usually winged forms, as for example, beetles, bugs, bees, flies, and other allied classes of anthropods whose members are wingless and usually have more than six legs, as for example, spiders, mites, ticks, centipedes, and wood lice.

insecticide
A pesticide compound specifically used to kill or control the growth of insects.

insert
To place one object inside another.

insertion mode
A data entry mode in which text or information entered by the user is placed in front of any existing text or information, shifting that existing text.

inshore traffic zone
A designated area between the landward boundary of a traffic separation scheme and the adjacent coast, intended for local traffic.

inside-out display
Any display which uses a vehicle as the frame of reference such that the display reflects the way the operator would see the external environment from inside.

inside work
(1) That work performed inside some structure which shields the worker at least in part from the atmosphere elements. Also referred to as *indoor work*. (2) *See* ***internal work***.

insidious
Spreading in a subtle manner.

insolation
The incoming solar radiation that reaches the earth and the atmosphere.

insoluble
Incapable of being dissolved.

insoluble wastes
Solid wastes consisting of inert natural minerals or inert synthetic materials.

insomnia
An extended period in which sleep is disturbed, not resulting from immediate external stimuli. The causes of insomnia may be physical or psychological or, most often, a combination of both. Some persons are more sensitive to conditions around them than others, and may be kept awake by slight noises, light, or the sharing of their bed. Beverages that contain caffeine, such as coffee, tea, and cola drinks, keep people awake. A heavy meal shortly before bedtime may prevent sleep. Drinking large quantities of fluids may cause an uncomfortable feeling of distention of the bladder.

insorption
Movement of a substance into the blood, especially from the gastrointestinal tract into the circulating blood.

inspect
To verify quality, integrity, and/or safety through testing, observation, or other processes.

inspection
(2) *General.* Visual examination for detection of features or qualities perceptible to the eye. (2) *OSHA.* Any investigation of an employer's factory, plant, establishment, construction site, or other area, workplace, or environment where work is performed by an employee of an employer, and includes re-inspection, follow-up inspection, accident investigation, or other inspection conducted under Section 8(a) of the Occupational Safety and Health Act. (3) *Law.* To examine; scrutinize; investigate; look into; check over; or view for the purpose of ascertaining the quality, authenticity, or conditions of an item, product, document, residence, business, etc.

inspection and maintenance
(1) Activities to assure proper emissions-related operation of mobile sources of air pollutants, particularly automobile emissions controls. (2) Also applies to wastewater treatments plants and other anti-pollution facilities and processes.

inspection and weighing services for motor vehicle transport facility
Establishments primarily engaged in the operation of fixed facilities for motor vehicle transportation, such as toll roads, highway bridges, and other fixed facilities, except terminals.

inspection error
Any incorrect reading, action, or other error of either omission or commission in the inspection process.

Inspection Register
A registration containing the name of each establishment scheduled for inspection and the order in which these establishments will be inspected. Compiled from the establishment list and the industry rank report. *See also* ***establishment list*** *and* ***industry rank report.***

inspection searches
Administrative searches conducted by local or state authorities for health or building law enforcement must be based on a warrant issued on probable cause. A warrant is likewise required for inspection of business premises by OSHA inspectors. An exception to the warrant requirement is in cases involving closely regulated industries where the commercial operator's privacy interest is adequately protected by detailed regulatory schemes authorizing inspections without warrants.

Inspirable Particulate Mass
Particulates that are hazardous when deposited anywhere in the respiratory tract.

Inspirable Particulate Mass TLVs (IPM-TLVs)
Exposure limits that are applied to those materials that are hazardous when deposited anywhere in the respiratory tract.

inspiration
The process of drawing air into the lungs.

inspiratory capacity (IC)
The maximum volume of air which can be inhaled after a normal expiration.

inspiratory reserve volume (IRV)
The maximum volume of air that can be forcibly expired following a normal inspiration.

inspire
Breathe or take air into the lungs, using one's own resources.

instant start fluorescent lamp
A fluorescent lamp which does not require preheating of the electrodes. Also referred to as cold start fluorescent lamp.

instantaneous acceleration
The rate of velocity change with time at any instant.

instantaneous velocity
The rate of change of displacement with time at any instant.

instep
The arch on the medial side of the foot.

instep circumstance
The surface distance around the foot in a coronal/frontal plane at the anterior junction of the leg and foot. Measured with the individual standing erect, having his weight distributed equally on both feet on the floor, and without any unnecessary leg or foot muscle tension.

instep length
The linear distance from the plane of the most posterior aspect of the heel to the point of maximum medial protuberance of the foot. Measured with the individual standing erect and the body weight equally distributed between both feet on the floor.

instinct
A genetically based or natural motivation or behavior.

institutional solid waste
Solid waste generated by educational, health care, correctional, and other institutional facilities.

instruction
One item of a set of procedures, standard practices, or steps for accomplishing a given task or job.

instruction aid
A job aid containing written instructions on a card or sheet of paper.

instructional flying
(1) *FAA.* Any use of an aircraft for the purpose of formal instruction with the flying instructor aboard, or with the maneuvers on the particular flight(s) specified by the flight instructor; excludes proficiency flying. (2) *NTSB.* Flying accompanied in supervised training under the direction of an accredited instructor.

instrument
(1) *General.* Any device for measuring, recording, and/or controlling the value of one or more variables. (2) *Aviation.* A device using an internal mechanism to show visually or aurally the attitude, altitude, or operation of an aircraft or aircraft part. It includes electronic devices for automatically controlling an aircraft in flight. (3) *Law.* A formal or legal document in writing, such as a contract, deed, will, bond, or lease.

instrument approach
Aviation. (1) An approach to an airport, with intent to land, by an aircraft flying in accordance with an Instrument Flight Rules (IFR) flight plan, when the visibility is less than 3 miles and/or when the ceiling is at or below the minimum initial altitude. (2) A series of predetermined maneuvers for the orderly transfer of an aircraft under instrument flight conditions from the beginning of the initial approach to a landing, or to a point from which a landing may be made visually. An instrument approach is prescribed and approved for a specific airport by competent authority Federal Aviation Regulation (FAR) Part 91.

instrument approach procedure
A series of predetermined maneuvers by reference to flight instruments with specified protection from obstacles from the initial approach fix, or where applicable, from the beginning of a defined arrival route to a point from which a landing can be completed and thereafter, if a landing is not completed, to a position at which holding or en route obstacle clearance criteria apply.

instrument approach procedures charts
Portrays the aeronautical data which are required to execute an instrument approach to an airport. These charts depict the procedures, including all related data, and the airport diagram. Each procedure is designated

for use with a specific type of electronic navigation system including nondirectional beacon (NDB), tactical aircraft control and navigation (TACAN), very high frequency omnidirectional range (VOR), instrument landing system/microwave landing system (ILS/MLS), and area navigation (RNAV). These charts are identified by the type of navigational aid(s) which provide final approach guidance.

instrument error
Any error made by an instrument. Such errors, if not discovered, can foul an experiment or test.

instrument flight rules (IFR)
Rules governing the procedures for conducting instrument flight. Also a term used by pilots and controllers to indicate type of flight plan.

instrument flight rules (IFR) aircraft
An aircraft conducting flight in accordance with instrument flight rules (IFR).

instrument flight rules (IFR) aircraft handled
The number of instrument flight rules (IFR) departures multiplied by two plus the number of IFR overs. This definition assumes that the number of departures (acceptances, extensions, and originations of IFR flight plans) is equal to the number of landings (IFR flight plans closed).

instrument flight rules (IFR) conditions
Weather conditions below the minimum for flight under visual flight rules (VFR).

instrument flight rules (IFR) departure
An instrument flight rules (IFR) departure includes IFR flights originating in the center's area, accepted by the center under SOLE EN ROUTE clearance procedures, and extended by the center.

instrument flight rules (IFR) over
An instrument flight rules (IFR) flight that originates outside the Air Route Traffic Control Center (ARTCC) area and passes through the area without landing.

instrument flight rules (IFR) over-the-top
With respect to the operation of aircraft, means the operation of an aircraft over-the-top on an instrument flight rules (IFR) flight plan when cleared by air traffic control to maintain "visual flight rules (VFR) conditions" or "VFR conditions on top."

instrument flight rules (IFR) takeoff minimums and departure procedure
Federal Aviation Regulations, Part 91, prescribes standard takeoff rules for certain civil users. At some airports, obstructions or other factors require the establishment of nonstandard takeoff minimums, departure procedures, or both to assist pilots in avoiding obstacles during climb to the minimum en route altitude. Those airports are listed in National Airspace System (NAS) Department of Defense (DOD) Instrument Approach Plate (IAP) Charts (Ws) under a section entitled "IFR Takeoff Minimums and Departure Procedures." The IAP chart legend illustrates the symbol used to alert the pilot to nonstandard takeoff minimums and departure procedures. When departing Instrument Flight Rules (IFR) from such airports or from any airports where there are no departure procedures, standard instrument departures (SIDs), or Air Traffic Control (ATC) facilities available, pilots should advise ATC of any departure limitations. Controllers may query a pilot to determine acceptable departure directions, turns, or headings after takeoff. Pilots should be familiar with the departure procedures and must assure that their aircraft can meet or exceed any specified climb gradients.

Instrument Flight Service Station (IFSS)
A central operations facility in the flight advisory system, staffed and equipped to control aeronautical point-to-point telecommunications, and air-ground telecommunications with pilots operating over international territory or waters, which provides flight plan following, weather information, search and rescue action, and other flight assistance operations.

Instrument Landing System (ILS)
A precision instrument approach system which normally consists of the following electronic and visual aids a) localizer provides course guidance to the runway; b) glide slope provides vertical guidance during approach; c) marker beacon provides aural and/or visual identification of a specific position along an instrument approach landing.

instrument landing system category
ILS Category I. An ILS approach procedure which provides for approach to a height above touchdown of not less than 200 feet and with

runway visual range of not less than 1,800 feet. *ILS Category II.* An ILS approach procedure which provides for approach to a height above touchdown of not less than 100 feet and with runway visual range of not less than 1,200 feet. *ILS Category III.* a) IIIA: An ILS approach procedure which provides for approach without a decision height minimum and with runway visual range of not less than 700 feet. b) IIIB: An ILS approach procedure which provides for approach without a decision height minimum and with runway visual range of not less than 150 feet. c) IIIC: An ILS approach procedure which provides for approach without a decision height minimum and without runway visual range minimum.

instrument meteorological conditions (IMC)
Meteorological conditions expressed in terms of visibility, distance from cloud, and ceiling less than the minima specified for visual meteorological conditions.

instrument operation
Arrivals or departures of an aircraft in accordance with an instrument flight rules (IFR) flight plan or special visual flight rules (SVFR) procedures or an operation where IFR separation between aircraft is provided by a terminal control facility. There are three kinds of instrument operations: a) *Primary Instrument Operations.* Arrivals and departures at the primary airport which is normally the airport at which the approach control facility is located. b) *Secondary Instrument Operations.* Arrivals and departures at all the secondary airports combined. c) *Overflights.* Operations in which an aircraft transits the area without intent to land.

instrument operations
Arrivals or departures of an aircraft in accordance with an IFR flight plan or special VFR procedures or an operation where IFR separation between aircraft is provided by a terminal control facility.

instrument practice approach
An instrument approach procedure conducted by a visual flight rules (VFR) or an instrument flight rules (IFR) aircraft for the purpose of pilot training or proficiency demonstrations.

instrument runway
A runway equipped with electronic and visual navigation aids for which a precision or non-precision approach procedure having straight-in landing minimums has been approved.

instrument shelter
A box-like wooden structure designed to protect weather instruments from direct sunshine and precipitation.

Instrument Society of America (ISA)
A group that sets standards of performance for instruments made and used in the United States.

instrumental activities of daily living (IADL)
Those functions likely to be carried out on a daily basis which involve the use of equipment or instrumentation for sustenance of the individual or a normally habitable environment. *See also* ***activities of daily living*** *and* ***daily living tasks***.

instruments of international traffic
Lift vans, cargo vans, shipping tanks, skids, pallets, caul boards, and cores for textile fabrics, arriving (whether loaded or empty) in use or to be used in the shipment of merchandise in international traffic.

insubordination
The state of being insubordinate; disobedience to constituted authority. Refusal to obey some order which a superior officer is entitled to give and have obeyed. The term implies willful or intentional disregard of the lawful and reasonable instructions of an employer.

insulated body
Transit. Truck or trailer designed for transportation of commodities at controlled temperatures. It may be equipped for refrigeration or heating.

insulated conductor
See ***conductor***.

insulated rail joint
Rail Operations. A joint in which electrical insulation is provided between adjoining rails.

insulation value of clothing
See ***thermal insulation value of clothing***.

insulin
A sulfur-containing hormone produced by the pancreas of vertebrates. This hormone stimulates the conversion of glucose to glycogen and fat. An insulin deficiency results in excess blood sugar and causes the condition *diabetes mellitus.* Various preparations of in-

sulin are used in the treatment of this illness. Types of insulin vary in the rapidity of action and the duration of effectiveness. Regular insulin is effective almost immediately after injection and reaches its peak of action within 2 hours. It is used most often in diabetic emergencies and in regulating dosage for a patient when diabetes is first diagnosed. Crystalline insulin is made of zinc-insulin crystals and is usually given to patients who are allergic to regular insulin. Other types of insulin contain substances that prolong the action of insulin. Protamine zinc insulin (PZI), isophane insulin (NPH), globin zinc insulin, and insulin lente are examples of long-acting preparations of insulin.

insurance

(1) A contract whereby, for a stipulated consideration, one party undertakes to compensate the other for loss on a specified subject by specified perils. The party agreeing to make the compensation is usually called the *insurer* or *underwriter;* the other, the *insured* or *assured;* the agreed consideration is the *premium;* the written contract, a *policy;* the events insured against, *risks* or *perils;* and the subject, right, or interest to be protected, the *insurable interest.* (2) A contract whereby one undertakes to indemnify another against loss, damage, or liability arising from an unknown or contingent event and is applicable only to some contingency or act to occur in the future. (3) An agreement by which one party for a consideration promises to pay money or its equivalent or to do an act valuable to another party upon destruction, loss, or injury of something in which the other party has an interest.

insured and principal

Transportation. The motor carrier named in the policy of insurance, surety bond, endorsement, or notice of cancellation, and also the fiduciary of such motor carrier.

intake

A measure of exposure expressed as the mass of substance in contact with the exchange boundary per unity body weight per unit time (e.g., mg/kg-day). Also referred to as the *normalized exposure rate.*

intangible risk

A risk involving unwanted consequences which are primarily nonphysical, such as public opinion, employee morale, etc., but may still have adverse effects.

integral absorbed dose

The energy imparted to matter by ionizing particles. The unit of measure is the gram-rad and is equal to 100 ergs.

integral mode controller

A type of controller whose output signal is proportional to the integral of the error signal.

integrate

(1) To compute the area under a curve. (2) To combine activities, information, or objects in a meaningful way for some purpose.

integrated carriers

Carriers that have both air and ground fleets; or other combinations, such as sea, rail, and truck. Since they usually handle thousands of small parcels an hour, they are less expensive and offer more diverse services than regular carriers.

integrated circuit

A small chip of silicon on which miniaturized circuits have been etched.

integrated controller

A device which coordinates the control of more than one aspect of some operation.

integrated electromyogram (IEMG)

The computed area under the curve of an electromyographic signal.

integrated error

The sum of the errors accumulated over a given task.

integrated gas company

A company that obtains a significant portion of its gas operating revenues from the operations of both a retail gas distribution system and gas transmission system. An integrated company obtains less than 90 percent but more than 10 percent of its gas operating revenues from either its retail or transmission operations or does not meet the classification of mains established for distribution.

Integrated Noise Model (INM)

Aviation. A computer modeling system used by the Federal Aviation Administration (FAA) to develop noise contours for airports and surrounding areas.

integrated pest management (IPM)
A mixture of pesticide and non-pesticide methods to control pests.

Integrated Risk Information System (IRIS)
An EPA database containing verified RfDs, slope factors, health risks, and EPA regulatory information for numerous chemicals. IRIS is EPA's preferred source for toxicity information for Superfund.

integrated tow
Barges designed to fit together so the underwater configuration is the equivalent of a single hull of a motorized vessel. This eliminates water turbulence and increases efficiency.

integrated transportation
See ***intermodalism (3)***.

intellect
The capacity for understanding and reasoning.

intelligence
The ability to recognize, learn, understand, reason, create, and react appropriately to a given set of living conditions. It is a general term for the practical functioning of the mind. It is basically a combination of reasoning, memory, imagination, and judgement. Each of these faculties relies upon the others. The brain may store up many memories, but they are useful only when brought to surface consciousness at the right time and in the right connection. Imagination is the faculty of associating several memories (e.g., facts, images, sensations, etc.) to produce another fact or image. In general, the more efficiently the brain combines memories in an orderly fashion, the greater the intelligence. Imagination, however, must be governed by reason and judgement. Reason is the ability to draw logical conclusions by relating memories and observations. Judgement relies on experience to choose between different forms of reasoning. All these factors are controlled by the cerebral cortex.

intelligence quotient (IQ)
A numerical score attributed to be one's intelligence level, typically the value of the ratio of mental age to chronological age, multiplied by 100. (Generally of limited value.) Expressed as:

$$IQ = \frac{mental\ age}{chronological\ age} \times 100$$

intelligence test
Any of a set of standardized tests which purport to measure an individual's intelligence.

Intelligent Vehicle Highway System
A planned passenger car highway system in which the routine driving, safety, and navigation functions are assumed by integrated computer systems.

intensity
(1) *General.* A measure of the strength or amount of some entity or sensation. (2) *Radiation.* The energy of any radiation incident upon (or flowing through) a unit area, perpendicular to the radiation beam, in a unit of time. (3) *Seismology.* A measure of the effects at a particular place by shaking during an earthquake (not to be confused with *magnitude*). It is a measure of the "strength" of shaking experienced in an earthquake. The Modified Mercalli Scale represents the local effect or damage caused by an earthquake; the "intensity" reported at different points generally decreases away from the earthquake epicenter. The intensity range, from I to XII, is expressed in Roman numerals. For example, an earthquake of intensity II barely would be felt by people favorably situated, while intensity X would produce heavy damage, especially to unreinforced masonry. Local geological conditions strongly influence the intensity of an earthquake. Commonly, sites on soft ground or alluvium have intensities 2 to 3 units higher than sites on bedrock. *See also* ***magnitude***.

intentionally
To do something purposely, and not accidentally.

interaction
The result from a particular combination of events, due solely to the combination and not any particular individual event.

interaction effect
That experimental or statistical result attributable solely to a particular combination of variables and beyond that which can be predicted from the variables independently.

interactive
Having the capability for one or more cycles of human input with rapid display feedback.

interactive corporate compliance
The theory that self-regulation can supplement, and even replace some of the command and control style of corporate regulation by government. Involves using the competitive forces of the economy to produce a set of co-operative policies which can produce acceptable forms of business self-regulation to prevent industrial abuses before they require control by the government. *See also* ***positive incentives, carrot and stick approach*** *and* ***environmental leadership program.***

interactive window
An active window which is receptive to user input.

Interagency Fleet Management System (IFMS)
The organizational title assigned to the General Services Administrations (GSA) interagency fleet operation which encompasses the Central Office, Regional Offices, and all Fleet Management Centers and Fleet Management Subcenters.

interaural phase
The apparent relative phase difference of a tone between the left and right ears.

intercept glideslope altitude
The minimum altitude to intercept the glideslope or path on a precision approach. The intersection of the published intercept altitude with the glideslope/path, designated on Government charts by the lightning bolt symbol, is the precision final approach fix (FAF). However, when Air Traffic Control (ATC) directs a lower altitude, the resultant lower intercept position is then the FAF. *See also* ***glideslope.***

interceptor sewer
Large sewer lines that, in a combined system, control the flow of the sewage to the treatment plant. In a storm, they allow some of the sewage to flow directly into a receiving stream, thus preventing an overload by a sudden surge of water into the sewers. They are also used in separate systems to collect the flows from main and trunk sewers and carry them to treatment points.

interchange
An area designated to provide traffic access between roadways of differing levels.

intercity and rural bus transportation
Establishments primarily engaged in furnishing bus transportation, over regular routes and on regular schedules, the operations of which are principally outside a single municipality and its suburban areas.

intercity bus
(1) A bus with front doors only, high-backed seats, separate luggage compartments, and usually with restroom facilities for use in high-speed long-distance service. (2) A standard size bus equipped with front doors only, high-backed seats, luggage compartments separate from the passenger compartment and usually with rest room facilities, for high-speed long distance service.

intercity passenger mile
The distance generated by moving one passenger one mile on a trip between two cities.

intercity rail passenger
A rail car, intended for use by revenue passengers, obtained by the National Railroad Passenger Corporation (currently Amtrak®) for use in intercity rail transportation.

intercity rail transportation
Transportation provided by Amtrak®.

intercity transportation
(1) Transportation between cities. (2) Transportation service provided between cities by certified carriers, usually on a fixed route with a fixed schedule.

intercity trucking
Trucking operations which carry freight beyond the local areas and commercial zones.

intercostal
Between the ribs.

interdigital crotch
That region of soft tissue between each pair of digits on the hand or foot. The pair of digits being referred to should be specified when using this term.

interest, long-term debt and capital leases
Interest on all classes of debt, both short-term and long-term, as well as the amortization of premium, discount and expense connected

with the issuance of such debt and interest expense on capital leases.

interested party
For the purposes of administrative hearings, those who have a legally recognized private interest, and not simply a possible pecuniary benefit.

interface
A common boundary or point of connection between two or more parts of a system or between systems, whether physical or perceptual.

interfacility
Aviation. Between adjacent facilities; between Air Control Facility (ACF) and ACF, or between ACF and Air Traffic Control Tower (ATCT), as contrasted with intrafacility.

interference
An undesired positive or negative response caused by a substance other than the one being monitored. Substances that may be present in the atmosphere along with the contaminant of interest, which, when sampled, affect the reading of an instrument, detector tube, or in the analysis of the sample. Interference can be positive or negative, significant or insignificant, accounted for or unaccounted for, and generally must be considered when assessing an exposure situation.

interference allowance
That time compensation given a worker for lost production due to interference time.

interference equivalent
Mass or concentration of an interfering substance which gives the same measurement reading as a unit mass or concentration of the substance being measured.

interference time
(1) That machine idle time which results from an operator's inability to service one or more machines due to other assignments. (2) That worker idle time when working as a member of a team in which one or more members of the team are required to wait while some task is carried out by another member.

interferon
Low-molecular-weight protein produced by cells infected with viruses. It will block viral infection of healthy cells and suppress viral multiplication in cells already infected.

interflection
The multiple reflections of light from an enclosed volume other than the luminaire prior to reaching the surface of interest.

Interglacial period
A time interval of relatively mild climate during the Ice Age when continental ice sheets were absent or limited in extent to Greenland and the Antarctic.

intergovernmental revenue
Amounts received from other governments as fiscal aid in the form of shared revenues and grants-in-aid, as reimbursements for performance of general government functions and specific services for the paying government, or in lieu of taxes. This revenue excludes amounts received from other governments for sale of property, commodities, and utility services.

interim (permit) status
Period during which treatment, storage, and disposal facilities coming under RCRA in 1980 are temporarily permitted to operate while awaiting denial or issuance of a permanent permit. Permits issued under these circumstances are usually called Part A or Part B permits.

inter-individual variation
The differences between individuals on the same or equivalent aspect or variable.

interior compartment door
Any door in the interior of the vehicle installed by the manufacturer as a cover for storage space normally used for personal effects.

interior hung scaffold
A work scaffold suspended from the ceiling or roof structure.

interior structural fire fighting
The physical activity of fire suppression, rescue, or both inside of buildings or enclosed structures which are involved in a fire situation beyond the incipient stage.

interlace
Scan across a display screen or other medium such that the distance from line to line in a

field is approximately twice the line width, and adjacent lines belong to different fields.

interlaced display
A display which uses an interlaced scanning format such that two fields must be written to completely update the display.

interlock
An electrical or mechanical device for preventing the continued operation of an instrument if the interlock is not working, or the inactivation of an instrument/appliance, until a condition has been corrected to enable its safe operation.

interlocked route
DOT. A route within interlocking limits.

interlocked switch
A switch within the interlocking limits the control of which is interlocked with other functions of the interlocking.

interlocking limits
The tracks between the opposing home signals of an interlocking.

interlocking machine
An assemblage of manually operated levers or other devices for the control of signals, switches, or other units.

interlocking signal
A roadway signal which governs movements into or within interlocking limits.

intermediary
An arbitrator or mediator. A broker; one who is employed to negotiate a matter between two parties, and who for that purpose may be an agent of both.

intermediate
A chemical formed as a middle step in a series of chemical reactions, especially in the formation of organic compounds.

intermediate approach segment
Aviation. That segment of an instrument approach procedure between either the intermediate approach fix and the final approach fix or point, or between the end of a reversal, race track, or dead reckoning track procedure and the final approach fix or point, as appropriate.

intermediate cuneiform bone
One of the distal group of foot bones in the tarsus.

intermediate fix (IF)
Aviation. The fix that identifies the beginning of the intermediate approach segment of an instrument approach procedure. The fix is not normally identified on the instrument approach chart as an IF.

intermediate grade gasoline
An increasingly common grade of unleaded gasoline with an octane rating intermediate between "regular" and "premium." Octane boosters are added to gasoline to control engine pre-ignition or "knocking" by slowing combustion rates.

intermediate infrared
That portion of the infrared spectrum from about 1400 to 5000 nm.

intermediate landing
Aviation. On the rare occasion that this option is requested, it should be approved. The departure center, however, must advise the Air Traffic Control Command Center (ATCCC) so that the appropriate delay is carried over and assigned at the intermediate airport. An intermediate landing airport within the arrival center will not be accepted without coordination with and the approval of the ATCCC.

intermediate product
Under ISO 14000, input or output from a unit process which requires further transformation.

intermediate type road surface
Mixed bituminous and bituminous penetration (Surface/Pavement Type Codes 52 and 53).

intermittent, casual, or occasional driver
A driver who in any period of 7 consecutive days is employed or used as a driver by more than a single motor carrier. The qualification of such a driver shall be determined and recorded in accordance with the provisions of 49 CFR 391.63 or 391.65 as applicable.

intermittent noise
Noise which occurs intermittently or falls below the audible or measurable level one or more times over a given period.

intermittent stream
A stream, the flow of which in the state of nature is interrupted either from time to time during the year or at various places along its course, or both.

intermittent work
That work, often physically demanding, which is performed only at certain points in time, not on a continuous basis.

intermodal
Used to denote movements of cargo containers interchangeably between transport modes, i.e., motor, water, and air carriers, and where the equipment is compatible within the multiple systems.

intermodal container
A freight container designed and constructed to permit it to be used interchangeably in two or more modes of transport.

intermodal passenger terminal
An existing railroad passenger terminal which has been or may be modified as necessary to accommodate several modes of transportation, including intercity rail service and some or all of the following: intercity bus, commuter rail, intracity rail transit and bus transportation, airport limousine service and airline ticket offices, rent-a-car facilities, taxis, private parking, and other transportation services.

intermodal portable tank
A specific class of portable tanks designed primarily for international intermodal use.

intermodal transport
Enables cargo to be consolidated into economically large units (e.g., containers, bulk grain railcars) optimizing use of specialized intermodal handling equipment to effect high-speed cargo transfer between ships, barges, railcars, and truck chassis using a minimum of labor to increase logistic flexibility, reduce consignment delivery times, and minimize operating costs.

intermodal transportation
Use of more than one type of transportation; e.g., transporting a commodity by barge to an intermediate point and by truck to destination.

intermodalism
Typically used in three contexts: a) Most narrowly, it refers to containerization, piggyback service, or other technologies that provide the seamless movement of goods and people by more than one mode of transport. b) More broadly, intermodalism refers to the provision of connections between different modes, such as adequate highways to ports or bus feeder services to rail transit. c) In its broadest interpretation, intermodalism refers to a holistic view of transportation in which individual modes work together or within their own niches to provide the user with the best choices of service, and in which the consequences on all modes of policies for a single mode are considered. This view has been called balanced, integrated, or comprehensive transportation in the past.

internal
(1) Within or beneath the surface of a body part or other structure. (2) *See* ***medial***.

internal audit
Audit performed by personnel of the company being audited to assure that internal procedures, operations, and accounting practices are in proper order, in contrast to an audit by outside, independent agencies.

internal biomechanical environment
The mechanical forces to which bodily tissues, particularly the musculoskeletal system, are subjected when executing motions or being acted upon by outside forces.

internal canthus
See ***endocanthus***.

internal clock
A hypothetical internal bodily mechanism responsible for maintaining biological rhythms. *See also* ***circadian pacemaker***.

internal combustion engine
An engine in which the power is developed through the expansive force of fuel that is fired or discharged within a closed chamber or cylinder.

internal consistency
Having data within an experiment, analysis, or test which are repeatable across subjects or which have logical relationships within a subject.

internal contamination
As pertains to ionizing radiation, radioactive contamination within a person's body as a result of inhaling, swallowing, or skin puncture by radioactive materials.

internal conversion
A mechanism of radioactive decay in which transition energy is transferred to an orbital electron, causing its ejection from the atom.

internal desynchronization
The loss of normal phase relationships between biological rhythms within a single entity.

internal ear
*See **inner ear**.*

internal injury
Any injury to organs lying within the thoracic or abdominal cavities.

internal naris
The junction of the posterior nasal cavity with the nasopharynx.

internal pacing
Pertaining to self-paced work.

internal radiation
Nuclear radiation (alpha and beta particles and gamma radiation) resulting from radioactive substances inside the body. Important sources are iodine-131 in the thyroid gland, and strontium-90 and plutonium-239 in bone.

internal reporting
Reporting of a violation of a law or corporate policy to upper-level management within an organization. Part of a Self-Reporting or Violation-Reporting System under the Federal Sentencing Guidelines.

internal traffic
*See **internal water transportation**.*

internal water transportation
Includes all local (intraport) traffic and traffic between ports or landings wherein the entire movement takes place on inland waterways. Also termed internal are movements involving carriage on both inland waterways and the water of the Great Lakes, and inland movements that cross short stretches of open water that link inland systems.

internal work
That manual work done by an operator during the operation of a machine or process he/she is supervising. Also referred to as *fill up work* and *inside work*.

internally paced work
*See **self-paced work**.*

international
Air Commerce. Traffic (passengers and freight) performed between the designated airport and an airport in another country or territory.

international air operator
Commercial air transportation outside the territory of the United States, including operations between the U.S. and foreign countries and between the U.S. and its territories and possessions.

International Air Transportation Association (IATA)
Established in 1945, a trade association serving airlines, passengers, shippers, travel agents, and governments. The association promotes safety, standardization in forms (baggage checks, tickets, weight bills), and aids in establishing international airfares. IATA headquarters are in Geneva, Switzerland.

international airport
(1) Any airport designated by the Contracting State in whose territory it is situated as an airport of entry and departure for international air traffic. (2) An airport of entry which has been designated by the Secretary of Treasury or Commissioner of Customs as an international airport for customs service. (3) A landing rights airport at which specific permission to land must be obtained from customs authorities in advance of contemplated use. (4) Airport designated under the Convention on International Civil Aviation as an airport for use by international commercial air transport and/or international general aviation.

international and territorial operations
Aviation. The operation of aircraft flying between the 50 United States and foreign points, between the 50 United States and U.S. possessions and territories, and between two foreign points. Includes both the combination passenger/cargo and the all-cargo carriers engaged in international and territorial operations.

international bunkers
Storage compartments on vessels and aircraft engaged in international commerce, where fuel to be used by the vessel or aircraft is stored.

International Cargo Handling Coordination Association (ICHCA)
a) Collects, edits, and disseminates technical information relating to cargo handling by all modes of transport. b) Maintains consultative status with the International Standards Or-

ganization for the development of standards relating to cargo handling equipment (such as hooks, containers, wire slings, spreaders, and pallets). c) Maintains a library for members' use. d) Represents members' interests on an international basis.

International Civil Aviation Organization (ICAO)
A specialized agency of the United Nations whose objective is to develop the principles and techniques of international air navigation and to foster planning and development of international civil air transport. ICAO Regions include (AFI) African Indian Ocean Region, (CAR) Caribbean Region, (EUR) European Region, (MID/ASIA) Middle East/Asia Region, (NAM) North American Region, (NAT) North Atlantic Region, (PAC) Pacific Region, (SAM) South American Region.

International Civil Aviation Organization Broadcast
A transmission of information relating to air navigation that is not addressed to a specific station or stations.

International Committee on Radiation Protection (ICRP)
An international group of scientists that develops recommendations on ionizing radiation dose limits and other radiation protection measures.

International Court of Justice
The judicial arm of the United Nations. It has jurisdiction to give advisory opinions on matters of law and treaty construction when requested by the General Assembly, Security Council, or any other international agency authorized by the General Assembly to petition for such opinion. It also has jurisdiction to settle legal disputes between nations when voluntarily submitted to it. Its judgements may be enforced by the Security Council, Its jurisdiction and powers are defined by statute, to which all member states of the U.N. are parties. Judges of this Court are elected by the General Assembly and Security Council of the U.N.

international flight information manual
A publication designed primarily as a pilot's preflight planning guide for flights into foreign airspace and for flights returning to the U.S. from foreign locations.

International Flight Service Station (IFSS)
A central operations facility in the flight advisory system, manned and equipped to control aeronautical point-to-point telecommunications, and air/ground telecommunications with pilots operating over international territory or waters, providing flight plan filing, weather information, search and rescue action, and other flight assistance operations.

international freight forwarder
A person, duly registered with the Federal Maritime Board, engaged in the business of dispatching shipments on behalf of other persons, for a consideration, by ocean going vessels in commerce from the United States, its territories, or possessions, and handling the formalities incident to such shipments.

international inland waterways transport
Inland waterways transport between two places (a place of loading/embarkment and a place of unloading/disembarking) located in two different countries. It may involve transit through one or more additional countries.

international jurisdiction
Power of a court or other organization to hear and determine matters between different countries or persons of different countries or foreign states.

international law
Those laws governing the legal relations between nations.

International Maritime Organization (IMO)
Established as a specialized agency of the United Nations in 1948. The International Maritime Organization facilitates cooperation on technical matters affecting merchant shipping and traffic, including improved maritime safety and prevention of marine pollution. Headquarters are in London, England.

International Maritime Satellite Organization (INMARSAT)
An international partnership of signatories from 67 nations. The partnership provides mobile satellite capacity to its signatories, who, in turn, use the capacity to provide worldwide mobile satellite services to their maritime, aeronautical, and land-mobile customers including shipping, cruise, fishing, research and offshore exploration industries, and airlines. INMARSAT began service in 1976.

international operations
In general, operations outside the territory of the U.S., including operations between the U.S. and foreign countries, and the U.S. and its territories or possessions. Includes both the combination passenger/cargo carrier and the all-cargo carriers engaged in international and territorial operations.

International Organization for Standardization (ISO)
Founded as a worldwide federation to promote the development of international manufacturing, trade, and communication standards, thereby facilitating the international exchange of goods and services. ISO has promulgated more than 8,000 international standards.

international passenger
Any person traveling on any type of public conveyance (e.g., waterborne, airborne, etc.) between the United States and foreign countries and between Puerto Rico and the Virgin Islands and foreign countries.

International System of Units (SI)
See ***basic units***.

international transportation
Transportation between any place in the United States and any place in a foreign country; between places in the United States through a foreign country; or between places in one or more foreign countries through the United States.

Internet
A network of computers.

Internet Service Provider (ISP)
Provides Internet access to people or corporations.

interoceptor
Any sensory receptor sensitive to changes within the viscera and blood vessels

interocular breadth
See ***endocanthic breadth***.

interocular distance
See ***interpupillary breadth***.

interphalangeal
Between the phalanges of the hand or foot.

interpolate
To estimate one or more unknown values within a range of known values using some predictor.

interpupillary
Pertaining to the region between the eye pupils.

interpupillary breadth
The horizontal linear distance between the centers of the pupils of the eyes. Measured with the individual's scalp muscles relaxed, the eyes open, and looking straight ahead.

interrogator
Aviation. The ground-based surveillance radar beacon transmitter-receiver, which normally scans in synchronism with a primary radar, transmitting discrete radio signals which repetitiously request all transponders on the mote being used to reply. The replies received are mixed with the primary radar returns and displayed on the same plan position indicator (radar scope). Also, applied to the airborne element of the TACAN/DME system.

interrogatories
Part of the pre-trial discovery process. A formal set of questions, usually written, specific to the case, that must be answered by the party served, usually in writing and before the trial date.

interrupter
A mechanical barrier in a fuse that prevents transmission of an explosive effect to some elements beyond the interrupter.

interrupter switch
As pertains to systems over 600 volts (nominal), a switch capable of making, carrying, and interrupting specified currents.

interscapulae
Pertaining to the region of the back between the two scapular bones.

interscye, bent torso
The surface distance across the back between the scye points. Measured with the individual standing, the torso bent forward from the waist at an angle of about 90°, and the arms hanging relaxed.

interscye, seated forward reach
The surface distance across the back between the scye points. Measured with the individual sitting erect with his/her arms extended forward horizontally.

interscye, seated leaning
The surface distance across the back between the posterior axillary folds at the lower level of the armpits. Measured with the individual seated and leaning forward with his/her hands on his/her knees.

interscye, standing erect
The surface distance across the back between the posterior axillary folds at the lower level of the armpits. Measured with the individual standing erect and his/her body weight distributed equally between the two feet.

intersecting runway
Two or more runways which cross or meet within their lengths.

intersection
Aviation. (1) A point defined by any combination of courses, radials, or bearings of two or more navigational aids. (2) Used to describe the point where two runways, a runway and a taxiway, or two taxiways cross or meet.

intersection departure
Aviation. A departure from any runway intersection except the end of the runway.

Inter-Society Color Council - National Bureau of Standards color system
See ***color ordering system***.

interstate
(1) Limited access divided facility of at least four lanes designated by the Federal Highway Administration as part of the Interstate System. (2) Meaning between states or involving more than one state.

interstate air commerce
The carriage by aircraft of persons or property for compensation or hire, or the carriage of mail by aircraft, or the operation or navigation of aircraft in the conduct or furtherance of a business or vocation, in commerce between a place in any State of the United States, or the District of Columbia, and a place in any other State of the United States, or the District of Columbia; or between places in the same State of the United States through the airspace over any place outside thereof; or between places in the same territory or possession of the United States, or the District of Columbia.

Interstate Air Pollution Control Agency
Under the Clean Air Act: An air pollution control agency established by two or more states, or an air pollution control agency of two or more municipalities located in different states.

interstate air transportation
The carriage by aircraft of persons or property as a common carrier for compensation or hire, or the carriage of mail by aircraft in commerce: a) between a place in a State or the District of Columbia and another place in another state or the District of Columbia; b) between places in the same state through the airspace over any place outside that state; or c) between places in the same possession of the United States; whether that commerce moves wholly by aircraft or partly by aircraft and partly by other forms of transportation.

interstate carrier water supply
A source of water for drinking and sanitary use on planes, buses, trains, and ships operating in more than one state. These sources are federally regulated.

interstate commerce
Trade, traffic, or transportation in the United States which is between a place in a state and a place outside of such state (including a place outside of the United States) or is between two places in a state through another state or a place outside of the United States.

Interstate Commerce Act
The act of Congress of February 4, 1887 (49 U.S.C.A. § 10101 et seq.), designed to regulate commerce between the states, and particularly the transportation of persons and property, by carriers, between interstate points.

Interstate Commerce Commission (ICC)
The federal body charged with enforcing Acts of Congress affecting interstate commerce.

Interstate Commerce Commission authorized carrier
A for-hire motor carrier engaged in interstate or foreign commerce, subject to economic regulation by the ICC.

Interstate Commerce Commission exempt carrier
A for-hire motor carrier transporting commodities or conducting operations not subject to economic regulation by the ICC.

interstate highway (freeway or expressway)
A divided arterial highway for through traffic with full or partial control of access and grade separations at major intersections.

interstate highway system
This system is part of the Federal Aid Primary system. It is a system of freeways connecting and serving the principal cities of the continental United States.

interstate pipeline
(1) A pipeline or that part of a pipeline that is used in the transportation of hazardous liquids or carbon dioxide in interstate or foreign commerce. (2) A natural gas pipeline company that is engaged in the transportation, by pipeline, of natural gas across state boundaries, and is subject to the jurisdiction of the Federal Energy Regulatory Commission (FERC) under the Natural Gas Act. *See also* ***intrastate pipeline***.

interstate waters
Waters that flow across or form part of state or international boundaries, e.g., the Great Lakes, the Mississippi River, or coastal waters.

interstimulus-onset interval
The length of time between the onset of one stimulus and the onset of a second stimulus.

interstitial
The space between cellular components or parts of a structure or organ.

interstitial monitoring
A technique for monitoring the integrity of the area between the primary and secondary containment systems of underground storage tanks (USTs).

Intertropical Convergence Zone (ITCZ)
The boundary zone separating the northeast trade winds of the Northern Hemisphere from the southeast trade winds of the Southern Hemisphere.

interval scale
See ***equal-interval scale***.

intervertebral disk
A circular mass of fibrous cartilage located between adjacent vertebrae in the spine.

interview
A spontaneous or organized sequence of questions and discussion to exchange information relevant to a particular situation between two or more individuals.

intervocalic
Occurring between vowels.

intestine
The membranous tube extending from the pylorus of the stomach to the anus, consisting of the small intestine and the large intestine.

intort
To rotate a structure toward the midline, especially the eye.

intorter
A muscle which intorts.

intoxication
(1) A state of having been poisoned by any toxic substance, whether unknowingly or due to one's own voluntary actions. Intoxication in the sense of poisoning can be caused by carbon monoxide, lead, or other toxic agents. Some medications can be poisonous in excessive doses. Intoxication can also occur in persons who have an allergy to medications such as penicillin, to various serums, and to other substances. Any type of drug addiction is medically recognized as a state of intoxication. In addition to those mentioned, there are the commonly recognized types of poisoning such as those caused by chemicals and food contaminants. Acid intoxication and alkaline intoxication are acidosis and alkalosis, respectively, of a severe grade. Intoxication in the sense of drunkenness occurs when the concentration of alcohol in the blood reaches about one-tenth of 1 percent. (2) A state of intense mental or emotional excitement.

intra-abdominal pressure (IAP)
That pressure exerted on the internal walls by gravity, the abdominal viscera, arterial supply, and the musculature.

intra-cellular water
That water contained within the cells of the body. One of two components of total body water.

intrafacility
Aviation. Within a single facility; for example, between two sectors within the same Area Control Facility (ACF), as contrasted with interfacility.

intrafusal fiber
The small muscles fibers within a muscle spindle that are involved in sensing length changes. *See also* ***extrafusal fiber***.

intra-individual variation
That variation which occurs within a single person over time on the same or similar testing or observation.

intransit deliveries
Redeliveries to a foreign country of foreign gas received for transportation across U.S. territory, and deliveries of U.S. gas to a foreign country for transportation across its territory and redelivery to the United States.

intransit passengers
Aviation. Revenue passengers onboard international flights that transit an airport for non-traffic purposes in the 50 states.

intransit receipts
Receipts of foreign gas for transportation across United States (U.S.) territory and redelivery to a foreign country, and redeliveries to the U.S. of U.S. gas transported across foreign territory.

intra-ocular muscle
An involuntary, intrinsic, smooth muscle within the eye, specifically the ciliary and pupillary muscles.

intra-ocular pressure
That fluid pressure within the eyeball.

intraperitoneal
Within the abdominal/pelvic cavity.

intrapleural
Within the chest cavity.

intrasensory matching
A procedure in which a subject matches the magnitude of a stimulus in a sensory modality with the magnitude of another stimulus using the same modality.

intrastate
Travel or movement or events occurring within the same state.

intrastate air transportation
The carriage of persons or property as a common carrier for compensation or hire, by turbojet powered aircraft capable of carrying thirty or more persons, wholly within the same state of the United States.

intrastate commerce
Any trade, traffic, or transportation in any state which is not described in the term "interstate commerce."

intrastate pipeline
(1) A pipeline or that part of a pipeline to which 49 CFR 195.2 applies that is not an interstate pipeline. (2) A natural gas pipeline company engaged in the transportation, by pipeline, of natural gas not subject to the jurisdiction of the Federal Energy Regulatory Commission (FERC) under the Natural Gas Act. *See also* ***interstate pipeline***.

intraterritorial traffic
Traffic between ports in Puerto Rico and the Virgin Islands, which are considered as a single unit.

intratracheal
Endotracheal or within or throughout the trachea.

intravehicular activity (IVA)
Any activity occurring within a vehicle, especially referring to a space vehicle.

intravenous
Within a vein.

intravenous infusion
Administration of fluids through a vein. Also referred to as venoclysis and intravenous feeding. This method of feeding is used most often when a patient is suffering from severe dehydration and is unable to drink fluids because he/she is unconscious, recovering from an operation, unable to swallow normally, or vomiting persistently.

intrinsic
Pertaining to a structure or mechanism which originates within a structure and acts on itself.

intrinsic muscle
Any muscle having both its origin and insertion located within a given structure and which is involved in the function of that structure.

intrinsically safe
Incapable or producing sufficient energy to ignite an explosive atmosphere and two-fault tolerant against failure with single fault tolerance at 1.5 times the maximum voltage or energy.

intubation
Insertion of a tube. The purpose varies with the location and type of tube inserted; generally the procedure is done to allow for drainage, to maintain an open airway or for the

administration of anesthetics or oxygen. Intubation in the stomach or intestine is done to remove gastric or intestinal contents for the relief or prevention of distention, or to obtain a specimen for analysis. A tube may be inserted in the common bile duct to allow for drainage of bile from the ducts that drain the liver after surgery on the gallbladder or the common bile duct. Tracheal intubation can be achieved by the insertion of an endotracheal tube into the trachea via the nose or mouth.

inundation area
An area of land subject to flooding.

invasive
Pertaining to a procedure which requires breaking the skin, insertion of any object into any body cavity except the mouth, or which causes extreme discomfort.

inventory
(1) *General.* Materials on hand, or, a physical count of the materials on hand. (2) *TSCA.* An inventory of chemicals produced pursuant to Section 8(b) of the Toxic Substances Control Act.

inventory management
Those techniques involved in maintaining the desired inventory levels, including planning, tracking, distribution, providing storage, and purchasing.

inverse power function
An exponential mathematical relationship involving a negative exponent or where the variable would be represented in the denominator with a positive exponent.

inverse square law
Doubling the distance from a source of light, noise, ionizing radiation, etc. reduces the intensity of the exposure to the source at that point by one-fourth (i.e., the intensity varies inversely with the square of the distance from the source).

inversion
An atmospheric condition caused by a layer of warm air preventing the rise of cooling air trapped beneath it. This condition prevents the rise of pollutants that might otherwise be dispersed and can result in an air pollution episode.

invert
To turn inward.

invertebrate
(1) Having no vertebral column. (2) An animal organism that has no vertebral column.

inverted image
An image which has been rotated within its plane by 180°.

inverted U function
*See **concave function**.*

invertor
Any muscle which turns the sole of the foot inward.

investigation parameters
As defined by management, the specific considerations which must be evaluated to determine the focus of the accident investigation process. Examples include such aspects as the types of occurrences that will require reporting and investigating, the elements of a business operation or function to be investigated and to what extent, how accidents and incidents shall be formally reported, and what use shall be made of the information reported.

investigatory interrogation
An investigatory interrogation outside the scope of the *Miranda Rule* is the questioning of persons by law enforcement officers in a routine manner in an investigation which has not reached an accusatory stage and where such persons are not in legal custody or deprived of their freedom of action in any significant way.

investments and special funds
Investments and advances to investor-controlled and other associated companies, notes and receivables not due within one year, investment in securities issued by others, allowance for unrealized gain or loss on noncurrent marketable equity securities, funds not available for current operations, investments in leveraged leases, and net investments in direct financing and sales-type leases which are not reasonably expected to be amortized within one year.

involuntary muscle
Those muscles not normally under conscious control, such as the smooth muscles.

involution
The process of decline or decay in human processes later in life.

iodine
A chemical element, atomic number 53, atomic weight 126.904, symbol I. Salts of iodine and tincture of iodine were once used as antiseptics. Iodine is a strong poison, however, and has largely been replaced by other antiseptics that are less irritating to the tissues and equally effective. Since iodine salts are opaque to x-rays, they can be combined with other compounds and used as contrast media in diagnostic x-ray examinations of the gallbladder and kidneys.

IOHA
International Occupational Health Association.

ion
An atom or chemical radical (group of chemically combine atoms) bearing a positive or negative electrical charge caused by a deficiency or excess of electrons.

ion exchange
The reversible interchange of ions of like charge between an insoluble solid and a surrounding liquid phase in which there is no permanent change in the structure of the solid.

ion exchange resin
Synthetic resins which contain active groups enabling the resin to combine with, or exchange ions between it and those in another substance.

ion exchange treatment
A common water softening method often found on a large scale at water purification plants that removes some organics and radium by adding calcium oxide or calcium hydroxide to increase the pH to a level where the metals will precipitate out.

ion pair
Two particles of opposite charge. One method by which ionizing radiation gives up its energy is by the production of ion pairs. *See also* ***ionization***.

ionization
The separation of a normally electrically neutral atom or molecule into electrically charged components. The term may also be used to describe the degree or extent to which this separation occurs. Ionization is the removal of a negatively charged electron from the atom or molecule (either directly or indirectly) leaving a positively charged ion. The separated electron and ion are then referred to as an *ion pair*. *See also* ***ionizing radiation***.

ionization chamber
A device consisting of two electrically opposed, charged plates used to measure radioactivity.

ionization density
Number of ion pairs per unit volume.

ionization track
The detectable path of an ionizing photon or particle following passage through tissue or another substance.

ionizing radiation
Electromagnetic radiation (x-ray or gamma ray photons or quanta) or corpuscular radiation (alpha or beta particles, electrons, positrons protons, neutrons, or heavy particles) capable of producing ions by direct or secondary processes as it passes through matter.

ionometer
An instrument for measuring the intensity or quantity of x-rays.

ionosphere
An electrified region of the upper atmosphere where fairly large concentrations of ions and free electrons exist.

iontophoresis
The process of transferring ions across some barrier with direct current.

IPM
See ***integrated pest management***. *See also* ***Inspirable Particulate Mass***.

IPM-TLVs
See ***Inspirable Particulate Mass TLVs***.

IQ
See ***intelligence quotient***.

IR
Infrared.

iridescent
Pertaining to the optical interference effects in thin films or of reflected diffracted light from ribbed surfaces.

iridescent clouds
Clouds that exhibit brilliant spots or borders of colors, most often red and green. Observed up to about 30° from the sun.

IRIS

See ***Integrated Risk Information System***.

iris

(1) The colored circular structure in the aqueous humor of the eye which encircles the pupil between the cornea and the lens and regulates the amount of light reaching the retina. (2) An arrangement of flat leaf-like structures which provides an approximately circular opening on retraction.

iris reflex

The adjustment of muscle fiber length in the iris to accommodate light levels to which the eye is exposed.

iritis

Inflammation of the iris. The condition may be acute, occurring suddenly with pronounced symptoms, or chronic, with less severe but longer-lasting symptoms. The cause is often obscure. Frequently, the condition is associated with rheumatic diseases, particularly rheumatoid arthritis, and with diabetes mellitus, syphilis, diseased teeth, tonsillitis, and other infections. It may also be caused by injury. Iritis is characterized by severe pain, usually radiating to the forehead and becoming worse at night. The eye is usually red and the pupil contracts and may be irregular in shape. There is extreme sensitivity to light, together with blurring of vision and tenderness of the eyeball. The iris becomes swollen and discolored. If not treated promptly, iritis can be dangerous because of scarring and adhesions that may cause impaired vision and possibly blindness.

iron

(1) *General.* A chemical element, atomic number 26, atomic weight 55.847, symbol Fe. (2) *Human Physiology.* Iron is chiefly important to the human body because it is the main constituent of hemoglobin, and a constant although small intake of iron in food is needed to replace erythrocytes that are destroyed in the body's processes. Most iron reaches the body in food, where it occurs naturally in the form of iron compounds. These are converted for use in the body by the action of the hydrochloric acid produced in the stomach. This acid separates the iron from the food and combines with it in a form that is readily assimilated by the body. Vitamin C enhances the absorption of food iron. The administration of alkalis hampers iron absorption. The amount of new iron needed every day by the adult is approximately 15 mg. (3) *Transit (slang).* An old model truck.

iron bacteria

Bacteria capable of metabolizing reduced iron. Also called *crenothrix.*

iron lunger

Transit (slang). The conventional 220 or 250 horsepower engine.

IRPA

International Radiation Protection Association.

irradiance

The density of radiant flux per unit area on a specified surface.

irradiate

To expose to some form of directed energy.

irradiated food

Food that has been subject to brief radioactivity, usually by gamma rays, to kill insects, bacteria, and mold, and preserve it without refrigeration or freezing.

irradiation

Exposure to radiation of wavelengths shorter than those of visible light (gamma, x-ray, or ultraviolet) for medical purposes, the destruction of bacteria in milk or other foodstuffs, or for inducing polymerization of monomers or vulcanization of rubber.

irregular element

A work element occurring at other than regular intervals, but which may be statistically predicted.

irregular shift

A variable work schedule, set by the employer for his/her convenience, usually to accommodate anticipated workloads.

irreparable damages

In the law pertaining to injunctions, damages for which no certain pecuniary standard exists for measurement. Damages not easily ascertainable at law. With reference to public nuisances which a private party may enjoin, the term includes wrongs of a repeated and continuing character, or which occasion damages estimatable only by conjecture, and not by any accurate standard.

irrespirable
Unfit for breathing.

irreversible effect
An effect that is not reversible once the exposure has terminated.

irreversible injury
An injury that is neither repairable nor can be expected to heal.

irrigation
(1) Washing a body cavity or wound by a stream of water or other fluid. (2) Technique for applying water or wastewater to land areas to supply the water and nutrient needs of plants.

irrigation machine
An electrically driven or controlled machine, with one or more motors, not hand portable, and used primarily to transport and distribute water for agricultural purposes.

irritability
(1) Ability of an organism or a specific tissue to react to the environment. (2) The state of being abnormally responsive to slight stimuli or unduly sensitive.

irritant
Substance that induces local inflammation of normal tissues on immediate, prolonged, or repeated contact.

irritant smoke
A smoke-like material that is used in determining whether a mechanical air-purifying respirator wearer achieves a good fit in a qualitative fit test. Stannic oxychloride or titanium tetrachloride are used as the source of irritant smoke. This smoke is also used in ventilation system evaluations (i.e., smoke tubes).

irritation
A reaction of tissues to an injury that results in an inflammation; the response or reaction by tissues to the application of a stimulus.

IRV
See ***inspiratory reserve volume***.

ISA
See ***Instrument Society of America***.

isallobar
A line of equal change in atmospheric pressure during a specified time interval.

ischemia
A condition in which there is an insufficient amount of blood to a part of the body, due to a functional constriction or blockage of a blood vessel, that can result in damage to the affected area.

ischemic hypoxia
A form of hypokinetic hypoxia in which arterial blood flow is reduced.

ischial tuberosity
A projection at the base of the ischium which can become a pressure point when sitting on a hard surface.

ischium
The inferior and posterior portion of each coxal bone.

Ishihara test
A commonly used color deficiency test using plates on which numbers of a given color are embedded in a variety of hues. One number is seen by those with normal vision, another number by those with a color vision deficiency or color blindness.

ISO
See ***International Organization for Standardization***.

ISO 14000
Established in 1996 by the International Organization for Standardization, a worldwide federation funded to promote international standards in many areas. It is a set of voluntary standards in the area of environmental management, auditing, performance evaluation, and life cycle analysis. The standards consist of guidelines or principles, systems, and supporting techniques with two primary components: establishment of an environmental management system, and development of the practice of environmental auditing. Certification often qualifies companies for a decrease in insurance premium levels and an acknowledgment worldwide of their commitment to maintaining and improving the environment.

ISO 14010
International Standard under ISO 14000 which includes general principles of environmental auditing.

ISO 14011/1
International Standard under ISO 14000 which includes auditing of environmental management systems.

ISO 14012
International Standard under ISO 14000 which includes qualification criteria for environmental auditors.

ISO 14040
International Standard under ISO 14000 to address the life cycle assessment of products.

isoacceleration
Having a constant acceleration.

isobar
(1) One of several nuclides having the same number of nucleons, but different combinations of protons and neutrons (i.e., the same mass number, but different atomic numbers). (2) A line connecting points of equal pressure. A series of points on a map or chart that when connected depicts a line of constant atmospheric pressure.

isobaric chart
See ***constant pressure chart.***

isobaric surface
A surface along which the atmospheric pressure is equal everywhere. See also ***constant pressure chart.***

isocandela diagram
A set of plotted isocandela lines on a coordinate system to show lighting intensity spatial relationships.

isocandela line
A contour line representing an area of equal lighting intensity.

isocyanate asthma
Bronchial asthma as a result of an allergy to toluene diisocyanate and similar cyanate compounds.

isoforce
See ***isotonic.***

isoinertial
Pertaining to the force applied to a constant, moving mass.

isoinertial action
That dynamic muscle action involved in moving a constant mass. Also referred to as *isoinertial concentration.*

isoinertial concentration
See ***isoinertial action.***

isokinetic
Pertaining to movement at a constant velocity.

isokinetic action
A dynamic muscle action in which muscle contraction occurs at a constant velocity and maximal tension is maintained during the entire movement sequence. Also referred to as *isokinetic contraction.*

isokinetic contraction
See ***isokinetic action.***

isokinetic sampling
An air sampling technique used to measure particulates and other contaminants in exhaust stacks.

isolated
Not readily accessible to persons unless special means for access are used.

isolated power system
A system comprising an isolating transformer or its equivalent, a line isolation monitor, and its ungrounded circuit conductors.

isolated word recognition
See ***word recognition.***

isolating switch
See ***disconnecting switch*** *and* ***switch (3).***

isolation
(1) *General.* Any spatial or physical separation from other humans or certain individuals. (2) *Power Systems.* The process by which a permit space is removed from service and completely protected against the release of energy and material into the space by such means as blanking or binding; misaligning or removing sections of lines, pipes, or ducts; a double block and bleed system; lockout or tagout of all sources of energy; or blocking or disconnecting all mechanical linkages. (3) *Acoustics.* The use of materials or construction around a noise source to limit the transmission of sound from that source.

isomer
Chemical substances with the same number of characteristic atoms which are electronically arranged differently producing compounds with distinguishable physical and chemical properties.

isomerization
The process whereby any isomer is converted into another, usually requiring special conditions of temperature, pressure, or catalysts.

isometric
Having or maintaining the same dimension. May refer to either a muscle length or an engineering display/drawing.

isometric action
A muscular process in which tension increases in one or more muscles, but the muscles retain approximately the same length and essentially no movement of the body link(s) occurs for prolonged periods of time. Also referred to as *isometric contraction*.

isometric joystick
A non-moving joystick which provides a directional output proportional to the force applied by the user. Also referred to as *force joystick* and *pressure joystick*.

isometric strength test
A test to determine the safe static load handling capabilities for workers.

isometric view
A three-dimensional appearing view of an object on a display or drawing that has been constructed so that perspective has been ignored.

isopleth
See ***sound level contour***.

isoseismal lines
Seismology. A line connecting points of identified intensity for a given earthquake. Also referred to as *isoseisms*.

isoseismal map
Seismology. A map showing the distribution of intensity across a region for a particular earthquake using *isoseismal lines* to connect points of equal intensity.

isoseisms
See ***isoseismal lines***.

isotach
A line connecting points of equal wind speed.

isotherm
A line connecting points of equal temperature.

isotonic
Having uniform tension or force. Also known as *isoforce*.

isotonic action
A dynamic muscle action in which the muscle length of one or more muscles shortens and movement of one or more body links occurs, with constant muscle tension throughout the movement.

isotonic contraction
See ***isotonic action***.

isotonic hypoxemia
A hypoxemic condition with a normal partial pressure of oxygen, generally due to decreased hemoglobin or toxin/drug effects.

isotonic joystick
A joystick whose output is proportional to and in the same direction as the displacement of the joystick from its null point. Sometimes referred to as *displacement joystick*.

isotope
Forms of the same element having nearly identical chemical properties but differing in their atomic masses (due to different numbers of neutrons in their respective nuclei) and in nuclear properties such as radioactivity or fission.

isotropic
Having an equal spatial distribution or growth in all directions.

ISP
See ***Internet Service Provider***.

Itai Itai disease
Name given to a disease that was considered to be a result of eating rice that had been contaminated with cadmium from industrial emissions.

itch
(1) Any skin disease attended by itching. (2) Scabies.

ITCZ
See ***Intertropical Convergence Zone***.

iterate
To calculate desired result using repeated operations.

itinerant aircraft operations
All aircraft operations other than local operations.

IVA
See ***intravehicular activity***.

IWT
See ***inland waterway transport***.

J

J route

See ***jet route***.

jack staff

A vertical pole erected on the lead barge of a tow used by the pilot for aligning the heading of the tow.

jacket

(1) A short, lightweight coat. (2) A thermal blanket or insulating material placed around certain pipes or piping systems to ensure protection from ambient temperatures that may fall below the freezing point.

jackhammer

A hand-controlled chisel device operated by compressed air.

Workers using a jackhammer to break rock at a demolition site

jacking it around

Transit (slang). Backing a semitrailer around a very sharp curve.

jackknife

Transit (trucking). A jackknife can occur at any time during a crash sequence. Jackknifing is usually restricted to truck tractors pulling a trailing unit in which the trailing unit and the pulling vehicle rotate with respect to each other. *See also* ***accident*** *and* ***rollover***.

jacksonian epilepsy

A progression of involuntary clonic movement or sensation, with retention of consciousness.

jacob's ladder

A rope ladder suspended from the side of a vessel and used for boarding.

jamming

Aviation. Electronic or mechanical interference which may disrupt the display of aircraft on radar or the transmission/reception of radio communications/navigation.

jaundice

An abnormal physical condition caused by bile pigments in the blood and characterized by yellowing of the skin and sclera of the eye and by lassitude and loss of appetite. It is usually first noticeable in the eyes, although it may come on so gradually that it is not immediately noticed by those in daily contact with the jaundiced person. Jaundice is not a disease. It is a symptom of one of a number of different diseases and disorders of the liver, gallbladder, and blood. One such disorder is the presence of a gallstone in the common bile duct, which carries bile from the liver to the intestine. This may obstruct the flow of bile, causing it to accumulate and enter the bloodstream. The obstruction of bile flow may cause bile to enter the urine, making it dark in color, and also decrease the bile in the stool, making it light and clay-colored. This condition requires surgery to remove the gallstone before it causes serious liver injury. Jaundice may also be a symptom of infectious hepatitis. This very infectious disease may result in damage to the liver if not treated. Certain diseases of the blood, such as hemolytic anemia, increase the amount of yellow pigment in the bile, causing jaundice. The pigment that causes jaundice is called bilirubin. It is derived from hemoglobin that is released when erythrocytes are hemolyzed and therefore is constantly being formed and introduced into the blood as worn-out or defective erythrocytes are destroyed by the body. Normally the liver cells absorb the bilirubin and secrete it along with other bile constituents. If the liver is diseased, or if the flow of bile is obstructed, or if destruction of erythocytes is excessive, the bilirubin accumulates in the blood and eventually will produce jaundice. A diagnostic test for determination of the level of bilirubin in the blood (called the van den Bergh test) is of value in detecting elevated bilirubin levels at the earliest stages before jaundice appears, when liver disease or hemolytic anemia is suspected.

jaw
The two bones forming the skeletal framework for the mouth, the maxilla for the upper jaw bone, the mandible for the lower.

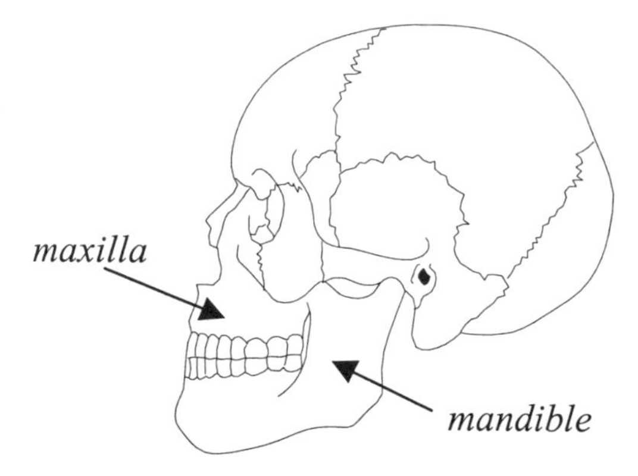

The human skull showing the bones of the jaw

jerk
(1) The rate of change of acceleration with time. (2) A sudden, spasmodic or reflex body movement.

jet blast
Jet engine exhaust (thrust stream turbulence).

jet down
To sink an object, generally a buoy sinker, deep into the mud below the river bottom by the use of high pressure water jet.

jet fuel
The term includes kerosene-type jet fuel and naphtha-type jet fuel. Kerosene-type jet fuel is a kerosene quality product used primarily for commercial turbojet and turboprop aircraft engines. Naphtha-type jet fuel is a fuel in the heavy naphtha boiling range used primarily for military turbojet and turboprop aircraft engines.

jet lag
A slang term describing the state of general discomfort from crossing one or more time zones rapidly, due to circadian rhythm desynchronization. Also referred to as *desynchronosis*.

jet route
A route designed to serve aircraft operations from 18,000 feet mean sea level (MSL) up to and including flight level 450. The routes are referred to as "J" routes with numbering to identify the designated route, e.g., J105.

jet stream
(1) Relatively strong winds concentrated within a narrow band in the atmosphere. (2) A migrating stream of high-speed winds present at high altitudes.

jettisoning of external stores
Airborne release of external stores, e.g., tip tanks, ordnance. *See also* ***fuel dumping***.

JHA
See ***Job Hazard Analysis***.

jig
Any precision mechanical device used to support or hold parts in position or act as a guide.

jitney
(1) Privately owned, small or medium-sized vehicle usually operated on a fixed route but not on a fixed schedule. (2) Passenger cars or vans operating on fixed routes (sometimes with minor deviations) as demand warrants without fixed schedules or fixed stops.

jitter
A periodic jumping of a target or small structure on a display.

JND
See ***just noticeable difference***.

job
(1) The sum of all the tasks and duties assigned to and carried out by one or more workers toward the completion of some goal. (2) That work specified in a contract work order, usually to be performed by several people.

job aid
See ***work aid***.

job analysis
An evaluation of job requirements through an evaluation of the duties and tasks, facilities and working conditions, and worker qualifications and responsibilities necessary to perform a job.

job breakdown
A division of a job into its elements; a listing of the elements comprising a job.

job class
A job classification level in which jobs involve similar types of work, difficulty, and/or pay.

job classification
The arrangement of jobs by job class.

job content
The total makeup of a job, including the physical tasks and the psychological factors of challenge, variety, and feeling of worth.

job costing
A cost determination in which manufacturing costs are attributed to individual items.

job demand
The combined physiological, sensory-perceptual, and psychological requirements for or loads experienced by a worker performing a particular job.

job description
A written general statement of the scope, duties, and responsibilities of a particular job.

job design
The process of determining what the job content should be for a set of tasks, how the tasks should be organized, and what linkage should exist between jobs. *See also* ***work design***.

job dimension
Any of the primary quantifiable aspects of a job for evaluation purposes.

job element
Some distinct portion of a specified job.

job enlargement
An increase in job scope with the intent to make jobs more interesting, through the addition of more tasks of a similar nature to the duties or tasks currently being performed

job enrichment
An increase in the scope of a worker's job, with the intent of increasing variety and significance by adding additional duties such as planning, greater control over operations, and more interaction with others.

job evaluation
The process of determining the relative worth or utility of a job.

job factor
An essential element of a job which gives management some basis for setting a wage range for the job, as well as the selection and training of workers.

job hazard analysis (JHA)
See ***job safety analysis***.

job modification
A minor change to a job. *See also* ***job redesign***.

job plan
An organized approach or document by management showing detailed procedures for each job.

job redesign
A significant, intentional change in job design.

job restriction
A condition in which an individual returning to the workforce following an illness or occupational injury is not permitted to perform certain tasks which might aggravate that illness or injury.

job rotation
The assignment to or performance of different activities by a group of workers on a periodic basis.

job safety analysis (JSA)
A generalized examination of the tasks associated with the performance of a given job and an evaluation of the individual hazards associated with each step required to properly complete the job. The JSA also considers the adequacy of the controls used to prevent or reduce exposure to those hazards. Usually performed by the responsible supervisor for that job and used primarily to train new employees, the JSA is also an excellent source of *paper evidence* during an accident investigation. Also known as *job hazard analysis*.

job satisfaction
The degree to which the work environment provides such qualities as variety, comfort, compensation, and social expression to make a job meaningful in meeting an individual's goals.

Job Severity Index (JSI)
A guideline for matching job design and employee placement such that an acceptable risk of injury potential is present.

job sharing

A work schedule in which two part-time workers perform the duties which would normally be assigned to one full-time person.

job shop

A company whose primary function is to produce small quantities of specialized parts or components for customers who will integrate them into larger products.

job skill

The combination of physical and mental abilities, experience, and training which enable a worker to perform a given task.

job standardization

Having or implementing a standard practice or method for some job.

jockey line

Maritime. Lashing used to prevent lateral movement between barges connected in tandem.

jockeying

That customer or user behavior in which he/she has the option of using several queues or lines, possibly even changing lines while waiting.

jog

An intermediate gait between walking and running, or an alternating combination of walking and running which is used as a form of exercise.

johnboat

A flat-bottomed skiff type boat with a square bow and stern.

joint

(1) Articulation between two bones that may permit motion and flexibility in one or more planes. They may become sites of concern with certain cumulative trauma disorders. Some joints are actually immovable, such as certain fixed joints where segments of bone are fused together in the skull. Other joints, such as those between the vertebrae, have extremely limited motion. However, most joints allow considerable motion. Many joints have an extremely complex internal structure. They are not only composed of the ends of bones but also of ligaments (tough whitish fibers binding the bones together); cartilage (connective tissue covering and cushioning the bone ends); the articulating capsule (a fibrous tissue that encloses the ends of the bones); the synovial membrane (lines the capsule and secretes lubricating fluid, or synovia); and some times bursae (fluid-filled sacs that cushion the movements of muscles and tendons). Joints are classified by variations in structure that make different kinds of movement possible. The movable joints are usually subdivided into hinge (e.g., elbow joint), pivot (e.g., cervical vertebrae joint); gliding (e.g., intervertebral joints), ball-and-socket (e.g., shoulder joint), condyloid (e.g., wrist joint), and saddle (e.g., base of the thumb). (2) Pertaining to a coordinated action between two or more groups. (3) Slang term for a cigarette filled with marijuana instead of tobacco.

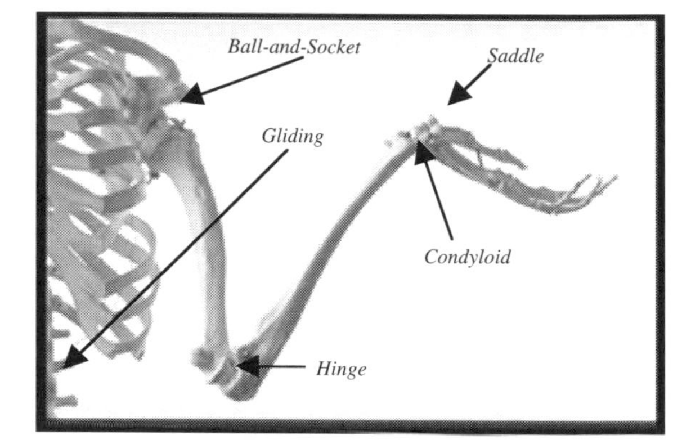

Types of anatomical joints

joint action

An action brought by two or more as plaintiffs or against two or more as defendants.

joint and several contracts

Contracts in which the parties bind themselves both individually and as a unit (jointly).

joint and several liability

A legal doctrine holding the parties involved equally responsible, each of which is 100% liable. Imposed in cases where the harm caused is indivisible (i.e., where individual or joint parties are potentially responsible for the harm but it cannot be determined with any degree of certainty which parties or defendants are responsible for which aspect of the damage). Parties can be held responsible independently or mutually.

joint capsule

The connective tissue and membrane surrounding a synovial joint cavity.

joint operations
Rail Operations. Operations conducted on a track used jointly or in common by two or more railroads subject to 49 CFR 225 or operation of a train, locomotive, car, or other on-track equipment by one railroad over the track of another railroad.

joint range of motion
The angle through which a single joint can be flexed, extended, rotated, or otherwise normally moved without discomfort, pain, or injury.

joint stability
A measure of the rigidity of a joint.

joint-use airport
A military installation at which the Department of Defense permits some degree of civil aviation use. Degrees of civil aviation use include a) open to all civil aviation under a joint-use agreement; b) joint-use agreement for limited use. Strictly military airports do not qualify for primary or commercial service status.

joint venture
A legal entity in the nature of a partnership engaged in the joint undertaking of a particular transaction for mutual profit.

Jones Act
Federal statute passed in 1920 which provides that a seaman injured in the course of his employment by the negligence of the owner, master, or fellow crew members can recover damages for his injuries. Similar remedies are available under the Act to personal representatives of a seaman killed in the course of his employment.

joule
The amount of energy provided by one watt flowing for one second.

journaling
The recording and storage within a computer of the keystrokes input by a user.

joystick
A lever control or computer input device having at least 2 degrees of freedom and with which an operator may control an electromechanical system or a cursor or other activity on a display.

JSA
See ***Job Safety Analysis***.

JSI
See ***Job Severity Index***.

judge
An officer so named in his/her commission, who presides in some court. A public officer, appointed to preside and to administer the law in a court of justice.

judicial branch of government
That branch of government which consists of the nation's court system, from the Supreme Court down.

judicial district
One of the circuits or precincts into which a state is commonly divided for judicial purposes.

judicial review
A type of civil litigation challenging a government decision, usually the propriety of some regulatory decision such as promulgation, interpretation, or application of regulations.

jugular
Pertaining to the neck.

jugular vein
Large veins that return blood to the heart from the head and neck. Each side of the neck has two sets of jugular veins, external and internal. The external jugular carries blood from the face, neck, and scalp and has two branches, posterior and anterior. The internal jugular vein receives blood from the brain, the deeper tissues of the neck, and the interior of the skull. The external jugular vein empties into the subclavian vein, and the internal jugular vein joins it to form the brachiocephalic vein, which carries the blood to the superior vena cava, where it connects to the heart.

Julian date
A number representing the current day within a given year. The range is from 1 through 365 (366 in a leap year).

jumbo barge
A barge 35 feet wide by 195 or 200 feet long that may be either a hopper or cover-type barge.

jumped the pin
Transit (slang). Missing the fifth wheel pin on the trailer when coupling tractor to trailer.

jumper's knee
Pain at proximal end of the patellar tendon.

junction
Area formed by the connection of two roadways, including intersections, interchange areas, and entrance/exit ramps.

jurisdiction
Generally, the authority of the court to hear and decide a case; or, the authority of a governing or responsible body or agency to create, decide, interpret, and implement policies as in "the authority having jurisdiction."

jury
A certain number of men and women selected according to law, and sworn to inquire of certain matters of fact, and declare the truth upon evidence to be laid before them.

jury instructions
A direction given by the judge to the jury concerning the law of the case. A statement made by the judge to the jury informing them of the law applicable to the case in general or some aspect of it.

just cause
(1) A cause outside the legal cause, which must be based on reasonable grounds, and there must be a fair and honest cause or reason, regulated by good faith. As used in the statutory sense, it is that which, to an ordinary intelligent person, is justifiable reason for doing or not doing a particular act. (2) Having good and fair reason(s) for taking disciplinary action.

just compensation
Compensation which is fair to both the owner and the public when property is taken for public use through condemnation (eminent domain). Consideration is taken for such criteria as the cost of reproducing the property, its market value, and the resulting damage to the remaining property of the owner. The Fifth Amendment to the U.S. Constitution provides that no private property shall be taken for public use without "just compensation."

just in time
Warehousing Industry. In this method of inventory control, warehousing is minimal or nonexistent; the container is the movable warehouse and must arrive "just in time," that is not too early or too late.

just noticeable difference (JND)
The smallest amount of change from a reference stimulus which an observer will report as a difference on a given trial.

just tolerable limit
The maximal level of short-term exposure to an agent which will prevent the average individual from developing either acute or chronic symptoms caused by that agent.

Justice Department
One of the executive departments of the federal government, headed by the Attorney General. The chief purposes of the Department of Justice are to enforce the federal laws, to furnish legal counsel in federal cases, and to construe the laws under which other departments act. It conducts all suits in the Supreme Court in which the United States is concerned, supervises the federal penal institutions, and investigates and detects violations against federal laws. It represents the government in legal matters generally rendering legal advice and opinions, upon request, to the President and to the heads of the executive departments. The Attorney General supervises and directs the activities of the U.S. attorneys and marshals in the various judicial districts.

justifiable cause
Justifiable cause for prosecution is the well-founded belief of a person of ordinary caution, prudence, and judgement in the existence of facts essential to prosecution.

justifiable homicide
The killing of another person in self-defense when the danger of death or serious bodily injury exists. Such a homicide generally connotes only the use of force which is necessary to resist the other party's misconduct, and use of excessive force destroys the jurisdiction.

justify
(1) To defend or vindicate. (2) To arrange text, graphics, or other material to be formatted such that it is aligned along the left and/or right margins of the page.

juxta nipple skinfold
The thickness of skinfold just superior to the nipple and parallel to the lateral margin of the pectoral muscle.

juxtaposition
A placing or being placed in nearness or contiguity (side by side).

K

K
See ***Kelvin scale.***

Kahler's disease
Multiple myeloma.

kakidrosis
The excretion of foul-smelling perspiration.

kala-azar
A fatal epidemic fever of tropical Asia, resembling malaria, caused by *Leishmania donovani,* a protozoan parasite. The sand fly is the vector for this disease. Symptoms are usually vague, resembling those of incipient pulmonary tuberculosis. The disease is often confused with malaria. There may be fever, chills, malaise, cough, anorexia, and loss of weight. The Leishmania organisms multiply in the cells of the reticuloendothelial system, eventually causing hyperplasia of the cells, especially those of the liver and spleen. Diagnosis is confirmed by demonstration of the parasite.

Kanawha River ratchet
Maritime Safety. A term for the placing of a "toothpick" or bar between a doubled-up line of barges to bring two barges together by twisting the bar around and around. A very dangerous practice.

Kan-sei engineering
A system for developing consumer products in which the product is designed to have sensory and emotional appeal.

kaolinosis
A pneumoconiosis resulting from the inhalation of kaolin clay dust.

Kaposi's disease
(1) Xeroderma pigmentosum. (2) Kaposi's varicelliform eruption. (3) *See* ***Kaposi's sarcoma.***

Kaposi's sarcoma
An opportunistic neoplasm associated with acquired immune deficiency syndrome (AIDS).

karst
Terrain with the characteristics of relief and drainage arising from a high degree of rock solubility in natural waters. The majority of karst occurs in limestone, but karst may also form in dolomite, gypsum, and salt deposits. Features associated with karst terrain typically include irregular topography, sinkholes, vertical shafts, abrupt ridges, caverns, abundant springs, and/or disappearing streams. Karst aquifers are associated with karst terrain.

karyotype
The chromosomal elements typical of a cell, arranged according to the Denver classification and drawn in their true proportions, based on the average of measurements determined in a number of cells.

Kata thermometer
An alcohol-based thermometer used for determining low air currents/velocities, in which the time required to cool from 100°F to 95°F corresponds to air velocity at that location.

katabatic wind
Any wind blowing downslope, usually a cold wind.

kb
Kilobyte(s). Approximately one thousand bytes.

kcal
Kilocalorie(s).

kCi
Kilocurie(s).

keelboat
A long, flat-bottomed boat with a keel used to haul freight and passengers before the appearance of steamboats on the western rivers.

Kefauver-Cellar Act
Federal anti-merger statute enacted in 1950 prohibiting the acquisition of assets of one company by another (generally in the same line of business) when the effect is to lessen competition.

K-electron capture
The process wherein the electron in the K shell of an atom is captured by the nucleus during a nuclear reaction. In this process, a characteristic x-ray is emitted.

keloid
(1) A mass of fibrous connective tissue, usually at the location of a scar. (2) A scar-like growth that rises above the skin surface, and is rounded, hard, shiny, and white, or some-

times pink. A keloid is a benign tumor that has its origin usually in a scar from surgery or a burn or other injury. Keloids are generally considered harmless and non-cancerous although they may produce contractures. Ordinarily they cause no trouble beyond an occasional itching sensation. Surgical removal is not usually effective because it results in a high rate of recurrence. However, radium and x-ray therapy often are of substantial help, provided care is taken not to destroy the surrounding healthy tissue.

Kelvin scale
A temperature scale with zero degrees equal to the theoretical temperature at which all molecular motion ceases. Also called the *absolute scale.*

Kendall's coefficient of concordance (W)
A measure of the degree of similarity in the rankings of a set of entities across two or more independent rank orderings of that set. Also referred to as *coefficient of concordance.*

Keogh Plan
A designation for retirement plans available to self-employed taxpayers (also referred to as H.R.10 plans). Such plans extend to the self-employed tax benefits similar to those available to employees under qualified pension and profit sharing plans. Yearly contributions to the plan (up to a certain amount) are tax deductible.

keratin
Tough, fibrous protein containing sulfur and forming the outer layer of epidermal structures, such as hair, nails, etc.

keratitis
Inflammation of the cornea. Keratitis may be deep, when the infection causing it is carried in the blood or spreads to the cornea from other parts of the eye, or superficial, caused by bacteria or virus infection or by allergic reaction. Microorganisms causing the inflammation can be introduced into the cornea during the removal of foreign bodies from the eye. All infections of the eye are potentially serious because opaque fibrous tissue or scar tissue may form on the cornea during the healing process and cause partial or total loss of vision. There are several kinds of keratitis. Dendritic keratitis is a viral form caused by the herpes simplex virus; it usually affects only one eye. A bacterial form, acute serpiginous keratitis, may result from infection by pneumococci, streptococci, or staphylococci. Some kinds of keratitis (e.g., dendritic keratitis) may follow symptoms of upper respiratory tract infection, such as fever. Burns of the cornea, such as those produced by chemicals, or ultraviolet rays, also give rise to a form of keratitis. In trachoma, a contagious disease of the conjunctiva, the eyes become inflamed, and small, gritty particles develop on the cornea. Herpetic keratitis may accompany herpes zoster. Interstitial keratitis is often caused by congenital syphilis, although occasionally, it may also result from acquired syphilis. When caused by congenital syphilis, the disease usually appears when the child is between the ages of 5 and 15. In rare cases, interstitial keratitis may also stem from tuberculosis or rheumatic infection in other parts of the body. Symptoms vary somewhat among the different forms of keratitis, but pain, which may be severe, and inability to tolerate light (photophobia) are usual. There may be considerable effusion of tears and conjunctival discharge.

keratoprotein
The protein of the horny tissues of the body, such as the hair, nails, and epidermis.

keratosis
Any horny growth on the skin, such as a wart.

kerogen
The organic component of oil shale.

kerosene
A petroleum distillate that boils at a temperature between 300 and 550 degrees Fahrenheit, that has a flash point higher than 100 degrees Fahrenheit by ASTM Method D 56, that has a gravity range from 40 to 46 degrees API, and that has a burning point in the range of 150 degrees to 175 degrees Fahrenheit. Kerosene is used in space heaters, cook stoves, and water heaters and is suitable for use as an illuminate when burned in wick lamps. *See also* ***fuel*** *and* ***gasoline***.

kerosene-type jet fuel
A quality kerosene product with an average gravity of 40.7 degrees API, and a 10 percent distillation temperature of 400 degrees Fahrenheit. It is covered by American Society of Testing Materials (ASTM) Specification

D1655 and Military Specification MIL-T-5624L (Grades JP-5 and JP-8). A relatively low-freezing point distillate of the kerosene type; it is used primarily for commercial turbojet and turboprop aircraft engines.

ketone

(1) A class of liquid organic compounds that is derived by the oxidation of secondary alcohols. They are used as solvents in paints and explosives. (2) A chemical compound characterized by the presence of the bivalent carbonyl group (>C:O).

ketone bodies

Substances synthesized by the liver as a step in the combustion of fats; they are betaoxybutyric acid, acetoacetic acid, and acetone. Initially, the combustion of fatty acids produces ketones, which eventually are broken down into carbon dioxide and water by the liver and other tissues of the body. Under abnormal conditions, such as uncontrolled diabetes mellitus, starvation, or the intake of a diet composed almost entirely of fat, the breakdown of fatty acids may be halted at the ketone stage, causing increasing levels of ketone bodies in the blood. This condition is called *ketosis* and is directly related to improper utilization or inadequate supplies of carbohydrates, which are necessary for proper combustion of fats.

ketosis

The accumulation of large quantities of ketone bodies in the body tissues and fluids. Ketosis is the result of incomplete combustion of fatty acids, which in turn is the result of improper utilization of, or lack of availability of, carbohydrates. When carbohydrates cannot be used as the source of energy, the body draws on its supply of fats. Deficiency of carbohydrates triggers several hormonal responses and greatly increases the removal of fatty acids from fatty tissues. As a result, large quantities of fatty acids must be oxidized, more in fact than the body cells can handle; thus their oxidation is incomplete and ketones accumulate in the blood and tissues. Ketosis may lead to severe acidosis because the ketone bodies beta-oxybutyric acid and acetoacetic acid decrease the blood pH and, more importantly, because when the ketone acids are excreted in the urine they take with them large quantities of sodium. The result is a depletion of the alkaline part of the body's buffer system, so that the acid-base balance is upset in favor of acidosis. Ketosis occurs in uncontrolled diabetes mellitus because carbohydrates are not properly utilized, and in starvation because carbohydrates simply are not available for utilization. Ketosis is sometimes produced intentionally in the treatment of epilepsy by means of the ketogenic diet, which contains large amounts of fat and little carbohydrate or protein.

keV

An abbreviation for kilo-electron volt and equal to 1000 electron volts.

kevel

Colloquial term used for a large steel cleat secured to the deck of a boat or a barge, used for securing, mooring, and towing lines. It is provided with two prongs called "horn." Also spelled "cavil," "cavel," "caval."

key click

An audible click which is presented whenever a keystroke is performed on a keyboard or keypad. Provides feedback to the user that a keystroke was made.

key event

One incident which is primarily responsible for the time, place, and severity of an accident or other significant happening.

key job

A job which has been evaluated itself and may be used as a benchmark for evaluating other similar, non-key jobs or work classes for evaluation, classification, and/or wage establishment purposes in the same company or industry.

key repeat rate

The repeat rate with a keyboard key when continuously depressed, providing the number of characters input per second.

keyboard

A computer input device or typewriter keying mechanism consisting of a panel containing alphanumeric, grammatical, function, and/or other keys for typing information/data entry to a computer or onto hardcopy.

The standard computer keyboard

keypad
A data entry pad consisting of the numeric keys 0-9, simple arithmetic function keys, a decimal key, and an enter key.

keypunch
An electromechanical device with a keyboard which punches holes in a car or onto tape (an older term).

keystroke
The depression of a key on a keyboard with a force greater than the actuation force for that key.

kg
1 kilogram (1000 g or 1,000,000 mg).

kgf
See ***kilogram force***.

kick down
Transit (slang). Shift down to lower gear.

kick line
Maritime. A line used to hold a towboat while the stern is being backed in so that the head will swing out into the stream.

kick the donuts
Transit (slang). To check the tires of a truck or other vehicle for proper inflation.

kickback
(1) The reaction of a piece of material back toward the operator as it is being fed into a mechanical processing device and meets the cutting or processing tool. A serious and potentially dangerous phenomenon. (2) The unlawful receipt of funds or other items of value as a type of payment or reward for work performed.

kickplate
Any vertical structure or covering device on or near the floor which protects the surface it covers from impact or which prevents accidental entry of the shoe/toes into a region which might be hazardous.

kidney
One of two glandular organs, almost bean-shaped located in the lumbar region, that secrete urine. Their function is to regulate the content of water and other substances in the blood, and to remove from the blood various wastes. In an average adult, each kidney is about 4 inches long, 2 inches wide, and 1 inch thick, and weighs 4 to 6 ounces. In this small area, the kidney contains over a million microscopic filtering units (called nephrons). Blood arrives at the kidney by way of the renal artery, and is distributed through arterioles into many millions of capillaries which lead to the nephrons. Fluids and dissolved salts in the blood pass through the walls of the capillaries, and are collected within the central capsule of each nephron, the malpighian capsule. The glomerulus, a tuft of capillaries within the capsule, acts as a semipermeable membrane permitting a protein-free ultrafiltrate of plasma to pass through. This filtrate is forced into hairpin-shaped collecting channels in the nephrons, called tubules. Capillaries in the walls of the tubules reabsorb the water and the salts required by the body and deliver them to a system in the kidney veins which, in turn, carry them into the renal veins and return them to the general circulation. Excess water and other waste materials remain in the tubules as urine. The urine contains, besides water, a quantity of urea, uric acid, yellow pigments, amino acids, and trace metals. The urine moves through a system of ducts into a collecting funnel (renal pelvis) in each kidney, where it is then lead into the two ureters. About 1.5 quarts (1500 cc) of urine are excreted daily by the average adult. The urine efficiency of the normal kidney is one of the most remarkable aspects of the body. It has a filtering capacity of a quart of blood per minute (or 15 gallons per hour, or 360 gallons per day). Ordinarily, it draws off from the blood about 180 quarts of fluid daily, and returns usually 98 to 99 percent of the water plus the useful dissolved salts, according to the body's changing needs.

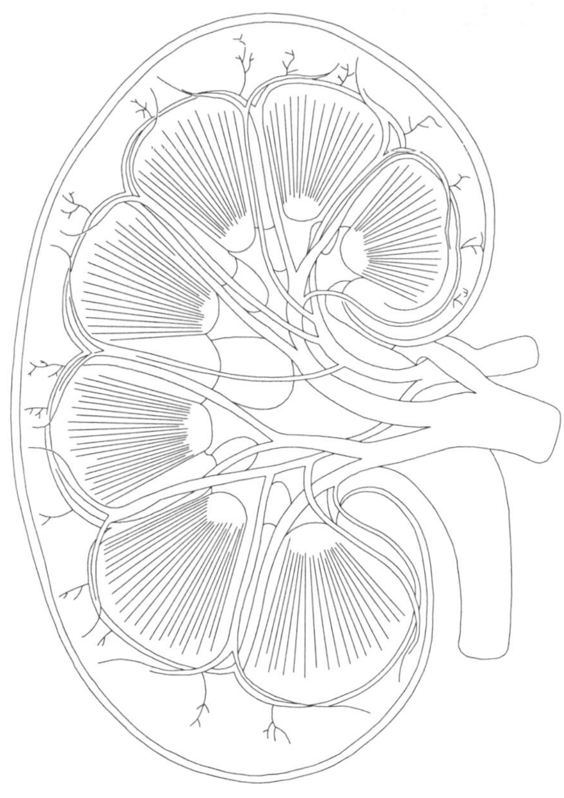
Cross section of the human kidney

kidney buster
Transit (slang). A hard riding truck.

kilo-
(prefix) One thousand or 10^3 times a base unit.

kilo electron volt
One thousand electron volts (KeV).

kilocalorie (kcal)
See ***Calorie***.

kilogram
(1) One thousand grams. (2) An international standard unit of mass in the SI/MKS system, corresponding to a specific platinum-iridium alloy mass.

kilogram force (kgf)
A force equivalent to that which the earth's gravity exerts on a one kilogram mass at the earth's surface.

kilogram-meter
A unit of work, representing the energy required to raise 1 kg. of weight 1 meter vertically against gravitational force.

kiloliter
One thousand liters (264 gallons).

kilopascal
A unit of pressure equal to one thousand pascals. One pound per square inch (psi) of pressure is equivalent to 6.894757 kilopascals.

kilovolt (kV)
The unit of electrical potential equal to 1000 volts.

kilowatt (kW)
One thousand watts.

kilowatt electric
One thousand watts of electric capacity.

kilowatt hour
One thousand watt/hours.

kinanthropometry
The study of human nutrition, growth, development, maturation, size, shape, proportion, body function, and body composition to understand and improve upon health and performance.

kinase
(1) An enzyme that catalyzes the transfer of a high-energy group of donors to an acceptor. (2) An enzyme that activates a zymogen.

kinematic chain
An open series of links or body segments, where the dimensions of each link are determined by the linear distance from one joint axis of rotation to another, with muscle mass and the type of articulation generally ignored.

kinematics
(1) The study of the geometry of motion without consideration of causal factors. (2) A technique which allows a computer graphics system to stimulate the movement of part or all of an image.

kinesimeter
A device which makes it possible to obtain quantitative measures of body motion, including displacement, velocity, and acceleration.

kinesiology
The study of movement of the human musculature.

kinesis
Objective physical body movement.

kinesthesia
That sense which originates in the stimulation of mechanoreceptors in joints, muscles, and/or

tendons and leads to awareness of position, movement, weight, and/or resistance of the limbs or other body parts.

kinesthesiometer
A device for measuring an individual's ability to sense body part position or movement.

kinetic
Pertaining to movement or motion.

kinetic art
The use of objects in motion as an expression of creativity.

kinetic energy
(1) The energy that a body possesses by virtue of its mass and velocity, or the energy of motion. (2) That portion of the energy of an object resulting from its motion. Expressed as:

$$KE = \frac{mv^2}{2}$$

where:
m = mass
v = velocity

kinetic friction
That friction between two surfaces in contact where there is relative motion between them.

kinetic rate coefficient
A number that describes the rate at which a water constituent, such as a biochemical oxygen demand or dissolved oxygen, increases or decreases.

kinetics
The study or use of the effects of mechanical forces and moments on material objects or to produce motion, especially of the human body.

kinetosphere
A reach envelope for the hand/arm combination or the leg/foot combination in which only translational motion of the limb is permitted, with the terminal segment (hand or foot) held in a constant position.

kingdom
One of the three major categories into which natural objects are usually classified: the animal (including all animals), the plant (including all plants) and the mineral (including all substances and objects without life).

king-pin saddle-mount
Transportation. That device which is used to connect the "upper-half" to the "lower-half" [of a "saddle-mount"] in such manner as to permit relative movement in a horizontal plane between the towed and towing vehicles.

Kirschner wire
A steel wire for skeletal transfixion of fractured bones and for obtaining skeletal traction in fractures. It is inserted through the soft parts and the bone and held tight in a clamp.

kiss and ride
Transit (slang). A place where commuters are driven and dropped off at a station to board a public transportation vehicle.

kitchen
A location in restaurants, homes, and some vehicles in which food is prepared for consumption. *See also* ***galley***.

knee
The junction of the femur, tibia, fibula, and patella, including all surrounding tissues. The knee is a complex hinge joint, one of the largest in the body, and one that sustains great pressure.

knee breadth
The horizontal linear distance between the most medial and lateral projections of the femoral epicondyles. Measured using firm pressure with the individual standing erect and with no excessive leg muscle tension. Also referred to as *femoral breadth*.

knee cap
See ***patella***.

knee circumference, fully bent
The distance around the maximum knee prominence and through the crease behind the knee. Measured with the individual in a squatting position with the knee joint maximally flexed.

knee circumference, sitting
The maximum surface distance around the knee, under the popliteal area and over the kneecap at an angle of 45° to the floor. Measured with the individual sitting erect, the upper leg horizontal, the lower leg vertical, and the foot flat on the floor.

knee circumference, standing
The surface distance around the knee measured at the level of the midpoint of the patella. Measured with the individual standing erect and his/her weight evenly distributed on both feet.

knee height, recumbent
The horizontal linear distance from the base of the heel to the anterior surface of the thigh at the femoral condyle. Measured with the knee flexed 90° and the longitudinal axis of the foot perpendicular to the longitudinal axis of the lower leg.

knee height, sitting
The vertical distance from the floor or other reference surface to the most superior part of the quadriceps musculature above the knee. Measured with the individual sitting, the knee flexed 90°, the foot flat on the floor/reference surface, and the lower leg vertical.

knee – knee breadth, sitting
The maximum horizontal linear distance between the lateral surface from one knee to the lateral surface of the other knee. Measured with the individual sitting erect, the knees flexed at right angles, and both knees touching but without significant tissue compression.

knee pad
A cushion for placement over the patella, usually having a strap or other attachment device around the knee, to protect against injury from kneeling or impacts between the anterior knee and other objects.

knee switch
An uniaxial control device which is operated by a lateral movement of the knee.

knee well
That region from the edge and extending under a table, desk, or other seated workstation which accommodates the legs in a seated posture, usually with the knees flexed.

knee well depth
The horizontal distance from the user's edge of a table, desk, or other seated workstation platform to a terminus against a wall or vertical panel on the opposite side.

knee well height
The vertical height from the floor or other reference level to the underside or lower surface of the structure forming a knee well.

knee well width
The horizontal distance from one side of a knee well to the other.

knob
A cover for placement on a rotational device or mechanism which normally protrudes from a surface for easier gripping and turning.

knocked down
(slang). Unassembled freight or merchandise.

knockout
Maritime (slang). To release a towboat from tow.

knockout single
Maritime (slang). To uncouple the towboat and lay alongside the barges for single lockage. Also called *single set over.*

knot
(1) A unit of speed equal to 1 natural mile per hour, or 6,080.20 feet per hour or 1.85 kilometers per hour. 1 knot equals 1.15 mph. (2) A branch or limb, imbedded in a tree and cut through in the process of lumber manufacture, classified according to size, quality, and occurrence. The size of the knot is determined as the average diameter on the surface of the piece of wood.

knowing endangerment
Knowing that one is placing another person in imminent danger of death or serious bodily injury.

knowledge-based behavior
A cognitive operating mode in which the individual attempts to achieve a goal in a situation with no clearly pre-established rules.

knowledge engineering
The process of identifying what information must be gathered, obtaining that information from one or more recognized experts, and organizing it into a rule structure to be used in decision-making for a specific problem.

knuckle
The protuberance of the heads of the metacarpals when the hand is clenched into a fist, or the protuberance at the interphalangeal joints when the fingers are flexed.

knuckle height
The vertical distance from the floor to the point of maximum protrusion of the metacarpal III knuckle. Measured with the individual

standing erect, the arm adducted to the side of the body, the palm flat against the side of the thigh, and the fingers extended.

knurled
Pertaining to a surface texture with small ridges, generally for providing a more firm grip.

Koch's law
For a given organism to be established as the cause of a given disease, the following conditions must be fulfilled: a) the microorganism is present in every case of the disease; b) it is to be cultivated in pure culture; c) inoculation of such culture must produce the disease in susceptible animals; d) it must be obtained from such animals, and again grown in a pure culture.

Koehler illumination
A type of illumination used in microscopy in which the light source is imaged in the aperture of the system and the lamp condenser is imaged in the specimen plane in order to obtain even brightness in the field of view and optimum resolving power of the microscope system.

Kolmer test
A complement-fixation technique used in the diagnosis of syphilis or other infections.

konimeter
A device for sampling airborne dust.

Korsakoff's syndrome
A mental disorder associated with chronic alcoholism and caused in part by vitamin B_1 (thiamine) deficiency. Characteristics include disturbances of orientation, memory defect, susceptibility to external stimulation and suggestion and hallucinations. There is irreversible brain damage; confinement to an institution is a frequent outcome of the condition.

kort nozzles
Cylindrical devices which surround the towboat propellers to increase the maneuverability and efficiency of the towboat.

kPa
Kilopascal(s).

kPa abs
Absolute pressure in kilopascals.

Kraepelin's classification
A classification of the manic-depressive and schizophrenic groups of mental disease.

krebiozen
A substance alleged to be capable of curing cancer.

Kretschmer somatotype
A body structure classification system developed by Ernst Kretschmer, supposedly to represent human character traits, in which men are divided into three basic groups; pyknic, athletic, and asthenic. An old system, no longer in use. *See also asthenic, pyknic,* and *athletic.*

krypton
A chemical element, atomic number 36, atomic weight 83.80, symbol Kr.

Kümmel's disease
A form of spondylitis of unknown origin or occurring at a great interval after the injury causing it, with collapse of the vertebra and thinning of the intervertebral disks.

kurtosis
A measure of the peak of a distribution, based on the fourth moment about the mean.

kuru
Disease in humans caused by a virus that affects the central nervous system and can be transmitted to subhuman primates.

Kussmaul disease
An inflammatory disease of the coatings of the small and medium-sized arteries of the body with inflammatory changes around the vessels and marked symptoms of systemic infection.

kV
Kilovolt(s).

kVp
Kilovolt peak.

kW
Kilowatt(s).

kymograph
An electromechanical device consisting of a rotating smoked drum or paper-covered cylinder with one or more styli for recording time-based events.

kyphoscoliosis
Backward and lateral curvature of the spine, such as that seen in vertebral osteochondritis (Scheuermann's disease).

kyphosis
A posture of the lumbar spine caused by bending forward, such as over a work bench or poorly positioned computer terminal, and involving reverse curvature of the spine.

kyrtorrhachic
Having a vertebral column in which the lumbar curvature is anteriorly convex.

L

L

Liter(s). Also, see ***lambert***.

L-1 maneuver

An anti-g straining maneuver for preventing gravity-induced loss of consciousness during high positive acceleration forces in high performance spacecraft, in which the crew member strains his skeletal body muscles and closes the glottis for a few seconds, then inhales and exhales rapidly before repeating the process. Also know as the *Leverett technique*. See also ***anti-g straining maneuver***.

LAA

See ***local airport advisory***.

LAAS

See ***low altitude airway structure***. *See also* ***low altitude alert system***.

Laban notation

A systematic method for describing body position in the field of dance.

label

(1) A descriptor of the contents of some container, which may include such information as the product name, manufacturer, amount present, instructions, and any warning(s). (2) Under the Federal Food, Drug, and Cosmetic Act: The written, printed, or graphic matter upon the immediate container or wrappers of any article or accompanying such article. (3) Under noise abatement program requirements, that item which is inscribed on, affixed to, or appended to a product, its packaging, or both for the purpose of giving noise reduction effectiveness information appropriate to the product. (4) A descriptor which helps to identify displayed screen or control structures.

label coding

The use of text, numerals, symbols, or other means to identify a control, device, or system.

labeled

Equipment is considered "labeled" if there is attached to it a label, symbol, or other identifying mark of a nationally recognized testing laboratory which a) makes periodic inspections of the production of such equipment, and b) whose labeling indicates compliance with nationally recognized standards or tests to determine safe use in a specified manner.

labeled molecule

A molecule containing one or more atoms distinguished by non-natural isotopic composition (with radioactive or stable isotopes).

labeling

Under the Federal Insecticide, Fungicide, and Rodenticide Act: All labels and all other written, printed, or graphic matter accompanying the pesticide or device at any time; or to which reference is made on the label in literature accompanying the pesticide or device except to current official publications of the EPA, the United States Departments of Agriculture and Interior, the Department of Health and Human Services, state experiment stations, state agricultural colleges, and other similar federal or state institutions or agencies authorized by law to conduct research in the field of pesticides.

labial

Pertaining to a lip, or labium.

labialism

Defective speech with the use of labial sound.

labile

(1) Gliding; moving from point to point over the surface; unstable. (2) Chemically unstable.

lability

The quality of being labile. In psychiatry, emotional instability; a tendency to show alternating states of happiness and somberness.

labiodental

Articulated with the lower lip touching the upper central incisors.

labium

A lip or lip-shaped structure.

labor

(1) The process of doing work, especially that involving physical effort. (2) A group of individuals consisting of or representing those working for hourly wages. (3) Work; toil; service; mental or physical exertion. (4) The function of the female organism by which the product of conception is expelled from the uterus through the vagina to the outside world.

labor a jury
To tamper with a jury; to endeavor to influence them in their verdict, or their verdict generally (jury tampering is a crime under 18 U.S.C.A. §§ 1503, 1504).

labor agreement
See ***labor contract.***

labor contract
A contract between employer and employees (i.e., union) which governs working conditions, wages, fringe benefits, and grievances.

labor cost
The portion of an employer's total cost of doing business which is attributable to wages and salaries, benefits, and other aspects of employment practices.

labor dispute
Term generally includes any controversy between employers and employees concerning terms, tenure, hours, wages, fringe benefits, or conditions of employment, or concerning the association or representation of persons in negotiating, fixing, maintaining, changing, or seeking to arrange terms or conditions of employment. However, it should be noted that not every activity of a labor organization and not even every controversy in which it may become involved are considered a "labor dispute" within the National Labor Relations Act.

labor-management relations
General term to describe the broad spectrum of activities which concern the relationship of employees to employers, both union and non-union

Labor-Management Relations Act
Federal statute (formally: Taft-Hartley Act) which regulates certain union activities, permits suits against unions for proscribed acts, prohibits certain strikes and boycotts, and provides machinery for settling strikes which involve national emergencies.

labor organization
An organization of any kind, or any agency or employee representation committee, group, association, or plan so engaged in which employees participate and which exists for the purpose, in whole or in part, of dealing with employers concerning grievances, labor disputes, wages, rates of pay, hours, or other terms or conditions of employment, and any conference, general committee, joint system board, or joint council so engaged which is subordinate to a national or international labor organization, other than a state or local central body.

labor picketing
The act of patrolling in motion at or near the employer or customer entrance; usually carrying placards with a terse legend or message communicating the basic gist of the union's claims. Certain forms are prohibited.

labor relations acts
State and federal laws that regulate relations between employers and employees.

labor turnover
A measure of how many employees enter and leave a particular workplace within a specified interval.

labor union
A combination or association of workers organized for the purpose of securing favorable wages, improved labor conditions, better hours of labor, etc., and righting grievances against employers. Such unions normally represent trades, crafts, and other skilled workers (e.g., machinists, electricians, etc.).

laboratory
A place for making tests or doing experimental work.

laboratory blank
Reagent laboratory grade water which is analyzed in the same way as field samples.

laboratory duplicates
Unmarked samples whose results help to ensure quality control.

laboratory hood
A ventilated enclosure designed to capture, contain, control, and remove gases, vapors, and particles generated within the enclosure.

laboratory study
An experimental study conducted in an environment in which the experimenter(s) have some degree of control over the variables involved in the phenomenon of interest.

laboratory ventilation
Air-moving systems and equipment which serve laboratories.

laborer
The word ordinarily denotes one who subsists by physical labor. One who, as a means of livelihood, performs work and labor for another. Any person who follows any legitimate employment or discharges the duties of any office.

laboring
Maritime. The effect of shallow water on the sound or performance of the boat's engine.

labyrinth
(1) A complicate maze. (2) *See* ***inner ear***.

labyrinthine nystagmus
See ***vestibular nystagmus***.

laceration
A wound caused by the tearing of body tissues, as distinguished from a cut or incision. External lacerations may be small or large and may be caused in many ways. Some common causes of lacerations are a blow from a blunt instrument, a fall against a rough surface, and an accident with machinery. A laceration may be a ragged tear with many tag ends of skin or a torn flap of skin and flesh. Although the bleeding may be less than that caused by a cut, the danger of infection may be greater. In a laceration there is likely to be more damage to surrounding tissue, with a greater area exposed. Because of the danger of infection, cleaning the laceration is the first and most important step in treatment. Lacerations within the body occur when an organ is compressed or moved out of place by an external or internal force. This kind of laceration may result from a blow that does not penetrate the skin. Surgical repair is usually necessary for internal lacerations.

lachrymation
See ***lacrimation***.

lack of jurisdiction
The phrase may mean a lack of power of a court to act in a particular manner or to give certain kinds of relief.

lacrimal
Pertaining to tears.

lacrimal bone
A small bone making up part of the medial orbit of the skull.

lacrimation
The excessive secretion and discharge of tears. Also spelled *lachrymation*.

lacrimator
A substance, such as a gas, that increases the flow of tears.

lactase
An enzyme that catalyzes the conversion of lactose into glucose and galactose.

lactic acid
A three-carbon organic acid product of anaerobic metabolism in tissue, especially muscle tissue.

ladder
An appliance usually consisting of two side rails joined at regular intervals by cross-pieces called steps, rungs, or cleats, on which a person may step in ascending or descending.

ladder jack scaffold
A light duty scaffold supported by brackets attached to ladders.

ladder safety device
Any device, other than a cage or well, designed to eliminate or reduce the possibility of accidental falls and which may incorporate such features as life belts, friction brakes, and sliding attachments. *See also* ***cage*** *and* ***well***.

laden in bulk
A term of maritime law, applied to a vessel which is freighted with a cargo which is neither in casks, boxes, bales, nor cases, but lies loose in the hold, being defended from wet or moisture by a number of mats and a quantity of dunnage. Cargoes of corn, salt, etc. are usually shipped in this manner.

lading
Refers to the freight shipped; the contents of a shipment.

lag
(1) The period of time by which a second event trails the leading event. (2) That distance at which a second moving object trails a leading object.

lag time
See ***dead time***.

lagging
An acoustical treatment involving the encapsulation of vibrating structures or ducts con-

taining fluid-borne noise in order to reduce radiated noise.

lagoon

(1) A shallow pond where sunlight, bacterial action, and oxygen work to purify wastewater; also used for the storage of wastewater or spent nuclear fuel rods. (2) Shallow body of water, often separated from the sea by coral reefs or sandbars. The sheet of water between an offshore reef, especially of coral, and the mainland. The sheet of water within a ring or horseshoe-shaped atoll.

lake

(1) Any standing body of inland water. (2) A considerable body of standing water in a depression of land or expanded part of a river. (3) An inland body of water or naturally enclosed basin serving to drain the surrounding land; or a body of water of considerable size surrounded by land; a widened portion of a river or a lagoon. (4) A body of water, more or less stagnant, in which the water is supplied from drainage. (5) An inland body of water of considerable size occupying a natural basin or depression in the earth's surface below the ordinary drainage level of the region.

lake breeze

A wind blowing onshore from the surface of a lake.

lake-effect snow

Localized snowstorm that forms on the downward side of a lake. Such storms are common in late fall and early winter near the Great Lakes as cold, dry air picks up moisture and warmth from the unfrozen bodies of water.

lake/pond

A standing body of water with a predominantly natural shoreline surrounded by land.

lakewise or Great Lakes

These terms apply to traffic between U.S. ports of the Great Lakes system. The Great Lakes system is treated as a separate system rather than as a part of the inland system.

lambert (L)

A unit of luminance; equals $1/\pi$ candela per cm^2 (an older term).

lambert surface

A reflecting or emitting surface whose brightness appears to be the same regardless of the angle of observation.

Lambert's cosine law

A law providing that the luminous intensity from a perfectly diffusing surface varies with the cosine of the angle between the perpendicular and the direction of interest.

lame duck

(1) An elected officeholder who is to be succeeded by another, between the time of the election and the date that his/her successor is to take office. (2) A speculator in stock who has overbought and cannot meet his/her commitments.

Lame Duck Amendment

Popular name for the Twentieth Amendment to the U.S. Constitution, abolishing the short congressional term.

lame duck session

Legislative session conducted after election of new members but before they are installed, and hence one in which some participants are voting for the last time as elected officials because of their failure to become reelected or due to voluntary retirement.

laminar flow

Ideally, air flow in which air molecules travel parallel to all other molecules; flow characterized by the absence of turbulence. Also known as *streamline flow*.

laminar flow clean room

A room with laminar air flow and Class 10,000 Clean Room or better.

laminectomy

Surgical excision of the posterior arch of a vertebra. The procedure is most often performed to relieve the symptoms of a ruptured intervertebral disk (slipped disk). When several disks are involved, spinal fusion may be performed so that the vertebrae in the affected area will remain in a fixed position. Bone grafts, usually taken from the iliac crest, are applied to fuse the affected vertebrae permanently, resulting in limitation of movement of this portion of the spine. Laminectomy is also performed for the removal of an intervertebral or spinal cord tumor.

lamp
A device used to produce artificial heat or light.

lamp burnout (LBO)
The cessation of light output from an artificial source. A recoverable light loss factor.

lamp burnout factor
The proportional loss of illuminance from the non-replacement of burned out lamps.

LAN
*See **Local Area Network**.*

land
(1) *General.* Any ground, soil, or earth whatsoever; including fields, meadows, pastures, woods, moors, waters, marshes, and rock. (2) *Maritime.* To moor or bring a boat to the riverbank.

land application
Discharge of wastewater onto the ground for treatment or reuse. *See also **irrigation**.*

land area
Based on the U.S. Bureau of the Census definition, this includes dry land and land temporarily or partially covered by water, such as marshlands, swamps, and river flood plains, systems, sloughs, estuaries and canals less than 1/8 of a statute mile (0.2 kilometers) in width and lakes, reservoirs, and ponds less than 1/16 square mile (0.16 square kilometers) in area. [For Alaska, 1/2 mile (0.8 kilometers) and 1 square mile (2.60 square kilometers) are substituted for these values]. The net land area excludes areas of oceans, bays, sounds, etc. lying within the 3-mile (4.8 kilometers) U.S. jurisdiction as well as inland water areas larger than indicated above.

land-ban
Under RCRA, the mandated phasing out of land disposal of untreated hazardous waste.

land breeze
A coastal breeze that blows from land to sea, usually at night.

land damages
A term sometimes applied to the amount of compensation to be paid for land taken under the power of eminent domain or for injury to, or depreciation of, land adjoining that taken.

land disposal
According to the Federal Solid Waste Disposal Act, the term includes, but is not limited to, any placement of hazardous waste in a landfill, surface impoundment, waste pile, injection well, land treatment facility, salt dome formation, salt bed formation, or underground mine or cave. *See also **landfills**.*

land farming
With regard to waste, a disposal process in which hazardous waste deposited on or in the soil is naturally degraded by microbes.

land use
(1) Under the Federal Coastal Zone Management Act of 1972: Activities which are conducted in or on the shore lands within the coastal zone. (2) Designates whether the general area in which a vehicle crash occurred is urban or rural, based on 1990 Census Data.

land wall
The concrete wall that forms part of the lock and is nearest to the land on the shore on which a lock chamber is constructed.

landbridge
An intermodal connection between two ocean carriers separated by a land mass, linked together in a seamless transaction by a land carrier. *See also **intermodal** and **minibridge**.*

landed cost
The dollar per barrel price of crude oil at the port of discharge. Included are the charges associated with the purchase, transporting, and insuring of a cargo from the purchase point to the port of discharge. Not included are charges incurred at the discharge port (e.g., import tariffs or fees, wharfage charges, and demurrage charges).

landed weight
The weight of an aircraft providing scheduled and non-scheduled service of only property (including mail) in intrastate, interstate, and foreign air transportation.

landfills
(1) Sanitary landfills are land disposal sites for nonhazardous solid wastes at which the waste is spread in layers, compacted to the smallest practical volume, and cover material is applied at the end of each operating day. (2) Secure chemical landfills are disposal sites for hazardous waste. They are selected and

designed to minimize the chance of release of hazardous substances into the environment.

landing
The level region at the bottom of a stair.

landing area
(1) Any locality either on land, water, or structures, including airports/heliports and intermediate landing fields, which is used, or intended to be used, for the landing and takeoff of aircraft whether or not facilities are provided for the shelter, servicing, or for receiving or discharging passengers or cargo. (2) That part of a movement area intended for the landing or takeoff of aircraft.

landing direction indicator
A device installed on the airport property which visually indicates the direction in which landings and takeoffs should be made (e.g., wind sock).

landing distance available (LDA)
The runway length declared available for landing an airplane.

landing gear
(1) *Trucking.* Device that supports the front end of semitrailer when not attached to a tractor. (2) *Aviation.* The wheels that support the aircraft during landing and while moving on the ground. The term generally includes all the components that support the wheel structures, not just the wheels themselves.

landing gear extended speed
The maximum speed at which an aircraft can be safely flown with the landing gear extended.

landing gear operating speed
The maximum speed at which the landing gear can be safely extended or retracted.

landing minimums
Aviation. The minimum visibility prescribed for landing a civil aircraft while using an instrument approach procedure. The minimum applies with other limitations set forth in Federal Aviation Regulation Part 91 with respect to the Minimum Descent Altitude (MDA) or Decision Height (DH) prescribed in the instrument approach procedures as follows: a) *Straight-in landing minimums.* A statement of MDA and visibility, or DH and visibility, required for a straight-in landing on a specified runway; or b) *Circling minimums.* A statement of MDA and visibility required for the circle-to-land maneuver. Descent below the established MDA or DH is not authorized during an approach unless the aircraft is in a position from which a normal approach to the runway of intended landing can be made and adequate visual reference to required visual cues is maintained.

landing place
Maritime. A place for loading and unloading passengers or cargo to and from water vessels.

landing rights airports
Any aircraft may land at one of these airports after securing prior permission to land from U.S. Customs.

landing roll
The distance from the point of touchdown to the point where the aircraft can be brought to a stop or exit the runway.

landing sequence
The order in which aircraft are positioned for landing.

landing signal
Maritime. A prearranged signal which the towboats of some companies sound when approaching their dock.

landmark
(1) *General.* A fixed object serving as a boundary mark to a tract of land, as a guide to travelers, etc. A prominent object in the landscape. (2) *Ergonomics.* An easily located position on or near the body surface. Also known as *anatomical reference point.*

Landolt C
See ***Landolt ring***.

Landolt ring
A ring having a small gap at some orientation, both the width of the gap and the ring thickness being one-fifth the outer diameter of the ring. For use in vision testing, in which the observer is expected to report the orientation of the gap.

Landrum-Griffin Act
Federal statute enacted in 1959, known as the Labor-Management Reporting and Disclosure Act, designed to curb corruption in union leadership and undemocratic conduct of internal union affairs as well as to outlaw certain types of secondary boycotts and "hot cargo"

provisions in collective bargaining agreements.

landscaping
Colloquial term meaning to clear shore structure of brush and vegetation in order to obtain optimum range of visibility. *See also* ***brush out***.

Landsteiner's classification
A classification of blood types in which they are designated O, A, B, and AB, depending on the presence or absence of agglutinogens A and B in the erythrocytes. Also called international classification.

lane
(1) A prescribed course for ships or aircraft, or a strip delineated on a road to accommodate a single line of automobiles; not to be confused with the road itself. (2) A portion of a street or highway, usually indicated by pavement markings, that is intended for one line of vehicles.

Langerhans' islands
Masses in the pancreas composed of cells smaller than the ordinary cells; they produced the hormone insulin and their degeneration is one of the causes of diabetes mellitus.

Langer's line
See ***cleavage line***.

LANL
Los Alamos National Laboratory (previously referred to a Los Alamos Scientific Laboratory or LASL).

lanolin
Wool fat or wool grease that is refined and incorporated into many commercial preparations. Lanolin is a byproduct of the process that accompanies the removal of sheep's wool from the pelt. In its crude form, it is a greasy yellow wax of unpleasant odor. This odor disappears when the lanolin is emulsified and made into salves, creams, ointments, and cosmetics. Although lanolin is slightly antiseptic, it has no other medicinal benefits and is valuable principally because of the ease with which it penetrates the skin, and because it does not turn rancid.

lanthanum
A chemical element, atomic number 57, atomic weight 138.91, symbol La.

lap
That region formed by the upper thighs to the junction of the lower abdomen with the body in an erect sitting posture.

lap belt
See ***seat belt***.

lapse rate
The rate at which an atmospheric variable, usually temperature, decreases with height. *See also* ***environmental lapse rate***.

large air carrier
Scheduled and nonscheduled aircraft operating under Federal Aviation Regulation (FAR) Parts 121 or 127. *Note:* Part 129 operations (foreign air carriers) are not included in the National Transportation Safety Board (NTSB) accident database, nor are hour and departure data available for these air carriers.

large air traffic hub
A community enplaning 1.00 percent or more of the total enplaned passengers in all services and all operations for all communities within the 50 States, the District of Columbia, and other U.S. areas designated by the Federal Aviation Administration.

large aircraft
Aircraft of more than 12,500 pounds, maximum certificated takeoff weight.

large aircraft commercial operator
Commercial operator operating aircraft with 30 seats or more or a maximum payload capacity of 7,500 pounds or more. Also, a commercial operator aircraft of more than 12,500 pounds maximum certificated takeoff weight.

large calorie
See ***Calorie***.

large certificated air carrier
An air carrier holding a Certificate of Public Convenience and Necessity that a) operates aircraft designed to have a maximum passenger seating capacity of more than 60 seats, or b) maximum payload capacity of more than 18,000 pounds.

large fleet
A fleet of 2,000 or more reportable vehicles, domestic or foreign, for which accountability is held by a department, independent establishment, bureau, or a comparable organiza-

tional unit of that department or independent establishment.

large nuclei
See ***condensation nuclei***.

large quantity generator
Generators producing more than 1000 kilograms per month of hazardous real estate.

large regional carrier group
Air carrier groups with operating revenues between $20,000,000 and $99,000,000.

large truck
Trucks over 10,000 pounds gross vehicle weight rating, including single unit trucks and truck tractors.

larva
The first or worm-like stage of an insect on issuing from the egg.

laryngectomy
Partial or total removal of the larynx by surgery. It is usually performed as treatment for cancer of the larynx. The person learns afterward to speak without his/her voice box.

laryngitis
Inflammation of the mucous membrane of the larynx affecting the voice and breathing. Laryngitis may be acute or chronic, or may occur in other forms. Acute laryngitis may be caused by overuse of the voice, allergies, irritating dust or smoke, hot or corrosive liquids, or even violent weeping. It also occurs in viral or bacterial infections, and is frequently associated with other diseases of the respiratory tract. In adults, a mild case of acute laryngitis begins with a dry, tickling sensation in the larynx, followed quickly by partial or complete loss of the voice. There may be a slight fever, minor discomfort, and poor appetite, with recovery after a few days. Other and more uncomfortable symptoms can include a feeling of heat and pain in the throat, difficulty in swallowing, and dry cough followed by expectoration; the voice may be either painful to use or absent. Swelling of the larynx and epiglottis may impair breathing. Increasing difficulty in breathing may be a sign of edematous laryngitis, or croup. After repeated attacks of acute laryngitis, chronic laryngitis may develop. This is caused mostly by continual irritation from overuse of the voice, tobacco smoke, dust, or chemical vapors, or by a chronic nasal or sinus disorder. Often, the moist mucous membrane lining the larynx becomes granulated. The granulation can proceed to thickening and hardening of the mucous membrane, which changes the voice or makes it hoarse. There is little or no pain, though there may be tickling in the throat and a slight cough. Chronic laryngitis that has persisted for a number of years may result in chronic hypertrophic laryngitis, a condition in which there is a permanent change in the voice because of hypertrophy of the membrane lining the larynx.

laryngopharynx
The lowest portion of the pharynx which extends from the level of the hyoid bone to the junction of the esophagus and larynx.

laryngoscopy
Direct visual examination of the larynx with a laryngoscope.

larynx
The essential sphincter guarding the entrance into the trachea and functioning secondarily as the organ of voice. The anterior protruding part forms the Adam's Apple. The larynx is a muscular and cartilaginous structure, lined with mucous membrane, situated at the top of the trachea and below the root of the tongue and the hyoid bone. The larynx contains the vocal cords, and is the source of the sound heard in speech (for this reason, it is also called the voice box). It is part of the respiratory system, and air passes through the larynx as it travels from the pharynx to the trachea and back again on its way to and from the lungs.

larynx to wall
The horizontal linear distance from the wall to the most anterior portion of the tissue overlying the thyroid cartilage. Measured with the individual standing erect, and with buttocks, shoulders, and occiput against the wall.

laser
Acronym for "light amplification by stimulated emission of radiation." Lasers are devices which convert electromagnetic radiation of mixed frequencies to one or more discrete frequencies of highly amplified and coherent visible radiation.

LASH
See ***lighter-aboard-ship***.

lashing
Maritime. A comparatively short manila line with an eye spliced in one end, used to moor barges and tows when passing through locks. Its average length is about 60 feet with sizes varying from 1 3/4 to 3 inches in circumference. The line is thrown somewhat in the manner of a lasso (hence the eye spliced) to catch a wall pin or bollard so as to snub the movement of barges and then moor them in the lock chamber. Also, any short length of line used to secure two barges end to end or side by side.

lasing medium
With regard to lasers, the material which absorbs and emits laser radiation. Lasers can be classified according to the state of their lasing media (e.g., gas, liquid, solid, and semiconductor).

LASL
*See **LANL**.*

lassitude
Weakness or exhaustion.

last assigned altitude
The last altitude/flight level assigned by Air Traffic Control (ATC) and acknowledged by the pilot.

last clear chance doctrine
This doctrine permits a plaintiff in a negligence action to recover, notwithstanding his/her own negligence, on showing that the defendant had the last chance to avoid the accident. The doctrine imposes upon a person the duty to exercise ordinary care to avoid injury to another who has negligently put himself/herself in a position of peril, and who he/she can reasonably assume is unconscious of or inattentive to peril or unable to avoid imminent harm.

latch block
Transit. The lower extremity of a latch rod which engages with a square shoulder of the segment or quadrant to hold the lever in position.

latch shoe
Transit. The casting by means of which the latch rod and the latch block are held to a lever of a mechanical interlocking machine.

Late Quaternary
*See **Quaternary**.*

late radiation effects
Those ionizing radiation effects which have a long latency.

latency period
The time interval between exposure to toxic chemical agents and the onset of signs and symptoms of illness.

latent
Present or potential, but not manifest.

latent defect
A hidden or concealed defect. One which could not be discovered by reasonable and customary observation or inspection; one not apparent on the face of goods, products, documents, etc.

latent heat of fusion
The amount of heat required to convert a unit mass of solid to liquid at the melting point.

latent heat of vaporization
The amount of heat required to convert a unit mass of substance from a liquid to a gas at a certain temperature.

latent period
The period of time between exposure to an injurious agent (chemical, physical, or biological) and the observation of an effect. It is the incubation period of an infectious disease. Also referred to as the latency period of a disease.

lateral
(1) Pertaining to, near, or toward the sides of the body or a symmetrical structure. (2) A consonant produced by closing off the midline of the mouth with the tongue, but allowing passage of air around one or both sides.

lateral bending moment
Those torques acting on the spine which result from sideways motion.

lateral canthus
*See **ectocanthus**.*

lateral cricoarytenoid
A skeletal muscle in the larynx which, on contraction, causes the glottis to close.

lateral cuneiform bone
One of the distal group of foot bones making up the tarsus.

lateral dam
Usually a rock and brush structure constructed parallel to normal stream flow to train or confine the current to a definite channel.

lateral disparity
*See **binocular disparity**.*

lateral displacement
*See **abduct**.*

lateral exhaust hood
A slot hood typically used to exhaust air contaminants from an open surface tank, and requiring full access to the top of the tank. The slots are narrow rectangular openings, usually located in a plenum at the rear of the tank opening. Also known simply as a *slot hood.*

lateral fault
A fault that slips in such a way that the two sides move with a predominantly lateral motion (with respect to each other). There are two kinds of lateral slip: right-lateral and left-lateral. They can be distinguished by standing on one side of the fault, facing the fault (and, of course, the other side), and noting which way the objects across the fault have moved with respect to you. If they have moved to your right, the fault is right-lateral. If the motion is to the left, then the fault is left-lateral.

lateral inhibition
A phenomenon in which neurons in the vicinity of a stimulation point, especially in sensory pathways, show reduced reactivity compared to those at the stimulation point.

lateral malleolus
The lateral protrusion of the fibula at the ankle.

lateral malleolus height
The vertical linear distance from the floor or other reference surface to the most lateral point of the lateral malleolus. Measured with the individual standing erect and his/her weight distributed equally on both feet.

lateral railroad
A lateral road is one which proceeds from some point on the main trunk between its terminal. An offshoot from the main line of the railroad.

lateral rectus muscle
A voluntary extraocular muscle with an anterior-posterior extent parallel to the optical axis along the lateral eyeball for rotating the anterior portion of the eyeball to the side.

lateral retinal image disparity
*See **binocular disparity**.*

lateral separation
Aviation. The lateral spacing of aircraft at the same altitude by requiring operation on different routes or in different geographical locations.

lateral sewers
Pipes that run under city streets and receive the sewage from homes and businesses.

lateral transfer
A personnel reassignment to another position at the same or approximately the same level of salary or responsibility.

laterality
A concept that different functions and modes of operation are allocated to different sides of the brain.

lateralization
The localization of a dichotically presented sound via earphones in apparent space along an imaginary line connecting the two ears.

launch vehicle
A vehicle that carries and/or delivers a payload to a desired location. This is the generic term. It includes, but is not limited to, airplanes, all types of space launch vehicles, manned launch vehicles, missiles and rockets and their stages, probes, aerostats and balloons, drones, remotely piloted vehicles, projectiles, torpedoes, and air-dropped bodies.

launching ramp
A transportation structure used for launching boats.

laundry booster
Any substance or combination of substances intended to aid detergents in the removal of certain stains from fabrics.

lavage
Irrigation or washing out an organ or cavity, especially the stomach or intestine. *Gastric lavage,* or irrigation of the stomach, is usually done to remove ingested poisons. It also may be employed as an emergency operation if there is danger of vomiting and aspiration during anesthesia, or in cases of persistent vomiting.

law
That which is laid down, ordained, or established. Law, in its generic sense, is a body of rules of action or conduct prescribed by a controlling authority, and having binding legal force.

law of inertia
See ***Newton's first law of motion***.

law of reflection
A physical law that an energy wave is reflected from a surface at an equal angle from the perpendicular as the incident wave, and both are in the same plane.

lawful arrest
The taking of a person into legal custody either under a valid warrant or on probable cause, believing that he/she has committed a crime, or under civil process which permits his/her arrest.

lawful authorities
Those persons who have the right to exercise public power, to require obedience to their lawful commands, to command or act in the public name.

lawful cause
Legitimate reason for acting, based on the law or on the evidence in a particular case as contrasted with acting on a whim or out of prejudice, or for a reason not recognized by the law.

lawrencium
A chemical element, atomic number 103, atomic weight 257, symbol Lw.

laws
Rules promulgated by the government as a means to an orderly society.

lawsuit
A vernacular term for a suit, action, or cause instituted or pending between two private persons in the courts of law. *See also* ***suit***.

lawyer
A person learned in the law; as an attorney, counsel, or solicitor; a person licensed to practice law.

laxative
A medicine that loosens the bowel contents and encourages evacuation. A laxative with a mild or gentle effect on the bowels is also known as an aperient; one with a strong effect is referred to as a cathartic or a purgative.

lay on the air
Transit (slang). Apply brakes.

lay witness
A person called to give testimony who does not possess any expertise in the matters about which he/she testifies.

layer
A cross-section through a three-dimensional object or computer model.

layering
The use of multiple display windows, allowing them to overlap and partially or completely hide the contents of the covered windows.

layoff
A termination of employment at the will of the employer. Layoffs can be temporary or permanent.

layover
Eight hours or more rest before continuing a trip or any off-duty period away from home.

lazy foot rule
A workplace design guideline that guards and lock out switches should be easily removable and replaceable so that workers will replace them.

lb
Pound(s).

lb/ft^3
Pounds per cubic foot.

LBB
See ***leak before burst***.

LBO
See ***lamp burnout***.

LBP
See ***lead-based paint***.

LC
(1) Lethal concentration. (2) Liquid chromatography.

LC_{50}
A standard measure of acute toxicity, normally applied to inhalation hazards but may also be applied in some cases to concentrations in water (or solution), designating the median *lethal concentration* of a chemical

that is estimated to kill 50% of the exposed organisms in a specific period of time and under a specific set of conditions. It is typically represented as micrograms per cubic meter ($\mu g/m^3$). LC_{50} is used in the Hazard Ranking System in assessing acute toxicity.

LCG
See ***liquefied compressed gas***.

LCL
See ***lifting condensation level***.

LD
(1) Lethal dose. (2) Legionnaire's disease.

LDA
See ***landing distance available***.

LDAR
Leak detection and repair program for fugitive emission sources.

Ldn
Average day-night sound level.

LD 0
The highest concentration and dosage of a toxic substance that kills test organisms.

LD_{50}
A standard measure of acute toxicity, normally applied to ingestion and/or absorption hazards, designating the median *lethal dose* of radiation or chemical applied directly to experimental organisms that will kill 50% of the exposed population within a specific period of time and conditions.

LD LO
The lowest concentration and dosage of a toxic substance that kills test organisms.

LDL
See ***lower detectable limit***. Also, *low density lipoprotein*.

leachate
A liquid that results from water collecting contaminants as it trickles through wastes, agricultural pesticides, or fertilizers. Leaching may occur in farming areas, feedlots, and landfills, and may result in hazardous substances entering surface water, groundwater, or soil.

leachate collection system
A system that gathers leachate and pumps it to the surface for treatment.

leaching
The process by which soluble constituents are dissolved and carried down through the soil by a percolating fluid. *See also* ***leachate***.

lead
(1) A chemical element, atomic weight number 82, atomic weight 207.19, symbol Pb. (2) A heavy metal that is hazardous to health if breathed or swallowed. Its use in paints, gasoline, and plumbing compounds has been sharply restricted or eliminated by federal laws and regulations. *See also* ***heavy metal***.

lead angle
An angle in which the load line is pulled during hoisting. Commonly used to refer to an angle in line with grooves in the drum or sheaves.

lead barge
The head, or first, barge of a tow generally with a rake.

lead-based paint (LBP)
A paint or other surface-coating product which has a lead content of 0.06% by weight in the total nonvolatile content of the paint, or by weight in the dried paint film. Also referred to as *lead-containing paint*.

lead free
According to the Federal Public Health Service Act: With respect to a drinking water cooler, that each part or component of the cooler which may come in contact with drinking water contains not more than 8 percent lead, except that no drinking water cooler which contains any solder, flux, or storage tank interior surface which may come in contact with drinking water shall be considered lead free if the solder, flux, or storage tank interior surface contains more than 0.2 percent lead. The EPA Administrator may establish more stringent requirements for treating any part or component of a drinking water cooler as lead free for purposes of this part whenever he determines that any such part may constitute an important source of lead in drinking water.

lead intoxication
Exhibiting of any neural, anemic, or colic symptoms resulting from lead absorption into body tissues.

lead line
A symptom of lead poisoning. A blue line on the gums as a result of excessive exposure to lead.

lead poisoning
A form of poisoning caused by the presence of lead or lead salts in the body. Lead poisoning affects the brain, nervous system, blood, and digestive system. It can be either chronic or acute. Chronic lead poisoning (plumbism) was once fairly common among painters, and was called "painter's colic." It became less frequent as paints composed of other chemicals were substituted for lead-based paints and as plastic toys replaced lead ones. Symptoms include weight loss, anemia, stomach cramp (lead colic), a bluish black line in the gums, and constipation. Other symptoms may be mental depression and, in children, irritability and convulsions. In addition to poisoning, the anemia and weight loss must also be treated, usually by providing an adequate diet. Acute lead poisoning, which is rare, can be caused in two ways. Lead may accumulate in the bones, liver, kidneys, brain, and muscles and then be released suddenly to produce an acute condition; or large amounts of lead may be inhaled or ingested at one time. Symptoms are metallic taste in the mouth, vomiting, bloody or black diarrhea, and muscle cramps. Diagnosis is made by examination of the blood and urine.

lead wall
The long wall of a lock, also known as a "guided wall," outside the confines of the lock chamber, usually the land wall in the case of older locks. *See also* ***guide wall***.

leaded gasoline
Gasoline to which lead has been added to raise the octane level. Contains more than 0.05 grams of lead per gallon or more than 0.005 grams of phosphorus per gallon. The actual lead content of any given gallon may vary. Premium and regular grades are included, and depending on the octane rating, also leaded gasohol. Blendstock is excluded from the definition until blending has been completed. Alcohol that is to be used in the blending of gasohol is also excluded. *See also* ***gasoline***.

leading question
One which instructs a witness how to answer, or "puts words into his/her mouth" to be echoed back. Leading questions are usually deemed improper on direct examination during litigation (except as may be necessary to develop the witnesses' testimony. Ordinarily, leading questions are permitted on cross-examination. However, it should be noted that, in some cases, leading questions may be permitted on direct examination (e.g., if the witness is very young, mentally disabled, or unfamiliar with the language spoken).

leak before burst (LBB)
A failure mode in which it can be shown that any initial flaw will grow through the wall of a pressure vessel or pressurized structure and cause leakage rather than brittle fracture/burst before leak. Normally determined at or below the maximum expected operating pressure (MEOP).

leak test
As pertains to ionizing radiation, a type of test for determining if a radioactive material is effectively contained or has escaped or leaked from a sealed source. It involves wiping surfaces on which the material would collect if it was released from the sealed source.

leakage
(1) *General.* The waste or diminution of all liquid caused by its leaking from a cask, barrel, or other vessel in which it was placed. (2) *Radiation.* Ionizing radiation, other than the useful beam, that is emitted from radiation producing equipment. Leakage from a sealed source of ionizing radiation (e.g., radioisotope) is the radioactive contamination that results outside the sealed source if the integrity of the seal fails to contain the material.

lean body mass
That mass of the body, including bones, muscles, and other tissues except for body fat. Also referred to as *fat-free mass*.

lean body weight
The lean body mass acted on by the acceleration due to gravitational or other forces according to Newton's second law. Also referred to as *fat-free weight*.

leaning
Pertaining to a posture in which the body longitudinal axis is away from vertical.

learn
Change behavior as a result of formal education, training, practice, or experience.

learning allowance
That time allowance given to a trainee or new worker while their skills are developed on a new job or task.

learning control
Having a control system with adequate memory and computing power to be able to modify its own operation in concert with newly acquired knowledge.

learning curve
A concept, mathematical function, or graphical representation of performance versus time in which performance improves with time as a result of learning/feedback.

learning hierarchy
A set of behavioral objectives, concepts, and principles arranged in the order in which they should be learned for optimum performance.

learning hierarchy analysis
A determination of the order in which the learning hierarchy should be taught.

lease
(1) A type of contract between parties (landlord and tenant) dealing with the use and occupancy of real estate or the use of property (e.g., automobile, office equipment, etc.). (2) Acquisition of a vehicle by an agency from a commercial firm, in lieu of government ownership, for a period of 60 continuous days or more.

leased property
Under Capital Leases: The total cost for all property obtained under leases that meet one or more of the following criteria: a) the lease transfers ownership of the property to the lessee by the end of the lease term; b) the lease contains a bargain purchase option; c) the lease term is equal to 75 percent or more of the estimated economic life of the leased property; or d) the present value at the beginning of the lease term of the minimum lease payments, excluding the portion of the payments representing executory costs such as insurance, maintenance, and taxes to be paid by the lessor, including any profit thereon, equals or exceeds 90 percent of the excess of the fair value of the lease property to the lessor at the inception of the lease over any related investment tax credit retained by the lessor and expected to be realized by him/her.

leasor
A person or firm that grants a lease.

least squares method
A mathematical technique for fitting a straight line or curve to a set of data points where the sum of the squares of the perpendicular distances from each data point to the line or curve is minimized.

leave of absence
A temporary absence from employment or duty with the intention to return during which time remuneration and seniority are not normally affected.

LED
See ***light-emitting diode***.

ledger
A horizontal scaffold member which extends from post to post and which supports the putlogs or bearer forming a tie between the posts. Sometimes called a *stringer*.

leeside low
Storm systems (extratropical cyclones) that form on the downward (lee) side of a mountain chain. In the United States, for example, leeside lows frequently form on the eastern side of the Rocky and Sierra Nevada mountain ranges.

leeward side
The side of an object away from the direction in which the wind is blowing.

LEF
See ***lighting effectiveness factor***.

left bank
The left descending bank of a river. The side of the river marked by red buoys, white or red lights, and red reflective material. *See also* ***right bank***.

left-hand draft
Current which pulls tow to the left.

left-hand draft in this set of marks
Maritime. Communication Protocol. Channel report term meaning that one should expect

the tow to drift to the left while running this course.

left-hand reef makes well in toward channel
Maritime. Communication Protocol. Term indicating that an underwater sandbar is building in toward the channel. A condition requiring extra caution on the part of the pilot and possibly the need for a flanking maneuver if the channel is considerably constricted.

leg
(1) *Anatomy.* The femur, tibia, fibula, and their surrounding associated and supporting soft tissues. (2) *Aviation.* Any portion of a flight plan from one point to another.

leg clearance
See ***knee well height.***

leg-foot
Involving both the leg and the foot, generally referring to internally generated or motor activities. *See also* ***foot-leg.***

leg inseam
The inside length of a trouser leg from the pubic crotch to approximately the dorsal/superior surface of the foot.

leg room
A measure of that usable volume beneath some table, platform, or other structure which the legs would normally occupy when in a seated posture.

legal auditing
An in-depth review of all phases of a company's operations to determine whether the company is fulfilling its obligations to laws and regulations, its permits, and any agreements with government agencies. An audit often includes an assessment of the company's management systems in order to identify ways in which the company might alter its existing structure or procedures to foster compliance.

legal cause
Proximate cause. Substantial factor in bringing about harm. In conflicts, denotes fact that the manner in which the actor's tortious conduct resulted in another's injury is such that the law holds the actor responsible unless there is some defense to liability.

legal weight
The weight of the goods plus any immediate wrappings which are sold along with the goods, e.g., the weight of a tin can as well as its contents.

legal willfulness
Intentional disregard of a known duty necessary to ensure the safety of persons or the property of another and the entire absence of care for life, persons, or the property of others.

legally liable
Liable under the law as interpreted by the courts.

legend
An explanatory symbol on a display or control, or on a drawing or blueprint.

legend switch
A labeled switch.

Legionnaire's Disease (LD)
Pneumonia caused by a bacterium, *Legionella pneumophila.* It has occurred among occupants of buildings in which this organism is present in the air at high concentrations. *See also* ***building-related illness.***

Legionella
The bacterium that is the causative agent of Legionnaire's disease and Pontiac fever.

legionellosis
Diseases caused by *Legionella* bacteria.

legislate
To enact laws or pass resolutions via legislation, in contrast to court-made law.

legislation
The act of giving or enacting laws; the power to make laws; the act of legislating; preparation and enactment of laws; the making of laws via legislation, in contrast to court-made laws.

legislative
Making or giving laws. Pertaining to the function of law-making or the process of enactment of laws.

legislative act
The enactment of laws. Law passed by legislature in contrast to court-made law. One which prescribes what the law shall be in future cases arising under its provision.

legislative branch of government

That branch of government which consists of this nation's law-making bodies, primarily the houses of Congress.

legislative immunity

The Constitution grants two immunities to members of Congress. First, that except for treason, felony, and a breach of the peace, they are "privileged from Arrest during their Attendance" at sessions of their legislative body. Second, that "for any Speech or Debate in either House, they shall not be questioned in any other Place." The first immunity is of little practical value, for its exceptions withdraw all criminal offenses and arrests therefore from the privilege, and it does not apply to the service of any process in a civil or criminal matter. The second immunity is liberally construed and includes not only opinion, speeches, debates, or other oral matter, but also voting, making a written report, or presenting a resolution, and in general to whatever member of Congress feels necessary to transact the legislative functions and business. Even a claim of a bad motive does not destroy the immunity, for it is the public good which is thereby served.

legislature

The department, assembly, or body or persons that makes statutory laws for a state or nation. At the federal level, and in most states, the legislature is bicameral in structure, usually consisting of two branches (upper house or Senate and the lower house or House of Representatives). Legislative bodies at the local levels are variously called city councils, boards of aldermen, etc.

LEL

See ***lower explosive limit***.

Lemnaceae

Floating aquatic plants that provide a habitat for aquatic organisms capable of metabolizing wastewater organics. Also commonly referred to as *duckweed*.

length

(1) The extent or distance from one end of an object to the other, or a distance in space from one clearly identified point to any other such point. In the International System of Units, the basic unit of length is the *meter,* which has been defined as the length of path traveled by light in a vacuum during a time interval of 1/299,792,458,458 of a second. In the MKS System, the basic unit of length is the *meter*. In the CGS System, the basic unit of length is the *centimeter*. In the English System, the basic units of length can be either the *foot,* the *inch,* or the *yard*. (2) An open anthropometric measurement from one point on the body to another which contains as a major portion a relatively straight line, but may also contain some brief curvature, such as a round a flexed joint.

length-tension curve

An inverted-U-shaped function which indicates that muscle tension capability falls off to either side of an optimum length.

lens

(1) A transparent device for refracting or otherwise directing electromagnetic radiation. (2) A glass for converging or scattering rays of light. (3) The crystalline lens, a transparent organ lying behind the pupil and iris and in front of the large vitreous-filled cavity of the eye. The crystalline lens refracts (bends) light rays so that they are focused on the retina. For the eye to see objects close at hand, light rays from the objects must be bent more sharply to bring them to focus on the retina; light rays from distinct objects require much less refraction. It is the function of the lens to accommodate or make some adjustment for viewing near objects and objects at a distance. To accomplish this, the lens must be highly elastic so that its shape can be changed and made more or less convex. The more convex the lens, the greater the refraction. Small ciliary muscles create tension on the lens, making it less convex; as the tension is relaxed, the lens becomes more spherical in shape and hence more convex. With increasing age, the lenses lose their elasticity. Thus, their ability to focus light rays in the retina becomes impaired. This condition is referred to as presbyopia. In farsightedness (hyperopia), the image is focused behind the retina because the refractive power of the lens is too weak, or the eyeball axis is too short. Nearsightedness (myopia) occurs when the refractive power of the lens is too strong or the eyeball is too long, so that the image is focused in front of the retina. *See also* ***contact lens***.

lenticular
(1) Pertaining to or shaped like a lens. (2) Pertaining to the crystalline lens. (3) Pertaining to the lenticular nucleus.

lenticular cloud
A cloud in the shape of a lens.

LEPC
*See **Local Emergency Planning Committee**.*

leprosy
A chronic communicable disease characterized by the production of granulomatous lesions of the skin, upper respiratory and ocular mucous membranes, peripheral nerves, and the testes; also called Hansen's disease. Not really contagious, it often results in severe disability but is rarely fatal. The cause of leprosy is believed to be a species of bacteria, *mycobacterium leprae* or *Hansen's bacillus,* which usually attacks the skin and nerves, but not the brain. Although it is not inherited, the actual means of transmission is not fully understood.

leptokurtic
Pertaining to a highly peaked normal distribution.

leptospirosis
An infection transmitted to man by dogs, swine, and rodents or by contact with contaminated water.

lesion
An abnormal localized change in the structure of an organ or tissue resulting from disease or injury. Lesion is a broad term, including wounds, sores, ulcers, tumors, cataracts, and any other tissue damage. Lesions range from skin sores associated with eczema to the changes in lung tissue that occur in tuberculosis.

less than truckload (LTL)
Transit. A quantity of freight less than that required for the application of a truckload rate. Usually less than 10,000 pounds and generally involves the use of terminal facilities to break and consolidate shipments.

lessons learned
(1) *General.* A formal, documented account or report of both the positive and negative aspects of operational or task experience which is compiled after the conclusion of the task. Used generally to highlight those actions which should or should not be allowed to occur during any subsequent performance of like or similar tasks. (2) *Accident Investigation.* A formal, documented account or report of both the positive and negative aspects of the operation or task involved in the accident. Intended for use as a tool for the prevention of accident recurrence.

lessor trochanter
A rounded projection on the medial proximal femur.

LET
*See **linear energy transfer**.*

lethal
Sufficient to cause, or capable of causing death.

lethal concentration (LC)
That quantity of an agent which is sufficient to cause death. The term "concentration" generally refers to a substance inhaled.

lethal concentration median (LC_{50})
That which causes death of 50 percent of the test population within 24 hours of exposure.

lethal dose (LD)
That quantity of an agent which is sufficient to cause death. The term "dose" generally refers to a substance ingested.

lethal dose median (LD_{50})
That which causes death of 50 percent of the test population within 24 hours of exposure.

letter of intent
A written promise to carry out a specified action at some point in the future

leucine
A naturally occurring amino acid, one of those essential for human metabolism.

leukemia
A disease of unknown specific cause characterized by an overproduction of leukocytes and their precursors, and enlargement of the spleen. The disease is variable, at times running a more chronic course in adults than in children. Exposure to low intensities of ionizing radiation is thought to be one possible cause. Leukemia is classified clinically on the basis of 1) the duration and character of the disease – acute or chronic; 2) the type of cell involved – myeloid (myelogenous), lymphoid (lymphogenous), or monocytic; and 3) in-

crease or no increase in the number of abnormal cells in the blood – leukemic or aleukemic (subleukemic). In acute leukemia, the white cells resemble precursor, or immature, cells. They are larger than normal cells, and they accumulate much more rapidly than in chronic leukemia. They are incapable of performing their normal function of combating infection. In chronic leukemia, the white cells are more mature, resembling normal cells and having some limited capacity to oppose invading organisms. *See also* ***leukocyte***.

leukemogenic
A substance that can cause leukemia. Also referred to as a *leukomogen*.

leukocyte
A white (colorless) blood corpuscle in the blood, lymph, or tissues that plays a major role in the body's defense against disease. There are five types: lymphocytes, monocytes, neutrophils, eosinophils, and basophils; the last three are often referred to as granulocytes. The leukocytes act by moving through blood vessel walls to reach a site of injury. Foreign particles such as bacteria may be engulfed or phagocytosed by the leukocytes, especially the neutrophils and monocytes. It is this process that causes the increase in the number of leukocytes in the blood during infection, and one of the laboratory determinations to diagnose infectious states is based on it. The leukocytes also play some role in the repair of injured tissue, though their function here is not clear.

leukocytosis
A transient increase in the number of white cells in the blood as a result of fever, infection, inflammation, etc.

leukopenia
A reduction, to below the normal level, of the number of white cells in the blood.

levator
Any muscle producing an upward movement.

levee
A built-up embankment on or back from the riverbank for the purpose of containing floodwater.

level A
The level of protection the EPA considers necessary for work in or entry into hazardous environments or contaminated sites where the potential for serious adverse occupational health effects are present. This includes sites contaminated with unknown materials or sites or areas where materials are known to exist that could cause both respiratory and dermal exposure effects. Level A protective equipment includes supplied-air positive pressure respirator and does not require skin protection.

level B
The level of protection described by the EPA that includes the maximum degree of respiratory protection but a lesser degree of full body and skin protection. Examples include airborne respirable contaminants that are very toxic but not toxic through skin absorption.

level C
The protection level described and required by the EPA where known concentrations of airborne contaminants exist but are suitably protected against by air-purifying respirators and do not require skin protection.

level D
The protection level required and described by the EPA where nuisance respiratory exposures and non-absorbing skin contaminants exist. Respirators would be required only when air sampling and monitoring determine that hazards exist above the nuisance level.

level of concern (LOC)
The concentration in air of an extremely hazardous substance above which there may be serious immediate health effects to anyone exposed to it for short periods of time.

level of effort
(1) A type of contract or agreement in which a certain number of people are supported to perform specified tasks. (2) The amount of physical or mental activity exerted or required to perform at a certain level.

level of service
Transportation. (1) A set of characteristics that indicate the quality and quantity of transportation service provided, including characteristics that are quantifiable and those that are difficult to quantify. (2) For highway systems, a qualitative rating of the effectiveness of a highway or highway facility in serving traffic, in terms of operating conditions. (3)

For paratransit, a variety of measures meant to denote the quality of service provided, generally in terms of total travel time or a specific component of total travel time. (4) For pedestrians, sets of area occupancy classifications to connect the design of pedestrian facilities with levels of service.

leveled time
*See **normal time**.*

leveling
A performance rating method in which an observer adjusts a worker's time to compare with normal time. *See also **performance rating**.*

lever
A rigid linear structure which is capable of movement and exerting force about a fulcrum.

lever arm
(1) The distance from a joint axis to the point of a muscle attachment. (2) The distance from the fulcrum to the point of effort or resistance on a lever.

lever switch
A type of toggle switch in which the activating mechanism is a manually operated lever.

leverage
That mechanical advantage achieved by using a lever.

Leverett technique
*See **L-1 maneuver**.*

lexical decision task
The process in which a judgement is made as to whether or not a letter string is a word.

LFL
*See **lower flammable limit**.*

LIA
Laser Institute of America.

liability
Being bound or obligated by law to do, pay, or make good something.

liability insurance
Insurance that covers suits against the insured for such damages as injury or death to other drivers or passengers, property damage, and the like. It is insurance for those damages for which the driver can be held liable. Liability insurance is that form of insurance which indemnifies against liabilities on account of injuries to the person or property of another. It is distinguished from *indemnity insurance* and may be issued to cover the liability of, for example, carriers, contractors, employers, landlords, manufacturers, and drivers. *See also **insurance** and **indemnity insurance**.*

liable
Bound or obligated in law or equity. Responsible, chargeable, answerable, and/or compelled to make satisfaction, compensation, or restitution. The condition of being bound to respond because a wrong has occurred.

liable parties
Under the Comprehensive Emergency Response, Compensation, and Liabilities Act (CERCLA), any person who by contract, agreement, or otherwise, arranged for disposal or treatment of hazardous substances owned or processed by such person.

libel
A method of defamation expressed by print, writing, pictures, or signs. In its most general sense, any publication that is injurious to the reputation of another. A false and unprivileged (i.e., without legal precedence) publication in writing of defamatory material.

license
(1) *General.* An authorization granted by a government agency to conduct an activity under the conditions specified in the license. (2) *Radiation.* The company or person authorized to use a radioactive material obtained under a license issued by the NRC or an Agreement State.

license plate lamp
A lamp used to illuminate the license plate on the rear of a motor vehicle.

licensed driver
Any person who holds a valid driver's license from any state.

licensed material
Radiation. Source material, special nuclear material or byproduct material that is received, possessed, used, or transferred under a special license issued by the licensing agency (e.g., NRC, Energy Research and Development Administration, or an Agreement State).

lichen
(1) A name applied to many different kinds of papular skin diseases. (2) Any species or

plant of a group believed to be composed of symbiotic algae and fungi.

lidar

An instrument that uses a laser to generate intense pulses that are reflected from atmospheric particles of dust and smoke. Lidars have been used to determine the amount of particles in the atmosphere as well as particle movement that has been converted into wind speed. Lidar means *light* detection and ranging.

lie sheet

Transit (slang). A driver's log book.

lien

A legal restriction imposed on a piece of equipment or real estate, usually by permission of a court, to secure the payment of money under a contract or if damages are awarded in litigation.

life cycle

(1) *System Safety.* A phased concept to explain the various stages of product or system progression consisting of the concept phase, design phase, production phase, operational phase, and disposal phase. In system safety, the product or system life cycle is often used to indicate the timing of certain types of analytical evaluations. (2) *Environmental.* Consecutive and interlinked stages of a product or service system, from the production and delivery of raw material or the generation of natural resources to the final disposal.

life cycle characteristic curve

A graph curve used to describe the expected phases over the lifetime of a machine or electromechanical system or process. It consists of a steeply declining initial segment (run-in, or infant mortality phase), a relatively flat middle segment (the useful phase), and a moderately increasing terminal segment (the wear-out or disposal phase). Synonymous with *bathtub curve*.

life cycle cost

The total cost of an item over its useful life, including purchase, maintenance, and operations.

life expectancy

The number of years a person may be expected to live from a given age, based on the mean length of life of persons of a similar age.

life jacket

A personal flotation device worn about the torso and normally secured with straps across the torso and through the pelvic crotch. U.S. Coast Guard requires a life jacket for each occupant of a floating craft. Normally, children should be wearing theirs at all times while in a floating craft (in actual practice, however, this is rarely seen).

Life jackets are required for all occupants of a floating craft

life performance curve

A functional relationship between some particular characteristic of a lamp and its age.

life support

That function which addresses the sustenance, health promotion, and protection of personnel under all reasonably expected conditions for a specified activity.

life support system

Any system which provides life support.

lifeline

A rope or other type of cable intended to save an individual's life should an accident occur under hazardous working conditions. May function to break fall before striking an object, to keep from drifting off, to keep from being washed overboard, etc.

lift
In a sanitary landfill, a compacted layer of solid waste and the top layer of cover material.

lift vessel
A vessel designed to be loaded or unloaded by moving the containers with a heavy crane.

lifting condensation level (LCL)
The level or altitude at which a parcel of air, when lifted dry adiabatically, would become saturated.

lifting station
*See **pumping station**.*

lifting task
Any task which involves manually changing the location of an object without external mechanical assistance and applies a force and/or torque to the vertebral column.

lifting technique
A procedure recommended or used by an individual to perform a particular lifting task.

lifting torque
The product of the load and distance of the load from a fulcrum in the vertebral column which is created by a lifting task.

ligament
A band of dense fibrous connective tissue which interconnects the articular aspects of bones.

ligature
A thread or wire used in surgery to tie off blood vessels to prevent bleeding, or to treat abnormalities in other parts of the body by constricting the tissues.

light
(1) Radiation from a region of the electromagnetic spectrum of which an organism becomes aware through stimulation of the retina or other visual receptor; that stimulation which excites visual receptors. (2) Not heavy.

light activity
That level of physical activity which requires/consumes 60 – 100 calories per square meter of skin surface per hour, including the basal metabolic rate.

light adaptation
An adjustment within the visual system making it more or less sensitive to light by adjusting the threshold.

light boat
A towboat without a tow.

light density railroad
Railroads with 1200 or less train-miles per road mile.

light duty
A work classification in which an individual is not permitted to do heavy lifting for health or other reasons.

light duty scaffold
A scaffold designed and constructed to carry a working load not to exceed 25 pounds per square foot.

light duty vehicle
Automobiles and light trucks combined.

light effort
The level of physical work which can be maintained for a work shift without undue fatigue.

light-emitting diode (LED)
Any semiconductor diode which emits light when current is applied.

light-emitting diode display
Any display using light-emitting diodes as a radiant source.

light field microscopy
Microscopic technique that relies on the amplitude modulation of light to make specimens visible. Different portions of the specimen absorb light to a differing degree, thereby providing specimen details as differences in the intensity of light reaching the eye.

light gun
Aviation. A handheld directional light signaling device which emits a brilliant narrow beam of white, green, or red light as selected by the tower controller. The color and type of light transmitted can be used to approve or disapprove anticipated pilot actions where radio communication is not available. The light gun is used for controlling traffic operating in the vicinity of the airport and on the airport movement area.

light loss factor (LLF)
Any of a set of possible variables which may contribute to a decrease in available luminance for a given location.

light mixing
*See **additive color mixing.***

light pen
A pen-shaped interactive device which emits a light beam for striking a certain region of a display to initiate a certain system action.

light quantity
A measure of the amount of light used, equal to the product of the luminous flux and the time duration for which it is sustained.

light rail
(1) *DOT.* A streetcar type vehicle operated on city streets, semi-exclusive rights-of-way, or exclusive rights-of-way. Service may be provided by step entry vehicles or by level boarding. (2) *APTA.* An electric railway with a "light volume" traffic capacity compared to "heavy rail." Light rail may use shared or exclusive rights-of-way, high or low platform loading, and multi-car trains or single cars. (3) *FTA.* Lightweight passenger rail cars operating singly (or in short, usually two-car, trains) on fixed rails in a right-of-way that is not separated from other traffic for much of the way. Light rail vehicles are driven electrically with power being drawn from an overhead electric line via a trolley or a pantograph. Also known as *streetcar, troller car,* and *tramway.*

light rail (streetcar)
Urban transit which uses predominantly reserved but not always grade-separated rights-of-way. Electrically powered rail vehicles operate alone or in trains.

Street cars are used extensively throughout Europe as well as in many cities in the United States

light rail vehicles (streetcars)
Rail cars with motive capability, usually driven by electric power taken from overhead lines, configured for passenger traffic and usually operating on non-exclusive right-of-way.

light scatter fraction
The ratio of scattered light to specularly reflected light.

light stand
Maritime Navigation. Colloquial term meaning the position or location of a shore-lighted aid to navigation.

light truck
(1) An automobile other than a passenger automobile which is either designed for off-highway operation or designed to perform at least one of the following functions: a) transport more than 10 persons; b) provide temporary living quarters; c) transport property on an open bed; d) provide greater cargo-carrying than passenger-carrying volume; or e) permit expanded use of the automobile for cargo-carrying purposes or other nonpassenger-carrying purposes through the removal of seats by means installed for that purpose by the automobile's manufacturer or with simple tools, such as a screwdriver and wrenches, so as to create a flat, floor level, surface extending from the forward most point of installation of those seats to the rear of the automobile's interior. An automobile capable of off-highway operation is an automobile a) that has 4-wheel drive, or is rated at more than 6,000 pounds gross vehicle weight; and b) that has at least four of the following characteristics calculated when the automobile is at curb weight, on a level surface, with the front wheels parallel to the automobile's longitudinal centerline, and the tires inflated to the manufacturer's recommended pressure: (i) approach angle of not less than 28 degrees; (ii) break over angle of not less than 14 degrees; (iii) departure angle of not less than 20 degrees; (iv) running clearance of not less than 20 centimeters; (v) front and rear axle clearances of not less than 18 centimeters each. (2) Trucks of 10,000 pounds gross vehicle weight rating or less, including pickups, vans, truck-based station wagons, and utility vehicles. (3) Two-axle, four-tire trucks.

lighted airport
An airport where runway and obstruction lighting are available.

lighter
A barge used in off-loading an oceangoing vessel.

lighter-aboard-ship (LASH)
A type of barge-carrying vessel equipped with an overhead crane capable of lifting barges of a common size and stowing them into cellular slots in a thwart ship position. Lighter-aboard ship is an all-water technology analogous to containerization.

lighter-than-air aircraft
Aircraft that can rise and remain suspended by using contained gas weighing less than the air that is displaced by the gas.

The zeppelin is the best example of a lighter-than-air aircraft. The modern-day blimp is in this category of aircraft.

lighting
The collective sensation or description of the light being input to the visual environment.

lighting effectiveness factor (LEF)
The ratio of equivalent sphere illumination to calculated illumination or illumination measured with a meter.

lighting fixture
Any structure designed and built specifically for the installation of light-producing devices and to direct illumination.

lighting outlet
An outlet intended for the direct connection of a lamp holder, a lighting fixture, or a pendant cord terminating in a lamp holder.

lightness
(1) A judgement as to the weight of an object, on a scale from light to heavy. (2) That apparent degree to which something is judged as lighter or darker compared to a similarly reflecting or transmitting white or achromatic reference.

lightning
A visible electrical discharge produced by thunderstorms. It may take place within a cloud, from one cloud to another, from a cloud to the surrounding air, or from a cloud to the ground. For lightning to occur, separate regions containing opposite electrical charges must exist within a cumulonimbus cloud. Exactly how this charge separation comes about is not totally understood. However, it is believed that, because unlike charges attract one another, the negative charge at the bottom of the cloud causes a region of the ground beneath it to become positively charged. As the thunderstorm moves along, this region of positive charge follows the cloud like a shadow. The positive charge is most dense on protruding objects, such as trees, poles, and buildings. The difference in charges causes an electrical potential between the cloud and ground, which may be 10,000 volts per meter. In dry air, however, a flow of current does not occur because the air is a good electrical insulator. Gradually, the electrical potential builds and when the electrical field associated with it exceeds about 3 million volts per meter, the insulating properties of the air break down, a current flows, and lighting occurs. *Cloud-to-ground lightning* begins as a flow of electrons from the middle of the cloud rushes toward the base. This discharge of electrons proceeds toward the ground in a series of steps. Each discharge covers approximately 50 meters, then stops for about 50 millionths of a second, then occurs again over another 50 meters. This *stepped leader* is very faint and is usually invisible to the human eye. As the top of the stepped leader approaches the ground, a current of positive charge starts upward from the ground to meet it. After they meet, large numbers of electrons flow to the ground and a much larger, brighter *return stroke* surges upward to the cloud along the path followed by the stepped leader. Even though the bright return stroke travels from the ground upward to the cloud, it happens so

quickly (in 1/10,000 of a second) that the human eye cannot resolve the motion and we see what appears to be a continuous bright flash of light. Sometimes there is only one lightning stroke, but more often each flash is actually a series of very rapid strokes that travel between the cloud and ground. A lightning flash consisting of many strokes usually lasts less than a second. During this short period of time the human eye may not be able to perceive the individual strokes and the flash appears to flicker. The lightning stroke can heat the air through which it travels to an incredible 30,000°C (54,000°F), which is approximately five times the temperature of the sun's surface. This extreme heating causes the air to expand explosively, thus initiating a booming sound wave called *thunder* that travels outward in all directions from the flash. Because light travels so fast, it reaches the eye nearly instantly. But the sound, traveling at only 330 meters/second (1100 feet/second), takes much longer to reach the ear. Hence, by counting the seconds from the moment the lightning is seen until the thunder is heard, one can approximate their distance from the lightning stroke. Because it takes sound about 3 seconds to travel one kilometer (5 seconds for each mile), thunder that is heard 15 seconds after the lighting was seen is approximately 5 km (3 miles) away.

Lightning burst, a violent display of nature's power

Likert scale
A technique for rating surveys on a discrete, integer-based scale having an odd number of discrete options and consisting of a range, generally from 1 to 5, from strongly disagree to strongly agree, respectively. (May occasionally see scales to 7 or 9 options).

limb
(1) An arm or leg, including all its component parts. (2) A structure or part resembling an arm or leg.

limb coordination
A measure of the degree of integrated functioning of the limbs in performing some activity.

limb-load aggregate
The combined mass/torque from the working load plus the mass/torque from the limb(s) involved in a lifting or movement task.

limb movement velocity
The rate at which a single movement of a limb can be accomplished, without regard for accuracy or coordination.

limestone scrubbing
Process in which sulfur gases moving toward a smokestack are passed through a limestone and water solution to remove sulfur before it reaches the atmosphere.

liminal contrast
See ***contrast threshold***.

liminal contrast threshold
See ***contrast threshold***.

limit load
The maximum load, or combination of loads, a part or structure is expected to experience at any time during its intended operation and expected environment, as follows:

limit load = (load factor) x (rated load)

limit of detection (LOD)
The smallest amount of an analyte that can be distinguished from background or the lowest concentration that can be determined to be statistically different from a blank. Typically, it is that amount of analyte which is three standard deviations above the background response. *See also* ***lower detectable limit*** *and* ***detection limit***.

limit of quantitation (LOQ)
The amount of analyte above which quantitative results may be reported with a specific degree of confidence. Typically, this value is 10 times the standard deviation of concentrations very near the limit of detection. *See also* ***limit of detection***.

limit stop
Any device or mechanism which prevents further movement of a control, door, drawer, or other object at a certain point when motion beyond that point might have undesirable consequences. May be accomplished by audible click or tactile sensation.

limit switch
An electrical switch which is capable of cutting the power supply if the device being monitored goes beyond a specified range.

limitation of damages
Provision in a contract or agreement by which parties agree in advance as to the amount or limit of damages for breach.

limited radar airport traffic control tower
Airport traffic control tower at which air traffic control specialists are permitted to provide radar approach control service that requires only limited vectoring, as well as to handle takeoffs and landings.

limiting factor
A condition, whose absence or excessive concentration, is incompatible with the needs or tolerance of a species or population and which may have a negative influence on their ability to grow or even survive.

limnology
The study of the physical, chemical, meteorological, and biological aspects of fresh water.

limousine or auto rental with driver
Establishments primarily engaged in furnishing limousines or auto rentals with drivers, where such operations are principally within a single municipality, contiguous municipalities, or a municipality and its suburban areas e.g., automobile rental with driver, limousine rental with driver, hearse rental with driver, passenger automobile rental with driver.

limp
(1) A type of gait in which steps are halting and the time spent on one leg is shorter than the other. *See also* ***gait***. (2) Flaccid; having less than normal tonicity.

line
Rail Operations. One or more running tracks, each kilometer of line counting as one, however many tracks there may be. The total length of line operated is the length operated for passenger or goods transport, or both. Where a section of network comprises two or more lines running alongside one another, there are as many lines as routes to which tracks are allotted exclusively.

line and staff organization
In the structure of an organization, those members who are directly accountable and responsible for the daily operations of the enterprise are considered *line* management with the authority to implement or change company policy and operating procedures. Those who serve as advisors to the line and can only recommend changes are considered *staff* management.

line breaking
The intentional opening of a pipe, line, or duct that is or has been carrying flammable, corrosive, or toxic material, an inert gas, or any fluid at a volume, pressure, or temperature capable of causing injury.

line-clearance tree trimming
The pruning, removing, trimming, maintaining, repairing, or clearing of trees or cutting of brush that is within 10 feet (305 cm) of electric supply line or equipment.

line-haul
Rail Operations. Transportation from one city to another as differentiated from local switching service. *See also* ***linehaul***.

line-haul operation railroads
Establishments primarily engaged in line-haul railroad passenger and freight operations.

line miles
The sum of the actual physical length (measured in only one direction) of all streets, highways, or rights-of-way traversed by a transportation system (including exclusive rights-of-way and specially controlled facilities) regardless of the number of routes or vehicles that pass over any of the sections.

line of flow
See ***flow path***.

line of sight (LOS)
That path from the lateral and vertical center of the eye pupil to an object being fixated or direction being viewed.

line section
A continuous run of pipe that is contained between adjacent pressure pump stations,

between a pressure pump station and a terminal or breakout tank, between a pressure pump station and a block valve, or between adjacent block valves.

line spar
Maritime. Line used to secure spar to deck.

line spectrum
A frequency spectrum in which the components are shown as lines at discrete frequencies.

line through
Maritime. To pull a boat through a swift shallow channel by means of lines placed on the bank. A term seldom used today.

line width
The width of a line on a display or hardcopy.

line worker
A worker employed by a utility company (e.g., electrical, telephone, etc.) who performs the majority of his/her assigned duties out "on the line," that is, working on the electrical lines or telephone lines, to ensure uninterrupted service and/or to make improvements to existing service.

Line workers are exposed to a number of hazards on a daily basis such as working at heights and around high voltage

linear
(1) Pertaining to a linear function. (2) Measured in a straight line.

linear algebra
The study and/or use of simultaneous linear equations, as used in vectors and linear transformations.

linear correlation
A relationship between two variables which may be represented graphically by a straight line or by a linear function.

linear energy transfer (LET)
The linear rate of energy loss locally absorbed by an ionizing radiation particle passing through a material medium.

linear equation
See ***linear function***.

linear function
A mathematical function which may be represented by a straight line, having an equation of the form below.

$$y = mx + b$$

linear momentum
The tendency for an object to continue moving in a straight line.

linear movement control
A control device which moves in a straight line when force is applied.

linear programming
A technique for determining a solution to a problem using the assumptions a) that the function is linear and b) that the process involved can be represented as a set of linear equations or inequalities.

linear range
Instruments. The ratio of the largest concentration to the smallest concentration within which the detector response is linear. It is also expressed as the range (i.e., lower value to upper value) over which the detector response is linear.

linear referencing system (LRS)
The total set of procedures for determining and retaining a record of specific points along a highway. Typical systems used are mile point, milepost, reference point, and link-node.

linear service
International water carriers that ply fixed routes on published schedules.

linear system
A system in which output varies according to some proportionality constant and the input.

linearity
(1) The straightness of a line, or column, or row on a display. (2) That property between two variables in which a change in one variable results in a directly proportional change in the other.

linehaul
Rail Operations. The movement of trains between terminals and stations on the main or branch lines of the road, exclusive of switching movements. *See also* ***line-haul***.

liner
(1) A relatively impermeable barrier designed to prevent leachate from leaking out of a landfill. Liner materials include plastic and dense clay. (2) An insert or sleeve for sewer pipes to prevent leakage or infiltration. (3) A word derived from the term "line traffic," which denotes operation along definite routes on the basis of definite, fixed schedules; a "line" thus is a vessel that engages in this kind of transportation, which generally involves the haulage of general cargo as distinct from bulk cargo. (4) A vessel sailing between specified ports on a regular basis.

liner terms
An expression covering assessment of ocean freight rates generally implying that loading and discharging expenses will be for the ship owner's account, and usually apply from the end of ship's tackle in port of loading to the end of ship's tackle in port of discharge.

lines
Maritime. The various types used with regard to towing: back line, backing line, breast line, check line, dropping out line, face line, fore and aft lines, handy line, head line, jockey line, lashing, lead line, lock line, monkey line, quarter line, side line, spar line, stem line, tow line, spring line, peg line.

linguadental
Articulated with the tip of the tongue placed on the upper front teeth.

liniment
An oily, soapy, or alcoholic preparation to be rubbed on the skin.

lining
A material permanently attached to the inside of the outer shell of a garment for the purposes of thermal protection and padding.

link
(1) Any interface between the human operator and a machine, at which movement in one produces movement in the other. May be referred to as *fixed linkage mechanism*. (2) A straight line representing a body segment, terminating at pivot points on the body. (3) Any interface, interaction, or bond between individuals. Also referred to as *linkage*.

link analysis
(1) An examination and study of the biomechanical link actions or positions of the body. (2) An identification and examination of the sensorimotor and mechanical and/or electrical interfaces between individuals, machines, or human and machine in a system.

linkage
(1) *Chemistry.* The connection between different atoms in a chemical compound, or the symbol representing it in structural formulae. (2) *Genetics.* The tendency for a group of genes in a chromosome to remain in continuous association from generation to generation. (3) *Psychology.* The connection between a stimulus and its response. (4) *See* ***link (3)***.

linked passenger trip
A trip from origin to destination on the transit system. Even if a passenger must make several transfers during a journey, the trip is counted as one linked trip on the system.

lip breadth
The maximum horizontal linear distance between the most lateral point of the junction of the upper and lower lips on each side of the mouth opening. Measured with the facial muscles relaxed.

lip breadth, smiling
The maximum horizontal linear distance between the corners of the mouth opening. Measured with the individual smiling broadly.

lip – lip length
The vertical distance, in the midsagittal plane, from the lower margin of the lower lip to the upper margin of the upper lip. Measured with the facial muscle relaxed and the lips together.

lip protrusion
The most anterior point of either the upper or lower lip, whichever is more anterior. Must specify which lip, if different.

lip protrusion to wall
The horizontal linear distance from a wall to the most anterior point of the lips. Measured with the individual standing or sitting erect with the back of the head against the wall.

lip-reading
Perception of speech through the sense of sight, by recognition of the words formed from movement of the lips.

lipid solubility
The maximum concentration of a chemical that will dissolve in fatty substances; lipid soluble substances are insoluble in water. If a substance is lipid soluble it will very selectively disperse through the environment via living tissue.

lipids
A comprehensive term for fats and fat-derived materials that denotes substances extracted from animal or vegetable cells by non-polar or fat-soluble solvents. Lipids are among the chief structural components of living cells.

lipochrome
Any one of a group of fat-soluble hydrocarbon pigments, such as lutein, chromophane, and the natural yellow coloring material of butter, egg yolk, and yellow corn. They are also known as carotenoids.

lipoprotein
A combination of lipid and protein, having the general properties (e.g., solubility) of proteins. Practically all of the lipids of the plasma and lipoprotein complexes (alpha- and beta-lipoproteins) can be distinguished by electrophoresis. Elevated levels of *low density lipoprotein (LDL)* are generally considered harmful and may be a major contributory factor in arterial sclerosis and other heart-related diseases. Whereas, elevated levels of *high density lipoprotein (HDL)* are generally considered beneficial in the overall prevention of heart disease and related ailments.

liposome
One of the particles of lipoid matter held emulsified in the tissues in the form of invisible fat.

lipotropism
Affinity for fat or fatty tissue, especially that of certain agents that are capable of decreasing the deposits of fat in the liver. Also called *lipotropy*.

liquefaction
Changing a solid into a liquid.

liquefied compressed gas (LCG)
A compressed gas which is partially liquid at the cylinder pressure and a temperature of 70°F (21°C).

liquefied natural gas (LNG)
(1) Natural gas or synthetic gas having methane as its major constituent and which has been changed to a liquid or semisolid. (2) Natural gas (primarily methane) that has been liquefied by reducing its temperature to -260°F at atmospheric pressure.

liquefied natural gas (LNG) facility
A pipeline facility that is used for liquefying or solidifying natural gas or synthetic gas or transferring, storing, or vaporizing liquefied natural gas.

liquefied petroleum gas (LPG)
Ethane, ethylene, propane, propylene, normal butane, butylene, and isobutane produced at refineries or natural gas processing plants, including plants that fractionate new natural gas plant liquids.

liquid
A material that has a vertical flow of over 2 inches (50 mm) within a three minute period, or a material having one gram or more liquid separation, when determined in accordance with the procedures specified in American Society for Testing and Materials (ASTM) specification D4359-84, "Standard Test Method for Determining whether a Material is a Liquid or Solid," 1984 edition.

liquid phase
DOT. A material that meets the definition of liquid when evaluated at the higher of the temperature at which it is offered for transportation or at which it is transported, not at the 37.8°C (100°F) temperature specified in American Society for Testing and Materials (ASTM) specification D4359-84.

liquid spiking
Introducing a solvent-containing analyte of interest directly onto a sorbent media. Subsequent desorption and analysis of the liquid spike should have a recovery of greater than or equal to 75%.

liquidated damages and penalties
The term is applicable when the amount of the damages has been ascertained by the judgment in the action, or when a specific sum of

money has been expressly stipulated by the parties to a bond or other contract as the amount of damages to be recovered by either party for a breach of the agreement by the other. The purpose of a penalty is to secure performance, while the purpose of stipulating damages is to fix the amount to be paid in lieu of performance. Liquidated damages is the sum which a party to a contract agrees to pay if he/she breaks some promise and, which having been arrived at by a good faith effort to estimate actual damage that will probably ensue from a breach, is recoverable as agreed damages if a breach does in fact occur.

LIRS
See ***low impact resistant supports***.

list
Shorthand term for the EPA list of violating facilities, or lists of firms debarred from obtaining government contracts because they violated certain sections of the Clean Air Act or Clean Water Act. The list is maintained by the Office of Enforcement and Compliance Monitoring.

listed
With regard to equipment, it is considered "listed" if it is of a kind mentioned in a list which a) is published by a nationally recognized laboratory which makes periodic inspections of the production of such equipment, and b) states such equipment meets nationally recognized standards or has been tested and found safe for use in a specified manner.

listed waste
Wastes listed as hazardous under RCRA but which have not been subjected to the Toxic Characteristics Listing Process because the dangers they present are considered self-evident.

lite locomotive
A locomotive or a consist of locomotives not attached to any piece of equipment or attached only to a caboose.

liter
In the metric system, a unit of measurements equivalent to 1.0567 quarts.

lithium
A chemical element, atomic number 3, atomic weight 6.939, symbol Li.

lithosphere
The solid part of the Earth below the surface, including any groundwater contained within it.

litigant
A party to a lawsuit (i.e., plaintiff or defendant); one engaged in litigation; usually spoken of active parties, not of nominal ones.

litigate
To dispute or contend in form of law; to settle a dispute or seek relief in a court of law; to carry on a lawsuit.

litigation
A lawsuit. Legal action, including all proceedings therein.

litigious
That which is the subject of a lawsuit or action; that which is contested in a court of law.

litmus
A blue stain prepared by enzymatic fermentation of coarsely powdered lichens.

litmus paper
Absorbent paper impregnated with a solution of litmus, dried and cut into strips. It is used to indicate the acidity or alkalinity of solutions. If dipped into alkaline solution, it remains blue; acid solution turns it red.

Little Ice Age
The period from about 1550 to 1850 when average global temperatures were about 1.5°C cooler, and alpine glaciers increased in size and advanced down mountain canyons.

little league elbow
An overuse injury caused by stress on the muscles, tendons, epiphyses, and articular surface of the elbow joint.

little league shoulder
A condition of tendonitis or metaphysical fracture causing pain from excessive internal and rotational stresses around the shoulder.

live room
A room characterized by a small amount of sound absorption. *See also* ***reverberation***.

liver
A large gland of red color located in the upper right portion of the abdomen. It has many functions concerned with the process of digestion and with the development of the erythrocytes. It produces bile, helps detoxify harmful substances in the blood, and stores food. It is the largest internal organ. The

liver can store up to 20 percent of its weight in glycogen and up to 40 percent of its weight in fats. The basic fuel of the body is a simple form of sugar called glucose. This comes to the liver as one of the products of digestion, and is converted into glycogen for storage. It is reconverted to glucose, when necessary, to keep up a steady level of sugar in the blood. This is normally a slow, continuous process, but in emergency conditions, the liver, responding to epinephrine in the blood, releases large quantities of this fuel into the blood for use by the muscles. As the chief supplier of glucose in the body, the liver is sometimes called upon to convert other substances into sugar. The liver cells can make glucose out of protein and fat. This may also work in reverse: the liver cells can convert excess sugar into fat and send it for storage to other parts of the body. In addition to these functions, the liver builds many essential proteins and stores up certain necessary vitamins until they are needed by other organs in the body. The liver disposes of worn-out blood cells by breaking them down into their different elements, storing some, and sending others to the kidneys for disposal in the urine. It filters and destroys bacteria and also neutralizes poisons. The liver also helps to maintain the balance of sex hormones in the body. A certain amount of female hormone is normally produced in males, and male hormone in females. When the level of this opposite sex hormone rises above a certain point, the liver takes up the excess and disposes of it. Finally, the liver polices the proteins that have been passed through the digestive system. Some of the amino acids derived from protein metabolism cannot be used by the body: the liver rejects and neutralizes these acids and sends them to the kidneys for disposal.

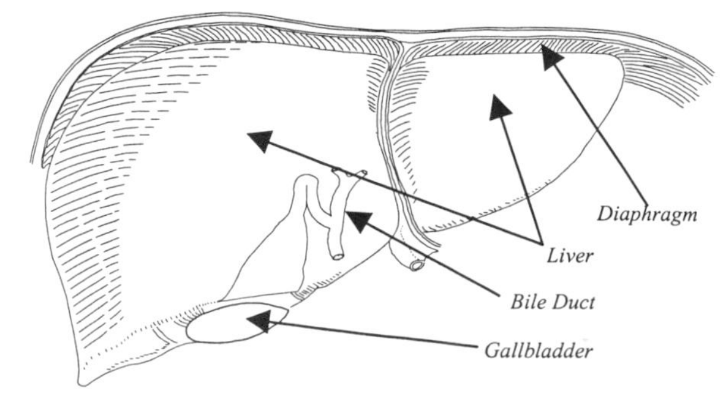

The liver, gallbladder, bile duct (which leads to the small intestine), and their relative position under the diaphragm

lives lost

U.S. Coast Guard. Those persons who perished as a direct result of the distress incident to which the Coast Guard was responding. *Lives lost before* refers to those persons who were considered lost prior to Coast Guard notification. *Lives lost after* refers to those persons who were alive at the time of Coast Guard notification, but who subsequently died.

lives saved

U.S. Coast Guard. Those persons who would have been lost without Coast Guard assistance.

livestock body

Truck or trailer designed for the transportation of farm animals.

LLF

*See **light loss factor**.*

LLNL

Lawrence Livermore National Laboratory.

ln

Natural logarithm.

LNG

*See **liquefied natural gas**.*

load

The performance demands required from a system or individual at any given time.

load cell

A strain gauge-based device for measuring the amount of force applied to an object.

load-endurance curve

A graphical curve illustrating the relationship between the percentage of maximum load and the length of time which that load will be voluntarily held.

load factor

(1) *General.* A factor that accounts for unavoidable deviations of the actual load from the nominal value. Examples of load factors include wind, shock, seismic, and dynamic load factors. (2) *Aviation.* The ratio of a specified load to the total weight of the aircraft. The specified load is expressed in terms of any of the following: aerodynamic forces, inertia forces, or ground or water reactions. (3) *Transportation.* The percentage of seating or freight capacity which is utilized. Also, a term relating the potential capacity of a sys-

tem relative to its actual performance. It is often calculated as total passenger miles divided by total vehicle miles. (4) *Performance.* That proportion of the work cycle time required for a worker to perform the necessary work at standard performance during a machine-paced cycle.

load limit
The maximum weight or stress which an individual, floor, vehicle, or other structure can safely support. Also referred to as *allowable load.*

load ratio
Transportation (Trucking). The ratio of loaded miles to empty miles per tractor.

load stress
That type of sensory overload caused by having too many channels of information to process effectively.

load weight
The maximum weight which a vehicle can safely carry.

loaded car mile
Transportation. A loaded car mile is a mile run by a freight car with a load. In the case of intermodal movements, the car miles generated will be loaded or empty depending on whether the trailers/containers are moved with or without a waybill, respectively.

loading island
Transit. (1) A pedestrian refuge within the right-of-way and traffic lanes of a highway or street. It is provided at designated transit stops for the protection of passengers from traffic while they wait for and board or alight from transit vehicles. (2) A protected spot for the loading and unloading of passengers.

loading secondary task
A secondary task which must be constantly attended to.

loading spectrum
A representation of the cumulative loading anticipated for the structure under all expected operating environments; significant transportation and handling loads are included.

loading tramway
Maritime. A pair of rails running down the riverbank upon which a cart rides for the purpose of loading buoys and other equipment aboard a tender.

LOAEL
See ***lowest observable adverse effect level***.

loaned employee
Loaned servant is an employee who is loaned or hired out to another employer for some specific service or particular transaction and who is under exclusive control of that employer who may then be held vicariously liable for acts of the employee under ordinary principles of respondeat superior.

loaned servant doctrine
When one employer lends an employee to another for a particular employment, the employee, for anything done in that employment, must be dealt with as an employee of the one to whom he/she has been lent.

lobe
A more or less well-defined portion of an organ or gland.

lobectomy
Excision of a lobe, as of the lung, brain, or liver.

lobotomy
The cutting of nerve fibers connecting a lobe of the brain with the thalamus. In most cases the effected parts are the prefrontal or frontal lobes, the areas of the brain involved with emotion; thus, the operation is referred to as prefrontal or frontal lobotomy. A lobotomy is a form of psychosurgery, a field in which the purpose of an operation is not to remove a growth or repair an injury to the body but to change the patient's mental and emotional state. In modern medical practice, physicians usually regard lobotomy as a last resort. The operation is rarely performed anymore (except in extremely violent cases when all other treatments and modalities fail. Certain drugs have been developed to treat mental illnesses which have all but eliminated the potential for lobotomy in most cases.

local aircraft operations
Performed by aircraft that a) operate in the local traffic pattern or within sight of the airport; b) are known to be departing for, or arriving from flight in local practice areas located within a 20-mile radius of the airport;

c) execute simulated instrument approaches or low passes at the airport.

local airport advisory (LAA)
A service provided by flight service stations or the military at airports not serviced by an operating control tower. This service consists of providing information to arriving and departing aircraft concerning wind direction and speed, favored runway, altimeter setting, pertinent known traffic, pertinent known field conditions, airport taxi routes and traffic patterns, and authorized instrument approach procedures. This information is advisory in nature and does not constitute an Air Traffic Control (ATC) clearance.

local and suburban and interurban passenger transportation transit
Includes establishments that provide local and suburban passenger transportation, such as those providing passenger transportation within a single municipality, contiguous municipalities, or a municipality and its suburban areas by bus, rail, car subway, either separately or in combination. Also included are sightseeing, charter, intercity passenger operations, and establishments providing passenger terminal and maintenance facilities.

local and suburban transit
Establishments primarily engaged in furnishing local and suburban mass passenger transportation over regular routes and on regular schedules, with operations confined principally to a municipality, contiguous municipalities, or a municipality and its suburban areas. Also included in this industry are establishments primarily engaged in furnishing passenger transportation by automobile, bus, or rail to, from, or between airports or rail terminals over regular routes and those providing bus and rail commuter services.

local application system
A fixed fire suppression system which has a supply of extinguishing agent, with nozzles arranged to automatically discharge the extinguishing agent directly on the burning material to extinguish or control a fire.

Local Area Network (LAN)
A communication link over which computers and peripherals may be connected within a limited geographical region.

local bus charter service
Establishments primarily engaged in furnishing local bus charter service where such operations are principally within a single municipality, contiguous municipalities, or a municipality and its suburban areas.

local courier service
Establishments primarily engaged in the delivery of individually addressed letters, parcels, or packages (generally under 100 pounds), except by means of air transportation or by the United States Postal Service. Delivery is usually made by street or highway within a local area or between cities.

local courts
Courts whose jurisdiction is limited to a particular territory or district. The term usually refers to the courts of the state, as opposed to the United States courts, or to municipal or county courts in contrast to courts with statewide jurisdiction.

local definition
An elaboration on a more generic definition by providing additional details to suit the purpose of a specialized condition or location.

local effect
An effect which occurs to a localized part of the body, such as irritation of the respiratory tract or eyes.

Local Emergency Planning Committee (LEPC)
A committee appointed by the state emergency response commission, as required by SARA, Title III to formulate a comprehensive emergency plan for its jurisdiction.

local exhaust system
A system composed of an exhaust opening, such as a hood, ductwork to transport exhausted air to a source of suction (fan, eductor, etc.), and frequently, an air cleaner to remove contaminants from the exhaust air before discharge to the environment. The air cleaner is typically positioned before the fan in the system to prevent fan wear.

local exhaust ventilation system
An air-handling system designed to capture and remove process emissions before they can escape into the workplace or the environment, generally consisting of a hood, conveying ductwork, an air-handling device, a fan, and an exhaust stack.

local freight
Maritime. Freight movements within the confines of a port, whether the port has only one or several arms or channels (except car ferry and general ferry). The term is also applied to marine products, sand, and gravel taken directly from the Great Lakes.

local government
City, county, or other governing body at a level smaller than a state. Local government has the greatest control over real property, zoning, and other local matters.

local horizontal
Pertaining to a region within a larger coordinate system in which a secondary, smaller coordinate system defines a horizontal axis or plane.

local lighting
That lighting intended to provide illumination only for a small region.

local magnitude (M_L)
A measure of the strain energy released by an earthquake within 100 kilometers of its epicenter. Strictly defined by Charles Richter as the base-10 logarithm of the amplitude, in microns, of the largest trace deflection that would be observed on a standard torsion seismograph at a distance of 100 km from the epicenter. *See also* ***surface-wave magnitude, moment magnitude,*** *and* ***Richter scale****.*

local minimum
The smallest value within a restricted range of values.

local operations
Aviation. Performed by aircraft which: a) operate in the local traffic pattern or within sight of the airport; b) are known to be departing for, or arriving from, flight in local practice areas within a 20 mile radius of the airport; c) execute simulated instrument approaches or low passes at the airport. Itinerant operations are all airport operations other than local operations.

local passenger (not elsewhere classified) transportation
Establishments primarily engaged in furnishing miscellaneous passenger transportation, where such operations are principally within a single municipality, contiguous municipalities, or a municipality and its suburban areas.

local roads
Those roads and streets whose principal function is to provide direct access to abutting land.

local streets and roads
Streets whose primary purpose is feeding higher order systems, providing direct access with little or no through traffic.

local toxic effect
An effect that is observed at the site of contact. For example, a skin burn from a corrosive substance.

local traffic
Aviation. (1) Aircraft operating in the traffic pattern or within sight of the tower. (2) Aircraft known to be departing or arriving from flight in local practice areas. (3) Aircraft executing practice instrument approaches at the airport.

local trip
An intracity or short mileage trip by a commercial motor vehicle.

local trucking (with storage)
Establishments primarily engaged in furnishing both trucking and storage services, including household goods, within a single municipality, contiguous municipalities, or a municipality and its suburban areas.

local trucking (without storage)
Establishments primarily engaged in furnishing trucking or transfer services without storage for freight generally weighing more than 100 pounds, in a single municipality, contiguous municipalities, or a municipality and its suburban areas.

local vertical
Pertaining to a region within a larger coordinate system in which a secondary, smaller coordinate system defines a vertical axis.

local winds
Winds that tend to blow over a relatively small area. Often due to regional effects, such as mountain barriers, large bodies of water, local pressure differences, and other influences.

localize
Determine the source of a stimulus or signal in space and/or time.

localizer
Aviation. The component of an instrument landing system (ILS) which provides the aircraft with course guidance to the runway. *See also* ***glideslope, instrument landing system, middle marker,*** *and* ***outer marker.***

localizer course
Aviation. The locus of points, in any given horizontal plane, at which the difference in depth of modulation (DDM) is zero.

localizer type directional aid
Aviation. A Navigational Aid (NAVAID) used for nonprecision instrument approaches with utility and accuracy comparable to a localizer but which is not a part of a complete instrument landing system and is not aligned with the runway.

localizer usable distance
Aviation. The maximum distance from the localizer transmitter at a specified altitude, as verified by flight inspection, at which reliable course information is continuously received.

location
(1) *Damp Location.* Partially protected locations under canopies, marquees, roofed open porches, and like locations, and interior locations subject to moderate degrees of moisture, such as some basements, some barns, and some cold-storage warehouses. (2) *Dry Location.* A location not normally subject to dampness or wetness. A location classified as dry may be temporarily subject to dampness or wetness, as in the case of a building under construction. (3) *Wet Location.* Installations underground or in concrete slabs or masonry in direct contact with the earth, and locations subject to saturation with water or other liquids, such as vehicle-washing areas, and locations exposed to weather or otherwise unprotected.

location coding
The identification of controls, devices, or systems through their placement on some panel or other structure.

location identifier (LOCID)
Aviation. A unique code which is assigned by the Federal Aviation Administration (FAA) to identify each airport.

locator
Aviation. A low/medium frequency (LM/MF) nondirectional beacon (NDB) used as an aid to final approach. Note: A locator usually has an average radius of rated coverage of between 18.6 and 46.3 miles (10 and 26 NM).

LOCID
See ***location identifier.***

lock
(1) *General.* A security device used to prevent access or secure property from theft and/or unauthorized use. (2) *Maritime.* An enclosure in a water body with gates at each end to raise or lower water vessels as they pass from one level to another.

lock cell
Maritime. The chamber of a lock.

lock gate
A movable, structural barrier to hold back the water in a lock chamber.

lock line
A long line leading from the bow and the stern of the tow to the lock wall.

lock rod
Rail Operations. A rod, attached to the front rod or lug of a switch, movable-point frog or derail, through which a locking plunger may extend when the switch points or derail are in the normal or reverse position.

lock traffic lights
Maritime. Red, yellow, and green lights displayed at the entrances of the lock, both up bound and down bound, for the purpose of controlling traffic.

locking bar
Rail Operations. A bar in an interlocking machine to which the locking dogs are attached.

locking bed
Rail Operations. That part of an interlocking machine that contains or holds the tappets, locking bars, cross-locking, dogs, and other apparatus used to interlock the levers.

locking dog
Rail Operations. A steel block attached to a locking bar or tappet of an interlocking machine, by means of which locking between levers is accomplished. *See also* ***dog chart.***

locking face
Rail Operations. The locking surface of a locking dog, tappet, or cross-locking of an interlocking machine.

locking sheet
Rail Operations. A description in tabular form of the locking operations in an interlocking machine.

locking time
Maritime. The total time required for a tow to pass through a locking procedure. This includes approach time, chamber time, and time to clear the lock.

lockjaw
*See **tetanus**.*

lockout device
A device that uses a lock and key to hold an energy-isolating device in the safe position for the purpose of protecting personnel.

lockout/tagout
A formal procedure for isolating equipment, machinery, or a process to prevent unintentional operation during maintenance, servicing, or for other reasons. The energized equipment, machinery, etc. is first put into an energy-isolated state and each individual who will work on the device places his/her lock and/or tag on the electrical switch or other startup means to keep it in a zero-energy state until the work is completed by each individual who has affixed a lock and/or tag to it. The policy and procedure related to this practice are to clearly and specifically outline the purpose, responsibility, scope, authorization, rules, definitions, and measures to enforce compliance.

locomotion
The active movement of the body from one place to another.

locomotive
(1) A self-propelled unit of equipment designed for moving other railroad rolling equipment in revenue service including a self-propelled unit designed to carry freight or passenger traffic, or both, and may consist of one or more units operated from a single control. (2) A self-propelled unit of equipment designed primarily for moving other equipment. It does not include self-propelled passenger cars. (3) A piece of on-track equipment other than hi-rail, specialized maintenance, or other similar equipment: a) with one or more propelling motors designed for moving other equipment; b) with one or more propelling motors designed to carry freight or passenger traffic or both; or c) without propelling motors but with one or more control stands. (4) A self-propelled unit of equipment which can be used in train service.

locomotive cab
That portion of the superstructure designed to be occupied by the crew while operating the locomotive.

locomotive mile
The movement of a locomotive under its own power the distance of one mile.

locomotive unit mile
The movement of a locomotive unit one mile under its own power. Miles of locomotives in helper service are computed on the basis of actual distance run in such service. Locomotive unit miles in road service are based on the actual distance run between terminals and/or stations. Train switching locomotive unit miles are computed at the rate of six miles per hour for the time actually engaged in such service.

locomotor system
The various bodily systems, structures, and tissues used in locomotion.

LOD
*See **limit of detection**.*

loft
The trapped air in clothing.

log
Abbreviation for logarithm.

$\log_{10}$
Logarithm to the base 10.

log body
Truck or trailer designed for the transportation of logs or other loads which may be boomed or chained in place.

logarithm
A function represented by the real-valued exponent of some base number.

logarithmic interval scale
An alternative to the basic measurement scales in which the magnitudes corresponding to points are given by:

$$\log x_n - \log x_{n+1} - \log x_{n+2}, \text{ etc.}$$

logic gate
As pertains to the system safety applications of fault tree analysis (FTA) and/or the man-

agement oversight and risk tree (MORT), a symbol used to identify the association between events on a logic tree.

log-normal distribution
The distribution of the logarithms of a random variable that has the property that the logarithms are normally distributed.

log-normally distributed variable
A variable is considered to be log-normally distributed if the logarithms of the variable are normally distributed.

long bone
Any bone whose length greatly exceeds its width.

long range navigation (LRNAV)
Aviation. A method of navigation that permits navigation over long distances. This is in contrast to the relatively short range navigation provided by the Very high frequency Omni-directional Range (VOR) radio system.

long term
Pertaining to events or conditions which develop or are maintained for an extended period of time, typically on the order of years.

long-term exposure
Continuous or repeated exposure of an individual to a substance or agent over a period of several years or working lifetime.

long-term memory
A coded form of memory which apparently exists indefinitely.

long ton
A unit of mass in the English system equal to 2,240 pounds.

long-waisted
(slang) Having a longer than normal trunk for the total stature.

long wavelength infrared
See ***far infrared***.

long waves in the westerlies
A wave in the major belt of westerlies characterized by a long length (thousands of kilometers) and significant amplitude. Also called *Rossby waves*.

longer combination vehicles
Any combination of truck tractor and two or more trailers or semitrailers which operates on the Interstate System at a gross vehicle weight greater than 80,000 pounds.

longitudinal
Transit. Parallel to the longitudinal centerline of the vehicle.

longitudinal axis
An approximate centerline of a body segment which is parallel to the length dimension of that segment.

longitudinal design
Any research methodology in which data are collected from the same individual(s) over a long period of time.

longitudinal separation
Aviation. The longitudinal spacing of aircraft at the same altitude by a minimum distance expressed in units of time or miles.

longitudinal study
An experiment or observation using a longitudinal design.

longitudinal wave
A waveform in which the direction of propagation and displacement is the same.

Longshore and Harbor Workers' Compensation Act
Federal Act (33 U.S.C.A § 901 et seq.) designed to provide workers' compensation benefits to employees, other than seamen, or private employers any of whose employees work in maritime employment upon the navigable waters of the United States (including any adjoining pier, wharf, dry dock, terminal, building way, marine railway, or other adjoining area customarily used by an employer in loading, unloading, repairing, or building a vessel). The primary occupations subject to the Act are stevedoring and ship service operations. The Act is administered by the Office of Workers' Compensation Programs.

longshoreman
A maritime laborer, such as a stevedore or loader, who works about wharves of a port. A person who loads and unloads ships.

longshoring operations
The loading, unloading, moving, or handling of cargo, ship's stores, gear, etc. into, in, on, or out of any vessel on the navigable waters of the United States.

longwave radiation
A term most often used to describe the infrared energy emitted by the earth and the atmosphere.

LOQ
See ***limit of quantitation***.

Loran
An electronic navigational system by which hyperbolic lines of position are determined by measuring the difference in the time of reception of synchronized pulse signals from two fixed transmitters. Loran A operates in the 1750-1950 kHz frequency band. Loran C and D operate in the 100-110 kHz frequency band.

lordosis
A curving of the cervical-lumbar regions of the spine in the sagittal plane to yield an anterior convexity.

LOS
See ***line of sight***.

loss
(1) *General.* A generic and relative term that signifies the act of losing or the thing lost; it is not a word of limited, hard, and fast meaning and has been held synonymous with, or equivalent to, *damage, damages, deprivation, detriment, injury,* and *privation*. (2) *Finance.* Expenses exceeding costs, or, actual losses. Bad and uncollectable accounts, damage; a decrease in value of resources or an increase in liabilities; depletion or depreciation or destruction of value. (3) *Insurance.* A state of fact of being lost or destroyed; ruin, or destruction. (4) *System Safety.* Anything that increases costs or reduces productivity and has any adverse effect on the organization or society resulting from either normal operations or unplanned events. (5) *Ventilation.* Usually refers to the conversion of static pressure to heat, noise, or vibration in components of the ventilation system (e.g., the hood entry loss). (6) *Law.* Loss is a generic and relative term. It signifies the act of losing or the thing lost. It is not a word of limited, hard and fast meaning and has been held synonymous with, or equivalent to damage, deprivation, detriment, injury, and privation.

loss control
The overall objective of accident investigation. A management responsibility to prevent or control the occurrence of those events which downgrade performance, negatively impact productivity, or otherwise result in a loss of some nature and degree.

loss ratio
Term used in the insurance industry. A ratio calculated by dividing the amount of loss(es) by the amount of premium(s). Normally expressed as a percentage of the premiums.

lost communications
Aviation. Loss of the ability to communicate by radio. Aircraft are sometimes referred to as NORDO (No Radio). Standard pilot procedures are specified in Federal Aviation Regulation (FAR) Part 91. Radar controllers issue procedures for pilots to follow in the event of lost communications during a radar approach when weather reports indicate that an aircraft will likely encounter instrument flight rules (IFR) weather conditions during the approach.

lost time
(1) That time which an individual would normally be at his/her workplace but is not due to an occupational illness or injury. (2) *See* ***delay time***.

lost time accident
An accident which results in a significant period of time away from the job.

lost time illness
An occupational illness which results in more than one day off from work, usually referring to something more serious than a minor illness.

lost-time injury
A work injury resulting in death or disability and in which the injured person is not able to work the next regularly scheduled shift.

lost workdays
(1) *General.* Under OSHA 29 CFR 1904.12(f), the number of days (consecutive or not) after, but not including, the day of injury or illness during which the employee would have worked but could not do so; that is, could not perform all or any part of his/her normal assignment during all or any part of the workday or shift, because of the occupational injury or illness. (2) *Lost Workday - Away from Work.* A day on which the employee would have worked but could not because of occupational injury or illness. This does NOT include the day of the injury or on-

set of illness or any days on which the employee would not have worked anyway (such as a weekend or holiday). (3) *Lost Workday - Restricted Work Activity.* A day on which, because of injury or illness, the employee was assigned to another job on a temporary basis; or, worked at a permanent job less than full time; or worked at a permanently assigned job but could not perform all duties normally connected with that job. This does not include the day of the injury or onset of illness or any days on which the employee would not have worked anyway (such as a weekend or holiday). (4) *Rail Operations.* Any full day or part of a day (consecutive or not) other than the day of injury, that a railroad employee is away from work because of injury or occupational illness.

loudness
An observer's impression of a sound's amplitude; a purely subjective assessment. The intensive attribute of an auditory sensation, in terms of which sounds may be arranged on a scale extending from soft to loud. It depends primarily on the sound pressure of the stimulus, as well as on its frequency and wave form.

loudness contour
A curve of sound pressure level values plotted against frequency which are required to produce a given loudness sensation for a normal listener.

loudness level
The loudness level of a sound, in phons, is numerically equal to the median sound pressure level, in decibels, relative to 2 E-4 microbar, of a free progressive wave of 1000 hertz presented to listeners facing the source, which in a number of trials, is judged by the listeners to be equally loud.

louse
(plural is *lice*) A general name for various parasitic insects. The true lice, which infest mammals, belong to the suborder *Anoplura*. They are grayish, wingless insects that vary in length from on-sixth to one-sixteenth of an inch.

louver
Panels used in hoods for distributing airflow at the hood face.

low
*See **extratropical cyclone**.*

low altitude airway structure (LAAS)
The network of airways serving aircraft operations up to but not including 18,000 feet mean sea level (MSL).

low altitude alert system (LAAS)
An automated function of the TPX42 that alerts the controller when a Mode C transponder-equipped aircraft on an instrument flight rules (IFR) flight plan is below a predetermined minimum safe altitude. If requested by the pilot, low altitude alert system monitoring is also available to visual flight rules (VFR) Mode C transponder-equipped aircraft.

low approach
An approach over an airport or runway following an instrument approach or a visual flight rules (VFR) approach including the go-around maneuver where the pilot intentionally does not make contact with the runway.

low boy
A low trailer for hauling heavy machinery.

low density lipoprotein (LDL)
A substance present in the blood which carries high levels of cholesterol, occasionally depositing it on arterial walls as plaque. *See also **high density lipoprotein** and **lipoprotein**.*

low density wood
That wood which is exceptionally light in weight and usually deficient in strength properties for the species.

low emission vehicle
A clean fuel vehicle meeting the low-emission vehicle standards.

low frequency
The frequency band between 30 and 300 kHz.

low-hazard contents
Those contents of such low combustibility that no self-propagating fire therein can occur and that consequently the only probable danger requiring the use of emergency exits will be from panic, fumes, or smoke, or fire from some external source.

low-hazard permit space
A confined space in which there is an extremely low likelihood that an immediately dangerous to life and health (IDLH) or en-

gulfment hazard could be present and in which all other potentially serious hazards have been controlled.

low head
Vertical difference of 100 feet or less in the upstream surface water elevation (headwater) and the downstream surface water elevation (tailwater) at a dam.

low impact resistant supports (LIRS)
Aviation. Supports designed to resist operational and environmental static loads and fail when subjected to a shock load such as that from a colliding aircraft.

low-level jet stream
Jet streams that typically form near the earth's surface below an altitude of about 2 km and usually attain speeds of less than 60 knots.

low-level radioactive waste
Under the Federal Nuclear Waste Policy Act of 1982: Radioactive material that is not high-level radioactive waste, spent nuclear fuel, transuranic waste, or byproduct material.

low-noise emission product determination
An EPA determination of whether or not a product, for which a properly filed application has been received, meets the low-noise emission product criterion.

low-order detonation
See ***detonation***.

low type road surface
Bituminous surface-treated Surface/Pavement Type Code 51.

low water dam
(1) A low-level dam designed to hold back a head of water so as to maintain project depth in a certain area. The dam may be visible at the low water stage. (2) A dam that is more effective at low water; at high water the dam becomes a weir. *See also* ***weir***.

low water datum
A term used by the Army Corps of Engineers to define their originating point of elevation in determining stages of water when erecting various gauges along a river.

lower detectable limit (LDL)
Instruments. The smallest concentration of the substance of interest that produces an output change in a reading of at least twice the noise level.

lower explosive limit (LEL)
The concentration of a compound in air below which a flame will not propagate if the mixture is ignited. Also referred to as *lower flammable limit*.

lower flammable limit (LFL)
Often referred to as the Lower Explosive Limit (LEL), it is the lowest concentration of gas or vapor in the air that will propagate a flame if a spark or heat source is present. *See also* ***lower explosive limit***.

lower gauge
A gauge located in the tailwater of a dam (downstream side); colloquially called *tailgate*.

lower-half of saddle-mount
Transit. That part of the device which is securely attached to the towing vehicle and maintains a fixed position relative thereto but does not include the "king-pin."

lowest achievable emission rate
According to the Clean Air Act, this is the rate of emissions which reflects either the most stringent emission limitation which is contained in the implementation plan of any state for such source (unless the owner or operator of the proposed source demonstrates such limitations are not achievable), or the most stringent emissions limitation achieved in practice, whichever is more stringent. Application of this term does not permit a proposed new or modified source to emit pollutants in excess of existing new source standards.

lowest observed adverse effect level (LOAEL)
In dose-response experiments, the experimental exposure level representing the lowest tested at which adverse effects were demonstrated.

lowpast filter
A device which allows frequencies lower than the cutoff frequency to exit from the device unattenuated, while the intensity of frequencies higher than the cutoff frequency is attenuated.

LPG
See ***liquid petroleum gas***.

LPM
Liter(s) per minute.

LRNAV
See ***long range navigation.***

LRS
See ***linear referencing system.***

LSD
See ***lysergic acid diethylamide.***

LSO
Laser safety officer.

LTL
See ***less than truckload.***

lubricating oil
Under the Federal Solid Waste Disposal Act: The fraction of crude oil which is sold for purposes of reducing friction in any industrial or mechanical device. Such term includes re-refined oil.

lumbago
A low level of pain in the lumbar region of the back.

lumbar
Refers to the five vertebrae of the lower back between the thorax and the pelvis.

lumbar disk
An intervertebral disk separating the lumbar vertebrae in the spine.

lumbar vertebra
Any of the vertebral bones in the lumbar spine, L1–L5.

lumber body
Platform truck or trailer body with traverse rollers designed for the transportation of sawed lumber.

lumbosacral angle
The angle between the posterior of the lumbar spine and the sacrum.

lumen
(1) Used to represent total light output; the unit of luminous flux emitted through a unit solid angle from a uniform point source of one candela. (2) A hole or passage in a tube-like structure within the body.

lumen depreciation
That decrease in luminous flux emitted by certain types of light sources over time.

lumen-hour (lm-hr)
A unit for that amount of light delivered by a luminous flux in one hour.

lumen-second (lm-sec)
See ***talbot.***

luminaire
A complete light fixture including the lamp, parts to distribute the light, position the fixture, and connect the lamp to the power supply.

luminaire dirt depreciation
The loss of luminous flux from lighting due to dirt collection on the luminaire or particulates in the atmosphere. A recoverable lighting loss factor.

luminaire surface depreciation
Any reduction in luminous output due to physical or chemical changes in materials associated with a luminaire, such as transmittance through or reflections from enclosing materials. A non-recoverable light loss factor.

luminance
A physical measure of the luminous flux per unit solid angle incident on a surface. An older term. Synonymous with *photometric brightness*. *See also* ***brightness.***

luminance contrast
A measure of the physical relationship in luminance between two adjacent, non-specular surfaces under the same general illumination and immediate surroundings, generally defined by an equation similar to the form below. *See also* ***brightness contrast.***

$$C_L = \frac{\Delta L}{L}$$

luminance ratio
The value of the ratio between the luminance of any two surfaces or objects in the visual field.

luminescence
The emitting of light due to some mechanism other than high temperatures.

luminosity
A measure of the relative efficiency of various wavelengths of visible light for exciting the retina.

luminosity function
See ***spectral luminous efficiency function.***

luminous efficacy, flux
The value of the ratio of the total luminous flux to the total radiant flux encompassing all wavelengths.

luminous efficacy, source
The value of the ratio of the total luminous flux emitted by a lamp to the total electrical power input.

luminous efficiency function
See ***spectral luminous efficiency function.***

luminous environment
That portion of the visual environment generated by the luminaire type, luminous intensity, direction, and hues.

luminous flux (Φ)
The rate of visible light energy emitted from a source over time.

luminous intensity
A measure of the power of a light source in terms of luminous flux per unit solid angle.

lunate bone
One of the proximal bones of the wrist.

lung
One of the asymmetrical bilateral organs within the chest which is involved in gaseous respiration. The lungs supply the blood with oxygen inhaled from the outside air, and they dispose of waste carbon dioxide in the exhaled air, as part of the respiration process. The lungs are made of elastic tissue filled with interlacing networks of tubes and sacs carrying air, and with blood vessels carrying blood. The bronchi, which bring air to the lungs, branch out within the lungs into many smaller tubes, the bronchioles, which culminate in clusters of tiny air sacs called alveoli, whose total runs into the millions. The aveoli are surrounded by a network of capillaries. Through the thin membranes of the capillaries, the air and blood make their exchange of oxygen and carbon dioxide. The lungs are divided into lobes, the left lung having two lobes and the right lung having three. The lungs are inflated and deflated via the action of the diaphragm and the intercostal muscles.

lung diffusing capacity
A measure of the amount of gas at standard temperature and pressure (STP) which diffuses across the pulmonary membrane in the alveolus.

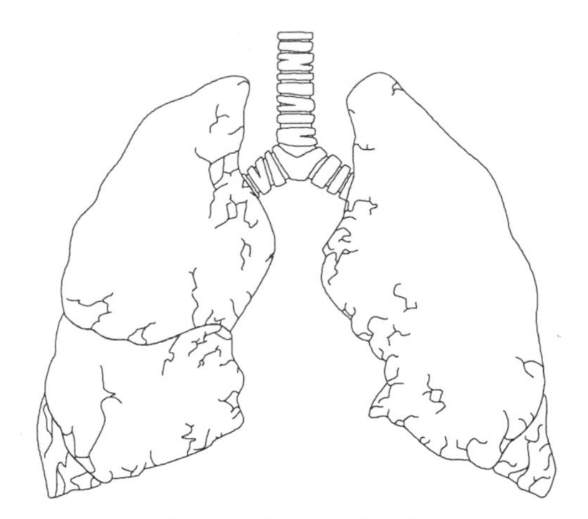

The human lungs and trachea

lung expiratory reserve volume
See ***expiratory reserve volume.***

lung function test
A test, usually employing a spirometer, that measures an individual's breathing capacity and, indirectly, their ability to wear a respirator.

lung functional residual capacity
See ***functional residual capacity.***

lung inspiratory capacity
See ***inspiratory capacity.***

lung vital capacity
See ***vital capacity.***

lung volume
The volume of measurable gas in the lungs under specified conditions.

lunula
The lighter-colored portion of the nail body near the nail root of the hand or foot.

lupus
Tuberculosis of the skin marked by the formation of brownish nodules on the corium, called *lupus vulgaris*.

lupus erythematosus
An inflammatory disease that takes two forms. One, the systemic or disseminated form, causes deterioration of the connective tissues in various parts of the body. This disease may attack the soft internal organs as well as the bones and muscles, and is often fatal. In its other form, the discoid type, it is a fairly mild skin disorder. Symptoms of the more serious form vary widely, but may include fever, abdominal pains, and pains in the muscles and joints. Often the symptoms

come and go over a long period of time. Diagnosis of the disease is difficult. The cause is unknown, but the disease is believed not to be infectious and possibly to be related to allergies. Lupus erythematosus is one of a group of similar disorders known as the collagen diseases. There is no specific treatment, though corticosteriods may be used to control symptoms.

LUST
Leaking underground storage tank.

lustermeter
A device developed by Hunter to measure contrast gloss and compute luster.

lux (lx)
Metric unit of illuminance equal to one lumen per square meter. One foot-candle is equal to 10.76 lux. Dividing the lux value by 10 provides the approximate equivalent foot-candle value.

LV%
Liquid volume percent.

lx
*See **lux**.*

lying
Pertaining to a posture in which an individual's torso is horizontal to a reference surface, but not prone, with possible flexion of the hips and knees.

Lyme disease
A bacterial disease transmitted to man by a tick bite. Symptoms of Lyme disease, including rash, headache, fever, tiredness, numbness, and others, mimic those of many other diseases and, therefore, Lyme disease may often be initially misdiagnosed. It was first identified in the community of Lyme, Connecticut and, hence, the name.

lymph
The water and various dissolved substances and particulates which enter the lymphatic system from the interstitial fluid. It is a colorless, odorless fluid, slightly alkaline, and has a salty taste. Lymph is approximately 95% water; the remainder consists of plasma proteins and other chemical substances contained in the blood plasma, but in a slightly smaller percentage than in plasma. In addition, the lymph contains a high concentration of lymphocytes. The body contains three main kinds of fluid: blood, tissue fluid, and lymph. The blood consists of the blood cells and platelets, the plasma, or fluid portion, and a variety of chemical substances dissolved in the plasma. When the plasma, without its solid particles and some of its dissolved substances, seeps through the capillary walls and circulates among the body tissues, it is known as tissue fluid. When this fluid is drained from the tissues and collected by the lymphatic system, it is called lymph. The lymphatic system eventually returns the lymph to the blood, where it again becomes plasma.

lymph gland
*See **lymph node**.*

lymph node
An ovoid-shaped structure occurring in lymph vessels which serves as a collection and filtration point for lymph in fighting infection. Lymph nodes filter and destroy invading bacteria and are the site of production of lymphocytes and certain antibodies. The main lymph nodes are in the neck, axillae, and groin. Sometimes called, incorrectly, *lymph gland*.

lymph vessel
Any of a range of diameters of tubular structures from capillary size to those resembling moderate-sized veins in the cardiovascular system which carry lymph.

lymphatic
Pertaining to lymph, lymph vessels, or lymph nodes.

lymphatic ducts
The two larger vessels into which all lymphatic vessels converge. The right lymphatic duct joins the venous system at the junction of the right internal jugular and subclavian veins and carries lymph from the upper right side of the body. The left lymphatic dust, or thoracic duct, enters the circulatory system at the junction of the left internal jugular and subclavian veins; it returns lymph from the upper left side of the body and from below the diaphragm.

lymphatic system
The fluid and the various structures involved in collecting interstitial fluid, removing foreign particles, and returning the fluid ultimately to the cardiovascular system.

lymphocyte
A white blood cell found in lymphatic tissues (e.g., lymph nodes, spleen, thymus) that is immunologically important and attacks invading pathogens.

lymphocytopenia
Reduction in the number of lymphocytes in the blood.

lymphogranuloma
See ***Hodgkin's disease***.

lymphoma
(1) A primary tumor of lymphoid tissue. (2) Any one of various conditions of unknown etiology chiefly affecting lymph nodes, considered to be neoplastic. (3) Various abnormally proliferative diseases of the lymphoid tissue of the lymphatic system. A tumor of lymphoid tissue.Also called *malignant lymphoma*.

lysergic acid diethylamide (LSD)
A hallucinogenic compound chemically related to ergot, having consciousness-expanding effects and capable of producing a state of mind in which there is false sense perception (hallucination). The perceptual changes brought about by LSD in normal persons are extremely variable and depend on factors such as age, personality, education, physical make-up, and state of health. The danger of the drug lies in the fact that it loosens control over impulsive behavior and may lead to a full-blown psychosis or less serious mental disorder in persons with latent mental illness. LSD was first developed in 1938 and was believed to be potentially useful in the treatment of mental illness. This theory was based on the belief that the drug could produce a schizophrenic syndrome and that psychiatrists and other persons concerned with mental illness could observe the manifestations of a psychosis under controlled conditions. However, competent investigators have shown that the effect of LSD is more closely related to a toxic psychosis such as that produced by fever, stress, or drugs of many kinds and is of doubtful use in understanding the mechanism of true psychosis resulting from severe mental disorder.

lysin
A substance that causes lysis; an antibody that causes dissolution of cells or other material.

lysine
A naturally occurring amino acid, one of those essential for human metabolism.

lysis
(1) Destruction or decomposition, as of a cell or other substance, under the influence of a specific agent. (2) Solution or separation, as of adhesions binding different anatomic structures. (3) Gradual abatement of the symptoms of a disease.

lysoenzyme
A crystalline, basic protein, which is present in saliva, tears, egg white, and many animal fluids and which functions as an antibacterial enzyme.

lysosome
A minute body occurring in a cell and containing various enzymes, mainly hydrolytic.

lysotype
(1) The type of microorganism as determined by its reactions to specific bacteriophages. (2) A taxonomic subdivision of bacteria based on their reactions to specific bacteriophages, or a formula expressing the reactions on which such a subdivision is based.

M

M
Molar (solution).

M-1 maneuver
A technique for air crew personnel to prevent gravity-induced loss of consciousness due to high positive acceleration maneuvers in aircraft, in which the crew member generally grunts with the glottis partially closed to increase intrathoracic pressure, thereby increasing blood pressure and blood flow to the brain. *See also* ***anti-g straining maneuver***.

m^3
Cubic meter (1000 liters (L) or 1.000,000 milliliters.

mA
Milliamp.

maceration
The softening of a solid by soaking; wasting away, softening and fraying, as if by action of soaking.

mach
A unit representing the velocity of sound, usually in air.

mach indicator
An aircraft or spacecraft display which provides the vehicle's velocity as a ratio to velocity of sound.

mach number
The ratio of true airspeed to the speed of sound.

mach technique
Aviation. Describes a control technique used by air traffic control whereby turbojet aircraft operating successively along suitable routes are cleared to maintain appropriate mach numbers for a relevant portion of the en route phase of flight. The principal objective is to achieve improved utilization of the airspace and to ensure that separation between successive aircraft does not decrease below the established minima.

machine
A mechanically or electromechanically powered device consisting of both fixed and moving parts and having one or more specific functions.

machine ancillary time
That time during which a machine is unavailable for use due to calibration, changeover, cleaning, or other related causes.

machine assignment
That function or operator to which a machine is assigned.

machine attention time
That time during which an operator must observe a machine's operations in the event intervention or servicing is required. Does not involve actually operating for production or servicing the machine.

machine available time
That time during which a machine is performing or could perform work.

machine capacity
Some measure of the normally expected output from a machine.

machine-controlled time
That time in a given work cycle which a machine requires to perform its portion of a task or process, independent of an operator.

machine-controlled time allowance
The expected or scheduled time given a worker for a machine to perform its portion of a given task.

machine cycle time
That time required for a machine to perform one complete cycle of a process.

machine downtime
That amount of time during which a machine is not able to perform its designated function due to a breakdown, routine servicing, or a materials shortage.

machine effective utilization index (MEUI)
The value of the ratio of the time which a machine is running under standard conditions compared to the time which the machine is available.

machine efficiency index (MEI)
The value of the ratio of the machine standard running time to the machine running time.

machine element
A work element performed entirely by a machine.

machine guard
Any piece of equipment or device on a machine intended to reduce or eliminate the chance of injury through the use of that machine.

machine hour
A unit of measure for the utilization of machines, corresponding to one machine working for one hour.

machine idle time
The amount of time a machine is available but not productive due to the operator performing other work, due to a shortage of materials, etc.

machine interference
A situation in which a demand for simultaneous operator attention by two or more machines results in machine idle time.

machine load
The proportion or percentage of scheduled or actual usage of machine available time during a given time interval.

machine maximum time
The total time in a day, week, or other time period during which one or more machines could work.

machine-paced work
That restricted or externally paced work in which machinery controls the rate at which the work cycle progresses.

machine running time
The actual operating, productive time by a machine.

machine standard running time
That time at which a machine operates at optimum capacity.

machine utilization index
The value of the ratio of the amount of time a machine is running compared to the time it is available.

Mackinaw boat
A crudely built flatboat used on the Ohio and Upper Mississippi Rivers during the 18th century.

macro-
(1) *Prefix.* Large, large-scale, or long-length. (2) *Computing.* A set of keystrokes or computer instructions which may be executed with a single command.

macro-command
That command which initiates a macro-execution.

macrobiota
The macroscopic living organisms of a region.

macroblast
An abnormally large, nucleated erythrocyte.

macrochemistry
Chemistry in which the reactions may be seen with the naked eye.

macroclimate
The general climate of a large area, such as a country.

macrocrania
Abnormal increase in the size of the skull in relation to the face.

macroelement
A work element which is of sufficiently long duration to permit observation and timing with a manually operated stopwatch or stopclock.

macrometeorology
Meteorological characteristics of a regional area, such as part of a province, region, state, or of a larger area.

macrophage
A typically large and long-lived cell originating in the bone marrow and distributed throughout the body that plays an important role in immunity by either engulfing invading microorganisms, through the production of antibodies, or by presenting antigens to lymphocytes for destruction.

macroscale
The normal meteorological synoptic scale for obtaining weather information. It can cover an area ranging from the size of a continent to the entire globe.

macroscopic
Visible to the eye without the aid of a microscope.

macroskelic
Having long legs relative to the torso length.

MACT
See maximum achievable control technology.

macula
(1) The area of the eye that is most responsive for color vision. Also, an opacity of the cornea. (2) A stain or spot, especially a discolored spot on the skin that is not elevated above the surface.

macula lutea
The yellow-colored central region of the fovea, at which visual acuity is greatest.

macular degeneration
A disease of the macula, the center of the retina, which is responsible for detailed vision. The most common form is characterized by the thinning of the yellowish macular pigment, scarring, and accumulation of oxidized fats called lipofuscin or drusen.

MADD
Mothers Against Drunk Driving.

maduromycosis
A chronic fungus affecting various body tissues, including the hands, legs, and feet. The most common form affects the foot (Madura foot) and is characterized by sinus formation, necrosis, and swelling. Also called *mycetoma*.

magazine
Any building or structure, other than an explosives manufacturing building, used for the storage of explosives.

magazine vessel
A vessel used for the receiving, storing, or dispensing of explosives.

maggot
The soft-bodied larva of an insect, especially one living in decaying flesh.

magma
(1) A suspension of finely divided material in a small amount of water. (2) A thin, paste-like substance composed of organic material.

magnesium
A chemical element, atomic number 12, atomic weight 24.312, symbol Mg.

magnet
An object having polarity and capable of attracting iron.

magnetic field
That vector field generated by a magnetic substance or which exists in conjunction with an electric field via current condition or electromagnetic radiation.

magnetic levitation (MAGLEV)
A rail transportation system with exclusive right-of-way which is propelled along a fixed guideway system by the attraction or repulsion of magnets on the rails and under the rail cars.

Magnetic Resonance Imaging (MRI)
The use of a combined static and radio frequency electromagnetic field to measure energy absorption by certain atoms, which can be processed and presented as an image cross-section of the body, a body segment, or any other object transparent to the electromagnetic field.

magnetic storm
A worldwide disturbance of the earth's magnetic field caused by solar disturbances.

magnetosphere
The region around the earth in which the earth's magnetic field plays a dominant part in controlling the physical processes that take place.

magnetotherapy
Treatment of disease by the use of magnetic currents.

magnitude
(1) *General.* The numerical absolute value of a vector. (2) *Seismology.* A general term for a measure of the strength or energy of an earthquake as determined from seismographic information. More specifically, it is a measure of the size of an earthquake. The Richter Scale, named after Charles F. Richter of the California Institute of Technology, is the best known scale for the measuring of magnitude (M) of earthquakes. The scale is logarithmic; a recording of 7, for example, signifies a disturbance with ground motion 10 times as large as a recording of 6. The energy released by an earthquake of M 7, however, is approximately 30 times that released by an earthquake of M 6; an earthquake of M 8 releases 900 times (30x30) the energy of an earthquake of M 6. An earthquake of magnitude 2 is the smallest earthquake normally felt by humans. Earthquakes with a Richter value of 5 or higher are potentially damaging. Some of the world's largest recorded earthquakes include one on January 31, 1906, off

the coast of Colombia and Ecuador, and on March 2, 1933, off the east coast of Honshu, Japan, had magnitudes of 8.9 on this scale, which is open ended. As the Richter scale does not adequately differentiate between the largest earthquakes, a new "moment magnitude" scale is being used by seismologists to provide a better measure. On the moment magnitude scale, the San Francisco earthquake is estimated at magnitude 7.7 compared to an estimated Richter magnitude of 8.3.

mail revenue
Revenues from the carriage of mail bearing postage for air transportation. Both U.S. and foreign mail that go by air on priority and nonpriority bases.

main
(1) *General.* The primary thing of consideration. (2) *Ventilation.* A duct or pipe connecting two or more branches of an exhaust system to the exhauster or air-cleaning equipment. (3) *Gas Industry.* A distribution line that serves as a common source of supply for more than one gas service line.

main deck
Maritime. The lowest deck on a river steamboat. The main deck supports the vessel's engines and boilers and has space for fuel and cargo.

main event
See ***contributory event***.

main heating fuel
Fuel that powers the main heating equipment.

main menu
The top-level menu within a software package.

main rotor
The rotor that supplies the principal lift to a rotorcraft.

main stem
The main portion of navigable channel of a river where more than one channel exists.

main track
A track, other than an auxiliary track, extending through yards or between stations, upon which trains are operated by timetable or train order or both, or the use of which is governed by a signal system.

mainshock
Seismology. The largest earthquake in any series of earthquakes. To be definitively called a mainshock, it should generally be at least half a magnitude unit larger than the next largest quake in the series. Otherwise, the series of quakes may more accurately fit the definition of a swarm. *See also* ***swarm***.

maintain
Aviation. (1) Concerning altitude flight level, the term means to remain at the altitude flight level specified. The phrase "climb and" or "descend and" normally precedes "maintain" and the altitude assignment, e.g., "descend and maintain 5,000." (2) Concerning other Air Traffic Control (ATC) instructions, the term is used in its literal sense, e.g., maintain visual flight rules (VFR).

maintainability
An expression of the ability of a given product or system to be maintained (with minimum maintenance and repair) and remain in intended service throughout the operational phase of the product life cycle.

maintained illuminance
That proportion of initial illuminance which a light or luminaire retains over some specified period of time.

maintenance
(1) The performance of those functions necessary to keep a machine, process, or system in or return it to a proper state of repair for safe and/or efficient operation. (2) Inspection, overhaul, repair, preservation, and the replacement of parts, but excludes preventive maintenance. (3) All expenses, both direct and indirect, specifically identifiable with the repair and upkeep of property and equipment. (4) The performance of services on fire protection equipment and systems to assure that they will perform as expected in the event of a fire. Maintenance differs from inspection in that maintenance requires the checking of internal fittings, devices, and agent supplies. *See also* ***preventive maintenance***.

maintenance control center (MCC)
U.S. Government. Responsible for the oversight of authorization for vehicle repair and authorization and certification of maintenance and repair invoices for Interagency Fleet Management System (IFMS) vehicles within

the specified region(s). The MCC also contacts vendors to schedule vehicle services. *See also* ***preventive maintenance***.

maintenance management
The process of deciding what type of maintenance will be used for systems under an individual's or organization's control, which may include a) the conducting of tradeoff studies, b) a decision as to what risks are acceptable and what are not, c) the scheduling and implementation of maintenance, and d) the development of maintenance procedures.

maintenance time
That time estimated, allowed, used, or required to perform some act of maintenance on a system.

major alteration
Aviation. An alteration not listed in the aircraft, aircraft engine, or propeller specifications that a) might appreciably affect weight, balance, structural strength, performance, powerplant operation, flight characteristics, or other qualities affecting airworthiness; or b) might be done according to accepted practices cannot be done by elementary operations.

major axis
The longer axis in defining an ellipse.

major carrier group
Air carrier groups with annual operating revenues exceeding $1,000,000,000.

major crimes
A loose classification of serious crimes such as murder, rape, armed robbery, etc.

major defect
A defect which results in a serious malfunction of a product.

major emitting facility
Under the Federal Clean Air Act, any of the following stationary sources of air pollutants which emit, or have the potential to emit, one hundred tons per year or more of any air pollutant from the following types of stationary sources: fossil-fuel fired steam electric plants of more than two hundred and fifty million British thermal units per hour heat input; coal cleaning plants (thermal dryers); kraft pulp mills; Portland Cement plants; primary zinc smelters, iron and steel mill plants; primary aluminum ore reduction plants; primary copper smelters; municipal incinerators capable of charging more than fifty tons of refuse per day; hydrofluoric; sulfuric; and nitric acid plants; petroleum refineries; lime plants; phosphate rock processing plants; coke oven batteries; sulfur recovery plants; carbon black plants (furnace process); primary lead smelters; fuel conversion plants; sintering plants; secondary metal production facilities; chemical process plants; fossil-fuel boilers of more than two hundred and fifty million British thermal units per hour heat input; petroleum storage and transfer facilities with a capacity exceeding three hundred thousand barrels; taconite ore processing facilities; glass fiber processing plants; charcoal production facilities. Such term also includes any other source with the potential to emit two hundred and fifty tons or more per year of any air pollutant. This term shall not include new or modified facilities which are nonprofit health or education institutions which have been exempted by the state.

major fuel
Fuels or energy sources such as electricity, fuel oil, liquefied petroleum gases, natural gas, district steam, district hot water, and district chilled water.

major injury
An occupational or other injury which results in a loss of time to the injured person and a medical expense.

major interstate pipeline company
A company whose combined sales for resale, including gas transported interstate or stored for a fee, exceeded 50 million thousand cubic feet in the previous year.

major mishap
An event or incident that has the potential of resulting in a fatality or major damages, such as the loss of a facility.

major modification
This term is used to define modifications with respect to prevention of significant deterioration (PSD) and new source review under the Clean Air Act and refers to modifications to major stationary sources of emissions and provides significant pollutant increase levels below which a modification is not considered major. *See also* ***prevention of significant deterioration, new source****, and* ***major stationary source***.

major repair

Aviation. A repair that, if improperly done, might appreciably affect weight, balance, structural strength, performance, powerplant operation, flight characteristics, or other qualities affecting airworthiness; or that is not done according to accepted practices or cannot be done by elementary operations.

major river

A river that, because of its velocity and vessel traffic, would require a more rapid response in case of a worst case discharge.

major source

According to the Federal Clean Air Act: Any stationary source or group of stationary sources located within a contiguous area and under common control that emits or has the potential to emit, in the aggregate, 10 tons per year or more of any hazardous air pollutant or 25 tons per year or more of any combination of hazardous air pollutants. The EPA Administrator may establish a lesser quantity, or in the case of radionuclides different criteria, for a major source than that specified in the previous sentence, on the basis of the potency of the air pollutant, persistence, potential for bioaccumulation, other characteristics of the air pollutant, or other relevant factors.

Major stationary sources are very closely regulated

major stationary sources

Term used to determine the applicability of prevention of significant deterioration (PSD) and new source regulations. In a nonattainment area, any stationary pollutant source that has a potential to emit more than 100 tons per year is considered a major source. In PSD areas, the cutoff level may be either 100 or 250 tons, depending upon the type of source. *See also* ***prevention of significant deterioration***.

make-or-buy analysis

A study to determine whether it is more advantageous to develop and produce an item in-house or purchase the item from outside sources.

make-ready allowance

See ***setup allowance***.

make short approach

Aviation. Communication Protocol. Used by Air Traffic Control (ATC) to inform a pilot to alter his/her traffic pattern 80 degrees to make a short final approach.

make-up air

Air brought into a building from outdoors through the ventilation system and that has not been previously circulated through the system. *See also* ***replacement air*** (the two terms are synonymous).

make up tow

To assemble barges into a tow.

malaise

A general feeling of fatigue or exhaustion; a lack of energy and/or desire to do anything; can appear as a symptom of disease and illness.

malaria

A serious infectious illness characterized by periodic chills and high fever. It responds well to modern drugs but can be chronic. Malaria is primarily found in tropical and subtropical climates. Malaria is caused by a protozoan parasite, the Plasmodium, which is carried by the Anopheles mosquito. When the mosquito bites an infected person, it sucks in the parasites, which reside in the blood. In the mosquito, the plasmodia multiply and travel to the salivary glands from which they are transmitted to the human bloodstream by the mosquito bite. Inside the human host, they penetrate the erythrocytes where they mature, reproduce, and at complete maturity, burst out of the blood cell. The life cycle varies according to the species of Plasmodium. There are usually no symptoms until several cycles have been completed. Then there is a simultaneous rupturing of cells by the entire blood supply, causing the characteristic chills followed in a few hours by fever. The temperature may rise to 104°or 105°F. As it subsides, there is profuse perspiring. Other symptoms are headache, nausea, body pains and, after the attack, exhaustion. The symptoms last from 4 to 6 hours and recur at regular intervals, depending upon the parasitic species and its cycle. If the attack occurs every other day, the disease is called tertian malaria; if it occurs at three-day intervals, it is quartan malaria. As the disease progresses, the attacks occur less frequently. Bouts of malaria last from 1 to 4 weeks but usually about 2 weeks. Relapses are common, with attacks ceasing and recurring at irregular intervals for several years, especially if untreated. Malaria is not usually fatal; when it is, it is almost always caused by the falciparum species.

Malcolm Baldrige National Quality Award
A nationally based award which is presented annually to an organization judged best in several categories such as human resource utilization, quality assurance, and leadership.

malfeasance
Evil doing, ill conduct, or the commission of some act which is positively unlawful. The doing of an act which is wholly unlawful and wrong.

malic acid
A crystalline acid from juices of many fruits and plants, and an intermediary product of carbohydrate metabolism in the body.

malice
The intentional doing of a wrongful act without just cause or excuse, with an intent to inflict an injury or under circumstances that the law will imply as evil intent.

malicious injury
An injury committed against a person at the prompting of malice or hatred toward that person, or done spitefully or wantonly.

malignancy
Cells having the ability to invade surrounding tissue and spread to distant sites (i.e., cancerous).

malignant
Pertaining to continuing abnormal tissue growth, possibly with eventual metastasis, culminating in death unless successfully treated.

malignant granuloma
*See **Hodgkin's disease**.*

malignant tumor
A tumor capable of metastasizing or spreading cancerous cells from one part of the body to another.

malingerer
An individual who feigns illness or another problem to get out of work or responsibility.

malleolus
A rounded bony projection at the ankle.

malnutrition
A condition in which there is an inadequate nutritional intake or an inability to utilize ingested nutrients. Extreme malnutrition may lead to starvation.

malodorant
Any odorant having a strong or offensive odor.

malpractice
Misconduct or lack of proper professional skill on the part of a professionally trained person, such as an engineer, physician, dentist, attorney, or other professional in doing his or her work.

malpractice insurance
Type of liability insurance which protects professional people (e.g., lawyers, doctors, accountants) against claims of negligence brought against them. *See also **insurance**.*

maltodextrin
A complex carbohydrate that is commercially manufactured by the enzymatic treatment of corn.

maltreatment
In reference to the treatment of a patient by a surgeon, this term signifies improper or unskillful treatment. It may result either from ignorance, neglect, or willfulness, but the word does not necessarily imply that the con-

duct of the surgeon is either willfully or grossly careless.

mammalian diving response
A physiological response to high environmental pressure in which the peripheral arteries contract and the heart rate slows due to the body's attempt to preserve oxygen flow to the brain and other vital organs.

mammatus clouds
Clouds that look like pouches hanging from the underside of a cloud.

mammography
Roentgenography of the breast with or without injection of an opaque substance into its ducts. Simple mammography, without the use of a contrast medium, is sometimes used in the diagnosis of cancer and other disorders of the breast.

man-amplifier
The concept of a human using an exoskeleton or other device which enables him/her to perform feats requiring much greater strength or other capabilities than would be normally humanly possible without such a device.

man-computer dialogue
See ***human-computer dialogue***.

man-computer interaction
See ***human-computer interaction***.

man-computer interface
See ***human-computer interface***.

man-hour
An industrial unit of production reflecting paid labor hours.

man-machine chart
A multiple activity process chart in which both personnel and machines are used.

man-minute
A unit of measure of work equivalent to the utilization, scheduling, or availability of one person working for one minute.

man-multiplier
A concept in which one person controls many machines, all performing the same tasks.

man-paced work
See ***self-paced work***.

man-tool interface
Any portion of a tool where a person might grasp, carry, and/or hold a tool for performing manipulations on other objects.

manage
To organize and direct human, economic, and material resources toward developing and accomplishing one or more specified objectives.

management
The group of people within an organization who manage.

management buy-in
With regard to compliance, a declaration (often in the Compliance Procedures Manual) of the organization's commitment and support to the compliance program, both in words and resources, by those who control the organization. That support includes the implementation of a process to achieve compliance.

management of migration
Actions that are taken to minimize and mitigate the migration of hazardous substances or pollutants or contaminants and the effects of such migration. Measures may include, but are not limited to, management of a plume of contamination, restoration of a drinking water aquifer, or surface water restoration.

mandamus
A court order compelling a government agency to do a duty expressly provided in some statute or regulation. Also, the civil cause of action against an agency to seek to compel compliance, as in filing a complaint in the nature of mandamus.

mandate
A command, order, or direction, written or oral, which a court (or other regulatory body) is authorized to give and a person is bound to obey.

mandatory altitude
An altitude depicted on an instrument approach procedure chart requiring the aircraft to maintain altitude at the depicted value.

mandatory standard
A procedural, performance, or other type of standard which is regulated by law via one or more governmental agencies.

mandatory statutes
Generic term describing statutes which require and not merely permit a course of action. They are characterized by such directives as "shall" and not "should."

mandatory use seat belt law
A law requiring some adult occupants of some traffic vehicles to use available restraint systems. *See also* ***manual restraint system, restraint usage***.

mandible
The lower, horseshoe-shaped jawbone of the skull. It consists of a central portion, which forms the chin and supports the lower teeth, and two perpendicular portions, or *rami,* which point upward from the back of the chin on either side. *See also* ***jaw*** *and* ***jaw bone***.

maneuver boat
Boat used by the Corps of Engineers in raising and lowering movable wickets of dams on the Ohio River.

maneuvering
Maritime Navigation. Changing course, speed, or similar boat handling action during which a high degree of alertness is required or the boat is imperiled because of the operation, i.e., docking, mooring, undocking, etc.

manganese
A chemical element, atomic number 25, atomic weight 54.938, symbol Mn.

mania
A disordered mental state of extreme excitement.

manic-depressive
A psychosis marked by alternating periods of elation and depression.

manifest
(1) *Human Perception.* Something that becomes evident to the senses, especially to sight. Obvious to the understanding, evident to the mind, not obscure or hidden, and is synonymous with open, clear, and self-evident. (2) *Shipping-General.* A document used in shipping and warehousing containing a list of the contents, value, origin, carrier, and destination of the goods to be shipped or warehoused. (3) *Shipping-EPA.* The uniform shipping document required by the EPA and established as a tracking mechanism by the Resource Conservation and Recovery Act (RCRA). This tracking document follows the hazardous waste from point of generation to its final destination. Copies are maintained by the state where the wastes are generated, the destination state, the transporting company, and by the generator. Proper DOT shipping names, EPA waste codes and other specific information are required. The manifest certifies everything has been done within the generators power to reduce the volume and/or toxicity of its hazardous wastes. (4) *Law.* In evidence, that which is clear and requires no proof.

manifold pressure
Absolute pressure as measured at the appropriate point in the induction system and usually expressed in inches of mercury.

manipulate
To handle, move, or operate on one or more objects or controls using the hands or other dexterous controlling device(s) in conjunction with a vision or other sensory system.

manipulation
Skillful or dexterous treatment by the hands. In physical therapy, the forceful passive movement of a joint beyond its active limit of motion.

manipulative dexterity
A measure of the skill which an individual or robotic device possesses for the coordinated use of fingers/hands/wrists or their robotic analogies for fine tasks.

manipulator
Any non-mobile mechanical device for handling, moving, or controlling operations at a distance.

manmade air pollution
Air pollution which results directly or indirectly from human activities.

manmade fiber
Any textile fiber made from synthetic or natural chemical substances.

manmade ionizing radiation
Ionizing radiation produced by a manmade source, such as an x-ray machine.

manmade mineral fiber (MMMF)
A fibrous material that is manmade as opposed to a naturally occurring fibrous material like asbestos. Manmade mineral fibers are used as substitutes for asbestos-containing materials. They include fibrous glass, mineral wool, refractory ceramic fibers, etc.

manmade noise
Any electrical or acoustic noise having a human source or resulting from manmade equipment.

manmade vitreous fiber (MMVF)
Fibrous, amorphous, inorganic substances that are made primarily from rock, clay, slag, or sand. They include fibrous glass, mineral wood (rock and slag), and refractory ceramic fibers.

Mann-Whitney U test
A nonparametric statistical test using rank-ordered data for comparing two independent groups.

mannequin
An anthropomorphic figure which has joints or other superficial human physical characteristics and which is used in modeling, clothing, display, training, or art.

manner of collision
A classification for crashes in which the first harmful event was a collision between two motor vehicles in transport.

manometer
Instrument for measuring the pressure of any fluid or the difference in the pressure between fluids, whether liquid or gas.

manoptoscope
A device for determining which eye is dominant.

MANOVA
See ***multivariate analysis of variance***.

Manpower and Personnel Integration (MANPRINT)
A U.S. Army management and technical human factors program for improving weapon-soldier system performance.

manslaughter
The unjustifiable, inexcusable, and intentional killing of a human being without deliberation, premeditation, and malice. Criminal homicide constitutes manslaughter when a) it is committed recklessly, or b) a homicide which would otherwise be murder is committed under the influence of extreme mental or emotional disturbance for which there is reasonable explanation or excuse.

Mansonia
A genus of mosquitoes comprising some 55 species, distributed primarily in tropical regions, important as vectors of microfilariae and viruses.

manual
(1) Pertaining to an operation or set of operations performed solely by humans, rather than by machines or with machine assistance. (2) A document which provides instructions or other information for operation of some equipment.

manual control
(1) Any control mechanism intended for manipulation by humans. The individual is the feedback element. (2) A discipline which studies and incorporates the human operator as a feedback element within a closed-loop system.

manual dexterity
A measure of the ability to make rapid, coordinated, fine, or gross movements of the fingers, hand(s), and/or arm(s) for handling independent objects.

manual element
A work element performed by a worker using no more than simple tools, and not involving machines.

manual input
The use of a human operator to input data to a computer via some computer input device.

manual interlocking
Rail Operations. An arrangement of signals and signal appliances operated from an interlocking machine and so interconnected by means of mechanical and/or electric locking that their movements must succeed each other in proper sequential, train movements over all routes being governed by signal indication.

manual labor
Literally, work done with the hands. Generally, it refers to labor performed by hand or by the exercise of physical force, with or without the aid of tools, machinery, or equipment, but depending for its effectiveness chiefly upon personal muscular exertion rather than upon skill, intelligence, or adroitness.

manual materials handling (MMH)
The non-equipment-aided human act of relocating an object, consisting of approximately the following stages: approach, grasp, pickup, move or carry, putdown, adjust.

manual rating insurance
Type of insurance in which the premium is set from a manual classifying types of risk on a

general basis such as a particular industry without reference to the individual case. *See also* ***insurance***.

manual restraint system
Occupant restraints that require some action, usually buckling, before they are effective. They include shoulder belt, lap belt, lap and shoulder belt, infant carrier, or child safety seat. *See also* ***mandatory use seat belt law, restraint usage***.

manual steadiness
See ***hand steadiness***.

manual time
The amount of time required to execute a manual element.

manually propelled mobile scaffold
A portable rolling scaffold support by casters.

manubrium
The triangular-shaped superior segment of the sternum.

manufacture
(1) *General.* The process of making products by hand, machinery, or other automated means. (2) *TSCA.* To import into the customs territory of the United States or to produce or manufacture chemical substances. (3) *EPCRA.* To produce, prepare, import, or compound a toxic chemical. Manufacture also applies to a toxic chemical that is produced coincidentally during the manufacture, processing, use, or disposal of another chemical or mixture of chemicals, including a toxic chemical that is separated from that other chemical or mixture of chemicals as a by-product, and a toxic chemical that remains in that other chemical or mixture of chemicals as an impurity.

manufacturer
(1) *General.* One who by labor, art, or skill transforms raw material into some kind of a finished product or article of trade. (2) *U.S. Coast Guard.* a) Any person engaged in the manufacture, construction, or assembly of boats or associated equipment. b) The manufacture or construction of components for boats and associated equipment. c) Equipment to be sold for subsequent assembly. d) The importation into the United States for sale of boats, associated equipment, or components thereof.

manufacturer identification code (MIC)
U.S. Coast Guard. Three-character identifier assigned by Coast Guard Headquarters on request to those manufacturers and importers defined under 33 CFR 181.31.

manufacturer's formulation
A list of substances or component parts as described by the maker of a coating or pesticide.

Manufacturing Automation Protocol (MAP)
A set of communication standards for use in automated manufacturing.

manufacturing cost
The total cost of manufacturing an item, including materials, direct labor, overhead, and depreciation.

manufacturing engineering
That field of engineering specializing in the research, planning, design, integration, and development of the methods, facilities, tools, and processes involved in the production of goods.

manufacturing progress function
The improvement in production efficiency with time.

manuometer
A spring device for measuring static strength of the finger flexor muscles.

map
A drawing used to illustrate the physical relationships between the elements of people, equipment, materials, and environmental structures associated with an accident or incident.

MAP
See ***Model Accreditation Plan***. *See also* ***Manufacturing Automation Protocol***. *See also* ***Michigan Anthropometric Processor***.

MARAD
See ***Maritime Administration***.

margin
(1) A distance, setting, or other limit which should not be exceeded under normal circumstances. (2) That region, typically without printing, which separates printed text and/or graphics from the paper or other material edge on a hardcopy.

margin of safety

The percentage by which the allowable load (stress) exceeds the limit load (stress) for specific design conditions, represented as follows:

$$\text{Margin of Safety} = \left[\left(\frac{\text{Yield Strength}}{\text{Limit Load Stress}}\right) x \left(\text{Yield Factor of Safety}\right)\right] - 1$$

or:

$$\text{Ultimate Margin of Safety} = \left[\left(\frac{\text{Ultimate Strength}}{\text{Limit Load Strength}}\right) x \left(\text{Ultimate Factor of Safety}\right)\right] - 1$$

marginal cost

That cost incurred for an additional unit of output.

marginal event

An occurrence, subsequent to the introduction of a hazard or set of hazards into a system, that results in a level of injury, damage, or loss of minimal consequences. Quick recovery would be possible and probable. The parameters for this categorization are usually established by management in the System Safety Program Plan, or other policy-making documentation.

marginal product

That additional unit of output which is obtained by adding an extra unit of some factor.

marginal revenue

That additional income realized by selling one additional product unit.

Marie-Tooth disease

Progressive neuropathic (peroneal) muscular atrophy.

marihuana

An annual herb, cannabis sativa, having angular rough stem and deeply lobed leaves. The bast fibers of cannabis are the hemp of commerce. A drug prepared from cannabis sativa, designated in technical dictionaries as cannabis and commonly known as *marijuana, marajuana,* or *maraguana*. Marihuana means all parts of the plant *cannabis sativa L*, whether growing or not, the seeds thereof, the resin extracted from any part of the plant; and every compound, manufacture, salt, derivative, mixture, or preparation of the plant, its seeds, or its resin. It does not include the mature stalks of the plant, fiber produced from the stalks, oil, or cake made from the seeds of the plant, any other compound, manufacture, salt, derivative, mixture, or preparation of the mature stalks (except the resin extracted therefrom), or sterilized seed of the plant which is incapable of germination. Marihuana is also commonly referred to as *pot, grass, tea, weed,* or *Mary-Jane;* and in cigarette form as a *joint* or *reefer*.

marina

Establishments primarily engaged in operating marinas. These establishments rent boat slips and store boats, and generally perform a range of other services including cleaning and incidental boat repair. They frequently sell food, fuel, and fishing supplies, and may sell boats. *See also* ***dock, pier, harbor,*** *and* ***wharf***.

marine cargo handling

Establishments primarily engaged in activities directly related to marine cargo handling from the time cargo, for or from a vessel, arrives at ship side, dock, pier, terminal, staging area, or in-transit area until cargo loading or unloading operations are completed. Included in this industry are establishments primarily engaged in the transfer of cargo between ships and barges, trucks, trains, pipelines, and wharves. Cargo handling operations carried on by transportation companies and separately reported are classified here. This industry includes the operation and maintenance of piers, docks, and associated buildings and facilities.

marine insurance

A contract whereby one party, for a stipulated premium, undertakes to indemnify the other against certain perils or sea risks to which his/her ship, freight, and cargo, or some of them, may be exposed during a certain voyage, or a fixed period of time. An insurance against risks connected with navigation, to which a ship, cargo, freightage, profits, or other insurable interest in movable property may be exposed during a certain voyage or a fixed period of time. *See also* ***insurance***.

marine mammal

According to the Federal Marine Mammal Protection Act of 1972: Any mammal which is morphologically adapted to the marine environment (including sea otters and members of the orders *Sirenia, Pinnipedia* and *Cetacea*), or primarily inhabits the marine environment (such as a polar bear).

marine mammal product
According to the Federal Marine Mammal Protection Act of 1972: Any item of merchandise which consists, or is composed in whole or in part, of any marine mammal.

marine pollutant
A hazardous material which is listed in Appendix B to CFR 172.101 and, when in a solution or mixture of one or more marine pollutants, is packaged in a concentration which equals or exceeds a) ten percent by weight of the solution or mixture for materials listed in the appendix; or b) one percent by weight of the solution or mixture for materials that are identified as severe marine pollutants in the appendix. *See also* ***hazardous material***.

marine sanitation device
Any equipment or device installed on board a vessel to receive, retain, treat, or discharge sewage and any process to treat such sewage.

marine terminal
A designated area of a port, which includes but is not limited to wharves, warehouses, covered and/or open storage spaces, cold storage plants, grain elevators and/or bulk cargo loading and/or unloading structures, landings, and receiving stations, used for the transmission, care, and convenience of cargo and/or passengers in the interchange of same between land and water carriers or between two water carriers.

marine terminal operator
Person or entity that operates the various marine terminals at ports, usually under long-term lease agreements with local or state governments or port authorities. The marine terminal operator provides receiving and delivery, and other terminal services for the cargoes moving through these facilities.

maritime
Business pertaining to commerce or navigation transacted upon the sea or in seaports in such matters as the court of admiralty has jurisdiction.

Maritime Administration (MARAD)
The Maritime Administration was established by Reorganization Plan No. 21 of 1950 (5 U.S.C. app.) effective May 24, 1950. The Maritime Act of 1981 (46 U.S.C. 1601) transferred the Maritime Administration to the Department of Transportation, effective August 6, 1981. The Administration administers programs to aid in the development, promotion, and operation of the U.S. Merchant Marine. It is also charged with organizing and directing emergency merchant ship operations. It administers subsidy programs, provides financing guarantees for the construction, reconstruction, and reconditioning of ships; and enters into capital construction fund agreements that grant tax deferrals on monies to be used for the acquisition, construction, or reconstruction of ships. The Administration constructs or supervises the construction of merchant type ships for the Federal Government, helps industry generate increased business for U.S. ships, and conducts programs to develop ports, facilities, and intermodal transport, and to promote domestic shipping. The Administration conducts program and technical studies and administers a War Risk Insurance Program, and under emergency conditions the Maritime Administration charters government-owned ships to U.S. operators, requisitions or procures ships owned by U.S. citizens, and allocates them to meet defense needs. It maintains a National Defense Reserve Fleet of government-owned ships, regulates sales to aliens and transfers to foreign registry of ships that are fully or partially owned by U.S. citizens. It also operates the U.S. Merchant Marine Academy, Kings Point.

maritime air
Moist air whose characteristics were developed over an extensive body of water.

maritime carrier
Carriers which operate on the open sea, i.e., their operations must include a foreign or international component and may include a domestic component.

Maritime Commission
The Federal Maritime Commission regulates the waterborne foreign and domestic offshore commerce of the United States, assures that the United States international trade is open to all nations on fair and equitable terms, and guards against unauthorized monopolies in the waterborne commerce of the United States. This is accomplished through maintaining surveillance over steamship confer-

ences and common carriers by water; assuring that only the rates on file with the Commission are charged; approving agreements between persons subject to the Shipping Acts of 1916 and 1984; guaranteeing equal treatment to shippers and carriers by terminal operators, freight forwarders, and other persons subject to the shipping statutes; and ensuring that adequate levels of financial responsibility are maintained for the indemnification of passengers or oil spill cleanup.

maritime revenue
Revenue received for operations in international or foreign shipping.

maritime tort
Civil wrongs committed on navigable water.

Mark Twain
Maritime Navigation (slang). Colloquial term for 12-foot depth or mark 2 on the lead line.

marked channel
A channel marked by buoys.

marker
To monitor for a unique component of a mixture and use its result as an indicator of the presence of the mixture.

marker beacon
Aviation. An electronic navigation facility transmitting a 76-mHz vertical fan or bone-shaped radiation pattern. Marker beacons are identified by their modulation frequency and keying code, and when received by compatible airborne equipment, indicate to the pilot, both aurally and visually, that he is passing over the facility.

market analysis
A study involving the collection of data to determine information for a product or service such as the identification of potential customers, trends in the marketplace, why a consumer might purchase it, etc.

market research
The process of gathering and analyzing data regarding the potential sale of goods or services to a consumer.

marketable title
A title which is free from encumbrances and any reasonable doubt as to its validity, and such as a reasonably intelligent person, who is well-informed as to the facts and their legal bearings, and ready and willing to perform his/her contract, would be willing to accept in exercise of ordinary business prudence.

marketed production
Gross withdrawals less gas used for repressurization, quantities vented and flared, and nonhydrocarbon gases removed in treating or processing operations. Includes all quantities of gas used in field and processing operations.

marketing policy
That guideline which determines what products will be offered, what types of markets will be approached, what selling and promotional techniques will be used, what process will be charged, etc.

marking
A descriptive name, identification number, instruction, caution, weight, specification, or combinations thereof, on outer packagings of hazardous materials.

marrow
A soft tissue material in the interior of many bones. Bone marrow is a network of blood vessels and special connective tissue fibers that hold together a composite of fat and blood-producing cells. The chief function of marrow is to manufacture erythrocytes, leukocytes, and platelets. These blood cells normally do not enter the bloodstream until they are fully developed, so that the marrow contains cells in all stages of growth. If the body's demand for white blood cells is increased because of infection, the marrow responds immediately by stepping up production. The same is true if more red blood cells are needed, as in hemorrhage or some other types of anemia. There are two types of marrow: red and yellow. Red marrow produces blood cells while yellow marrow, which is mainly composed of fatty tissue, normally has no blood-producing function. During infancy and early childhood, most bone marrow is red. But gradually, as one gets older and less blood-cell production is needed, the fat content of the marrow increases to turn some of the marrow from red to yellow. Red marrow continues to be present in adulthood only in the flat bones of the skull, the sternum, ribs, vertebral column, clavicle, humerus, and part of the femur. However, under certain conditions, as after hemorrhage, yellow marrow in other bones may again be converted to red

and resume its cell-producing functions. The marrow is occasionally subject to disease, as in aplastic anemia, which may be caused by destruction of the marrow by chemical agents or excessive x-ray exposure. Other diseases that affect the bone marrow are leukemia, pernicious anemia, myeloma, and metastatic tumors.

marsh
A type of wetland that does not accumulate appreciable peat deposits and is dominated by herbaceous vegetation. Marshes may be either fresh or saltwater and tidal or non-tidal. *See also* ***wetlands***.

marshal
The President is required to appoint a U.S. marshal to each judicial district. It is the responsibility of U.S. marshals to execute all lawful writs, processes and orders issued under authority of the United States. In executing the laws of the United States within a state, the marshal may exercise the same powers which a sheriff of the state may exercise in executing the laws thereof.

marstochron
See ***chronograph***.

marstograph
See ***chronograph***.

Martin's diameter
Length of the line which divides a particle into two equal areas.

maser
Acronym for *microwave amplification by stimulated emission of radiation*. A device that produces an extremely intense, small, and nearly nondivergent beam of monochromatic radiation in the microwave region, with all the waves in phase.

mask
(1) To cover or conceal. (2) In audiometry, to obscure or diminish a sound by the presence of another sound of different frequency. (3) An appliance for shading, protecting, or medicating the face. (4) To increase the threshold level of a stimulus or condition by presenting a second (masking) stimulus simultaneously or in close time or space proximity.

masking
In acoustics, a process by which the threshold of audibility for one sound is raised by the presence of another (masking) sound.

masking level
The difference in original stimulus intensity required to reach a reported threshold due to a masking stimulus.

mason's adjustable multiple-point suspension scaffold
A scaffold having a continuous platform supported by bearers suspended by wire rope from overhead supports, so arranged and operated as to permit the raising or lowering of the platform to desired working positions.

mass
(1) *General.* The fundamental measure of the quantity of matter. Mass is different from weight in that it does not depend upon gravitational force. (2) *Physics.* The measure of a body's resistance to acceleration. The mass of any object is different than, but proportional to, its weight which is the force of attraction that exists between the object being considered and any other proximate massive object (i.e., the earth). In the International System of Units, the basic unit of mass is the *kilogram,* which has been defined as being equal to the mass of the international prototype of the kilogram. In the MKS System, the basic unit of mass is the *kilogram*. In the CGS System, the basic unit of mass is the *gram*. In the English System, the basic unit of mass is the *slug*. (3) A lump or collection of cohering particles.

mass flow meter
An electrically heated tube and an arrangement of thermocouples to measure the differential cooling caused by a gas (e.g., air) passing through the tube. The thermoelectric elements generate a voltage proportional to the rate of gas flow through the tube.

mass media
Those forms of the media which typically reach large numbers of people, especially newspapers, television, and radio.

mass median aerodynamic diameter (mmad)
The mass median diameter of spherical particles of unit density which have the same falling velocity in air as the particle in question.

mass median size
The mass median size of a particle in a distribution of particles such that the mass of all

particles larger than the median is equal to the mass of all smaller particles.

mass number
The number of protons and neutrons in the nucleus of an atom. *See also* ***atomic weight***.

mass psychogenic illness (MPI)
Term used in describing illnesses experienced by workers and for which no definitive cause/source can be identified. Often, it is more commonly referred to as the "I'm sick, you're sick syndrome."

mass spectrography
An instrumental analytical method for identifying substances from their mass spectra.

mass spectrometer
An electronic instrument used for the separation of electrically charged particles by mass.

mass transit
Another name for public transportation.

mass transportation
Another name for public transportation.

Mass Transportation Agency
An agency authorized to transport people by bus, rail, or other conveyance, either publicly or privately owned, and providing to the public general or special service (but not including school, charter, or sightseeing service) on a regular basis.

massage
To rub, stroke, knead, or impact the superficial muscles of the body, either by hand or with some instrument for therapeutic or other purposes.

massed practice
Continuous, repeated, or extended training, without time for rest periods.

masseter muscle
The muscle that closes the jaws.

mast cell
A connective tissue cell whose specific physiological function is unknown. It elaborates granules that contain histamine, heparin, and, in the rat and mouse, serotonin.

mastectomy
Surgical removal of breast tissue. Mastectomy is usually performed to treat malignant breast tumors, although rarely it may be advisable to use the procedure for benign tumors and for other diseases of the breast, such as chronic cystic mastitis. *See also* ***mastitis***.

master agreement
The omnibus labor agreement reached between a union and the leaders of the industry or trade association. It becomes the pattern for labor agreements between the union and individual employers.

master plan
Term used in land use control law, zoning, and urban redevelopment to describe the omnibus plan of a city or town for housing, industry, recreational facilities, and their impact on environmental factors.

master-servant rule
Under this rule of law, the master (employer) is liable for the conduct of the servant (employee) which occurs while the servant is acting within the scope of his/her employment or within the scope of his/her authority.

master-slave manipulator
Pertaining to any device in which the remote operator is intended to follow either exactly or proportionately the motions and forces of the input controller.

Master Standard Data (MSD)
A universal predetermined motion time system.

Master's two-step test
The simple exercise of repeatedly ascending over two nine-inch steps to test cardiovascular function.

mastitis
Inflammation of the breast, occurring in a variety of forms and in varying degrees of severity. Chronic cystic mastitis is the most common disorder of the breast resulting from hormonal imbalance. This condition generally occurs in women between the ages of 30 and 50. It is probably related to the activity of the ovaries and is rare after menopause. The disease is characterized by the formation of cysts which give a lumpy appearance to the breast. Symptoms may include pain and tenderness, which are usually aggravated before the menstrual period, at which time the cysts tend to enlarge. There may also be discharge from the nipple. Periodic change in the size of a lump or its rapid appearance and disappearance is common in cystic mastitis. Since

there are times when it is difficult to distinguish this condition from cancer of the breast, biopsy may be necessary. Treatment may involve removing fluid from the cysts.

mastoid
(1) Nipple shaped. (2) The portion of the temporal bone lying behind the meatus of the ear (pars mastoidea), or more specifically, the conical projection from it (mastoid process). *See also* ***mastoid process***.

mastoid process
The bony projection on the inferior lateral surface of the temporal bone.

mastoiditis
Inflammation of the mastoid antrum and cells, usually the result of an infection of the middle ear with which the mastoid cells communicate. Mastoiditis most commonly follows sore throat and respiratory infection, but it can also be caused by such diseases as diphtheria, measles, and scarlet fever. Symptoms include earache and a ringing in the ears. The mastoid process may become painful and swollen.

MATC
See ***maximum acceptable toxicant concentration***.

matched groups design
An experimental design in which group selection is made by matching individuals across those groups based on one or more variables which are to be manipulated or controlled during the experiment. Also referred to as *equivalent groups method*.

matched pairs design
An experimental methodology in which assignment to groups is not strictly random, but based on one or more pairing criteria on which individuals are paired.

matching
Comparison for the purpose of selecting objects having similar or identical characteristics.

matching individual
An individual acting as a control for another individual in a matched pair.

material damping
Sound attenuation due to energy loss in the substance through which the energy is being transmitted.

material evidence
That quality of evidence which tends to influence the trier of fact because of its logical connection with the issue.

material handling
Short-distance movement of goods within a storage area.

material handling equipment (MHE)
A broad term used to delineate any equipment used to handle, lift, support, or manipulate hardware, materials, or other such equipment. MHE includes, but is not limited to, cranes, hoists, sling assemblies, load cells, forklifts, handling structures, and personnel work platforms.

material requirements planning (MRP)
The process of reducing each final product to its elementary parts, forecasting the product output required, and coordinating the production quantities of elementary parts.

material safety data sheet (MSDS)
A compilation of data required under OSHA's Hazard Communication Standard on the identity of hazardous chemicals, their health and physical hazards, exposure limits, and precautions. Section 311 of SARA requires facilities covered by the OSHA standard to submit MSDSs under certain circumstances.

material-type flow process chart
A flow process chart which indicates material usage.

material witness
A person who can give testimony relating to a particular matter no one else, or at least very few, can give. In an important criminal case, a material witness may sometimes be held by the government against his or her will. This witness may be the victim or an eyewitness.

maternity leave
That leave of absence, granted either with or without pay, for a female employee to give birth to a child and recover before returning to work.

mathematical reasoning
The ability to understand and organize a mathematical problem, then select a method to find a solution to the problem. Excludes the actual numerical manipulation.

matrix
(1) A rectangular array of numbers with a designated rows-by-columns structure. (2) A generative, or basic, structure from which a tissue or organ develops, such as the organs from which the hair and nail grow.

matrix spikes
Duplicate field samples that are spiked in the laboratory with measured quantities of contaminant; the volume of contamination in a matrix spike can be subtracted from the overall quantity of contaminant in the pure sample to determine the contamination level in the original soil sample.

matte
Having or pertaining to a surface with a dull appearance, exhibiting primarily or only diffuse reflections.

matter
(1) Anything that has mass or occupies space. (2) Physical material having form and weight under ordinary conditions of gravity.

matter of law
Whatsoever is to be ascertained or decided by the application of statutory rules or the principles and determinations of the law, as distinguished from the investigation of particular facts.

mature thunderstorm
The second stage in the three-stage cycle of an air mass thunderstorm. This stage is characterized by heavy showers, lightning, thunder, and violent vertical motions inside cumulonimbus clouds.

MAWP
Maximum allowable working pressure.

maxi-cube vehicle
A combination vehicle consisting of a power unit and a trailing unit, both of which are designed to carry cargo. The power unit is a non-articulated truck with one or more drive axles that carries either a detachable or a permanently attached cargo box. The trailing unit is a trailer or semitrailer with a cargo box so designed that the power unit may be loaded and unloaded through the trailing unit.

maxilla
A bilaterally fused bone making up much of the anterior portion of the face, including the upper part of the mouth/jaw, part of the nasal cavities, and the floor of the orbits. *See also **jaw**.*

maximal aerobic capacity
The level at which oxygen uptake during performance of a task reaches a steady state and no additional oxygen can be used by the muscles involved in the task. Also referred to as *aerobic capacity, aerobic endurance capacity, aerobic work capacity, maximal oxygen uptake/consumption, maximal aerobic power, maximum aerobic work capacity,* and *maximum oxygen uptake*.

maximal aerobic power
*See **maximal aerobic capacity.***

maximal isometric force
The maximum force generated during an isometric contraction for a specified muscle or muscle group.

maximal oxygen uptake/consumption
*See **maximal aerobic capacity.***

maximal voluntary contraction (MVC)
The greatest force which a muscle or muscle groups involved can develop under voluntary control when contracting against a resistance under specified conditions.

maximally exposed individual
The individual with the highest exposure in a given population.

maximum
The largest measured, existing, or permissible value of a set.

maximum acceptable toxicant concentration (MATC)
The geometric mean of the highest tested concentration which did not cause the occurrence of a specified adverse effect and the lowest concentration which did cause the specified adverse effect.

maximum achievable control technology (MACT)
The level of air pollution control technology required by the Clean Air Act.

maximum aerobic work capacity
*See **maximal aerobic capacity.***

maximum allowable flight duty period
The greatest number of hours an air crew can fly in an aircraft in any 24-hour period.

maximum allowable slope
That ratio of the horizontal distance from the edge of an excavation to the depth which must be provided for the existing soil or rock conditions.

maximum breathing capacity
See ***maximum voluntary ventilation***.

maximum contaminant level (MCL)
The maximum permissible level of a contaminant in water delivered to any user of a public water system. MCLs are enforceable standards.

maximum contaminant level goal (MCLG)
Under the Federal Safe Drinking Water Act, as amended, a non-enforceable concentration for a substance in drinking water that is protective of adverse human health effects and allows for an adequate margin of safety.

maximum design pressure (MDP)
See ***maximum expected operating pressure (MEOP)***.

maximum detection limit (MDL)
The lowest concentration of analyte that a method can detect reliably in either a sample or a blank.

maximum evaporative capacity
The maximum amount of sweat that can be evaporated from the body's surface under the environmental conditions that exist. The evaporation of sweat is limited by the moisture content of the air.

maximum expected operating pressure (MEOP)
The highest pressure that a pressure vessel, pressurized structure, or pressure component is expected to experience during its service life and retain its functionality, in association with its applicable operating environments. It includes the effect of temperature, pressure transients and oscillations, vehicle quasi-steady and dynamic accelerations and relief valve operating variability. Synonymous with *maximum operating pressure (MOP)* or *maximum design pressure (MDP)*.

maximum extent practicable
The limits of available technology and the practical and technical limits on a pipeline operator in planning the response resources required to provide the on-water recovery capability and the shoreline protection and cleanup capability to conduct response activities for a worst case discharge from a pipeline in adverse weather.

maximum high water elevation
The highest water level reached during the past 200 years of recordkeeping.

maximum intended load
In scaffolding, the total of all loads including the working load, the weight of the scaffold, and such other loads as may be reasonably anticipated.

maximum metabolic rate
The highest metabolic rate consistent with sustained aerobic metabolism.

maximum operating pressure (MOP)
The maximum operating pressure a system will be subjected to during planned static and dynamic conditions. *See also* ***maximum expected operating pressure (MEOP)***.

maximum oxygen uptake
See ***maximal aerobic capacity.***

maximum performance
The performance level which results in the highest possible production.

maximum permissible concentration (MPC)
The amount of radioactive material that can be tolerated in the environment or in the body without producing a significant injury. It is the recommended maximum average concentration of radionuclides in air or water to which a person (radiation worker or member of the general public) may be exposed, assuming 40 hours per week exposure for the worker and 168 hours per week for the public. It is the amount of radiation per unit volume of air or water which, if inhaled or ingested over a period of time, would result in a body burden that is believed will not produce significant injury.

maximum permissible dose (MPD)
That amount of ionizing radiation which can be absorbed per unit mass of irradiated material at a specific location without being expected to cause radiation injury to a person during one's lifetime.

maximum permissible exposure (MPE)
See ***maximum permissible dose*** *and* ***radiation protection guide***.

maximum permissible lift (MPL)
Three times the acceptable lift in kilograms or pounds.

maximum permissible limit (MPL)
A NIOSH guideline for manual lifting under specified conditions, above which musculoskeletal injury is a high probability.

maximum sound level
The greatest A-weighted sound level in decibels measured during the designated time interval or during the event, with either fast meter response or slow meter response. It is abbreviated as L_{max}.

maximum use concentration (MUC)
In radiation protection, the maximum concentration that can exist for which a specific type of respiratory protection can be used. It is equal to the permissible exposure limit for the substance to which exposure occurs times the assigned protection factor.

maximum voluntary ventilation
The volume of air breathed with maximum voluntary effort by an individual for a given period of time, usually 10-15 seconds, corrected to one minute. Also known as *maximum breathing capacity*.

maximum working area
That portion of the working surface which is easily accessible to the operator's hands with the elbow and shoulder fully extended in the normal working posture.

maximum working volume
That maximal region within which an operator can be expected to reach via any combination of shoulder, elbow, and wrist motions.

Maxwell
A unit of magnetic flux in the meter-kilogram-second electromagnetic system.

may
If a discretionary right, privilege, or power is abridged or if an obligation to abstain from acting is imposed, the word "may" is used with a restrictive "no," "not," or "only" (for example, no employer may…; an employer may not…; only qualified persons may…).

May Day
The international radiotelephony distress signal. When repeated three times, it indicates imminent and grave danger and that immediate assistance is requested. *See also* ***Pan Pan***.

Maynard Operation Sequence Technique (MOST)
A predetermined motion time system.

Mb
Megabyte, or one million bytes.

MBO
Management by Objective. An approach to organizational management characterized by the establishment of specific objectives or goals and requirements for every element of the organizational structure to work toward the achievement of that goal(s). While each element cannot possibly be responsible for the accomplishment of these objectives individually, it is their collective contributions that allow the organization to achieve its goals.

MCA
Manufacturing Chemists Association.

MCC
See ***maintenance control center***.

MCH
See ***mean corpuscular hemoglobin***.

MCHC
See ***mean corpuscular hemoglobin concentration***.

mCi
Millicurie(s).

MCL
See ***maximum containment level***. *See also* ***middle compass locator***.

MCLG
See ***maximum contaminant level goal***.

MCS
Multiple Chemical Sensitivity.

MCV
(1) *See* ***mean corpuscular volume***. (2) *See* ***mean clinical value***.

MDA
See ***minimum descent altitude***.

MDC
Mild detonating cord.

MDF
Mild detonating fuse.

MDL
See ***maximum detection limit***.

MDP
Maximum design pressure. *See* ***maximum expected operating pressure (MEOP).***

MEA
See ***minimum en route altitude.***

meal break
That segment of the work shift, typically about mid-shift, which an employee is allotted for eating a meal. May be compensable or not, depending on whether primarily for the benefit of the employer or the employee.

mean
In statistical analysis, the arithmetic average derived from the addition of all value points in the sample, divided by the total number of points in the sample. *See also* ***estimate ratio*** *and* ***ratio estimate.***

mean annual temperature
The average temperature at any given location for the entire year.

mean body temperature
An estimated value of the average body temperature based on skin and core temperature measurements, usually as a function of the weighted mean skin temperature and the rectal temperature.

mean clinical value (MCV)
Obtained by assigning a numerical value to the response noted in a number of patients receiving a specific treatment, adding these numbers, and dividing by the number of patients treated.

mean corpuscular hemoglobin (MCH)
An expression of the average hemoglobin content of a single cell in micromicrograms, obtained by multiplying the hemoglobin in grams by 10 and dividing by the number of erythrocytes (in millions).

mean corpuscular hemoglobin concentration (MCHC)
An expression of the average hemoglobin concentration in percent, obtained by multiplying the hemoglobin in grams by 100 and dividing by the hematocrit determination.

mean corpuscular volume (MCV)
An expression of the average volume of individual cells in cubic microns, obtained by multiplying the hematocrit determination by 10 and dividing by the number of erythrocytes (in millions).

mean deviation
The average of the absolute deviations of values in a distribution from the mean.

mean radiant temperature
The temperature of a black body which would exchange the same amount of radiant heat as a worker would at the same location in a hot environment.

mean skin temperature
A measure intended to represent the average temperature of the skin over its total body surface.

mean time between failures (MTBF)
The average time expected between failures of a system or piece of equipment.

mean time to failure (MTTF)
The average time to the first failure of a component or system.

means of egress
A continuous and unobstructed way of exit travel from any point in a building or structure to a public way and consisting of three separate and distinct parts: the way of exit access, the exit, and the way of exit discharge. A means of egress comprises the vertical and horizontal ways of travel and shall include intervening room spaces, doorways, hallways, corridors, passageways, balconies, ramps, stairs, enclosures, lobbies, escalators, horizontal exits, courts, and yards.

means of transportation
A mode used for going from one place (origin) to another (destination). Includes private and public modes, as well as walking. For all travel day trips, each change of mode constitutes a separate trip. *See also* ***mode.***

measles
Also called *rubeola* (differentiated from *rubella,* or German measles). A highly contagious illness caused by a virus. Measles is a childhood disease but it can be contracted at any age. Epidemics of measles usually recur every 2 or 3 years and are most common in the winter and spring. The virus that causes measles is spread by droplet infection. The virus can also be picked up by touching an article, such as a handkerchief, that an infected person has recently used. The incuba-

tion period is usually 11 days, although it may be as few as 9 or as many as 14. The infected person can transmit the disease from 3 or 4 days before the rash appears until the rash begins to fade, a total of about 7 or 8 days. One attack of the measles usually provides a lifetime of immunity to rubeola, but not to German measles (rubella), which is somewhat similar to ordinary measles. Symptoms generally appear in two stages. In the first stage the patient feels tired and uncomfortable, and may have a running nose, a cough, a slight fever, and pains in the head and back. The eyes may become reddened and sensitive to light. The fever rises a little each day. The second stage begins at the end of the third or beginning of the fourth day. The person's temperature is generally between 103° and 104°F. Koplik's spots, small white dots like grains of salt surrounded by inflamed areas, can often be seen on the gums and the inside of the cheeks. A rash appears starting at the hairline and behind the ears and spreading downward, covering the body in about 36 hours. At first the rash appears as separate pink spots, about a quarter of an inch in diameter, but later some of these spots may run together, giving the person a blotchy look. The fever usually subsides after the rash has spread. The rash turns brown in color and fades after 3 or 4 days.

measure
(1) To read or otherwise obtain one or more numerical values from observations for analysis according to certain rules. (2) An aspect or dimension.

measure of availability
See ***availability.***

measure of central tendency
Any variable or value which is used to represent the central tendency of a distribution, such as the mean, mode, or median.

measure of dispersion
Any value which is an indicator of the spread of a distribution, such as the range, variance, or standard deviation. Also known as *measure of variability* and *dispersion*.

measure of variability
See ***measure of dispersion.***

measured daywork
That work performed at standard levels for an established hourly, nonincentive wage.

measured work
That work for which performance standards have been set using some form of work measurement technique.

measurement
The taking of data or the data resulting from a measure.

measurement error
The difference between the true value and the value initially obtained by the measuring device.

measurement error standard deviation
The square root of the within-subject variance when a group of individuals has each been measured more than once.

measurement ton
40 cubic feet.

measuring and monitoring
With regard to environmental performance, a system in place within an organization to measure and monitor actual performance against the organization's environmental objectives and targets in the areas of management systems and operational processes. This includes evaluation of compliance with relevant environmental legislation and regulations. The results should be analyzed and used to determine areas of success and to identify activities requiring corrective action and improvement.

meat wrapper's asthma
The respiratory response that may occur among meat-packaging personnel as a result of their exposure to contaminants emitted during the cutting and heat-sealing of the polyvinyl chloride plastic wrap used to package meat products.

mechanical advantage
The value of the ratio of force output by a mechanical device to the force applied to it.

mechanical aeration
Use of mechanical energy to inject air into water to cause a waste stream to absorb oxygen.

mechanical efficiency
The value of the ratio of external work performed to physiological energy production.

mechanical filter respirator
A respiratory protective device which provides protection from airborne particulates, such as dusts, mists, fumes, fibers, and other particulate type contamination.

mechanical hazard
Any unsafe situation due to machinery, equipment, tools, and/or physical structures.

mechanical impedance (Z_m)
The complex ratio of force to velocity during simple harmonic motion.

mechanical noise
Noise due to impact, friction, or vibration.

mechanical ohm
A unit for mechanical resistance, reactance, and impedance.

mechanical reactance (X_m)
The imaginary portion of mechanical impedance.

mechanical resistance (R_m)
(1) The real portion of the mechanical impedance; the opposition of a structure or object to a mechanical force either to change or to deform the structure. (2) A qualitative indication of the mechanical forces which must be overcome to move an object, control, or other mechanism.

mechanical shock
A relatively rapid transmission of mechanical energy into or out of a system.

mechanical turbulence
Random irregularities of fluid motion in air caused by buildings or non-thermal mechanical processes.

mechanical ventilation
Air movement caused by a fan or other type of air moving device.

mechanics
That field which studies the mechanical environmental effects on physical systems.

mechanize
To introduce machinery to carry out certain functions previously performed by humans.

mechanoreceptor
Any sensory receptor which is stimulated by a local change in mechanical pressure, force, or tension due to some type of movement.

MED
Minimal erythermal dose.

media
(1) The news press; those individuals representing public and/or private news gathering and reporting organizations. (2) Specific environments, such as air, water and soil, which are the subject of regulatory concern and activities. (3) General term referring to the substance or material on or in which a contaminant is collected. The media can be a liquid absorbent, solid adsorbent, filter, or other material. Typically referred to as the sampling media.

medial
Lying near or toward the midsagittal plane of the body or other approximately symmetrical structure.

medial calf skinfold
The thickness of a vertical skinfold on the medial surface of the calf at the level of the calf circumference point midway along the antero-posterior direction. Measured with the individual standing, the knee flexed 90° and the foot resting flat on an elevated platform.

medial canthus
See ***endocanthus***.

medial cuneiform bone
One of the distal group of foot bones of the tarsus.

medial malleolus height
The vertical distance from the floor or other reference surface to the most medially projecting point of the medial ankle bone. Measured with the individual standing erect and his/her weight evenly distributed on both feet.

medial rectus muscle
A voluntary extraocular muscle located parallel to the optical axis along the medial side of the eyeball. Involved in rotating the anterior portion of the eyeball toward the body midline.

median
(1) In statistical analysis, that value point which is precisely in the center (i.e., half the

value points fall below the median and half lie above the median). (2) Situated in the midline, or in the *median plane,* of a body or structure.

median category
Transit. Inclusion of a median within a single instance of the road.

median included
Transit. Median is included within the instance of the road.

median lethal concentration
The concentration of a substance in air which is lethal to 50% or more of those exposed to it. *See also* $\boldsymbol{LC_{50}}$.

median lethal dose
The dose of a material/agent necessary to kill 50% of those receiving it. *See also* $\boldsymbol{LD_{50}}$.

median lethal time
That time required for 50% of the organisms to die following a given dose of a drug, radiation, biological agent, or other agent.

median nerve
A major nerve that controls the flexor muscles of the wrist and hand. Its location in the carpal tunnel of the wrist makes it susceptible to injury or trauma as a result of overuse of tendons that pass through the same area. When the tendons swell, the nerve may be pinched causing severe pain in an illness known as carpal tunnel syndrome (CTS).

median not included
Transit. Median is not included because there is no median or the median is wide enough to cause separate instances of road.

median particle diameter
The particle size, in micrometers, about which an equal number of particles are smaller or larger in size. *See also* ***median particle size***.

median particle size
The median size of a particle in a distribution of particles by their size in microns. *See also* ***median particle diameter***.

mediastinum
(1) A medium septum or partition. (2) The mass of tissues and organs separating the sternum in front and the vertebral column behind, commonly considered to have three divisions (anterior, middle, and superior).

mediation
A proceeding involving a disinterested party who hears a dispute and recommends a resolution. If all interested parties agree to abide by these recommendations, then the dispute will be considered resolved without having had to go before the courts.

medical
Pertaining, relating, or belonging to the study and practice of medicine, or the science and art of the investigation, prevention, cure, and alleviation of diseases.

medical evidence
Evidence furnished by doctors, nurses, and other medical personnel testifying in their professional capacity as experts, or by standard treatises on medicine or surgery.

medical examiner
Public officer charged with the responsibility of investigating all sudden, unexplained, unnatural, or suspicious deaths reported to him/her, including the performance of autopsies and assisting the state in criminal homicide cases.

medical expert
Any licensed physician found qualified to give testimony as an expert witness by a court.

medical jurisprudence
The science which applies the principles and practice of the different branches of medicine to the elucidation of doubtful questions in a court of law.

medical pathology
A disorder or disease.

medical radiation
Any ultrasound, electromagnetic, or particulate radiation emitted by or received from diagnostic or therapeutic radiological procedures.

medical surveillance program
A medical program that calls for detailed physical examinations for a specific or purpose.

medical treatment
According to OSHA 29 CFR 1904.12(d), includes treatment administered by a physician or by registered professional personnel under the standing orders of a physician. Medical

treatment does not include first aid treatment even though provided by a physician or registered professional personnel.

medical waste
(1) Isolation wastes, infectious agents, human blood and blood products, pathological wastes; sharps, body parts, contaminated bedding, surgical wastes and potentially contaminated laboratory wastes, dialysis wastes, and such additional medical items as the EPA Administrator shall prescribe by regulation (Federal Marine Protection, Research and Sanctuaries Act of 1972, Federal Water Pollution Control Act). (2) Any solid waste which is generated in the diagnosis, treatment, or immunization of human beings or animals, in research pertaining thereto, or in the production or testing of biologicals (Federal Solid Waste Disposal Act). Also referred to as *medically contaminated waste*.

medically contaminated waste
See ***medical waste***.

medicine
(1) A drug or remedy. (2) The art or practice of healing.

medium
(1) An agent by which something is accomplished or impulse is transmitted. (2) A substance providing the proper nutritional environment for the growth of microorganisms.

medium air traffic hub
Aviation. A community enplaning from 0.25 to 0.99 percent of the total enplaned passengers in all services and all operations for all communities within the 50 States, the District of Columbia, and other U.S. areas designated by the Federal Aviation Administration.

medium duty scaffold
A scaffold designed and constructed to carry a working load not to exceed 50 pounds per square foot.

medium frequency
That portion of the electromagnetic spectrum consisting of radiation frequencies between 300 kHz and 3 mHz.

medium or heavy trucks
A motor vehicle with a Gross Vehicle Weight Rating (GVWR) greater than 10,000 pounds (buses, motor homes, and farm and construction equipment other than trucks are excluded).

medium regional carrier group
Carrier groups with annual operating revenues less than $19,999,999 or that operate only aircraft with 60 seats or less (or 18,000 lbs. maximum payload).

medium size bus
A bus from 29 to 34 feet in length.

medium speed
A speed not exceeding 40 miles per hour.

medium voltage (MV) cable
Type MV medium voltage cable is a single- or multi-conductor solid dielectric insulated cable rated 2000 volts or higher.

medulla
The central or inner portion of an organ.

medulla oblongata
That part of the hind brain lying between the pons above and the spinal cord below. It houses nerve centers for both motor and sensory nerves, where such functions as breathing and the beating of the heart are controlled.

meets intent certification
A certification used to indicate an equivalent level of safety is maintained despite not meeting the exact requirements stated in the document.

mega-
Prefix indicating 1 E+6.

megacurie
One million curies.

megahertz (mHz)
One million hertz.

megawatt electric (MWE)
One million watts of electric capacity.

megger
A high voltage resistance meter.

MEI
See ***machine efficiency index***.

meibomian cyst
See ***chalazion***.

mel
A 1,000 hertz tone, 40 decibels above a listener's threshold, produces a pitch of 1,000 mels.

melanin
Dark, sulfur-containing pigment normally found in the skin, ciliary body, choroid of the eye, pigment layer of the retina, certain nerve cells, and hair. It occurs abnormally in certain tumors, known as *melanomas,* and is sometimes excreted in the urine when such tumors are present.

melanoma
A malignant tumor containing dark pigment.

melatonin
A hormone produced in the pineal gland with a circadian cycle, and believed to have a relationship to circadian rhythms.

melting point (mp)
The temperature at which a solid changes to a liquid phase.

membrane potential
That voltage difference measured across the membrane between the interior and exterior of a cell or across an artificial membrane.

membranous labyrinth
A collection of soft-tissue ducts containing endolymph within the osseous labyrinth of the inner ear comprising the semicircular ducts, the cochlear duct, saccule, and utricle.

memomotion
A method of visually sampling work activities at specified periods of time using time-lapse photography or videography.

memomotion study
The use of memomotion for the analysis of long-duration events or processes. Also referred to as *camera study*.

memorandum of understanding (MOU)
An agreement in the nature of a contract between government agencies about how to interpret laws or regulations or implement policies or programs. Sometimes called a *memorandum of agreement (MOA).*

memorize
To absorb information with perfect recall, usually in text, numeric, or pictorial form.

memory
(1) The capacity for mental storage of feelings, sensations, information, movement patterns, and events. The ability of the brain to retain and to use knowledge gained from past experience is essential to the process of learning. Although the exact way in which the brain remembers is not completely understood, it is believed that a portion of the temporal lobe of the brain, lying in part under the temples, acts as a kind of memory center, drawing on memories stored in other parts of the brain. (2) Any of several types of storage means for bits in a computer.

menarche
That phase in a female's life when menstruation begins.

Menière's disease
A disorder of the labyrinth of the inner ear. It is believed to result from dilation of the lymphatic channels in the cochlea. In about 90 percent of cases, only one ear is affected. The usual symptoms are tinnitus, heightened sensitivity to loud sounds, progressive loss of hearing, headache, and dizziness. In the acute stage there may be severe nausea with vomiting, profuse sweating, disabling dizziness, and nystagmus. Some attacks last only minutes, and others continue for hours; they may occur frequently or only several weeks apart. The disease usually lasts a few years, with progressive loss of hearing in the affected ear. Sometimes the symptoms subside before all hearing is lost. If loss of hearing in the affected ear does become complete, nausea symptoms are likely to disappear. The disease sometimes develops after an injury to the head or an infection of the middle ear. Many cases, however, have no apparent cause. The disorder is most common among men between the ages of 40 and 60.

meninges
See ***meninx***.

meningitis
Inflammation of the meninges, the membranes that cover the brain and spinal cord. There are several varieties of meningitis. The two most important are *meningococcal meningitis* (the most common) and *tuberculous meningitis*. Others include *aseptic meningitis* and *viral meningitis*. Meningococcal meningitis is caused by meningococci. It is generally the epidemic type and is very contagious because the bacteria are present in the throat as well as in the cerebrospinal fluid. It is transmitted by contact and by droplet infection. The incubation period for epidemic

meningitis is 2 to 10 days. Also called *cerebrospinal meningitis*.

meninx
A layer of tissue which covers the brain and spinal cord (plural is *meninges*).

meniscus
Something of crescent shape, as the concave or convex surface of a column of liquid in a pipette or burette, or a crescent-shaped fibrocartilage (semi-lunar cartilage) in the knee joint.

menopause
That phase of a women's life at which the menstrual cycle terminates.

menses
The time of menstruation.

menstrual cycle
The hormonally regulated period of approximately 28 days during which a woman normally undergoes ovulation and menses.

menstruate
To pass blood and other tissues from the uterus via the vaginal orifice during part of the menstrual cycle.

mental
Of or pertaining to the mind or intellectual/cognitive activities or functions.

mental age
The mental competence of an individual relative to the chronological age of an average individual with equivalent mental competence. *See also* ***chronological age*** *and* ***developmental age***.

mental basic element
Any work element which involves some form of mental activity.

mental health
A state in which an individual or population has accomplished a high degree of self-realization and integrated its own desires while successfully adapting to its environment.

mental hygiene
That field of study and practice for the development and/or preservation of mental and emotional health.

mental retardation
A mental handicap in which less than normal intellectual functioning is exhibited.

mental work
Any work done by an individual primarily using perceptual and cognitive abilities, especially those involving such activities as calculating, reasoning, monitoring, decision-making, and verbal/image processing.

mental workload
Any measure of the amount of mental effort required to perform a task.

menthol
An alcohol from various mint oils or produced synthetically, used locally to relieve itching.

menton
The point at the tip of the chin in the midsagittal plane. Typically represented by the most anterior point in anterior-posterior measures, by the most inferior point in vertical measures.

menton – crinion length
The vertical distance from the bottom surface of the tip of the chin to the hairline in the midsagittal plane. Measured with the individual standing or sitting erect, looking straight ahead with the facial muscles relaxed (not applicable on bald or balding persons).

menton projection
The horizontal linear distance in the midsagittal plane from the most anterior point of the chin to the junction of the neck and the bottom of the jaw. Measured with the facial musculature relaxed.

menton – sellion length
The vertical linear distance from the inferior tip of the chin to the deepest point of the nasal root depression. Measured with the individual sitting or standing erect, looking straight ahead with the facial muscles relaxed.

menton – subnasale length
The vertical linear distance between the junction of the base of the nasal septum and the superior philtrum to the base of the chin in the midsagittal plane. Measured with the individual sitting or standing erect, looking straight ahead with the facial muscles relaxed.

menton to back of head
The horizontal linear distance from inion to the most anterior portion of the chin. Measured with the individual standing erect and looking straight ahead.

menton to vertex
The vertical linear distance from the inferior tip of the chin to the vertex plane level. Measured with the individual standing erect and looking straight ahead with the facial muscles relaxed.

menton to wall
The horizontal linear distance from a wall to the most anterior portion of the chin. Measured with the individual standing erect with his/her back and head against the wall.

menu
A display of the possible options available to the user from a given command location in a software package.

menu bar
A function area within a screen display which contains a menu.

menu hierarchy
The structure with which a menu is organized, generally with higher level menus providing access to other comparable level menus as well as lower level menus under a given level.

MEOP
*See **maximum expected operating pressure**.*

meq
Milliequivalent.

mercers
Pertaining to mercury in its lower valence; containing monovalent mercury.

mercury
A chemical element, atomic number 80, atomic weight 200.59, symbol Hg. A heavy metal, existing in liquid state under standard conditions, that can accumulate in the environment and is highly toxic if breathed or swallowed. *See also **heavy metal**.*

mercury-fluorescent lamp
A relatively high-intensity discharge lamp using high pressure mercury enclosed within a tube whose interior is coated with phosphorus to convert the ultraviolet light into visible light.

mercury lamp
An illumination source which operates by passing an electrical current between two electrodes in an ionized mercury vapor atmosphere, giving off a bluish-green light with a significant amount of ultraviolet light.

meridian
An imaginary line on the surface of a globe or sphere, connecting the opposite ends of its axis.

meridional flow
A type of atmosphere circulation pattern in which the north-south component of the wind is pronounced.

merit rating
(1) The process of assessing, or the result of an assessment of, performance regarding an employee in a job, usually according to some periodic interval and some specified group of factors such as dependability and work quality or quantity. (2) The process of determining or the resulting determination of tax or insurance premium rates based on an employer's record for disabling injuries and layoffs. Also referred to as *experience rating*.

mescaline
A poisonous alkaloid derived from a Mexican cactus, which produces hallucinations of sound and color.

mesh
That latticework in computer modeling which divides a large object into finite elements.

mesocyclone
A vertical column of cyclonically rotating air within a severe thunderstorm.

mesomorph
A Sheldon somatotype denoted by prominent muscular tissue, heavy bones, broad shoulders, and a flat abdomen.

meson
A short-lived unstable particle with or without electric charge which generally weighs less than a proton and more than an electron.

mesopic vision
That vision using both the rods and cones at moderate luminous intensities; an intermediary between photopic and scotopic vision.

mesoscale
The scale of meteorological phenomena that

ranges in size from a few kilometers to about 100 kilometers. It includes local winds, thunderstorms, and tornadoes.

mesosphere
The atmospheric layer between the stratosphere and the thermosphere. Located at an average elevation between 50 and 80 kilometers above the earth's surface.

mesothelioma
A rare neoplasm that grows as a thick sheet in the pleura of the lungs and in the peritoneum. This condition has been demonstrated in workers who have had extensive exposures to asbestos. *See also* ***asbestos***.

mesothelium
A layer of flat cells, derived from the mesoderm, that lines the coelom of the body cavity of the embryo. In the adult it forms the simple squamous-celled layer of the epithelium that covers the surface of all true serous membranes (peritoneum, pericardium, and pleura).

message area
A function area for the system or other users to communicate with a user.

message line
A single line within a message area.

met
A unit of physiological workload; the metabolic thermal output of an average, sitting, resting individual under conditions of thermal comfort.

metabolic gradient
A difference in degree of metabolic activity from one region of the body to another.

metabolic heat production
The transformation of chemical energy into heat energy by the body.

metabolic rate
The calories (or BTUs) required by the body to sustain vital functions, such as the action of the heart and breathing. The rate depends on the physical activity of the individual and physiological factors. Also referred to as *energy expenditure*. *See also* ***basal metabolic rate***.

metabolic reaction
These include hydrolysis, oxidation, reduction, and conjugation (alkylation, esterification, and acylation).

metabolic reserves
The potential chemical energy source, stored primarily as glycogen, which can be rapidly mobilized for use by the body, especially for muscular activity involving effort beyond one's normal level of activity.

metabolism
(1) The set of biochemical transformations that a chemical undergoes in the body by which energy is made available for use by the organism. (2) The sum total of the physical and chemical processes and reactions taking place among the ions, atoms, and molecules of the body. Essentially these processes are concerned with the disposition of the nutrients absorbed into the blood following digestion. There are two phases of metabolism: the anabolic and catabolic. The anabolic, or constructive, phase is concerned with the conversion of simpler compounds derived from nutrients into living, organized substances that the body cells can use. In the catabolic, or destructive, phase these organized substances are reconverted into simpler compounds, with the release of energy necessary for the proper functioning of the body cells. The rate of metabolism can be increased by exercise, by elevated body temperature (as in high fever, which can more than double the metabolic rate), by hormonal activity (such as that produced by thyroxine, insulin, and epinephrine), and by specific dynamic action that occurs following the ingestion of a meal. The *basal metabolic rate* refers to the lowest rate obtained while an individual is at complete physical and mental rest. This rate is frequently used in the diagnosis of various diseases, especially in malfunctioning of the thyroid gland.

metabolite
Any product (foodstuff, intermediate, waste product) of metabolism.

metacarpal bone
One of the bones in the hand between the wrist bones and the phalanges which make the rigid structure of the palm and back of the hand. *See also* ***hand***.

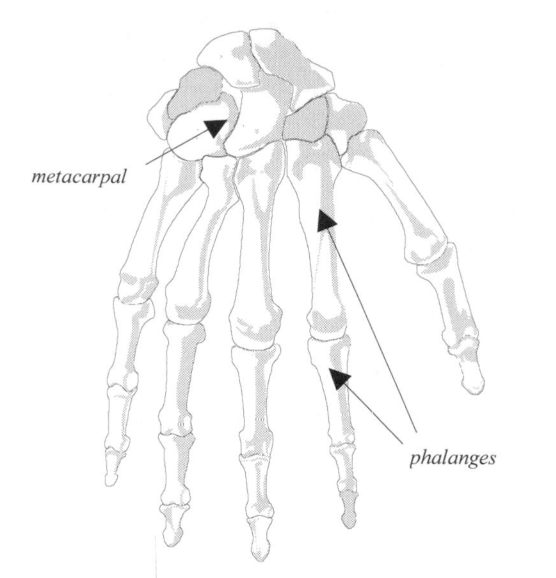

Metacarpal bones and their relationship to the phalanges

metal-clad (MC) cable
Type MC cable is a factory assembly of one or more conductors, each individually insulated and enclosed in a metallic sheath of interlocking tape, or a smooth or corrugated tube.

metal fume fever
An acute condition, usually of short duration, caused by the inhalation of finely divided fumes of zinc, magnesium, copper, or their oxides, and possible others produced during hot work (such as welding). Symptoms can appear 4 to 12 hours after exposure and consist of fever and chills. Most cases are the result of inhalation of zinc oxide from the welding of galvanized steel.

metal halide lamp
A high-intensity discharge lamp in which the primary light is produced from metal halide radiation and its dissociation products.

metalizing
An industrial process involving the coating of parts with molten metal, usually aluminum, by means of vacuum deposition. The process often presents occupational health hazards from metal fumes, dust, heat, and nonionizing radiation.

metallic
Possessing a brilliant luster, characteristic of most metals.

metamer
A visual stimulus which is perceptually indistinguishable from another visual stimulus under one given type of illumination, but which has a different spectral composition and may be distinguishable under another type of illumination.

metameric pair
Two colored visual stimuli which appear identical to the eye, but which consist of different spectral compositions.

metamerism
A condition in which two colored stimuli appear the same under one illuminant but different under another illuminant.

metaphysis
The region of bone growth near the ends of a long bone.

metastasis
(1) Transmission of disease from the original site to one or more other sites elsewhere in the body. (2) Spread by malignancy from the site of primary cancer to a secondary site by transfer through the lymphatic or blood system. (3) The transfer of disease from one organ or part to another not connected with it. Metastasis may be due to the transfer of pathogenic microorganisms or to the transfer of cells, as occurs in malignant tumors. (4) The transfer of malignant neoplastic cells from the original or parent site to a more distant one, with the resultant appearance of a neoplasm.

metastasize
To be transferred, transmitted, or transformed by metastasis. *See also **metastasis***.

metatarsal bone
One of the bones of the foot anterior to the tarsus.

metencephalon
(1) The part of the central nervous system comprising the pons and cerebellum. (2) The anterior of two brain vesicles formed by specialization of the rhombencephalon in the developing embryo.

Meteorological Impact Statement
An unscheduled planning forecast describing conditions expected to begin within 4 to 12 hours which may impact the flow of air traffic in a specific Air Route Traffic Control Center's (ARTCC) area.

meteorology
The study of the atmosphere and atmospheric phenomenon as well as the atmosphere's in-

teraction with the earth's surface, oceans, and life in general.

meter (m)
An SI/MKS unit of length; a distance equal to 1,650,763.73 wavelengths of the radiation emitted corresponding to the Krypton-86 atom transition between levels $2p_{10}$ and $5d_5$ in vacuum.

meter-angle
The amount of eye convergence when the visual axis of each eye is centered on an object one meter distant from the cornea of each eye.

meter fix/slot time
Aviation. A calculated time to depart the meter fix to cross the vertex at the Actual Calculated Landing Time (ACLT). This time reflects descent speed adjustment and any applicable time that must be absorbed prior to crossing the meter fix.

metered data
End-use data obtained through the direct measurement of the total energy consumed for specific uses within the individual household. Individual appliances can be sub-metered by connecting the recording meters directly to individual appliances.

metering
Aviation. A method of time regulating arrival traffic flow into a terminal area so as not to exceed a predetermined terminal acceptance rate.

metering airports
Airports adapted for metering and for which optimum flight paths are defined. A maximum of 16 airports may be adapted.

metering fix
Aviation. A fix along an established route from over which aircraft will be metered prior to entering terminal airspace. Normally, this fix should be established at a distance from the airport which will facilitate a profile descent 10,000 feet above airport elevation or above.

meters, kilograms, and seconds (MKS) system
A metric system of measurement. *See also* ***basic units***.

methadone
A synthetic compound with pharmacological properties qualitatively similar to those of morphine.

methamphetamine
An adrenergic, central nervous system stimulant used in the treatment of narcolepsy, chronic fatigue states, alcoholism, and depression. Since it depresses the appetite, it is also used in the control of obesity. Its actions are similar to those of amphetamine and so it may produce insomnia, excitement, and elevation of blood pressure. Prolonged use can lead to dependence.

methane
A colorless, nonpoisonous, flammable gas created by anaerobic decomposition of organic compounds.

methanol
(1) A light, volatile alcohol (CH_3OH) eligible for motor gasoline blending. (2) A colorless poisonous liquid with essentially no odor and very little taste. It is the simplest alcohol and boils at 64.7 degrees Celsius. In transportation, methanol is used as a vehicle fuel by itself (M100), or blended with gasoline (M85).

methemoglobin
(1) An oxidized form of hemoglobin containing iron in the ferric state. (2) A compound formed with hemoglobin as a result of the oxidation of iron present in hemoglobin, from the ferrous to ferric state. This form of hemoglobin does not combine with and transport oxygen to tissues.

methemoglobinemia
A blood disease caused by the accumulation of excessive methemoglobin in the blood, usually due to toxic action of drugs or other agents, or to the hemolytic process. In infants, it may be caused by the ingestion of water high in nitrates and may also be referred to as *blue baby syndrome* since the excessive concentration of reduced hemoglobin in the blood causes the surface of the skin to appear as a tinted blue, especially the fingers, hands, feet, lips, and ears. See also ***cyanosis***.

method
A technique, orderly sequence of steps, or set of operations used to perform some given task.

method 18
An EPA test method which uses gas chromatographic techniques to measure the concentration of individual volatile organic compounds in a gas stream.

method 24
An EPA reference method to determine density, water content, and total volatile content (water and VOC) of coatings.

method 25
An EPA reference method to determine the VOC concentration in gas stream.

method blank
Used to calibrate the instrument chosen to test a sample.

method of adjustment
A psychophysical methodology in which the subject actively varies some aspect of a stimulus until the variable stimulus appears either to match or be just noticeably different from a fixed reference stimulus, as specified for the test.

method of constant stimuli
A psychophysical methodology in which stimuli are presented to the subject who is to make judgements about how they differ from a standard stimulus, whether greater or lesser along some dimension.

method of equal-appearing intervals
A psychophysical methodology in which the subject adjusts a set of stimuli until the elements of the set appear equidistant from each other along some dimension in an attempt to establish interval level data.

method of limits
A psychophysical methodology in which some dimension of a stimulus is changed in small increments in an ascending/descending manner until the subject either ceases responding or changes his/her response.

method of loci
A visualization type of mnemonic in which a sequence of locations is used to remember a sequence of events.

method of magnitude estimation
A psychophysical methodology in which a subject assigns relative quantitative values to stimuli based on their intensity compared to a reference value.

method of paired comparisons
A psychophysical methodology in which all possible pairs of stimuli are presented to a subject for comparison along one or more dimensions.

method of rank order
An ordinal-level psychophysical methodology in which stimuli are presented to a subject for ranking along a specified dimension.

method of ratio estimation
A psychophysical methodology in which a subject is instructed to adjust or rate a stimulus along some dimension such that it is a specified ratio of a reference stimulus.

methodology
The standard technique used to accomplish different tasks.

methods design
The process of developing improved work methods to improve job performance.

methods engineering
The analysis, design, and implementation of improved work methods and systems where human effort is used.

methods of self-reporting
Reporting of suspected violations of the law or corporate policy vary according to the size of the organization. They may include direct communication with a supervisor, suggestion boxes, toll-free numbers or hotlines. Such systems allow reports to be made without fear of retribution and maximum confidentiality. *See also* ***Self-Reporting System***.

methods study
A systematic examination of the techniques, factors, and resources involved in the component parts of one or more operations. With the intent of improving techniques and productivity, while reducing costs.

Methods Time Measurement (MTM)
A predetermined motion time system (exists in several versions).

methylate
A compound of methyl alcohol and a base.

metric
Refers to the modernized metric system known as the International System.

metric system
International decimal system of weights and measures based on the meter and kilogram.

metric ton (MT)
(1) A unit of weight equal to 2,204.6 pounds.
(2) 1000 kilograms; equal to 2204.6 pounds avoirdupois or 2679.23 pounds troy.

metrocyte
(1) A mother cell. (2) A large uninuclear cell containing hemoglobin; supposed to be the mother cell of the red corpuscles of the blood.

Metropolitan Planning Area
The geographic area in which the metropolitan transportation planning process required by 23 U.S.C. 134 and section 8 of the Federal Transit Act (49 U.S.C. app. 1607) must be carried out.

Metropolitan Planning Organization (MPO)
The forum for cooperative transportation decision making for a metropolitan planning area. Formed in cooperation with the state, develops transportation plans and programs for the metropolitan area. For each urbanized area, a Metropolitan Planning Organization (MPO) must be designated by agreement between the Governor and local units of government representing 75% of the affected population (in the metropolitan area), including the central cities or cities as defined by the Bureau of the Census, or in accordance with procedures established by applicable State or local law (23 U.S.C. 134(b)(1) of the Federal Transit Act of 1991, Sec. 8(b)(1)).

metropolitan railway
Another name for "heavy rail."

Metropolitan Statistical Area (MSA)
Areas defined by the U.S. Office of Management and Budget. A Metropolitan Statistical Area (MSA) is a) a county or a group of contiguous counties that contains at least one city of 50,000 inhabitants or more; or b) an urbanized area of at least 50,000 inhabitants and a total MSA population of at least 100,000 (75,000 in New England). The contiguous counties are included in an MSA if, according to certain criteria, they are essentially metropolitan in character and are socially and economically integrated with the central city. In New England, MSAs consist of towns and cities rather than counties. *See also* ***central city*** *and* ***standard metropolitan statistical area***.

metropolitan status
A building classification referring to the location of the building either located within a Metropolitan Statistical Area (MSA) or outside a MSA.

MEUI
See ***machine effective utilization index***.

MeV
Mega electron volts or million electron volts. A unit of energy commonly used in nuclear physics equivalent to 1.6 x 10^{-6} ergs. Approximately 200 MeV are produced for every nucleus that undergoes fission.

mfpcf
See ***million fibers per cubic foot***. The former method to express airborne asbestos fiber concentration.

mg
Milligram (0.001 gram (g) or 1000 micrograms (L)].

mG
Milligauss.

mg/kg
Milligrams per kilogram.

mg/m^3
Milligram(s) per cubic meter.

mgd
See ***million gallons per day***.

mGy
Milligray(s).

mho
A unit of conductance equal to the reciprocal of the ohm.

mHz
See ***megahertz***.

MIC
See ***manufacturer identification code***.

mica pneumoconiosis
A disease of the lung caused by excessive inhalation of mica dust usually over a period of many years.

Michigan Anthropometric Processor (MAP)
A software program developed by the University of Michigan for analog-to-digital acquisition and real-time checking of anthropometric data.

micro-
Prefix indicating one-millionth.

microaerophilic
Refers to organisms that are aerobic but require reduced concentrations of oxygen (pressures lower than 0.2 atmosphere) and elevated levels of carbon dioxide to grow.

microbar
A unit of pressure commonly used in acoustics and equal to one dyne per square centimeter.

microbes
Microscopic organisms, such as algae, viruses, rickettsia, bacteria, fungi, and protozoa, some of which cause disease in humans, plants, and/or animals.

microbial
See ***biological contaminants***.

microbial growth inhibition
Control of growth of microorganisms by various methods (such as heat, dehydration, refrigeration, chemicals, etc.), that can also be used to destroy them in some cases.

microbial pesticide
A microorganism that is used to control a pest. They are of low toxicity to man.

microbiologicals
See ***biological contaminants***.

microbiology
The science concerned with the study of microscopic and ultramicroscopic organisms (protistology).

microbridge
A cargo movement in which the water carrier provides a through service between an inland point and the port of load/discharge.

microburst
A small downburst with outbursts of damaging winds extending 2.5 miles or less. In spite of its small horizontal scale, an intense microburst could induce wind speeds as high as 150 knots.

microchronometer
A large-faced electric clock with marked time units in decimal minutes with rapidly moving hands used for noting the time in micromotion studies.

microclimate
The climate structure of the airspace near the surface of the earth.

microcurie
One-millionth of a curie.

microearthquake
A term used to describe earthquakes under Richter magnitude 2, and occasionally, slightly larger quakes, especially those not felt by people nearby.

microelement
An element of work which occurs in an interval of time too short to allow it to be adequately observed with the unassisted capacity of the human eye/perceptual system.

microgravity
Any environment in which objects of significant size and mass appear to remain suspended indefinitely, usually due to hypogravitational conditions on the order of 10^{-6} g. Commonly referred to as *zero gravity*.

microgravity growth factor
That proportional increase in body height, primarily within the torso due to release from intervertebral disk compression, which an individual experiences when exposed to microgravity conditions. Generally about 3%.

Micro-Matic Methods and Measurement
A computerized predetermined motion time system.

micrometeorology
The meteorological characteristics of a local area that is usually small in size (e.g., acres or several square miles) and is often limited to a shallow layer of the atmosphere near the ground.

micrometer (µm)
A unit of length equal to one-millionth of a meter.

micromicrocurie
One-trillionth of a curie.

micromicrogram
One-millionth (10^{-6}) microgram, or 10^{-12} gram.

micromole
One millionth of a mole.

micromotion data
See ***simultaneous motion chart***.

micromotion study
The use of normal or high-speed photographic or videographic frame rates for the frame-by-frame analysis of events or processes which

occur too rapidly for adequate real-time observation by the eye.

micron
A one-millionth part of a meter (i.e., 10^{-6} meter or 10^{-4} centimeter). It is roughly four one-hundred thousandth (4×10^{-5}) of an inch.

microorganism
Any microscopic of submicroscopic organism, especially any of the viruses, rickettsia, bacteria, or protozoa.

microphone
An electroacoustic transducer that responds to sound waves and delivers essentially equivalent electric waves. A conduit for producing amplified sound.

micropredator
An organism that derives elements for its existence from other species of organisms larger than itself, without destroying them.

microreciprocal Kelvin (mirek)
A unit of reciprocal color temperature equal to $10^{-6}/T_k$.

microscope
An instrument used to obtain an enlarged image of small objects and reveal details of structure not otherwise distinguishable.

microscopy
By use of a microscope; investigation by means of a microscope.

microsleep
A very brief sleep period, usually on the order of a few seconds to minutes, in an individual who is not permitted to go to sleep but is too fatigued to remain awake.

microspectrophotometer
A device for measuring the wavelength and intensity of light absorbed as it passes through a transparent substance.

microwave
Nonionizing electromagnetic radiation that has a wavelength of 3 meters to 3 millimeters (a frequency of 30 mHz to 300 GHz), often used in cooking, radar, communications, and dosimetry.

microwave cataract
A partial or complete opacity of the eye lens due to microwave radiation exposure.

microwave dosimetry
The study or measurement of the amount of microwave energy to which a system is exposed.

microwave hearing effect
An auditory sensation apparent as a clicking or buzzing sound in humans exposed to pulsed microwave energy.

microwave landing system (MLS)
(1) An instrument landing system operating in the microwave spectrum which provides lateral and vertical guidance to aircraft having compatible avionics equipment. (2) A precision instrument approach system operating in the microwave spectrum which normally consists of the following components: a) azimuth station, b) elevation station, c) precision distance measuring equipment.

microwave nonthermal effects
The presumed non-heating effects from exposure to low power microwave energy.

microwave oven
Oven which is designed to heat, cook, or dry food through the application of electromagnetic energy, and which is designed to operate at a frequency of 916 megahertz (mHz) or 2.45 gigahertz (GHz).

microwave radiation
Electromagnetic radiation in the frequency range from 30 mHz to 300 GHz.

microwave thermal effects
An alteration in biological systems due to the heat produced by absorbed microwave energy.

micturition
The process of urine secretion.

mid-
Located at approximately the axial center of some entity.

midaxillary line
An imaginary or marked vertical line passing through the antero-posterior center of the axilla and down the side of the trunk.

midaxillary line at umbilicus level skinfold
The thickness of a skinfold at the umbilicus level in the midaxillary line. Measured with the individual standing comfortably erect and arms hanging naturally at the sides.

midaxillary plane
A vertical plane extending through the centers of the armpits which divides the body into anterior and posterior segments.

midbrain
The short part of the brain stem just above the pons. It contains the nerve pathways between the cerebral hemisphere and the medulla oblongata, and also contains nuclei (relay stations or centers) of the third and fourth cranial nerves. The center for visual reflexes, such as moving the head and eyes, is located in the midbrain.

midday period
The period between the end of the A.M. peak and the beginning of the P.M. peak.

middle bar
A bar in the middle of a river.

middle compass locator (MCL)
Aviation. A compass locator installed at the site of the middle marker of an instrument landing system.

middle ear
That portion of the ear within the temporal bone in which the auditory ossicles are located, between the tympanic membrane and the oval and round windows.

middle latitude cyclone
See ***extratropical cyclone.***

middle marker (MM)
Aviation. A marker beacon that defines a point along the glideslope of an instrument landing system (ILS) normally located at or near the point of decision height (ILS Category I). It is keyed to transmit alternate dots and dashes, with the alternate dots and dashes keyed at the rate of 95 dot/dash combinations per minute on a 1300 Hz tone, which is received aurally and visually by compatible airborne equipment. *See also* ***compass locator, glideslope, instrument landing system, localizer,*** *and* ***outer marker.***

middle ultraviolet
That portion of the ultraviolet radiation spectrum from about 200 to 300 nm.

midget
A normal dwarf. An individual who is undersized but perfectly proportioned.

midget impinger
A sampling device which can be used to collect dusts for concentration determinations by the light field microscopic technique or to collect materials (gases, mists, or vapors) by absorption in a liquid absorbent material. The sampling rate for dusts by this method is 0.1 cubic feet per minute and the result is expressed as millions of particles per cubic foot of air.

midgrade unleaded gasoline
Gasoline having an antiknock index (R+M/2) greater than or equal to 88, or less than or equal to 90, and containing not more than 0.05 grams of lead or 0.005 grams of phosphorus per gallon. *See also* ***fuel, gasohol, gasoline,*** *and* ***kerosene.***

midi
See ***musical instrument digital interface.***

midline
An imaginary line or plane which divides a structure into two approximately symmetrical parts.

midpatella
A point on the patella which is midway between the superior and the inferior margins of the patella.

midpatella height
The vertical distance from the floor or other reference surface to midpatella. Measured with the individual standing erect, but with relaxed leg musculature, and his/her weight evenly distributed between both feet.

mid-rail
A rail approximately midway between a guardrail and a platform, used when required, and secured to the uprights erected along the exposed sides and ends of platforms.

midsagittal
Pertaining to the midsagittal plane or a point on it.

midsagittal plane
The imaginary plane which divides the body or other (approximately) symmetrical structure into right and left sections.

midshoulder
A point half way between the neck-shoulder junction and acromion.

midshoulder height, sitting
The vertical distance from the seat upper surface to midshoulder. Measured with the individual sitting erect, with the head and back against a wall.

midthigh
A position midway between the inguinal crease and superior aspect of the patella along the midline of the leg as described in the thigh length measure.

MIG welding
Metal inert gas welding.

migraine
A headache, usually severe, often limited to one side of the head, and sometimes accompanied by nausea and vomiting. Although the cause is not completely understood, migraine is thought to be associated with constriction and then dilation of the cerebral arteries. It is also thought to have a psychological aspect, since it occurs most often in persons with particular types of personalities and often follows emotional disturbances. Migraine tends to run in families. In women, the headaches often occur during menstrual periods. The symptoms of migraine vary greatly not only from person to person but also from time to time in the same person. The headaches are usually intense and they frequently occur on one or the other side of the head. They are often accompanied by nausea and vomiting. A typical migraine attack begins with changes in vision, such as a flickering before the eyes, flashes of light, or a blacking out of part of the sight.

migration
In air sampling, the undesired transfer of an absorbed material from the front section of a solid sorbent tube to the back-up section.

mil
Unit of length equal to one-thousandth of an inch. Also, one-thousandth of a radian.

Milankovitch theory
A theory proposed by Milutin Milankovitch in the 1930s suggesting that changes in the earth's orbit were responsible for climatic changes and the ice ages.

miliaria
A cutaneous condition with retention of sweat, which is extravasated at different levels in the skin. Also called *prickly heat* or *heat rash*.

mile
A statute mile (5,280 feet). Most mileage computations are based on statute miles.

mile board
A 12 by 36 inch board mounted horizontally above a shore to aid in navigation and labeled with the river mileage at that point.

mile marker
A point on a feature indicating the distance, in miles, measured along the course or path of the feature from an established origin point on the feature.

mileage death rate
See ***motor vehicle incidence rate***.

miles in trail
A specified distance between aircraft, normally in the same stratum associated with the same destination or route of flight.

miles of road operated
The single or first main track, measured by the distance between terminals, over which railway transportation service is conducted.

miles of track
The number of tracks per one mile segment of right-of-way. Miles of track are measured without regard to whether or not rail traffic can flow in only one direction on the track. All track is counted, including yard track.

miles of track operated
Total track mileage consisting of first, second, and other main tracks, and of yard tracks and sidings over which railway transportation service is conducted. *See also* ***track mile***.

miles per gallon (MPG)
A measure of vehicle fuel efficiency. Miles per gallon (MPG) represents "fleet miles per gallon." For each subgroup or "table cell," MPG is computed as the ratio of the total number of miles traveled by all vehicles in the subgroup to the total number of gallons consumed. MPGs are assigned to each vehicle using the Environmental Protection Agency (EPA) certification files and adjusted for on-road driving.

miles per gallon (MPG) shortfall
The difference between actual on-road miles per gallon (MPG) and Environmental Protec-

tion Agency (EPA) laboratory test MPG. Miles per gallon (MPG) shortfall is expressed as gallons per mile ratio (GPMR).

milieu
Environment or surroundings.

milinch
One-thousandth of an inch.

military approach controls
Aviation. Military approach control facilities include Army Radar Approach Controls (ARACs), Radar Air Traffic Control Facilities (RATCFs), and Radar Approach Controls (RAPCONs).

military authority assumes responsibility for aircraft separation
A condition whereby the military services involved assume responsibility for separation between participating military aircraft in the Air Traffic Control (ATC) system. It is used only for required instrument flight rules (IFR) operations which are specified in letters of agreement or other appropriate Force Module (FM) or military documents.

military base
An area owned and operated by the government in which various military activities take place.

military fuel
Kerosene-type jet fuel intended for military use.

military instrument flight rules (IFR) training route
Routes used by the Department of Defense and associated Reserve and Air Guard units for the purpose of conducting low altitude navigation and tactical training in both instrument flight rules (IFR) and visual flight rules (VFR) weather conditions below 10,000 feet mean sea level (MSL) at airspeeds in excess of 250 knots indicated airspeed (IAS).

military operations
(1) *General.* Refers to all classes of military operations. (2) *Aviation.* Arrivals and departures of aircraft not classified as civil.

military operations area (MOA)
Aviation. An airspace assignment of defined vertical and lateral dimensions established outside positive control areas to separate/segregate certain military activities from intermediate fix (IF) traffic and to identify for visual flight rules (VFR) traffic where these activities are conducted.

Military Standard (MIL-STD)
A mandatory standard issued by the U.S. Department of Defense for use by contractors or others in manufacturing items for DOD use. *See also **Department of Defense Standard**.*

military training route
Aviation. Airspace of defined vertical and lateral dimensions established for the conduct of military flight training at airspeeds in excess of 250 knots IAS.

milk run
Transit (slang). An easy trip.

mill capital
Cost for transportation and equipping a plant for processing ore or other feed materials.

milli-
Prefix indicating one-thousandth or 10^{-3} of the basic unit.

milliampere
One-thousandth of an ampere.

millibar
One-thousandth of the standard barometric pressure, 1 E+2 newtons per square meter, or 9.87 E-4 bar.

millicurie
One-thousandth of a curie.

millicurie-of-intensity-hour
*See **sievert**.*

milliequivalent
One-thousandth of an equivalent weight of a substance.

milligram
One-thousandth of a gram.

milligram-hour
A unit of radiation dose, equivalent to 1 milligram of radium for one hour.

milliliter
One-thousandth of a liter.

millimeter
One-thousandth of a meter.

millimeter of mercury (mm Hg)
A unit of pressure equal to that exerted by a column of liquid mercury one millimeter high at standard temperature.

millimicron
Unit of length equal to one-thousandth of a micron.

millimole
One-thousandth of a mole.

million-gallons per day (mgd)
A measure of water flow.

millions of fibers per cubic foot of air (mfpcf)
Former unit of expressing the airborne concentration of asbestos fibers in air.

millions of particles per cubic foot of air (mppcf)
Former unit for expressing the airborne concentration of dusts, such as coal dust.

millirem
One-thousandth of a rem. *See also* ***rem***.

milliroentgen
One-thousandth of a roentgen. *See also* ***roentgen***.

Minamata disease
A neurological disorder caused by alkyl mercury poisoning, typically characterized by peripheral and circumoral parasthesia, ataxia, dysarthria, and loss of peripheral vision and leading to permanent neurological and mental disability or death.

Mine Safety and Health Administration (MSHA)
A United States federal agency which regulates matters pertaining to health and safety issues regarding mining operations and the mineral industry. It carries out inspections, investigations, enforces regulations, provides technical support, develops relevant training programs, and assesses penalties for violations of regulations.

mineral
A naturally occurring inorganic homogeneous substance. There are 19 or more minerals forming the mineral composition of the human body, at least 13 are essential to health.

mineral-insulated (MI) metal-sheathed cable
Type MI metal-sheathed cable is a factory assembly of one or more conductors insulated with a highly compressed refractory mineral insulation and enclosed in a liquid-tight and gas-tight continuous copper sheath.

mineral resources
All nonliving, natural, nonrenewable resources, including fossil fuels, minerals, whether metallic or nonmetallic, but not including ice, water, or snow.

mineral wood
A manmade mineral fiber material that is made from various types of silicate rock. Also referred to as rock wool or slag wool, depending upon the source.

miner's asthma
Asthma associated with anthracosis.

miner's cramp
See ***heat cramps***.

miner's helmet
A safety helmet with an attached lamp.

mini-Gym
See ***MK-1, II***.

mini landbridge
Transportation. An intermodal system for transporting containers first by ocean and then by rail or motor to a port for additional transport over water.

mini service
Service station attendants pump vehicle fuel but do not provide other services, such as checking oil and tire pressure or washing windshields.

miniature railway
Small-scale railway used for amusement.

minibridge
A joint water, rail, or truck container move on a single Bill of Lading for a through route from a foreign port to a U.S. port destination through an intermediate U.S. port or the reverse. *See also* ***intermodal*** *and* ***landbridge***.

minimal passageway
That minimal height and width of a corridor which allows an individual clothed for specified working conditions to pass without conflict with boundaries or other persons.

minimal perceptible erythema
See ***erythemal threshold***.

minimal weight
The least amount a person can weigh without endangering lean body mass and essential fat storage.

minimum
The lowest active, existing, or permissible value.

minimum altitude
Minimum altitudes for instrument flight rules (IFR) operations as prescribed in Federal Aviation Regulation (FAR) Part 91. These altitudes are published on aeronautical charts and prescribed in FAR Part 96 for airways and routes, and in FAR Part 97 for standard instrument approach procedures. If no applicable minimum altitude is prescribed in FAR Part 96 or FAR Part 97, the following minimum IFR altitude applies: a) in designated mountainous areas, 2,000 feet above the highest obstacle within a horizontal distance of 4 nautical miles from the course to be flown; b) other than mountainous areas, 1,000 feet above the highest obstacle within a horizontal distance of 4 nautical miles from the course to be flown; or c) as otherwise authorized by the [Federal Aviation Administration (FAA)] Administrator or assigned by Air Traffic Control (ATC).

minimum angle of resolution
See ***minimum resolution angle***.

minimum cost life
See ***economic life***.

minimum crossing altitude
The lowest altitude at certain fixes at which an aircraft must cross when proceeding in the direction of a higher minimum en route instrument flight rules (IFR) altitude (MEA).

minimum descent altitude (MDA)
The lowest altitude, expressed in feet above mean sea level, to which descent is authorized on final approach or during circle-to-land maneuvering in execution of a standard instrument approach procedure where no electronic glide slope is provided. *See also* ***height above airport***.

minimum design duct velocity
See ***transport velocity***.

minimum detectable quantity
In instrumentation, the amount of material (e.g., micrograms) which gives a response equal to twice the detector noise level.

minimum detectable sensitivity
In instrumentation, the smallest amount of input concentration that can be detected as the concentration approaches zero.

minimum detection limit
The lowest concentration or weight of a substance which an instrument can reliably quantify.

minimum dose
The smallest quantity of an agent which will produce a physiological effect.

minimum en route instrument flight rules (IFR) altitude (MEA)
The lowest published altitude between radio fixes which assures acceptable navigational signal coverage and meets obstacle clearance requirements between those fixes. The minimum en route altitude (MEA) prescribed for a federal airway, or segment thereof, area navigation low or high route, or other direct route applies to the entire width of the airway, segment, or route between the radio fixes defining the airway, segment, or route.

minimum erythemal dose
(1) The amount of energy (usually ultraviolet) expressed in microwatt-seconds per square centimeter of skin to which skin can be safely exposed. (2) The smallest radiant exposure (e.g., UV radiation) that produces a barely perceptible reddening of the skin that disappears after 24 hours. Also referred to as *minimal erythemal dose*.

minimum fuel
Aviation. Indicates that an aircraft's fuel supply has reached a state where, upon reaching the destination, it can accept little or no delay. This is not an emergency situation but merely indicates an emergency situation is possible should any undue delay occur. *See also* ***fuel remaining***.

minimum holding altitude
The lowest altitude prescribed for a holding pattern which assures navigational signal coverage, communications, and meets obstacle clearance requirements.

minimum instrument flight rule (IFR) altitude (MIA)
Minimum altitudes for Instrument Flight Rules (IFR) operations as prescribed in Federal Aviation Regulations (FAR) Part 91.

minimum lethal dose (MLD)
The smallest dose which kills one of a group of test animals (an older term no longer used).

minimum/minimal separable acuity
See ***minimum resolution angle***.

minimum navigation performance specifications (MNPS)
A set of standards which requires aircraft to have a minimum navigation performance capability in order to operate in MNPS-designated airspace. In addition, aircraft must be certified by their State of Registry for MNPS operation.

minimum navigation performance specifications airspace (MNPSA)
Designated airspace in which minimum navigation performance specifications (MNPS) procedures are applied between MNPS certified and equipped aircraft. Under certain conditions, non-MNPS aircraft can operate in Minimum Performance Specifications Airspace (MNPSA). However, standard oceanic separation minima are provided between the non-MNPS aircraft and other traffic. Currently, the only designated MNPSA is described as follows: a) between Flight Levels 275 and 400; b) between latitudes 27-N. and the North Pole; c) in the east, the eastern boundaries of the Control Areas (CTA) Santa Maria Oceanic, Shanwick Oceanic, and Reykjavik; d) in the west, the western boundaries of CTA's Reykjavik and Gander Oceanic and New York Oceanic excluding the area west of 60-W and south of 38-30'N.

minimum obstruction clearance altitude
The lowest published altitude in effect between radio fixes on very high frequency (VHF) omnidirectional range (VOR) airways, off-airway routes or route segments which meets obstacle clearance requirements for the entire route segment and which assures acceptable navigational signal coverage only within 25 statute (22 nautical) miles of a VOR.

minimum pool elevation
Maritime. The least depth to which a pool is permitted to go and still maintain project channel depth.

minimum population estimate
An estimate of the number of animals in a stock that is based on the best available scientific information on abundance, incorporating the precision and variability associated with such information and provides reasonable assurance that the stock size is equal to or greater than the estimate.

minimum resolution angle
The smallest angular or linear separation at which an individual can resolve two visual objects as separate under a specified set of conditions. Also referred to as *angle of resolution, minimum angle of resolution, resolution angle,* and *minimum/minimal separable acuity.*

minimum safe altitude (MSA)
Aviation. (1) The minimum altitude specified in Federal Aviation Regulation (FAR) Part 91 for various aircraft operations. (2) Altitudes depicted on approach charts which provide at least 1,000 feet of obstacle clearance for emergency use within a specified distance from the navigation facility upon which a procedure is predicated. These altitudes will be identified as Minimum Sector Altitudes or Emergency Safe Altitudes and are established as follows: a) Minimum Sector Altitudes. Altitudes depicted on approach charts which provide at least 1,000 feet of obstacle clearance within a 25-mile radius of the navigation facility upon which the procedure is predicated. Sectors depicted on approach charts must be at least 90 degrees in scope. These altitudes are for emergency use only and do not necessarily assure acceptable navigational signal coverage; b) Emergency Safe Altitudes. Altitudes depicted on approach charts which provide at least 1,000 feet of obstacle clearance in non-mountainous areas and 2,000 feet of obstacle clearance in designated mountainous areas within a 100-mile radius of the navigation facility upon which the procedure is predicated and normally used only in military procedures. These altitudes are identified on published procedures as "Emergency Safe Altitudes."

minimum safe altitude warning (MSAW)
A function of the Automated Radar Terminal System (ARTS) III computer that aids the controller by alerting him when a tracked Mode C equipped aircraft is below or is predicted by the computer to go below a predetermined minimum safe altitude.

minimum sector altitude
The lowest altitude which may be used under emergency conditions which will provide a minimum clearance of 300 m (1,000 feet)

above all obstacles located in an area contained within a sector of a circle of 46 km (25 nautical miles) radius centered on a radio aid to navigation.

minimum transport velocity
The minimum velocity necessary to transport particulates through a ventilation system without their settling out. *See also* ***transport velocity***.

minimum vectoring altitude (MVA)
The lowest mean sea level (MSL) altitude at which an instrument flight rule (IFR) aircraft will be vectored by a radar controller, except as otherwise authorized for radar approaches, departures, and missed approaches.

minimum wage
The minimum hourly rate of compensation for labor, as established by federal statute and required of employers engaged in businesses which affect interstate commerce.

minimums
Aviation. Term used to describe the weather condition requirements established for a particular operation or type of operation, e.g., instrument flight rules (IFR) takeoff or landing, alternate airport for IFR flight plans, visual flight rules (VFR) flights, etc.

mining
The process or business of extracting from the earth the precious or valuable metals, either in their native state or in their ores.

mining danger area
An area identified as a danger to maritime navigation due to unexploded ordinances.

mining waste
Residues which result from the extraction of raw materials from the earth.

minivan
A type of small van that first appeared with that designation in 1984. Any of the smaller vans built on an automobile-type frame. Earlier models such as the Volkswagen van are now included in this category. *See also* ***automobile, bus, car,*** *and* ***motor vehicle***.

minnie
Transit (slang). Less than 100 pound shipment.

minometer
An instrument for measuring stray radiation from radioactive sources.

minor
An individual less than 18 years of age.

minor arterial
Transit. Streets and highways linking cities and larger towns in rural areas in distributing trips to small geographic areas in urban areas (not penetrating identifiable neighborhoods). *See also* ***arterial highway*** *and* ***principal arterial***.

minor axis
The shorter axis defining an ellipse.

minor defect
A defect which may affect appearance, slightly reduce functionality, or other characteristics, but which causes no serious malfunction.

minor injury
An occupational or other injury in which no significant amount of time from work is lost and no major medical costs are incurred.

Minor's sweat test
An examination to measure possible damage to the sympathetic nervous system by determining which dermatomes of the body do not perspire.

minute respiratory volume
The total volume of air moved into and out of the respiratory system per minute.

minutes in trail
Aviation. A specified interval between aircraft expressed in time. This method would more likely be utilized regardless of altitude.

miosis
Contraction of the pupil of the eye.

mirage
A refraction phenomenon that makes an object appear to be displaced from its true position. When an object appears higher than it actually is, it is called a *superior mirage*. When an object appears lower than it actually is, it is an *inferior mirage*.

mirek
See ***microreciprocal Kelvin***.

mirror

(1) A highly specularly reflecting surface. (2) To create a mirror image in computer modeling or graphics.

mirror image

A structure which would correspond at least in part to the reflection of another part of an original object about some plane.

mirror stereoscope

A laboratory device used to present separate images of a scene to each of the eyes by a system of mirrors. Also referred to as *Wheatstone stereoscope.*

misbehavior

Ill conduct; improper or unlawful behavior.

misbranded

According to the Federal Insecticide, Fungicide, and Rodenticide Act: A pesticide is misbranded if (1) its labeling bears any statement, design, graphic representation relative thereto or to its ingredients which is false or misleading in any particular; (2) it is contained in a package or other container or wrapping which does not conform to the standards established by the Administrator of the EPA pursuant to Title 7 Agriculture (Environmental Pesticide Control); (3) it is an imitation of, or is offered for sale under the name of, another pesticide; (4) its label does not bear the registration number assigned under Title 7 to each establishment in which it was produced; (5) any word, statement, or other information required by or under authority of Title 7 to appear on the label or labeling is not prominently placed thereon with such conspicuousness (as compared with other words, statements, designs, or graphic matter in the labeling) and in such terms as to render it likely to be read and understood by the ordinary individual under customary conditions of purchase and use; (6) the labeling accompanying it does not contain directions for use which are necessary for effecting the purpose for which the product is intended and if complied with, together with any requirements imposed under Title 7, are adequate to protect health and the environment; (7) the label does not contain a warning or caution statement which may be necessary and if complied with, together with any requirements imposed under Title 7, is adequate to protect health and the environment; or (8) in the case of a pesticide not registered in accordance with Title 7 and intended for export, the label does not contain, in words prominently placed thereon with such conspicuousness (as compared with other words, statements, designs, or graphic matter in the labeling) as to render it likely to be noted by the ordinary individual under customary conditions of purchase and use, the following: "Not Registered for Use in the United States of America".

A pesticide is also misbranded if the label does not bear an ingredient statement on that part of the immediate container (and on the outside container or wrapper of the retail package, if there is one, through which the ingredient statement on the immediate container cannot be clearly read) which is presented or displayed under customary conditions or purchase, except that a pesticide is not misbranded under this subparagraph if (1) the size or form of the immediate container, or the outside container or wrapper of the retail package, makes it impracticable to place the ingredient statement on the part which is presented or displayed under customary conditions of purchase; and (2) the ingredient statement appears prominently on another part of the immediate container, or outside container or wrapper, permitted by the EPA Administrator.

A pesticide is also misbranded if the labeling does not contain a statement of the use classification under which the product is registered.

A pesticide is also misbranded if there is not affixed to its container, and to the outside container or wrapper of the retail package, if there is one, through which the required information on the immediate container cannot be clearly read, a label bearing (1) the name and address of the producer, registrant, or person for whom produced; (2) the name, brand, or trademark under which the pesticide is sold; (3) the net weight or measure of the content, except that the EPA Administrator may permit reasonable variations; and (4) when required by regulation of the EPA Administrator to effectuate the purposes of Title 7, the registration number assigned to the pesticide under Title 7 and the use classification.

The pesticide is also misbranded if it contains any substance or substances in quantities highly toxic to man, unless the label shall bear, in addition to any other matter required by Title 7 (1) the skull and cross-bones; (2) the word "poison" prominently in red on a background of distinctly contrasting color; (3) and a statement of a practical treatment (first aid or otherwise) in case of poisoning by the pesticide.

miscellaneous transport revenue
Aviation. Other revenues associated with air transportation performed by air carriers, such as transportation fees collected from those traveling on free or reduced transportation and processing service charges such as lost tickets.

miscible
Capable of being mixed in any concentration without separation of phrases.

misconduct
A transgression of some established and definite rule of action; a forbidden act; a dereliction from duty; unlawful behavior; willful in character; improper or wrong behavior; delinquency; impropriety; mismanagement; offense; but not negligence or carelessness.

misdemeanor
Offenses lower than felonies and generally those punishable by fine, penalty, forfeiture, or imprisonment.

misfeasance
The improper performance of some act which a person may lawfully do.

mishap
An occurrence that results in some degree of injury, property damage, or both.

misrepresentation
Any manifestation by words or other conduct by one person to another that, under the circumstances, amounts to an assertion not in accordance with the facts. An untrue statement of fact.

missed approach
Aviation. (1) A maneuver conducted by a pilot when an instrument approach cannot be completed to a landing. The route of flight and altitude are shown on instrument approach procedure charts. A pilot executing a missed approach prior to the missed approach point (MAP) must continue along the final approach to the MAP. The pilot may climb immediately to the altitude specified in the missed approach procedure. (2) A term used by the pilot to inform Air Traffic Control (ATC) that he is executing the missed approach. (3) At locations where ATC radar service is provided, the pilot should conform to radar vectors when provided by ATC in lieu of the published missed approach procedure. *See also* ***go around***.

missed approach point
Aviation. A point prescribed in each instrument approach procedure at which a missed approach procedure shall be executed if the required visual reference does not exist.

missed approach procedure
Aviation. The procedure to be followed if the approach cannot be continued.

missed executed approach
Aviation. Instructions issued to a pilot making an instrument approach which means continue inbound to the missed approach point and execute the missed approach procedure as described on the Instrument Approach Procedure Chart or as previously assigned by Air Traffic Control (ATC). The pilot may climb immediately to the altitude specified in the missed approach procedure upon making a missed approach. No turns should be initiated prior to reaching the missed approach point. When conducting an airport surveillance radar (ASR) or precision approach radar (PAR) approach, execute the assigned missed approach procedure immediately upon receiving instructions to "execute missed approach."

missile
Sometimes applied to space launch vehicles, but more properly connotes automated weapons of warfare, i.e., a weapon which has an integral system of guidance, as opposed to the unguided rocket.

mission
That designated activity at a particular location which a system is intended to accomplish.

mission reliability
The probability that a given product or system will complete a specified mission.

Mississippi River System
Includes the Mississippi River from the head of navigation to its mouth, and navigable tributaries including the Illinois Waterway, Missouri River, Ohio River, Tennessee River, Allegheny River, Cumberland River, Green River, Kanawha River, Monongahela River, and such others to which barge operations extend.

mist
Liquid particles, measuring 40 to 50 microns, that are generated by condensation from the gaseous state to the liquid state, or by the break up of a liquid into a dispersed state (splashing, foaming, or atomizing). In contrast, *fog* particles are smaller than 40 microns.

mistake
Some unintentional act, omission, or error arising from ignorance, surprise, imposition, or misplaced confidence. A state of mind not in accord with reality.

mistrial
An erroneous, invalid, or nugatory trial. A trial of an action which cannot stand in law because of want of jurisdiction, or a wrong drawing of jurors, or disregard of some other fundamental that expenditure of further time and expense would be wasteful if not futile. It is a trial which has been terminated prior to its normal conclusion. The judge may declare a mistrial because of some extraordinary event (e.g., death of a juror or attorney), for prejudicial error that cannot be corrected at trial, or because of a deadlocked jury.

miter gates
Maritime Navigation. Vertical gates which form the openings of navigation locks; these gates consist of two swinging leaves and close at the center.

miter sill
Maritime Navigation. The underwater concrete sill across the openings in the upper and lower lock chamber that the movable lock gates close on. The depth over these cells exceeds project depth and is registered on the several gauges within the lock chamber.

mitigating factors
Under the Federal Sentencing Guidelines (FSGs): A reduction in the amount of damages or penalties. According to the FSGs, penalties may be reduced if an organization a) has an effective compliance program to prevent and detect criminal conduct, and b) is willing to self-report violations and is cooperative and accepting of responsibility for any violations. *See also* ***Federal Sentencing Guidelines***.

mitigation
Measures taken to reduce adverse impacts on the environment, the workplace, or both.

mitigation of damages
Although the law of damages contemplates full and just compensation for negligently inflicting injuries, the law also prescribes, as a reciprocal principle, that a tort feasor should not sustain liability for those damages not attributable to the injury producing event. Consequently, a plaintiff may not recover damages for the effects of an injury which reasonably could have been avoided or substantially ameliorated. This limitation on recovery is generally referred to as *mitigation of damages* or *avoidance of consequences.* Mitigation of damages or avoidance of consequences arises only after the injury-producing event has occurred.

mitosis
Nuclear cell division in which the resulting nuclei have the same number and kind of chromosomes as the original cell. The first step in mitosis is duplication of all genes and chromosomes. To accomplish this, the cell must double its content of DNA. Chromosomes are composed of the DNA molecule loosely bound with protein; genes are segments of the DNA molecule. Since the DNA molecule has the ability to duplicate itself (replication), it is possible for the cell to form two identical sets of chromosomes and genes. After they are duplicated, they divide between the two separate nuclei that have formed. The final step in mitosis is the splitting of the parent cell into two identical daughter cells, each with a full complement of genes and chromosomes. Most cells of the body are continually growing and reproducing, so that when the old cells die the new ones take their place. Thus, mitosis is a continuous process. It is obvious that this reproduction must take place in an orderly manner, but the exact way in which cell growth and reproduction are regulated is not completely understood. Although

certain cells such as blood-forming cells of the skin grow and reproduce continually, other cells such as neurons (nerve cells) do not reproduce during a person's lifetime. Neoplastic disorders such as cancer are a result of the abnormal and unrestricted growth and reproduction of certain body cells.

mitten
A type of fitted hand wear for covering the hand which has a slot for the thumb, but does contain separate finger slots.

mixed cargo
Indicates that a vessel carries any combination of grains, government aid, containers, general or bulk cargoes.

mixed cloud
A cloud containing both water drops and ice crystals.

mixed liquor
A mixture of activated sludge and water containing organic matter undergoing activated sludge treatment in an aeration tank.

mixed radioactive and other hazardous substances
Material containing both radioactive hazardous substances and non-radioactive hazardous substances, regardless of whether these types of substances are physically separated, combined chemically, or simply mixed together.

mixed trains
Mixed trains are passenger-carrying trains consisting of both passenger and freight cars. Freight cars, such as baggage cars, that are equipped with passenger-type braking and suspension systems, are considered to be passenger cars when utilized in passenger service.

mixed waste
Under the Federal Solid Waste Disposal Act: Waste that contains both hazardous waste and source, special nuclear, or byproduct material subject to the Atomic Energy Act of 1954.

mixing depth
The unstable atmospheric layer that extends from the surface up to the base of an inversion.

mixing zone
A term used to represent the volume of receiving water (e.g., river, stream) which is permitted for mixing of the discharge with the receiving water.

mixture
(1) *General.* A heterogeneous association of substances which cannot be represented by a chemical formula. (2) *Federal Toxic Substances Control Act (TSCA).* Any combination of two or more chemical substances if the combination does not occur in nature and is not, in whole or in part, the result of a chemical reaction. Such term does include any combination which occurs, in whole or in part, as a result of a chemical reaction if none of the chemical substances comprising the combination is a new chemical substance and if the combination could have been manufactured for commercial purposes without a chemical reaction at the time the chemical substances comprising the combination were combined.

mixture rule
Used to determine the hazardous nature of a waste product. Although the EPA has specifically excluded numerous chemical mixtures from this rule, it is still generally true that any mixture of a listed hazardous waste with another non-hazardous waste will render the entire volume of the waste product hazardous and subjected to regulation.

MK-I, II
A small commercial exercise device flown in earth orbit on Skylab for exercising arm and back muscles. Also called *mini-Gym.*

MKS System
See ***meters, kilograms and seconds system.*** *See also* ***basic units.***

mL
Milliliter (0.001 liter or 1000 microliters (L).

M_L
See ***local magnitude.***

MLD
See ***minimum lethal dose.***

MLS
See ***microwave landing system.***

mm
Millimeter(s).

mm^2
Square millimeters.

mm^3
Cubic millimeters.

mm Hg
See ***millimeter of mercury***.

MM
See ***middle marker***.

MMA welding
Manual metal arc welding.

mmad
See ***mass median aerodynamic diameter***.

mmcf
Million cubic feet.

MMH
See ***manual materials handling***.

MMMF
See ***manmade mineral fibers***.

mmol
Millimole.

MMPA
Marine Mammal Protection Act of 1972.

MMVF
See ***manmade vitreous fiber***.

mmx
See ***multimedia extension***.

M'Naghten rule
Rule of law that states: "To establish a defense on the ground of insanity, it must be clearly proved that at the time of committing the act the party accused was laboring under such a defect of reason from disease of the mind as not to know the nature or quality of the act he was doing, or, if he did know it, that he did not know he was doing what was wrong."

mnemonic
Any formal technique for aiding in memory storage or recall.

MNPS
See ***minimum navigation performance specifications***.

MOA
Memorandum of agreement. *See* ***memorandum of understanding***. *See also* ***military operations area***.

mobile
Having the freedom or ability to physically move about from one location to another through relatively independent means.

mobile home
A housing unit built on a movable chassis and moved to the site. It may be placed on a permanent or temporary foundation and may contain one room or more. If rooms are added to the structure, it is considered a single-family housing unit. A manufactured house assembled on site is a single-family housing unit, not a mobile home.

mobile home park
An area maintained for the parking of inhabited mobile homes.

mobile source
A moving producer of air pollution, mainly forms of transportation such as cars, trucks, motorcycles, airplanes, etc.

mobile x-ray
X-ray equipment mounted on a permanent base with wheels and/or casters for moving while completely assembled.

mobility aid
Any physical device which enhances one's mobility, especially with regard to the handicapped.

mobility analysis
A determination of which employees have the skills, training, experience, or other capability to move to other jobs if it becomes necessary.

mockup
A full-scale, representative physical layout of a workstation, equipment, or situation used for training or as a design tool.

modal
See ***mode***.

modal share
The percentage of total freight moved by a particular type of transportation.

modal split
(1) The proportion of total person trips that uses each of various specified modes of transportation. (2) The process of separating total person trips into the modes of travel used. (3) A term that describes how many people use alternative forms of transportation. It is frequently used to describe the percentage of people who use private automobiles, as op-

posed to the percentage who user public transportation.

modal time
That element time which occurs with the highest frequency during a time study.

modality
Any sense, such as vision or hearing.

mode
(1) *Statistical Analysis.* The most common or most frequent value that appears during evaluation or observation of a sample population of values. (2) *Transportation.* Any of the following transportation methods: rail, highway, air, or water. Also, transportation planners, analysts, and decision makers refer to the means of transportation as a mode. (3) *Transit.* Service operated in a particular format. There are two types: fixed-route and non-fixed route. (4) *Aviation.* The letter or number assigned to a specific pulse spacing of radio signals transmitted or received by ground interrogator or airborne transponder components of the Air Traffic Control Radar Beacon System (ATCRBS) Mode A (military Mode 3) and Mode C (altitude reporting) are used in air traffic control.

mode C intruder alert
Aviation. A function of certain air traffic control automated systems designed to alert radar controllers to existing or pending situations between a tracked target (known instrument flight rules (IFR) or visual flight rules (VFR) aircraft) and an untracked target (unknown IFR or VFR aircraft) that require immediate attention/action. *See also* ***conflict alert***.

mode S
Aviation. A secondary surveillance radar and communication system in which each aircraft is assigned a unique address code. Using this unique code, interrogations and other messages can be directed to a particular aircraft, and replies can be unambiguously identified.

model
A mathematical representation of real phenomena. It serves as a pattern from which interrelationships can be identified, analyzed, altered, or synthesized without distributing the real world situation. A mathematical and/or physical representation of real world phenomena which serves as a plan or pattern from which interrelationships can be identified, analyzed, synthesized, and altered without disturbing real world processes.

Model Accreditation Plan (MAP)
Related to the accreditation of persons who inspect for the presence of asbestos, develop asbestos management programs, etc. under AHERA and ASHARA as they relate to public buildings.

model bow
Maritime. A shaped, pointed bow.

model plant
A description of a typical but theoretical plant used for developing economic, environmental impact, and energy impact analyses as support for regulations or regulatory guidelines. It is an imaginary plant, with features of existing or future plants used to estimate the first step in exploring the economic impact of a potential NSPS.

Model Rules of Professional Conduct
Rules that were adopted by the American Bar Association in 1983, with technical amendments adopted in 1987, which provide comprehensive treatment of professional conduct in the form of rules as to what an attorney may and may not do in dealing with the court, opposing counsel, his/her client, and third persons. These Rules, which replace the former American Bar Association (ABA) Code of Professional Responsibility, have been adopted by many states (usually by the state supreme court) to govern the conduct of attorneys admitted to practice in the state.

model year
The year in which the particular style or design of vehicle was introduced or manufactured.

modeling
An investigative technique using a mathematical or physical representation of a system or theory that accounts for all or some of its known properties. Models are often used to test the effect of changes of system components on the overall performance of the system.

modem
Modulator/demodulator. A device employed to transform signals for transmission of information and data by telephone lines. A

communication device that allows information to be exchanged between computers via telephone lines.

moderate work
That level of work activity which has a gross metabolic cost of 180–280 calories per square meter of skin surface per hour.

moderator
(1) *General.* In arbitration, or during formal proceedings of any nature, an unbiased person responsible for ensuring the proceedings adhere to a pre-established schedule or agenda. (2) *Nuclear.* A material, such as beryllium, graphite (carbon), or water, which is capable of reducing the speed of neutrons, thereby increasing the likelihood for them to produce fission in the nuclear reactor.

modification
According to the Federal Clean Air Act: Any physical change in, or change in the method of operation of, a stationary source which increases the amount of any air pollutant emitted by such source or which results in the emission of any air pollutant not previously emitted. Any change in, or change in the method of operation of, a major source which increases the actual emissions of any hazardous air pollutant emitted by such source by more than a de minimis amount or which results in the emission of any hazardous air pollutant not previously emitted by more than a de minimis amount.

modified Cooper-Harper Scale
See ***Cooper-Harper Scale, modified.***

Modified Mercalli Intensity Scale
An earthquake intensity scale adopted in 1931 that divides the effects of an earthquake into twelve categories, from I (not felt by people) to XII (damage total).

Modified Rhyme Test
A multiple choice test in which an individual is to select the word he/she believes he/she heard spoken from a selection of rhyming alternatives.

modular
See ***module.***

modular design
Consisting of modules.

modular workstation
A workstation which may be assembled from modular components in a variety of different configurations. *See also* ***cluster workstations.***

modulation
The variation in value of some parameter characterizing a periodic oscillation.

module
A standard unit which may serve as a building block for larger structures.

modulus
The numerical value assigned to a standard stimulus, against which other stimuli are judged and assigned relative values.

modus operandi
Method of operating or doing things (M.O.). The term used by the police and criminal investigators to describe the particular method of a criminal's activity. It refers to the pattern of criminal behavior so distinct that separate crimes or wrongful conduct are recognized as the work of the same person.

moist adiabatic rate
The rate of change of temperature in a rising or descending saturated air parcel. The rate of cooling or warming varies but a common value of 6°C per 100 meters (3.3°F per 1000 feet) is used.

moisture vapor transmission rate
The mass of water vapor passing through a specified area of one or more fabrics per unit time.

Mojave Block
The tectonic region located between the Garlock fault and the San Andreas fault, and extending eastward roughly to the California-Arizona (and California-Nevada) border.

mol
Molecular weight expressed in grams.

molal
A solution containing one mole of solute per liter of solution.

molar volume
The volume occupied by a gram mole of a substance in its gaseous state. This is equal to 22.414 liters at standard conditions (temperature of 0°C and 760 mm Hg pressure) and to 24.465 liters at normal temperature and pres-

sure (25°C and 760 mm Hg) in industrial hygiene work.

mold
*See **fungus**.*

mole
The basic measure of the amount of any substance. The mole has been defined to be the precise number of elementary entities, as there are atoms in exactly 0.012 kilograms (12.0 grams) of $_6C^{12}$. When the mole is used, the specific elementary entities must be specified; however, they may be atoms, molecules, ions, electrons, protons, neutrons, other particles, or any specified groupings of such particles. In general, one mole of any substance will contain Avogadro's Number, N_A, of atoms, molecules, or particles of some sort. Avogadro's Number is 6.022 x 10^{23}. *See also **gram-mole** or **gram molecular weight**.*

mole percent
The ratio of the number of moles of one substance to the total number of moles in a mixture of substances, multiplied by 100.

molecular viscosity
The small-scale internal fluid friction that is due to the random motion of the molecules within a smooth-flowing fluid, such as air.

molecular weight
The relative weight of a molecule of any substance as compared to the weight of an atom of carbon-12 (12.00000).

molecule
Ultimate unit of quantity of a chemical compound that can exist by itself and retain all the properties of the original substance.

moment
(1) A statistic measure, represented by the sum of the deviations from the mean, raised to some power, and divided by the number of terms used in accumulating the sum. (2) The tendency of a force to generate rotation in a body or torsion about an origin.

moment arm
That component of the vector representing the distance from a point of rotation which is perpendicular to the line of action of a force creating a torque.

moment concept
The idea that lifting stress is also a function of the bending moments at the spine, not just of the weight lifted.

moment magnitude (M_W)
The seismic moment of an earthquake, converted to a magnitude scale that roughly parallels the original Richter magnitude scale. However, since it is not based on the same measurement as Richter (local or surface-wave) magnitudes, the different magnitudes do not always agree, particularly for very large quakes. Because it relates directly to the energy released by an earthquake, it has become the standard in modern seismology. *See also **local magnitude, surface-wave magnitude, seismic moment,** and **Richter scale**.*

moment of force
*See **torque**.*

moment of inertia
The tendency of an object to retain its current rotational motion about an axis.

$$I = \sum m_i r_i^2 = \int r^2 dm$$

where:
m = mass element
r = distance from the axis of rotation

momentary hold
The maintenance of some position for a brief period of time (may be planned or unplanned).

momentum
The product of the mass of a body and its velocity, expressed in units of g-cm/s.

monaural
Indicating sound reception by only one ear.

Monday morning heart attack
Term used to describe heart attacks observed among dynamite workers. The effect is believed to be the result of the vasodilatory effect of ethylene glycol dinitrate and nitroglycerine which are used in dynamite manufacture.

monel
Term for a large group of corrosion-resistant alloys of predominantly nickel and copper with very small percentages of carbon, manganese, sulfur, and silicon. Some may contain aluminum, titanium, and cobalt.

Monge's disease
*See **altitude sickness**.*

mongolism
A congenital condition involving some degree of mental retardation and various physical malformations. The name is based on characteristic facial traits resembling somewhat those of persons of the Mongolian race. The term mongolism is now considered to be inaccurate and undesirable and has been replaced by the term Down's syndrome, or trisomy 21. The latter name refers to the presence of three twenty-first chromosomes, found in those with Down's syndrome, instead of the usual pair. *See also **Down's syndrome**.*

monitor
(1) *General.* To observe, listen to keep track, or exercise surveillance of ongoing progress, events, or situations by any appropriate means. (2) *Computing.* A peripheral device that allows the user of a CPU to directly view information and processing data. (3) *Aviation.* When used with communication transfer, "monitor" means to listen on a specific frequency and stand by for instructions. Under normal circumstances, communications are not to be established.

monitor circuit
A circuit used to verify the status of a system, such as an inhibit directly; control circuits can be monitored but they cannot serve as a monitor circuit.

monitoring
(1) *General.* Periodic or continuous surveillance or testing to determine the level of compliance with statutory requirements and/or pollutant levels in various media or in humans, animals, and other living things. (2) *Health Physics.* Periodic or continuous determination of the amount of ionizing radiation or radioactive contamination present in an occupied space.

monitoring strategy
The plan for implementing and carrying out a monitoring campaign to determine worker exposure to a contaminant, physical agent, etc.

monitoring wells
Wells drilled at a hazardous waste management facility or Superfund site to collect groundwater samples for the purpose of physical, chemical, or biological analysis to determine the amounts, types, and distribution of contaminants in the groundwater beneath the site.

monkey line
Maritime (slang). Small hand line used by a lockman to throw down or bring up the lockline.

mono
Prefix denoting one, or single.

monochromasia
Total color blindness, in which all the red, green, and blue cones are missing or nonfunctional. The individual sees only shades of gray, lightness. Also known as *monochromasy*.

monochromat
An individual having monochromasia.

monochromatic
Having only one color, or producing light of only one wavelength.

monochromatic radiation
Electromagnetic radiation of a single wave length, or in which all the photons have the same energy (e.g., lasers).

monochrome
Pertaining to a screen display or hardcopy having a single color image against a background.

monoclonal antibodies
Molecules of living organisms that selectively find and attach to other molecules to which their structure conforms exactly. This could also apply to equivalent activity by chemical molecules.

monocular
Pertaining to only one eye, or vision using one eye.

monocular visual field
That part of the visual environment which can be seen by a single eye at any given instant with the head and eye stationary. *See also **binocular visual field**.*

monodisperse aerosol
A uniform aerosol with a standard deviation of 1.0. That is, the aerosol is all of one size.

monoenergetic radiation
Particulate radiation of a given type (alpha, beta, neutrons, etc.) in which all particles have the same energy.

monomer
A compound of relatively low molecular weight which, under certain conditions, either alone or with another monomer, forms various types and lengths of molecular chains called polymers or copolymers of high molecular weight. For example, styrene is a monomer that polymerizes readily to form polystyrene.

mononucleosis
Excess of mononuclear leukocytes in the blood. Infectious mononucleosis is an acute disease that causes changes in the leukocytes. The exact cause is not clearly understood, but it is widely considered to be a viral infection. Transmission of the disease is also not clearly understood. It occurs more frequently in the spring and affects primarily children and young adults. Generally, after an incubation period of uncertain duration (1 week to several weeks), headache, sore throat, mental and physical fatigue, severe weakness, and symptoms typical of influenza develop. Skin rashes may also occur.

monoplegia
The paralysis of a single limb, or a single muscle group.

monorail
(1) An electrical railway in which a rail car or train of cars is suspended from or straddles a guideway formed by a single beam or rail. Most monorails are either heavy rail or automated guideway systems. (2) A single rail on which a vehicle or train of cars travels.

monorail vehicles
Guided transit passenger vehicles operating on or suspended from a single rail, beam, or tube.

monotone
*See **monotonic**.*

monotonic
Pertaining to a function in which the dependent variable either continuously increases or continuously decreases in magnitude with an increase in the independent variable throughout the range of values under consideration, such that each point for either function uniquely defines one point for the other.

monotonic-decreasing
Pertaining to a function in which the dependent variable continuously decreases in magnitude with an increase in the independent variable.

monotonic-increasing
Pertaining to a function in which the dependent variable continuously increases in magnitude with an increase in the independent variable.

monotony
The psychological state created by the lack of variety due to the repeated performance of a non-challenging task or long-duration task.

monsoon depression
A weak low-pressure area that tends to form in response to divergence in an upper-level jet stream. The circulation around the low strengthens the monsoon wind system and enhances precipitation during summer.

monsoon wind system
A wind system that reverses direction between winter and summer. Usually the wind blows from land to sea in winter and from sea to land in summer.

Monte Carlo Method
A probabilistic technique for obtaining solutions to problems by statistical sampling methods.

Montreal Protocol
The Montreal Protocol on Substances that Deplete the Ozone Layer, a protocol to the Vienna Convention for the Protection of the Ozone Layer, including adjustments adopted by Parties thereto and amendments that have entered into force.

moonlight
(1) The nighttime luminance created as a result of the sun's reflection on the lunar surface. (2) To work a second job.

mooring
(1) A floating ball, can, or other structure, which is permanently secured to the harbor bottom by means of a heavy chain and anchor system and to which vessels are made fast, but able to swing in the wind and/or current. (2) The place where a craft may be secured to the ground, wharf, pier, post, or buoy.

mooring cell
A river-front structure generally composed of steel piling or a cluster of wooden piles used for securing barges along the bank at loading facilities. *See also* ***dolphin***.

MOP
See ***maximum operating pressure***.

moped
Includes motorized bicycles equipped with a small engine, typically 2 horsepower or less. Mini-bikes, dirt bikes, and trail bikes are excluded. Note that a motorized bicycle may or may not be licensed for highway use. *See also* ***motorcycle*** *and* ***motor-driven cycle***.

morale
A measure of the level of confidence and enthusiasm of an individual or group.

morbid
Diseased.

morbidity
The condition of being sick or morbid. The ratio of sick to well persons in a population.

morbidity rate
The number of cases of a specific disease occurring in a population within a specified time interval.

mordant
A substance that is capable of binding a dye to a textile fiber.

morning person
(slang) An individual who typically wakes up easily in the morning, ready for the day, and has trouble staying up late at night.

morphine
An opium alkaloid, a narcotic analgesic and respiratory depressant, usually used as morphine sulfate.

morphology
Structural configuration. The science of the forms and structure of organized beings and other materials (e.g., objects).

mortal
Subject to death.

mortality rate
The number of deaths occurring per 1000 population in a specified time period. Also referred to as *death rate*.

mortality tables
A means of ascertaining the probable number of years any man or women of a given age and of ordinary health will live. A mortality table expresses, on the basis of the group studied, the probability that, of a number of persons of equal expectations of life who are living at the beginning of any year, a certain number of deaths will occur within that year. These tables are used by insurance companies to determined the premium to be charged for those in the respective age groups.

MOST
See ***Maynard Operation Sequence Technique***.

most harmful event
Transportation. The event during a crash for a particular vehicle that is judged to have produced the greatest personal injury or property damage.

most restrictive state
Rail Operations. The mode of an electric or electronic device that is equivalent to a track relay in its deenergized position.

mother-of-pearl clouds
See ***nacreous clouds***.

motile
Moving or having the power to move spontaneously.

motion aftereffect
Any illusion of continuing motion which begins on cessation or change of a particular motion.

motion analysis
The acquisition, processing, organization, and use of data obtained from human physical activity, whether of certain specific joints, body segments, or the body as a whole.

motion cycle
The entire set of physical activities required to perform a given work cycle one time.

motion efficiency
The concept that body motions in performing a task should be reduced to the minimal, simplest, least fatiguing possible set.

motion efficiency principles
A set of some common sense or empirically determined concepts dealing with human movements for the industrial/manufacturing

workplace to simplify and improve the effectiveness of manual work and minimize fatigue. General principles include

a) use natural, rhythmic, easy movements
b) establish habitual movements
c) use both hands simultaneously in parallel motions, not sequential
d) minimize movements
e) involve the fewest body segments possible in performing the work
f) distribute actions among the various muscles of the body
g) use ballistic movements rather than slower, controlled movements
h) use momentum to aid performance
i) minimize momentum when muscular effort must be used to overcome it
j) use continuous, curved movements, not straight lines involving rapid changes in direction

Also referred to as *principles of motion economy, principles for motion improvement,* and *characteristics of easy movement.* *See also* ***workplace design*** and ***display-control layout***.

motion sickness
A condition in which the signs or symptoms of nausea, vomiting, and/or physiological effects are produced by either real or perceived motion of the body or its surroundings. The discomfort is caused by irregular and abnormal motion that disturbs the organs of balance located in the inner ear.

motion study
See ***motion analysis***.

Motion Time Analysis (MTA)
A predetermined motion time system

motions inventory
The nature and quality of possible motions within the capabilities of an individual under specified circumstances.

motivation
A psychophysiological construct which is involved in the initiation, direction, and sustenance of behavior by an individual or group toward accomplishing some goal.

motor
Pertaining to the activation of muscles by efferent neurons or nerves.

motor activity
Any pattern of muscular activity concerned with locomotion or the moving of a limb or body part.

motor bus
(1) A rubber-tired, self-propelled, manually steered vehicle with fuel supply carried on board the vehicle. (2) Rubber-tired passenger vehicles that operate on roadways. Motor bus service implies fixed routes and schedules. *See also* ***intercity bus*** *and* ***motorbus***.

motor bus, class A (>35 seats)
Rubber-tired passenger vehicles powered by diesel, gasoline, battery, or alternative fuel engines contained within the vehicle. Class A motor buses are equipped with more than 35 seats.

motor bus, class B (25-35 seats)
Rubber-tired passenger vehicles powered by diesel, gasoline, battery, or alternative fuel engines contained within the vehicle. Class B motor buses are equipped with 25 to 35 seats.

motor carrier
(1) A for-hire motor carrier or a private motor carrier of property. The term "motor carrier" includes a motor carrier's agents, officers, and representatives as well as employees responsible for hiring, supervising, training, assigning, or dispatching of drivers and employees concerned with the installation, inspection, and maintenance of motor vehicle equipment and/or accessories. (2) An employer firm that is primarily engaged in providing commercial motor freight or long distance trucking or transfer services.

motor carrier of passengers
A common, contract, or private carrier using a bus to provide commercial transportation of passengers.

motor-circuit switch
See ***switch (4)***.

motor-driven cycle
A motorcycle with a motor that produces 5 brake horsepower or less. *See also* ***moped*** *and* ***motorcycle***.

motor end plate
See ***end plate (1)***.

motor fitness
A measure of an individual's physical suitability for a particular task.

motor freight transportation warehousing and stockyards
Includes establishments that provide local or long-distance trucking or transfer services, warehousing and storage of farm products, furniture or other household goods, and commercial goods of a general nature. The operation of terminal facilities for handling freight, with or without maintenance facilities is also included. Stockyards, establishments that provide holding pens for livestock in transit, are included in this major group. These stock yards do not sell or auction livestock.

motor gasoline
A complex mixture of relatively volatile hydrocarbons, with or without small quantities of additives, obtained by blending appropriate refinery streams to form a fuel suitable for use in spark ignition engines. Motor gasoline includes both leaded and unleaded grades of finished motor gasoline, blending components, and gasohol. *See also **fuel** and **gasohol**.*

motor home
Includes self-powered recreational vehicles (RV) that are operated as a unit without being towed by another vehicle (e.g., a Winnebago motor home).

motor homunculus
A representation of the human body on the surface of the motor cortex, whose distribution is proportional to the density of innervation in various parts of the body.

motor learning
Any form of learning involving the coordinated activities of muscles.

motor nerve
An efferent nerve which provides motor innervation to a muscle.

motor neuron
An efferent neuron which sends or carries information toward a neuromuscular junction.

motor point
A location on the skin at which electrical stimulation will cause contraction of the underlying muscle.

motor skill
The ability to move some or all parts of the body in a coordinated fashion toward the performance of some task.

motor unit
The combination of a motor neuron, its axon, the neuronal terminal branches, and the muscle fibers they innervate.

motor vehicle
(1) Any self-propelled vehicle, truck, tractor, semitrailer, or truck-full trailers used for the transportation of freight over public highways. (2) A vehicle, machine, tractor, trailer, or semitrailer, or any combination thereof, propelled or drawn by mechanical power and used upon the highways in the transportation of passengers or property. It does not include a vehicle, locomotive, or car operated exclusively on a rail or rails, or a trolley bus operated by electric power derived from a fixed overhead wire, furnishing local passenger transportation similar to street-railway service. (3) Any mechanically or electrically powered device not operated on rails, upon which or by which any person or property may be transported upon a land highway. The load on a motor vehicle or trailer attached to it is considered part of the vehicle. *See also **automobile, bus, car, minivan, semitrailer, taxi, tractor (or truck tractor), tractor-semitrailer, truck, van,** and **vehicle**.*

motor vehicle accident
An unstable situation that includes at least one harmful event (injury or property damage) involving a motor vehicle in transport (in motion, in readiness for motion or on a roadway, but not parked in a designated parking area) that does not result from discharge of a firearm or explosive device and does not directly result from a cataclysm.

motor vehicle chassis
The basic operative motor vehicle, including engine, frame, and other essential structures and mechanical parts, but excluding body and all accessories and auxiliary equipment.

motor vehicle in transport
A motor vehicle in motion on the trafficway or any other motor vehicle on the roadway, including stalled, disabled, or abandoned vehicles.

motor vehicle incidence rate
A factor for rating the number of deaths from motor vehicular accidents by miles of vehicular travel.

motor vehicle traffic accident
An accident involving a motor vehicle in use within the right-of-way or other boundaries of a trafficway open for the use of the public.

motor vessel
Any vessel equipped with propulsion machinery (other than steam) more than sixty-five feet long.

motorboat
Any vessel equipped with propulsion machinery, not more than sixty-five feet in length.

motorbus
(1) Rubber-tired, self-propelled, manually steered bus with fuel supply on board the vehicle. Motor bus types include intercity, school, and transit. (2) Rubber-tired passenger vehicles which operate on roadways. Motorbus service implies fixed routes and schedules. *See also* ***intercity bus, motor bus, school and other nonrevenue buses,*** *and* ***transit bus***.

motorcycle
(1) All two- or three-wheeled motorized vehicles. Typical vehicles in this category have saddle type seats and are steered by handle bars rather than a wheel. This category includes motorcycles, motor scooters, mopeds, motor powered bicycles, and three-wheeled motorcycles. (2) A two- or three-wheeled motor vehicle designed to transport one or two people, including motor scooters, minibikes, and mopeds. *See also* ***moped*** *and* ***motor-driven cycle***.

motorized vehicle
Includes all vehicles that are licensed for highway driving. Specifically excluded are snow mobiles and mini-bikes.

mottled
Covered with spots or streaks of different shades or colors.

MOU
See ***memorandum of understanding***.

mountain and valley breeze
A local wind system of a mountain valley that blows downhill (*mountain breeze*) at night and uphill (*valley breeze*) during the day.

mountain sickness
See ***altitude sickness***.

mounting height
That vertical height above the floor, table, ground, or other surface at which an illumination source is located.

mouse
A computer input device having one or more buttons and capable of two-dimensional rolling motion which can drive a cursor on the display and perform a variety of selection options or commands.

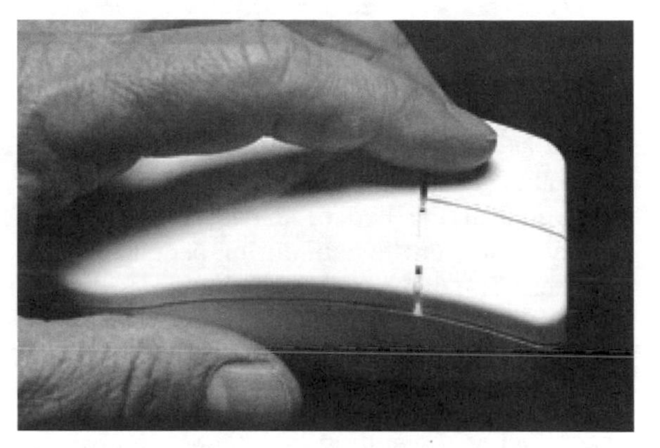

The typical (two-button) computer mouse

mouse keys
An interactive feature for handicapped individuals which will allow them to use certain keys on a computer keyboard to control a cursor normally operated by a mouse.

mouth
(1) *Anatomy*. An opening, especially the oral cavity, forming the beginning of the digestive system in which the chewing of food takes place. The mouth is also the site of the organs of taste and the teeth, tongue and lips. Not only is the mouth the entrance to the body for food and sometimes air, but it is a major organ of speech and emotional expression. (2) *Geography*. The exit or point of discharge of a stream into another stream, lake, or sea.

mouth stick
A rod for allowing a quadriplegic or other handicapped individual to operate various forms of equipment by holding the device in his/her mouth and using pressure to operate the equipment.

mouth-to-mouth resuscitation
A method of artificial respiration in which the rescuer covers the victim's mouth with his/her

own and breaths out vigorously in an attempt to resuscitate the person.

movable bridge
That section of a structure bridging a navigable waterway so designed that it may be displaced to permit passage of traffic on the waterway.

movable bridge locking
The rail locks, bridge locks, bolt locks, circuit controllers, and electric locks used in providing interlocking protection at a movable bridge.

movable dam
A dam that is predominantly constructed of a series of wickets which may be raised or lowered as water stages dictate for passing water through the dam. These wickets may all be lowered at the bed of the river and vessels may pass over the dam during periods of high water. The dam and/or river is then said to be "open."

move
(1) To execute one or more isotonic muscular contractions, resulting in a change in position of one or more parts of the body. (2) To transfer (cut and paste) a segment of text, graphics, or other material in a computer system from one location to another. (3) A physical basic work element involving motion of the hand carrying one or more objects.

movement area
Aviation. (1) The runways, taxiways, and other areas of an airport/heliport which are utilized for taxiing/hover taxiing, air taxiing, takeoff, and landing of aircraft, exclusive of loading ramps and parking areas. At those airports/heliports with a tower, specific approval for entry onto the movement area must be obtained from ATC. (2) That part of an aerodrome to be used for the takeoff, landing and taxiing of aircraft, consisting of the maneuvering area and the apron(s).

movement disorder
Any pathological condition which results in an abnormal deviation from an intended movement, an inability to execute a desired movement, or an undesired involuntary movement.

moving average
An arithmetic mean based on a fixed number of samples over time, in which as each new sample is added, the oldest sample is dropped.

moving target indicator
Aviation. An electronic device which will permit radar scope presentation only from targets which are in motion. A partial remedy for ground clutter.

mp
*See **melting point**.*

MPC
*See **maximum permissible concentration**.*

MPD
*See **maximum permissible dose**.*

MPE
*See **maximum permissible exposure**.*

MPG
*See **miles per gallon**.*

MPI
*See **mass psychogenic illness**.*

MPL
*See **maximum permissible lift**.*

MPO
*See **Metropolitan Planning Organization**.*

mppcf
Millions of particles per cubic foot (of air) (mppcf x 35.3 = million particles per cubic meter = particles per cubic centimeter).

MPRSA
Marine Protection, Research, and Sanctuaries Act of 1972 (Federal).

mps
Meter(s) per second.

mR
Milliroentgen(s).

mrad
Millirad(s).

mrem
Millirem(s).

mrem/h
Millirem(s) per hour.

MRI
*See **Magnetic Resonance Imaging**.*

MRP
See ***material requirements planning.***

M_S
See ***surface-wave magnitude.***

MS
Mass spectrometer. *See also* ***multiple sclerosis.***

MSA
See ***Metropolitan Statistical Area.*** *See also* ***minimum safe altitude.***

MSAW
See ***minimum safe altitude warning.***

MSD
Musculoskeletal disorder. *See also* ***Master Standard Data.***

MSDS
See ***material safety data sheet.***

MSHA
See ***Mine Safety and Health Administration.***

mSv
Millisievert(s).

mT
Millitesla.

MT
See ***metric ton.***

MTA
See ***Motion Time Analysis.***

MTBF
See ***mean time between failures.***

MTD
Maximum tolerated dose.

MTM
See ***Methods Time Measurement.***

MTTF
See ***mean time to failure.***

mu locomotive
A multiple-operated electric locomotive described in 49 CFR 229.4 paragraph (i)(2) or (3).

MUC
See ***maximum use concentration.***

muck soils
Earth made from decaying plant materials.

mucociliary clearance
Removal of materials from the respiratory tract via cilia action.

mucosa
See ***mucous membrane.***

mucous membrane
Membrane lining all channels in the body that communicate with the air, such as the respiratory tract, stomach, urinary tract, intestines, and the alimentary canal (digestive tract), the glands of which secrete mucus. Also referred to as *mucosa*. *See also* ***mucus.***

mucus
The viscous suspension of mucin, water, cells, and inorganic salts secreted as a protective lubricant coating by glands in the mucous membranes. *See also* ***mucous membrane.***

muffler
In acoustics, a device for reducing noise emissions from engine exhausts, vents, etc. Two types of mufflers, namely, the *dissipative* and *reactive,* are available.

mulch
A layer of material (wood chips, straw, leaves, etc.) placed around plants to hold moisture, prevent weed growth, protect the plants, and enrich the soil.

mule
Small tractor used in warehouse to pull two-axle dollies, also yard tractor.

mule train
The maneuver of towboats in ice-choked channels whereby the tow is strung out single file, the barges fitted with loose couplings or lashings, and the tow pulled behind the towboat. Also known as *string out.*

multicell storm
Thunderstorms in a line, each of which may be in a different stage of development.

multicom
Aviation. A mobile service not open to public correspondence used to provide communications essential to conduct the activities being performed by or directed from private aircraft.

multi-factor plan
An incentive plan in which employee awards are based on more than one factor.

multilevel sampling
The selection of a primary, large or high level unit, followed by secondary, tertiary, et cetera units, each selected from within the next higher level unit.

multilimb coordination
The ability to meaningfully integrate the movements of two or more limbs to fulfill some purpose such as manipulating a control, an object, or locomotion.

multimedia
Two or more elements, such as sound and animation, or video in a computer program.

multimedia extension (mmx)
In computing, advanced processor functionality for running multimedia programs.

multimedia inspection
Environmental. Inspection strategy sometimes employed by the Federal Environmental Protection Agency (EPA), as well as many state agencies, whereby a number of inspectors each with an expertise in a given environmental subject area (e.g., CAA, CWA, RCRA, etc.) conduct an inspection simultaneously at the same location, thereby effectively covering all applicable aspects of environmental compliance during a single inspection.

multimodal transportation
Often used as a synonym for intermodalism. Congress and others frequently use the term intermodalism in its broadest interpretation as a synonym for multimodal transportation. Most precisely, multimodal transportation covers all modes without necessarily including a holistic or integrated approach. *See also* ***intermodal***.

multiple activity process chart
A process chart showing the chronological activities involving a work system, with each component of the system allocated a separate vertical column to show relative or coordinated activities.

multiple chemical sensitivity
Term used by some people to refer to a condition in which a person is considered to be sensitive to a number of chemicals at very low concentrations. There are a number of views about the existence, potential causes, and possible remedial actions regarding this phenomenon.

multiple correlation
The degree of relationship between a criterion variable and two or more predictor variables.

multiple correlation coefficient (R)
A numerical value representing the correlation between a set of two or more predictor variables and one criterion variable. Synonymous with *coefficient of multiple correlation* and *multiple R.*

multiple machine work
A work assignment which has a worker attending to two or more machines.

multiple myeloma
A malignant neoplasm of plasma cells usually arising in the bone marrow and manifested by skeletal destruction, pathologic fractures, and bone pain.

multiple R
See ***multiple correlation coefficient***.

multiple receptacle
See ***receptacle (3)***.

multiple regression
The analysis or use of the combined and individual contributions from more than one predictor variable for predicting the value of a single criterion variable.

multiple runway
The utilization of a dedicated arrival runway(s) for departures and a dedicated departure runway(s) for arrivals when feasible to reduce delays and enhance capacity.

multiple sclerosis (MS)
A disease resulting in demyelination within the CNS and the corresponding movement, speech, and other difficulties.

multiple use
Under the Federal Land Policy and Management Act of 1976: The management of the public lands and their various resource values so that they are utilized in the combination that will best meet the present and future needs of the American people. Making the most judicious use of the land for some or all of these resources or related services over areas large enough to provide sufficient latitude for periodic adjustments in use to conform to changing needs and conditions. The use of some land for less than all of the resources. A combination of balanced and diverse resource uses that takes into account the long-term needs of future generations for renewable and nonrenewable resources, including, but not limited to, recreation, range, timber, minerals, watershed, wildlife and fish, and natural scenic, scientific and historical values, and harmonious and

coordinated management of the various resources without permanent impairment of the productivity of the land and the quality of the environment with consideration being given to the relative values of the resources and not necessarily to the combination of uses that will give the greatest economic return or the greatest unit output.

multipurpose dry chemical
A dry chemical which is approved for use on Class A, Class B and Class C fires.

multipurpose passenger vehicle
A motor vehicle with motive power, except a trailer, designed to carry 10 persons or less which is constructed either on a truck chassis or with special features for occasional off-road operation.

multisensory
Combining or related to more than one sensory modality.

multistop body
Fully enclosed truck body with driver's compartment designed for quick, easy entrance and exit.

multitasking
(1) The processing of more than one dataset or application at a time, usually with the operator working directly only on one application. (2) The assignment of a worker to more than one task or job.

multi-trailer five or less axles truck
All vehicles with five or less axles consisting of three or more units, one of which is a tractor or straight truck power unit.

multi-trailer seven or more axles truck
All vehicles with seven or more axles consisting of three or more units, one of which is a tractor or straight truck power.

multi-trailer six axle truck
All six-axle vehicles consisting of three or more units, one of which is a tractor or straight truck power-unit.

multivariate
(1) Having more than one dependent variable. (2) Pertaining to more than one variable.

multivariate analysis
Any statistical analysis involving more than one independent variable and/or more than one dependent variable.

multivariate analysis of variance (MANOVA)
An analysis involving two or more of both independent and dependent variables.

mumps
A communicable viral disease that attacks one or both of the parotid glands, the largest of the three pairs of salivary glands. Occasionally, the submaxillary glands are also affected. Although older people may contract the disease, mumps usually strikes children between the ages of 5 and 15. Mumps is spread by droplet infection. The disease is contagious in the infected person from 1 to 2 days before symptoms appear until 1 or 2 days after they disappear. The incubation period is usually 18 days, although it may vary from 12 to 26 days. One attack usually gives immunity. Often, the first noticeable symptoms of mumps is a swelling of one of the parotid glands. The swelling is frequently accompanied by pain and tenderness. Occasionally, acid foods and beverages may cause an increase in the pain. In the first stage of mumps, the person may have a fever of 100° to 104°F. Other common symptoms include loss of appetite, headache, and back pain. Also called *epidemic parotitis*.

municipality
A legally incorporated or duly authorized association of inhabitants of a limited area for local governmental or other public purposes.

Munsell chromas
Saturation in the Munsell color system.

Munsell color system
A color ordering system for surfaces which divides colors into perceptually uniform segments for ordering and specifying with regard to hue, chroma (saturation), and value (lightness).

Munsell value
A measure of lightness in the Munsell color system, on a scale ranging from 1 (black) to 10 (white).

murder
The unlawful killing of a human being by another with malice aforethought, whether express or implied (the crime is defined by statute in most states).

muscle
(1) A structure composed of a mass of muscle tissue, usually enclosed by some type of

sheath, and forming a distinct unit. Muscles are responsible for locomotion and play an important part in performing vital body functions. They also protect the contents of the abdomen against injury and help support the body. Muscle fibers range in length from a few hundred thousandths of an inch to several inches. They also vary in shape, and in color from white to deep red. Each muscle fiber receives its own nerve impulses, so that fine and varied motions are possible. Each has its small stored supply of glycogen which it uses as fuel for energy. Muscles, especially the heart, also use free fatty acids as fuel. At the signal of an impulse travelling down a nerve, the muscle fiber changes chemical energy into mechanical energy, and the result is muscle contraction. Some muscles are attached to bone by tendons. Other are attached to other muscles, and to the skin – producing the smile, the wink, and other facial expressions, for example. All or part of the walls of hollow internal organs, such as the heart, stomach, intestines, and blood vessels, are composed of muscles. The last stages of swallowing and of peristalsis are actually series of contractions by the muscles in the walls of the organ involved. There are three types of muscle: involuntary, voluntary, and cardiac. They are composed, respectively, of smooth, striated (or striped), and mixed smooth and striated. Muscles that are not under the control of the conscious part of the brain are called involuntary muscles. They respond to nerve impulses of the autonomic nervous system. These involuntary muscles are countless short-fibered, or smooth, muscles of the internal organs. They power the digestive tract, the pupils of the eyes, and all other involuntary mechanisms. The muscles controlled by the conscious part of the brain are called voluntary muscles, and are striated. These are skeletal muscles that enable the body to move, and there are more than 600 of them in the human body. The fibers of voluntary muscles are grouped together in a sheath of muscle cells. Groups of fibers are bundled together into fascicles and the bundles are surrounded by a tough sheet of connective tissue to form a muscle group like the biceps. Unlike the voluntary muscles, which can remain in a state of contraction for long periods without tiring and are capable of sustained rhythmic contractions, the voluntary muscles are readily subject to fatigue. They also differ from the involuntary muscles in their need for regular and proper exercise. The third kind of muscle, cardiac muscle, or the muscle of the heart, is involuntary and consists of striated fibers different from voluntary muscle fibers. The contraction and relaxation of cardiac muscle continue at a rhythmic pace until death, unless the muscle is injured in some way. No muscle stays completely relaxed, and as long as a person is conscious, it remains slightly contracted. This condition is called *tonus,* or tone. It keeps bones in place and enables a posture to be maintained. It allows a person to remain standing, sitting up straight, kneeling, or in any other natural position. Muscles also have elasticity. They are capable of being stretched and of performing reflex actions. This is made possible by the motor and sensory nerves which serve the muscles. (2) *See **muscle tissue**.*

muscle action
Any muscle activity which results in a change in length or in an increase in tension in the muscle.

muscle capacity
*See **muscular endurance**.*

muscle fatigue
*See **muscular fatigue**.*

muscle fiber
A muscle cell. *See also **intrafusal fiber** and **extrafusal fiber**.*

muscle group
A collection of individual skeletal muscles which have similar innervation and perform a similar/common/related function.

muscle hemoglobin
*See **myoglobin**.*

muscle testing
Any procedure intended to measure the performance of a restricted number of muscles on some graded basis.

muscle tissue
An irritable, contractile, extensible elastic tissue composed of long tubular or spindle-shaped cells.

muscle tone
A state of continuous mild muscle contraction.

muscular endurance
The maximum time under stated conditions which a muscle or muscle group can maintain a given measure of external force.

muscular fatigue
The buildup of lactic acid in muscle tissue due to prolonged heavy exertion.

musculoskeletal system
Pertaining to or comprising the skeleton and the muscles.

musculospiral nerve
See ***radial nerve***.

mustache
That long-term accumulation of hair growth which originates on the face generally above the upper lip, medial to the lip margins, and beneath the nasal septum base.

MUSYA
Multiple-Use Sustained-Yield Act of 1960 (Federal).

mutagen
A chemical substance that has the ability to produce a change *(mutation)* in the genetic composition of the DNA in a cell. The change is capable of being passed on to succeeding generations. Mutations can also be brought about by radiation exposures.

mutagenesis
The process in which normal cells are converted into genetically abnormal cells.

mutagenic
An agent that induces genetic mutation.

mutagenicity
The property of being able to induce genetic mutation.

mutant
An individual who has been altered as a result of mutation (i.e., from a change in the character of a gene that is perpetuated in the subsequent division of the cell in which it occurs).

mutate
To bring about a change in the genetic constitution of a cell by altering its DNA. In turn, "mutageneis" is any process by which cells are mutated.

mutation
A change in the characteristics of an organism produced by an alteration of the DNA of living cells.

mutism
Inability or refusal to speak. In almost all cases, mutes are unable to speak because their deafness has prevented them from hearing the spoken word.

mutuality of obligation
Mutuality of obligation requires that unless both parties to a contract are bound, neither is bound.

mV
Millivolt(s).

MVA
See ***minimum vectoring altitude***.

MVC
See ***maximal voluntary contraction***.

mw
Molecular weight.

M_W
See ***moment magnitude***.

MWE

See ***megawatt electric***.

myalgia
Pain in a muscle or muscles.

myasthenia
Muscular debility or weakness.

mycetoma
See ***maduromycosis***.

mycotoxin
A toxin produced by a mold growing on a specific substrate, many of which are known to be potent carcinogens.

myelin
The white, fatty substance that forms a sheath around certain nerve fibers.

myelin sheath
The collective concentric wrapping of the membranes of many neural support cells around an axon at intervals along its length. Each support cell forms one internode and permits saltatory conduction.

myelogenous
Produced in the bone marrow.

myelogenous leukemia
Leukemia arising from myeloid tissue.

myeloid tissue
Tissue pertaining to, derived from, or resembling bone marrow.

myeloma
A tumor composed of cells of the type normally found in the bone marrow.

myelotoxicity
Deterioration of the bone marrow structure that results in dangerous changes in blood composition.

myocarditis
Inflammation of the muscular walls of the heart. The condition may result from bacterial or viral infections or it may be a toxic inflammation caused by drugs or toxins from infectious agents. Other systemic diseases that may be accompanied by myocarditis are trichinosis, serum sickness, rheumatic fever, and collagen diseases. In many cases the etiology is unknown. The most common symptoms of acute myocarditis are pain in the epigastric region or under the sternum, dyspnea, and cardiac arrhythmia. If the condition persists and becomes chronic, there is pain in the right upper quadrant of the abdomen, owing to hepatic congestion. The latter symptom is a sign of left ventricular failure and often is accompanied by edema and other signs of congestive heart failure.

myocardium
The muscular substance of the heart.

myoclonus
An isolated involuntary contraction of one or more muscles.

myoelectric limb
A limb prosthesis which senses muscle electrical activity in the proximal remaining portion of the limb or the trunk region and uses those signals to drive one or more motors to operate the prosthesis.

myofibril
The basic unit of contractile structure in skeletal muscle cells.

myoglobin
A protein in muscle which may function as an oxygen carrier. Also called *muscle hemoglobin.*

myography
*See **electromyography**.*

myoma
A tumor formed of muscular tissue. Myomas are often multiple, although a single tumor may occur They are usually small but may grow quite large and occupy most of the uterine wall.

myoneural junction
*See **neuromuscular junction**.*

myopia
A refraction error in which parallel light rays from a distant object are focused anterior to the retina under relaxed accommodation. The error of refraction is caused by rays of light entering the eye parallel to the optic axis and brought to a focus in front of the retina as a result of the eyeball being too long from front to back. Hence, vision for near objects is better than for far. Also know as *nearsightedness* for this reason.

myositis
Inflammation of a voluntary muscle.

N

N

(1) *See* ***Newton***. (2) Normal (solution).

N589

A classical acoustics modeling software package.

NAAQS

See ***National Ambient Air Quality Standards***.

NACE

National Association of Corrosion Engineers.

NACOSH

See ***National Advisory Committee on Occupational Safety and Health***.

nacreous clouds

Clouds of unknown composition that have a soft, pearly luster and that form at latitudes about 25 to 30 km above the earth's surface. They are also called *mother-of-pearl clouds*.

nail

(1) A slender piece of metal, usually pointed at one end and broadened at the other, used for driving into or through wood or other materials so as to hold or fasten one piece to another or to project a peg. (2) A rod of metal, bone, or other material used for fixation of the ends of fractured bones. (3) The elastic protein tissue covering the dorsal portion of the terminal phalanges of the hand and foot. The nails are part of the outer layers of the skin. They are composed of keratin, the substance that gives the skin its toughness. The appearance of the nails can sometimes indicate general physical health. For example, any change in the basic structure, shape, or appearance of the nails (such as softness, brittleness, furrowing, or speckling) may be a symptom of a disease affecting the whole body. Marked pallor of the nails may suggest anemia. In certain cases of hemiplegia and poliomyelitis the nails cease to grow. Curing the disease will cure the condition. Certain disorders affect the nails themselves. They are readily exposed to outside sources of infection and are particularly vulnerable to injury in the course of daily life. Many of the diseases that afflict the skin may also affect the nail bed and be aggravated by the confining presence of the nail. Congenital defects and metabolic disturbances may affect the nails. Most infections involving the nails originate in the folds of tissue around them. Inflammation of this area is called paronychia. It is a fairly common infection by staphylococci, streptococci, or other bacteria or fungi, and causes painful swelling around the nail, with red, shiny skin. If untreated, paronychia may spread to the nail bed and cause inflammation there. This condition is known as onychia, and is more serious. The bacteria grow under the nail and can cause severe inflammation and pain. Onychia may also arise when the nail is injured and bacteria or fungi gain entrance to the tissue underneath. If the organisms that penetrate the nail produce pigments, the nail may change color as a result. In extreme cases onychia may also cause the nail to separate from its bed. Among the diseases from which paronychia and onychia may result are tuberculosis, diphtheria, and syphilis, and also skin diseases such as psoriasis, fungus diseases, and contact dermatitis. Dermatitis is the most common disorder to involve the nails and often leads to the complete loss of the nail. After treatment the nail will generally grow back, but if the matrix is severely damaged a new nail may be deformed or may fail to grow. *See also* ***finger nail*** *and* ***toe nail***.

nail body

The exposed portion of the nail.

nail fold

The rounded skin at the lateral and proximal portions of a nail.

nail groove

The depressed region between the nail and the nail fold.

nail matrix

That structure beneath the skin from which nail tissue is formed.

nail root

That portion of the nail which lies beneath the skin between the lunula and the nail matrix.

NAM

National Association of Manufacturers.

named insured

In insurance, the person specifically designated in the policy as the one protected and, commonly, it is the person with whom the contract of insurance has been made.

NAMS
National Air Monitoring Station.

nano-
Prefix indicating one-billionth or 10^{-9} of the basic unit.

nanogram
One billionth of a gram.

nanometer
The billionth part of a meter.

NANPCA
Nonindigenous Aquatic Nuisance Prevention and Control Act of 1990 (Federal).

nap
(1) A brief period of sleep. (2) The short, small fibers on a fabric surface.

nape
The back of the neck. Technically referred to as the *nucha*.

napestrap
A strap-like device extending from a piece of headgear over the nape of the neck to assist in headgear retention.

naphtha
A generic term applied to a petroleum fraction with an approximate boiling range between 122 and 400 degrees Fahrenheit.

naphtha-type jet fuel
American Petroleum Institute (API). A fuel in the heavy naphtha boiling range with an average gravity of 52.8 degrees and 20 to 90 percent distillation temperatures of 290 degrees to 470 degrees Fahrenheit, meeting Military Specification MIL-T-5624L (Grade JP-4). JP-4 is used for turbojet and turboprop aircraft engines, primarily by the military. Excludes ram-jet and petroleum rocket fuels.

napier
*See **néper**.*

narcoanalysis
Process whereby a subject is put to sleep, or into a semisomnolent state by means of chemical injections and then interrogated while in this dreamlike state.

narcolepsy
A disorder in which an individual experiences numerous severe occasions of sleepiness during the day.

narcosis
A reversible stupor or state of unconsciousness that may be produced by some chemical substances.

narcotic
Compound that produces stupor. Many opium derivatives are examples of strong narcotics. Narcotics can affect the central nervous system (CNS) and the gastrointestinal tract. The CNS effects include analgesia, euphoria, sedation, respiratory depression, and antitussive action. Chronic use of narcotics develops tolerance to the compounds and physical dependence. Medically, the term narcotic includes any drug that has this effect. By legal definition, however, the term refers to habit-forming drugs, for example, opiates such as morphine and heroine and synthetic drugs such as mepredine (Demerol). Narcotics can be legally obtained only with a doctor's prescription. The sale or possession of narcotics for other than medical purposes is strictly prohibited by federal, state, and local laws.

naris
(1) The passage at either the anterior or posterior nasal cavity. (2) An opening into the nasal cavity on the exterior of the body (anterior or external naris) or into the nasopharynx (posterior naris).

narrative evidence
Testimony from a witness which he/she is permitted to give without the customary questions and answers (e.g., when a witness explains in detail what happened without interruption).

narrow band
Pertaining to a frequency band consisting of a few hertz on either side of a center frequency.

narrow band analysis
A type of frequency analysis in which sound intensity level measurements are restricted to a few hertz on either side of a center frequency.

narrow channel
A channel with very little room to spare.

narrow railway gauge
Distance between the rails of a track less than 4 ft 8.5 inches.

NAS
National Academy of Sciences. *See also* ***National Airspace System***.

NASA
National Aeronautics and Space Administration.

nasal
(1) *See* ***nose***. (2) Pertaining to the sound produced when the velum is lowered to allow air passage through the nasal cavity.

nasal bone
The bone forming part of the bridge of the nose and extending in an anterior-inferior direction to form the base for the protruding portion of the nose.

nasal breadth
See ***nose breadth***.

nasal cavity
The region between the external nares and the nasopharynx.

nasal field
The medial portion of the eye's field of view.

nasal height
See nose ***height***.

nasal reflex
The induction of sneezing due to stimulation of the nasal mucous membranes.

nasal root
The junction of the nasal bone with the frontal bone.

nasal root breadth
The minimum horizontal linear distance across the base of the nose between the eyes. Measured with the facial muscles relaxed.

nasal root depression
The concave region where the bridge of the nose meets the forehead between the eyes.

nasal septum
The collective tissues separating the right nostril from the left.

nasalize
To produce a sound with the nasal portion of the vocal tract open.

NASC
National Aeronautics and Space Council.

nascent
Coming into existence or in the process of emerging.

nasion
The horizontal and vertical midpoint of the nasofrontal suture on the skull.

nasogastric tube
A tube of soft rubber or plastic that is inserted through a nostril and into the stomach. The tube may be inserted for the purpose of instilling liquid foods or other substances, or as a means of withdrawing gastric contents.

nasolacrimal
Pertaining to the nose and lacrimal apparatus.

nasolacrimal duct
The tubular structure interconnecting the medial portion of the eye to the nasal cavity for drainage of tears.

nasopharyngitis
Inflammation of the nasopharynx which is situated above the soft palate at the roof of the mouth.

nasopharynx
The uppermost cavity of the pharynx, lying behind the internal nasal cavity and the soft palate.

National Advisory Committee on Occupational Safety and Health (NACOSH)
Committee established to advise, consult, and make recommendations to the Secretary of Health and Human Services on matters regarding administration of the Department of Labor's Occupational Safety and Health Act.

National Aeronautics and Space Administration Standard (NASA-STD)
A document containing standards published by NASA for use in the U.S. space program and related aerospace or medical work.

National Airspace System (NAS)
The common network of U.S. airspace; air navigation facilities, equipment, and services; airports or landing areas; aeronautical charts, information, and services; rules, regulations, and procedures; technical information, manpower, and material. Included are system components shared jointly with the military.

National Airspace System (NAS) Stage A
The en route Air Traffic Control (ATC) system's radar, computers and computer programs, controller plain view displays (Plain View Displays (PVD)/Radar Scopes), input/output devices, and the related communi-

cations equipment which are integrated to form the heart of the automated instrument flight rules (IFR) air traffic control system. This equipment performs flight data processing (FDP) and radar data processing (RDP). It interfaces with automated terminal system and is used in the control of en route IFR aircraft.

National Ambient Air Quality Standards (NAAQS)
Air quality criteria established by the EPA that apply to outside air. Established under the Clean Air Act of 1970, NAAQS set the maximum concentration levels for various pollutants. NAAQS are promulgated on a pollutant-by-pollutant basis. Originally, the EPA applied NAAQS to seven specific pollutants, which became known as *criteria pollutants*. These pollutants are carbon monoxide, hydrocarbons, lead, nitrogen dioxide, sulfur dioxide, ozone, and suspended particulates. *See also* ***criteria pollutants***.

National Beacon Code Allocation Plan Airspace
Airspace over United States territory located within the North American continent between Canada and Mexico, including adjacent territorial waters outward to about boundaries of oceanic control areas (CTA)/flight information regions (FIR).

National Boating Safety Advisory Council (NBSAC)
A 21-member council, equally represented by industry, the public, and State Boating Law Administrators, with expertise, knowledge, and experience in boating safety. The Council acts in an advisory or consulting capacity to the Commandant – U.S. Coast Guard.

National Bridge Inspection Standards (NBIS)
Federal regulations establishing requirements for inspection procedures, frequency of inspections, qualifications of personnel, inspection reports, and preparation and maintenance of a state's bridge inventory.

National Bridge Inventory (NBI)
The aggregation of structure inventory and appraisal data collected to fulfill the requirements of the National Bridge Inspection Standards that each state shall prepare and maintain an inventory of all bridges subject to the National Bridge Inspection Standards.

National Cancer Institute (NCI)
Supports research and the dissemination of information related to occupational cancer hazards, as well as for other causes of cancer.

National Carrier Group
Air carrier groups with annual operating revenues between one hundred million and one billion dollars.

National Contingency Plan
See ***National Oil and Hazardous Substances Contingency Plan***.

National Cooperative Highway Research Program (NCHRP)
The cooperative research, development, and technology transfer (RD&T) program directed toward solving problems of national or regional significance identified by states and the FHWA, and administered by the Transportation Research Board, National Academy of Sciences.

National Cooperative Transit Research and Development Program
A program established under Section 6a) of the Urban Mass Transportation Act of 1964, as amended, to provide a mechanism by which the principal client groups of the Urban Mass Transportation Administration can join cooperatively in an attempt to resolve near-term public transportation problems through applied research, development, testing, and evaluation. NCTRP is administered by the Transportation Research Board.

National Council on Radiation Protection (NCRP)
An advisory group, chartered by the U.S. government to develop and make recommendations on ionizing radiation protection in the United States.

national emergency
A state of national crisis; a situation demanding immediate and extraordinary national or federal action. Congress has made little or no distinction between a "state of national emergency" and a "state of war."

National Emission Standards for Hazardous Air Pollutants (NESHAPS)
These emission standards are set by the EPA for an air pollutant not covered by NAAQS that may cause an increase in deaths, or irreversible or incapacitating illness. Primary

standards are designed to protect human health, secondary standards to protect public welfare. *See also* ***National Ambient Air Quality Standards***.

National Environmental Policy Act (NEPA) of 1969
The Act emphasizes the need for a national environmental policy for public awareness and national response.

National Fire Protection Association (NFPA)
A voluntary, nonprofit association committed to making both the home and the workplace more fire-safe. Members promote scientific research into the development and updating of fire safety awareness and produce information and practical publications on fire safety that are of interest to all concerned with the preservation of life and property from fire. The NFPA is a non-regulatory agency (i.e., carries no force of law). However, many of its codes have been adopted by regulatory agencies and, therefore, compliance is mandatory. Examples include NFPA 70E (National Electrical Code®) and NFPA 101 (Life Safety Code®), which have been adopted by OSHA.

National Flight Data Center
A facility in Washington DC, established by the Federal Aviation Administration (FAA) to operate a central aeronautical information service for the collection, validation, and dissemination of aeronautical data in support of the activities of government, industry, and the aviation community. The information is published in the National Flight Data Digest.

National Flight Data Digest
A daily (except weekends and federal holidays) publication of flight information appropriate to aeronautical charts, aeronautical publications, Notices to Airmen, or other media serving the purpose of providing operational flight data essential to safe and efficient aircraft operations.

National Health Interview Survey
A survey conducted by the National Center for Health Statistics via interviews of household samples of the U.S. civilian, non-institutionalized population on various health and health status issues.

National Highway System (NHS)
This system of highways designated and approved in accordance with the provisions of 23 U.S.C. 103b).

National Highway Traffic Safety Administration (NHTSA)
The Administration was established by the Highway Safety Act of 1970 (23 U.S.C. 401 note). The NHTSA was established to carry out a congressional mandate to reduce the mounting number of deaths, injuries, and economic losses resulting from motor vehicle crashes on the nation's highways and to provide motor vehicle damage susceptibility and ease of repair information, motor vehicle inspection demonstrations and protection of purchasers of motor vehicles having altered odometers, and to provide average standards for greater vehicle mileage per gallon of fuel for vehicles under 10,000 pounds (gross vehicle weight).

National Income
The aggregate earnings of labor and property which arise in the current production of goods and services by the nation's economy.

National Inland Waterways Transport
Inland waterways transport between two places (a place of IWT loading/embarkment and a place of unloading/disembarking) located in the same country irrespective of the country in which the IWT vessel is registered. It may involve transit through a second country.

National Inland Waterways Transport Vessel
Inland waterways transport (IWT) vessel which is registered at a given date in the reporting country.

National Institute for Occupational Safety and Health (NIOSH)
That part of the U.S. Department of Health and Human Services that is responsible for investigating the occurrence and causes of occupational diseases and for recommending appropriate standards to the Occupational Safety and Health Administration.

National Institutes of Health (NIH)
A section of the Public Health Service that conducts research related to diseases and body injuries and helps establish burn treatment centers.

National Labor Relations Act
A federal statute known as the Wagner Act of 1935 and amended by the Taft-Hartley Act of 1947. It is a comprehensive legislation regulating the relations between employers and employees, including supervised elections, and establishing the National Labor Relations Board.

National Labor Relations Board (NLRB)
The NLRB is an independent agency created by the National Labor Relations Act of 1935 (Wagner Act), as amended by the acts of 1947 (Taft-Hartley Act) and 1959 (Landrum-Griffin Act). The Board has two principal functions under the Act: preventing and remedying unfair labor practices by employers and labor organizations or their agents, and conducting secret ballot elections among employees in appropriate collective bargaining units to determine whether or not they desire to be represented by a labor organization. The Board also conducts secret ballot elections among employees who have been covered by a union-shop agreement to determine whether or not they wish to revoke their union's authority to make such agreements. In jurisdictional disputes, the NLRB decides and determines which competing group of workers is entitled to perform the work involved. The Board also conducts secrete ballot elections among employees concerning the employers' final settlement offers in national emergency labor disputes.

National Mediation Board
Created on June 21, 1934 by an Act of Congress amending the Railway Labor Act, the Board's major responsibilities are a) the mediation of disputes over wages, hours, and working conditions which arise between rail and air carriers and organizations representing their employees; and b) the investigation of representation disputes and certification of employee organizations as representatives of crafts or classes of carrier employees.

National Oil and Hazardous Substances Contingency Plan
The federal regulation that guides determination of the sites to be corrected under the Superfund program and the program to prevent or control spills into surface waters or other portions of the environment. Also known as the *National Contingency Plan* or *NCP*.

National Pollution Discharge Elimination System (NPDES)
A provision of the Clean Water Act which prohibits discharge of pollutants into waters of the United States without a special permit issued by the EPA, a state, or (where delegated) a tribal government on an Indian Reservation.

National Priorities List (NPL)
The EPA's list of the most serious uncontrolled or abandoned hazardous waste sites identified for possible long-term remedial action under Superfund. A site must be on the NPL to receive money from the Trust Fund for remedial actions. The list is based primarily on the score a site receives from the Hazard Ranking System (HRS). The EPA is required to update the NPL at least annually.

National Response Center (NRC)
The federal operations center that receives notifications of all releases of oil and hazardous substances into the environment. The NRC is operated 24 hours per day by the United States Coast Guard, which evaluates all reports and notifies the appropriate agencies.

National Response Team (NRT)
Representatives of thirteen federal agencies that, as a team, coordinate federal responses to nationally significant incidents of pollution and provide advice and technical assistance to the responding agencies before and during a response action.

National Safety Council (NSC)
Independent nonprofit organization that provides information, literature, training, and support for occupational safety and health related programs and issues with the goal of reducing the number and severity of accidents/occupational diseases in the U.S. and finding ways to prevent their occurrence.

National Search and Rescue Plan
An interagency agreement which provides for the effective utilization of all available facilities in all types of search and rescue missions.

National Toxicology Program (NTP)
Established to determine the toxic effects of chemicals and to develop more effective and less expensive toxicity test methods.

National Transit Database
(Formerly Section 15) A reporting system, by uniform categories, to accumulate mass transportation financial and operating information and a uniform system of accounts and records. The reporting and uniform systems shall contain appropriate information to help any level of government make a public sector investment decision. The Secretary [of Transportation] may request and receive appropriate information from any source.

National Transportation System (NTS)
An intermodal system consisting of all forms of transportation in a unified, interconnected manner to reduce energy consumption and air pollution while promoting economic development and supporting the nation's preeminent position in international commerce. The NTS includes the National Highway System (NHS), public transportation, and access to ports and airports.

Nationally Recognized Testing Laboratory (NRTL)
A laboratory that has been accredited (for a minimum of five years) by OSHA to test and certify products that require certification under OSHA's safety and health standards.

nation's freight bill
The amount spent annually on freight transportation by the nation's shippers; also represents the total revenue of all carriers operating in the nation.

Nationwide Personal Transportation Survey (NPTS)
(1) A nationwide home interview survey of households that provides information on the characteristics and personal travel patterns of the U.S. population. Surveys were conducted in 1969, 1977, 1983, and 1990 by the U.S. Bureau of Census for the U.S. Department of Transportation. (2) A periodic national survey that provides comprehensive information on travel by the U.S. population, along with related socioeconomic characteristics of the trip maker. The NPTS is designed to allow an analysis of travel by characteristics of the trip (e.g., length, purpose, mode), the trip maker (e.g., age, sex, household income), and the vehicle used (e.g., model year, vehicle type, make, and model). NPTS surveys were conducted in 1969, 1977 and 1983 by the Bureau of Census (BOC) for the Department of Transportation (DOT). The 1990 NPTS was sponsored by a group of DOT agencies, specifically the Federal Highway Administration (FHWA), Federal Railroad Administration (FRA), National Highway Traffic Safety Administration (NHTSA), Office of the Secretary (OST), and the Federal Transit Administration (FTA). The survey was conducted for DOT by Research Triangle Institute. Information was collected on all trips taken by each household member age 5 and older during a designated 24-hour period, known as a "travel day," and on trips of 75 miles or more taken during the preceding 14-day period, known as the "travel period." The trip information was expanded to annual estimates of trips and travel. The survey encompassed trips on all modes of transportation for all trip purposes and all lengths.

natural and probable consequences
Those consequences that a person, by prudent human foresight, can anticipate as likely to result from an act because they happen so frequently from the commission of such an act that in the field of human experience they may be expected to happen again.

Natural Color System
See ***color ordering system***.

natural draft
The negative pressure created by the height of a stack or chimney and the temperature difference between the flue gas and the outside.

natural environment
That environment relatively unaffected by man.

natural fiber
Any fiber having a plant or animal origin.

natural flood channel
A channel beginning at some point on the banks of a stream and ending at some other point lower downstream, through which flood waters naturally flow at times of high water.

natural frequency
As pertains to vibration, the frequency at which an undamped system will oscillate when momentarily displaced from its rest position.

natural gas
A mixture of naturally occurring gases of hydrocarbon and nonhydrocarbon components

(the main component is methane) usually associated with petroleum deposits.

natural gas marketed production
Gross withdrawals of natural gas from production reservoirs, less gas used for reservoir repressuring; nonhydrocarbon gases removed in treating and processing operations; and quantities vented and flared.

Natural Gas Policy Act (NGPA) of 1978
Section 311, Construction, allows an interstate pipeline company to transport gas "on behalf of" any intrastate pipeline or local distribution company. Pipeline companies may expand or construct facilities used solely to enable this transportation service, subject to certain conditions and reporting requirements.

natural gas transmission
Establishments engaged in the transmission and/or storage of natural gas for sale.

natural gas transmission and distribution
Establishments engaged in both the transmission and distribution of natural gas for sale.

natural language
A computer language in which the rules approximate those of the user's normally written language.

natural radioactivity
The property of radioactivity exhibited by more than fifty naturally occurring radionuclides.

natural resources
Land, fish, wildlife, biota, air, water, groundwater, drinking water supplies, and other such resources belonging to, managed by, held in trust by, appertaining to, or otherwise controlled by the United States, any state or local government, any foreign government, any Indian tribe, or, if such resources are subject to a trust restriction on alienation, any member of an Indian tribe.

natural selection
The process of survival of the fittest, by which organisms that adapt to their environment survive and those that do not disappear.

natural ventilation
Air movement created by wind, a temperature difference, or other non-mechanical means. The movement of outdoor air into a space through intentionally provided openings, such as doors or windows, as well as by infiltration.

natural wet-bulb temperature
The temperature indicated by a wetted thermometer bulb that is exposed to and cooled by the movement of the surrounding air. *See also* ***wet-bulb temperature***.

naturally occurring background levels
Ambient concentrations of chemicals that are present in the environment and have not been influenced by humans (e.g., aluminum, magnesium).

naturally occurring radioactive material (NORM)
Any nuclide which is radioactive in its natural physical state but does not include source material or special nuclear material (i.e., plutonium, uranium-233, or uranium enriched in the isotopes uranium-233 or uranium-235).

naturopathy
A drugless system of healing by the use of physical methods, such as light, air, water, etc.

nausea
An unpleasant physical sensation, often culminating in vomiting. Nausea may be a symptom of a variety of disorders, some minor and some more serious. Nausea is usually felt when nerve endings in the stomach and other parts of the body are irritated. The irritated nerves send messages to the center of the brain that controls the vomiting reflex. When the nerve irritation becomes intense, vomiting results. Nausea and vomiting may be set off by nerve signals from many other parts of the body besides the stomach. For example, intense pain in almost any part of the body can produce nausea. The reason is that the nausea-vomiting mechanism is part of the involuntary autonomic nervous system. Nausea can also be precipitated by strong emotions.

NAVAID
See ***navigation aid***.

navel
See ***umbilicus***.

navicular bone
(1) *Foot.* A bone in the posterior portion of the foot. (2) *Wrist.* The largest bone in the

wrist, located in the proximal row of bones on the thumb side. Also called *scaphiod bone*.

navigable airspace
Airspace at and above the minimum flight altitudes prescribed in the Federal Aviation Regulations (FARs) including airspace needed for safe takeoff and landing.

navigable canal
Waterway built primarily for navigation.

navigable inland waterway
A stretch of water, not part of the sea, over which vessels of a carrying capacity of not less than 50 tons can navigate when normally loaded. This term covers both navigable rivers and lakes and navigable canals.

navigable lake
Natural expanse of water open for navigation.

navigable pass
The water pass through which vessels may pass over a movable dam during periods of high water. The wickets of the dam are lowered to the riverbed and the water flows with little or no obstruction. Navigable passes are usually from 600 feet to 900 feet in width when the dam is lowered. These are found only in the Ohio and Illinois rivers.

navigable river
Natural waterway open for navigation irrespective of whether it has been improved for that purpose.

navigable waters
(1) Traditionally, waters sufficiently deep and wide for navigation by all, or specified sizes of vessels; such waters in the United States come under federal jurisdiction and are included in certain provisions of the Clean Water Act. (2) The waters of the United States, including the territorial sea and such waters as lakes, rivers, streams; waters which are used for recreation; and waters from which fish or shellfish are taken and sold in interstate or foreign commerce.

navigable waters of the U.S.
Waters of the United States that are subject to the ebb and flow of the tide, and/or are presently used, or have been used in the past, or may be susceptible to use to transport interstate or foreign commerce.

navigation
Those activities involved in directing the movement of a vehicle toward its intended destination.

navigation aid (NAVAID)
Includes electrical and visual air navigation aids, lights, signs, and their supporting equipment.

navigation aid classes
Very high frequency omnidirectional radio range (VOR), combined very high frequency omnidirectional radio range (VOR) and tactical aircraft control and navigation (TACAN) navigational facility (VORTAC), and TACAN aids are classed according to their operational use. The three classes of NAVAIDs are T-Terminal, L-Low altitude, H-High altitude.

navigation bulletin
See ***public notice***.

navigational aid
Any visual or electronic device airborne or on the surface which provides point-to-point guidance information or position data to aircraft in flight. *See also* ***air navigation facility***.

NBI
See ***National Bridge Inventory***.

NBIS
See ***National Bridge Inspection Standards***.

NBS
National Bureau of Standards.

NBSAC
See ***National Boating Safety Advisory Council***.

NCC
See ***noise criterion curve***.

NCHRP
See ***National Cooperative Highway Research Program***.

NCI
See ***National Cancer Institute***.

NCP
National Contingency Plan. *See* ***National Oil and Hazardous Substances Contingency Plan***.

NCRP
See ***National Council on Radiation Protection***.

NDE
See ***non-destructive evaluation***.

NDI
Nondestructive inspection. *See discussion of* ***nondestructive evaluation (NDE)***.

NDIR
See ***nondispersive infrared***.

NDT
Nondestructive testing. *See discussion of* ***nondestructive evaluation (NDE)***.

near-accident
See ***near-miss***.

near field
(1). *Acoustics*. The area close to a sound source within which the sound pressure level does not obey the inverse square law concept (i.e., reduction of 6 dBA for each doubling of distance from the noise source). (2). *Electromagnetic Radiation.* Region near a radiating electromagnetic source or structure in which the electric and magnetic fields do not have a substantially planewave character, but vary considerably from point to point. Typically the near field extends out to at least five wavelengths from the radiating device.

near infrared
That portion of the infrared radiation spectrum just beyond the visual range, from about 750 nm to 1400 nm.

near midair collision
An incident associated with the operation of an aircraft in which a possibility of collision occurs as a result of proximity of less than 500 feet to another aircraft, or a report is received from a pilot or flight crew member stating that a collision hazard existed between two or more aircraft. Near midair collisions are categorized based upon the *degree of hazard,* as follows:

Critical: A situation in which collision avoidance was due to chance rather than an act on the part of the pilot. Less than 100 feet of aircraft separation would be considered critical.

Potential: An incident which would probably have resulted in a collision if no action had been taken by either pilot. Closest proximity of less than 500 feet would usually be required in this case.

No Hazard: A situation in which direction and altitude would have made a midair collision improbable regardless of evasive action taken.

near-miss
An occurrence or happening that had the potential to result in some degree of injury, property damage, or both, but did not. Also referred to as a *near-accident.*

near ultraviolet
Pertaining to that portion of the ultraviolet radiation spectrum having wavelengths ranging from about 300 to 380 nm.

near vision
The ability to see the close physical environment.

near vision chart
Any of a number of cards with letters, words, or paragraphs for determining the smallest font size which can be easily read under given conditions.

nearsightedness
A condition in which vision for near objects is better than for distant ones. *See* ***myopia***.

nebula
(1) Slight corneal opacity. (2) Cloudiness in urine. (3) A liquid substance prepared for use as a spray. (4) A cloud of interstellar gas and dust.

NEC
National Electrical Code.

necessary damages
A term said to be much wider in scope in the law of damages than pecuniary. It embraces all those consequences of an injury usually referred to as *general* damages, as distinguished from *special* damages; whereas the phrase *pecuniary damages* covers a smaller class of damages within the larger class of general damages.

neck
The region of the body comprised of those tissues which connect the trunk with the head, including the cervical vertebrae.

neck breadth
The horizontal linear distance from one side of the neck to the other at the vertical mid-

point between otobasion inferior and the shoulder. Measured with the individual standing erect, the facial and neck musculature relaxed, and without flesh compression.

neck – bustpoint length
The surface distance from the neck-shoulder junction to the tip of the bra. Measured with the individual standing erect, the facial, neck, and torso musculature relaxed.

neck – cervical length
The surface distance from the cervicale to the point at which the neck-shoulder junction becomes the vertical portion of the neck. Measured with the individual standing erect, the neck and scalp musculature relaxed.

neck circumference, maximum
The maximum surface distance around the neck, including the thyroid cartilage in the male. Measured without flesh compression, with the individual standing erect, looking straight ahead, and the neck musculature relaxed.

neck circumference, minimum
The minimum surface distance around the neck inferior to the laryngeal prominence. Measured without flesh compression, with the individual standing erect, looking straight ahead, and the neck musculature relaxed.

neck depth
The horizontal linear distance from the anterior protrusion of the neck to the nape of the neck. Measured with the individual standing erect, the facial and neck musculature relaxed, and without flesh compression.

neck – shoulder junction
The level of the lateral point at which the shoulder and neck meet and the angle of the surface arc is 45° above horizontal.

neck – waist length
The surface distance from the superior point of the neck-shoulder junction over the front midline of the body to the midsagittal waist level.

neckrest
Any padded structure which provides support to the neck, especially when sitting.

necropsy
Examination of the body after death.

necrosis
The death of one or more cells or a portion of a tissue or organ, usually resulting from irreversible damage to the affected area.

needle
Maritime. A long stick of timber placed between the wickets of a movable dam to stop the leakage of water between the gates. A needle flat is a small barge used in transporting these timbers.

needle beam scaffold
A light duty scaffold consisting of needle beams supporting a platform.

needs analysis
The breakdown of identified needs into their component parts to determine the causes or reasons for the needs.

needs assessment
The determination and identification of the knowledge, skills, abilities, or other characteristics that are required for a task, job, or operation.

negative
Having a value of less than zero; including lack or absence. Characterized by denial or opposition.

negative acceleration
See ***deceleration***.

negative afterimage
An image seen on a bright background following the removal of a stimulus, and which is approximately the complementary color of the original stimulus.

negative air machine
A device consisting of a fan and ductwork and often a high-efficiency particulate air (HEPA) filter. It is used to move air from one area to another to maintain a state of negative air pressure inside a contaminated area and thus prevent leakage of the contaminants into other areas.

negative air pressure
(1) Air pressure in a room or duct that is less than the pressure in adjacent areas. (2) Condition that exists when less air is supplied to a space than is exhausted from the space, so the air pressure within that space is less than that in surrounding areas.

negative contact
Aviation. Term used by pilots to inform Air Traffic Control (ATC) that a) previously issued traffic is not in sight. It may be followed by the pilot's request for the controller to provide assistance in avoiding the traffic; b) they were unable to contact ATC on a particular frequency.

negative feedback
A signal which tends to decrease the output of a system.

negative feedback mechanism
*See **feedback mechanism**.*

negative g
An acceleration acting along the body's longitudinal axis in a superior direction.

negative pressure
Condition that exists when less air is supplied to a space than is exhausted from the space so the air pressure within that space is less than that in the surrounding area.

negative pressure respirator
A respirator in which the pressure inside the respirator is negative during inspiration relative to the pressure outside, and positive inside the respirator relative to the pressure outside during exhalation.

negative reinforcement
That which causes a weakening or a decrease in the frequency or size of a response as a result of contingent reinforcement.

negative skew
Having a distribution curve with the mean less than the mode.

negative transfer
A condition in which previous experience causes interference with the learning of a new task, usually due to conflicting stimuli or response requirements.

negative work
That dynamic work done by a person using external forces and eccentric muscle contractions.

negatron
A negatively charged electron.

neglect
May mean to omit, fail, or forbear to do a thing that can be done, or that is required to be done, but it may also import an absence of care or attention in the doing or omission of a given act.

negligence
Failure to take reasonable care to avoid causing foreseeable harm to another and which failure caused the harm.

negligible event
An occurrence, subsequent to the introduction of a hazard or set of hazards into a system, that results in a level of injury, damage, or loss of such insignificant consequence that quick or total recovery would be highly probable and possible. The parameters for this categorization are usually established by management in the System Safety Program Plan, or other policy-making documentation.

negotiation
Discussion among the interested parties in a dispute to seek a resolution.

NEI
Non-explosive initiator.

nem
A nutrition unit, based on the caloric content of one gram of standard composition breast milk. Equal to about 0.6 calorie.

NEMA
National Electrical Manufacturers Association.

nematocide
A chemical agent which is destructive to nematodes (roundworms or threadworms). *See also **nematode**.*

nematode
Invertebrate animals of the phylum nemathelminthes and class nematoda, that is, unsegmented round worms with elongated, fusiform, or sac-like bodies covered with cuticle, and inhabiting soil, water, plants, or plant parts; may also be called nemas or eelworms.

neonate
A newborn infant.

neoplasia
A condition characterized by the presence of new growths (tumors).

neoplasm
(1) Term commonly used to describe a cancer, but technically means any new growth of cells that is more or less unrestrained and not governed by the usual limitations of normal

growth. The growth is *benign* if there is some degree of growth restraint and no spread to distant parts. It is *malignant* if it invades other tissues of the host, spreads to distant parts, or both. (2) A mass of new, abnormal tissue; a new growth or tumor. *See also* ***tumor***.

neoplastic
Of or pertaining to a *neoplasm* or a cancer.

NEPA
See ***National Environmental Policy Act***.

néper (Np)
A unit of absorption/attenuation for sound waves, based on the natural logarithm of the ratio of two quantities. Also *napier*.

nephelometer
An instrument which measures the scattering of light due to particles suspended in a medium, such as water.

nephelometry
Photometric analytical technique for measuring the light scattered by finely divided particles of a substance in suspension.

nephrectomy
Surgical removal of a kidney. The procedure is indicated when chronic disease or severe injury produces irreparable damage to the renal cells. Tumors, multiple cysts, and congenital anomalies may also necessitate removal of a kidney. A single kidney can carry on function formerly done by both kidneys, and thus a patient can survive nephrectomy in good health.

nephritis
Inflammation of the kidneys. The most usual form is glomerulonephritis, that is, inflammation of the glomeruli (clusters of renal capillaries). Damage to the membranes of the glomeruli results in impairment of the filtering process, so that blood and proteins such as albumin pass out into the urine. Depending on the symptoms it produces, nephritis is classified as acute nephritis, chronic nephritis, or *nephrosis*. *Acute nephritis* occurs most frequently in children and young people. The disease seems to strike those who have recently suffered from sore throat, scarlet fever, and other infections that are caused by streptococci, and it is believed to originate as an immune response on the part of the kidney. An attack of acute nephritis may produce no symptoms. More often, however, there are headaches, a malaise, back pain, and perhaps slight fever. The urine may look smoky, bloody, or wine colored. Analysis of the urine shows the presence of erythrocytes, albumin, and casts. Another symptom is edema. If this occurs, the face or ankles are swollen, more so in the morning than in the evening. The blood pressure usually rises during acute nephritis and, in severe cases, hypertension may be accompanied by convulsions. *Chronic nephritis* may follow a case of acute nephritis immediately or it may develop after a long interval during which no symptoms have been present. Many cases of chronic nephritis occur in people who have never had the acute form of the disease. The symptoms of chronic nephritis are often unpredictable, with great variations in different cases. But in almost every case of the disease there is steady, progressive, permanent damage to the kidneys. Chronic nephritis generally moves through three stages. In the first stage, the latent stage, there are few outward symptoms. There may be slight malaise, but often the only indication of the disease is the presence of albumin and other abnormal substances in the urine. If a blood count is performed during this stage, anemia may be found. There is no special treatment during the latent stage of chronic nephritis. The person can lead a perfectly normal life but should avoid extremes of fatigue and exposure, and should eat a well-balanced diet. There may be a second stage of chronic nephritis in which edema occurs. Excess body fluids collect in the face, legs, or arms. It is particularly important, at any stage of chronic nephritis to avoid other infections, which will aggravate the condition. The final stage of chronic nephritis is uremia. At this point, damage to the kidneys is so extensive that they begin to fail. Also call *Bright's disease*. *See also* ***nephrosis***.

nephroptosis
Downward displacement of a kidney, also called *floating* or *dropped kidney*. Displacement can occur when the kidney supports are weakened by a sudden strain or blow, or are congenitally defective.

nephrosclerosis
Hardening of the kidney associated with hypertension and disease of the renal arterioles. It is characterized as benign or malignant depending on the severity and rapidity of the hypertension and arteriolar changes.

nephrosis
A disease of the kidneys in which there is malformation of the kidney tissue without inflammation. It probably represents one stage of nephritis and is marked by excessive accumulation of fluid in the body, apparently due to the inability of the kidneys to regulate the body's water content properly. It is further characterized by a great loss of protein in the urine and decreased serum albumin. The exact cause is unknown but the disease may follow acute nephritis.

nephrotoxic agent
*See **nephrotoxin**.*

nephrotoxin
A toxin known to have deleterious effects on the kidney tissue. Also referred to as a *nephrotoxic agent*.

nerve
(1) A collection of one or more axons bound together by connective tissue and having a defined origin and termination. (2) A cordlike structure of the body, composed of highly specialized tissue, by which impulses are conveyed from one region of the body to another. Depending on their function, nerves are known as sensory, motor, or mixed. Sensory nerves, sometimes called afferent nerves, carry information from the outside world to the brain and spinal cord. Sensation of heat, cold, and pain are conveyed by the sensory nerves. Motor nerves, or efferent nerves, transmit impulses from the brain and spinal cord to the muscles. Mixed nerves are composed of both motor and sensory fibers, and transmit messages in both directions at once. Together, the nerves make up the peripheral nervous system, as distinguished from the central nervous system, which consists of the brain and spinal cord. There are twelve pairs of cranial nerves, which carry messages to and from the brain. Spinal nerves arise from the spinal cord and pass out between the vertebrae. There are 31 pairs: 8 cervical, 12 thoracic, 5 lumbar, 5 sacral, and 1 coccygeal. The various nerve fibers and cells that make up the autonomic nervous system serve the glands, heart, blood vessels and involuntary muscles of the internal organs.

nerve cell
The basic unit of the nervous system. The highly specialized cell has many fibers extending from it which carry messages in the form of electrical charges and chemical changes. The fibers of some cells are only a fraction of an inch long, but those of others (for example, the sciatic nerve) extend for 2 or 3 feet. These fibers reach into muscles and organs throughout the body, to the ends of the fibers and toes, and cluster by the thousands in areas of the skin no larger than the head of a pin. *See **neuron**.*

nerve deafness
A hearing impairment due to some abnormality in the auditory nerve.

nervous breakdown
A popular term for any type of mental illness that interferes with a person's normal activities. The term does not refer to a specific disturbance. The so-called "nervous breakdown" can include any of the mental disorders, including neurosis, psychosis, or depression.

nervous system
A system comprised of neural and various supporting tissues which is capable of taking input, integrating that input, and providing motor output.

nervousness
Morbid or undue excitability.

NESHAPS
*See **National Emission Standards for Hazardous Air Pollutants**.*

nested
Located within some larger structure.

net energy
*See **delivered energy**.*

net horsepower
The usable power output of an engine "as installed." Net horsepower is the gross horsepower minus the horsepower used to drive the alternator, water pump, fan, etc. at a specified rpm.

net income or loss before income taxes
The operating profit (or loss) which is operating revenues less operating expenses less non-

operating income and expenses produces the net income, but before "nonrecurring items."

net instrument response
The gross instrument response for the sample, minus the sample blank.

net maximum dependable capacity
The gross electrical output measured at the output terminals of the turbine generator(s) during the most restrictive seasonal conditions, less the station service load.

net metabolic cost
That metabolic activity incurred only from a particular activity, with the basal metabolic rate subtracted from the gross metabolic cost.

net module shipments
Represents the difference between module shipments and module purchases. When exported, incomplete modules and unencapsulated cells are also included.

net productivity rate
The annual per capita rate of increase in a stock resulting from additions due to reproduction, less losses due to mortality.

net receipts
Shipping. The difference between total movements into and total movements out of each PAD District by pipeline, tanker, and barge.

net tare weight
The weight of an empty cargo-carrying piece of equipment plus any fixtures permanently attached.

net ton mile
Rail Operations. The movement of revenue and/or nonrevenue freight a distance of one mile. Includes a reasonable portion of the weight of exclusive work equipment and motorcar trains moved one mile.

net tonnage
The net or register tonnage of a vessel is the remainder after deducting from the gross tonnage of the vessel the tonnage of crew spaces, master's accommodations, navigation spaces, allowance for propelling power, etc. It is expressed in tons of 100 cubic feet.

net weight
Shipping. Weight of the goods alone without any immediate wrappings, (e.g., the weight of the contents of a tin can without the weight of the can). *See also* ***gross weight***.

network
(1) *General.* To meet or otherwise correspond with a select group of individuals who have something in common with one another.
(2) *Computing.* The purposeful grouping of computer terminals for the purpose of sharing resources (peripherals, software, files, applications, etc.), usually in a business environment.

network interface card (NIC)
A device for transferring data over a network. *See also* ***network (2)***.

network navigable inland waterway
All navigable inland waterways open to public navigation in a given area.

neural
Pertaining to a nerve, the nerves, or the nervous system.

neural loss
Hearing loss due to nerve damage.

neuralgia
Sudden pain in a nerve or along the course of one or more nerves. Neuralgia is usually a sharp, spasm-like pain that may recur at intervals. It is caused by inflammation of or injury to a nerve or group of nerves. Inflammation of a nerve (neuritis) may affect different parts of the body, depending upon the location of the nerve. A commonly encountered form of neuralgia is sciatica, or pain occurring along the sciatic nerve. This pain is felt in the back and down the thigh to the ankle. It may result from inflammation of or injury to the sciatic nerve, and is often associated with conditions such as arthritis of the spine, slipped intervertebral disk, diabetes mellitus, and gout. *See also* ***neuritis***.

neurasthenia
Neuroses marked by a lack of energy, depression, loss of appetite, insomnia, and inability to concentrate resulting from a functional disorder of the nervous system, due usually to prolonged and excessive expenditure of energy. Commonly referred to as nervous prostration.

neuritis
Inflammation of a nerve. There are many forms with different effects. Some increase or

decrease the sensitivity of the body part served by the nerve; others produce paralysis; some cause pain and inflammation. The cases in which pain is the chief symptom are generally called *neuralgia*. Neuritis and neuralgia attack the peripheral nervous system (i.e., the nerves that link the brain and spinal cord with the muscles, skin, organs, and all other parts of the body). These nerves usually carry both sensory and motor fibers; hence both pain and some paralysis may result. Certain toxic substances such as lead, arsenic, and mercury may produce a generalized poisoning of the peripheral nerves, with tenderness, pain, and paralysis of the limbs. Other causes of *generalized neuritis* include alcoholism, vitamin-deficiency diseases (such as beriberi and diabetes mellitus), thallium poisoning, some types of allergy, and some viral and bacteria infections (such as diphtheria, syphilis, and mumps). Some attacks of generalized neuritis begin with fever and other symptoms of an acute illness. However, neuritis caused by lead or alcohol poisoning comes on very slowly over the course of weeks or months. Usually an attack of generalized neuritis will subside by itself when the toxic substance is eliminated. Frequently, instead of a generalized irritation of the nerves, only one nerve is affected. For example, Bell's palsy (or facial paralysis) results when the facial nerve is affected. It usually lasts only a few days or weeks. Sometimes, however, the cause is a tumor pressing on the nerve, or injury to the nerve by a blow, cut, or even a bullet. In that event, recovery depends on the success in treating the tumor or the injury.

neuroanatomy
The study of nervous system structure.

neuroceptor
One of the terminal elements of a dendrite that receives a stimulus from the neuromittor of the adjoining neuron.

neurochemistry
The scientific study of the chemical processes taking place in the nervous system.

neurocirculatory
Pertaining to all or part of both the nervous and circulatory systems or the intersection between them.

neurology
Branch of medicine dealing with the nervous system and its disorders.

neuromuscular
Pertaining to all or part of both the motor aspects of the nervous system, the muscular system, or the interaction between them.

neuromuscular junction
That point of interface between the motor neuron and muscle tissue at which the synapse occurs. Also called *myoneural junction*.

neuromuscular spindle
A capsular proprioceptive sensory structure located within skeletal muscles which contains several intrafusal fibers and is responsive to stretch for providing nervous system feedback to prevent damage by overstretching.

neuromuscular stimulation
The stimulation of nervous and/or muscle tissue(s) via electrical, chemical, or other means.

neuron
A nerve cell, with its processes, collaterals, and terminations, regarded as the basic structural unit of the nervous system. Neurons are highly specialized cells having two characteristic properties: irritability, which means they are capable of being stimulated; and conductivity, which means they are able to conduct impulses. They are composed of a nerve cell body (neurosome), one or more processes (nerve fibers) extending from the body, dendrites, and an axon.

The processes or nerve fibers are actually extensions of the cytoplasm surrounding the nucleus of the neuron. A nerve cell may have only one such slender fiber extending from its body, in which case it is classified as unipolar. A neuron having two processes is bipolar, and one with three or more processes is considered multipolar. Most neurons are multipolar, this type of neuron being widely distributed throughout the central nervous system and autonomic ganglia. The multipolar neurons have a single process called an axon and several branched extensions which are the dendrites. The dendrites receive stimuli from other nerves or from a receptor organ, such as the skin or ear, and transmit them through the neuron to the axon. The axon conducts the impulses to the dendrite of another neuron or to an effector organ that is thereby stimulated to action. Many process are covered with a layer of lipid material called myelin. Peripheral nerve fibers have a thin outer covering called neurolemma.

Neurons that receive stimuli from the outside environment and transmit them toward the brain are called afferent or sensory neurons. Neurons that carry impulses in the opposite direction, away from the brain and other nerve center muscles, are called efferent or motor neurons. Another type of nerve cell, the association of internuncial neuron, or interneuron, is found in the brain and spinal cord. These neurons conduct impulses from afferent to efferent neurons. The point at which an impulse is transmitted from one neuron to another is called a synapse. The transmission is chemical in nature (i.e., there is no direct contact between the axon of one neuron and the dendrites of another). The cholinergic nerves (parasympathetic nervous system) liberate at their axon endings a substance called acetylcholine, which acts as a stimulant to the dendrites of adjacent neurons. In a similar manner, the adrenergic nerves (sympathetic nervous system) liberate sympathin, a substance that closely resembles epinephrine and probably is identical to norepinephrine.

The synapse may involve one neuron in chemical contact with many adjacent neurons, or it may involve the axon terminals of one neuron and the dendrites of a succeeding neuron in a nerve pathway. There are many different patterns of synapses.

The dendrites of the sensory neurons are designed to receive stimuli from various parts of the body. These dendrites are called receptor end-organs and are of three general types: exteroceptors, interoceptors, and proprioceptors. Their names give a clue to their specific function. The exteroceptors are located near the external surface of the body and receive impulses from the skin. They transmit information about the senses of touch, heat, cold, and other factors in the external environment. The interoceptors are located in the internal organs and receive information from the viscera (e.g., pressure, tension, and pain). The proprioceptors are found in muscles, tendons, and joints and transmit "muscle sense," by which one is aware of the position of his/her body in space.

The axons of motor neurons form synapses with skeletal muscle fibers to produce motion. These junctions are called motor end-plates or myoneural junctions. The axon of a motor neuron divides just before it enters the muscle fibers. These motor neurons are called somatic efferent neurons. Visceral efferent neurons form synapses with smooth muscle, cardiac muscle, and glands.

neuropathy

A general term denoting functional disturbances and/or pathological changes in the peripheral nervous system.

neurophysiology

The basic physiology of neurons and the nervous system in general, from simple metabolism to the generation and conduction of impulses.

neurosis

An emotional disorder that can interfere with a person's ability to lead a normal, useful life, or can impair his/her physical health. A neurosis is generally a milder form of mental illness than a psychosis. Those persons with neurotic symptoms are usually in contact with reality; they are able to function in society even though they may feel uncomfortable or their efficiency may be impaired. By contrast, psychotic persons tend to withdraw from the real world into one of their own, or to act in strange, even bizarre, ways, and are often not aware of their illness. Sometimes referred to as *psychoneurosis*.

neurotendinous spindle

*See **Golgi tendon organ**.*

neurotic

(1) Pertaining to or affected with a neurosis. (2) Pertaining to the nerves. (3) A nervous person in whom emotions predominate over reason.

neurotoxic agent

*See **neurotoxin**.*

neurotoxicity

Toxic effects on the central or peripheral nervous system causing behavioral or neurological abnormalities.

neurotoxicology
The study of the effects of toxins on nerve tissue.

neurotoxin
A substance that is poisonous or destructive to nerve tissue. Also referred to as a *neurotoxic agent.*

neurotransmitter
A chemical which is released from one neuron at a chemical synapse and for which a receptor is located nearby on the same or another neuron.

neutral body posture
That posture which the body tends to assume when relaxed with the eyes closed or covered in a microgravity environment; the arms lie in front of the body with the elbows, the neck, the hips, and knees all somewhat flexed (given sufficient volume/space to assume this position under microgravity conditions).

neutral body posture stature
The maximum perpendicular linear distance from a plane at the most distal part of the feet to a plane at the highest point on the head when the subject is in the neutral body posture.

neutral density filter
An optical filter which reduces the intensity of light without appreciably changing the relative spectral distribution.

neutral stability
An atmospheric condition that exists in dry air when the environmental lapse rate equals the dry adiabatic rate. In saturated air the environmental lapse rate equals the moist adiabatic rate. Also referred to *neutrally stable air.*

neutralization
Decreasing the acidity or alkalinity of a substance by adding to it alkaline or acidic materials, respectively, as required.

neutralizer
A muscle which functions to prevent some undesired action of another muscle.

neutrally stable air
See ***neutral stability***.

neutrino
A particle resulting from a nuclear reaction which carries energy away from the system but has no mass or charge.

neutron
A neutral particle (i.e., one without an electrical charge) of approximate unit mass present in all atomic nuclei, except those of ordinary (or light) hydrogen. Neutrons are used to initiate the fission and fusion process.

neutron chain reaction
A process in which some of the neutrons released in one fission event cause other fissions to occur.

nevus
A small, flat, elevated or pedunculated lesion of the skin, pigmented or nonpigmented, and with or without hair growth, characterized by a specific type of cell. More commonly referred to as a *mole.* Most moles are either brown, black, or flesh-colored. They may appear on any part of the skin. They vary in size and thickness, and occur in groups or singly. Usually they are not disfiguring. A nevus is usually not troublesome unless it is unsightly or unless it becomes inflamed or cancerous. Fortunately, nevi seldom become cancerous and, if they do, the cause is often constant irritation. Any change in size, color, or texture of a mole, or any excessive itching or any bleeding, should be reported to a physician. Moles can be removed by surgery or by one of several other methods, such as the application of solid carbon dioxide, injections, and radium treatment.

new animal drug
Any drug intended for use for animals other than man, including any drug intended for use in animal feed but not including such animal feed. The composition is a) such that such drug is not generally recognized, among experts qualified by scientific training and experience to evaluate the safety and effectiveness of animal drugs, as safe and effective for use under the conditions prescribed, recommended, or suggested in the labeling thereof; except that such a drug not so recognized shall not be deemed to be a "new animal drug" if at any time prior to June 25, 1938, it was subject to the Food and Drug Act of June 30, 1906, as amended, and if at such time its labeling contained the same representations concerning the conditions of its use; or b) the composition is such that the drug, as a result of investigations to determine its safety and effectiveness for use under such conditions,

has become so recognized but which has not, otherwise than in such investigations, been used to a material extent or for a material time under such conditions.

new candle
See ***candela***.

new drug
According to the Federal Food, Drug, and Cosmetic Act: Any drug (except a new animal drug or an animal feed bearing or containing a new animal drug) the composition of which is such that such drug is not generally recognized, among experts qualified by scientific training and experience to evaluate the safety and effectiveness of drugs, as safe and effective for use under the conditions prescribed, recommended, or suggested in the labeling thereof, except that such a drug not so recognized shall not be deemed to be a "new drug" if at any time prior to the enactment of the Federal Food, Drug and Cosmetic Act it was subject to the Food and Drugs Act of June 30, 1906, as amended, and if at such time its labeling contained the same representations concerning the conditions of its use.

A new drug is also any drug (except a new animal drug or an animal feed bearing or containing a new animal drug) the composition of which is such that such drug, as a result of investigations to determine its safety and effectiveness for use under such conditions, has become so recognized, but which has not, otherwise than in such investigations, been used to a material extent or for a material time under such conditions.

new look bus
A bus with the predominant styling and mechanical equipment common to buses manufactured between 1959 and 1978.

new source
Any stationary source which is built or modified after publication of final or proposed regulations that prescribe a standard of performance which is intended to apply to that type of emission source.

New Source Performance Standards (NSPS)
Uniform national EPA air emissions and water effluent standards which limit the amount of pollution allowed from new sources, or from existing sources that have been modified.

new underground storage tanks
Tanks used to contain regulated substances and installed after December 22, 1988.

new vehicle
A vehicle which is offered for sale or lease after manufacture without any prior use.

new vehicle storage
A Fleet Management System (FMS) inventory status indicating vehicles that are placed in storage when first received and are awaiting assignment.

Newton (N)
A unit of force which, when applied to a mass of one kilogram, will give it an acceleration of one meter per second.

newton-meter
A unit of torque in the SI/MKS system, equal to a 1 N force acting perpendicularly at 1 meter from a point of rotation.

Newton's first law of motion
Every mass maintains its current state of motion unless acted on by one or more non-equilibrating forces. Also called *law of inertia*.

Newton's laws of motion
Three physical laws which govern the basic interactions of physical objects and forces. *See* ***Newton's first law of motion, Newton's second law of motion,*** *and* ***Newton's third law of motion***.

Newton's second law of motion
The force required to impart a given acceleration to an object is proportional to the mass of that object.

Newton's third law of motion
For every action, there is an equal and opposing reaction.

NFPA
See ***National Fire Protection Association***.

ng
Nanogram [0.001 g or 1000 picograms (pg)].

NGPA
See ***Natural Gas Policy Act of 1978***.

NGT
See ***nominal group technique***.

NHS
See ***National Highway System***.

NHTSA
See ***National Highway Traffic Safety Administration.***

NIBS
National Institute of Business Sciences.

NIC
See ***network interface card.***

nickel itch
A type of dermatitis seen in some workers who are exposed to nickel.

nicotine
An alkaloid that in its pure state is a colorless, pungent, oily, and highly poisonous liquid, having an acrid burning taste. It is a constituent of tobacco. In water solution, it is sometimes used as an insecticide and plant spray and is generally highly effective because of its systemic poisoning properties. Although nicotine is highly toxic, the amount inhaled while smoking tobacco is too small to cause immediate death. The nicotine in tobacco can, however, cause indigestion and increase blood pressure, and dull the appetite. It also acts as a vasoconstrictor.

NIEHS
National Institute of Environmental Health Sciences.

night
(1) The hours between the end of evening civil twilight and the beginning of morning civil twilight or such other period between sunset and sunrise as may be specified by the appropriate authority. (Note: Civil twilight ends in the evening when the center of the sun's disk is 6 degrees below the horizon and begins in the morning when the center of the sun's disk is 6 degrees below the horizon). (2) The hours from 6:00 p.m. to 5:59 a.m.

night blindness
The inability of the eyes to quickly adjust to changes in light intensity. For example, people whose eyes cannot quickly adjust to the darkness of a movie theater or are temporarily blinded by the glare of headlights at night probably have night blindness.

night shift
See ***third shift.***

night vision
See ***scotopic vision.***

night vision goggles (NVG)
A light image intensifying device for enabling an individual to see terrain, objects, or other items of interest at very low light levels.

NIH
See ***National Institutes of Health.***

NIHL
See ***noise-induced hearing loss.***

nimbostratus
A dark, gray cloud characterized by more or less continuously falling precipitation. It is not accompanied by lightning, thunder, or hail.

nine-foot contour line
A meandering line not necessarily on the channel edge at which a depth of 9 feet is obtained at river stage low water reference plane.

Nineteenth Amendment
Known as the women's suffrage amendment to the U.S. Constitution, it provides that the right of citizens of the U.S. to vote shall not be denied or abridged by the U.S. or by any state on account of sex. The 19th Amendment was ratified in 1920.

Ninth Amendment
This amendment to the U.S. Constitution provides that the enumeration in the Constitution of certain rights, shall not be construed to deny or disparage others retained by the people.

NIOSH
See ***National Institute for Occupational Safety and Health.***

nip point
The nearest point of intersection or near contact of two oppositely rotating circular surfaces or a rotating circular surface and a planar surface.

nipple
(1) A projection from the proximal center of the breast, usually of different hue and texture than normal flesh and which, in the female, provides the terminal milk duct outlets. (2) Any structure resembling or serving a function similar to the human nipple.

NIPTS
See ***noise-induced permanent threshold shift.***

NIST
National Institute of Standards and Technology (formerly the National Bureau of Standards).

nit
A unit of luminance, equal to 1 candela/m^2.

nitrate
A compound containing nitrogen which can exist in the atmosphere or as a dissolved gas in water and which can have harmful effects on humans and animals. Nitrates in water can cause severe illness in infants and cows.

nitric oxide (NO)
A gas formed by combustion under high temperature and high pressure in an internal combustion engine. It changes to nitrogen dioxide (NO_2) in the ambient air and contributes to photochemical smog.

nitrification
The process whereby ammonia in wastewater is oxidized to nitrite and then to nitrate by bacterial or chemical reactions.

nitrilotriacetic acid (NTA)
A compound being used to replace phosphates in detergents.

nitrite
(1) An intermediate in the process of nitrification. (2) Nitrous oxide salts used in food preservation.

nitrogen (N_2)
A chemical element, atomic number 7, atomic weight 14.007, symbol N. A colorless, odorless, and chemically reactively inert gas. Makes up approximately 78-79% of the earth's atmosphere. Also, *gaseous nitrogen* or *GN_2*.

nitrogen dioxide (NO_2)
The result of nitric oxide combining with oxygen in the atmosphere. A major component of photochemical smog.

nitrogen fixation
The utilization of atmospheric nitrogen to form chemical compounds. In nature this is accomplished by bacteria resulting in the ability of plants to synthesize proteins.

nitrogen narcosis
A condition, due to breathing of nitrogen gas under high pressures, whose symptoms range from joviality and lack of concern, to drowsiness and weakness, to unconsciousness and death, depending on the pressure and duration of exposure. Sometimes referred to as *rapture of the deep/depths*. *See also* ***inert gas narcosis***.

nitrogen oxide (NO_X)
Product of combustion from transportation and stationary sources and a major contributor to the formation of ozone in the troposphere and acid rain deposition.

nitrogenous wastes
Animal or vegetable residues that contain significant amounts of nitrogen.

NITTS
See ***noise-induced temporary threshold shift***.

NLRB
See ***National Labor Relations Board***.

nm
Nanometer (common unit for absorbance used in spectroscopy).

N/m^2
Newtons per square meter.

NMOG
See ***nonmethane organic gas***.

NMR
Nuclear magnetic resonance.

NNI
See ***noise and number index***.

NOAA
National Oceanic and Atmospheric Administration.

NOAEL
See ***nonobserved-adverse-effect-level***.

no brain rule
(slang) A task design guideline that the workplace should prevent the worker from getting hurt even if he/she doesn't think before acting.

no eyewitness rule
The principle by which one who is charged with the burden of showing freedom from contributory negligence is assumed to have acted with due care for his/her own safety in the absence of eyewitnesses or of any obtainable evidence to the contrary.

no-fault auto insurance
Type of insurance in which claims for personal injury (and sometimes property dam-

age) are made against the claimant's own insurance company (no matter who was at fault) rather than against the insurer of the party at fault. Under such state "no-fault" statutes only in cases of serious personal injuries and high medical costs may the injured bring an action against the other party of his/her insurer. No-fault statutes vary from state to state in terms of scope of coverage, threshold amounts, threshold types (e.g., monetary or verbal), etc.

no-fire level
The maximum direct current or RF energy at which an electroexplosive initiator shall not fire with a reliability of 0.999 at a confidence level of 95 percent as determined by a Bruceton test method and shall be capable of subsequent firing within the requirements of performance specifications. *See also* ***Bruceton test method***.

no gyro approach
Aviation. A radar approach vector provided in case of a malfunctioning gyro-compass or directional gyro. Instead of providing the pilot with headings to be flown, the controller observes the radar track and issues control instructions "turn right/left" or "stop turn" as appropriate.

no-observed-effect level
In dose-response experiments, the experimental exposure level representing the highest level tested at which no effects at all were demonstrated.

noble gas
A gas that is either completely nonreactive or reacts only to a very limited extent with other elements. The noble gases are helium, argon, neon, krypton, xenon, and radon.

NOC
See ***not otherwise classified***.

noctilucent cloud
A wavy, thin, blue-to-white cloud that is best seen at twilight in polar latitudes. It forms at altitudes about 80 to 90 kilometers above the surface.

nocturnal
Pertaining to a species or individual who prefers to be active at night.

nocturnal inversion
See ***radiation inversion***.

nodal point (of the eye)
An imaginary midpoint in the eyeball at which light rays from any point in the visual field will intersect the visual axis.

node
(1) A junction. (2) A point or region of minimum or zero amplitude in a periodic system.

node of Ranvier
A gap in the myelin sheath of a nerve fiber in which the axon membrane is exposed. Enables the neural impulse to jump from node to node, providing for faster transmission.

NOEL
No-observable-effect level.

noise
(1) *Acoustics.* Commonly defined as unwanted sound and is usually expressed in decibels on the A scale (dBA), which is the scale thought to most approximate human hearing. Noise is characterized by both frequency (pitch) and pressure (intensity). (2) *Instrument.* Any unwanted electrical disturbance or spurious signal which modifies the transmission, measurement, or recording of desired data. An output signal of an instrument that does not represent the variable being measured or the variation in the signal from an instrument that is not caused by variations in the concentration of the material being measured.

noise and number index (NNI)
An index used for rating the noise environment near airports and the noise associated with aircraft flyby.

noise cancellation
An active noise reduction technique using a device which monitors an incoming signal with noise and produces an opposing signal prior to passing the signal to the observer.

noise contour
A continuous line on a plot plan or map which connects all points of a specified noise level (such as 85 dBA).

noise control
The process of achieving a more nearly acceptable environment through the use of any noise reducing techniques.

noise criterion curve (NCC)
Any of several sets of criteria for providing a single number rating the acceptability of continuous environmental noise, based on curves of noise intensity or sound pressure level vs. frequency. Ratings are given for the noise criterion curve which is not exceeded. Each curve is named for the dB level where the curve crosses the 2-kHz point.

noise exposure
(1) The cumulative amount of acoustic stimulation which reaches the ear of an individual over some specified period of time. (2) Exposure to any unwanted sound. Overexposure to occupational noise in the United States is considered to be 90 dBA over an 8-hour time-weighted average (TWA).

noise-induced hearing loss (NIHL)
A progressive hearing loss that is the result of exposure to noise, generally of the continuous type over a long period of time, as opposed to acoustic trauma, which results in immediate hearing loss.

noise-induced permanent threshold shift (NIPTS)
A permanent hearing loss due to extremely high noise exposure levels.

noise-induced temporary threshold shift (NITTS)
A temporary hearing loss due to high noise exposure levels.

noise level
For airborne sound, unless otherwise specified to the contrary, noise level is the weighted sound pressure level, called sound level, the weighting of which must be indicated (e.g., A, B, or C weighting). *See also **sound level**.*

noise margin
The margin between the worst case noise level and logic circuitry threshold.

noise meter
*See **sound level meter**.*

noise nuisance
An environmental problem consisting of human activities that arise from unreasonable or unlawful use by a person of his/her own property, obstructing or injuring others' rights, and producing material nuisance or discomfort presumed by the law to result in damage.

noise pollution
An amount of noise in the environment considered to be excessive by the majority of the population.

noise rating number
The perceived noise level of specified acoustic conditions that is tolerable.

noise reduction
The reduction in the sound pressure level of a noise, or the attenuation of unwanted sound by any means.

noise reduction coefficient (NRC)
The average sound absorption coefficient for a material over the logarithm of frequency in the range from 256 to 2048Hz.

noise reduction rating (NRR)
(1) A measure of the effectiveness of a given hearing protector, usually expressed in decibels. Assuming a complete and perfect fit, the NRR is the difference between the sound pressure levels outside the ear and those inside the ear. (2) A single number noise reduction factor in decibels, determined by an empirically derived technique which takes into account performance variation of protectors in noise reducing effectiveness due to differing noise spectra, fit variability, and the mean attenuation of test stimuli at the one-third octave band test frequencies.

noise suppressor
An electronic circuit which is capable of automatically inhibiting the amplifier of a radio receiver to eliminate background noise when no signal is being received.

noisy shoulder
(slang) Emitting a grating noise on elevation or depression of the shoulder. Often due to a snapping tendon over the scapula.

nolo contendere
Latin phrase meaning "I will not contest it." A plea in a criminal case which has a similar legal effect as pleading guilty.

Nomex®
A fire-resistant material used for clothing and other industrial applications.

nominal bandwidth
The range between the specified upper and lower cutoff frequencies of a system.

nominal damages
A trifling sum awarded to a plaintiff in an action, where there is no substantial loss or injury to be compensated, but still the law recognizes a technical invasion of his/her rights or a breach of the defendant's duty, or in cases where there has been a real injury, the plaintiff's evidence entirely fails to show its amount.

nominal group technique (NGT)
A method for generating innovative product ideas in which the individuals within the group communicate verbally with each other only at specified periods of time, using their individual creativity the remainder of the time.

nominal scale
A basic measurement scale in which items are categorized or classified using only labeling methods.

nomograph
A chart in the form of linear scales which represents an equation containing a number of variables so that a straight line can be placed across them, cutting the scales at values of the variables satisfying the equation and yielding an answer to that for which one is solving.

nomological validity
An aspect of construct validity concerned with the fit between theoretical postulates and empirical data.

NON
Notice of noncompliance. *See* ***violation notice***.

nonaccidental injury
Any injury which cannot be traced to a specific accident.

nonadaptive response
A reaction to a situation which does not support continued survival.

nonasbestiform fiber
A fibrous material which contains no asbestos.

nonattainment area
Geographic area which does not meet one or more of the National Ambient Air Quality Standards (NAAQS) for the any of the EPA's listed criteria pollutants designated under the Clean Air Act. *See also* ***NAAQS*** *and* ***criteria pollutants***.

nonauditory effects of noise
(1) Effects from exposure to noise, such as stress, fatigue, reduction in work efficiency, etc. (2) Any physiological or psychological effect of noise other than via the auditory system.

nonballistic movement
See ***controlled movement***.

non-blackbody
Any surface which reflects at least some of the radiation impinging upon it.

non-bulk packaging
A packaging which has a) a maximum capacity of 450 L (119 gallons) or less as a receptacle for a liquid; b) a maximum net mass of 400 kg (882 pounds) or less and a maximum capacity of 450 L (119 gallons) or less as a receptacle for a solid; or c) a water capacity of 454 kg (1000 pounds) or less as a receptacle for a gas as defined in 49 CFR 173.115.

non-causal association
A statistical association in which no cause-and-effect relationship is apparent between two variables.

noncertified color additive
Any of a category of substances which are approved by the FDA for cosmetic use without special safety testing.

noncoincidental peak-day flow
Gas Industry. The largest volume of gas delivered to a particular customer by a pipeline company in a single day during the year.

noncollision accident
A motor vehicle accident which does not involve a collision. Non-collision accidents include jackknifes, overturns, fires, cargo shifts and spills, and incidents in which trucks run off the road.

noncollision crash
A class of crash in which the first harmful event does not involve a collision with a fixed object, non-fixed object, or a motor vehicle. This includes overturn, fire/explosion, falls from a vehicle, and injuries in a vehicle.

noncombustible
Pertaining to a substance which is essentially incapable of burning or supporting a fire.

noncommunity water system
A public water system that is not a community water system, such as the water supply at a camp site or national park.

noncompliance
(1) *General.* Not in compliance with a specific or given requirement; or, to be found not in compliance with a specific or given requirement. Noncompliance may lead to fines and penalties under certain circumstances. (2) *U.S. Coast Guard.* Failure to comply with a standard or regulation issued under 46 U.S.C. Chapter 43, or with a section of the statutes.

nonconventional pollutant
Any pollutant that is not statutorily listed or which is poorly understood by the scientific community.

noncurrent liabilities
Noncurrent portion of long-term debt and of capital leases, advances to associated companies and other liabilities not due during the normal business cycle.

noncyclic element
A segment or step of a process or operation which doesn't occur within each cycle.

nondestructive evaluation (NDE)
Any testing, inspection, or evaluation that does not cause harm or impair the usefulness of the object being tested. *Nondestructive testing (NDT)* often refers just to the test methods and test equipment with only a general reference to materials and/or parts. *Nondestructive inspection (NDI)* relates to specific written requirements, procedures, personnel, standards, and controls for the testing of a particular material of a specific part. *Nondestructive evaluation (NDE)* is concerned with the decision-making process, the determination of the meanings of the results, of the final acceptance or rejection of the material or part, and may be qualitative or quantitative.

nondestructive inspection (NDI)
See ***nondestructive evaluation***.

nondestructive testing (NDT)
See ***nondestructive evaluation***.

nondetects
Chemicals that are not detected in a particular sample above a certain limit. This limit usually will be the quantitation limit for the chemical in that sample. (Note, however, that it is possible to detect and estimate concentrations of chemicals below the quantitation limit but above the detection limit.)

nondeterministic
Pertaining to any event or condition which cannot be reliably predicted given certain prior events and currently known laws.

nondisabling injury
Any injury not resulting in death, permanent disability, or temporary total disability.

nondispersive infrared (NDIR)
A measurement principal that can be employed to measure the airborne concentration of some materials (e.g., CO, CO_2, etc.) using an infrared source and photocell to determine the absorption of the IR radiation, which is dependent on contaminant concentration in the sample.

nondisruptive
Pertaining to an activity which does not interfere with any other ongoing activity.

nonearthen shore
A structure built of stone, brick, concrete, or other building materials, that borders a body of water and that is not otherwise classified.

nonfatal accident
A motor vehicle traffic accident that results in one or more injuries, but no fatal injuries.

nonfatal alcohol involvement crash
Alcohol-related or alcohol-involved if police indicate on the police accident report that there is evidence of alcohol present. The code does not necessarily mean that a driver, passenger, or nonoccupant was tested for alcohol.

nonfatal casualty
Rail Operations. Injuries and occupational illnesses incurred during railroad operations and maintenance procedures.

nonfatal injury
A nonfatal injury is any traffic accident injury other than a fatal injury.

nonfatal injury accident
(1) A nonfatal injury accident is a traffic accident that results in nonfatal injuries. (2) Accident in which at least one person is injured, and no injury results in death.

nonfatal injury accident rate
The nonfatal injury accident rate is the number of nonfatal injury accidents per 100 million vehicle miles of travel.

nonfatal (most serious) injured
Nonfatally injured persons whose injuries are classified as incapacitating (as defined in the "Manual On Classification of Motor Vehicle Traffic Accidents," American National Standards Institute (ANSI) D16.1-1989). States may receive information about these injuries on the accident report forms as incapacitating, incapacitating injury, incapacitated, disabled, carried from scene, severe injury, critical nonfatal, major injury, or other similar terms.

nonfatally injured person
A nonfatally injured person is one who suffers a nonfatal injury in either a fatal or nonfatal injury traffic accident.

nonfatally injured persons rate
The nonfatally injured persons rate is the number of nonfatally injured persons per 100 million vehicle miles of travel.

nonfeasance
Nonperformance of some act which a person is obligated or has the responsibility to perform.

non-fixed guideway directional route miles
Rail Operations. The mileage operated by non-rail modes on fixed routes and schedules in mixed traffic rights-of-way.

non-fixed route
Service not provided on a repetitive, fixed-schedule basis along a specific route to specific locations. Demand response is the only non-fixed route mode.

nonflammable
A material or substance that will not burn readily or quickly.

nonfriable
With regard to asbestos, a material which contains more than 1% asbestos (by weight) and which cannot be crumbled by hand pressure when dry.

nonhub
Aviation. A community enplaning less than 0.05 percent of the total enplaned passengers in all services and all operations in all communities within the 50 States, and District of Columbia, and other U.S. areas designated by the Federal Aviation Administration.

nonincendive
A device that will not ignite the group of gases or vapors for which it is rated. It is similar to *intrinsically safe,* but does not include failure tolerance ratings. It is typically used in rating electrical products for use in Class I, Division 2 locations only. *See also **intrinsically safe** and **Class I, Division 2 location**.*

nonindigenous species
Any species or other viable biological material that enters an ecosystem beyond its historic range, including any such organism transferred from one country into another.

noninteractive window
An active window which is not receptive to user input.

noninterlaced display
A visual display in which the entire display is not receptive to user input.

noninvasive
Pertaining to those clinical or experimental procedures which do not require breaking the skin, insertion into any body cavity except the mouth, and which do not cause extreme discomfort.

nonionic detergent
Any detergent with molecules which don't ionize in water.

nonionizing radiation
Electromagnetic radiation, such as ultraviolet, laser, infrared, microwave, and radiofrequency radiation, that does not cause ionization.

nonisolated intermediates
TSCA. (1) Chemicals that are both manufactured and partially or totally consumed in the chemical reaction process. (2) Chemicals intentionally present in order to affect the rate of chemical reactions by which other chemical substances or mixtures are being manufactured.

nonlinear correlation
A correlation which does not follow the linear relationship $Y = a + bX$. Also referred to as *curvilinear correlation.*

nonlinear damping
Damping due to a force that is not proportional to velocity.

nonlinear function
Any function which can't be expressed in the form $y = a_1x_1 + a_2x_2 + \ldots + a_nx_n$.

nonlinear regression
Any type of regression involving a function or curve which has other than a directly or inversely proportional relationship between variables. Also referred to as *curvilinear regression*.

nonloading secondary task
A secondary task which may be attended to when the operator's primary task does not require attention.

nonmetallic sheathed cable
A factory assembly of two or more insulated conductors having an outer sheath of moisture-resistant, flame-retardant, nonmetallic material. Nonmetallic sheathed cable is manufactured in the following types: a) *Type NM.* The overall covering has a flame-retardant and moisture-resistant finish. b) *Type NMC.* The overall covering is flame-retardant, moisture-resistant, fungus-resistance, and corrosion-resistant.

nonmethane organic gas (NMOG)
According to the Clean Air Act: The sum of nonoxygenated and oxygenated hydrocarbons contained in a gas sample, including at a minimum, all oxygenated organic gases containing 5 or fewer carbon atoms (i.e., aldehydes, ketones, alcohols, ethers, etc.) and all known alkanes, alkenes, alkynes, and aromatics containing 12 or fewer carbon atoms. To demonstrate compliance with a NMOG standard, NMOG emissions shall be measured in accordance with the "California Non Methane Organic Gas Test Procedures". In the case of vehicles using fuels other than base gasoline, the level of NMOG emissions shall be adjusted based on the reactivity of the emissions relative to vehicles using base gasoline.

nonmetropolitan
Households not located within Metropolitan Statistical Areas (MSA) as defined by the U.S. Office of Management and Budget.

nonmonetary incentive
Any incentive plan not involving monetary compensation, such as improved working conditions and social benefits.

nonmonotonic
Pertaining to a function which contains cyclic or both increasing and decreasing aspects within the region of interest.

nonmotorist
Any person who is not an occupant of a motor vehicle in transport and includes the following: a) pedestrians, b) pedal cyclists, c) occupants of parked motor vehicles, d) others such as joggers, skateboard riders, people riding on animals, and persons riding in animal-drawn conveyances.

nonmotorist location
The location of nonmotorists at time of impact. Intersection locations are coded only if nonmotorists were struck in the area formed by a junction of two or more traffic ways. Nonintersection location may include nonmotorists struck on a junction of a driveway/alley access and a named traffic way. Nonmotorists who are occupants of motor vehicles not in transport are coded with respect to the location of the vehicle.

nonobserved-adverse-effect-level (NOAEL)
In dose-response experiments, the experimental exposure level representing the highest level tested at which no adverse effects were demonstrated.

nonoccupant
Any person who is not an occupant of a motor vehicle (e.g., pedestrian or pedal cyclist), or who is an occupant of a motor vehicle which is not in transport.

nonoperating income and expense
Interest income and expense, unusual foreign exchange gains or losses, and capital gains or losses in disposition of property and equipment.

nonparametric
Statistical methods that do not assume a particular distribution for the population under consideration.

nonperformance
Neglect, failure, or refusal to do or perform an act stipulated or contracted to be done.

nonpermit confined space
A confined space that does not contain or, with respect to atmospheric hazards, have the potential to contain any hazard capable of causing death or serious physical harm.

nonpilot personnel
May include any of the following personnel: maintenance, servicing, inspection, rules, regulations, standards, weather service, airport management, production design, ground signalman, passenger, driver of vehicle, flight engineer, radio operator, flight instructor on ground, operational supervisor, air traffic control, airway facilities, pilot of another aircraft, ground crewman, spectator, third pilot, navigator, flight attendant, dispatching and other flight personnel.

nonpoint source
Pollution sources which are diffuse and do not have a single point of origin or are not introduced into a receiving stream from a specific outlet. The pollutants are generally carried off the land by stormwater runoff. The commonly used categories for nonpoint sources are agriculture, forestry, urban, mining, construction, dams and channels, land disposal, and salt water intrusion.

nonpolar compound
A compound for which the positive and negative electrical charges coincide and the molecules do not ionize in solution and impart electrical conductivity.

nonpolar solvents
The aromatic and petroleum hydrocarbon group of compounds.

nonport terminal
Waterfront terminals individually located along navigable rivers, having place identity but not otherwise classifiable as a regional distribution or subregional port.

nonprecision approach procedure
A standard instrument approach procedure in which no electronic glideslope is provided.

nonprecision instrument runway
A runway with an approved or planned straight-in instrument approach procedure which has no existing or planned precision instrument approach procedure.

nonpriority U.S. mail
Mail bearing postage for surface transportation that goes by air on a space available basis at rates lower than those fixed for priority (i.e., air) mail.

nonradar approach control tower
An airport traffic control tower (ATCT) providing approach control service without the use of radar. In other words, an ATCT at which air traffic control specialists are permitted to provide manual approach control service, as well as to handle takeoffs and landings.

nonrandom sample
Any sample taken in such a manner that some members of the defined population are more likely to be sampled than others.

nonrapid eye movement sleep
Any phase of sleep in which rapid eye movements are not present. Also called *non-REM sleep*.

nonrecoverable light loss factor
Any light loss factor due to equipment or other conditions which cannot be remedied through normal maintenance, specifically including temperature, lamp position/tilt, equipment operation, luminaire surface deterioration, line voltage, and ballast characteristics.

nonrecurring items
Discontinued operations, extraordinary items, and accounting changes in income or loss.

nonregulated trucking
A carrier which is exempt from economic regulation, e.g., exempt agricultural shipments and private trucking operations.

nonrepetitive
Pertaining to an operation, process, or job which is frequently changed or altered in some way.

nonresident commercial driver's license
A CDL (commercial driver's license) issued by a state to an individual domiciled in a foreign country.

nonrevenue freight
Company materials and supplies transported without charge in freight trains of a particular railroad for its own use.

nonroad engine
Under the Clean Air Act: An internal combustion engine (including the fuel system) that is not used in a motor vehicle or a vehicle used solely for competition.

nonroad vehicle
Under the Clean Air Act: A vehicle that is powered by a nonroad engine and that is not a motor vehicle or a vehicle used solely for competition.

nonroutine respirator use
The wearing of a respirator when carrying out a special task that occurs infrequently.

nonscheduled air transportation
Establishments primarily engaged in furnishing nonscheduled air transportation. Also included in this industry are establishments primarily engaged in furnishing airplane sightseeing services, air taxi services and helicopter passenger transportation services to, from, or between local airports, whether or not scheduled.

nonscheduled service
Revenue flights, such as charter flights, that are not operated in regular scheduled service, and all nonrevenue flights incident to such flights.

non-self-propelled
Vessels not containing within themselves the means for their own propulsion.

nonserious violation citation
Under OSHA citation criteria, a category of citation issued when a situation would affect worker safety or health but would not cause death or serious physical harm.

nonstandard
Differing from established specifications, conditions, or requirements.

nonstationary
Pertaining to a condition or function where the mean, spectral density, and probability distribution vary with time.

nonstationary time series
A stochastic time series whose characteristics change with an integral increase in the time axis.

nontoxic
A material is nontoxic when experience and/or experiments have failed to cause physiological, morphological, or functional changes which adversely affect the health of man or animal.

non-train incident
An event arising from railroad operations but not from the movement of on-track equipment, which does not exceed the reporting threshold, and results in a death, a reportable injury, or a reportable occupational illness.

nontrespassers
Rail Operations. A person who is lawfully on any part of railroad property which is used in railroad operations or a person who is adjacent to railroad premises when injured as a result of railroad operations.

nontrespassers (Class D)
Rail Operations. Persons lawfully on that part of railroad property that is used in railroad operation (other than those herein defined as employees, passengers, trespassers, or contractor employees) and persons adjacent to railroad premises when they are injured as the result of the operation of a railroad. This class also includes other persons on vessels or buses, whose use arises from the operation of a railroad.

nonutility unit
Under the Clean Air Act, a unit other than a utility unit.

nonvehicle maintenance
All activities associated with facility maintenance, including administration; repair of buildings, grounds, and equipment as a result of accidents or vandalism; operation of electric power facilities; and maintenance of vehicle movement control systems; fare collection and counting equipment; structures, tunnels, and subways; roadway and track; passenger stations, operating station buildings, grounds and equipment; communication systems; general administration buildings, grounds, and equipment; and electric power facilities.

nonvessel operating common carrier (NVOCC)
(1) A company operating as a freight forwarder involved in ocean-going vessel freight only. (2) A form of freight consolidation for the selling of space on ocean-going vessels.

nonvessel owning carrier (NVOC)
A firm which consolidates and disperses international containers that originate at, or are bound for, inland ports.

nonvolatile
Material that does not evaporate at ordinary temperature.

nonwoven fabric
Any type of cloth produced by a semirandom arrangement of fibers, whether synthetic or

natural, held together by adhesives or needling.

NOPPA
National Ocean Pollution Planning Act of 1978.

noradrenalin
See ***norepinephrine***.

nordo
See ***lost communications***.

norepinephrine
A catecholamine which serves both as a hormone and a neurotransmitter. Also referred to as *noradrenalin* and *arterenol*. *See also* ***catecholamine***.

NORM
See ***naturally occurring radioactive material***.

normal (N)
(1) That which conforms to some standard; typical or commonplace. (2) A solution containing one gram-equivalent weight per liter of solution. (3) Perpendicular to a vector, plane, or other entity. (4) Pertaining to or having a Gaussian (normal) distribution. (5) The moving average of temperature which is recognized as a standard for a given location.

normal distribution
(1) In statistical analysis, that distribution of events which occurs most often and is typically represented graphically as a bell-shaped curve. (2) If the mean, median, and mode are the same in a set of data, the data assume a completely symmetrical, bell-shaped distribution which is called a normal distribution. This distribution is characterized by a maximum number of occurrences at the center or mean point, a progressive decrease in the frequency of occurrences with distance from the center, and a symmetry of distribution on either side of the center. Also called the *Gaussian distribution*.

normal effort
That amount of effort required or expended in manual work by an average operator with average skill and attention to the task.

normal element time
A statistically determined element time based on the expected or required performance by an average qualified worker working at a normal pace.

normal event
As pertains to fault tree analysis (FTA) and/or the Management Oversight and Risk Tree (MORT), an event which occurs as a normal function in system operation that may or may not present a risk of hazard to that system. Represented graphically by a house shape in FTA and a scroll shape in MORT.

normal fault
Seismology. A fault characterized by predominantly vertical displacement in which the hanging wall is moved downward with respect to the footwall of the fault. Generally, this kind of fault is a sign of tectonic extension.

normal humidity
A range of 40 to 80% relative humidity.

normal line of sight
That line of sight which is assumed by an individual in a relaxed posture and is typically directed approximately 10° to 15° below the horizontal plane.

normal operator
An operator who is adapted to his/her position and attains normal performance when using prescribed methods and working at a normal pace.

normal pace
The manual productivity level required or achieved by a normal effort.

normal performance
That output expected from an average qualified operator working with prescribed methods at an average pace.

normal pool
The stage of an artificially impounded water body that prevails for the greater part of the year.

normal pool elevation
Height in feet above sea level at which a section of the river is to be maintained behind the dam.

normal pool stage
That level of the river maintained by the desired dam operations.

normal range
In biological testing, the range of values of a biological analyte that would be expected without exposure to the environmental contaminant in the workplace.

normal saline
A solution of 0.9 g of sodium chloride in 100 ml of water, which is isotonic with body fluids.

normal temperature and pressure
In the practice of industrial hygiene, normal conditions are considered to be 25°C (77°F) and 760 mm pressure.

normal time
That temporal period required for a qualified worker to perform some task or operation while working at a normal pace without personal, fatigue, or other allowances. Also referred to as *base time* or *leveled time*.

normal working area
The approximately planar region of a work surface which is bounded distally by the arc swept out by a worker's extended fingertips and proximally by the body while pivoting the shoulder laterally (lateral rotation) in the normal working position. *See also* ***normal working area, one-handed; normal working area, two-handed; normal working volume;*** *and* ***normal working posture***.

normal working area, one-handed
The normal working area for only the right or only the left arm.

normal working area, two-handed
The normal working area for that overlapping area between the two individual one-handed working areas.

normal working position
See ***normal working posture***.

normal working posture
The typical posture assumed by a worker for a given task, generally taken as a standing or sitting position, with the upper arm hanging in a relatively stationary position close to the body and the elbow flexed at about 90º.

normal working volume
The three-dimensional region bounded proximally by a worker's body and distally by the arc swept by the fingertips of one or both hands with a range of elbow flexion and/or body rotation about its vertical axis.

normalize
To carry out a transformation on a variable to obtain a linear function, a normal distribution, or a desired range, or to have the sum of the parts equal 1.0.

normalized exposure rate
See ***intake***.

normative
Pertaining to or the establishment of a norm or standard for evaluation.

normotonic
Having normal muscle tone.

normoxic
Having a normal oxygen level.

northeaster
A name given to a strong, steady wind from the northeast that is accompanied by rain and inclement weather. It often develops when a storm system moves northeastward along the coast of North America.

northern lights
The aurora borealis. *See* ***aurora***.

nose
(1) The fleshy protuberance in the center of the face which contains the nostrils and associated airway passages through which breathing occurs. Air breathed in through the nose is warmed and filtered, as opposed to that breathed in through the mouth. The nostrils, which form the external entrance of the nose, lead into two nasal cavities, which are separated from each other by a partition (the nasal septum) formed of cartilage and bone. Three bony ridges project from the outer wall of each nasal cavity and partially divide the cavity into three air passages. At the back of the nose these passages lead into the pharynx. The passages also are connected by openings with the paranasal sinuses. One of the functions of the nose is to drain fluids discharged from the sinuses. The nasal cavities also have a connection with the ears by the eustachian tubes, and with the region of the eyes by the nasolacrimal ducts. The interior of the nose is lined with mucous membrane. Most of this membrane is covered with minute hairlike projections called cilia. Moving in waves, these cilia sweep out from the nasal passages the nasal mucus, which may contain pollen, dust, and bacteria from the air. The mucous membrane also acts to warm and moisten the inhaled air. High in the interior of each nasal cavity is a small area of mucous membrane

that is not covered with cilia. In this pea-sized area are located the endings of the nerves of smell, commonly called the olfactory receptors. These receptors sort out odors. Unlike the taste buds of the tongue, which distinguish only between four different tastes (sweet, sour, salt, and bitter), the olfactory receptors can detect innumerable different odors. This ability to smell contributes greatly to what we usually think of as taste because much of what we consider flavor is really odor. (2) That portion of a tread of a stairway projecting beyond the face of the riser immediately below. (3) The front or beginning of an air frame.

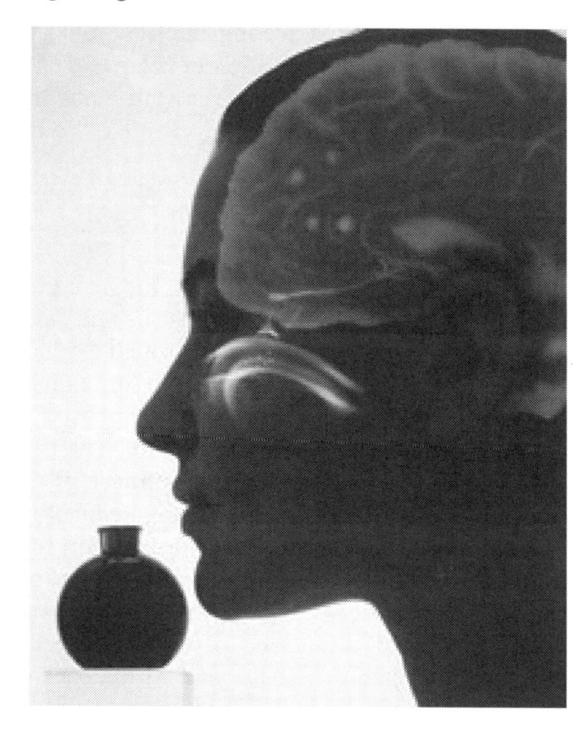

The nose not only allows us to breath cleaned and warmed air, it also provides us with the ability to sense an infinite number of differing odors.

nose breadth
The maximum horizontal linear distance across the nose, at whatever level it occurs. Measured with the facial muscles relaxed and without compressing tissue.

nose clip
Any spring device which pinches off the nostrils to prevent entry of water or air.

nose dive
Transit (slang). Trailer tipped forward on its nose.

The nose section of a Boeing 747 aircraft

nose height
The linear vertical distance from subnasale to sellion.

nose height – breadth index
The percentage value of the ratio between the nose height and the nasal breadth.

nose length
The linear distance from sellion to pronasale. Measured parallel to the ridge of the nose.

nose protrusion
The linear horizontal distance from subnasale to pronasale. Measured with the individual standing or sitting erect, with the facial muscles relaxed.

nosebleed
Bleeding from the nose for any number of reasons (e.g., injury, irritation, etc.). Also called *epistaxis*.

nose'er in
Maritime Navigation (slang). To land by putting the bow of the boat into the bank.

nosocomial disease
A disease with its source in a hospital and which is contracted as a result of being there.

nostril
See ***external naris***.

not otherwise classified (NOC)
A category of items including relatively infrequent dissimilar items.

not paved surface
All surfaces other than asphalt or concrete.

NOTAM
See ***notice to airmen***.

notary public
(1) A public officer whose function is to administer oaths; to attest and certify, by his/her hand and official seal, certain classes of documents, in order to give them credit and authenticity in foreign jurisdictions; to take acknowledgements of deeds and other conveyances, and certify the same; and to perform certain official acts, chiefly in commercial matters, such as the protesting of notes and bills, the noting of foreign drafts, and marine protests in cases of loss or damage. (2) One who is authorized by the state or federal government to administer oaths, and to attest to the authenticity of signatures.

notch
Maritime Navigation. A void or opening or any place where barge head logs do not meet, where they are not even with each other, or where no other barge is faced to a barge.

notes and accounts receivable
Aviation. Current notes and accounts receivable which are reasonably expected to be amortized within one year. These receivables include passenger receivables for air travel to be performed both by the selling carrier and other airlines, for which the related liabilities (to passenger or to the performing air carrier) are included in the "air traffic liabilities" account.

notice and comment rulemaking
See ***informal rulemaking***.

notice letter
The EPA's formal notice to potentially responsible parties (PRPs) that CERCLA-related action will be taken at a site for which the PRP is considered responsible.

notice of noncompliance (NON)
See ***violation notice***.

notice of proposed rulemaking
A public notice of proposed regulations required by law, which allows for public comments and scheduling of public hearings. *See also* ***advanced notice of proposed rulemaking***.

notice of violation (NOV)
A formal document completed by regulatory agencies as a result of established violations at or by a hazardous waste facility, transporter, or generator. This official notification is a legal document directing the violator to correct violations of existing environmental law(s) and may include or be followed by fines. *See also* ***violation notice***.

notice to airmen (NOTAM)
Aviation. A notice containing information (not known sufficiently in advance to publicize by other means) concerning the establishment, condition, or change in any component (facility, service, or procedure of, or hazard to the national airspace system), the timely knowledge of which is essential to personnel concerned with flight operations.

notice to mariners
A bulletin or information to mariners issued by the Coast Guard. *See also* ***public notice***.

notices to airmen publication (NTAP)
A publication issued every 14 days, designed primarily for the pilot, which contains current notice to airmen (NOTAM) information considered essential to the safety of flight as well as supplemental data to other aeronautical publications. The contraction NTAP is used in NOTAM text.

NOV
See ***notice of violation***.

nox
A unit for measuring levels of illumination. Equals 10^{-3} lux.

NO_x
Oxides of nitrogen.

noxious
Pertaining to that which is harmful or poisonous.

noy
A unit used in the calculation of perceived noise level relative to the perceived noise level of random noise of bandwidth 1000 ± 90 Hz at a sound pressure level of 40 dB, referenced to 2×10^{-4} microbar.

Np
See ***néper***.

np chart
A graph or display, tracking over time, the number of nonconforming units in samples when the number of items in each sample is constant.

NPA
National Particleboard Association.

NPAR
Nonbinding Preliminary Allocation of Responsibility.

NPDES
See ***National Pollution Discharge Elimination System***.

NPL
See ***National Priorities List***.

NPRM
Noticed of proposed rulemaking. *See* ***advanced notice of proposed rulemaking***.

NPT
National pipe thread.

NPTS
See ***Nationwide Personal Transportation Survey***.

NRC
See ***National Response Center***. *See also* ***Nuclear Regulatory Commission***. *See also* ***noise reduction coefficient.*** Also, acronym for National Research Council.

NRDC
Natural Resources Defense Council.

NRR
See ***noise reduction rating***.

NRT
See ***National Response Team***.

NRTL
See ***Nationally Recognized Testing Laboratory***.

NSC
See ***National Safety Council***.

NSF
National Science Foundation.

NSPS
See ***New Source Performance Standard***.

NTA
See ***nitrilotriacetic acid***.

NTAP
See ***notices to airmen publication***.

NTIS
National Technical Information Service.

NTP
See ***National Toxicology Program***. *See also* ***normal temperature and pressure***.

NTS
See ***National Transportation System***.

NTSB
See ***National Transportation Safety Board***.

nub
An intentional knot or tangle in a fabric which gives it an irregular texture.

nucha
See ***nape***.

nuchale
The lowest point in the midsagittal plane of the occiput that can be palpated among the muscles in the posterior-superior part of the neck.

nuchale tubercle
See ***cervicale***.

nuclear disintegration
A process resulting in the change of a radioactive nucleus through the emission of alpha or beta particles.

nuclear energy
The energy released as the result of a nuclear reaction. The processes of fission or fusion are employed to create a nuclear reaction.

nuclear fission
A type of nuclear transformation characterized by the splitting of a nucleus into at least two other nuclei and the release of a relatively large amount of energy. *See also* ***fission***.

nuclear fuel cycle
The operations defined to be associated with the production of electrical power for public use by any fuel cycle through utilization of nuclear energy.

nuclear fusion
The joining of a hydrogen nucleus with another hydrogen or heavier nucleus in a thermonuclear reaction to form heavier nuclei,

with the release of energy. More commonly referred to simply as *fusion*. *See also* ***fission***.

nuclear incident
According to the Federal Atomic Energy Act of 1954: Any occurrence, including an extraordinary nuclear occurrence, within the United States causing, within or outside the United States, bodily injury, sickness, disease, or death, or loss of or damage to property, or loss of use of property, arising out of or resulting from the radioactive, toxic, explosive, or other hazardous properties of source, special nuclear, or byproduct material.

nuclear magnetic resonance (NMR)
See ***magnetic resonance imaging***.

nuclear power plant
A facility that converts atomic energy into usable power; heat produced by a reactor makes steam to drive turbines which produce electricity.

nuclear radiation spectrum
The frequency distribution of nuclear or ionizing radiation with respect to energy.

nuclear reaction
A reaction which alters the energy, composition, or structure of an atomic nucleus.

nuclear reactor
A device in which a chain reaction is initiated and controlled, with the consequent production of heat. Typically utilized for power generation.

Nuclear Regulatory Commission (NRC)
A U.S. federal agency which regulates all commercial uses of nuclear energy, including the construction and operation of nuclear power plants, nuclear fuel reprocessing, research applications of radioactive materials, etc.

nuclear winter
Prediction by some scientists that smoke debris rising from massive fires resulting from nuclear war could enter the atmosphere and block out sunlight for weeks or months. The scientists making this prediction project a cooling of the earth's surface, and changes in climate which could, for example, negatively affect world agricultural and weather patterns.

nucleic acids
(1) A polymer of purine and pyrimidine bases, each chemically combined with a five-carbon sugar and phosphoric acid. (2) Nucleic acids are found in the cells of all living tissue. They are extremely complex and of high molecular weight, containing phosphoric acid, sugars, and purine and pyrimidine bases. Two pentose sugars are involved as constituents of nucleic acids: ribose and deoxyribose. Thus are derived from the names of the nucleic acids ribonucleic acid (RNA) and deoxyribonucleic acid (DNA). The nucleic acids and their derivatives are of great importance in metabolism, and though all of their functions are not completely understood, they appear to be concerned with controlling the general pattern of metabolism and acting as catalysts in many chemical reactions within the cell. The synthesis of proteins by the nucleic acids are the intermediate steps in the metabolism of other foodstuffs. The nucleic acids are also of great biologic significance. For example, DNA and RNA are the chemical repositories of genetic information and therefore affect the transmission of individual characteristics and functions from cell to cell and also from individual persons to their offspring. *See also* ***deoxyribonucleic acid (DNA)***.

nucleon
Common name for a constituent particle of the nucleus. It is applied to protons, neutrons, and any other particle found to exist in the nucleus.

nucleotide
One of a group of compounds obtained by hydrolysis of nucleic acids consisting of purine or pyrimidine bases linked to sugars, which in turn are esterified with phosphoric acid.

nucleus
(1) *Radiation*. The atomic nucleus; the small, centrally located, positively charged region of the atom that carries essentially all the mass. Except for the nucleus of ordinary (or light) hydrogen, which is a single proton, all atomic nuclei contain both protons and neutrons. (2) *Biology*. The structure within cells that contains chromosomes and one or more nucleoli. The nucleus contains large quantities of deoxyribonucleic acid (DNA), a nucleic acid that controls the synthesis of protein enzymes of the cytoplasm and also cellular reproduction. Because of its DNA content, the nucleus

is considered to be the control center of the cell. (3) *Anatomy.* A mass of gray matter in the central nervous system, especially such a mass marking the central termination of a cranial nerve.

nucleus pulposus
The viscous fluid in the center of an intervertebral disk.

nuclide
A general term referring to any nuclear species, either stable (of which there are about 270 in number) and unstable (of which there are about 500), of the chemical elements.

nude
(1) Having a minimal amount of clothing (e.g., underwear), in which many anthropometric measurements are taken. (2) Without clothing. Synonymous with *naked*.

nude body dimensions
Anthropometric measures which have been taken with a nude subject.

nuisance
(1) That activity which arises from the unreasonable, unwarranted, or unlawful use by a person of his/her own property, working obstruction or injury to the right of another, or to the public, and producing such material annoyance, inconvenience, and discomfort that the law will presume resulting damage. (2) Engaging in an unreasonable use of land so as to materially and substantially interfere with the use and enjoyment of the land of another.

nuisance dust
Airborne particulates which neither alter the architecture of the airspaces of the lungs nor produce scar tissue to a significant extent, and the tissue reaction they do produce is reversible. They are not recognized as the direct cause of a serious pathological condition. *See also* ***inert dust***.

null
Having a quantity of zero; a nonexistent entity.

null gravity
See ***microgravity***.

null hypothesis
The hypothesis about a population parameter to be tested. A statement proposing that there is no statistically significant difference with respect to one or more given variables between two or more groups.

numb
Having an impaired ability or no ability to feel tactile sensations.

number
A symbol, as a figure or word, expressive of a certain value or a specified quantity.

number facility
The ability to perform basic arithmetic processing correctly within a reasonable time limit (e.g., add, subtract, multiply, and divide—individually or in combination).

numbered vessel
An undocumented vessel numbered by a state with an approved numbering system or by the Coast Guard under Chapter 123 of title 46, U.S.C.

numbering system
Any plan for the assignment of numeric values to items, cases, or events as a means for classification.

numbness
A paresthesia of touch insensibility in a part.

numerical analysis
The use of mathematical approximation techniques to solve problems.

numerical control
A method for precisely controlling the motions of a mechanical device, usually some type of machine tool, via a mathematical description of the object being manufactured.

numerical display
Any electrical display involving numbers, as on a panel or instrumentation. Also known as *digital display*.

numerical weather prediction (NWP)
Forecasting the weather based upon the solutions of mathematical equations by high-speed computers.

numerous targets vicinity (location)
Aviation. A traffic advisory issued by Air Traffic Control (ATC) to advise pilots that targets on the radar scope are too numerous to issue individually.

nurse
(1) A person who makes a profession of car-

ing for the sick, disabled, or enfeebled. (2) To care for a sick or disabled person or one unable to provide for his/her own needs. (3) To nourish at the breast.

nurture
A substance which nourishes.

nutrient
Any substance assimilated by living things that promotes growth. The term is generally applied to nitrogen and phosphorus in wastewater, but is also applied to other essential elements found in foodstuffs and soils.

nutrition
(1) The requirements and processes of the living body involved with activity, growth, maintenance, and repair. (2) The nourishment of the body by food. It includes all the processes by which the body uses food for energy, maintenance, and growth. Nutrition is particularly concerned with those properties of food that build sound bodies and promote health. In this sense, good nutrition means a balanced diet containing adequate amounts of the essential nutritional elements that the body must have to function normally. The essential ingredients of a balanced diet are proteins, vitamins, minerals, fats, and carbohydrates. The body can manufacture sugars from fats, and fats from sugars and protein, depending on the need. But it cannot manufacture proteins from sugars and fats. The most important constituents of proteins are the amino acids. These complex organic compounds of nitrogen play a vital role in nutrition. The best sources of complex proteins (e.g., proteins containing all the essential amino acids) are meat, fish, eggs, and dairy products. The amount of protein that the average person actually needs, however, is much smaller than that of popular opinion. Vitamins are special substances that are present, in varying amounts, in all food. Their absence from the diet can cause such diseases as beriberi (lack of vitamin B, or thiamine), pellagra (lack of the B vitamin niacin), and scurvy (lack of vitamin C, or ascorbic acid). The principal minerals needed by the body are calcium and phosphorus (to build bones and teeth) and iron (to assure a sufficient supply of erythrocytes). All three are plentiful in eggs, dairy products, lean meat, and enriched flour. The trace of iodine needed to prevent goiter is easily provided by iodized table salt. The minute amounts of magnesium, manganese, and copper that are necessary are found in any balanced diet. For quick energy, the body should have sugars (carbohydrates) and starches, which the body converts into sugars. Fats and proteins can also provide energy and can be stored for future use, whereas sugars and starches cannot. Since the body can manufacture most of its own fat, fats are of secondary importance in a balanced diet.

nutriture
The status of the body in relation to nutrition.

NVG
*See **night vision goggles**.*

NVLAP
National Voluntary Laboratory Accreditation Program (part of the National Institute of Standards and Technology, NIST).

NVOC
*See **nonvessel owning carrier**.*

NVOCC
*See **nonvessel operating common carrier**.*

NWP
*See **numerical weather prediction**.*

NWPA
Nuclear Waste Policy Act of 1982 (Federal).

nyctalgia
Pain that occurs only in sleep.

nyctalopia
*See **nightblindness**.*

nylon
Any of a set of long-chained amide polymers used in fabrics.

nystagmogram
A recording or display of nystagmus.

nystagmograph
An instrument for recording the movements of the eyeball in nystagmus.

nystagmus
Involuntary movement of the eyeballs often experienced by workers who continuously subject their eyes to abnormal or unaccustomed movements. The condition is often accompanied by headaches, dizziness, and fatigue. The most prevalent form of occupational nystagmus occurs in miners.

nytophilia
A preference for darkness or for night.

O

OA
See ***outdoor air***.

OA sound pressure level
See ***overall noise***.

oakie blower
An air scoop on the air intake used to increase power.

OALT
See ***operational acceptable level of traffic***.

OAPCA
Organotin Antifouling Paint Control Act of 1988 (Federal).

OBA
See ***octave band analyzer***.

obesity
That condition resulting from a prolonged condition in which caloric energy intake exceeds output, the excess being converted to fat and deposited within the body. This excessive accumulation of fat in the body leads to an increase in weight beyond that considered desirable with regard to age, height, and bone structure. Being "overweight" can affect physical and mental health. Too many extra pounds are a strain on the body, and can eventually shorten the span of life. Obesity is also unattractive, and this may create psychological problems. The overweight person is susceptible to a number of normally unnecessary complications. These include an overworked heart, shortness of breath, a tendency to arteriosclerosis and high blood pressure or to diabetes mellitus, chronic back and joint pains (from increased strain on joints and ligaments), a greater tendency to contract infectious diseases, and a reduced ability to exercise or enjoy sports.

object
(1) *General.* Any physical entity which can be viewed or manipulated. (2) *Aviation.* Includes, but is not limited to aboveground structures, people, equipment, vehicles, natural growth, terrain, and parked aircraft. (3) *Computing.* Any structure which can be displayed or manipulated by a computer system.

object class
Transit. As the term is used in expense classification, an object is an article or service obtained. An object class is a grouping of expenses on the basis of goods or services purchased. The object classes include salaries and wages, fringe benefits, services, materials and supplies, and other expenses as defined in Section 7.2, Volume II of the Uniform System of Accounts (USOA).

object free area (OFA)
Aviation. A two-dimensional ground area surrounding runways, taxiways, and taxi lanes which is clear of objects except for Navigation Aids (NAVAIDs) and objects whose location is fixed by function.

objection
Act of objecting. That which is, or may be presented in opposition. An adverse reason or argument. A reason for objecting or opposing. A feeling of disapproval.

objective
Pertaining to a measure or aspect which can be observed or evaluated by more than one person independently.

objective basic element
Any of a set of work elements which involve an observable element.

objective rating
A type of performance rating which has an objective, as opposed to subjective, basis.

objective symptom
Those which a surgeon or physician discovers from an examination of a patient (subjective symptoms being those which the surgeon learns from what a patient tells him/her).

objective tree
A qualitative form of relevance tree which may be used simply to place variables in perspective.

objects not fixed
Transit. Objects that are movable or moving but are not motor vehicles. Includes pedestrians, pedal cyclists, animals, or trains (e.g., spilled cargo in roadway).

obligate anaerobes
Microorganisms that are strictly intolerant of oxygen in their environment.

obligate parasites
Organisms which can only survive in living cells.

obligation
A very broad term but generally meaning that which a person is bound to do or forbear; any duty imposed by law, promise, contract, relations of society, courtesy, kindness, etc.

obligations under capital leases (current and noncurrent)
Liability applicable to property obtained under capital leases.

oblique fault
Seismology. Describing motion that is a combination of movement both perpendicular and parallel to the strike of a fault. A combination of strike-slip and dip-slip (whether normal or reverse). Also referred to as an *oblique slip*.

observation board
A clipboard or similar tool used to support the timing device and hold any forms in gathering time and motion data.

observation form
Any generic or specially designed form for recording the different work elements in a particular time study.

observe
View to acquire data for documentation or study.

observed rating
That rating applied to a worker's pace by the time and motion study individual relative to that individual's judgement of what the standard pace should be.

observer
An individual who makes the observations in a study.

2° observer
See ***CIE Standard Observer***.

10° observer
See ***CIE Supplementary Standard Observer***.

observer error
Any error due to intra- or interobserver unreliability or differences in judgement.

obsessive-compulsive
Marked by a compulsion to repeatedly perform certain acts or carry out certain rituals. Obsessive-compulsive reaction is a type of neurosis in which there is the intrusion of insistent, repetitious, and unwanted ideas or impulses to perform certain acts.

obstacle
Aviation. An existing object, object of natural growth, or terrain at a fixed geographical location or which may be expected at a fixed location within a prescribed area with reference to which vertical clearance is or must be provided during flight operation.

obstacle free zone (OFZ)
Aviation. The obstacle free zone is a three-dimensional volume of airspace which protects the transition of aircraft to and from the runway. The OFZ clearing standard precludes taxiing and parked airplanes and object penetrations, except for frangible navigation aid (NAVAID) locations that are fixed by function. Additionally, vehicles, equipment, and personnel may be authorized by air traffic control to enter the area using the provisions of Order 7110.65, Air Traffic Control, paragraph 3-5. The runway OFZ and when applicable, the inner-approach OFZ, and the inner-transitional OFZ, comprise the OFZ. *See also* ***inner-approach obstacle free zone, inner-transitional obstacle free zone,*** *and* ***runway obstacle free zone***.

obstructing justice
The act of impeding or obstructing those who seek justice in a court, or those who have duties or powers of administering justice therein. The term also applies to obstructing the administration of justice in any way (as by hindering witnesses from appearing, assaulting an officer of the court, influencing jurors, obstructing court orders or criminal investigations).

obstruction
Aviation. Any object/obstacle exceeding the obstruction standards specified by Federal Aviation Regulations (FAR) Part 77, Subpart G.

obstruction accident
Rail Operations. An accident/incident consisting of striking: a) a bumping post or a foreign object on the track right-of-way; b) a highway vehicle at a location other than a highway-rail crossing site; derailed equipment; or c) a track motorcar or similar work equipment not equipped with Association of

American Railroad couplers, and not operating under train rules.

obstruction light
Aviation. A light or one of a group of lights, usually red or white, frequently mounted on a surface structure or natural terrain to warn pilots of the presence of an obstruction.

obstruction to air navigation
Aviation. An object of greater height than any of the heights or surfaces presented in Subpart C of Federal Aviation Regulations (FAR) Part 77.

obvious danger
Danger or dangerous conditions that are apparent in the exercise of ordinary observation and disclosed by the use of the eyes and other senses. Danger that is plain and apparent to a reasonably observant person.

obvious risk
Risk so plain that it would be instantly recognized by a person of ordinary intelligence (it does not mean unnecessary risk).

Occam's razor
A rule that, given two theories which explain a phenomenon, the simpler is preferred.

occasional
In terms of probability of hazard or mishap occurrence, a hazard or event likely to occur sometime during the life of an item.

occasional element
A job element which occurs at irregular intervals, less than once in a given work cycle or operation.

occipital bone
A curved, flat bone forming a portion of the posterior and inferior skull.

occipital condyle
One of a pair of bilaterally distributed condyles at the base of the skull which articulate with the atlas bone.

occipital lobe
A pyramid-shaped structure at the posterior portion of the cerebrum whose primary function is visual processing.

occipital pole
The posterior tip of the occipital lobe of the brain.

occiput
The posterior portion of the head.

occluded
Closed, shut, or blocked.

occluded front
A complex system that ideally forms when a cold front overtakes a warm front. When the air behind the front is colder than the air ahead of it, the front is called a *cold occlusion*. When the air behind the front is milder than the air ahead of it, it is called a *warm occlusion*.

occluded gases
Those gases forced into a closed space or tunnel with blowers.

occlusion
(1) The act of closure or state of being closed. (2) The contact of the teeth of both jaws when closed or during the movements of the mandible in mastication.

occupancy
(1) *Life Safety.* The number of people permitted to occupy a building or region within a building. (2) *Transportation.* The number of persons, including driver and passenger(s) in a vehicle. Nationwide Personal Transportation Survey (NPTS) occupancy rates are generally calculated as person miles divided by vehicle miles.

occupant
Any person who is in or upon a motor vehicle in transport. Includes the driver, passengers, and persons riding on the exterior of a motor vehicle (e.g., a skateboard rider who is set in motion by holding onto a vehicle).

occupation
That trade, profession, or other activity which occupies one's time for compensation.

occupational acne
An occupational skin disorder involving acne resulting from regular exposure to acne-causing material(s) such as tar, wax, and chlorinated hydrocarbons, and which disappears on removal from those material(s). *See also* ***occupational dermatosis***.

occupational biomechanics
The study of the volitional acts of the individual in loading the musculoskeletal system in the working environment.

occupational contact dermatitis
See ***industrial dermatitis***.

occupational dermatosis
Any of a class of occupational skin disorders involving one or more regions of the skin, such as contact dermatitis, eczema, or rash. See also occupational acne.

occupational disease
(1) *General.* A disease which is a result of exposure to a hazardous material, physical agent, biological organism, or ergonomic stress in the course of one's work. (2) *Law.* A disease resulting from an exposure during employment to conditions or substances detrimental to health. Impairment of health not cause by accident but by exposure to conditions incidental to and arising out of or in the course of one's employment.

occupational dose
As pertains to ionizing radiation, the dose received by an individual in a restricted area or in the course of employment in which the individual's assigned duties involve exposure to radiation or to radioactive materials from licensed and unlicensed sources of radiation, whether in the possession of the licensee or other person.

occupational ecology
The study of the worker, the working environment, and the interaction between the two.

occupational ergonomics
The study and/or practice of human factors in the workplace.

occupational exposure
Exposure to a health hazard such as a chemical, physical, or biologic agent, or an ergonomic factor while carrying out work within the workplace.

occupational exposure limit (OEL)
A term indicating the concentration of an airborne contaminant or physical stress that is acceptable for exposure to it for a specified period of time.

occupational hazard
A risk of accident or disease which is peculiar to a particular calling or occupation.

occupational health
A subset of occupational medicine dealing with promoting the maintenance of worker mental and physical well-being, including means of disease prevention.

occupational illness
(1) *OSHA.* Any abnormal condition or disorder, other than one resulting from an occupational injury, caused by exposure to environmental factors associated with employment. This includes any acute or chronic illnesses or diseases that may be caused by inhalation, absorption, ingestion, or direct contact. (2) *Federal Railroad Administration.* Any abnormal condition or disorder of a railroad employee, other than one resulting from injury, caused by environmental factors associated with his or her railroad employment, including, but not limited to, acute or chronic illnesses or diseases which may be caused by inhalation, absorption, ingestion, or direct contact. Also, any abnormal condition or disorder caused by environmental factors associated with a worker's employment, but not the result of an injury.

occupational injury
Any injury that results from a work accident or from a single instantaneous exposure in the work environment.

occupational medicine
A branch of medicine dedicated to the appraisal, maintenance, restoration, and improvement of the health of workers through the scientific application of preventive medicine, emergency medical care, rehabilitation, epidemiology, and environmental medicine.

occupational neurosis
Any neuropsychological disorder, not caused directly by an individual's occupation, but which is characterized by symptoms such as pain or fatigue involving those parts of the body normally in his/her occupation.

occupational noise
That noise found in the workplace.

occupational nystagmus
An ocular nystagmus resulting from prolonged exposure to poor lighting conditions or retinal fatigue.

occupational paralysis
A muscular weakness or atrophy due to nerve compression resulting from the working environment.

occupational physiology
See ***work physiology***.

occupational psychiatry
A specialty within psychiatry concerned in business and industry with a) the promotion of mental health; b) diagnosis and treatment of mental illness; and c) dealing with the psychological aspects of personnel problems such as hiring, absenteeism, vocational adjustments, and retirement.

occupational safety
The study and/or implementation of principals intended to recognize hazards and prevent accidents in work-related situations.

Occupational Safety and Health Act
Federal law (1970) administered by the Occupational Safety and Health Administration (OSHA) enacted to reduce the incidence of injuries, illnesses, and deaths among working men and women in the United States which result from their employment.

Occupational Safety and Health Administration (OSHA)
A federal agency within the U.S. Department of Labor responsible for establishing and enforcing standards for the exposure of workers to safety hazards or harmful materials that they may encounter in the work environment, as well as other matters that may affect the safety and health of workers. The agency was established under the Occupational Safety and Health Act of 1970 along with the National Institute for Occupational Safety and Health (NIOSH) and the Occupational Safety and Health Review Commission (OSHRC).

Occupational Safety and Health Review Commission (OSHRC)
A commission that is independent of OSHA and has been established to review and rule on contested OSHA cases. It was created under the Occupational Safety and Health Act of 1970 along with the National Institute for Occupational Safety and Health (NIOSH) and the Occupational Safety and Health Administration (OSHA).

occupational skin disease
Any occupational disease involving the skin. Also referred to as *occupational skin disorder*. *See also* ***occupational dermatosis*** *and* ***occupational acne***.

occupational strain
The reaction of one or more parts of the body to occupational stressors.

occupational stress
An internal condition resulting from any forces exerted on the individual as a result of performing some task in the work environment. Also called *work stress*.

occupational stressor
Any stressor present in the workplace. Also called *work stressor*.

occupational therapist
One who is licensed or otherwise qualified to practice occupational therapy.

occupational therapy
The training or use of certain occupational skills for therapeutic or rehabilitation purposes.

occupied caboose
A rail car being used to transport non-passenger personnel.

occupied zone
In the study of indoor air quality, those locations/positions where the people work or occupy space within a building.

occurrence
An incident.

ocean
Any portion of the high seas beyond the contiguous zone.

ocean bill of lading
A receipt for the cargo and a contract for transportation between a shipper and the ocean carrier. It may also be used as an instrument of ownership which can be bought, sold, or traded while the goods are in transit.

ocean freight differential (OFD)
The amount by which the cost of the ocean freight bill for the portion of commodities required to be carried on U.S. flag vessels exceeds the cost of carrying the same amount on foreign flag vessels. When applied to agricultural commodities shipped under Food for Peace, OFD is the amount paid by the Commodity Credit Corporation.

ocean going container
Usually made of steel, it is a large rectangular box designed for easy lift on/off by cranes.

ocean waters

Those waters of the open seas lying seaward of the base line from which the territorial sea is measured, as provided for in the Convention on the Territorial Sea and the Contiguous Zone.

oceanic airspace

Airspace over the oceans of the world, considered international airspace, where oceanic separation and procedures per the International Civil Aviation Organization are applied. Responsibility for the provisions of air traffic control service in this airspace is delegated to various countries, based generally upon geographic proximity and the availability of the required resources.

oceanic display and planning system

Aviation. An automated digital display system which provides flight data processing, conflict probe, and situation display for oceanic air traffic control.

oceanic front

A boundary that separates masses of water with different temperatures and densities.

oceanic navigational error report (ONER)

Aviation. A report filed when an aircraft exiting oceanic airspace has been observed by radar to be off course. ONER-reporting parameters and procedures are contained in Order 7110.82, Monitoring of Navigational Performance in Oceanic Areas.

oceanic published route

Aviation. A route established in international airspace and charted or described in flight information publications, such as Route Charts, Department of Defense (DOD) En Route Charts, Chart Supplements, NOTAMs, and Track Messages.

oceanic transition route

Aviation. Route established for the purpose of transitioning aircraft to/from an organized track system.

OCL

See ***outer compass locator***.

OCMI

See ***Officer in Charge of Marine Inspection***.

OCR

See ***optical character recognition***.

octane number

A numerical rating used to grade the relative antiknock properties of gasoline. A high octane fuel (e.g., octane rating of 89 or more) has better antiknock properties than one with a lower number.

octanol-water partition coefficient (K_{ow})

Measure of the extent of partitioning of a substance between water and octanol at equilibrium. The K_{ow} is determined by the ratio between concentration in octanol divided by the concentration in water at equilibrium (unitless).

octave

The interval between two sounds having a frequency ratio of two to one.

octave band

As applied to noise, a bandwidth that has an upper band frequency that is twice its lower band frequency. The term is used to describe the separation of noise energy into frequency bands which cover a 2 to 1 range of frequencies. The center frequencies of these bands are 31.5, 63, 125, 250, 500, 1000, 2000, 4000, 8000, and 16,000 Hz. This separation is used to analyze noise. One-third-octave band and one-tenth-octave band analyses are also used to obtain a more detailed analysis of noise.

octave band analyzer (OBA)

A portable instrument used for characterizing the frequency and amplitude characteristics of a sound.

ocular

Pertaining to the eye.

ocular dominance

A condition in which one eye is subconsciously relied upon more than the other. Also referred to as *eye dominance*.

oculogram

A surface electrical recording of activity adjacent to the eye which indicates eye movement patterns.

oculogravic illusion

An illusion indicating a tilting of the visual field produced when a change in vertical gravity occurs, as in a centrifuge or other linear acceleration. Also referred to as *agravic illusion*.

oculogyral illusion
A visual illusion involving a sense of rotation in the opposite direction produced when an abrupt change in rotational velocity occurs.

oculomotor
Pertaining to eye movements.

oculomotor nerve
The third cranial nerve, which provides motor input to the intrinsic and some extrinsic eye muscles.

o.d.
Outside diameter.

OD
See ***optical density***.

ODC
(1) Other direct costs. (2) Ozone depleting chemical. *See* ***ozone level depleting substances (OLDS)***.

odontoma
A tumor derived from tissues involved in tooth formation.

odor
The characteristic of a substance that makes it perceptible to the sense of smell.

odor threshold
The minimum concentration of a substance that can be detected and identified by a majority of the exposed population.

odorant
Any relatively volatile substance which is added to an odorless or offensive material to give the latter a distinctive odor for safety, attractant, or other purposes. *See also* ***malodorant***.

odoriferous
Having an odor.

odorimetry
The study or measurement of the effects of odors on the olfactory sensory structures.

odorize
Add an odorant to another substance.

OEL
See ***occupational exposure limit***.

OFA
See ***object free area***.

OFD
See ***ocean freight differential***.

Office of Pollution Prevention and Toxics (OPPT)
A section of the U.S. Environmental Protection Agency.

off course
Aviation. A term used to describe a situation where an aircraft has reported a position fix or is observed on radar at a point not on the air traffic control (ATC) approved route of flight. *See also* ***on course*** *and* ***on-course indication***.

off-line
Pertaining to a terminal or other hardware not ready for access to a computer or network.

off peak period
See ***base period***.

off-road vehicular area
An area for the testing of, or use by, vehicles that are designed to travel across the terrain.

off route vector
Aviation. A vector by Air Traffic Control (ATC) which takes an aircraft off a previously assigned route. Altitudes assigned by ATC during such vectors provide required obstacle clearance.

off-site facility
A hazardous waste treatment, storage, or disposal (TSD) area that is located at a place away from the generating site.

off time
That period within a given day when an individual is not scheduled to be at work.

offense
A felony or misdemeanor. A breach of the criminal laws; a violation of the law for which a penalty is prescribed.

offgassing
The release of adsorbed or occluded substances from a solid or liquid material, often by exposure to heat. Synonymous with *outgassing*.

office
Any location in which management, supervision, and administrative support personnel are housed and their respective functions are performed.

office automation
The use of implementation of computers or electromechanical devices for communica-

tions or manipulating, storing, or sending documents.

office layout
The arrangement of desks, filing cabinets, photocopiers, other associated equipment, and the personnel who occupy an office.

Office of the Secretary of Transportation (OST)
The Department of Transportation is administered by the Secretary of Transportation, who is the principal adviser to the President in all matters relating to federal transportation programs. The Secretary is assisted in the administration of the Department by a Deputy Secretary of Transportation, a Associate Deputy Secretary, the Assistant Secretaries, a General Counsel, the Inspector General, and several Directors and Chairpersons.

Office on Environmental Policy (OEP)
Created by President Clinton in 1993 to replace the Council on Environmental Quality (CEQ). The OEP is responsible for coordinating environmental policy within the federal government. The level of OEP participation in the major governmental policy councils (the National Security Council, the National Economic Council, and the Domestic Policy Council) clearly establishes the OEP on a much higher, visible plane than the old CEQ. *See* ***Council on Environmental Quality***.

Officer in Charge, Marine Inspection (OCMI)
A person from the civilian or military branch of the Coast Guard designated as such by the Commandant and who under the supervision and direction of the Coast Guard District Commander is in charge of a designated inspection zone for the performance of duties with respect to the enforcement and administration of Title 52, Revised Statutes, acts amendatory thereof or supplemental thereto, rules and regulations thereunder, and the inspection required thereby.

official immunity doctrine
Doctrine of *official immunity* provides that government officials enjoy an absolute privilege from civil liability should the activity in question fall within the scope of their authority and if the action undertaken requires the exercise of discretion, and this rule of immunity is not limited to the highest executive officers of the government.

offset parallel runway
Runways that are staggered but have centerlines which are parallel to one another.

offshore
That geographic area that lies seaward of the coastline. In general, the coastline is the line of ordinary low water along with that portion of the coast that is in direct contact with the open sea or the line marking the seaward limit of inland water.

offshore breeze
A breeze that blows from the land out over the water. Opposite of *onshore breeze*.

offshore control area
Aviation. That portion of airspace between the U.S. 12-mile limit and the Oceanic Control Area/Flight Information Region (CTA/FIR) boundary within which air traffic control is exercised. These areas are established to permit the application of domestic procedures in the provision of air traffic control services. Offshore control area is generally synonymous with Federal Aviation Regulations, Part 71, Subpart E, "Control Areas and Control Area Extensions."

offshore facility
According to the Comprehensive Environmental Response, Compensation, and Liabilities Act: Any facility of any kind located in, on, or under, any of the navigable waters of the United States, and any facility of any kind which is subject to the jurisdiction of the United States and is located in, on, or under any other waters, other than a vessel or a public vessel.

offshore supply vessel
A cargo vessel of less than 500 gross tons that regularly transports goods, supplies, or equipment in support of exploration or production of offshore mineral or energy resources.

OFZ
See ***obstacle free zone***.

OGC
Office of General Counsel.

ogive
A cumulative distribution curve, generally resembling an "S" shape, depending on the distribution.

OHA
Operational hazard analysis. *See* ***operating and support hazard analysis***.

OH&S
Occupational Health and Safety.

ohm (Ω)
A unit of electrical resistance equal to the electrical resistance between two points of a conductor when a constant potential of 1 volt is applied between the two points and produces a current of one ampere.

Ohm's law
A law that is applied to the flow of electricity through a conductor. It states that the current flow in amperes is proportional to the voltage divided by the resistance in ohms.

oil
(1) Oil of any kind or in any form, including, but not limited to, petroleum, fuel oil, sludge, oil refuse, and oil mixed with wastes other than dredged spoil. (2) A mixture of hydrocarbons usually existing in the liquid state in natural underground pools or reservoirs. Gas is often found in association with oil.

oil acne
Acneform dermatitis resulting from the skin's allergic reaction to oil or oil products. Appears as red bumps, usually with pustules which progress to sores. *See also* ***industrial dermatitis***.

oil and gas production
The lifting of oil and gas to the surface and gathering, treating, field processing (as in the case of processing gas to extract liquid hydrocarbons), and field storage. The production function shall normally be regarded as terminating at the outlet valve on the lease or field production storage tank. If unusual physical or operational circumstances exist, it may be more appropriate to regard the production function as terminating at the first point at which oil, gas, or gas liquids are delivered to a main pipeline, a common carrier, a refinery, or a marine terminal.

oil field body
Heavily constructed platform-type truck body equipped with instruments for oil drilling.

oil (filled) cutout
As pertains to systems over 600 volts (nominal), a cutout in which all or part of the fuse support and its fuse link or disconnecting blade are mounted in oil with complete immersion of the contacts and the fusible portion of the conducting element (fuse link). Any arc interruption, caused by severing of the fuse link or by opening of the contacts, will occur under oil.

oil fingerprinting
A method that identifies sources of oil and allows spills to be traced back to their source.

oil folliculitis
Acne-like lesions resulting from repeated skin contact with some oil products, such as insoluble cutting oils.

oil mist
Aerosol produced when oil is forced through a small orifice, splashed or spun into the air during operations, or vaporized and then condensed in the atmosphere.

oil pipeline mode
Covers crude oil, petroleum product and gas trunk lines. The pipeline industry, which transports oil and petroleum products, is an important if specialized freight mode. *See also* ***pipeline***.

oil spill
An accidental or intentional discharge of oil which reaches bodies of water. Can be controlled by chemical dispersion, combustion, mechanical containment, and/or adsorption.

oil spill removal organization
An entity that provides response resources.

oilless compressor
An air compressor that is not lubricated with oil. Also referred to as a breathing air compressor. Thus, it does not generate carbon monoxide or oil mist when in operation.

ointment
A semisolid preparation for external application to the body. Official ointments consist of medicinal substances incorporated in suitable vehicles.

OJT
See ***on-the-job training***.

OKN
See ***optokinetic nystagmus***.

OLDS
See ***ozone level depleting substances***.

olecranon
The proximal end of the ulna which forms the elbow prominence. More accurately called the *olecranon process*.

olecranon fossa
A depression in the posterior distal end of the humerus, into which the olecranon process of the ulna fits when the elbow is extended.

olecranon height
The vertical distance from the floor or other reference surface to the underside of the elbow. Measured with the individual standing erect, the elbow flexed 90°, and the upper arm vertical.

OLF
A perceived air quality term which attempts to quantify the level of odorous pollutants in OLFs.

olfaction
The sense of smell.

olfactometer
Any device for measuring the sensitivity of smell.

olfactory
Pertaining to the sense of smell.

olfactory fatigue
Condition in which the sense of smell has been diminished to the extent that an odor cannot be detected.

olfactory nerve
The first cranial nerve, which conveys sensory information regarding smell to the brain.

oligotrophic lakes
Deep clear lakes with low nutrient supplies. They contain little organic matter and have a high dissolved-oxygen level.

OM
*See **outer marker**.*

OMB
Office of Management and Budget.

ombudsman
An individual with whom an employee or agent can discuss confidential, work-related concerns and receive a quick, neutral response. The employee or agent may also report violations of law or corporate policies to this individual without fear of retribution. The ombudsman is usually trusted and respected by employees, agents, and management of an organization and knows the organization's corporate culture. The ombudsman may be someone within the organization or knowledgeable individuals outside of the company.

omega
Aviation. An Area Navigation (RNAV) system designed for long-range navigation based upon ground-based electronic navigational aid signals.

omega high
A ridge in the middle or upper troposphere that has the shape of the Greek letter omega.

omission
Neglecting to perform what the law requires. The intentional or unintentional failure to act which may or may not impose criminal liability depending upon the existence of a duty to act under the circumstances.

omphalion height
The linear vertical distance from the floor or other reference surface to omphalion. Measured with the individual standing erect, with his/her body weight equally distributed on both feet.

on course
Aviation. (1) Used to indicate that an aircraft is established on the route centerline. (2) Used by Air Traffic Control (ATC) to advise a pilot making a radar approach that his aircraft is lined up on the final approach course. *See also **off-course** and **on-course indication**.*

on-course indication
An indication on an instrument, which provides the pilot a visual means of determining that the aircraft is located on the centerline of a given navigational track, or an indication on a radar scope that an aircraft is on a given track. *See also **off course** and **on course**.*

on-demand
Supplied as a result of a user-initiated response.

on-flight passenger trip length
Aviation. The average length of a passenger trip, calculated by dividing the number of revenue passenger-miles in scheduled service by the number of revenue passenger enplanements in scheduled service.

online
Pertaining to a fully connected, powered, and ready for operation terminal or other hardware access to a computer or network.

on-off control
Any simple control mechanism which has only two possible discrete outcomes, either full on or full off, with no intermediate state possible.

on-off switch
A type of on-off control which consists of a manual, remote, or automatic switch.

on-road mile per gallon (mpg)
A composite miles per gallon (mpg) that was adjusted to account for the difference between the test value and the fuel efficiency actually obtained on the road.

on-scene coordinator (OSC)
The predesignated EPA, Coast Guard, or Department of Defense Official who coordinates and directs Superfund removal actions or Clean Water Act oil (or hazardous) spill corrective actions.

on-site facility
A hazardous waste treatment, storage, or disposal (TSD) area that is located on the generating site.

on-system
Any point on or directly interconnected with a transportation, storage, or distribution system operated by a natural gas company.

on-the-job training (OJT)
The training of an employee by doing the tasks or job he/she will be expected to perform when training is completed, rather than by classroom or other training techniques.

on-time performance
The proportion of the time that a transit system adheres to its published schedule times within state tolerances.

on-track equipment
Railroad rolling stock used to transport freight or passengers; includes locomotives, railroad cars, maintenance equipment, and one or more locomotives coupled to one or more cars.

oncogene
A viral gene, found in some retroviruses, that may transform the host cell from normal to neoplastic. More than 30 oncogenes have been identified in humans.

oncogenesis
The production or causation of tumors.

oncogenic
A substance that causes tumors, whether benign or malignant.

oncogenicity
The quality or property of being able to cause tumor formation.

oncology
The study of tumors, including the study of causes, development, characteristics, and the treatment of the tumor.

one-hole test
A psychomotor skill test in which an individual is required to grasp, move, and position a small cylindrical object in a hole with close tolerances.

one-inch rule
*See **residue rule**.*

one-point discrimination
The ability to localize a point on the body surface where pressure is being applied.

one-tailed test
A test of statistical significance in which a directional hypothesis is used, stating that a sample value will be exclusively either less than or greater than some value.

one-tenth-octave band
A band-width equal to one-tenth of an octave. *See also **octave band**.*

one-third-octave band
A band-width equal to one-third of an octave. *See also **octave band**.*

ONER
*See **oceanic navigational error report**.*

onshore breeze
A breeze that blows from the water onto the land. Opposite of an *offshore breeze*.

onshore facility
According to the Comprehensive Environmental Response, Compensation, and Liabilities Act: Any facility (including, but not limited to, motor vehicles and rolling stock) of any kind located in, on, or under any land or nonnavigable waters within the United States.

onshore oil pipeline facilities
New and existing pipe, rights-of-way and any equipment, facility, or building used in the transportation of oil located in, on, or under any land within the United States other than submerged land.

ontogeny
The study of the origin and development of an individual organism, from the zygote to adult.

oocytes
Developing egg cell.

opacity
The amount of light obscured by particulate pollution in the air; clear window glass has a zero opacity, a brick wall has 100 percent opacity. Opacity is used as an indicator of changes in performance of particulate matter pollution control systems.

opalescence
The clouded, iridescent appearance of a translucent substance or material when illuminated by more than one frequency of visible light.

open-access transportation
The contract carriage delivery of nonsystem supply gas on a nondiscriminatory basis for a fee generally subject to transportation tariffs which are usually on an interruptible service basis on a first-come, first-serve capacity usage.

open-body type vehicle
A vehicle having no occupant compartment top or an occupant compartment top that can be installed or removed by the user at his convenience.

open burning
(1) Uncontrolled fires in an open dump. (2) In solid waste, the combustion of waste without the control of combustion air to maintain adequate temperature for efficient combustion, containment of the combustion reaction in an enclosed device to provide sufficient residence time and mixing for complete combustion, or control of the emission of the combustion products.

open channel
That portion of the river above pool water.

open-circuit SCBA
A type of self-contained respiratory protection device which exhausts exhaled air to the atmosphere rather than recirculating it.

open court
Common law requires a trial in open court, meaning a court to which the public will have a right to be admitted.

open cut
Rail transit way below surface in an excavated cut that has not had a covering constructed over it. Transition segments to open cut or subway-tunnel/tube segments are included.

open dump
An uncovered site used for disposal of waste without environmental controls. *See also* ***dump***.

open-face filter cassette
A cassette holding a filter that collects airborne particulate matter (usually fibers) on removal of the entire lid and not just the small inlet plug of the cassette.

open fracture
See ***compound fracture***.

open insurance policy
A marine insurance policy that applies on all shipments over a period of time rather than on a single shipment.

open loop system
Any system in which its own output provides insignificant or no input back to the system, with all or the remainder of the input coming from another source.

open motorboat
Craft of open construction specifically built for operating with a motor, including boats canopied or fitted with temporary partial shelters.

open path detectors
Line of sight contaminant detection systems that can cover a wide area. Detection is dependent on the contaminant crossing or breaking the detector line of sight measurement beam, such as an IR or UV source. Results are typically expressed in ppm-meters.

open riser
The airspace between the treads of stairways without upright members (risers).

open river
(1) Any river having no obstructions such as dams. (2) When the stage of a pooled river running through movable dams is high

enough for traffic to clear the dams, the river is said to be "open."

open shop
A facility in which employment is available to both labor union members and nonunion workers.

open stope
Pertaining to an underground workplace which is either unsupported or supported only by occasional timbers or rock pillars.

open system
A system in which the handling or transfer of a material occurs in a manner such that there is contact of the material with the atmosphere.

open timbering
A technique for supporting the soil or rock in a shaft or tunnel in which vertical supports are located some distance apart, with overhead horizontal struts between them.

open to public travel road
A road must be available, except during scheduled periods, extreme weather or other emergency conditions, and open to the general public for use by four-wheel, standard passenger cars without restrictive gates, prohibitive signs, or regulation other than restrictions based on size, weight, or class of registration. Toll plazas of public toll facilities are not considered restrictive gates.

open top
A trailer with sides but without a permanent top; often used for heavy equipment that must be lowered into place by crane.

open union
A labor union without restrictive membership provisions.

open window
A display window which is perceptually and functionally available to the user.

open wiring on insulators
An exposed wiring method using cleats, knobs, tubes, and flexible tubing for the protection and support of single insulated conductors run in or on buildings, and not concealed by the building structure.

operable unit
Term for each of the separate activities undertaken as part of a Superfund site cleanup. A typical operable unit would be removing drums and tanks from the surface of a site.

operant conditioning
A form of learning/training in which an organism provides a certain response to obtain a reward, which reinforces the occurrence of that response in the future.

operate
With respect to aircraft, means use, cause to use, or authorize to use aircraft, for the purpose (except as provided in 14 CFR 91.13) of air navigation including the piloting of aircraft, with or without the right of legal control (as owner, lessee, or otherwise).

Operating and Support Hazard Analysis (O&SHA)
A system safety analytical technique (also know as the *operational hazard analysis* or *OHA*) which focuses primarily on the hazards associated with or caused or enhanced by the human/task interface of system operations.

operating assistance
Financial assistance for transit operations (not capital expenditures). Such aid may originate with federal, local, or state governments.

operating cost
Transit. (1) Fixed operating cost, in reference to passenger car operating cost, refers to those expenditures that are independent of the amount of use of the car, such as insurance costs, fees for license and registration, depreciation and finance charges. (2) Variable operating cost, in reference to passenger car operating cost, expenditures which are dependent on the amount of use of the car, such as the cost of gas and oil, tires, and other maintenance.

operating employee
Transit. An employee involved with operation, maintenance, or administration of the transit system, excluding those involved in construction and capital procurement.

operating expenses
(1) Expenses of furnishing transportation service including the expense of maintenance and depreciation of the plant used in the service. (2) The costs of handling traffic, including both direct costs, (such as driver wages and fuel) and indirect costs (e.g., computer expenses and advertising) but excluding

interest expense. (3) Expenses incurred in the performance of air transportation, based on overall operating revenues and overall operating expenses. Does not include nonoperating income and expenses, nonrecurring items or income taxes.

operating life
The period of time in which prime power is applied to electrical or electronic components without maintenance or rework.

operating practice
Railroad employment performance and adherence to the established operating rules of a railroad company.

operating profit or loss
Aviation. Profit or loss from performance of air transportation, based on overall operating revenues and overall operating expenses. Does not include nonoperating income and expenses, nonrecurring items, or income taxes.

operating property and equipment
Aviation. Owned assets including capital leases and leaseholds which are used and useful to the air carrier's central business activity, excluding those assets held for resale, or inoperative or redundant to the air carrier's current operations. These assets include loans and units of tangible property and equipment that are used in air transportation services and services incidental thereto.

operating ratio
The ratio of operating expenses to operating revenues.

operating revenue
Transit. (1) The amount of money which a carrier receives from transportation operations. (2) Revenues from the performance of air transportation and related incidental services. Includes a) transport revenues from the carriage of all classes of traffic in scheduled and nonscheduled services and b) nontransport revenues consisting of federal subsidy (where applicable) and revenues for services related to air transportation.

operating system (OS)
A collection of computer programs that control how a computer works.

operation
The act of performing any planned job or task by one or more humans with or without machines/equipment in which value added to a product or information is input, processed, or output.

operation analysis
A systematic review and study of the purpose, procedures, time, and motions required, tools and equipment used, materials used, standards, workplace design, and working conditions for any operation.

operation analysis chart
A form which lists all relevant variables involved in an operation.

operation and maintenance
(1) Activities conducted at a site after a Superfund site action is completed to ensure that the action is effective and operating properly. (2) Actions taken after construction to assure that facilities constructed to treat wastewater will be properly operated, maintained, and managed to achieve efficiency levels and prescribed effluent limitations in an optimum manner.

operation of a railroad
Inclusive term used to describe all activities of a railroad related to the performance of its rail transportation business.

operation process chart
An abbreviated flow process chart consisting of a graphic/symbolic description providing a top-level view of the sequence for an entire operation, specifying such information as the actions and inspections involved, materials used, and pints of introduction, etc. Also called *outline process chart.*

operational
Ready for immediate use, or in the process of being used.

operational acceptable level of traffic (OALT)
An air traffic activity level associated with the designed capacity for a sector or airport. The OALT considers dynamic changes in staffing, personnel experience levels, equipment outages, operational configurations, weather, traffic complexity, aircraft performance mixtures, transitioning flights, adjacent airspace, handoff/point-out responsibilities, and other factors that may affect an air traffic opera-

tional position or system element. The OALT is normally considered to be the total number of aircraft that any air traffic functional position can accommodate for a defined period of time under a given set of circumstances.

operational containment
An active process for preventing an interface between entities which should be kept separate.

operational control
With respect to a flight, means the exercise of authority over initiating, conducting, or terminating a flight.

operational deviation
Aviation. An occurrence where applicable separation minima were maintained (as defined under *operational error*), but a) less than the applicable separation minima existed between an aircraft and protected airspace without prior approval; b) an aircraft penetrated airspace that was delegated to another position of operation or another facility without prior consideration and approval; c) an aircraft penetrated airspace that was delegated to another position of operation or another facility at an altitude or route contrary to the altitude or route requested and approved in direct coordination or as specified in a Letter of Agreement, pre-coordination, or internal procedure; d) an aircraft, vehicle, equipment, or personnel encroached upon landing area was delegated to another position of operation without prior coordination and approval. *See also* ***operational error***.

operational effectiveness
A measure of satisfaction of the work accomplished or the rate at which work is being done within a given total system environment.

operational error
Aviation. An occurrence attributable to an element of the air traffic control system which a) results in less than applicable separation minima between two or more aircraft, or between an aircraft and terrain or obstacles and obstructions. Obstacles include vehicles, equipment, personnel on runways; or b) aircraft lands or departs on a runway closed to aircraft operations after receiving air traffic authorization. *See also* ***operational deviation***.

operational maintenance
Any minor inspection, cleaning, servicing, adjustment, or parts replacement in equipment which can normally be performed by an operator without any specialized training or high-level technical skills.

operational readiness
A state or condition in which a system is not functioning due to scheduling or other reasons, but will perform its intended function when called upon to do so.

operational road
A usable road and intended for use roadway.

operational runway
A usable and intended for use runway.

operational suitability
A measure of the ease of use or usability of a manufactured product.

operations
The sum of all activities of an organization.

operations research
The application of scientific, statistical, and/or modeling methodology toward obtaining information for management to make objective, quantitatively based decisions using specified criteria regarding the men, machines, materials, and money under their control.

operative temperature
A measure of heat stress.

operator
(1) An individual or robot whose functions may include manipulating, supporting, and operational maintenance of a system or piece of equipment. (2) A person who controls the use of an aircraft, vessel, or vehicle. (3) A person who owns or operates a liquefied natural gas (LNG) facility. (4) A person who owns or operates onshore oil pipeline facilities. (5) The company responsible for the management and day-to-day operations of natural gas production, gathering, treating, processing, transportation, storage, and/or distribution facilities, and/or a synthetic natural gas plant.

operator error
See ***human error***.

operator input
That information or data presented to/received by an operator via instructions, displays, ob-

serving equipment/system operation, or the general working environment.

operator instruction sheet
Any form of written instructions for providing the operator with details for performing a given task or job.

operator output
Any physical or verbal action taken by an operator.

operator overload
A condition in which an operator is expected to do more than he/she is capable of performing effectively within the given workplace, environment, or other constraints.

operator performance
Any measure of the work output of an operator.

operator process chart
An operation process chart describing the activities of a single worker without differentiation between the two hands.

operator training
Instruction which is intended to enable or enhance an individual's performance on a job or task.

operator utilization
The ratio of actual working time to total clock time.

operator workload assessment
The use of any relevant physiological, cognitive, or other measure to determine operator workload.

operators
Maritime. The personnel (other than security agents) scheduled to be aboard vehicles in revenue operations including vehicle operators, conductors, and ticket collectors.

ophthalmic
Pertaining to the eye.

ophthalmologist
A physician who specializes in the structure, function, and diseases of the eye.

opiate
Any substance having an addiction-forming or addiction-sustaining capability similar to morphine or being capable of conversion into a drug having addiction-forming or addiction-sustaining capabilities.

opinion
(1) *General.* A view or belief based on a judgement about what is believed to be true regarding some issue, object, or event, but without absolute certainty or knowledge. (2) *Law.* A document prepared by an attorney for his/her client, embodying his/her understanding of the law as applicable to a state of facts submitted to him/her for that purpose (e.g., an opinion of an attorney as to the marketability of a land title as determined from a review of the abstract of title and other public records.

opinion evidence
Evidence (testimony) of what the witness thinks, believes, or infers in regard to the facts in dispute, as distinguished from his/her personal knowledge of the facts themselves. Opinions are not generally considered with the same regard as actual facts. An exception to this rule exists with regard to *expert witnesses.* These are witnesses who, by education and experience, have become an expert in some art, science, profession, or calling and, as such, may state their opinions as to relevant and material matter, in which they profess to be expert, and may also state their reasons for the opinion. In general, expert witness opinions shall be considered with the same regard as the actual facts of the case.

opisthocranion
That point on the occipital bone in the midsagittal plane which marks the posterior extremity of the largest skull diameter measure.

opponent color
One of a set of pairs of opposing colors. *See also **opponent process theory**.*

opponent colors system
A color ordering system in which specified color pairs are considered to be a t the ends of a single dimension: red vs. green, blue vs. yellow, and white vs. black.

opponent process theory
A theory that there are receptors in the eye for red or green for blue or yellow, and for white or black. Synonymous with *Hering's opponent process theory.*

opportunistic infection
An infection caused by a microorganism that does not ordinarily cause disease but can become pathogenic under some circumstances.

opposing signals
Rail Operations. Roadway signals which govern movements in opposite directions on the same track.

opposing train
A train, the movement of which is in a direction opposite to and toward another train on the same track.

opposite direction aircraft
Aircraft are operating in opposite directions when a) they are following the same track in reciprocal directions; or b) their tracks are parallel and the aircraft are flying in reciprocal directions; or c) their tracks intersect at an angle of more than 136 degrees.

OPPT
*See **Office of Pollution Prevention and Toxics**.*

optic
Pertaining to the eye.

optic chiasm
The point at which some of the neural fibers from the retina cross to the opposite side of the brain.

optic nerve
The third cranial nerve. It is purely sensory and is concerned with carrying visual information to the brain. The rods and cones of the retina are connected with the optic nerve which leaves the eye slightly to the nasal side of the center of the retina. The point at which the optic nerve leaves the eye is called the blind spot because there are no rods and cones in this area. The optic nerve passes through the optic foramen of the skull and into the cranial cavity. It then passes backward and undergoes a division; those nerve fibers leading from the nasal side of the retina cross to the opposite side while those from the temporal side continue to the thalamus uncrossed. After synapsing in the thalamus, the neurons convey visual impulses to the occipital lobe of the brain. Degenerative and inflammatory lesions of the optic nerve occur as a result of infections, toxic damage to the nerve, metabolic or nutritional disorders or trauma. Syphilis is the most frequent cause of infectious disorders of the optic nerve. Methanol (methyl alcohol) is highly toxic to the optic nerve and can cause total blindness. Diabetes mellitus and anemia are examples of metabolic and nutritional disorders that can lead to damage to the optic nerve and produce serious loss of vision.

optical axis
An imaginary straight line extending along a horizontal plane of the eye through the midpoint of the cornea, the pupil, and the retina. The optical axis is separated from the visual axis by about 4°.

optical brightener
*See **brightener**.*

optical cavity
A system of using mirrors to pass a light beam through a lasing medium several times, thereby amplifying the number of photons emitted.

optical character reader
A device having the capability to scan a single or limited type of standardized text.

optical character recognition (OCR)
The study or use of photoelectric methods to identify printed or handwritten characters.

optical density (OD)
A measure of the total luminous transmittance of an optical material. A logarithmic expression of the attenuation provided by a filter. The logarithmic value of the ratio between the intensity of transmitted light through a clean filter and a sample.

optical element
Any structure within an optical device involved in shaping or directing light passage through that device.

optical glass
A glass which meets certain standards in being free from imperfections which would adversely affect its light transmission (bubbles, seeds, haze, etc.).

optician
A person who measures and grinds eyeglasses to prescription.

optics
The study of the generation, transmission, refraction, reflection, and detection of electromagnetic radiation between x-rays and radio waves.

optimal
The most desirable.

optimal menu hierarchy
That hierarchy of menu structures which yields either the lowest average access times or fewest number of steps in getting to a specified point or in the most common uses.

optimistic time
The shortest possible time in which a given operation, task, or other activity could be completed.

optimum location principal
The concept in designing a man-machine interface that each display and control should be placed at the best site according to one or more criteria for its intended use.

optimum replacement interval
*See **economic life**.*

optimum sustainable population
With respect to any population stock, the number of animals which will result in the maximum productivity of the population or the species, keeping in mind the carrying capacity of the habitat and the health of the ecosystem of which they form a constituent element.

option approach
*See **cleared for the option**.*

optokinetic nystagmus (OKN)
That nystagmus in a normal individual caused by a succession of objects moving across the visual field. The movement is relative. It may be achieved with the individual stationary and moving objects or a moving individual passing a number of stationary objects.

optometer
An instrument for determining the visual capacities of the eyes.

optometrist
A professional person trained to examine the eyes and prescribe eyeglasses to correct irregularities in the vision. The optometrist is not a physician and is not qualified to diagnose or treat diseases or injuries of the eye, or perform surgery.

optometry
The study and/or measurement of the human eye's capabilities.

oral
(1) Pertaining to the mouth. (2) Spoken.

oral cavity
*See **mouth**.*

oral fissure
That approximately elliptical opening formed by the separation of the facial lips.

oral ingestion
The swallowing of a material.

oral verbal comprehension
The ability to understand spoken language.

oral verbal expression
The ability to use spoken language to communicate with others.

orbit
(1) *Anatomy.* Referring to the eye socket. More specifically, the bony cavity containing the eyeball and its associated muscles, vessels, and nerves. The ethmoid, frontal, lacrimal, nasal, palatine, sphenoid, and zygomatic bones and the maxilla contribute to its formation. (2) *Astronautics.* To maintain a roughly elliptical position in the space around a stationary body. (3) *Science.* The path of an electron around the nucleus of an atom.

orchard heater
An oil heater placed in orchards that generates heat and promotes convective circulation to protect fruit trees from damaging low temperatures.

order
(1) A written or verbal direction or command from someone in authority. (2) Having some systematic structure or pattern (e.g., a lack of chaos). (3) A request for a specific number and type of goods or services. Also called a *purchase order*. (4) *See **order of magnitude**.*

order entry
The process of inputting the information pertaining to a purchase order into a computer for processing.

order of magnitude
An integer value representing an exponent of some number or expression.

ordered metric scale
A basic measurement scale in which items can be classified by rank using some magnitude measure, but with no specification about the absolute magnitudes or magnitudes of differences between items.

ordinal scale
A basic measurement scale in which items can be classified by rank, using some magnitude measure, but with no specification about the absolute magnitudes or magnitudes of differences between items.

ordinance
A municipal statute or regulation.

ordinary care
Law. That degree of care which persons of ordinary care and prudence are accustomed to using and employing, under the same or similar circumstances. Or, it is that degree of care which may reasonably be expected from a person in the party's situation, that is, *reasonable care. See also* ***care*** *and* ***reasonable care****.*

ordinary hazard contents
Those contents which are liable to burn with moderate rapidity and to give off a considerable volume of smoke but from which neither poisonous fumes nor explosions are to be feared in the case of fire.

ordinary high water line
In nature, that water elevation below which aquatic vegetation will not grow. In practice, a water surface elevation arbitrarily fixed from past experience or the establishment of navigation pools.

ordinate
(1) The vertical or dependent axis on a two-dimensional graph, typically labeled the y axis. (2) A particular value on a graph, represented by the perpendicular distance from the abscissa.

ordnance
All ammunition, demolition material, solid rocket motors, liquid propellants, pyrotechnics, and explosives.

ordnance component
A component such as a squib, detonator, initiator, ignitor, or linear-shaped charge, in an ordnance system.

organ
Organized group of tissues that perform one or more definite functions in an organism.

organ of Corti
An organ, lying against the basilar membrane in the cochlear duct of the ear, which contains special sensory receptors for hearing.

organic
Of, pertaining to, or derived from living organisms. Also, in chemistry, refers to substances containing carbon compounds.

organic carbon partition coefficient (K_{oc})
Measure of the content of partitioning of a substance, at equilibrium, between organic carbon in geologic materials and water. The higher the K_{oc}, the more likely a substance is to bind to geologic materials than to remain in water.

organic chemicals/compounds
Animal- or plant-produced substances containing mainly carbon, hydrogen, and oxygen.

organic matter
Carbonaceous waste contained in plant or animal matter and originating from domestic or industrial sources.

organic peroxide
An organic compound containing the bivalent -O-O- structure and which may be considered a derivative of hydrogen peroxide where one or more of the hydrogen atoms have been replaced by organic radicals.

organism
Any living biological entity composed of one or more cells.

organization
That structure of people, concepts, or other entities which exist or are created to carry out or assist in one or more specific objectives.

organization chart
A graphic representation of the interrelationships within an organization, which may indicate lines of authority and areas of responsibility.

organizational climate
Those properties of the working environment which may have effects on employee productivity.

organizational psychology
That field of study and practice involving the structure and function of organizations.

organized track system
Aviation. A movable system of oceanic tracks that traverses the North Atlantic between Europe and North America the physical position of which is determined twice daily taking the best advantage of the winds aloft.

organo-
Having a carbon base.

organogenesis
The period in the development of a fetus during which the organs are developing.

organometallic compound
A chemical compound in which a metal is chemically bonded to an organic compound. Examples include organophosphate compounds, tetraethyl lead, manganese cyclopentadienyl tricarbonyl, etc.

organophosphates
Pesticide chemicals containing phosphorous that are used to control insects. They are usually short-lived, but some can be toxic when first applied. In sufficient quantities of exposure, there can be systemic poisoning effects.

organotin
As defined by the Federal Organotin Antifouling Paint Control Act of 1988: Any compound of tin used as a biocide in an antifouling paint.

orientation
The process of providing a new employee with some background information on the organization, its policies, and its procedures.

orientation reflex
See ***orienting response.***

orienting response
A mild psychophysiological response involving a sudden shift of attention to process information, associated with some sudden event. Also referred to as *orientation reflex*. *See also* ***startle response.***

orifice
An opening or hole of controlled size that can be used for the measurement of liquid or gas flow.

orifice meter
A device for determining flow rate. A flow meter employing, as the measure of flow, the pressure difference as measured on the upstream side of a specific type of restriction within a pipe or duct.

origin
(1) *Transportation.* Starting point of a trip. (2) *Transit.* The country in which the cargo was loaded and/or the transit originated. (3) *Anatomy.* The point or bone surface location from which skeletal muscle originates (opposite of *insertion*). (4) *Mathematics.* The null reference point for a coordinate system, at which all axes meet and are usually assigned values of zero.

original document rule
The best evidence of the contents of a document is the original of that document. The party bearing the burden of proving the contents of a document is required to introduce the original unless he/she is excused from its production because of its non-availability and in this instance, secondary evidence is admissible.

originality
The ability to produce new, unusual, or clever thoughts on a given topic.

originated carload
An originated carload is one which is loaded and begins its journey on a particular railroad.

originating statute
See ***statutory mandate.***

ORM
See ***other regulated material.***

ORNL
Oak Ridge National Laboratory.

orographic uplift
The lifting of air over a topographic barrier. Clouds that form in this uplifting process are called *orographic clouds.*

oropharynx
The middle region of the pharynx, from the level of the soft palate to the level of the hyoid bone, and from the posterior pharyngeal wall to the fauces.

Orsat apparatus
A device for measuring the percentage of carbon dioxide, oxygen, and carbon monoxide in flue gas.

orthoaxis
The true anatomical axis of rotation for a limb.

orthocenter
The instantaneous anatomical center of rotation for a joint.

orthogonal
(1) Being perpendicular or at right angles. (2) Completely independent or separable.